D1394223

Agricultural Innovation Systems

AN INVESTMENT SOURCEBOOK

AGRICULTURE AND RURAL DEVELOPMENT

Seventy-five percent of the world's poor live in rural areas and most are involved in agriculture. In the 21st century, agriculture remains fundamental to economic growth, poverty alleviation, and environmental sustainability. The World Bank's Agriculture and Rural Development publication series presents recent analyses of issues that affect the role of agriculture, including livestock, fisheries, and forestry, as a source of economic development, rural livelihoods, and environmental services. The series is intended for practical application, and we hope that it will serve to inform public discussion, policy formulation, and development planning.

Titles in this series:

Agribusiness and Innovation Systems in Africa

Agricultural Innovation Systems: An Investment Sourcebook

Agricultural Land Redistribution: Toward Greater Consensus

Agriculture Investment Sourcebook

Bioenergy Development: Issues and Impacts for Poverty and Natural Resource Management

Building Competitiveness in Africa's Agriculture: A Guide to Value Chain Concepts and Applications

Changing the Face of the Waters: The Promise and Challenge of Sustainable Aquaculture

Enhancing Agricultural Innovation: How to Go Beyond the Strengthening of Research Systems

Forests Sourcebook: Practical Guidance for Sustaining Forests in Development Cooperation

Gender and Governance in Rural Services: Insights from India, Ghana, and Ethiopia

Gender in Agriculture Sourcebook

The Land Governance Assessment Framework: Identifying and Monitoring Good Practice in the Land Sector

Organization and Performance of Cotton Sectors in Africa: Learning from Reform Experience

Reforming Agricultural Trade for Developing Countries, Volume 1: Key Issues for a Pro-Development Outcome of the Doha Round

Reforming Agricultural Trade for Developing Countries, Volume 2: Quantifying the Impact of Multilateral Trade Reform

Rising Global Interest in Farmland: Can It Yield Sustainable and Equitable Benefits?

Shaping the Future of Water for Agriculture: A Sourcebook for Investment in Agricultural Water Management

The Sunken Billions: The Economic Justification for Fisheries Reform

Sustainable Land Management: Challenges, Opportunities, and Trade-Offs

Sustainable Land Management Sourcebook

Sustaining Forests: A Development Strategy

Agricultural Innovation Systems

AN INVESTMENT SOURCEBOOK

THE WORLD BANK
Washington, D.C.

© 2012 International Bank for Reconstruction and Development / International Development Association or
The World Bank
1818 H Street NW
Washington DC 20433
Telephone: 202-473-1000
Internet: www.worldbank.org

1 2 3 4 15 14 13 12

This volume is a product of the staff of The World Bank with external contributions. The findings, interpretations, and conclusions expressed in this volume do not necessarily reflect the views of The World Bank, its Board of Executive Directors, or the governments they represent.

The World Bank does not guarantee the accuracy of the data included in this work. The boundaries, colors, denominations, and other information shown on any map in this work do not imply any judgment on the part of The World Bank concerning the legal status of any territory or the endorsement or acceptance of such boundaries.

Rights and Permissions

The material in this work is subject to copyright. Because The World Bank encourages dissemination of its knowledge, this work may be reproduced, in whole or in part, for noncommercial purposes as long as full attribution to the work is given.

For permission to reproduce any part of this work for commercial purposes, please send a request with complete information to the Copyright Clearance Center Inc., 222 Rosewood Drive, Danvers, MA 01923, USA; telephone: 978-750-8400; fax: 978-750-4470; Internet: www.copyright.com.

All other queries on rights and licenses, including subsidiary rights, should be addressed to the Office of the Publisher, The World Bank, 1818 H Street NW, Washington, DC 20433, USA; fax: 202-522-2422; e-mail: pubrights@worldbank.org.

ISBN (paper): 978-0-8213-8684-2
ISBN (electronic): 978-0-8213-8944-7
DOI: 10.1596/978-0-8213-8684-2

Library of Congress Cataloging-in-Publication Data

Agricultural innovation systems : an investment sourcebook.
    p. cm. — (Agriculture and rural development)
  Includes bibliographical references and index.
   ISBN 978-0-8213-8684-2 — ISBN 978-0-8213-8944-7 (electronic)
    1. Agricultural innovations—Economic aspects. 2. Agriculture—Economic aspects. 3. Rural development. I. World Bank. II. Series: Agriculture and rural development series.

  S494.5.I5A32213 2012
  338.1'6—dc23
         2011052338

Cover photograph: Anne Wamalwa/CIMMYT
Cover design: Critical Stages

CONTENTS

BOXES

TABLES

Agricultural Innovation Systems: An Investment Sourcebook was made possible by a number of key individuals, donors, and partner organizations. The AIS Sourcebook was part of the Finland / infoDev / Nokia program "Creating sustainable businesses for the knowledge economy," whose contributions together with World Bank funding served as a foundation for the production of the Sourcebook. Financing was also provided through a grant from the Gender Trust Fund and many in-kind contributions from the following institutions: *CATIE, CIAT, CRS, BASIX, FAO, ICRISAT, IFDC, KIT, LINK Ltd, PIPRA, Prolinnova, University of Guelph, University of Hohenheim, UNIDO, and USAID.* The overall task was managed and coordinated by Riikka Rajalahti and Andrea Pape-Christiansen (World Bank). They collectively assume responsibility for remaining errors and omissions.

The preparation of this Sourcebook involved many individuals within several units of the World Bank and a variety of partner organizations. As module coordinators the following individuals played a key role in the preparation of the individual modules: Javier Ekboir and Riikka Rajalahti (module 1); Charles Maguire (module 2); Kristin Davis and Willem Heemskerk (module 3); John Lynam (module 4); Josef Ernstberger (module 5); Han Roseboom (module 6); and Helen Hambly Odame, Andy Hall, and Kumuda Dorai (module 7).

We are grateful to the more than 70 individuals who made written contributions to module Overviews, Thematic Notes, or Innovative Activity Profiles. Contributors within the Bank included Seth Ayers (WBI, formerly infoDev), Marie-Hélène Collion (LCSAR), Luz Diaz Rios (ARD), Michelle Friedman (LCSAR), Indira Ekanayake (AFTAR), Willem Janssen (LCSAR), Gunnar Larson (ARD), Andrea Pape-Christiansen (ARD), Riikka Rajalahti (ARD), Florian Theus (WBI), Ariani Wartenberg (LCSAR), Melissa Williams (SASDA), and Douglas Zhihua Zeng (AFTFE).

Contributors of material for module Overviews, Thematic Notes, or Innovative Activity Profiles outside the Bank included Nuria Ackermann (UNIDO), John Allgood (IFDC), S. Aravazhi (ICRISAT), Peter Ballantyne (ILRI), Alan B. Bennett (PIPRA, University of California, Davis), Rupert Best (CRS), Regina Birner (University of Hohenheim, formerly IFPRI), Sara Boettiger (PIPRA, University of California, Berkeley), Arnoud R. Braun (Famer Field School Foundation Wageningen University), Alistair Brett (Consultant), Judy Chambers (IFPRI), Anthony Clayton (University of the West Indies), Kristin Davis (IFPRI and GFRAS), Kumuda Dorai (LINK Ltd), Deborah Duveskog (Consultant), Javier Ekboir (ILAC), Howard Elliott (Consultant), Wyn Ellis (Consultant), Josef Ernstberger (Consultant), Josef Geoola (GALVmed), Peter Gildemacher (KIT), María Verónica Gottret (CATIE), Andy Hall (LINK Ltd., Open University), Helen Hambly Odame (University of Guelph), Frank Hartwich (UNIDO), Willem Heemskerk (KIT), Douglas Horton (Consultant), Steen Joffee (Innodev), Trish Kammili (INRA), S.M. Karuppanchetty (ICRISAT), Godrick Khisa (Wageningen University Farmer Field School Foundation), Laurens Klerkx (Wageningen University), Stanley Kowalski (University of New Hampshire), Patti Kristjanson (CIFOR), Anton Krone (SaveAct, Prolinnova), Ninatubu Lema (NARS Tanzania), Tarmo Lemola (ADVANSIS Ltd.), David Lugg (FAO), Mark Lundy (CIAT), John Lynam (Consultant), Charles J. Maguire (Consultant), Vijay Mahajan (BASIX), Morven McLean (ILSI), Mohinder S. Mudahar

(Consultant), Bernardo Ospina Patiño (CLAYUCA), Christopher Palmberg (ADVANSIS Ltd.—now at Tekes), Robert Potter (Consultant), John Preissing (FAO), Ranjitha Puskur (ILRI), Catherine Ragasa (IFPRI), Johannes Roseboom (Consultant), Silvia Sarapura (University of Guelph), Eva Schiffer (Consultant), Kiran K. Sharma (ICRISAT), David J. Spielman (IFPRI), Rasheed Sulaiman V (CRISP), Bernard Triomphe (CIRAD), Klaus Urban (FAO), K. Vasumathi (BASIX), Rodrigo Vega Alarcón (Consultant, formerly FIA), Laurens van Veldhuizen (ETC Foundation), Ann Waters-Bayer (ETC Foundation), and Mariana Wongtschowski (KIT).

The Sourcebook was reviewed and refined by a large number of reviewers, whose work is gratefully acknowledged. The team appreciates the substantive comments, and suggestions from the following peer reviewers, including Adolfo Brizzi, Willem Janssen, and Alfred Watkins (World Bank); Manuel Lantin and Jonathan Wadsworth (CGIAR); Dr. Mruthyunjaya (ICAR); Tom Mwangi Anyonge, Maria Hartl, Shantanu Mathur, and Vineet Raswant (IFAD); Marco Ferroni and Yuan Zhou (Syngenta Foundation).

In addition to the peer review, several people provided an overall review of the concept note and module drafts: Gary Alex (USAID), Jock Anderson (Consultant), Regina Birner (University of Hohenheim), Andreas Blom (World Bank), Derek Byerlee (Consultant), Montague Demment (University of California, Davis, APLU), Grahame Dixie (World Bank), Kumuda Dorai (LINK Ltd.), Fionna Douglas (World Bank), Andy Hall (LINK Ltd.), Willem Heemskerk (KIT), Doug Horton (Consultant), Steven Jaffee (Innodev), Willem Janssen (World Bank), Nancy Johnson (ILRI), Trish Kammili (INRA), Manuel Lantin (CGIAR), Esparanza Lasagabaster (World Bank), Tarmo Lemola (Advansis Ltd.), John Lynam (Consultant), Charles Maguire (Consultant), Calvin Miller (FAO), Dr. Mruthyunjaya (ICAR), Remco Mur (KIT), Gordon Myers (IFC), Ajai Nair (World Bank), David Nielson (World Bank), Eija Pehu (World Bank), John Preissing (FAO), Ranjitha Puskur (ILRI), Luz Berania Diaz Rios (World Bank), William M. Rivera (Consultant), Bill Saint (Consultant), Iain Shuker (World Bank), Amy Stilwell (World Bank), Burt Swanson (University of Illinois), Jee-Peng Tan (World Bank), Jonathan Wadsworth (CGIAR, formerly DFID), and Johannes Woelcke (World Bank).

The Sourcebook team would like to acknowledge Juergen Voegele and Mark Cackler (World Bank) for their guidance and support to the Sourcebook throughout its preparation.

Technical editing and final editing were done by Kelly Cassaday (Consultant). Production of the Sourcebook was handled by Patricia Katayama, Aziz Gokdemir, and Nora Ridolfi (EXTOP). The Sourcebook team would also like to acknowledge the production assistance of Kaisa Antikainen, Fionna Douglas, Anthony Stilwell, and Sarian Akibo-Betts (ARD), as well as Felicitas Doroteo-Gomez (ARD), who provided administrative assistance.

ABBREVIATIONS

AACREA	Asociación Argentina de Consorcios Regionales de Experimentación Agrícola (Argentine Association of Regional Consortiums for Agricultural Experimentation)
ABI	agri-business incubation
AERI	Agricultural Exports and Rural Income Project
AFAAS	African Forum for Agricultural Advisory Services
AGLED	Agricultural, Livestock, and Enterprise Development
AIS	agricultural innovation system
ARDA	Agriculture and Rural Development, World Bank
ATMA	Agriculture Technology Management Agency (India)
ATS	agricultural technical school
ATVET	agricultural technical–vocational education and training
BACET	Building Agribusiness Capacity in East Timor
BDS	business development services
BRAC	Acronym is now official name
CAADP	Comprehensive African Agriculture Development Programme
CAC	Codex Alimentarius Commission
CATIE	Tropical Agriculture Research and Education Center
CBR	community-based research
CGIAR	Consultative Group on International Agricultural Research
CGPs	competitive grant programs
CIAT	Centro Internacional de Agricultura Tropical (International Center for Tropical Agriculture)
CILSS	Comité Inter-états de Lutte Contre la Sècheresse au Sahel (Committee for Drought Control in the Sahel)
CIMMYT	Centro Internacional de Mejoramiento de Maíz y Trigo (International Maize and Wheat Improvement Center)
CIP	Centro Internacional de la Papa (International Potato Center)
CIRAD	Centre de Coopération Internationale en Recherche Agronomique pour le Développement (Agricultural Research for Development, France)
CLAYUCA	Consorcio Latinoamericano y del Caribe de Apoyo a la Investigación y al Desarrollo de la Yuca (Latin American and Caribbean Consortium to Support Cassava Research and Development)
CMSA	community-managed sustainable agriculture
CNFA	Citizens Network for Foreign Affairs

COFUPRO	Coordinadora Nacional de las Fundaciones Produce (National Coordinating Agency of Produce Foundations), Mexico
CORAF	Conseil Ouest et Centre Africain pour la Recherche et le Développement Agricoles
CSO	civil society organization
DHRUVA	Dharampur Uththan Vahini (Vanguard of Awakening in Dharampur) (Gujarat)
DIPP	Disease Intellectual Property Plan
EARTH	Escuela de Agricultura de la Región Tropical Húmeda (EARTH University), Costa Rica
ECOWAS	Economic Community of West African States
EMBRAPA	Empresa Brasileira de Pesquisa Agropecuária (Brazilian Agricultural Research Corporation)
EU	European Union
F2F	Farmer to farmer
FAO	Food and Agriculture Organization of the United Nations
FBO	farmer-based organization
FFS	Farmer Field School
FIA	Fundación para la Innovación Agraria (Foundation for Agricultural Innovation), Chile
FIC	Fondo de Innovación para la Competitivad (Innovation Fund for Competitiveness), Chile
FIF	Farmer Innovation Fund
FOB	free on board
FPEAK	Fresh Produce Exporters Association of Kenya
GAP	good agricultural practice
GDP	gross domestic product
GE	genetically engineered
GFRAS	Global Forum for Rural Advisory Services
GIZ	Deutsche Gesellschaft für Internationale Zusammenarbeit (German International Cooperation Agency)
GLOBALG.A.P.	Global Good Agricultural Practices
GM	genetically modified
GMO	genetically modified organism
GPSG	gobal positioning systems
HACCP	Hazard Analysis and Critical Control Point
HRD	human resource development
HRM	human resource management
IAP	innovative activity profile
ICAR	Indian Council of Agricultural Research
ICI	Imperial Chemical Company
ICRISAT	International Crops Research Institute for the Semi-Arid Tropics
ICT	information and communication technology
IDRC	International Development Research Centre
IFAD	International Fund for Agricultural Development
IFDC	International Fertilizer Development Center
IFPRI	International Food Policy Research Institute
IIRR	International Institute of Rural Reconstruction
ILAC	Institutional Learning and Change Initiative (CGIAR)
ILO	International Labour Organisation
ILRI	International Livestock Research Institute
INCAGRO	Innovación y Competitividad para el Agro Peruano (Innovation and Competitiveness Program for Peruvian Agriculture)
INSAH	Institut du Sahel (Sahel Institute)

IP	intellectual property
IPO	initial public offering
IPR	intellectual property right(s)
IRR	internal rate of return
ISNAR	International Service for National Agricultural Research
ISO	International Organization of Standardization
IT	information technology
KfW	Kreditanstalt für Wiederaufbau (KfW Development Bank, Germany)
KHDP	Kerala Horticultural Development Programme
KIT	Koninklijk Instituut voor de Tropen (Royal Tropical Institute), the Netherlands
KTDA	Kenya Tea Development Agency
LBDSs	Local business development services
LCSAR	Latin America and Caribbean Region, Agriculture and Rural Development (World Bank)
LISFs	Local Innovation Support Funds
LSA	livelihood services adviser
LSP	livelihood services provider
M&E	monitoring and evaluation
MOFA	Ministry of Food and Agriculture, Ghana
MUCIA	Midwest Universities Consortium for International Activities
NAADS	National Agricultural Advisory Services, Uganda
NAIP	National Agricultural Innovation Project, India
NARI	national agricultural research institute
NARS	national agricultural research system
NDDB	National Dairy Development Board, India
NGO	nongovernmental organization
NIA	National Innovation Agency, Thailand
NIS	national innovation system
OECD	Organisation for Economic Co-operation and Development
OIE	Office International des Epizooties (World Organisation for Animal Health)
PASAOP	Programme d'Appui aux Services Agricoles et aux Organisations Paysannes (Agricultural Services and Farmer Organization Support Program), Mali
PIPRA	Public Intellectual Property Resource for Agriculture, University of California, Davis
PPP	public-private partnership
PRONEA	Programa Nacional de Extensão Agrária (National Agricultural Extension Program), Mozambique
PVP	plant variety protection
PVR	plant variety rights
R&D	research and development
RAS	rural advisory services
RD&D	research, development, and deployment
RUFORUM	Regional Universities Forum for Capacity Building in Agriculture
SACCO	Savings and Credit Cooperative
SAFE	Sasakawa Africa Fund for Extension Education
SAU	State Agricultural University, India
SDC	Swiss Agency for Development and Cooperation
SERP	Society for the Elimination of Rural Poverty
SHG	self-help group
SMART	specific, measurable, attainable, realistic, and timely
SMEs	small and medium enterprises

SNIC	Sistema Nacional de Innovación para la Competitivad (National System of Innovation for Competitiveness), Chile
SOCAD	State Office for Comprehensive Agricultural Development, China
SPS	sanitary and phyosanitary
SSA	Sasakawa Africa Association
SSTAB	Short-Term Technical Assistance in Biotechnology
STI	Science, Technology, and Innovation; Office of Science, Technology, and Innovation, Thailand
T&V	Training and Visit
TBT	Agreement on Technical Barriers to Trade
Tekes	Teknologian ja innovaatioiden kehittämiskeskus (Finnish Funding Agency for Technology and Innovation)
TIA	Technology Innovation Agency, South Africa
TN	thematic note
TRF	Thailand Research Fund
TRIPS	Agreement on Trade-Related Aspects of Intellectual Property Rights
TTO	Technology Transfer Office
TVET	technical–vocational education and training
UCC	University of Cape Coast (Ghana)
UNESCO	United Nations Educational, Scientific, and Cultural Organization
UNIDO	United Nations Industrial Development Organization
UNU	United Nations University
UPOV	Union Internationale pour la Protection des Obtentions Végétales (International Union for the Protection of New Varieties of Plants)
VET	vocational education and training
VFPCK	Vegetable and Fruit Promotion Council, Kerala
VO	village organization
WAEMU	West African Economic and Monetary Union
WECARD	West and Central African Council for Agricultural Research and Development
WHO	World Health Organization
WIPO	World Intellectual Property Organization
WTO	World Trade Organization

Sourcebook Overview and User Guide

Riikka Rajalahti, World Bank

ORIGINS AND PURPOSE OF THIS SOURCEBOOK

Consensus is developing about what is meant by "innovation" and "innovation system" (box O.1). The agricultural innovation system (AIS) approach has evolved from a concept into an entire subdiscipline, with principles of analysis and action; yet no detailed blueprint exists for making agricultural innovation happen at a given time, in a given place, for a given result. This sourcebook draws on the emerging principles of AIS analysis and action to help to identify, design, and implement the investments, approaches, and complementary interventions that appear most likely to strengthen innovation systems and promote agricultural innovation and equitable growth.

Although the sourcebook discusses *why* investments in AISs are becoming so important, it gives most of its attention to *how* specific approaches and practices can foster innovation in a range of contexts. Operationalizing an AIS approach requires a significant effort to collect and synthesize the diverse experiences with AISs. The information in this sourcebook derives from approaches that have been tested at different scales in different contexts. It emphasizes the lessons learned, benefits and impacts, implementation issues, and prospects for replicating or expanding successful practices. This information reflects the experiences and evolving understanding of numerous individuals and organizations concerned with agricultural innovation, including the World Bank. (For a complete list of the contributors, see the acknowledgments.)

The sourcebook is targeted to the key operational staff in international and regional development agencies and national governments who design and implement lending projects and to the practitioners who design thematic programs and technical assistance packages. The sourcebook is also an important resource for the research community and nongovernmental organizations (NGOs) and may be a useful reference for the private sector, farmer organizations, and individuals with an interest in agricultural innovation.

This overview begins with a discussion of why innovation is vital to agricultural development, how innovation occurs, and why complementary investments are needed to develop the capacity and enabling environment for agricultural innovation. It concludes with details on the sourcebook's structure, a summary of the themes covered in each module, and a discussion of the cross-cutting themes treated throughout the sourcebook.

INNOVATION AND AGRICULTURAL DEVELOPMENT

Agricultural development enables agriculture and people to adapt rapidly when challenges occur and to respond readily when opportunities arise—as they inevitably will, because agriculture's physical, social, and economic environment changes continually (box O.2). Some changes occur with unpredictable force and suddenness; since June 2010, for example, rapidly rising food prices have pushed about 44 million people into poverty, and another 10 percent rise in

---

Box O.1   Innovation and Innovation Systems Defined

Innovation is the process by which individuals or organizations master and implement the design and production of goods and services that are new to them, irrespective of whether they are new to their competitors, their country, or the world.

An *innovation system* is a network of organizations, enterprises, and individuals focused on bringing new products, new processes, and new forms of organization into economic use, together with the institutions and policies that affect their behavior and performance.

Source: World Bank 2006.

Box O.2   The World's Need for Agriculture, Agricultural Development, and Innovation

In one way or another, agriculture is integral to the physical and economic survival of every human being. The United Nations forecasts that the global population will reach more than 9 billion by 2050. To feed everyone, food production will have to increase 70 percent. Helping the world's farmers and fishers to achieve this target is challenging in itself, but beyond providing food, agriculture sustains the economies of most countries in significant ways, especially in the developing world. Across Sub-Saharan Africa, for example, agriculture accounts for three-quarters of employment and one-third of GDP; 75 percent of the world's poor live in rural areas and have an economic link to agriculture. For very

poor households, agricultural development not only is a defense against hunger but also can raise incomes nearly *four times more effectively* than growth in any other sector. These circumstances help to explain why agricultural development is such a powerful tool for reducing global poverty and eliciting economic development.

Agricultural development demands and depends on innovation and innovation systems. Innovation is widely recognized as a major source of improved productivity, competitiveness, and economic growth throughout advanced and emerging economies. Innovation also plays an important role in creating jobs, generating income, alleviating poverty, and driving social development.

Sources: Author; OECD 2009.

---

the food price index could impoverish 10 million more people. Food prices are expected to remain volatile for the foreseeable future.

Other changes emerge more gradually, but are no less significant. Agriculture is more vulnerable to the increasing effects of climate change than any other economic sector, and it uses almost 80 percent of the world's freshwater—a vanishing resource in some parts of the world. A changing, less predictable, and more variable environment makes it imperative for the world's farmers and fishers to adapt and experiment. They require more knowledge that contributes to sustainable, "green" growth—as well as a greater capacity to help develop such knowledge.

Like climatic variability, globalizing markets for agricultural products, far-reaching developments in technology, and equally transformative evolution in institutions (including new roles for the state, the private sector, and

civil society) have also been altering agriculture's social and economic landscape over the past few decades (World Bank 2007b). Agriculture increasingly occurs in a context where private entrepreneurs coordinate extensive value chains linking producers to consumers, sometimes across vast distances. A growing number of entrepreneurial smallholders are organizing to enter these value chains, but others struggle with the economic marginalization that comes from being excluded from such opportunities.

In this context, markets, urbanization, globalization, and a changing environment not only influence patterns of consumption, competition, and trade but also drive agricultural development and innovation far more than before. More providers of knowledge are on the scene, particularly from the private sector and civil society, and they interact in new ways to generate ideas or develop responses to changing agricultural conditions (World Bank 2006).

If farmers, agribusinesses, and even nations are to cope, compete, and thrive in the midst of changes of this magnitude, they must innovate continuously. Investments in public research and development (R&D), extension, education, and their links with one another have elicited high returns and pro-poor growth (World Bank 2007b), but these investments alone will not elicit innovation at the pace or on the scale required by the intensifying and proliferating challenges confronting agriculture.

HOW AGRICULTURAL INNOVATION OCCURS

Agricultural innovation typically arises through dynamic interaction among the multitude of actors involved in growing, processing, packaging, distributing, and consuming or otherwise using agricultural products. These actors represent quite disparate perspectives and skills, such as metrology, safety standards, molecular genetics, intellectual property, food chemistry, resource economics, logistics, slash-and-burn farming, land rights—the list is far too long to complete here.

For innovation to occur, interactions among these diverse stakeholders need to be open and to draw upon the most appropriate available knowledge. Aside from a strong capacity in R&D, the ability to innovate is often related to collective action, coordination, the exchange of knowledge among diverse actors, the incentives and resources available to form partnerships and develop businesses, and conditions that make it possible for farmers or entrepreneurs to use the innovations. Box O.3 provides examples of how innovation has occurred in agriculture.

Agricultural innovation systems

Research, education, and extension are usually not sufficient to bring knowledge, technologies, and services to farmers

---

Box O.3 Examples of Agricultural Innovation and Innovation Processes

The instances of agricultural innovation listed here came about in different ways. In some cases, markets heightened the pressure to innovate, and the private sector played a decisive role in driving the subsequent innovation. In others, public sector interventions, such as policy, R&D, and other incentives, drove the innovation process.

- *Cassava-processing innovation system, Ghana.* Research-led development and promotion of new cassava products with a private sector coalition.
- *Cut flower innovation system, Colombia.* Continuous innovation in response to changing markets, using licensed foreign technology and coordinated by an industry association.
- *Medicinal plants innovation system, India.* Mobilization of traditional and scientific knowledge for rural communities, coordinated by a foundation.
- *Small-scale irrigation innovation system, Bangladesh.* Promotion by a civil society organization of a low-cost pump to create markets; innovation by small-scale manufacturers with the design of pumps in response to local needs.
- *Golden rice innovation system, global.* Complex partnership of multinational companies, interna-

tional agricultural research organizations, universities, and development foundations; complex but creative institutional arrangements over ownership; innovation targeted to poor (nutrient-deficient) users.

- *Potato, Peru.* Facilitation by an international research center of the development of new indigenous potato products with a coalition of researchers, smallholders, and multiple private actors (including supermarkets, traders, and restaurants).

In each case, the drivers of innovation and growth were different and the role of research and extension varied, but in all cases the actors used similar approaches to address their challenges and innovate. The challenges included meeting stringent quality standards, remaining competitive, responding to changing consumer tastes, and addressing technological problems.

The actors' ability to improve their interactions and strengthen their links to one another proved crucial to their success. All of the cases illustrate the importance of taking collective action, having the benefit of facilitation and coordination by intermediaries, building a strong skill base, and creating an enabling environment for innovation to take place.

Sources: Adapted from Bernet, Thiele, and Zschocke 2006; Hall, Clark, and Naik 2007; World Bank 2006; A. Hall, personal communication; R. Rajalahti, personal communication.

and entrepreneurs and to get them to innovate. Innovation requires a much more interactive, dynamic, and ultimately flexible process in which the actors deal simultaneously with many conditions and complementary activities that go beyond the traditional domains of R&D and extension. These conditions and complementary interventions have not been consistently addressed to date; new, additional ways and means of doing so are needed.

An AIS approach looks at the multiple conditions and relationships that promote innovation in agriculture. It may offer a more flexible means of dealing with the varied conditions and contexts in which innovation must occur. It considers the diverse actors involved, their potential interactions, the role of informal practices in promoting innovation, and the agricultural policy context.

The AIS principles of analysis and action integrate the more traditional interventions (support for research, extension, and education and creation of links among research, extension, and farmers) with the other complementary interventions needed for innovation to take place. Such interventions include providing the professional skills, incentives, and resources to develop partnerships and businesses; improving knowledge flows; and ensuring that the conditions that enable actors to innovate are in place.

Figure O.1 presents a simplified conceptual framework for an AIS. The figure shows the main actors (typical agricultural knowledge and technology providers and users, as well as the bridging or intermediary institutions that facilitate their interaction); the potential interactions between actors; and the agricultural policies and informal institutions, attitudes, and practices that either support or hinder the process of innovation.

EFFORTS TO STRENGTHEN KEY COMPONENTS OF THE INNOVATION SYSTEM

Agricultural research, extension, education, and training are key components of an AIS. The following sections summarize approaches that have been used to strengthen these components, what they achieved, and continuing concerns.

Figure O.1   An Agricultural Innovation System

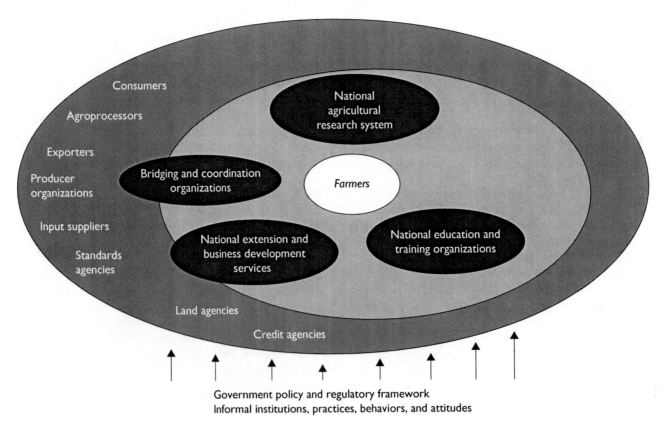

Source: Modified from Rivera et al. n.d.

Agricultural research

A strong science and technology system—encompassing basic, strategic, and adaptive agricultural science as well as sciences outside agriculture—is widely regarded as contributing to innovation and sustainable, equitable agricultural development. Development cannot occur without knowledge, much of which must be generated and applied nationally and often more locally. For this reason, sustaining food production and rural livelihoods while reducing poverty depends to a great extent on how successfully knowledge is generated and applied in agriculture and on whether the capacity to produce such knowledge is improved.

Aside from budgetary constraints (box O.4), many public research organizations face serious institutional constraints that inhibit their effectiveness, constrain their ability to attract funds, and ultimately prevent them from functioning as a major contributor to the innovation system. The main constraints associated with many national research organizations result from strong path-dependency in institutional development and slow institutional and policy change, such as the lack of consensus on a strategic vision, ineffective leadership and management, a continued emphasis on building centralized national agricultural research structures rather than on creating partnerships, the loss of highly qualified scientific staff, and weak links with and accountability to other actors involved in innovation processes (World Bank 2005).

Over the years, research organizations have attempted to address these various constraints. Most of these efforts have centered on shifting investments away from physical infrastructure, equipment, human resource development, and operating funds and toward improvements in the management of public research organizations—for example, through better planning, improved financial management, greater accountability, and more relevant programs for clients (developed with oversight from multistakeholder boards or through better research-extension linkages).

Box O.4   Trends in Financing Agricultural Science and Knowledge Systems

Global public investments in agricultural science, technology, and development have increased significantly over the years, rising from US$16 billion (reported in 1981) to US$23 billion in 2005 purchasing power parity dollars in 2000 (figures from Beintema and Elliott 2009; 2000 is the latest year for which comparable global data are available). The increase is somewhat deceptive, because it has been concentrated in just a handful of countries (Pardey et al. 2006). More recent data indicate that investments in science and technology continue to increase.

Government remains the largest contributor to public agricultural research, accounting for an average of 81 percent of funding (of more than 400 government agencies and nonprofit institutions in 53 developing countries sampled). Only 7 percent of funding is provided by donors as loans or grants. Funding supplied through internally generated funds, including contractual arrangements with private and public enterprises, on average accounts for 7 percent of the funding for public agricultural research. Nonprofit organizations, which collect about two-thirds of their funding from producer organizations and marketing boards, are also more active than government agencies at raising income from internally generated resources, which include contracts with private and public enterprises (26 percent).

The private sector spends an estimated US$16 billion (in 2005 purchasing power parity dollars) on agricultural research, equivalent to 41 percent of the global investment (public and private). Almost all of these private investments are made by companies pursuing agricultural R&D in high-income countries. In addition, several international research centers focus on agricultural R&D to produce international public goods.

Investments in R&D, including research and advisory services, have been the World Bank's major strategy to improve agricultural productivity and innovation (World Bank 2009b). The World Bank alone invested US$4.9 billion (US$5.4 billion in real million dollars, 2010 = 100) into agricultural R&D and advisory services over the 20 years from 1990 to 2010. The World Bank's annual commitments to agricultural research, extension, education, and training have ranged from US$100 million to US$800 million. The very low commitments by governments and donors to agricultural tertiary education since the early 1990s are an especially worrying trend (World Bank 2007a), because they imply that a capacity for innovation is not being sustained.

Sources: Author.

Much effort has focused on increasing client participation and on the financing and overall development of pluralistic agricultural knowledge and information systems (World Bank 2005). Table O.1 captures the main differences and changes in emphasis in World Bank investments to support innovation. Box O.5 describes recent reforms in agricultural research and extension organizations.

Approaches to international cooperation in agricultural R&D continue to change as well. Growing capacities in large national agricultural systems such as those of Brazil, China,

Table O.1 Defining Features of the Three Main Frameworks Used to Promote and Invest in Knowledge in the Agriculture Sector

Defining feature	National agricultural research systems	Agricultural knowledge and information systems	Agricultural innovation systems
Actors	Research organizations	Farmer, research, extension, and education	Wide spectrum of actors
Outcome	Technology invention and transfer	Technology adoption and innovation	Different types of innovation
Organizing principle	Using science to create new technologies	Accessing agricultural knowledge	New uses of knowledge for social and economic change
Mechanism for innovation	Technology transfer	Knowledge and information exchange	Interaction and innovation among stakeholders
Role of policy	Resource allocation, priority setting	Linking research, extension, and education	Enabling innovation
Nature of capacity strengthening	Strengthening infrastructure and human resources	Strengthening communication between actors in rural areas	Strengthening interactions between all actors; creating an enabling environment

Source: World Bank 2006.

Box O.5 Recent Reforms in Public Agricultural Research and Extension

- Increasing the participation of farmers, the private sector, and other stakeholders in research governing boards and advisory panels to attain real influence over research decisions and priorities. The participation of women farmers is particularly important, given their crucial role in rural production systems, the special constraints under which they operate (for example, time constraints), and their range of activities and enterprises, including marketing, processing, and food storage.
- Decentralizing research to bring scientists closer to clients and better focus research on local problems and opportunities.
- Decentralizing extension services to improve accountability to local users and facilitate clients' "purchase" of research services and products that respond better to their needs. Matching-grant programs for farmer and community groups allow them to test and disseminate new technologies.
- Establishing competitive funding mechanisms that involve key stakeholders, especially users, in promoting demand-driven research, setting priorities, formulating projects, and screening pro-

posals. Competitive funds have increased the role of universities in agricultural R&D in some countries. Continuing challenges include limited engagement with the private sector, sustainability of funding, the bias against strategic R&D, and the heavy transaction costs.
- Promoting producer organizations to reach economies of scale in services and market activities, increase farmers' ability to demand better services, and help producers to hold service providers accountable.
- Mixing public and private systems by enabling farmer organizations, NGOs, and public agencies to outsource advisory services, identify the "best fit" for the particular job, and recognize the private-good attributes of some extension services. For example, approaches based on public funding that involve local governments, the private sector, NGOs, and producer organizations in extension service delivery may be most relevant to subsistence farmers, whereas various forms of private cofinancing may be appropriate for commercial agriculture, extending to full privatization for some services.

Sources: World Bank 2005, 2007b, 2010.

India, and South Africa hold huge potential for increased South-South cooperation, especially given the number of smaller developing countries that lag behind these agricultural research giants. These realities, among others, have impelled the Consultative Group on International Agricultural Research (CGIAR)—a global partnership of organizations that fund and conduct research for sustainable agricultural development—to examine and revise its approach to agricultural R&D through a change management process initiated in 2008.

Besides giving high priority to effectiveness, accountability, cost-effectiveness, and staff quality, research supported by the CGIAR will be based on the development of results-oriented research agendas directed toward significant and compelling challenges.[1] The CGIAR will give particular attention to enabling effective partnerships, because the complexity of scientific advances, socioeconomic developments, and environmental impacts, along with the higher costs associated with new lines of research, make partnerships essential for producing and delivering international public goods in agriculture. The CGIAR's contribution to agricultural development through research and knowledge management must be integrated with the wider development goals and activities of other actors, notably countries, international and regional development organizations, multilateral organizations, advanced research institutes, and the private sector.

Agricultural extension and advisory services

Like R&D, agricultural extension and advisory services have passed through similar cycles of challenge and reform. The public services that dominate extension services are plagued by widespread problems: limited funding, insufficient technology, poorly trained staff, weak links to research, and limited farmer participation (World Bank 2005). Because previous approaches have been ineffective, most extension programs are moving away from centralized systems and trying to improve links with research and farmers (World Bank 2007b). Most programs widely acknowledge the need to build social capital among farmers, pay greater attention to the needs of women and youth, and facilitate better links to markets.

Despite widespread agreement on the need for change, it is clear that no single extension model is universally relevant. New models need to be developed, based not only on general principles but also on analyses of the specific farming systems and social conditions they are expected to address. New models are more important than ever, because extension services are shifting their focus and changing their roles to improve service provision and act as brokers to the more diverse set of clients seen in an AIS.

The role of information and communications technology (ICT) in producing and disseminating knowledge has expanded exponentially. ICTs offer striking opportunities to change how agricultural science, innovation, and development occur by enabling a variety of stakeholders to interact and collaborate in new ways to enhance the innovation process (box O.6).

Agricultural education and training

Education and training institutions are especially significant in an AIS because they develop human resources and at the same time serve as a source of knowledge and technology. The absence or decline of these institutions leaves a large gap in a country's innovation capacity. Even so, government and donor investments in agricultural education and training (AET) have dropped to almost nothing since the early 1990s (World Bank 2008).

For AET, the primary constraint (among many) is that institutions have not kept pace with the labor market's demand for knowledge and practical competencies, especially in agribusiness, business and program management, and the problem-solving and interpersonal skills crucial for actors to function in an AIS. Despite this poor performance, global experience shows that it is possible to build productive and financially sustainable education systems (World Bank 2007b). Besides the AET system in a number of developed countries (Denmark, Japan, the Netherlands, and the United States), developing countries such as India, Malaysia, Brazil, and the Philippines have established productive AET systems.

LIMITATIONS OF CURRENT INVESTMENTS FOR INCREASING INNOVATION IN AGRICULTURE

As shown in box O.4, investments in science and technology have been a steady component of most strategies to improve and maintain agricultural productivity. The high returns and pro-poor growth emerging from investments in public agricultural research, advisory services, and education reflect a growing spectrum of initiatives to improve the response to clients' demands, work with farmer groups, communicate better with partners, and collaborate with the private sector. Yet efforts to strengthen research systems and increase the availability of knowledge have not necessarily increased innovation or the use of knowledge in agriculture (Rajalahti, Woelcke, and Pehu 2005). As noted, complementary investments are needed to build the capacity for

Box O.6   The Role of Information and Communications Technology in Knowledge Exchange and Innovation

For innovation to take place, effective bridging mechanisms are often needed to facilitate communication, translation, and mediation across the boundaries among the various actors in agricultural research and development and between knowledge and action. Such facilitating and bridging mechanisms can include diverse innovation coordination mechanisms such as networks, associations, and extension services, but also ICT.

ICTs offer the opportunity to improve knowledge flows among knowledge producers, disseminators, and users and, for example, among network partners; support the opening up of the research process to interaction and more accessible knowledge use; and more cost-effectively widen the participation of stakeholders in the innovation and governance process. ICTs have more often been associated with providing advanced services to number crunching and data management, geospatial applications, knowledge-based systems and robotics, and improved farm equipment and processes, but less often been considered for connecting diverse innovation communities—whether at the local, sub-sectoral, and national level.

ICTs that serve as information "collectors," "analyzers," "sharers," and "disseminators" are already positively affecting agricultural interventions in developing countries. Affordable mobile applications, in particular, provide linkages to previously isolated actors: information on prices, good farming practices, soil fertility, pest or disease outbreaks, and extreme weather has expanded farmers' opportunities to capitalize on markets, react to unfavorable agricultural conditions more effectively, and better interact with public service agents.

Satellite imagery and aerial photography have increased the capacity of scientists, researchers, and even insurance providers to study farm conditions in remote areas and assess damage from climatic challenges like drought. Increasingly affordable technologies like radio frequency identification tags and other wireless devices are improving livestock management, allowing producers to monitor animal health and trace animal products through the supply chain. A persistent barrier to innovation, the lack of rural finance, is also lifted by digital tools.

Sources: Author, drawing on Manning-Thomas 2009; World Bank 2011.

innovation across the spectrum of actors in the AIS and to develop an enabling environment for innovation to occur.

This sourcebook reviews and assesses experiences with those complementary investments. It outlines the needs, opportunities, and priorities for such investments and offers specific tools and guidance to develop interventions in different contexts. As emphasized in the next section—which offers more detail on the sourcebook's contents and organization—this sourcebook reflects work in progress and an evolving knowledge base. The emerging principles it contains will change as practitioners learn and develop creative new approaches to innovation for agricultural development.

SOURCEBOOK MODULES

The content of this sourcebook is presented in line with the project cycle or phased approach that practitioners use (table O.2). Modules 1 through 4 discuss the main investments related to innovation capacity (coordination and

organization of stakeholders, agricultural education and training, and research and advisory services). Module 5 is concerned with the incentives and resources needed for innovative partnerships and business development, and module 6 describes complementary investments that create a supportive environment for innovation. Module 7 provides information on assessing the AIS and identifying and prioritizing prospective investments, based partly on what has been learned from monitoring and evaluating similar efforts. A glossary defines a range of terms related to agriculture, innovation, and development.

Each module generally has four parts:

1. *The module overview* introduces the theme (a particular area of investment), summarizes the major issues and investment options, and points readers to more detailed discussions and examples in the thematic notes and innovative activity profiles that follow the overview. The overview provides substantive contextual information for each topic, including lessons from earlier approaches in

Table O.2 Sourcebook Modules	
Module number	**Title**
	Sourcebook Overview and User Guide
1	Coordination and Collective Action for Agricultural Innovation
2	Agricultural Education and Training to Support Agricultural Innovation Systems
3	Investment in Extension and Advisory Services as Part of Agricultural Innovation Systems
4	Agricultural Research within an Agricultural Innovation System
5	Incentives and Resources for Innovation Partnerships and Business Development
6	Creating an Enabling Environment for Agricultural Innovation
7	Assessing, Prioritizing, Monitoring, and Evaluating Investments in Agricultural Innovation Systems

Source: Author.

national agricultural research systems and agricultural knowledge and information systems.

2. *Thematic notes* discuss technical and practical aspects of specific investment approaches and programs that have been tested and can be recommended (sometimes with provisos) for implementation and scaling up. The notes review the considerations, organizing principles, questions, performance indicators, and lessons that would guide the design and implementation of similar approaches or programs.

3. *Innovative activity profiles* describe the design and highlight innovative features of recent projects and activities related to the area of investment described in the module. The profiles pay close attention to features that contributed to success and that technical experts can adapt for their own operations. The activities and projects described here have not yet been sufficiently evaluated to be considered "good practice" in a range of settings, but they should be monitored closely for potential scaling up. Their purpose is to ignite the imagination of task managers and technical experts by providing possibilities to explore and adapt in projects.

4. *References and further reading* offer resources and additional information.

THEMES COVERED IN THE MODULES

Each sourcebook module covers a theme related to assessing and designing investments in a particular area integral to the AIS. The discussion that follows gives readers a broad idea of the content and concerns of the modules. The nonagricultural and cross-cutting issues treated in each module are presented as well.

Building the capacity to innovate (Modules 1–4)

For an innovation system to be effective, the capacity of its diverse actors must be built and strengthened; many actors will increasingly possess a special mix of skills that contribute to the AIS in particular ways. Stronger technical skills are very important, but they must be complemented with functional expertise, because the new ways of working within an AIS require a range of skills: scientific, technical, managerial, and entrepreneurial skills and skills and routines related to partnering, negotiating, building consensus, and learning.

COORDINATION AND COLLECTIVE ACTION FOR AGRICULTURAL INNOVATION (MODULE 1). Coordination and organization of stakeholders may serve many purposes, such as building coherence and setting consensus-based priorities, strengthening the sharing of knowledge and resources, strengthening collaboration through joint processes and products, and reducing transaction costs and reaching economies of scale in extension and market activities. Without organizations (or brokers) to address social and resource imbalances and transaction costs, prospects for participating in innovation processes and systems are limited, especially for poor people. Effective platforms help to organize stakeholders with different assets, knowledge, and experience.

Module 1 discusses the capacities and resources required to organize and coordinate stakeholders, providing examples and lessons from previous efforts. The corresponding areas of investment include innovation coordination bodies (which can be national, multisectoral, or specific to the agriculture sector), subsector or industry associations or networks, producer organizations, productive alliances, and self-help groups to foster innovation. A range of policies and institutions is also needed to support coordination and collective action at different levels of governance in the AIS.

AGRICULTURAL EDUCATION AND TRAINING TO SUPPORT AIS (MODULE 2). Agricultural innovation is a product of the capacity, resources, and interactions that are brought to bear by actors from the wide range of fields related to food and agriculture. The capacity to generate new ideas, knowledge, technologies, processes, and forms of collaboration depends on an extensive array of skills—not only the expected technical, fiscal, and managerial competencies but also complementary skills in such areas as entrepreneurship, facilitation, conflict resolution, communications, contractual arrangements, and intellectual property rights. Universities, research institutes, and other learning institutions will have to reposition themselves to acquire and inculcate these skills. The critical functions of research, teaching, extension, and commercialization must be far more closely integrated. Module 2 reviews approaches to reorienting agricultural education and training to better serve the needs of a diverse cadre of AIS actors. The examples and lessons describe long-term reform processes, curriculum reform, technician training approaches, as well as on-the-job training.

INVESTMENT IN EXTENSION AND ADVISORY SERVICES AS PART OF AIS (MODULE 3). Better knowledge sharing and better use of available information and knowledge for desired changes are at the center of innovation processes. Extension and advisory services can become nodes for exchanging information and services that help to put knowledge to use; then they become well positioned to facilitate and support multistakeholder processes. Module 3 describes key principles for developing demand-driven, pluralistic advisory services (including the technical, entrepreneurial, and organizational aspects of this process) capable of supporting the heterogeneous client base of an AIS. The module reviews investments to support the shifting roles of advisory systems within an AIS, including the provision of market-centered advisory services and brokering services.

AGRICULTURAL RESEARCH WITHIN AN AIS (MODULE 4). Agricultural research systems are the source of new knowledge and the resulting products, services, and management practices that enable productivity to grow. Locating agricultural research within an AIS is a means of heightening the performance of research systems through improved articulation with demand, more effective, better-differentiated international and national institutional partnerships, and better market integration. Module 4 provides guidance on how research systems in three innovation contexts (agriculture-based, transitional, and urbanized economies) can function more effectively. The module reviews investments in mechanisms for articulating

demand and developing interfaces with other actors (for example, through codesign, innovation platforms, alliances and consortia, and technology transfer and commercialization), giving special attention to the potential for public-private partnerships and regional innovation systems. The module also provides considerable detail on the incentives and instruments needed for research system reforms to succeed.

Incentives and resources for innovative partnerships and business development (Module 5)

Economic change entails the transformation of knowledge into goods and services through partnerships and business enterprises. Strong links between knowledge and business development are a good indication of the vitality of an AIS. Partnerships for business development often require appropriate incentives to create such links, particularly to engage the private sector in R&D, technology transfer, and joint business activities (with producers). Module 5 provides lessons and examples of many potential interventions that promote private sector contributions to innovation either through service provision, technology commercialization, or other business-related innovation (through business support). The module describes key instruments for supporting technology commercialization (the establishment of technology transfer offices, incubators, and science parks); for supporting business (the provision of innovation funds, risk capital, and other resources to initiate and sustain novel partnerships); and for forming clusters, which enable stakeholders from a particular subsector or value chain to benefit from economies of scale, geographic proximity, and complementary public investments.

Creating an enabling environment for agricultural innovation (Module 6)

Farmers and entrepreneurs will not take the risk of innovating in unfavorable conditions. Researchers will not engage in long-term activities that are not aligned with the regulatory system (for example, researchers will not develop innovative plant-breeding processes if they cannot protect the resulting intellectual property). In many instances, innovation and business development do not occur without complementary investments to create a supportive environment. Module 6 discusses the roles of innovation policy and governance mechanisms, regulatory frameworks (for quality and safety, intellectual property, and biosafety), and market development in fostering agricultural innovation. It also describes investments (in infrastructure or financial services, for example) that have synergistic effects with other instruments such as innovation funds. Given the resource limitations and numerous choices, investments in an enabling environment must be prioritized and sequenced with great care.

Assessing, prioritizing, monitoring, and evaluating investments in agricultural innovation systems (Module 7)

AIS investments must be specific to the context and respond to the stage of development in a particular country and agriculture sector, especially the AIS. Given that optimal human and financial resources are rarely available, an incremental approach is advisable. The scale of operations is also likely to vary from local or zonal to subsectoral or national. This variation requires investments to be assessed, prioritized, sequenced, and tailored to the needs, challenges, and resources that are present.

The identification and design of appropriate interventions begin with a good understanding of the level of development and the strengths and weaknesses of the AIS. The status of an AIS and its critical needs can be assessed in several ways and at several levels. Module 7 reviews and provides lessons and examples of tools for assessing AISs, such as AIS frameworks, organizational assessments, NetMap, and benchmarking.

Investments are prioritized based on the needs that are identified, but setting priorities is also a political process involving negotiation to build a consensus. Scoping and consultation can help stakeholders to develop a shared perspective on goals and challenges and to identify specific needs, opportunities, and priority interventions. These processes ideally engage a diverse group of stakeholders from within and outside the sector through platforms, committees, alliances, and scenario and foresight exercises.

Monitoring and evaluation (M&E) of an AIS are essential for assessment and prioritization, as they allow insights on impact and change. As technological, institutional, policy, and other innovations arise through interactions between networks of stakeholders in an innovation process, M&E should encompass quantifiable assessments (of economic benefits, productivity increases, and so on) and nonquantifiable assessments (of learning by doing, institutional reorganization, capacity building, and so on).

This module reviews traditional and other methods that help practitioners to assess and understand learning processes, institutional change, changes in capacities, and other outcomes and to include the various stakeholders in the M&E process.

Nonagricultural and cross-cutting issues

Although the sourcebook focuses on innovation in agriculture, it draws on experience and lessons from other sectors, not least because so many "nonagricultural" issues impinge on agriculture and innovation. Such issues include rural finance, business development, innovation policies, and the governance of innovation, among others.

The sourcebook addresses three major cross-cutting themes—the role of the public and private sector, climate change, and gender—as appropriate (and when examples have been identified). These issues are briefly introduced in box O.7.

THE SOURCEBOOK AS A LIVING DOCUMENT

To the extent possible, the modules in this sourcebook reflect current knowledge and guidance for investments to support innovation systems in agriculture. Their content is based on the expert judgment of the authors and thematic specialists, as well as reviews by experienced specialists. Yet important gaps in knowledge remain, and new knowledge will emerge from approaches that are just now being devised and tested. For example, impact assessment methods and good M&E practices for an AIS are two areas in which much more knowledge is needed. Future iterations of this sourcebook will also benefit from additional examples of integrated AIS investments to strengthen innovation capacity in related areas such as education, research, advisory services, and brokering, among others.

This sourcebook is intended to be a living document that remains open to dialogue and new, imaginative approaches to innovation for agricultural development. Its primary home is not on the bookshelf but online (www.worldbank.org/ard/ais), where it will be updated and expanded as new experience is gained and new approaches and initiatives arise. The authors strongly encourage readers to update, verify, and offer feedback on the information here. Readers are encouraged to adapt key principles and relevant guidelines to their individual agricultural projects and programs—and to share the results widely.

Box O.7   Cross-Cutting Themes Addressed in This Sourcebook

The role of the public and private sector. The public sector is expected to remain an important provider and/or funder of R&D, education, and extension services in developing countries, where 94 percent of the investment in agricultural R&D still comes from public coffers (World Bank 2007b). Yet if markets now drive much of the agenda for agricultural R&D and new actors are more prominent in agriculture, what is the proper role of the public sector? Each module examines the roles of the public and private sector with a view to answering that question.

Climate change and green growth. Climate change adaptation and mitigation are key goals of an agricultural knowledge system. Technical as well as organizational innovations (for example, the use of climate-smart and green technologies, coupled with inclusive and effec-tive approaches on knowledge dissemination and adoption) are required to identify and develop appropriate solutions that contribute to adaptation, mitigation, and green growth. The modules describe institutional approaches that lend themselves well to generating and adopting climate-smart solutions.

Gender. The AIS approach argues that diversity, inclusion, and participatory approaches are critical to building the quality of social capital needed for resilient and sustainable innovation systems. It takes into account the many actors along the value chain; diverse organizational forms to facilitate education, research, and extension systems; and the practices, attitudes, and policies that frame agricultural production and trade. Every module addresses gender issues through examples and/or policy interventions.

NOTE

1. In contrast to the CGIAR's origins in the 1960s and 1970s as a mechanism for funding research divided largely along commodity and geographic lines. For more information on the change management process and on how the CGIAR has changed as agriculture, approaches to R&D, and approaches to funding R&D have changed, see www .cgiar.org.

REFERENCES AND FURTHER READING

Beintema, N. M., and H. Elliott. 2009. "Setting Meaningful Investment Targets in Agricultural Research and Development: Challenges, Opportunities, and Fiscal Realities." Paper presented at the Food and Agriculture Organization Expert Meeting "How to Feed the World in 2050," Rome, June 24–26. ftp://ftp.fao.org/docrep/fao/012/ak978e/ak978e00.pdf.

Bernet, T., G. Thiele, and T. Zschocke. 2006. *Participatory Market Chain Approach (PMCA): User Guide.* Lima, Peru: CIP-Papa Andina. http://papandina.cip.cgiar.org/fileadmin/PMCA/User-Guide.pdf.

Freeman, C. 1987. *Technology Policy and Economic Performance: Lessons from Japan.* London: Frances Pinter.

Hall, A., N. Clark, and G. Naik. 2007. "Technology Supply Chain or Innovation Capacity? Contrasting Experiences of Promoting Small-Scale Irrigation Technology in South Asia." UNU Merit Working Paper 2007-014, United Nations University, New York; Maastricht Economic and Social Research and Training Centre on Innovation and Technology, Maastricht. www.merit .unu.edu/publications/wppdf/2007/wp2007-014.pdf.

ISNAR (International Service for National Agricultural Research). 1992. "Service through Partnership: ISNAR's Strategy for the 1990s." The Hague: ISNAR.

Lundvall, B.-Å., ed. 1992. *National Systems of Innovation: Towards a Theory of Innovation and Interactive Learning.* London: Frances Pinter.

Manning-Thomas, N. 2009. "Changing the Emperor: ICT-Enabled Collaboration Transforming Agricultural Science, Research, and Technology into an Effective Participatory and Innovations System Approach." Technical Centre for Agricultural and Rural Cooperation of the ACP Group of States and the European Union. www.web2fordev.net/component/content/article/1-latest-news/70-changing-the-emperor.

Mytelka, L. K. 2000. "Local Systems of Innovation in a Globalised World Economy." *Industry and Innovation* 7 (1): 15–32.

OECD (Organisation for Economic Co-operation and Development). 2009. "Growing Prosperity. Agriculture, Economic Renewal, and Development." Draft Outcome Document from the Experts Meeting "Innovating Out of Poverty," OECD, Paris, April 6–7.

Pardey, P. G., N. M. Beintema, S. Dehmer, and S. Wood. 2006. *Agricultural Research: A Growing Global Divide?*

IPFRI Food Policy Report 17. Washington, DC: International Food Policy Research Institute.

Rajalahti, R., W. Janssen, and E. Pehu. 2008. *Agricultural Innovation Systems: From Diagnostics to Operational Practices.* Agriculture and Rural Development Discussion Paper 38. Washington, DC: World Bank.

Rajalahti, R., J. Woelcke, and E. Pehu. 2005. *Monitoring and Evaluation for World Bank Agricultural Research and Extension Projects: A Good Practice Note.* Agriculture and Rural Development Discussion Paper 20. Washington, DC: World Bank.

Rivera, William M., Gary Alex, James Hanson, and Regina Birner. 2006. "Enabling Agriculture: The Evolution and Promise of Agricultural Knowledge Frameworks." Association for International Agricultural and Extension Education (AIAEE) 22nd Annual Conference Proceedings (unpublished), Clearwater Beach, FL.

Rygnestad, H., R. Rajalahti, K. Khanna, and E. Pehu. 2007. "Review of Agricultural Innovation System (AIS) Supporting Investments in the World Bank's Agricultural Knowledge and Information Systems (AKIS) FY90–06 Portfolio." Unpublished Agriculture and Rural Development Department Portfolio Review, World Bank, Washington, DC.

Spielman, D., and R. Birner. 2008. *How Innovative Is Your Agriculture? Using Innovation Indicators and Benchmarks to Strengthen National Agricultural Innovation Systems.* Agriculture and Rural Development Discussion Paper 41. Washington, DC: World Bank.

UNIDO (United Nations Industrial Development Organization). 2008. "Innovation Systems in Practice, Charting a New Course for UNIDO." UNIDO Policy Position Paper. UNIDO, Vienna.

World Bank. 2005. *Agriculture Investment Sourcebook.* Washington, DC: World Bank.

———. 2006. "Enhancing Agricultural Innovation: How to Go Beyond the Strengthening of Research Systems." World Bank, Washington, DC.

———. 2007a. "Cultivating Knowledge and Skills to Grow African Agriculture: A Synthesis of an Institutional, Regional, and International Review." World Bank, Washington, DC.

———. 2007b. *World Development Report 2008: Agriculture for Development.* Washington, DC: World Bank.

———. 2008. Annotated AIS Database FY90–08. Internal document. World Bank, Washington, DC.

———. 2009a. *Gender in Agriculture Sourcebook.* Washington, DC: World Bank.

———. 2009b. "Implementing Agriculture for Development: World Bank Group Agriculture Action Plan, FY2010–2012." World Bank, Washington, DC.

———. 2010. "Designing and Implementing Agricultural Innovation Funds: Lessons from Competitive and Matching Grant Projects." World Bank, Washington, DC.

———. 2011. *Information and Communication Technologies for Agriculture e-Sourcebook.* Washington, DC: World Bank.

Coordination and Collective Action for Agricultural Innovation

OVERVIEW

Javier Ekboir, Institutional Learning and Change Initiative (ILAC)
Riikka Rajalahti, World Bank

EXECUTIVE SUMMARY

Actors in an agricultural innovation system (AIS) innovate not in isolation, but through interacting with other actors—farmers, firms, farmer organizations, researchers, financial institutions, and public organizations—and the socioeconomic environment. In other words, agricultural innovation is an organizational phenomenon influenced by individual and collective behaviors, capabilities for innovation, and enabling conditions. Interaction, coordination, and collective action are based above all on the actors' capacity to identify opportunities for innovation, assess the challenges involved, and access the social, human, and capital resources required for innovating, learning, and sharing information. Better coordination can improve the design and implementation of innovation policies by allowing more actors to voice their needs and concerns, resulting in more inclusive policies and faster diffusion of innovations. Stronger interaction and coordination can also induce all actors in an innovation system, particularly public research and extension organizations, to be more aware of and responsive to the needs and concerns of other actors, especially resource-poor households. Despite such advantages, interaction and coordination have been difficult to achieve for the same reasons that hinder collective action: opportunistic behavior; lack of trust, incentives, and capacity; and difficulties in setting and enforcing rules. Interaction and coordination in the innovation systems of developing countries are hampered as well by segmented markets, different technological regimes, lack of collaboration cultures, inappropriate incentives, weak channels of communication, and insufficient innovation capabilities.

Effective interaction, coordination, and collective action are based on existing capabilities, appropriate incentives, and the empowerment of individuals; thus they rely on voluntary action. Coordination and interaction can emerge spontaneously or be induced by specific public or private programs. Effective coordination requires (1) a committed and capable leadership; (2) appropriate incentives; (3) an enabling environment, in which important stakeholders that coordinate their activities have the mandate, culture, and freedom to participate; (4) stable support programs; (5) efforts to strengthen the capabilities for innovation and collective action; and (6) adaptation of public organizations to participate more effectively in innovation processes. The organizational innovations (committees or councils, platforms or networks, and diverse associations) reviewed in this module show that many innovations are not planned in detail beforehand; instead, they result from the adaptation of organizational structures in response to emerging problems or opportunities. They also show that creative and committed individuals guide the adaptation and that an enabling environment allows the organizations to change.

RATIONALE FOR INVESTING IN COORDINATION AND COLLECTIVE ACTION FOR INNOVATION

Many different actors in an AIS generate and disseminate innovations, including farmers, firms, farmer organizations, researchers, financial institutions, and public organizations. These actors do not innovate in isolation; rather, they innovate through interacting with other actors and the socioeconomic environment. Their interactions take different forms, such as consultations to define innovation policies, joint research activities, or participation in or facilitation of innovation networks and value chains (box 1.1). Agricultural innovation is increasingly recognized as an organizational phenomenon influenced by individual and collective behaviors (World Bank 2006). These behaviors, in turn, depend on the individual and collective capabilities possessed by the actors, on culture,[1] incentives, routines, and the environment (Ekboir et al. 2009).

The importance of interaction, coordination, and collective action in innovation systems has been recognized for more than two decades (Freeman 1987; Lundvall 1992; Nelson 1993). Common reasons for AIS actors to interact and address issues collectively include improved identification of opportunities for and challenges with innovation; leveraging of human and capital resources; learning and information sharing; and (obviously) economic and/or social benefits. Interaction and coordination may also improve the design and implementation of innovation

policies by allowing more actors to voice their needs and concerns, resulting in more inclusive policies and faster diffusion of innovations.[2] Stronger interaction and coordination also induce all actors, especially public research and extension organizations, to be more aware of and responsive to the needs and concerns of other actors, especially resource-poor households. Box 1.2 illustrates the benefits arising from one type of interaction: learning alliances (see also IAP 4 in module 4). The more general economic, social, and environmental benefits of investing in coordination, interaction, and collective action for agricultural innovation and in building organizational capabilities are summarized in the sections that follow.

Fostering economic growth

Better-connected actors with stronger innovation capabilities help to solve coordination problems among potential partners, build trust for collaboration, build up innovation capabilities, and develop a better understanding of the needs and capabilities of other actors in the AIS, especially marginalized groups. Other economic benefits of coordinating more capable actors include the following:

- Lowering the cost of searching for and using technical and commercial information, easing the identification of emerging technical, social, and economic needs and opportunities, facilitating experimentation on alternative

---

Box 1.1 Main Terms Used in This Module

Innovation network. A diverse group of actors that voluntarily contribute knowledge and other resources (such as money, equipment, and land) to jointly develop or improve a social or economic process or product. These networks are also known as *innovation platforms.*

Farmer organization. An organization integrated only by farmers seeking solutions to production or commercial problems.

Value chain. The set of linked activities conducted by the different actors that a firm organizes to produce and market a product.

Organization. A group of actors that collaborate over a sustained period. An organization can be either

formal or informal. Collaboration may take different forms, including frequent exchanges of information, joint priority setting for policies and programs, and joint implementation of innovation projects.

Innovation brokers. Teams of specialists that combine a strong background in science with knowledge of business and commercialization and/or the creation of innovation networks. Innovation brokers are also known as *change agents* or *technology brokers.*

Innovation capabilities. The skills to build and integrate internal and external resources to address problems or take advantage of opportunities.

Source: Author.

Box 1.2 Role of Learning Alliances in Enhancing Interaction and Improving Innovation Capabilities in Central America

A learning alliance is a process-driven approach that facilitates the development of shared knowledge among different actors. Learning alliances contribute to improved development outcomes because lessons are identified and learned more quickly and because stronger links among research organizations and other actors in the AIS improve the focus on research and development practices.

Since 2003, international and local nongovernmental organizations (NGOs), a national university, the International Center for Tropical Agriculture (CIAT), and the International Development Research Centre (IDRC) have come together to explore how learning alliances can improve links between research and development actors. The learning alliance has worked with 25 agencies as direct partners, and through their networks it has influenced 116 additional organizations. Over the years, the alliance fostered change in

organizations working with 33,000 rural families in El Salvador, Guatemala, Honduras, and Nicaragua by significantly changing the partners' attitudes and practices. As they became better connected, organizations working on similar topics improved their access to information and knowledge of rural enterprise development and their access to improved methods and tools. Attitudes shifted from competition to collaboration. Partners experienced how working together enhanced their capacity to serve rural communities and attract donor funding. The increased effectiveness of the partners' projects and the development of more strategic new projects indicate how development practices and knowledge management improved. These shifts have contributed to a more efficient innovation system, as seen in the shared use and generation of information, joint capacity-building programs, and large-scale, collaborative projects.

Source: Author, based on CIAT 2010.
Note: CIAT = Centro Internacional de Agricultura Tropical.

solutions, opening market opportunities, and developing competitive capabilities.

- Integrating more effectively into innovation networks.
- Developing new skills and more effectively using human, social, physical, and financial resources, thus fostering economic growth.
- Participating in the development and diffusion of innovations, including action-research projects and new approaches to extension.

Producing public goods

Individuals and organizations with facilitation and/or brokering skills (in other words, with the capacity to promote interaction) produce three important public goods. First, by linking public, private, and nonprofit actors, they facilitate the identification of emerging trends and improve policy dialogues and the design and implementation of innovation and agricultural policies at the global and domestic level. Second, by interacting more actively with researchers, they help researchers to generate more relevant scientific information. Third, by interacting with extension organizations, they can help extension services to become more receptive

to the needs of nonpublic actors and help them to use technical and commercial information, thus strengthening their innovation capabilities.

Reducing poverty

Poor households usually have limited human, social, physical, and financial resources (Neven et al. 2009). Individuals and organizations with facilitation and/or brokering skills can help these households to pool their limited resources among themselves or with other actors (for example, NGOs or supermarkets) to achieve economies of scale, enter new markets, or access new resources, such as technical information or credit (World Bank 2006). Although the direct impact of farmer organizations on poverty seems relatively modest (see TNs 2 and 4), organizations can have important indirect effects on poverty by fostering economic growth, creating employment, preventing buyers from benefiting at the expense of suppliers, building innovation capabilities, and protecting marginal groups (such as women or landless farmers) from further marginalization (for example, see IAPs 4 and 6). They can also negotiate with authorities on behalf of their members, increasing the public resources invested in poverty alleviation and affirmative action programs.

Improving environmental outcomes

Innovations that improve the sustainable use of natural resources are usually developed and diffused by networks with a diverse set of partners and capabilities (IAP 1). Their diversity facilitates access to a large pool of technical information, the implementation of participatory and action-research programs, effective diffusion activities (such as farmer-to-farmer extension), and collective action for the management of common resources. Individuals and organizations with stronger capabilities and facilitation and/or brokering skills can also help to articulate environmental demands that are often excluded from national policies. These demands are often linked to poverty alleviation, because poor households tend to live in unfavorable and fragile environments.

CONTENT OF THIS MODULE

After looking in detail at past experiences with coordinating and fostering collective action for innovation (such as through formal and informal coordination agents and/or organizations), this module describes the conditions, instruments, and incentives for coordination and discusses ways of building innovation capabilities in groups of actors in an AIS. It also briefly reviews key policy issues; new directions, priorities, and indicators that can help to monitor progress and assess the results of investing in interaction and coordination; and the conditions and capabilities that improve the chances of success.

PAST EXPERIENCE

Despite their advantages, interaction and coordination have been difficult to achieve for the same reasons that hinder collective action: opportunistic behavior; lack of trust, incentives, and capacity; and difficulties in setting and enforcing rules. Interaction and coordination in the innovation systems of developing countries are also hampered by segmented markets, different technological regimes, lack of collaboration cultures, inappropriate incentives, weak channels of communication, and insufficient innovation capabilities.

Interacting in a sustained way has proven difficult, particularly among research and educational institutions, which have an important role to play in facilitating access to scientific information, generating information that is

needed but does not yet exist, and preparing professionals with strong skills in disciplines required by innovating actors.[3] To play this role, researchers must interact intensively with other actors in the innovation system. While in recent years public research institutes and universities in developing countries have been pressed to open up to the needs of other stakeholders in the innovation system, few have been able to adapt because they do not have the capabilities and incentives to interact with nonacademic agents, have weak research capabilities, and resist change (Ekboir et al. 2009; Davis, Ekboir, and Spielman 2008).[4,5] Similarly, most farmers and NGOs have weak links with public organizations and governments, including the traditional research, extension, and regulatory agencies as well as local authorities and financing organizations.

The sections that follow review the main elements of effective interaction, coordination, and collective action, starting with the conditions for effective interaction and coordination and the instruments and incentives needed. The main types of coordinating bodies and organizations are described, and the crucial need to build innovation and organizational capabilities in new or existing organizations is discussed.

Conditions that foster the effective interaction and coordination of actors for collective action and agricultural innovation

Interaction and coordination for collective action and agricultural innovation can be successful only if (1) the AIS actors perceive that the benefits of contributing to a common effort are bigger than the associated costs; (2) they have the appropriate human and social resources to participate in collective action; and (3) they feel that they can influence the processes in which they participate. Interaction and coordination may emerge *spontaneously*. An effective value chain, in which actors collaborate (often without formal contracts) to supply a particular product to a market, is an example of spontaneous coordination. Coordination may also arise from *deliberate interactions*, such as a multistakeholder forum where innovation policies are discussed.

Effective interaction, coordination, and collective action are based on existing capabilities and on appropriate incentives and empowerment of individuals, and thus they rely on *voluntary action*.[6] Because effective interactions and trust seldom emerge spontaneously, programs that support intermediaries and build innovation capabilities are often necessary to facilitate the process. Box 1.3 lists factors essential to effective interaction and coordination.

Every innovation is a new combination of resources, particularly ideas, skills, information, different types of capabilities, interorganizational learning and knowledge, and specialized assets. Organizational innovations are as important as product or process innovations.

Individuals or organizations may facilitate the exchange of resources and the coordination of actions to develop innovations. This process among heterogeneous actors may increase the diversity of resources and ideas that are available. The greater the variety of these factors available to innovators, the greater the scope for them to be combined in different ways, producing innovations that can be both more complex and more sophisticated. But variety is not enough for effective innovation. It is also necessary to have the incentives and capacity to search the pool of ideas and know how to combine them. In other words, incentives and innovation capabilities determine how actors innovate.

Effective coordination for innovation occurs when (1) a committed and capable leadership promotes the collaboration; (2) one organization offers appropriate (often new) positive and negative incentives to individuals from cooperating organizations (such as researchers or farmers); (3) important stakeholders that coordinate their activities have the mandate, culture, and freedom to participate; and (4) individuals participating in the collaboration do not change often (a relatively common problem with high-level civil servants).

Other factors are also essential to interaction and cooperation:

- Interactions and cooperation are not costless or easy to implement; therefore, they have to be managed with a clear view of the outcomes being sought and the associated costs.
- Interaction and collaboration thrive only if they are based on trust, which fosters greater commitment, more thorough knowledge sharing, and better conflict resolution.
- Motivated, capable, and autonomous facilitators are essential for inducing collaboration. Well-connected facilitators and collaborators pull promising new entrants into their networks and collaborate with a wide assortment of partners, exposing them to more experiences, different competencies, and added opportunities. In rapidly changing industries, facilitators lacking such connections fail to keep pace.

- Facilitators work effectively only when financing is sustainable and stable. Private actors will rarely pay for services that are initially difficult to define and whose real value can be determined only after the intervention finishes. For this reason, public funds should be made available to support coordinators of innovation processes and the implementation of institutional and organizational innovations.
- Inducing actors to alter their behaviors may require changes in laws and regulations. Given the uncertain nature of innovation processes, such changes should be introduced after they have been tried in pilot projects and after different actors (private firms, public research organizations, and relevant stakeholders such as the main ministries, regulators, and NGOs) have been involved in policy design, consultations, strategizing, and implementation. It is also important to invest in preparing high-quality information to support decision making, such as background studies commissioned from national or international think tanks and experts, sectoral dialogues between employers and employees' unions, and high-level steering groups (see module 6, TN 1 and TN 2).
- Interaction and coordination require adaptive management to help participating actors to change their behaviors as new actors join the informal organization and the innovation process matures. Adaptive management cannot be implemented well without monitoring and evaluation systems that focus more on processes than on outcomes.
- Many public research and higher education organizations in developing countries have incubator programs to foster the emergence of private firms and other organizations. Before expanding these programs, it is important to assess their capabilities to manage innovation programs and, if their capabilities are weak, to consider creating new, dedicated agencies, not necessarily within the public sector. Also consider divesting from obsolete schemes and institutions.
- "System failures" are prevalent, caused by weak incentives for collaboration, conservative organizational cultures, lack of trust among potential partners, regulations and programs that hamper interactions, and ineffective financing for innovation. By diminishing these failures, facilitators or collaborators promote collective action, the production of public goods for innovation, and the development of the innovations themselves.

Sources: Authors, based on Fountain 1999; Axelrod and Cohen 1999; Hakansson and Ford 2002; Fagerberg 2005; Powell and Grodal 2005; Klerkx and Leeuwis 2009; Klerkx, Aarts, and Leeuwis 2010.

Instruments and incentives for interaction, coordination, and organization of actors for agricultural innovation

Effective coordination and organization of actors for agricultural innovation can be supported with different instruments, including building capabilities for innovation (TN 4); joint priority setting or technology foresight exercises (TN 1), joint research and/or innovation programs, efforts to foster the emergence of innovation platforms (module 4, TNs 1 and 2) and value chains (TN 3), the creation of venture capital (module 5, TN 6), and support for establishing innovation brokers (module 3, TN 4). Innovation councils and advisory committees involving different ministries can coordinate policies, joint priority setting, and technology foresight exercises, which are often supplemented with temporary stakeholder consultation arrangements (see module 7, TN 3). Innovation forums and market and technology intelligence can create common visions among agents, thus fostering coordination.

If they do not provide proper *incentives* for organizations and especially individuals, coordination initiatives result only in formal interactions that have little effect on the AIS. Individuals respond to the incentives offered to them. When organizations do not introduce incentives to support external collaboration and coordination, their members (whether researchers, employees, or farmers) simply continue their normal activities. For example, it has been very difficult for research and education organizations in developing countries to participate in innovation processes.

Effective participation of nonacademic actors in the governing bodies of research organizations can induce research organizations to interact better with nonacademic agents. For this to happen, the nonacademic actors must have a good understanding of the dynamics of innovation and research (Ekboir et al. 2009). When the agricultural technology institutes are part of the agriculture ministry and the latter has little interaction and no formal links with the offices in charge of general scientific and innovation policies (research councils or other ministries), agricultural research and innovation policies can be divorced from other science and technology policies. This separation has two important consequences. First, the incentives offered to researchers often discourage interactions with farmers, especially when these incentives value indexed publications over participation in innovation processes. Second, incentives for innovation are often designed in ways that prevent farmers from using them. For example, it is common to offer tax incentives for private investments in research and innovation, but

for most farmers and small companies the tax deductions are too small to finance research or innovation projects.

At the macro level, policies can facilitate coordination by setting new incentives and rules by which agents operate. At other levels of the AIS, successful coordination of actors requires *innovation capabilities* (discussed later) and appropriate incentives, such as effective markets (which create opportunities to benefit from innovations) and innovation funds (see module 5, TN 2). At lower levels, actors in the AIS coordinate their actions in response to public incentives (such as social programs financed by local or international donors), market opportunities (such as high-value agriculture), or problems that affect whole communities (such as the management of natural resources). Programs to promote collaboration at the lowest levels of the AIS include funding innovation brokers (see module 3, TN 4), extension agents, and incubators; fostering the emergence of innovation networks (TN 2); providing resources for coordination activities (such as face-to-face meetings), and building actors' capabilities so that they can better search for and use technical and commercial information. Finally, because innovators and brokers cannot innovate in an unfavorable environment, physical and communications infrastructure should be developed, institutions must be strengthened, and regulations must be updated periodically to adapt to new technologies and market requirements (see module 6 on the enabling environment for AIS).

Types of coordinating bodies and organizations in the AIS

Aside from the traditional coordinating activities led by ministries of agriculture, a wide array of coordinating bodies contributes to agricultural innovation. Coordinating bodies and other forms of organizing actors (either individuals or organizations) are becoming increasingly important owing to the challenges imposed by globalization, emerging technologies, the increasing complexity of science, new forms of innovation, and global issues such as climate change, access to clean water, and poverty reduction. These bodies do not necessarily belong to the public sector but often have links to the highest levels of government (vertical coordination). Interaction and coordination increasingly are promoted and supported by different agents at the specific level in which they operate. Almost any agent can coordinate an innovation process if it has the personal, financial, and social resources to do so.

Over the years, formal and informal structures that coordinate actors within and between different levels of the

AIS have emerged in almost every country. Informal organizations (for example, innovation networks or value chains)[7] are increasingly recognized as important sources of innovation, because they complement and bond to formal organizations through a dense web of personal relations (Hakansson and Ford 2002; Robinson and Flora 2003; Christensen, Anthony, and Roth 2004; Vuylsteke and van Huylenbroeck 2008).

Formal organizations, such as a firm or a national research council, have a leadership structure defined in by-laws. Formal, dedicated structures with set agendas do not guarantee that actors are coordinated effectively, however. Informal organizations lack formal structures but possess all the other features of organizations, such as shared cultures and communication codes, governance structures, incentives, and routines. Informal organizations have coordinators or leaders (Simon 1981) who emerge from the organization's internal dynamics, the relative strength of the partners, and the socioeconomic environment. For example, individual smallholders negotiate with buyers from a disadvantaged position, but they may gain bargaining strength when they organize themselves, even if their organization does not have legal status. Without the organization, leadership in the value chain rests with the buyers; with the organization, farmers can better influence the chain (Hellin, Lundy, and Meijer 2009).

Interaction and coordination may be strengthened by creating new organizations or strengthening existing ones (see the discussion on building innovation capabilities). Both approaches have different challenges. Table 1.1 summarizes the main types of coordinating bodies and/or organizations that operate at different levels of the AIS. Given the diversity of coordinating bodies, organizations, instruments, and the roles they play[8] (discussed next), it is important that decision makers at the highest level, such as ministers or donors, avoid—to the extent possible—imposing particular approaches and allow the catalyzing actors to search for the best instruments for the particular processes they coordinate.

AT THE MACRO LEVEL (NATIONAL). Several countries have embarked on improving overall innovation system governance at the national level. Governance concerns the mechanisms by which decisions are made in an organization, whether public, private, or nonprofit. Governance has several dimensions, including power, culture, incentives, leadership, and coordination. In the governance of a national innovation system, special attention is given to the systems and practices for setting priorities and agendas, designing and implementing policies, and obtaining

knowledge about their impacts. Some developed countries (Finland and the Republic of Korea are examples) have set up national science and/or innovation councils at the highest possible level. These councils engage ministers (such as the prime minister and finance minister) and representatives from universities, public research organizations, and industry in developing policy guidelines and facilitate coherence, consensus building, and coordination throughout the system. Similarly, several countries (Chile, the Netherlands, South Africa, Thailand, and the United Kingdom are examples) have created science and technology councils or other organizations, consisting of public and private research organizations, private firms, funding agencies, sectoral organizations, and farmer organizations to proactively promote coordination in the innovation system.[9]

A thematic note (module 6, TN 2) discusses the role of innovation system governance in greater detail, offering examples and guidelines on actors, structures (such as councils), and capacities needed for good governance. Thailand's National Innovation Agency is discussed in box 1.4 as an example of the challenges, objectives, evolving functions, and learning associated with innovation system governance.

AT THE MESO LEVEL (SECTORAL, REGIONAL, OR PROVINCIAL). At the AIS level, governance has been improved by creating formal but effective spaces for dialogue at different levels of the AIS, building up the actors' organizations and their capabilities, and improving formal and informal regulations (intellectual property rights regimes, the judiciary system, customs, and markets) that reduce transaction costs associated with interaction and collaboration. These governance systems typically aim at improving participation of stakeholders from producers to consumers; improving transparency and openness associated with decision making on funding and priorities; improving responsiveness and accountability to stakeholders; facilitating consensus building and coherence of policies, strategies, and activities; and building a strategic vision (Hartwich, Alexaki, and Baptista 2007).

Most efforts to coordinate innovation at the higher levels of the agriculture sector have centered on establishing formal apex *research councils*. In several countries—examples include Australia, Brazil, Ghana, India (box 1.11 in TN 1), and Mexico—these formal research councils or forums assist the government in designing and implementing agricultural research policies, setting priorities and agendas, coordinating the division of labor; managing large parts of public budgets for research, monitoring and evaluating

Table 1.1 Examples of Coordinating Bodies, Potential Participants, and the Levels at Which They Operate

Structure	Actors that participate in the coordination	Type of coordination	Countries where it can be found	Examples of coordinating bodies
Macro (national) level				
Councils of ministers or advisory councils to the president or ministries (through policies and regulations)	Ministries, science councils, public bodies, private firms, farmers, farmer organizations, NGOs	Spontaneous when actors follow the rules; formal when ministries coordinate policy making	Canada, Finland, Rep. of Korea	Research and Innovation Council, Finland, http://www.aka.fi/en-gb/A/Science-in-society/Strategic-Centres-for-Science-Technology-and-Innovation/Background-to-CSTIs/; National Innovation Agency, Thailand, www.nia.org.th
Science and innovation councils	Public and private research, private firms, funding agencies, sectoral and farmer organizations	Formal and informal	Brazil, Chile, European Union, Finland, India, Mexico, South Africa, United States	Consejo Nacional de Ciencia y Tecnología, Mexico, http://www.conacyt.gob.mx/Paginas/default.aspx
Meso (sectoral, regional, or provincial) level				
Agricultural science/research councils	Mostly public, but the newer types with public, private, farmer, civil society participation	Formal and informal	Australia, Austria, Bangladesh, India, Netherlands, Norway	Indian Council of Agricultural Research, www.icar.org.in; Latin American and Caribbean Consortium to Support Cassava Research and Development; Australia Rural Research and Development Council, www.daff.gov.au/agriculture-food/innovation/council; Bioconnect, Netherlands, www.bioconnect.nl
Coordinating bodies (for several sectors or specialized in the agriculture sector)	Public and private research, private firms, funding agencies, sectoral and farmer organizations	Formal and informal	Argentina, Australia, Chile, Mexico, Netherlands, Thailand, United Kingdom	Asociación Argentina de Productores de Siembra Directa, www.aapresid.org.ar; Fundación para la Innovación Agraria, Chile, www.fia.cl; Medicinal Herbs Board/Association, India
Micro (farmer) level				
Innovation networks	Private firms, farmers, farmer organizations, NGOs, funding agencies, researchers	Informal	All countries	Papa Andina, International Potato Center, www.papandina.org; Bioconnect, Netherlands, www.bioconnect.nl
Value chains	Private firms, farmers, farmer organizations, NGOs	Informal	All countries	Numerous examples exist; no specific example is given here
Product marketing organizations	Private firms, farmers, farmer organizations	Formal	Colombia, Israel, Kenya, New Zealand	Federación Nacional de Cafeteros de Colombia, www.cafedecolombia.com; Fresh Produce Exporters Association of Kenya, www.fpeak.org
Public-private partnerships	Private firms, farmers, farmer organizations, research organizations	Formal	Almost all countries	Bioceres, Instituto Nacional de Tecnología Agropecuaria, Argentina; http://www.bioceres.com.ar/trigo_biointa/trigo_b_regalias.html
Innovation parks	Private firms, research organizations	Formal and informal	Almost all countries	Waikato Innovation Park, New Zealand, www.innovationwaikato.co.nz
Incubators	Private firms, farmers, farmer organizations, research organizations	Formal and informal	Almost all countries	International Crops Research Institute for the Semi-Arid Tropics, Agribusiness Incubator, India, www.agri-sciencepark.icrisat.org

Source: Authors.

Box 1.4 Thailand's National Innovation Agency

Thailand's National Innovation Agency (NIA), established in 2003, supports the development of innovations to enhance national competitiveness and gives significant attention to agriculture and other biological sciences. Operating under the overall policy guidance of the Ministry of Science and Technology, in 2009 NIA had a budget of about US$10.8 million. NIA is unusual in that it offers direct financial support to private companies for innovation-related projects. In 2009, it supported 98 "innovation projects" initiated by private companies. The agency essentially shares the investment risks associated with innovative, knowledge-driven businesses through technical and financial mechanisms. NIA's main strategies are the following:

- *Upgrade innovation capability,* with a focus on biobusiness, energy and environment, and design and branding. NIA encourages the development of startups and supports commercialization of research.
- *Promote innovation culture* within organizations of all types. NIA operates an innovation management course for executives, National Innovation Awards, an innovation ambassador scheme, an Innovation Acquisition Service, and a Technology Licensing Office.

- *Build up the national innovation system.* Although NIA is in an ideal position to propose measures to enhance policy coherence across ministries, its portfolio suggests that it focuses more on discrete and disguised subsidies for firm-level innovation.

One challenge is that NIA's definition of its role as "coordinating industrial clusters both at policy and operational levels, promoting innovation culture, and building up innovation systems, with a broader aim to transform Thailand into an innovation-driven economy" appears to overlap with the mandates of the newly established Office of Science, Technology, and Innovation (STI), the National Economic and Social Development Board, National Science and Technology Development Agency, and Office for SME Promotion. Of particular note is the government's assignment of the mandate to draft Thailand's 10-year science, technology, and innovation policy to the STI rather than the NIA. In 2010 the NIA was upgraded by government decree from a project within the Ministry of Science and Technology to a public organization. Its new board comprises representatives from key government agencies as well as the private sector, and it is currently chaired by the executive chairman of Bangkok Bank.

Sources: Wyn Ellis, personal communication; NIA, www.nia.or.th; Brimble and Doner 2007; Intarakumnerd, Chairatana, and Tangchitpiboon 2002.

---

research programs, and often coordinating and creating improved links between public agricultural research and extension organizations. Research may also be coordinated at the regional level (IAP 5 presents an example related to cassava). However, the effectiveness of these councils varies greatly. Many have a narrow representation of stakeholders, consisting primarily of ministerial representatives or researchers, and their research prioritization is not necessarily consultative or does not rely on rigorous evidence. They often have little influence on the policy process and how research is conducted. For further details, see TN 1.

Given the predominance of national innovation councils and agricultural research councils, there are few "true" *agricultural innovation councils,* mandated to coordinate and prioritize investments in agricultural innovation. Notable

exceptions include the Australia Agriculture and Rural Development Council (box 1.10 in TN 1) and Chile's Fundación para la Innovación Agraria (FIA, Agricultural Innovation Foundation; IAP 3).

As few countries have national councils specifically mandated to coordinate agents in the AIS, most agricultural innovation is coordinated at the subsectoral level via *product marketing companies and associations,* such as the Fresh Produce Exporters Association of Kenya (box 1.5) and the Colombian Coffee Growers' Federation, or *farmer-managed foundations,* such as Mexico's Produce Foundations (IAP 2).

Marketing or commodity boards were the first type of coordinating bodies to be created and have been common in both developed and developing countries for many

Box 1.5   Fresh Produce Exporters Association of Kenya: A Sectoral Coordinating Body

Horticulture is the fastest-growing agriculture subsector in Kenya, earning roughly US$1 billion in 2010. The Fresh Produce Exporters Association of Kenya (FPEAK), established in 1975, serves as the premier trade association representing growers, exporters, and service providers in the subsector (fresh cut flowers, fruits, and vegetables). FPEAK is part of a larger Kenya Horticulture Council formed in 2007 through a merger between the Kenya Flower Council and FPEAK. The Kenya Horticulture Council's role is to enhance the effectiveness and efficiency of resource use and service delivery to Kenya's horticultural industry.

FPEAK provides a focal and coordination point for the horticulture export industry. A recognized partner of the leading agricultural legislation, certification, and research bodies and development partners in Kenya, it provides technical and marketing information and training, acts as an information center, and runs active lobbying and advocacy programs to enhance the sector's competiveness.

FPEAK structure. FPEAK is registered as a company limited by guarantee. Its elected board of directors consists of members actively engaged in the export business. The FPEAK secretariat is responsible for administrative functions and providing services to members. FPEAK operates independently of and receives support from partners.

Who qualifies to be a member? To become an ordinary member, an exporter must have been in business for six months. Affiliate membership is open to firms and/or individuals serving the industry. These include airlines, consultants, certification bodies, input suppliers such as seed suppliers, packaging manufacturers, chemical companies, and clearing and forwarding firms.

What are FPEAK's strategic goals and activities?

- FPEAK has developed and implemented the protocol for Kenya Good Agricultural Practices (Kenya-GAP), against which growers can be audited and certified, a process that has involved wide stakeholder consultation.
- FPEAK's information service disseminates news on technical issues, trade, official regulations, and market requirements.
- FPEAK receives trade inquiries from overseas buyers and passes them on to members.
- FPEAK offers training programs by specialists. For example, farmers, including smallholders, are trained in GAP and standards compliance in partnership with exporters who are members of the association.
- FPEAK agronomists visit members' farms and production sites upon request to advise on readiness for compliance, in particular to Kenya-GAP and GLOBAL GAP (which sets voluntary standards for the certification of agricultural products).
- FPEAK coordinates members' participation in trade events and assesses emerging markets.

Source: FPEAK, www.fpeak.org.

---

decades. After the wave of deregulation in the 1980s, many developing countries abolished or privatized marketing boards, but several public marketing boards remain. There is no generally agreed definition of a commodity board. Usually they are formal bodies in which different actors involved in the production, transformation, and marketing of a product discuss issues of mutual interest and sometimes regulate their activities. They may also finance or implement supporting activities such as managing research (either in their own institutes or by contracting external researchers), implementing generic advertising campaigns, and proposing legislation related to a product. In some cases, they also regulate production. Commodity boards are a diverse group of organizations representing an array of goals, structures, and challenges. Examples include India's National Dairy Development Board (box 1.6), the New Zealand Dairy Board, Kenya Tea Board, Ghana Cocoa Board, and the Colombian Coffee Growers' Federation.

AT THE MICRO LEVEL (ORGANIZING FARMERS). At the micro level, farmers form local producer organizations or join private firms and other actors in innovation networks and value chains. Farmer organizations are joined only by producers who seek solutions to particular production or commercial problems (TN 4).[10] The creation of new organizations for small-scale farmers, especially cooperatives,

Box 1.6   Indian National Dairy Development Board

The Indian National Dairy Development Board (NDDB) was set up in 1965 by Parliament as a national institution governed by a board of directors to promote, finance, and support producer-owned and -controlled dairy organizations and support national policies favoring their growth. The board's work and scope expanded under Operation Flood, a program supported by the World Bank from 1970 to 1996. NDDB places dairy development in the hands of milk producers and the professionals they employ to manage their cooperatives. The board also promotes other commodity-based cooperatives and allied industries. The government tasked NDDB with preparing a National Dairy Plan for meeting a projected demand for about 180 million tons of milk by 2021–22. The plan is being implemented with World Bank support.

In 2009, India's 1.3 million village dairy cooperatives federated into 177 milk unions and 15 federations, which procured an average 25.1 million liters of milk every day. In 2010, 13.9 million farmers were members of village dairy cooperatives. The following are the NDDB's main areas of focus:

- Support cattle and buffalo breeding, animal health programs, and biotechnology research to improve milk productivity.
- Add value by testing and transferring product, process, and equipment technologies as well as services for analysis of dairy products and milk quality sampling.

- Create self-reliant and professionally managed cooperative institutions, responsive to members' economic and social expectations, through cooperative development and governance programs to strengthen capacities of the primary members, management committee members, staff of village Dairy Cooperative Societies, and the professionals and elected boards of Milk Producers' Cooperative Unions. Women's Development and Leadership Development Programs are a central activity.
- Technical and professional skills training at Regional Demonstration and Training Centers of NDDB, Union Training Centers, and the Mansingh Institute of Technology. Programs are designed for dairy cooperative boards, chief executives, managers, field staff, and workers.
- Technical assistance and engineering inputs for clients such as milk producers' and oilseed growers' cooperative unions and federations and central and state government. Services offered by NDDB-qualified engineers include setting up and standardizing dairy plants, chilling centers, automatic bulk milk vending systems, cattle feed plants, and infrastructure for agro-based industry projects.

NDDB's coordination role has come in for some criticism. Although the board is strengthening cooperative dairies, which supply about 70 percent of marketed processed milk, it has not improved the capacity of the informal dairy sector, based on village vendors, which produces some 80 percent of the milk in India.

Sources: http://www.nddb.org; Rasheed Sulaiman, personal communication.

---

has had mixed results (TN 2). Many were initiated by external agents such as NGOs or research institutes. When the new organizations ran into financial problems, the inclination was to provide them with additional support. This intervention isolated them from potential partners and, in some cases, from markets, creating a vicious cycle of dependence on funders (Hellin, Lundy, and Meijer 2009). Additionally, when organizations responded to the objectives of their funders or were captured by elites, they provided little benefit to the intended beneficiaries. New organizations tended to benefit their stakeholders mostly when the organizations could adapt their objectives and

operating routines in response to unforeseen needs and opportunities, especially unexpected market and social developments; could participate actively in networks of specialized actors; and could try several institutional arrangements and routines until they found a configuration that enabled them to fulfill their mission (Ekboir et al. 2009; TN 4 and IAP 2).

Coordination of actors in the AIS at the meso and micro levels can also be strengthened by creating self-help groups, such as farmer groups (IAPs 1 and 4). Through the creation of rural productive alliances, which bring commercial buyers together with producer organizations, small-scale producers tap into vital resources for reaching important markets (IAP 6).

Farmers may form regional associations to conduct applied research (like the regional associations of no-till farmers in Brazil, described in IAP 1), provide services, lobby decision makers, or influence the agenda of public research institutes (like the Mexican Produce Foundations described in IAP 2). Sectoral organizations that facilitate market access to large numbers of small-scale farmers by setting standards, providing technical and financial assistance, and consolidating their output are another common form of coordination at the subsectoral level (TN 1).

Innovation networks are groups of agents (including farmers, private firms, and possibly researchers and farmer organizations) that voluntarily coordinate their actions and contribute knowledge and other resources to develop jointly or improve a social or economic process or product. The membership of innovation networks changes often in response to new challenges or opportunities (see TN 2, particularly on the Papa Andina network). In contrast, *value chains* are networks with a commercial focus, one actor (a supermarket or broker, for example) that "organizes" and commands the chain, and a relatively narrow, stable membership. The strengths of value chains often result from the development of organizational innovations (especially the coordination of actors along the chain) that enable the creation of new business models (see TN 1 and IAP 6).

The importance of *innovation brokers* is increasingly recognized. Innovation brokers are teams of specialists that combine a strong background in science with knowledge of business, marketing, and/or the creation of innovation networks. Innovation brokers support linkages among actors in the AIS and help farmer organizations and private firms to manage research and innovation projects. They teach courses on the management of innovation, assess the actors' innovation capabilities, propose actions to strengthen them, and may accompany the implementation of the recommendations. Innovation brokers may also help governments and donors to develop their own innovation capabilities and to explore new instruments to foster innovation. NGOs, specialized service providers, or public organizations (including research or educational institutions) can play this role.

Building innovation capabilities in coordinating bodies and organizations

Given the challenges of developing new organizations, it may be more feasible to strengthen organizations that already exist. Collaboration can be reinforced by transforming the actors so that they can contribute better to innovation processes. For example, collaboration with foreign agroprocessing and trading companies to expose the agriculture sector to different business cultures and provide access to new markets has been very effective. Another effective strategy is to visit other innovators (whether local, domestic, or foreign), especially when the visitors have strong innovation capabilities and the visits are part of a program to share the information with other innovators at home. Virtual platforms have also been useful when actors in the AIS have the capabilities to use them. Innovation brokers can help marginalized groups to develop these capabilities.

Whether one establishes new or strengthens existing organizations to support coordination, *innovation capabilities* are essential (box 1.1). Innovation capabilities depend both on individual traits (creativity, for example) and on collective factors, such as collective learning mechanisms and organizational cultures. *In other words, innovation capabilities depend not only on innovative individuals but also on internal features of the organization, especially incentives, cultures, organizational spaces for experimentation, coordinating structures, and collective action* (box 1.7).

Innovation capabilities cannot be bought or built easily, and their development requires important investments and strong leadership over long periods, as exemplified by Whirlpool, a company that transformed itself from selling commoditized appliances in mature markets to generating a stream of breakthrough innovations that multiplied the company's revenue 20 times in just three years (box 1.7). Given the complexity and major investments required for such a large set of interventions, it is unlikely that many countries and donors will implement programs of similar scope, particularly in the case of agricultural innovations, which are often developed by networks of actors—that is, by organizations with very weak hierarchies.

Often, however, an external event or a few key interventions can trigger a virtuous cycle that builds up innovation capabilities (box 1.8). Innovation capabilities should be built within organizations (farmer organizations, civil society organizations, and private firms), in innovation brokers, in supporting organizations (such as research institutes and ministries), and in the enabling environment. Programs to broaden organizational capabilities should be adapted to particular configurations of actors, problems to be solved, and socioeconomic and institutional environments.

Building organizational capabilities for innovation is particularly challenging for several reasons. First, the main

Instilling innovation as a core competence at Whirlpool took a massive, broad-based effort over several years, involving major changes to leader accountability and development, cultural values, resource allocation, knowledge management, rewards and recognition systems, and a whole host of other management practices and policies.

Here are just a few examples of these changes:

- The appointment of vice presidents of innovation at both the global and regional level.
- The creation of large, cross-functional "innovation teams" in each region employed solely in the search for breakthrough ideas.
- The introduction of a companywide training program aimed at developing and distributing the mind-set and skills of innovation.
- The appointment of more than 600 part-time "innovation mentors" and 25 full-time "innovation consultants," who act as highly skilled advisers to new project development teams around the world.

- The creation of "innovation boards" in each region and each major business unit, made up of senior staff who meet monthly not just to review ideas and projects, set goals, and allocate resources but to oversee the continuing innovation capability-building process.
- The organization of big communication events called Innovation Days, where innovation teams showcase their ideas to other Whirlpool people, the media, and even Wall Street analysts. Sometimes these events are also held in suburban shopping malls as a way of collecting feedback and additional ideas from potential users.
- The creation of a comprehensive set of metrics to continually measure the company's innovation performance as well as its progress in embedding innovation as a core competence.
- The establishment of a sophisticated IT [information technology] infrastructure called Innovation E-Space, which integrates all of Whirlpool's people into the innovation effort and allows them to track progress on innovation activities across the corporation.

Source: Quoted directly from Skarzynski and Gibson 2008, 7.

Box 1.8   Actions to Build Organizational Capabilities

- Assess the main organizations in the AIS, analyzing at least three issues: whether each organization is necessary, what capabilities it needs to fulfill its mission, and how those capabilities can be built.
- Introduce new incentives so that existing organizations, especially public research institutes and universities, can better innovate and integrate into innovation networks (see the remaining points and modules 2 and 4).
- Create awareness among decision makers of the importance and nature of organizational capabilities and of the need for sustained efforts to build them.
- Implement training and mentoring programs on the management of agricultural innovation; tailor these programs to the specific needs of important

stakeholders, including top managers, directors, policy makers, funders, and field staff.
- Create multistakeholder forums to discuss innovation policies and programs.
- Promote the emergence of innovation brokers and new types of interactions for innovation, such as public-private partnerships or innovation networks (module 3).
- Explore new models of extension to promote organizational innovations (module 3).
- Support organizations that seek to provide services such as farmer-led research, extension, credit, and the provision of inputs in rural areas (module 3).
- Support exchanges between foreign and domestic organizations.

Source: Authors.

factors that influence organizational capabilities are not well understood. Second, organizations are strongly conservative (Christensen, Anthony, and Roth 2004), especially public organizations that operate under the rules of the civil service. Third, because interventions to build a capacity to innovate must suit the particular needs of each organization, the design and implementation of projects to achieve this goal require that the implementing agencies themselves possess strong capabilities to innovate.

KEY POLICY ISSUES

The key policy issues surrounding the organization of actors for innovation—the need for coordination, collective action, and stronger innovation capabilities—concern the kinds of institutions needed to foster innovation, the roles of the organizations involved (including reforms or actions that help organizations and marginalized groups to participate more fully), and the sustainability of innovation programs.

Institutional and enabling considerations

An enabling environment (module 6) is a prerequisite for effective coordination and is developed more easily when governments, donors, and the other actors in the AIS have a clear understanding of innovation processes. When such processes are financed by governments or donors, the latter often try to influence the process; governments may also resent having to negotiate policies and priorities with other actors. Finally, coordination is more effective when laws, regulations, and interventions by external stakeholders (especially governments and donors) facilitate transparency and accountability to all stakeholders in the innovation process.

Roles of the private and public sectors and civil society

The private sector has been and is expected to continue being the source of most innovations (Fagerberg 2005). For commercial agriculture in particular, the private sector will likely continue to lead innovation, including organizing value chains and developing agricultural equipment and inputs. To develop these innovations, private firms organize networks with farmers, traders, and eventually strong research teams (TN 1). Usually these teams have been located in developed countries, but a few strong teams from developing countries have also participated. Public research and extension agents have coordinated the emergence of innovation networks, but mostly in exceptional cases. The private sector or nimble NGOs are better equipped to coordinate the development and diffusion of technical and commercial innovations that adapt to rapidly changing technical or economic conditions, such as value chains for high-value products, or environmental innovations that require collective action, such as the management of water resources or forests.

The public sector (including the central, provincial, and local governments) can support innovation by (1) setting up an institutional environment conducive to innovation, including regulations, sanitary services, and intellectual property regimes; (2) financing programs to support innovation, including support for coordination of actors, support for venture and angel funds, financing research and extension embedded in innovation programs, and strengthening innovation capabilities; (3) allowing innovators to experiment with alternative approaches to achieve the project's goals; and (4) building up the infrastructure, especially transportation networks and public research and extension institutions.

The nonprofit sector should coordinate innovation processes that open opportunities for marginalized groups and represent their interests in policy dialogues. Public organizations may feel threatened when nonpublic actors assume a leading role in fostering innovation and try to change how public organizations interact with the AIS. Social responsibility may induce private actors to create organizations to develop innovations without commercial value, like projects with environmental or poverty alleviation goals, but the public sector will continue to have a major responsibility in these areas.

Reform of research and educational organizations

Given the organizational inertias that characterize public institutions, policy makers should carefully assess whether to invest in (1) transforming traditional organizations, (2) supporting existing actors, or (3) creating new institutions that complement traditional ones. The reform of public research and educational organizations should start by identifying the roles they should play, considering that others in the AIS also produce technical and scientific information. It is necessary to understand how public agricultural research institutions can complement private research and innovation (for example, a public institution can study relatively unknown plant varieties and develop new material with useful traits that private seed companies can then use in their breeding programs). Once the roles have been defined, the resources needed to fulfill the new mandates must be identified—especially investments in physical, human, and social capital—and plans must be prepared to attain those resources. Finally, new incentives must be defined for managers and researchers so that they can better integrate into innovation processes. For example, the incentives should not prioritize scientific publications over interactions with actors in the AIS or the generation of other types of scientific output, such as new agronomic recommendations; additionally, the incentives should allow researchers to develop long-term research programs. A complementary approach is to provide resources and incentives directly to innovative researchers and professors so that they can join innovation networks. This approach somewhat resembles the awarding of funds directly to researchers, a process that can have the unintended consequence of bypassing their institutes' formal resource allocation structures.

Social and local considerations, including gender and equity

Because innovation capabilities, physical assets, and power are not distributed equally, the best-endowed actors can benefit the most from emerging opportunities. In hierarchical societies in which coordination must include social leaders, greater coordination can award even more power to dominant local groups or individuals. Local conditions, especially cultural issues, similarly influence coordination. Some cultures forbid interaction between certain ethnic groups, but simply forcing them to interact directly may not be as effective as acting through intermediaries.

Even though social marginalization is a key aspect of poverty, it is difficult to create and sustain coordination organizations that include marginalized actors, especially women and landless farmers. Such organizations are often opposed by civil servants, politicians, middlemen, or wealthier farmers who see their power challenged (World Bank 2009). Affirmative action measures, reinforced by disincentives for wealthier actors, help to reduce gender and income disparities in coordination organizations and can include the following:

- Fostering the emergence of organizations of women and poor households, such as those used in microfinance programs.
- Setting aside seats on boards of organizations for representatives of marginalized groups and ensuring that

the representatives possess the skills (literacy, for example) to participate. Often capacity building is needed as well to prevent the most powerful stakeholders from capturing coordinating organizations (World Bank 2009).
- Fostering the emergence of networks that focus on innovations appropriate for marginalized populations, such as the no-till package developed in Ghana (IAP 1) (Ekboir, Boa, and Dankyi 2002).
- At a more general level, institutionalizing gender and pro-poor policies and planning functions in governments, projects, and organizations, and opening women and disadvantaged farmers' desks to guide practitioners in mainstreaming affirmative action in planning, budgeting, and implementation.

Sustainability of innovation programs

Actors can coordinate spontaneously in response to a need or opportunity, or they may be induced by specific public policies and programs. It is easier to strengthen forms of coordination that have survived for a certain period, because the actors involved have solved many of the barriers to collective action. Existing organizations, especially informal ones, can profit greatly from programs that build their capabilities and link them with other actors in the AIS. Forms of coordination imposed by external partners usually command more resources but often are less sustainable. They run a greater risk of becoming only formal structures and may lose their autonomy and effectiveness. Failure rates are high in new organizations because collectively agreed rules must be defined and effectively enforced. In some cases, the transaction costs of establishing an organization outweigh the benefits, especially in markets with low transaction costs, such as those for undifferentiated commodities (Hellin, Lundy, and Meijer 2009).

For an organization to be sustainable, its stakeholders must develop organizational capabilities and have incentives to contribute to the common effort. They must effectively influence the organizations they participate in, and they must also perceive the benefits of participation. When this happens, stakeholders often invest their financial and political capital to ensure the programs' continuity. These conditions apply not only to organizations at the "base" of the AIS (such as farmer organizations or value chains) but also to organizations seeking to coordinate the AIS at the "top," such as research councils or innovation brokers.

NEW DIRECTIONS, PRIORITIES, AND REQUIREMENTS FOR INVESTMENT

As discussed, agricultural innovation may be coordinated by strengthening existing individuals and organizations or by setting up entirely new actors and organizations. The common denominator of these approaches is the need to provide incentives, apply appropriate instruments, and build innovation and organizational capabilities. Because effective collaboration among innovators is so difficult to implement, however, it is likely that new types of organizations will need to be supported, as described next.

Improving governance of the AIS

A number of factors impinge on the efficiency of governance in a national innovation system[11] in general and an AIS in particular—in other words, on the extent to which policy processes have the greatest effect with a given use of resources (OECD 2005). The evidence indicates that efficient governance depends on certain qualities, which include the following:

- *Legitimacy.* The policy actors and approaches adopted in policy processes have to be appropriate and widely accepted for the tasks at hand.
- *Coherence.* The strands of innovation policy and associated policy instruments must fit together.
- *Stability.* Innovation requires sufficiently stable framework conditions, institutions, and policy.
- *Adaptive ability.* As the environment for innovation and innovation itself keeps evolving, governance actors need to be able to adapt.
- *Ability to steer and give direction.* A related capability is the governance system's ability to provide direction to actors and steer the innovation system as a whole. This capacity requires commitment and leadership by policy makers at the highest level.

Governance of the AIS can be improved by creating formal but effective spaces for dialogue at different levels of the AIS (local, sectoral, and national), building actors' organizations and their capabilities, and improving formal and informal regulations that reduce transaction costs (such as intellectual property rights regimes, the judiciary system, customs, and markets). At higher levels of the AIS, investment is needed to establish and strengthen effective and responsive coordinating bodies for agricultural innovation, such as innovation councils and subsector-specific bodies that can contribute to collective identification of opportunities and challenges and help to align and allocate limited resources to key innovation issues (see details in TN 1). IAP 5 describes a regional approach to coordinate innovation actors in the cassava subsector, particularly researchers. IAP 3 focuses on the Foundation for Agricultural Innovation (Fundación para la Innovación Agraria, FIA), a ministry-affiliated foundation with independent governance that coordinates and incentivizes agricultural innovation in Chile.

Farmer and nonprofit organizations respond mostly to their funders, whose interests may not coincide with the needs of local stakeholders, especially if they are marginalized groups like women or landless rural households. It is important to provide some means for these groups to influence the AIS. Possibilities include arrangements for ensuring good governance and accountability and for training managers and members about their respective roles in their organizations and in the AIS. It is also important to keep external interventions in NGOs and civil society organizations to a minimum, allowing them to evolve as needed.

The rise of networks

TN 2 in this module examines the nature and dynamics of innovation networks, which are becoming more prevalent as the complexity of innovations grows and rapid economic and technological change forces agents to innovate at a faster pace. Innovation networks have developed important technical, commercial, and organizational innovations that have had major economic and social impacts. IAP 1 reviews the emergence and evolution of the innovation networks that developed no-till technologies for small-scale farmers in South America and Ghana. Where they have been adopted widely, these technologies have increased farmers' incomes, reduced food insecurity, diminished labor requirements for the production of staples, allowed poor rural households to engage in new income-generating activities, and enhanced the sustainability of agriculture in marginal and well-endowed areas. Although programs to diffuse no-till have been documented in more than 60 countries, massive adoption has occurred only in the handful of regions where diverse actors formed innovation networks to develop organizational and technical innovations adapted to local conditions.

Innovation brokers: promising, but challenging to implement

Innovation brokers can play a valuable role in an AIS. Several types have evolved, but few have survived without support from governments or donors (Klerkx and Leeuwis 2008).[12] Innovation brokering services are affected by severe information uncertainties. Individual actors cannot know the commitment and the capabilities of potential partners in advance. A conflict of interest may exist if the funders' requirements do not coincide with the needs of the other actors in the AIS. Giving the funds to users to pay for the brokering services may reduce the conflict, but effective controls are needed to make sure that funds are used appropriately and that actors follow the brokers' recommendations. Similarly, users often do not know the nature and quality of services offered by innovation brokers. They are reluctant to pay for services that are difficult to define beforehand and highly uncertain.

Organizing around value chains

Small-scale farmers' access to modern marketing chains, often organized by supermarkets, is analyzed in TN 3. Chains provide more stable incomes and sometimes higher profits for their adherents, but participating in chains requires commercial and technical skills. Recent research has found that in the long term few smallholders can survive in these chains as suppliers; only the more affluent smallholders, better endowed with natural resources, infrastructure, access to credit, and social capital, tend to participate. Despite poor rural households' limited access to the markets supported by modern value chains, the chains can bring important benefits to rural economies by creating many permanent and temporary positions on the farm and in associated services such as input supply, sorting and packaging, and transport. More dynamic local economies also create small business opportunities, such as food stalls and professional services. As mentioned, rural productive alliances, which are economic agreements between commercial buyers and formally organized producer organizations, enable small-scale producers to reach those markets. The agreements create favorable conditions and incentives for buyers and smallholders to establish mutually beneficial and sustainable relationships (IAP 6).

Supporting farmer organizations and self-help groups

Farmer organizations can participate in the financing, development, and diffusion of innovations, manage public and private funds and programs for innovation, collaborate in the design of innovation policies, coordinate other actors in the AIS, and influence research and extension organizations. TN 4 reviews the conditions that help farmer organizations to emerge and consolidate. It gives particular attention to the technical and market conditions that make farmer organizations a more effective means of coordinating AIS actors than other institutional arrangements, such as value chains and innovation networks. It also reviews factors that reinforce farmers' ownership and sustain their organizations.

IAP 2 examines the emergence and evolution of Mexico's Produce Foundations, which are civil society organizations that influence the design and implementation of research and innovation policies and programs. Soon after the federal government created the foundations in 1996, a few farmers pushed for their foundations to be independent of the government; they succeeded, and the remainder eventually followed suit. The foundations created a coordinating office that interacts with the government and promotes organizational learning. The experience in Mexico shows how a number of interacting factors support successful organization building. IAP 2 also analyzes the interplay between deliberate strategies to build organizations and strategies that emerge organically from the innovation process.

IAP 4 describes another form of organization—highly federated self-help groups for the poor established by the Society for Elimination of Rural Poverty in Andhra Pradesh. Each tier in the organization of self-help groups functions as a financial intermediary and provides specialized services to members (and other stakeholder groups) in a variety of sectors.

Building capabilities for coordination among individuals and organizations

As noted, building capabilities for coordination requires strong leadership and sustained, major investments. Investments can encompass support for physical, human, and social capital (basic education; management and entrepreneurial skills; learning to participate in social, innovation, and economic networks; and development of financial capacity), short- and long-term consultancies, formal courses, long-term mentoring, support for innovation brokers, the creation of dedicated bodies (innovation councils or programs to fund innovation), and the transformation of public universities and research and extension organizations (discussed in box 1.8). Transforming public universities and research and extension organizations is particularly problematic but not impossible. Employees have job security, governments are reluctant to pay locally competitive academic salaries that exceed civil servants' salaries, and it is challenging to hire good, experienced professionals to work under difficult conditions (Davis, Ekboir, and Spielman 2008).[13] In response to these challenges, many countries have created new public organizations with different conditions and adequate working resources. Alternatively, investments could create the conditions to induce the best employees of existing organizations to participate in innovation networks regardless of their institutional association. Another major problem that hampers efforts to build capacity for coordination in developing countries is the frequent rotation of capable civil servants. A major effort should be made to build the capabilities of nonpublic organizations, including political parties, private firms, and civil society and farmer organizations, so that they can influence the policy dialogue despite changes in government.

Organizing around a common vision of major issues

Coordination is facilitated when potential partners share a common vision of their problems and opportunities. This vision can be created through foresight exercises, studies and consultancies, gathering and processing of technological and market intelligence, sectoral dialogues, and interministerial committees. Because the common vision needs to be updated in response to technical, social, and economic change, these activities must be implemented periodically and include a broad array of stakeholders, not just those directly linked to the processes being analyzed (Skarzynski and Gibson 2008).

Setting agendas and priorities

Priorities for collaboration should be defined with participatory approaches and updated periodically, but not too often, which would disrupt the development of trust required for effective collaboration and implementation of long-term activities. Investments should also build the capabilities of all stakeholders so that they can participate actively in innovation processes with which they may be unfamiliar, such as policy design or the boards of research organizations.

MONITORING AND EVALUATING INVESTMENTS AND SCALING UP

The impact of organizational innovations on the AIS is notoriously difficult to measure, owing to the complex dynamics of innovation. Not only does every AIS have many actors with differing goals, but the outcomes of their formal or informal cooperation may not appear until many years have passed. Another issue in assessing outcomes is that monitoring and evaluation can be used for different, often conflicting, purposes, like learning and accountability.

Although it is difficult to define valid indicators for monitoring organizational innovations and their impacts, two principles are valuable to consider. *First, the monitoring system should be a learning tool.* When a project is defined, several critical assumptions are made, representing an explicit or implicit theory of how stakeholders' behaviors are expected to change over the project's life. The monitoring system should be designed to (1) test these assumptions early in the project's life and adjust the interventions if the assumptions are proved wrong, (2) identify unforeseen problems or opportunities as early as possible in the project's life (Spitzer 2007), and (3) measure changes in stakeholders' behaviors and provide feedback to stakeholders so that they can learn faster.

Second, many indicators, especially those intended to measure ill-defined processes such as organizational innovations, cannot be measured quantitatively. In such cases, qualitative indicators, such as stakeholders' opinions, are appropriate measures that can be tracked with specially designed techniques (such as Likert or rating scales).

The monitoring system should also cover at least four areas:

- *Organizational strengthening.* Assess the organization's ability to improve its governance, culture, and finances and expand its membership.
- *Organizational learning.* Assess the organization's ability to perform its normal activities better.
- *Exploration of new instruments to fulfill the organization's mandate.* Assess the organization's ability to search for new ways to reach its objectives or to define new objectives.
- *Changes in the AIS according to the underlying theory of change.* Assess the responses of the actors involved directly or indirectly in the project, especially the evolution of their interactions.

Finally, some potential indicators of organizational innovation are listed in table 1.2, along with their corresponding sources of information.

Table 1.2 Indicators of Organizational Innovation

Indicator	Source of information or tools
Process indicators	
1. Leveraged investments: additional time and resources invested by the organizations' members in joint activities	Surveys of value chains Surveys of innovation networks
2. Consolidation of innovation networks and value chains	Case studies Surveys of partners in innovation networks and value chains
3. Improvements in the innovation capabilities of farmer organizations (for example, changes in governance, learning routines, and experimentation)	Case studies Surveys of members, stakeholder interviews
4. Changes in the resources invested in building organizational capabilities	Review of public programs, survey of funders
5. Number of programs and resources invested in organizational innovation	Review of public programs, survey of funders
6. Number and types of innovation programs targeted to marginalized groups	Review of public programs, survey of funders
Output indicators	
7. New products or processes introduced by actors in the organization (or by the organization itself)	Surveys of the organization's stakeholders
8. Strength of the value chains (volume marketed, additional income generated, number of farmers benefiting directly, number of rural jobs created)	Surveys of supermarkets, brokers, wholesalers, farmers Trade statistics
9. Changes in value chains, especially in the number of intermediaries, their relative strength, and other institutional changes (from stakeholder platforms and other methods)	Surveys of supermarkets, brokers, wholesalers, farmers Trade statistics
10. New partnerships created (number, diversity, types, goals, achievements)	Review of programs that foster the creation of organizations for innovation Participatory rural appraisal Surveys of actors in the AIS
11. Expansion of the networks (such as the number and type of partners, effectiveness, innovations adopted)	Surveys of innovation networks, stakeholder interviews
12. Changes in curricula that prepare professionals in organizational innovation	Case studies Surveys of educational organizations
Outcome indicators	
13. Number of partnerships that survived after three years	Surveys of partnerships
14. Changes in the participation of farmer organizations in innovation processes	Case studies Surveys of partners in decision-making processes Stakeholder interviews
15. Improvement in the condition of marginalized groups (women and landless farmers) thanks to their participation in innovation programs	Case studies Surveys of partners in decision-making processes Stakeholder interviews
16. Changes in the performance of value chains after three years	Surveys of supermarkets, brokers, wholesalers, farmers Trade statistics
17. Changes in asset ownership and market participation induced by organizational innovations after three years	Case studies Surveys of farmers and other stakeholders in innovation processes

Source: Authors.

National Coordination and Governance of Agricultural Innovation

Riikka Rajalahti, World Bank

SYNOPSIS

Many countries could benefit from better coordination of agricultural innovation at the national sectoral level, where broad science and innovation policies, strategies, and activities are defined. This note describes the potential benefits and elements of an organization that successfully coordinates national innovation in agriculture. Ideally, a nationally mandated but independently governed agricultural innovation council or committee (consisting of diverse stakeholders) coordinates the development of a strategic vision for agricultural innovation. It also coordinates and formulates the corresponding agricultural innovation policy (to be increasingly integrated into general science-innovation policy), designs agricultural innovation priorities and agendas, and monitors and evaluates innovation programs and their impact. In theory responsibilities for policy making, financing, and implementation should be separate, but experience varies in practice. Many innovation councils are advisory and policy-making bodies with no mandate to channel funds, whereas others have been more effective at inducing coordination of policy when they control innovation funds. This note reviews lessons emerging from the three commonly applied modes of coordinating innovation for agriculture at higher levels to date: the national agricultural innovation council or committee, competitive innovation/research funds, and coordination by theme or subsector. New approaches could make all three options more effective at contributing to coordination and implementation of agricultural innovation policies, strategies, and agendas.

BACKGROUND AND CONTEXT FOR INVESTMENT

Many countries are gradually addressing challenges to coordinating innovation within specific agricultural subsectors. Even so, far greater impetus is needed at the national level to address coordination and collective action for agricultural innovation as part of the wider sphere in which science and innovation policies, strategies, and activities are defined. Ideally, interventions to improve national coordination and governance of the AIS seek to improve the participation of stakeholders, including end users such as producers and enterprises; improve the transparency and openness of decisions related to funding and priorities; improve responsiveness and accountability to stakeholders; build consensus; and develop consensual, coherent policies, strategies, and activities that reflect a strategic vision of innovation for agriculture.

The lack of appropriate coordination and governance for agricultural innovation at the national level is a chronic problem for many countries. As pointed out in the module overview, most efforts to coordinate agricultural innovation at the national level have focused on establishing formal apex research councils, as in Australia, Brazil, Ghana, India, and Mexico, to govern multi-institutional national agricultural research systems. They develop national research strategies and plans; link research to broader agricultural policy discussions; channel funds to priority research areas and thus coordinate research across institutions; promote collaboration and exchanges among the various parts of the national agricultural research system; and coordinate external links (Byerlee and Alex 1998). In several cases, though, design of the councils has reduced their impact. For example, Brazil's national (and most important) agricultural research organization, EMBRAPA,[1] is also the formal head of the national agricultural research system and transfers resources from its budget to other research organizations. Because of these multiple roles, several stakeholders in the AIS do not see EMBRAPA as an unbiased coordinator. For many years Mexico's office for coordinating national agricultural research and extension systems[2] had no instruments to induce coordination, and its attributions and roles had not been defined clearly; not surprisingly, it was largely ignored by the actors it was supposed to coordinate.

Despite their presence, national innovation councils and agricultural research councils rarely operate as true *agricultural innovation organizations or councils,* with a mandate to coordinate and prioritize investments in agricultural innovation at the highest level. This thematic note discusses measures that enable coordination processes to improve, adapt to changing circumstances, and rely on the growing array of stakeholders to improve the governance and impact of agricultural research and innovation. It begins by reviewing the characteristics and norms shaping a "true" organization for national coordination and governance of research and innovation (mandate and management structure, resources, and operating practices and values). The potential benefits of such an organization, the policy issues that impinge on its successful operation, and the many lessons emerging from previous efforts are all discussed in the sections that follow.

INVESTMENT NEEDED

An organization to foster national (sectoral) coordination and governance of agricultural innovation would be nationally mandated but independently governed. The general outlines of the mandate, governance structure, activities, resources, and practices of an effective national coordinating organization are presented next to provide an idea of the kinds of investment needed.

Organization and mandate

The coordinating organization would be formal and independently governed and managed as defined by its bylaws. The effectiveness, legitimacy, relevance of, and confidence in a coordinating organization depend on how effectively it reaches out to stakeholders from diverse areas of the economy. The composition of the organization should reflect the diversity of its stakeholders. The range of stakeholders in matters of agricultural policy such as innovation is likely to be very wide, including farmers and other actors associated with agricultural innovation (research, education, extension, and farmer organizations; private firms; and NGOs), rural territories, and consumers. Representatives from outside the agricultural sector can add diversity and value to discussions. All should have a voice and be included in decision-making concerning agricultural innovation strategies and programs.

A skilled management team (a secretariat, for example) would execute the activities identified by the organization's board. A typical mandate and set of activities would be the following:

- Coordinate the development of a strategic vision for agricultural innovation.[3]
- Coordinate and formulate agricultural innovation policy, which will be increasingly integrated into general science-innovation policy.
- Link agricultural innovation to broader agricultural policy and science-innovation discussions.
- Continue to contribute to the development of a strategic vision of the agricultural sector.
- Coordinate and design agricultural innovation priorities and agendas.
- Coordinate the division of labor and channeling of funds[4] to priority innovation areas.
- Monitor and evaluate innovation programs and their impact.
- Promote collaboration and exchanges among the various parts of the innovation system, including external linkages.

Ideally, innovation policy making, innovation financing, and implementation are separate functions. Agricultural innovation councils or committees should not be responsible for executing innovation programs, which is the task of science, technology, and innovation organizations.[5]

Funding, infrastructure, and capacities

Formal, dedicated structures with set agendas do not in themselves guarantee effective coordination of—and action by—actors. At a minimum, coordinating organizations will need operating funds, physical infrastructure, and communications infrastructure (ICTs, for example) to enable transparent and open communication and support effective coordination and governance. Coordinating organizations will also need to build capacity among actors and encourage them to address issues collectively. Box 1.9 summarizes the capacities and skills needed for innovation coordination and governance to be effective at a higher level.

Operational practices

The practices of an organization that coordinates national agricultural innovation will be guided by such values as transparency, responsiveness, accountability, consensus, and coherence within and between the organization's activities.

Transparency is achieved if all information regarding decisions on funding, priorities, and operations is open and freely available. This openness implies that central and regional governments and agricultural development agents

Governance capabilities are defined as the ability to

- Recognize system characteristics (strengths, weaknesses, problems, development potential)—which requires facilitation and analytical skills.
- Define the focus and the topics for political action (agenda setting)—which requires skills in communication and consensus-building.
- Encourage diverse players (through consultation and participation) to coordinate their activities in

and beyond their policy field—which requires skills in facilitation, negotiation, and consensus-building.
- Implement these policies—which requires policy capacity.
- Learn from previous experience (such as evaluation results)—which requires learning, intelligence, and accountability.
- Make adjustments over the complete policy cycle.

Source: Adapted from Ohler et al. 2005.

---

will actively communicate and deliver key messages to stakeholders about what they do and the decisions they take. They should use language that is accessible to the general public. Openness also implies that potential beneficiaries such as farmers and processors have equal access to information and funding opportunities (Hartwich, Alexaki, and Baptista 2007).

Those who govern organizations and societies, as well as the institutions and processes they establish, must be *responsive* to stakeholders. They must prove in some way that they are acting in response to stakeholders' priorities and generating outputs that meet their needs. *Accountability* means that the organization will be fiscally responsible and use efficient mechanisms to avoid corruption. Regular reporting on the efficient and effective use of project and program funds, along with evaluations of outputs and impacts, will help to prove that funds are used in the best and most correct way. In complex innovation systems, many actors are involved in setting priorities and using funds, but generally the agents that disburse the funds are held accountable (Hartwich, Alexaki, and Baptista 2007).

Differing interests need to be taken into account to balance regional and subsectoral development, and the institutions involved in agricultural innovation need to reach a broad *consensus* on which tasks are of general interest and who will play what role in those tasks. Now that the range of tasks required to achieve innovation-led growth in agriculture has grown so large, the need for *coherence* among those tasks has become more pronounced. Coherence aids in coordinating and forming critical masses for innovation, and it also enables innovating agents to be competent—to

focus on their own clearly defined share of the work. Coherence requires strong and effective leadership to ensure that roles and tasks are articulated, understood, and shared as agreed (Hartwich, Alexaki, and Baptista 2007).

Australia has developed a multistakeholder, multidisciplinary agricultural research and development council with a focus on agricultural innovation (box 1.10). Chile's FIA plays the roles of innovation coordinator, promoter of innovation, and technological broker, interacting with several actors in the AIS, including farmers, private firms, financial institutions, technological institutes, and universities (IAP 3). Despite Chile's competitiveness and innovativeness, its agricultural sector currently is not fully aligned with the rest of the national innovation system and associated policy coordination.

POTENTIAL BENEFITS

The reasons commonly cited for AIS actors to interact and address issues collectively include improved identification of opportunities and challenges related to innovation, better leveraging of human and capital resources, better learning and information sharing, improved implementation and results, and economic benefits. Coordination may also improve the design and implementation of innovation policies. Stronger coordination induces all actors, especially public research and extension organizations, to be more aware of and responsive to the needs and concerns of other actors, especially resource-poor households. More actors in the innovation system can voice their needs and concerns in the process of designing and implementing innovation

Box 1.10   The Rural Research and Development Council of Australia

Since 1994, the Government of Australia has developed rural research and development priorities that reflect the national understanding of critical needs for investment in agricultural, fisheries, forestry, and food industry research and development. A shared approach to priority setting among state and territory governments, industry, and research funders and providers enables issues of common concern to be explored in a coordinated, cost-effective way.

The priorities complement national research priorities and guide Australia's Rural Research and Development Corporations and Companies (RDCs). As the government's primary vehicle for funding rural innovation, RDCs are partnerships between government and industry, created to share funding, strategic planning, and investment in research and development and the subsequent adoption of outputs (see TN1 in module 4). The RDCs significantly influence the work of research providers and investors in related fields.

The Rural Research and Development Council was appointed in 2009 by the Minister for Agriculture,

Fisheries, and Forestry to be the government's independent, strategic advisory body on rural research and development. Its principal goal is to provide high-level advice and coordination to improve the targeting and effectiveness of government investments in rural research and development. The Council's 10 members represent research, academia, farmers, and the private sector. The Council works closely with RDCs, industry, research providers, state and territory jurisdictions, and government agencies to strengthen rural research and development through improved collaboration, facilitation, and prioritization of investment and performance measurement and reporting.

Given its recent establishment, the council's effectiveness has yet to be rigorously evaluated. Even so, the enhanced focus on research and development at the national level is expected to improve the productivity, profitability, sustainability, and global competitiveness of Australia's agricultural, fisheries, forestry, and food industries, with benefits for individual rural businesses, the environment, and the wider community.

Source: Rural Research and Development Council (http://www.daff.gov.au/agriculture-food/innovation/council).

policies, resulting in more inclusive policies and faster diffusion of innovations.

POLICY ISSUES

The policy issues involved in developing an organization responsible for coordinating agricultural innovation on a national level resemble the policy issues that apply to coordination more generally. They include policies to promote the organization's sustainability and effectiveness and to reflect the needs of all stakeholders, including marginalized groups.

Beyond the creation of formal coordinating bodies, effective coordination of stakeholders depends on stable financing and incentives for participating organizations and employees. Careful attention must be paid to developing stakeholders' capacity to contribute to shared goals, influence the organizations they participate in, and benefit from their participation in ways that encourage them to continue participating. Effective and relevant coordinating organizations will include a wide swathe of stakeholders, differing by geographical location, subsector, gender, and type of institution (public, private, and civil society).

To enhance transparency, it is better for the functions of innovation policy making, financing, and program implementation to reside in different organizations. Policy making is ideally the mandate of coordination and policy-making bodies, such as councils, whereas financing is the mandate of ministries and special agencies, leaving the implementation of innovation policy programs to research-innovation entities such as research organizations, universities, and private companies, among others.

Formal coordinating organizations at the national level may be set up or facilitated by public, private, or civil society actors. In practice, the public sector often takes the lead. It is essentially in the public interest to guarantee that society's resources are allocated to priority issues, identified collectively.

LESSONS LEARNED AND RECOMMENDATIONS FOR PRACTITIONERS

The lessons and recommendations emerging to date center on three commonly applied modes of coordinating innovation for agriculture at higher levels. The first is the national agricultural research council or committee. The second is the establishment of competitive or noncompetitive project-based innovation/research funds, with their associated governance and management structures. The third involves coordination by theme or subsector, which can be effective at generating innovations that reflect users' needs and government policies and strategies.

Research councils: too many roles, too few stakeholders, and stark funding and capacity needs

Most research councils or forums play an important role in research, development, and policy design, and they often collect large parts of public budgets for research. It is important to separate the design of policy from the implementation of agricultural research and innovation, but experiences and practices related to funding vary among research and innovation councils. In practice few apex councils function independently of the ministry in charge of agriculture and the agricultural research system. Many have become large research institutes in themselves. In still other cases, apex councils are advisory bodies without a mandate to channel funds, following the principle that ministries and agencies allocate funds (box 1.10). Both arrangements typically lead to failure (Byerlee and Alex 1998). However, some organizations have been more effective at inducing collaboration and coordination when equipped with control over incentives, such as innovation funds (IAP 3 and module 5, TN 2). The Indian Council of Agricultural Research (ICAR) functions as a national agricultural research council and executes research as well, as outlined in box 1.11. Although it has operated as a typical research-oriented council, it is transforming itself to operate more effectively as part of the wider Indian agricultural innovation system. With the state agricultural universities, it is implementing the National Agricultural Innovation Project (NAIP) and pursuing specific activities to catalyze and manage change in India's agricultural research system:

■ Strengthening communications ability and information capacity (enhancing dialogue and interaction with the public at large, the farming community, and the private sector, as well as among all key functionaries in the ICAR system).

■ Forming business development units and offering intellectual property rights (IPR) management.

■ Undertaking systemwide organizational and management reforms, including improvements in monitoring, evaluation, and financial management.

■ Undertaking visioning, technological foresight, and policy analyses.

■ Renewing links with the state agricultural universities, which have tended to be isolated from ICAR's research.

Box 1.11   Structure and Mandate of the Indian Council of Agricultural Research

The Indian Council of Agricultural Research (ICAR), established in 1929, is an autonomous organization under the Ministry of Agriculture. With 97 institutes and 47 agricultural universities, ICAR is the national apex body for coordinating, guiding, and managing research and education in agriculture. The members come entirely from the public sector. ICAR has two bodies: (1) the General Body, the supreme authority of ICAR, headed by the Minister of Agriculture and (2) the Governing Body, the chief executive and decision-making authority of ICAR, headed by the Director-General, ICAR. The correspondingly broad mandate of ICAR includes:

■ Plan, undertake, aid, promote, and coordinate education, research and their application in agriculture, agroforestry, animal husbandry, fisheries, home science, and allied sciences.

■ Act as a clearinghouse of research and general information in its areas of competence through its publications and information system.

■ Institute and promote transfer of technology programs.

■ Provide, undertake, and promote consultancy services.

■ Look into the problems relating to broader areas of rural development concerning agriculture, including postharvest technology, by developing cooperative programs with other organizations such as the Indian Council of Social Science Research, Council of Scientific and Industrial Research, Bhabha Atomic Research Centre, and the universities.

Source: http://www.icar.org.in/en/aboutus.htm; World Bank 2006c.

Many councils consist of representatives from government or research agencies, with few other stakeholders. It has been challenging to ensure representation of the wide spectrum of developing country farmers, which includes large-scale farmers or plantations producing traditional export products, small-scale farmers that supply supermarkets and fast-food restaurants (usually the better-off members of this group), large-scale farmers who sell through local or wholesale markets, and small-scale farmers who produce for their own consumption or sell a small surplus in local markets. *These groups operate in innovation systems that barely intersect.*

This lesson is being learned, however. A movement is underway to broaden the representativeness of governance bodies (by stakeholder group, geographical location, and discipline), improve the transparency of decision making, reduce bureaucracy, and use more rigorous and diverse prioritization tools. Box 1.12 describes mechanisms and processes to integrate additional stakeholders, particularly producers, in research priority setting and planning, based on an example from Uruguay. A related lesson is that effective agricultural research councils recognize that they are major stakeholders in, and must form an integral part of, national innovation councils.

The effectiveness of many councils is limited not only by a narrow constituency but by a lack of resources. Scarce financial resources restrict councils' influence on stakeholders, the policy process, and the way that research is conducted. For councils to be more than formal constructs, they must command resources, have continuity, be seen by other actors in the AIS as honest brokers, and those actors, especially top government officials, must be willing to listen to their advice. Coordination at the highest levels of the AIS has been hampered by the fact that formal coordinating structures (research and innovation councils) generally operate in an environment that gives other organizations in the AIS no complementary incentives to foster innovation. Government directives and agreements reached among organizations often cannot be

---

Box 1.12   Mechanisms to Articulate Producers' Needs in Uruguay

Producers have a significant role in the financing, governance, and research priority setting of Uruguay's main agricultural research institute, the Instituto Nacional de Investigación Agropecuaria (INIA). Formal mechanisms foster producers' participation in identifying, prioritizing, and planning research: Producers are active members of the board of directors, regional advisory councils, and working groups. They are involved in planning primarily through wide participation in identifying and analyzing problems. Specific mechanisms to articulate producers' demands and transfer technology have been developed as well, such as experimental units for validation and demonstration.

The five regional advisory councils, set up in 1990, act as "antennae" for capturing local demands in the area served by each experiment station. The councils are an important forum for regular exchanges of views and close contacts between producers and INIA staff. INIA also created working groups for major commodities at each regional experiment station to strengthen farmers' role in guiding commodity research. In these groups, INIA staff and farmers discuss research plans and results for specific commodities and production

systems. Meetings are open to all producers interested in attending, as well as other stakeholders (extensionists, representatives from industry, and policy makers). The working groups have become a very useful mechanism for formally incorporating inputs for research planning, monitoring, and evaluation.

INIA also has close links with its clients through a specialized unit for diffusing technology. For instance, INIA has a long-term agreement with a producer association to demonstrate new intensive cropping and livestock production technologies. The arrangement has validated promising technologies at the commercial level, facilitating their transfer and providing feedback to reorient research, and it has been expanded to support new technological developments in extensive livestock production.

Round tables are a third means of incorporating producers' demands into national research programs. Composed of specialists from INIA, other public research institutes, the university faculty of agronomy, and representatives from different stages in the agroindustrial chain (from producers to consumers), the round tables operate as self-directed work groups to identify relevant vertically integrated problems.

Source: Allegri 2002.

implemented because the representatives of national organizations cannot force provincial or local chapters to change their behavior when the only tool at their disposal is reasoning.

Another vital lesson is that capacity-building programs are often needed to build skills in collaboration, competition, and negotiation, particularly among individuals who lead the coordination and governance process. For details, see the module overview.

Project-based research and innovation funds: governance and management issues

The introduction of project-based (including competitive) funding schemes for agricultural research and technology transfer has in many instances been associated with changes in the governance of national agricultural innovation systems. At one time, national agricultural research agencies received an all-inclusive lump sum based on a broad research mandate and could set research priorities quite independently. More recently, priority-setting responsibilities have shifted owing to major changes in how society views science and to the introduction of competitive science and technology funding schemes.

The extent to which a project-based fund influences priority setting differs greatly between agriculture-specific and general project-based funding schemes. Specific, project-based science and technology funds tend to define agricultural research needs up front, before calling for proposals, whereas general science and technology funds are somewhat more open. For general science and technology funds, it is usually academic relevance that matters most in the selection procedure; in specific, project-based schemes, it is economic relevance (World Bank 2006b). Project-based funds can contribute positively as well as negatively to the governance of agricultural research (box 1.13).

The good practice for governing and managing project-based research funds is to maintain separate units for policy setting, technical evaluation, management, and governance. The main governing responsibility in project-based research funds should reside with a governing board that ideally consists of a distinguished group of senior decision makers. The good practice for appointing members to the board is to strike a balance among the stakeholder groups pertinent to the grant scheme and the wider innovation system. At a minimum, the board should consist of representatives from government, research organizations (including universities), extension, farmers, and the private sector. All too often, boards can be taken hostage by one interest group that dominates discussions and skews decisions in favor of its constituency at the expense of others. Similarly, boards can be held hostage by politicians. This practice is particularly damaging given the complex and multisectoral nature of funding for agricultural innovation systems. A governing board is typically supported by a technical advisory committee mandated to provide technical input for planning programs and setting priorities and a technical review panel mandated to evaluate, score, and rank proposals and make funding recommendations. A secretariat manages programs and carries out daily operations. Table 1.3 summarizes the principal characteristics of project-based competitive science and technology funds (many of which are closed) in five countries. For further details on innovation funds, see module 5, TN 2.

In Latin America, many science and technology funds specific to agriculture—especially funds focusing on adaptive agricultural research and technology transfer—have acknowledged the need to improve client orientation and participation. These funds have adopted strategies that involve farmers in identifying and prioritizing innovation needs and in developing, selecting, implementing, and funding subprojects. To reach their clients, funds have adopted decentralized strategies or are in the process of doing so. Although stronger client participation and orientation are generally considered positive, they may also have drawbacks (such as a bias toward short-term research, a lack of equity, and significant transaction costs).

Thematic or subsectoral coordination: the value of delegating to networks

Owing to the difficulties involved in establishing effective national coordination of agricultural innovation, many countries have implemented formal structures to coordinate actors at different levels of the AIS. The module overview mentions several examples, including associations, commodity boards, and networks. The delegation of research governance to networks is increasingly seen as a means of resolving problems endemic to traditional research funding processes. It can reduce the state's direct influence on funding policies, respect the independence of scientific institutions, foster vigorous scientific institutions, and ensure scientists' strong commitment to users' interests. One example of this new type of research-innovation council is Bioconnect, a research-oriented, multistakeholder network for organic agriculture in the Netherlands (box 1.14).

Competitive funds can contribute positively to governance, leading to

- *Improved identification and prioritization of agricultural research needs*, particularly with the more specific, competitive funds. (Note that general science and technology funds, in contrast, usually do not prioritize research needs beforehand. Because these funds leave much of the initiative of selecting research topics to individual researchers, the agricultural research agenda may not take into account the needs of AIS actors or emerging opportunities for innovation.)
- *Improved formulation of research project proposals.* The introduction of competitive funding schemes requires a strong project culture within agricultural research and technology transfer agencies; some take time to adjust to the new rules and regulations. Universities seem to have more experience with competitive funding schemes.
- *More transparent selection of agricultural research projects.* All project proposals are reviewed by external reviewers. Research projects that are approved and selected for financing are listed publicly. Many competitive funds have project databases that can be consulted online.
- *Improved monitoring and evaluation (M&E) of project implementation.* In most agricultural research and technology transfer agencies, M&E has long been a weak spot. Internal reporting mechanisms are rare, and sanctions are seldom applied for failure to report on progress. Competitive funding schemes, with their more stringent reporting requirements, have introduced innovations in M&E.

Competitive funds can also lead to governance problems:

- *By lacking sufficient objectivity.* Particularly in small science communities, it can be very difficult to organize impartial reviews of research project proposals. A solution may be to mobilize foreign reviewers, but this alternative remains quite costly to organize and manage without good access to ICTs.
- *By using a limited time horizon and funding only operational costs.* Competitive science and technology funding schemes, which usually fund only short-term projects (two to four years), are not necessarily the best instrument for funding long-term agricultural research activities, such as plant breeding and strategic research. The same problem arises because most project-based funds do not finance researchers' salaries or investments in equipment and infrastructure.
- *By being relatively inflexible in adjusting project proposals and implementation.* Strict adherence to selection transparency and procedures can jeopardize efficient selection and implementation of agricultural innovation projects. Simple mistakes in budgets or incomplete documentation sometimes result in outright rejection of project proposals.
- *By not fitting within existing bureaucratic procedures.* Government agencies often find it extremely difficult to administer a research grant within their bureaucratic procedures. For example, resources often cannot be transferred to the next financial year. A way to avoid this problem is to administer the research grant through a nonprofit foundation. Although this procedure adds to the overall administrative costs of an innovation project, timely and undisrupted disbursement of research resources may create some savings as well. In most countries, this legal construct is widely accepted.

Source: World Bank 2006b.

Through Bioconnect, all actors in the organic agriculture value chain, organized by product workgroups, have decision-making authority in research funding, utilizing public funds from the Ministry of Agriculture. Early results suggest that this model can induce more interactive and inclusive ways of working, given that the interaction of researchers and users is built into the system and is a prerequisite for obtaining funding.

Table 1.3 Client Orientation and Participation in Science and Technology Funds That Are Competitive and Specific to Agriculture in Select Latin American Countries

Country and competitive fund	Client orientation and participation
Brazil	
PRODETAB	The fund's steering committee (mainly government), in consultation with stakeholders, formulates and prioritizes innovation needs. Private sector involvement in project development and implementation is an important criterion for funding. The selection of projects for funding is a centralized process managed by experts. Special attention is given to ensuring geographic equity.
FNDCT Agribusiness	The private sector has a majority vote on the board. A consultancy firm produced the priority-setting document. Project selection is centralized and managed by experts.
Chile	
Several funds managed by FIA	FIA recently initiated regional consultation of farmers and other stakeholders to formulate regional priorities. It also started to issue regional calls for proposals in addition to a national call. A few small information offices have been opened to improve FIA's regional presence. Project selection is centralized and managed by experts. Most projects, depending on their particular objective, involve farmers.
Colombia	
PRONATTA	PRONATTA has about 20 local nodes throughout Colombia to assemble local researchers, extension agents, farmer representatives, government officials, and other interested stakeholders. The nodes identify and prioritize local research needs and develop project profiles, which are submitted to one of five regional coordination units. Projects are selected in two stages, first by a regional panel and ultimately by a national panel (consisting of the chairs of the regional panels). In both cases, a scoring method is used. Only the highest-scoring projects are funded. PRONATTA strongly favors farmer participation in the implementation of projects.
Ecuador	
PROMSA	Research priorities are based on past studies and refined at a workshop, where a scoring approach is used to develop priorities in a three-way matrix of commodities, agroclimatic regions, and thematic areas. Farmers' participation in priority setting has been low. Project selection is centralized and managed by experts, but each project has a reference group consisting of direct beneficiaries (farmers) and other stakeholders (other researchers, extension staff, agribusiness, and so on). Ideally the reference group participates in project design, planning, implementation, monitoring, and evaluation.
Mexico	
CONAFOR/CONACYT	The forestry sector is asked to submit its research needs, which form the basis for the call for proposals. Project selection is centralized and managed by experts.
SAGARPA/CONACYT Produce Foundations	Produce Foundations, set up in all 32 states, strongly involve farmers at all levels. Farmers have a majority vote on the boards of the foundations and provide the board chair. The identification of innovation needs and selection of projects are decentralized to production chains at the state level. Farmers are required to cofinance (usually in kind) technology transfer projects. Project selection is centralized and managed by experts.

Source: World Bank 2006b.

Note: CONACYT = Consejo Nacional de Ciencia y Tecnología (National Council of Science and Technology); CONAFOR = Comisión Nacional Forestal (National Forest Commission); FIA = Fundación para la Innovación Agraria (Agricultural Innovation Foundation); FNDCT = Fundo Nacional de Desenvolvimento Científico e Tecnológico (National Fund for Scientific and Technological Development); PRODETAB = Projeto de Apoio ao Desenvolvimento de Tecnologias Agropecuárias para o Brasil (Agricultural Technology Development Project for Brazil); PROMSA = Programa de Modernización de los Servicios Agropecuarios (Agricultural Services Modernizatiopn Program); PRONATTA = Programa Nacional de Transferencia de Tecnología Agropecuaria (National Agricultural Technology Transfer Program); and SAGARPA = Secretaria de Agricultura, Ganaderia, Desarrollo Rural, Pesca y Alimentacion (Ministry of Agriculture, Livestock, Rural Development, Fisheries, and Food).

Box 1.14   Bioconnect: A Networked Research Council in the Netherlands

Bioconnect is a new type of research "council" for organic agriculture, one that is able to induce an inclusive way of working.

Supportive government policy. To support knowledge development and exchange in organic agriculture, the government allocates the subsector 10 percent of its budget for policy support research and statutory research tasks (€9.6 million in 2008). In 2005, the government delegated responsibility for setting the research agenda for organic agriculture to stakeholders by creating Bioconnect. The goal is to determine whether delegating research governance to networks in which users allocate research funding can work for other areas of publicly funded, applied agricultural research. Farmer-driven research planning exists in the Netherlands but has not always forged broader linkages within the innovation system.

Bioconnect. Through product workgroups (dairy, glass house horticulture, and so on), the users of research (farmers, agrifood supply and processing companies, civic advocacy organizations representing consumers) unite with researchers, consultants, and policy makers to determine how to use public research funding. Working within themes established by the government, workgroups propose topics based on broad demand among their constituencies. They discuss and prioritize the topics with research coordinators to align research with sector needs. Research is contracted on the basis of proposals, which are selected through a review by the users and funding body and not solely through peer reviews of scientific merit. Bioconnect facilitates the participatory research arising from the priority setting and links it to legislative and market developments in an effort to ensure that research innovations have an impact.

A *knowledge manager* is the group's facilitator, streamlining information flows and mediating between actor groups. A *theme coordinator* informs researchers about workgroup results to ensure that proposals correspond to government funding guidelines. The knowledge managers embody the management of the network; a Knowledge Committee oversees the broad research themes and seeks to maintain consistency throughout the program.

Early results and challenges. Despite the reluctance of some researchers, the model does initiate learning about more interactive ways of working. Similarly, despite their strong commitment to users' interests, delegation systems also allow government to determine the macro priorities within which users can maneuver. As an intermediary in all aspects of the research process, from priority setting to disseminating results, Bioconnect occupies a pivotal but neutral position—one that is not easy to maintain. The position of such an intermediary depends on whether it promotes institutional learning with regard to the new roles of the actors involved and helps their goals to converge. Bioconnect must constantly balance the interests and gain the trust of the range of stakeholders for whom it mediates and on whom it depends for its resources (social and operating capital). As part of this balancing act, Bioconnect has to convince research contractors of the value of setting research priorities through a single, multiactor platform; balance the strategic interests of the research contractors; and urge government to achieve cohesion across ministry directorates with respect to budgets, macro priorities, and processes for monitoring the network.

Source: Klerkx, Hall, and Leeuwis 2009; Klerkx and Leeuwis 2008.

How to Build Innovation Networks

Javier Ekboir, Institutional Learning and Change Initiative (ILAC)

SYNOPSIS

The success of spontaneously emerging innovation networks has sparked interest in deliberate efforts to build new networks and strengthen existing ones. This note describes both spontaneous and deliberately created innovation networks, discusses the factors related to success, and offers lessons and guidelines for creating new networks or supporting existing ones. To support the emergence and consolidation of innovation networks, avoid imposing formal organizational arrangements. Promote policies that provide incentives (beyond simple subsidies for participation) and an enabling environment for innovation (especially, creating market and social opportunities for innovation). Consolidate the innovation capabilities of networks through policies that support action-research and decentralized experimentation with centralized learning, the construction of local and distant interactions, and resources for participatory assessment of innovation capabilities. Finally, build capabilities of members of innovation networks by strengthening innovation capabilities in private firms and civil society organizations, building the capacity of actors willing to be catalytic agents, and promoting cultural changes in public institutions that foster their participation in innovation networks.

BACKGROUND AND CONTEXT FOR INVESTMENT

Innovation networks emerge (and are becoming more prevalent) because no single actor commands all the resources required to innovate at the pace demanded by modern markets (Powell and Grodal 2005). To access the resources and information they lack, actors establish informal collaborations. Innovation networks emerge when the problems or opportunities affecting actors in the innovation system are not clearly defined. This uncertainty prevents effective contracting and structured partnerships, and it forces partners to use flexible approaches to explore potential solutions. Innovation networks can emerge from the deliberate actions of actors, as in Brazil with no-till agriculture (IAP 1), or they can emerge inadvertently as actors perform tasks that require collaboration.

In this module, an innovation network is defined as a diverse group of agents who voluntarily contribute knowledge and other resources (money, equipment, and land, for example) to jointly develop or improve a social or economic process or product. Innovation networks are a special form of organization with a nonhierarchical structure, a collaboration-based culture, consensus-based coordination (because members are free to leave the network at any time), usually no legal personality (especially in their early stages), and often relatively fuzzy objectives (such as improving the management of natural resources). They evolve with market opportunities and the technologies they develop. Different types of actors participate in innovation networks; in fact, membership changes in response to emerging problems and opportunities or the development and diffusion of innovations (IAP 1).

Innovation networks have supported the development of important innovations, both in agriculture and agroindustries, such as no-till agriculture and Chile's wine industry. The networks' flexibility allows groups of actors to minimize transaction costs and minimize risk by adding flexibility.

More than any other feature, it is diversity that characterizes innovation networks. Networks' goals can differ. Some develop technical solutions to specific problems (like the no-till networks), create new business models and new products (like Papa Andina, described in box 1.15), or reduce poverty. Others empower marginalized groups by fostering innovations appropriate to their resources and needs—an example is the International Livestock Research Institute's (ILRI's) innovation platforms to overcome fodder scarcity[1]—or develop improved crop varieties. Network membership can encompass farmers, farmer organizations,

Box 1.15   The Creation and Consolidation of Papa Andina

Since 1998, Papa Andina has fostered agronomic, technical, and commercial innovations in Andean potato-based food systems to improve farmers' access to more dynamic and lucrative markets. The network, which reaches about 4,000 poor rural households and includes about 30 partners in Bolivia, Ecuador, and Peru, pursues several strategies: increasing demand for native and commercial potato varieties, adding value to potatoes, improving contractual arrangements, and facilitating access to commercial information. Financed by the Swiss Agency for Development and Cooperation and other donors, Papa Andina is hosted by the International Potato Center (CIP).

Based on a participatory method for stimulating agricultural innovation (Rapid Appraisal of Agricultural Knowledge Systems), in 2000 CIP researchers started to enhance innovation through a participatory market chain approach and stakeholder platforms. These efforts bring researchers together with small-scale farmers, agricultural service providers, and market chain actors (including chefs, supermarkets, and potato processors). In each participating country, Papa Andina coordinates its activities with a "strategic partner" that assumes a leadership and coordinating role in market chain innovation: PROINPA Foundation in Bolivia, the INCOPA Project in Peru, and the National Potato Program of INIAP in Ecuador. Although the approaches to developing the networks are common (the participatory market chain approach and stakeholder platforms), different organizational arrangements, involving different partners and interaction patterns, emerged in each project implemented by Papa Andina.

Papa Andina's success resulted largely from the exploration of alternatives to reaching its goal (poverty alleviation), the involvement of different actors in developing and testing innovations, and the continued support of its funders.

Source: Author, based on Devaux et al. 2009.
Note: CIP = Centro Internacional de la Papa; INCOPA = Innovación, tecnológica y competitividad de la papa en Perú; INIAP = Instituto Nacional Autónomo de Investigaciones Agropecuarias; PROINPA = Promoción e Investigación de Productos Andinos.

private and public firms, researchers, extension agents, government agencies, and donors. Innovation networks differ in their origins and their approaches to developing and diffusing innovations. For example, private firms or farmers catalyzed no-till networks in South America and relied on farmer associations to diffuse the technology; in contrast, a private firm and NGO catalyzed Ghana's no-till network but relied heavily on a PhD student and traditional public extension methods (Ekboir 2002).

Innovation networks differ from farmer organizations in that farmer organizations have a homogeneous membership and more formal, stable relations. Innovation networks differ from value chains in that the latter are more stable, are focused on delivering a product or service, and are coordinated by a central actor (such as a supermarket) that organizes the operation (TN 3).

This thematic note describes strategies to foster the emergence of innovation networks or strengthen existing networks. It discusses the benefits and policy issues involved, particularly the need to strike a balance in how the public and private sector participate. The examples highlight the many forms and operating modes that innovation networks can adopt and the potential problems that can render networks ineffective or dependent on external funding.

INVESTMENT NEEDED

Because of their informal structure and frequent changes, innovation networks need flexible and sustained support, often from innovation brokers. This flexibility does not fit easily into the usual requirements of publicly funded projects, especially because it is difficult at first to define clear objectives and the steps that will deliver the innovation. In contrast, the private sector has long recognized the special nature of innovation and created flexible approaches to support it, including venture and angel funds and actions to develop capabilities that favor innovation. In recent years, some governments, international donors, and multilateral funding agencies have started to support innovation projects that include financing for innovation networks—either to support the emergence of new networks or to strengthen existing ones.

Fostering the emergence of innovation networks

Actions to foster the emergence of innovation networks seek to create trust among potential partners, identify common goals, establish the bases of collaboration, and develop innovation capabilities. Few developing countries have programs to foster the emergence of innovation networks; on the other hand, many developed countries support such programs, such as Canada's Agricultural Biorefinery Innovation Network for Green Energy, Fuels, and Chemicals and the Dutch InnovationNetwork.

More specifically, the following investments foster the emergence of innovation networks:

- Financing catalytic agents—innovation brokers, business incubators, NGOs, researchers, extension agents, and groups of farmers—that assemble potential partners. Their remuneration should be linked to measures of the consolidation of the network. The milestones should not be imposed by the financing institution but negotiated between potential partners, funders, and the innovation broker. When the milestones are defined by the funders alone, innovation brokers tend to respond to their interests instead of those of potential network partners. The catalytic agents should be trained in the different methodologies that have been developed to foster the emergence of partnerships (see, for example, Hartwich et al. 2007; USAID 2004).
- Giving small, short-term grants to potential catalytic agents of networks to facilitate interactions with potential partners, such as organizing meetings or establishing electronic platforms for communication.
- Broadening the mandate of research and extension institutions to include promoting innovation networks. Appropriate incentives should be introduced and resources made available.
- Establishing a team to develop capabilities and a monitoring and evaluation system to assess the different methods used to promote innovation networks and catalytic agents.

Supporting existing innovation networks

Given their voluntary nature, innovation networks survive when they can implement collective action. Support for existing networks should seek to (1) strengthen their ability to assess their strengths and weaknesses and gain access to needed resources; (2) build the capacity of network leaders to steer nonhierarchical organizations; (3) implement strategies to gain access to needed resources; (4) develop capabilities to explore new instruments to develop and foster innovations; and (5) facilitate the implementation of joint activities. Investments can include:

- Resources to consolidate innovation networks, including the implementation of collective action (hire facilitators, build human resources with formal and informal training, and support travel, meetings, communications and communications products, and experimentation).
- Strengthening innovation capabilities of nonpublic actors in the AIS, for example, with consulting services, extension activities, technological interchanges, seminars, and workshops on the dynamics of innovation networks.
- Creating venture funds to finance the development of innovations, similar to those used in the high-tech industries.
- Fostering the transformation of public research and extension institutions so that they can better integrate into innovation networks.
- Building the understanding of the main actors in the AIS, especially senior civil servants, of the dynamics of innovation and the nature of innovation networks.
- The no-till networks in South America are examples of consolidated innovation networks. Initially, they were supported by agrochemical companies that provided funds and expertise; once they matured, the networks were managed by farmers and supported by a large number of companies and farmers (IAP 1).

POTENTIAL BENEFITS

Innovation networks have many potential benefits:

- They can spur the development of innovations thanks to increased collaboration and coordination among diverse actors in the AIS; more effective identification of organizational, commercial, technical, and institutional opportunities; better exploration of alternative solutions to reach the network's goals; lower cost of searching for technical and commercial information; easier experimentation with alternative solutions; and better development of new ideas and skills.
- More rapid development of social and economic innovations can result in stronger economic growth and more sustainable use of natural resources. Strong networks are particularly necessary for innovation when public organizations, especially research and extension, are weak.
- Human, social, physical, and financial resources are used more effectively for innovation.

- Innovation programs become more efficient as all actors in the AIS, especially governments and donors, expand their innovation capabilities.
- Innovations diffuse faster because the participation of users in the network increases the odds that the results will be useful to farmers and other actors in the AIS.
- Public institutions become more effective when they participate actively in innovation efforts.
- Development is more inclusive when innovations meet the needs of marginalized groups.
- Institutional innovations become more inclusive and effective because marginalized groups gain a stronger influence on the design and implementation of innovation policies.

Box 1.16 illustrates some of the benefits that an innovation network can yield, using the example of local and foreign participation in developing small-scale equipment for no-till agriculture in South Asia.

POLICY ISSUES

Key policy issues related to the emergence and continued success of innovation networks include their consolidation and sustainability; social considerations (when networks form spontaneously, based on the resources that each actor contributes, the poorest and most marginalized groups may not be included); and the changing roles of public and private actors in the network.

Sustainability

The sustainability of innovation networks depends on many internal and external factors. Internal factors include the

---

Box 1.16   Benefits of Local and Foreign Collaboration to Develop Equipment for No-Till Agriculture in South Asia

No-till agriculture is thought to offer environmental and economic advantages for rice-wheat production systems in South Asia's Indo-Gangetic Plains. Farmers in Pakistan, India, Nepal, and Bangladesh have rapidly adopted the practices since 2000. No-till approaches used in other parts of the world were tested and modified to suit local conditions by a research consortium led by the International Maize and Wheat Improvement Center (CIMMYT) and the Indian Council on Agricultural Research. The technology did not take hold until researchers and agricultural engineers from abroad began working with local, small-scale manufacturers to design prototype no-till seeders. Particularly important were several exchanges of prototypes between small-scale manufacturers from Bolivia and India. The exchanges were mediated by CIMMYT, which worked in both countries with local artisans and handled the logistics of importing the prototypes. Several modifications were made to the original design, and manufacturers now produce and distribute a wide array of the new seeders. Results of the interaction between local and foreign actors included:

- No-till methods were adopted rapidly in irrigated rice-wheat systems. No-till was used on about 820,000 hectares by 2003–04.
- The number of small factories making no-till equipment grew. By 2003, an estimated 15,700 seeders had been sold in Haryana and Punjab, India.
- Profits for small-scale farmers increased, pollution decreased, and water savings increased. Seasonal savings in diesel for land preparation were estimated in the range of 15–60 liters per hectare, representing a 60–90 percent savings. Water savings in wheat production were estimated at 20–35 percent and profitability increased by 46 percent.
- Local manufacturers gained access to information from different countries in the form of original, nonadapted equipment.
- Equipment was more rapidly adapted to smallholders' conditions.
- Technical and scientific information was generated.
- Networks of farmers, equipment manufacturers, and researchers from international centers and state universities were consolidated.
- The impact of CIMMYT's actions was magnified.
- A market for planting services emerged.

Source: Ekboir 2002; World Bank 2006; Laxmi, Erenstein, and Gupta 2007.
Note: CIMMYT = Centro Internacional de Mejoramiento de Maíz y Trigo.

presence of effective leadership that works by consensus, the development of collective learning routines and trust, effective governance mechanisms for collective action, and a culture that respects the different partners. Support to sustain innovation networks should help build trust among potential partners and encourage a culture of collaboration. External factors influencing networks' sustainability include the emergence of commercial and technological opportunities, timely access to key inputs (particularly to specialized information and qualified professionals), the general socioeconomic climate, and noninterference from governments.

Because of their voluntary nature and low barriers to exit, innovation networks have no formal hierarchy and operate by consensus. Although some actors have more influence than others, they still have to consider the interests of other partners if they want them to remain in the network. The more focused a network, the better the chances of its consolidation. It is important to remember, however, that motives for joining the network may differ (profit, advancement of science, and so on) and could eventually conflict.

Social considerations

By definition, marginalized populations (poor households, women, and other groups) have few financial, human, and social resources to contribute to innovation networks (Spielman et al. 2008). Their participation in innovation processes often depends on funds and expertise provided by other actors (often NGOs or donors). Innovation networks can cater to the needs of marginalized populations in two ways: by organizing them to improve their access to markets and their influence on innovation processes and policies, and by developing innovations that address their specific needs.

The impacts of innovation networks on poverty are not easy to identify. For example, no-till practices caused land ownership to become more concentrated in southern Brazil. Many smallholders sold their farms and moved to newly developed areas in central and northern Brazil, where land prices were one-tenth of those in the south. Working on much larger areas, the former smallholders became middle-class farmers. Smallholders who remained in the south also benefited from no-till, but only after other actors formed innovation networks to develop special no-till packages for them (Ekboir 2003).

The role of public researchers in innovation networks

The participation of public sector researchers is not a condition for successful innovation networks, because most innovations are developed by networks of productive or social agents. Yet researchers are necessary when problems require science-based solutions, such as understanding the changes in soil dynamics induced by conservation practices or managing native forests that have life cycles that may last for over a century. In any event, the incentives and cultures within most public offices in developing countries, including research and extension institutes, hamper their participation in innovation networks. These restrictive environments have not prevented many researchers from making important contributions to innovation networks, however, as illustrated in box 1.17.

Public and private sector roles

Membership in innovation networks usually is varied and may include partners from the public and private sectors, civil society, and international organizations. Generally, networks formed only by private partners seek to develop products with commercial value. As some of the examples in this thematic note indicate, however, it is not uncommon for private firms to team with public or international research institutes to develop commercial innovations, such as improved plant varieties, or innovations with social content, such as no-till packages for small-scale farmers in southern Brazil (Ekboir 2002). The network partners may have had different interests, but they could still collaborate toward a common goal. For example, the private firms may have wanted to develop a new market, whereas the public partners wanted to achieve social objectives. Private firms may participate in innovation networks without a commercial focus when they have a social responsibility policy. Civil society organizations have also participated in networks to develop innovations that address their members' needs. In a few cases, these organizations were the catalytic agents for innovation networks.

As with other aspects of innovation, the public sector can play several roles in supporting innovation networks. It can provide funds to support innovation activities, implement programs to foster the emergence of networks (particularly by supporting innovation brokers), or provide specialized assets, such as research capabilities. The public sector can offer incentives for private firms to participate in innovation networks, such as tax rebates and matching funds, and it can promote reform in public research and educational institutions so that they can participate more effectively in innovation networks. Some public research institutions have taken steps to foster the emergence of

Box 1.17   The Roles of Individuals and Organizational Culture in the Development
            of Innovation Networks: A Mexican Example

In the early 1980s, a researcher from Mexico's national agricultural research organization crossbred Zebu cattle with European breeds in his ranch. The neighbors, interested in the new animals, soon organized a group to discuss technology issues, such as breeding techniques and pasture management. Based on experiences in Argentina with AACREA, an organization of farmer groups that conduct on-farm applied research and share the results with one another, a few researchers promoted the creation of similar groups of ranchers in the same region of Mexico. The groups started to exchange information and work on validating technologies. The research organization's authorities strongly objected to these actions, claiming that they were extension activities and thus beyond the mandate for national agricultural research. Researchers responded by meeting on Saturdays to avoid being accused of using their working hours for unauthorized activities. By the late 1990s, the groups—GGAVATTS—had developed and diffused several important innovations in their region. In the early 2000s, the federal authorities acknowledged their performance and man-

dated that their methodology be used in all publicly supported extension activities. They also decided that GGAVATTs should have priority in accessing support programs. The number of GGAVATTs exploded. Most of the new groups were created by technicians hired to do so, and they did not respond to farmers' goals as the original groups had done. Farmers joined chiefly to gain easier access to public resources. Only a few of the new GGAVATTs improved members' ranching technologies. Most faded away when government support ended.

This experience contains two important lessons for supporting innovation networks. First, innovation networks emerge when different actors find a common interest in exploring ways to improve a product or a process. The spontaneous convergence of interests is difficult, however, and facilitating convergence can greatly accelerate the emergence of networks. Second, facilitation should not be coupled with the disbursement of other subsidies. Otherwise farmers join to receive the subsidies and not because they are truly interested in innovating.

Source: Ekboir et al. 2009.
Note: AACREA = Asociación Argentina de Consorcios Regionales de Experimentación Agrícola (Argentine Association of Regional Consortiums for Agricultural Experimentation); GGVATTS = Grupos Ganaderos de Validación y Transferencia de Tecnología (Livestock Producer Groups for Technology Validation and Transfer).

innovation networks, but often their approach follows a traditional, linear vision of science.

LESSONS LEARNED

Experiences with innovation networks that develop spontaneously and those that are developed deliberately offer useful lessons about when networks are most effective, who joins, and what they accomplish. Other lessons relate to how networks can be catalyzed and otherwise motivated, how they can be consolidated, and how they can avoid depending on external funding agencies.

When are innovation networks effective?

Innovation networks are particularly effective for developing and diffusing technical and commercial innovations that deal with ill-defined or complex issues, like new

approaches for natural resource management (such as water catchment in arid regions) or the development of market niches for poor households (TN 3). The problems and the potential solutions that innovation processes will explore are rarely well defined at the beginning. Because of their voluntary nature, innovation networks have the flexibility to adapt to emerging needs and opportunities, but this responsiveness can depend on the availability of flexible funding and on enabling all actors in the AIS (especially public researchers and extension agents) to participate.

A network's origins greatly influence who joins and how it innovates

Innovation networks can emerge because of the interests of one person—a farmer, a private firm, or a researcher (as in box 1.17)—or because it is part of a research program, like the Ghanaian no-till equipment network (box 1.23 in IAP 1).

It can originate with field staff and then be integrated with an organization's work plan, or it can originate with the top management. When traditional research organizations attempt to create innovation networks, they usually emphasize the use of scientific experimentation. Other partners are sought to contribute complementary scientific capabilities, provide farmers' perspectives, or participate in diffusion. This kind of network includes the scientific network that developed Golden Rice (a more nutritious type of rice)[2] and demand-driven innovation networks in which farmers define the problems and researchers work on solutions (replicating the traditional, linear research pipeline).[3] When farmers, private firms, or NGOs create innovation networks, they emphasize the social and organizational dimensions. They often adapt techniques without the collaboration of scientists. They establish new kinds of organizations to diffuse them (like some of the no-till networks) (Ekboir 2002) or rely on farmer-to-farmer communication. More recently, social scientists in research institutes (such as CIP, ILRI, and Wageningen University) have helped to create innovation networks involving social and agronomic researchers and other actors from the AIS. As all innovation networks mature, the importance and roles of the actors change (see, for example, IAP 1).

The emergence of innovation networks requires two types of partners to collaborate: a catalytic agent and a funder. Committed and innovative catalytic agents are vital to the emergence of innovation networks, because by definition only they can induce other partners to invest time and resources in the network, and they also seek partners to contribute the resources needed by the network (Ekboir et al. 2009). Catalytic agents can have different organizational affiliations. For example, they can belong to an organization willing to start an innovation network in pursuit of its own interests or can be hired by a project as an innovation broker. Regardless of organizational affiliation, to start an innovation network, a catalytic agent must seek partners willing to contribute to the common effort, identify which capabilities the network needs, and look for new partners that can contribute those capabilities until at least one is found. In searching for partners to initiate an innovation network, it is important to focus both on individuals and institutions. Support from top management is of little help if the people who must participate in field activities are not motivated. For example, the first no-till network in Brazil was catalyzed by a researcher from a private firm. Realizing that no-till required new planters, he contacted all the manufacturers in the area until one agreed to participate in the emerging network. He also contacted several public researchers and research institutes until he gathered sufficient expertise in agronomy and soil science. The motivation of these researchers was critical to the network's success (IAP 1).

Funders, on the other hand, provide the resources that allow the catalytic agent to operate. Once the network has emerged, it is essential for it to establish effective links with other networks that can provide information, share experiences, and provide access to critical assets the network lacks. The links are often not provided by the catalytic agent but by other well-connected partners ("central nodes," in the terminology of Social Network Analysis). In addition to these partners (catalytic agents, central nodes, and funders), innovation networks are populated by individuals and organizations. Like any other organization, each innovation network develops its own culture, learning routines, heuristics, and modes of interaction.

Central nodes facilitate information flows within and between networks

Innovation networks facilitate the exchange of knowledge, abilities, and resources among their members, but effective networks also interact with other networks and sources of information (Ekboir et al. 2009) through a few central nodes.[4] Innovation brokers are particularly prepared to become central nodes. By linking clusters of network actors, the central nodes facilitate flows of information and resources, as exemplified by CIMMYT's role in developing no-till equipment in South Asia, discussed earlier. Securing funding to build these connections has been difficult, because the benefits of innovation networks have been identified so recently.

Supporting innovation networks without creating dependence

When donors or governments have tried to use innovation networks to distribute resources, the number of formal groups in the networks has surged, but most were less innovative and had weaker internal cohesion than groups formed without the incentive. As a result, most disappeared when the public program ended, as with the Mexican livestock groups described previously. Subsidies to create innovation networks have had similar results; partners did not develop the social capital to keep the network alive once the subsidies ended (TN 1). Supporting networks that already exist or supporting innovation brokers that foster the creation of social capital among partners are more sustainable alternatives.

Incentives and structure for innovation networks

Innovation networks are made up of individuals, even if they represent an organization. Their contribution to the collective effort depends on the personal benefits they gain from participation, the incentives offered by their organizations, and their organizations' cultures. It is relatively common for projects in developing countries to start to build a network by signing memoranda of understanding between public and private organizations. The public organizations often have weak accountability and lack the incentives and culture to effectively induce their members to collaborate in external networks (Ekboir et al. 2009). Despite these misaligned incentives, many researchers have participated in innovation networks out of personal interest.

Governance and accountability mechanisms are essential for innovation networks to survive, because one of their major existential threats is opportunistic behavior by members. Governance and accountability mechanisms cannot be imposed, however. All members must see such mechanisms as reasonable and practical. They must be negotiated clearly and with care, with all involved.

Innovation networks should not be pushed to adopt a formal structure. Given the uncertain nature of innovation, formal contracts and intellectual property rights are seldom important for the development of innovations that are a little more complex than incremental improvements (Rycroft and Kash 1999).

RECOMMENDATIONS FOR PRACTITIONERS

To support the emergence and consolidation of innovation networks, recognize their informal nature. *Avoid imposing formal organizational arrangements, and promote policies that*

- Provide incentives for innovation by creating market opportunities, helping to access crucial inputs (credit is a common need), and facilitating the flow of commercial and technical information.

- Create an environment conducive to innovation. Eliminate excessive bureaucratic requirements for business, develop basic infrastructure, and facilitate the formation of partnerships by, for example, training human resources, supporting exchange visits, holding meetings, and developing communications facilities and material.
- Strengthen analytical capacities in the public sector so that it can provide better support to innovation processes.

Consolidate the innovation capabilities of networks through policies that

- Support action-research projects and decentralized experimentation with centralized learning to identify new instruments to foster innovation networks and diffuse best practices.
- Support the construction of local and distant interactions. When distant ties are missing, create them by linking local networks to international sources of information and resources.
- Provide resources for participatory assessment of innovation capabilities.

Build capabilities of members of innovation networks by

- Providing resources to strengthen innovation capabilities in private firms and civil society organizations. For example, use consulting services, innovation brokers, innovation incubators, extension activities, technological interchanges, and seminars and workshops on the dynamics of innovation networks.
- Building the capacity of actors willing to be catalytic agents through courses and mentoring. These actors may be dedicated organizations that foster the exchange of ideas and shape the public debate on particular topics.
- Promoting a change of mentality, management (including incentives), and culture in public institutions, including research institutes, so that they are more willing to participate in innovation networks.

Facilitating Smallholders' Access to Modern Marketing Chains

Javier Ekboir, Institutional Learning and Change Initiative (ILAC)

SYNOPSIS

Modern marketing chains can heavily influence rural employment, poverty, and asset ownership, especially the concentration of land, physical assets, credit, and social capital in the hands of farmers with access to those chains. Only farmers with strong capabilities for innovation can survive in these highly competitive marketing chains, in which a dominant actor close to consumers (usually a supermarket or broker) organizes many producers and intermediaries and coordinates their interactions. The private sector or nongovernmental organizations (NGOs) will lead the development of modern marketing chains, but the public sector should play an important supporting role by improving infrastructure, facilitating access to input markets and financial services, modernizing traditional wholesale markets and linking them with modern marketing chains, improving rural education, and supporting extension and advisory services. Project managers, field staff, donors, and policy makers require sound knowledge of the dynamics of modern marketing chains, niche markets, and innovation processes, along with the opportunities and challenges they pose for small-scale farmers. For a group to identify and maintain market links, strong leadership with the appropriate experience is the most important factor. External facilitation can help to access modern marketing chains, but only if it uses appropriate approaches that target farmers with strong innovation and managerial capabilities. Building social capital should be an important component of projects to develop marketing chains, and social capital should be an important criterion in allocating funds to participating groups. Projects should not impose a model of interaction. Interventions generally should not organize the poorest farmers to sell directly to modern chains but rather to develop niche markets, sell to wholesalers, or obtain training for specialized off-farm employment.

BACKGROUND AND CONTEXT FOR INVESTMENT

Modern marketing chains can foster economic growth and alleviate poverty directly and indirectly. Direct effects include higher incomes for farmers who participate in the marketing chain; the development of innovation capabilities (through interactions with other actors in the chain); the diffusion of technical, organizational, and institutional innovations; and the creation of employment in rural areas. Indirect effects include expansion of local economies, exposure of local entrepreneurs outside the chain to the operation of modern markets, the development of links with new actors in the AIS, and the modernization of traditional wet and wholesale markets. Given these benefits, donors, multilateral organizations, and governments have assigned high priority to improving poor farmers' access to high-value agriculture and modern marketing chains, yet only a fraction of smallholders have managed to succeed, either individually or through organizations (Hellin, Lundy, and Meijer 2009; Wang et al. 2009).

What differentiates modern marketing chains from farmer organizations and innovation networks? *Modern marketing chains* have many features of buyer-driven value chains: an actor close to consumers (usually a supermarket or a broker) dominates, organizing many producers and intermediaries, deciding who participates in the chain, overseeing all the links from the farm to the shelf, defining the nature of the interactions and commercial conditions, and setting quality and safety standards. *Farmer organizations*, on the other hand, are member organizations that include only one type of actor; *innovation networks* are open organizations formed voluntarily by different types of actors. Other important features of modern marketing chains are that they focus on marketing specific products (vegetables, fruits, meat, and so on), access to the chain is highly restricted, verbal contracts based on trust are common but informal transactions rare, and technologies are generated mostly in developed countries and imposed by the leading agent.

High-value products are commercialized through four types of modern marketing chain. *The first type* includes traditional export crops produced on plantations, such as coffee, tea, and bananas, and marketed mainly by large companies and traditional agroprocessors. Most innovations in these marketing chains are technical, because the organizational aspects have already been resolved. In a few cases, such as the Kenya Tea Development Agency (TN 4, box 1.22) or the Colombian Coffee Growers' Federation,[1] small-scale farmers have created large, successful organizations that compete internationally.

The second type of modern marketing chain, which supplies developed countries with fresh and processed fruits and vegetables, fish and fish products, meats, nuts, spices, and flowers, started in the early 1980s following trade liberalization and improvements in logistics. Traders and agroprocessors usually work with large-scale farmers under different associative forms, although occasionally they provide financing and technical advice to smallholders. Both technical and organizational innovations are important in this chain, and local actors have to develop strong innovation capabilities to remain competitive in global markets.

The third type of modern marketing chain was triggered by the expansion of fast-food chains and supermarkets that supply mostly domestic markets, although increasingly they reach foreign markets. This type of chain is a product of the internationalization of wholesaling and logistics, consolidation of rural and urban wholesale markets, emergence of specialized and dedicated wholesalers who organize procurement, growth in vertical coordination, and the introduction of private grades and standards. The better-off smallholders tend to sell through this channel (sellers in traditional channels tend to be less well off); they have more capital (especially irrigation facilities), easier access to credit, and greater specialization in commercial horticulture (Reardon et al. 2009). Actors in these chains use sophisticated production packages, but the most important innovations are organizational: coordinating production by large numbers of farmers of products of consistently high quality (frequently highly perishable) and delivering them to numerous distant retail sales points (increasingly abroad). The same products that are sold in modern marketing chains are also sold in traditional wet and wholesale markets; the difference is that the latter usually handle products of mixed quality and operate with spot prices, whereas the modern chains must adhere to high quality standards, and deliveries and prices are set in advance. Box 1.18 describes the operation of these chains in three countries where they have attained different levels of development.

The fourth type of modern marketing chain comprises niche markets in their many forms. Examples include the following:

- Smallholders close to a large city, who sell directly to consumers in a process similar to the "locavore" movement in developed countries.
- Development projects that create new markets for traditional products or new products based on traditional crops (Papa Andina is an example for Andean potatoes; see module 7, IAP 3, and module 4, TN 4.
- The Fair Trade movement, which caters to specific population segments in developed countries.[2]

Some niches are more demanding than others in terms of quality and commercial requirements; for example, wealthy consumers in nearby cities are less demanding than consumers from developed countries who buy in Fair Trade shops. Participation in these markets depends on farmers' capacity to meet their requirements. Although niche markets have had important impacts on local communities, they cannot expand beyond a certain size without becoming commoditized. For this reason, they can make only a limited contribution to alleviating poverty.

INVESTMENT NEEDED

Investments to support modern marketing chains should support the development of organizational capabilities for innovation along the chain (especially among farmers), coordinate actors operating in the chain, and create the enabling conditions for innovation. It is important to recognize that although modern marketing chains are growing, traditional wholesale and wet markets will probably retain an important share of high-value agricultural produce and remain the main marketing option for most smallholders. As modern marketing chains become increasingly concentrated, it will be harder for smallholders and farmer organizations to access them directly. For this reason, support to modern marketing chains should be accompanied by programs to help the farmers who are excluded.

Supporting the expansion of modern marketing chains through targeted investments

Private actors make most of the investments to develop modern marketing chains, but targeted public investments

Mexico. Supermarket procurement systems have moved from traditional wholesale markets to (1) networks based on centralized distribution centers; (2) emerging regional networks; (3) sourcing from a combination of specialized wholesalers (which are increasingly backward integrated into production zones), wholesaler/growers, and direct relations with grower/packer/shippers; and (4) new institutions, such as emerging private quality standards and implicit preferred supplier (quasi-contractual) relations. These trends make it very difficult for most smallholder and even traditional large wholesalers and medium-scale growers to sell to supermarkets. Supplying directly is increasingly the domain of (1) the large grower/packer/shipper, (2) the backward-integrated, large, specialized wholesaler, and (3) to a certain extent the wholesaler and farm company managing contract farming.

Kenya. Supermarkets have a relatively small but growing share of urban markets for high-value products. The leading chains began to modernize procurement by creating systems of preferred suppliers (farmers and specialized wholesalers) for key products, centralizing procurement into distribution centers, and starting to use private quality standards. Smallholders have found it difficult to enter modern chains because most lack the physical, financial, human, and organizational capital to lower production

and transaction costs to a point where they become competitive. Modern marketing chains increasingly are supplied by a new type of "middle-class" farmer emerging between large-scale export farmers and small-scale, traditional market farmers. These middle-class farmers buy traditional farms and convert pasture and grain fields to horticultural crops grown with modern production techniques.

China. The expansion of supermarkets has been driven by factors common to other developing countries—urbanization, income growth, and liberalization of foreign direct investment in retailing—as well as a number of China-specific policies, such as government investment in the sector and policies promoting conversion of wet markets to supermarkets. A feature unique to China is that poorer households produce the largest share of horticultural crops; that share is growing, whereas the share of richer households is falling. At the same time, there is little evidence that modern wholesalers or supermarkets source produce directly in rural communities. Instead, they rely on urban wholesale markets dominated by small-scale traders; for this reason, little or no effort is made to impose or monitor quality or safety standards directly among producers. Some researchers have reported the incipient emergence of features common to more developed procurement systems, such as dedicated providers.

Sources: Author, based on Reardon et al. 2007; Neven et al. 2009; Wang et al. 2009.

in specific markets and chains can trigger virtuous cycles of investment (IAP 3). These investments may support the formation of farmer groups, finance infrastructure and specialized equipment, build capabilities for farmers and their organizations to meet the standards required by private companies, and help provide market intelligence. For example,

> Sam's Club required cooled, packed, and delivered raspberries by a group of small farmers in Michoacán (Mexico). SEDAGRO/Michoacán [the state department for rural development] discovered exactly what those requirements are in joint meetings with the chain and the group, and then provided efficient and targeted help to the group to make those investments. The group then entered the business relationship and that provided further capital to make investments on their own. [Reardon et al. 2007, 35]

Another type of investment is the facilitation of lead farmer–outgrower schemes. Lead farmers have proven their ability to produce to the standards required by modern chains. Supermarkets or specialized wholesalers encourage lead farmers to organize their neighbors to produce to the same standards; the only incentive the buyers offer is a guaranteed market opportunity. For a fee (usually a percentage of the final sales), lead farmers provide various services that may include production planning, technical assistance, access to inputs, market intelligence, sorting and packing, transportation to market, and financial administration. Lead farmer programs require less external support than support for farmer organizations, but the farmers themselves have to make higher investments (Reardon et al. 2009; Neven et al. 2009). Investments to support these programs include financing farm equipment and capacity building for the farmers willing to work with the lead farmer, including paying for the time of the lead farmer. Identifying opportunities for targeted investments requires fund managers to have the capabilities and incentives appropriate for operating flexible programs.

Fostering coordination among actors in modern marketing chains

Modern marketing chains usually use sophisticated production and marketing mechanisms to bring perishable products from rural areas to urban consumers in developed and developing countries. In addition, the technologies and market requirements (such as the varieties grown and packaging methods) change often, forcing farmers to adapt. Greater coordination along the marketing chain helps smallholders to adapt by facilitating access to up-to-date information and financial resources (IAP 6). Successful coordination, however, requires coordinating organizations to respond to the needs of actors in the marketing chain and adapt themselves to changing conditions. Otherwise, coordination results only in formal agreements with little impact on the ground (see, for example, World Bank 2006). Investments include market intelligence, development of market and farm infrastructure, facilitation of interactions (usually by the actor that dominates the chain), and the formation of farmer groups, assisted by innovation brokers.

Developing niche markets

Niche markets are a particular form of innovation network, and the investments resemble those described in TN 2. These investments include financing innovation brokers and other actors that bring together potential partners and strengthening the innovation capabilities of nonpublic actors in the innovation system—for example, through consulting services, extension activities, technological interchanges, or seminars and workshops on the dynamics of innovation networks. Investments also include small, short-term grants for potential catalytic agents of innovation networks (researchers, extension agents, and groups of farmers) to facilitate interactions with potential partners (through meetings or electronic communications platforms), build capacity, and facilitate collective action. Still other investments could create venture funds to finance the development of innovations, similar to those used in high-tech industries. Finally, it may be necessary to invest in transforming public research and extension institutions so that they can better integrate into innovation networks.

Improving traditional markets to benefit smallholders

Investments to modernize traditional markets and improve farmers' bargaining power include developing small-scale farmers' human and social capital by fostering the emergence of farmer organizations, providing technical and organizational support, and offering training in modern marketing methods. Another useful investment is to facilitate smallholders' access to modern infrastructure, especially ICTs, as well as to services for business registration and incorporation, which can facilitate contractual relationships with retailers. It is also important to build the capabilities of field staff, project managers, donors, and policy makers to support smallholders; these capabilities include a clear understanding of the dynamics of modern marketing chains, niche markets, and innovation processes, along with the opportunities and challenges they pose for small-scale farmers. Finally, another investment to benefit smallholders is to modernize wholesale and wet markets by improving buildings, storage facilities, communications facilities, and roads.

POTENTIAL BENEFITS

To recapitulate, modern marketing chains can offer the following potential benefits:

- More rapid organizational and economic innovation, resulting in stronger economic growth.
- Faster creation of wealth for the richer tier of small-scale farmers through direct participation in modern chains.
- Creation of jobs in rural areas, some exclusively for women (sorting and grading fruits and vegetables, for example).
- For farmers who sell in modern marketing chains and for smallholders who access niche markets, easier access to input and output markets and other resources for innovation, such as technical advice, innovation networks, and participation in action-research projects.
- Reduced commercial risk for farmers owing to steady demand, more formal transactions, and higher probability of being paid on time.
- More effective use of human, social, physical, and financial resources for innovation.
- Faster creation of wealth and empowerment among farmers who can access niche markets.
- Better interaction and coordination among actors, resulting in a more dynamic innovation system for agriculture.
- In the case of niche markets, more inclusive development.

- Cheaper food that often meets higher quality and safety standards for consumers.
- Faster expansion of exports, improving the country's balance of payments.

POLICY ISSUES

Key policy issues related to modern marketing chains involve identifying ways for the chains to spur growth and reduce poverty, ensuring that modern chains are sustainable (some may require more public support than others), fostering more equitable access, clarifying appropriate roles for public and private investors, and identifying appropriate public investments.

Can modern marketing chains spur growth and reduce poverty?

As noted, high-value agriculture is expected to continue growing and transforming developing country agriculture for the foreseeable future. Instead of trying to impose new directions on this process, policy makers should identify the main trends guiding the expansion of different markets and seek interventions that can steer the process in ways that spur economic growth and alleviate poverty. For example, buyers in modern marketing chains clearly prefer to operate with the better-off smallholders. They buy from the poorest only when they have no other option. Interventions should not induce the poorest farmers to organize to sell directly to

modern chains but should rather organize them to develop niche markets, sell to wholesalers, or obtain training leading to off-farm employment as specialized workers.

Sustainability and social considerations

As with most private firms, the consolidation of modern marketing chains depends on the central actors' capabilities and on socioeconomic conditions (discussed later). Given the international nature of supermarkets and traders, eventually all central actors develop their own capabilities to organize most modern marketing chains.

Niche markets are a different case, given that they are innovation networks. In addition to external factors, the sustainability of niche markets depends on the presence of effective leadership, on farmers developing strong innovation capabilities (especially collective learning routines and trust), on long-term support from donors, and on monitoring and evaluation systems that allow experimentation over several years (box 1.19).

The expansion of modern marketing chains has increased disparities in asset ownership but has also created employment in rural areas. Smallholders' access to modern chains depends not only on their assets and organization but on being located in the procurement area for a particular crop. Many programs have helped marginalized groups (small-scale farmers and/or women) to organize and access modern marketing chains, but the failure rate has been very high. In general, the participation of small-scale

---

Box 1.19   Sourcing Practices Used by Mexican Supermarkets Reveal Experimentation with Contractual Arrangements and Types of Growers

Mexican supermarkets have pursued a number of strategies to source frozen vegetables. Exporters have used backward integration, relying on their own farms, for "high care" products demanding traceability. Some supermarkets adopted intense, highly controlled outgrower schemes in which they provided resources under contracts with small-scale farmers. Others relied on small-scale outgrowers for less-demanding products and market segments requiring less traceability and less technology, with the company providing only extension services to its outgrowers. Other companies chose to contract only with large-scale farmers, with a combination of large- and small-scale farmers, or only with smallholders through resource-provision contracts. A multinational working with large- and small-scale farmers had seven contract types, ranging from contracts with large-scale farmers under which no resources were provided to contracts with the smallest-scale farmers under which the company provided "specialized inputs and equipment, credit, technical assistance, and insurance." These examples show that projects should allow considerable latitude in the types of contractual arrangements and organizations that can be financed.

Source: Author, drawing heavily on Reardon et al. 2009 (quoted text, p. 1722).

farmers has depended more on support and expertise from NGOs and donors than on private traders or supermarkets, and it has been somewhat easier for smallholders to participate in niche markets or sell to wholesalers who supply supermarkets.

Public and private sector roles

As indicated, different types of agents have facilitated smallholders' access to modern marketing chains. Private agents (brokers and supermarkets) sometimes provide credit, inputs, and technical assistance (Reardon et al. 2009), and research institutes, NGOs, civil society organizations, and farmer organizations have also managed successful programs.[3] The public sector has provided funds, developed critical market infrastructure (roads, electricity, water, and communications networks), and facilitated interactions among actors in modern marketing chains.

Projects led by nonprofit organizations work with small-scale farmers or community-based organizations to identify niche markets (see, for example, Poitevin and Hassan 2006; Vermeulen et al. 2008; Devaux et al. 2009). These programs usually use participatory methods and may involve other chain actors in the process. Components of such projects include stakeholder analysis to identify the key people and organizations that should be invited to participate, visits, skill development, organizational aid to small and micro enterprises and community-based organizations, business plan preparation, assessment of markets and of opportunities to network, and stakeholder workshops.

The private sector will continue to lead the development of modern marketing chains, but the public sector has an important supporting role to play by improving infrastructure, facilitating access to input markets and financial services, fostering the modernization of traditional wholesale markets and helping them to link with modern marketing chains, improving education in rural areas, and supporting extension and advisory services.[4]

LESSONS LEARNED

Because buyers establish different types of commercial relations, even with farmers in the same area, projects should not try to impose a model of interaction. Commercial chains use different suppliers according to the nature of the product and the type of farmers present in the procurement area. They generally buy from small-scale farmers only when smallholders dominate a particular market (tomatoes in Guatemala, guavas in Mexico); large-scale farmers have

several marketing options, increasing the risk of unstable supply;[5] small-scale farmers are more able and willing to follow highly labor-intensive field management practices needed by the companies; and small-scale farmers have reduced transaction costs by organizing (Reardon et al. 2009). If a buyer has only small-scale farmers in its procurement area, it usually assists them with various inputs, credit, and technical advice.

Social capital should be an important criterion to allocate funds. Social capital is a significant factor in a group's ability to sell in modern marketing chains. Social capital is strong within mature groups with strong internal institutions, intragroup trust, altruistic behavior, membership in other groups, and ties to external service providers. Older groups that have built substantial social capital should be ranked above newer groups in consideration for support. Additionally, building social capital should be an important component of projects that support the development of modern marketing chains. Such efforts will require sustained commitment and support from external agents, which may include NGOs and innovation brokers.

Capable and motivated program leaders are important. Strong leadership with the appropriate experience is regarded as the most important factor for a group to identify and maintain market links, followed by the quality of external facilitation. It is very difficult for one person to have all the required qualifications. Leaders should be supported by teams of specialists, not just in marketing and business management but also in the management of innovation. A similar approach has been used in several countries to support innovation in small firms (see, for example, Shapira 1999). It is important to offer program leaders incentives to experiment and allow them to change programs as needed. To this end, funders and supervising agencies should set up program committees to analyze potential changes or allow program leaders to use small amounts of money to experiment with new approaches, and they should review results annually. Firms use such approaches to develop strong innovation capabilities in traditional markets (Whirlpool is one example; see box 1.7 in the module 1 overview and Skarzynsky and Gibson 2008).

RECOMMENDATIONS FOR PRACTITIONERS

In addition to improving the infrastructure and socioeconomic policies that are conducive to the consolidation of private firms, facilitating small-scale farmers' access to modern marketing chains requires instruments appropriate for each of the four marketing channels identified at the beginning of this thematic note.

To foster the participation of better-off smallholders in modern marketing chains,

■ Support the development of innovation capabilities in large farmer organizations, as described in TN 3.
■ In conjunction with the actor dominating the marketing chain, define criteria to select farmers for support to enter into one of the several contractual modes used in modern marketing chains. Social capital should be one of the most important selection criteria.
■ Do not impose specific interaction patterns as a condition for participating in projects.

To foster the participation of poor farmers and marginalized groups,

■ Support the modernization of traditional wholesale markets and traditional retailers, and help them to link with modern marketing chains (one example is China's "200 Markets Upgrading Program").[6] Identify high-value products that modern marketing chains must source from smallholders, such as low-volume and niche products, and support associations of market agents and small-scale farmers to provide them.
■ Finance investments (cold storage, packaging, and so on) that target specific requirements of specific modern marketing chains.
■ Build the capabilities of field staff, project managers, donors, and policy makers to support small-scale farmers.
■ Develop market intelligence programs that provide detailed, current information on market developments and trends.

■ Implement "business linkage" programs, such as business round tables and conventions, and support travel to those programs by farmers and government officials.
■ Foster the emergence and consolidation of farmer organizations through technical and organizational support, as described in TN 4. These organizations, in turn, can work closely with specialized wholesalers.
■ Develop new financial instruments to take advantage of the stable relationship between small-scale farmers (individually or organized) and the main actors in modern marketing chains. These actors, operating as intermediaries between banks and farmers, can help farmers obtain credit at market rates. Several countries, including Mexico and Croatia, have implemented such programs.
■ Develop small-scale farmers' human and social capital through sustained programs that include facilitation to form associations and training in modern marketing.
■ Help small-scale farmers obtain bank accounts, register a business, and transact the other business required to incorporate as formal companies that can enter legal, contractual arrangements with retailers.
■ Facilitate access to modern infrastructure, especially ICTs.

To support the development of niche markets,

■ Conduct multistakeholder planning exercises to identify potential niches and define strategies to develop them.
■ Apply recommendations in TN 2 for the development of innovation networks.
■ Finance innovation brokers to support policy makers and all other participants in the development of niches.
■ Aim for sustainable results. If niche markets are not profitable, they can create dependency on donors.

Building Innovation Capabilities in Farmer Organizations

Javier Ekboir, Institutional Learning and Change Initiative (ILAC)

SYNOPSIS

Organizational capabilities for innovation encompass the abilities of the organization's members (such as specialized knowledge, creativity, and task commitment) and the organization's key characteristics (such as culture, governance, communications and learning routines, and resources). Interventions to build organizational capabilities for innovation involve building the individual and collective capabilities themselves and making the enabling environment more supportive of innovation. Instruments for capacity building have included fostering interactions among actors in the AIS, creating venture capital funds, consolidating innovation brokers, strengthening the policy framework for innovation, and setting up training and mentoring programs and multistakeholder platforms. The environment for innovation improves when governments and donors facilitate dialogues to develop consensual innovation policies, create market and social opportunities for innovation, facilitate access to crucial inputs, strengthen flows of commercial and technical information, offer public institutions incentives to interact with farmer organizations, and promote a legal environment that supports farmer organizations and encourages decentralization to empower local farmer organizations. The most innovative farmer organizations have a federated structure, i.e., are composed of small, linked farmer groups, thus combining the best features of large organizations and small groups. In designing projects to develop farmer organizations that can innovate, practitioners should select organizations that can learn; avoid overly strict criteria for participation; avoid requiring organizations to become formal entities; identify the networks in which the farmer organization participates, because its partners can indicate its own innovativeness; use external consultants and innovation brokers to design training programs; and develop a

monitoring and evaluation system that promotes exploration and learning.

BACKGROUND AND CONTEXT FOR INVESTMENT

The weakening of public research and extension organizations and wider recognition of the complexity of innovation processes are creating opportunities for farmer organizations to develop and diffuse innovations. By pooling resources for innovation, connecting with other partners in innovation processes, and building their individual and collective capacity, farmers who belong to organizations are participating more effectively in innovation networks and value chains. They are gaining better access to new markets and production methods, including methods to manage natural resources. Farmer organizations acquire new bargaining power in input and output markets and they can also coordinate other actors in the AIS. They increasingly participate in designing and implementing innovation and research policies and programs and join national innovation councils, sectoral coordinating bodies, and the boards of research institutes. They lobby public organizations such as research institutes and industry regulators (World Bank 2004). They manage research funding and share in the financing, development, and diffusion of technical, commercial, organizational, and institutional innovations.

Whether farmer organizations participate effectively in these processes is strongly determined by their origin, evolution, the enabling environment, and the capabilities they develop, as discussed in the module overview. Often farmer organizations are unskilled in identifying the specific capabilities they lack and in defining strategies to build them up. Organizations do not develop capabilities for innovation overnight; they must make a sustained effort, make major investments over the long term, have committed leadership, and work with specialists in organizational change and innovation processes.

For many years, governments and funders favored the creation of cooperatives, but their performance has been rather disappointing. Lately, farmers and rural households have sought alternative organizational arrangements. These arrangements have had different goals, operate at different levels (local, regional, and national), and include community organizations, self-help groups, associations to manage natural resources (such as water user associations), and lobbying associations. According to the capabilities they develop and the type of interactions they establish with other actors in the AIS, farmer organizations can be categorized into four types:

■ Traditional, commodity-based farmer organizations, such as the Colombian Coffee Growers' Federation, India's dairy cooperatives, or the Kenya Tea Development Agency. Input supply, output processing, and marketing can be done by the organization or outsourced to private firms. Less attention goes to facilitating interactions and cooperation with potential partners in innovation. While the declared goal is usually the diffusion of technical innovations, successful commodity-based farmer organizations have innovated to coordinate large numbers of farmers. These organizations often sponsor their own research teams.

■ Nontraditional, market-oriented farmer organizations seek to improve market access through collaboration with key actors in the marketing chain (supermarkets or brokers). Often this kind of farmer organization is created with assistance from NGOs and/or externally funded projects (Papa Andina is an example). Innovations are viewed as technical, commercial, and social processes to be addressed through participatory methods. Research capabilities reside in local and foreign universities or international research centers.

■ Innovation-oriented farmer organizations focus on developing technical innovations, but they can also develop commercial or organizational innovations or a combination of all three (a good example, discussed in IAP 1, is South America's no-till farmer associations). These organizations may be created by farmers, NGOs, or public programs, and they may use public or private funds. They usually become the coordinating agent of a diverse network that includes research institutes, private firms, and public programs. Some focus mainly on farmer-developed innovations and seek to improve

and/or diffuse them (the ecologically oriented Prolinnova network is one such organization).[1] Other farmer organizations concentrating on innovation include farmers and researchers as equal partners. These farmer organizations use participatory methods to manage the innovation process and may combine top-down and bottom-up approaches.

■ Farmer organizations that are service-oriented and networked, such as Mexico's Produce Foundations, promote the emergence of local farmer organizations that form part of larger networks (IAP 2). Through collective action and participation in local and national forums, they establish partnerships with other actors in the AIS for the provision of services, including research, extension, training, credit and savings schemes, lobbying (like the West Africa Network of Peasant and Agricultural Producers' Organizations),[2] or developing value chains (TN 3).

INVESTMENT NEEDED

No recipe exists to create farmer organizations for innovation (Rondot and Collion 2001). The organizations that carved a niche for themselves did so by developing: organizational innovations, such as a clearly articulated purpose, mission, and vision; organizational cultures that allowed change; effective learning routines and heuristics (especially the ability to analyze the organization's needs and to implement plans to address them); strong technical capabilities; skills to participate in innovation networks, interact with the political and economic environment, and mobilize the resources to accomplish the organization's goals; effective and transparent governance structures (including new incentives) and leadership that prevented them from being captured by donors, governments, or elites; and active participation of members (Ekboir et al. 2009).

Successful farmer organizations develop their own organizational models. Some emerge spontaneously, such as the innovation networks that developed no-till agriculture in Argentina, Bolivia, and Paraguay; others are created by a market actor (a supermarket leading a supply chain, for example), by a coalition of actors (as usually occurs in action-research projects), by outsiders in the framework of a project (Papa Andina), or by public programs (after which they assumed a life of their own, like Mexico's Produce Foundations). Independent of their origins, farmer organizations have succeeded when they could adapt their original "business plan" to unexpected problems and opportunities. Adaptation is possible only when farmer organizations have the organizational and innovation capabilities to create a shared understanding of the organization's goals, overcome institutional inertias and conservative cultures, train all members (board members, senior management, and field staff) on the nature of innovation processes, and muster the internal and external resources for innovation. Organizational and innovation capabilities cannot be built only with short courses. Nor can they be bought easily. They require a sustained effort to cultivate.

Relatively few publications describe steps for building innovation capabilities in nonprofit organizations, but many successful experiences have been documented in private firms (for example, see Skarzynsky and Gibson 2008; Davila, Epstein, and Shelton 2006; and Christensen, Anthony, and Roth 2004). From this literature, this note distills the most valuable recommendations for farmer organizations. Programs to build organizational capabilities for innovation feature a complex menu of interventions over a sustained period and usually include investments in physical capital, short- and long-term consultancies, courses, and long-term mentoring.[3] The interventions can be divided into two categories: building the capabilities themselves and making the enabling environment more supportive of innovation.

Building social capital

The lack of social cohesion and capital, including transparent governance mechanisms, often explains why farmer organizations fail or cannot be sustained for long. Capacity-building programs for any particular farmer organization should identify the extent to which the lack of social capital is an issue and determine the type of capacity building that can compensate for it. Otherwise, capacity-building investments may not yield the desired results.

Developing organizational capabilities for innovation

Organizational capabilities for innovation encompass the *abilities of the organization's members* and the *organization's key characteristics.* Organizational abilities for innovation include specialized knowledge, creativity, and commitment to the organization; developing a long-term vision for the organization; absorbing information generated by other agents (also called the absorptive capacity); creating new knowledge; and using this knowledge to develop innovations that address commercial, social, organizational, or technological needs or opportunities.[4] An organization's key characteristics include its culture, governance, and

communications routines (whether they are hierarchical or allow individual exploration of opportunities and horizontal communication); learning routines (the heuristics and methods used for collectively accepting new ideas and procedures); the propensity to interact and cooperate with other actors in the AIS; and resources available for the development of innovations (capital and specialized assets) (Davila, Epstein, and Shelton 2006; Ekboir et al. 2009). ICTs, which facilitate networking and information sharing, are one resource that appears inherently suited to improve the potential for innovation among farmer organizations (box 1.20).

A comprehensive program for capacity development contains a long list of measures, but often the implementation of a few critical actions is sufficient to trigger a virtuous cycle of autonomous capacity building. In fact, each organization has to develop its own menu of interventions. Possible investments include the following:

- An institutional assessment to identify the farmer organization's culture, learning routines, human capital, and leadership capabilities.[5] The assessment should be the basis for defining the investments required to develop organizational capabilities for innovation.
- Assess and reinforce the organization's social capital to prevent opportunistic behavior, strengthen governance structures, and establish links with external partners.
- Train and mentor an innovative leadership group.
- Develop learning mechanisms such as the one used by Mexico's Produce Foundations (IAP 2).
- Strengthen channels (including IT platforms) to exchange information on innovative activities.
- Implement budgeting procedures that allow experimentation.
- Create dedicated teams to explore potential innovations. These teams should include groups to design and try innovations; temporary teams of organization members, partners, and researchers to generate new ideas and proposals; an innovation board to screen and fund innovation proposals, both internal and with other actors in the AIS; and trained "innovation champions" to guide and mentor any member who has an innovative idea. If the farmer organization is too small to have these specialized structures, it should partner with others in the AIS that have similar attitudes toward innovation.

Farmer organizations cannot survive, let alone improve their capacity to innovate, unless they develop sustainable financing (box 1.21), based on successful experiences,

Box 1.20   ICTs Improve the Effectiveness of Farmer Organizations

In rural areas of many developing countries, farmer organizations may be the only organizations on the ground. Because information and communication technologies (ICTs) make it easier to speak for and to farmers, they can dramatically heighten the capacity for networking, good governance, collective action, and innovation in producer organizations and agricultural cooperatives:

- *ICTs enhance farmer organizations' connections and governance.* ICTs can facilitate the sharing of market and technical information, help organizations attract and retain members, and inform members of the organization's activities on their behalf. In Mali, Coprokazan (http://www.coprokazan.org), a cooperative for female producers of shea butter, computerized its operations to reach a wider market online, develop more effective training materials for coop members, and more generally raise its profile. An unexpected outcome of using ICTs was that members became more confident in the coop's governance after coop staff started using computers for routine administration and to develop visual overviews of yearly accounts and activities. The well-known capacity of ICTs to streamline administrative and accounting tasks of all kinds makes them highly useful for administering farmer organizations. In dairy and coffee cooperatives in India and Kenya, for example, farmers believe that automated measurement and record-keeping systems help ensure fair compensation from the cooperative. Automated systems have the added advantages of speeding transactions and reducing spoilage.

- *ICTs give organizations a stronger collective voice.* High-speed connectivity may still be out of reach in many parts of the world, but individual farmers can still use mobile phones and text messages to "have their say" on agricultural radio. Feedback about services offered by farmer organizations and local government, when expressed over the airwaves, has more influence than comments made in a less public forum. In Mali's Sikasso Province, an ambitious project brings farmer organizations into the national agricultural policy debate through telecenters in seven towns and villages, together with local radio stations. In locations throughout Africa, the Participatory Radio Campaigns of Farm Radio International (a nongovernmental organization) invite farmers' participation and respond to their feedback. Text messaging on cheap mobile phones is proving to be a highly adaptable medium for organizations to mobilize members around specific issues and concerns, acquire resources (information, credit, and even insurance) to facilitate production and marketing, and generally become more integrated and visible within the innovation system.

- *ICTs can foster innovation.* As the examples indicate, ICTs can spur farmer organizations to innovate in how and where they operate, with whom, and why. The use of ICTs can make organizations more effective lobbyists, better at acquiring information or services from public and private sources, more effective participants in markets and value chains, and more valuable partners in research and development initiatives (such as initiatives to provide early warning of plant and animal diseases).

Farmer organizations may be aware of the potential of ICTs but may not necessarily find it easy to acquire and learn to use them. Generally it is governments, donors, and nongovernmental organizations (NGOs) that have the funds to develop ICT solutions that benefit farmer organizations. (The software used in the Indian dairy cooperatives mentioned earlier was developed through public-private partnerships, for example; Farm Radio International is an NGO.) Another challenge is to sustain the use of ICTs after external support ends. Significant costs are usually associated with equipment, maintenance, training, and continuing system development. New technology must generate enough additional income for an organization to cover its ongoing costs, or it must generate enough additional benefits for individual users to be willing to pay for it. Finally, although ICTs facilitate communication, it is important to recognize that they cannot substitute for building social capital.

Source: World Bank 2011.

Box 1.21 Successful Financing of Farmer Organizations

Voluntary contributions. The Argentine Association of Regional Consortiums for Agricultural Experimentation (AACREA) is an apex organization of self-help groups of commercial farmers in Argentina. Each group has about 10 members who jointly hire a technical advisor and conduct adaptive research and validation. The group also contributes to the apex organization. The latter gathers the information developed by each group and shares it among its 1,880 members. This structure is an example of decentralized experimentation with centralized learning.

Levies. The Colombian Congress can tax specific agricultural products to finance programs to support them. A specific law must be enacted for each taxed product, and the proceeds can be used only to support the product's market chain. The central government collects the tax. The funds are administered by a contract between the Ministry of Agriculture and a farmer organization of national reach that represents producers of the taxed product. Colombia has 15 such taxes.

Sales of goods and services. Starting in 1970, India's National Dairy and Development Board imported food aid in the form of dairy products and marketed

them under its own brand name. The surplus from these sales was invested in the expansion of the cooperative movement in the dairy industry. Today farmers organize into village-level cooperatives, which in turn are organized into district-level cooperatives (comprising 400–1,000 primary village cooperatives). The district-level cooperatives federate into a state cooperative. At the apex is the National Cooperative Dairy Federation, which coordinates marketing for all state cooperatives. The cooperatives currently supply about 70 percent of the processed milk marketed in India and provide a wide range of services to members.

Government programs. The Mexican government created a program in 1996 to finance research and extension projects through Produce Foundations, civil society organizations in each state that currently also finance and implement innovation activities. The presence of a foundation in each state makes it possible to clearly identify local farmers' needs and foster the emergence of local innovation networks. The Foundations set the priorities, decide which projects are funded, and administer the projects; the federal and state governments audit the use of the funds. See IAP 2 for a detailed discussion.

Source: Author.

Note: AACREA = Asociación Argentina de Consorcios Regionales de Experimentación Agrícola.

voluntary contributions (such as those used in Argentina by AACREA),[6] levies (used in many countries, including Australia and Colombia), sales of goods and services (Kenya Tea Development Agency and the Indian dairy cooperatives), and partnerships between the government and civil society organizations (such as the Mexican Produce Foundations) (box 1.21).

Using new and more sophisticated instruments to support innovation

In the past, most projects that sought to build capabilities in organizations supported professional education, short courses, or consultancies on specific topics. Current projects also support interactions among different actors in the AIS (including public-private partnerships), creation of venture capital funds, consolidation of innovation brokers and "deal flow" promoters, and strengthening the policy framework for innovation. Other instruments introduced recently include mentoring programs, multistakeholder platforms, knowledge-exchange fairs, and IT platforms to facilitate communications and coordination.

Adding flexibility to the enabling environment

Farmer organizations' ability to change depends on whether the broad policy and institutional environment enables them to change. For this reason, policies that seek to foster innovation should also consider promoting changes in the enabling environment. Because so many actors participate and because the enabling environment is loosely governed, however, it is very difficult to initiate changes and, once they are introduced, to predict the effect of particular actions (Ekboir et al. 2009). Given this complexity, governments should use adaptive approaches to induce changes in the enabling environment. Various instruments can be used to this end:

- Devising capacity-building programs to strengthen the understanding that different actors in the AIS, especially governments, donors, extension agents, and research organizations, have of innovation processes.
- Building confidence and promoting coordination among potential partners, including financing for multiactor programs, innovation brokers, and sustainable extension programs.
- Strengthening research capabilities, especially by focusing on the quality of researchers and changing the culture and incentives of research organizations.
- Certifying innovation brokers.
- Financing programs for innovation, including venture and angel capital funds and innovation brokers; flexible financing of action-research projects to explore new instruments; and institutional arrangements to foster innovation. These programs could be organized as stakeholder-driven and client-controlled mechanisms, in which farmer organizations have a financial stake.

- Creating a committee to coordinate the AIS and facilitate multistakeholder dialogues on innovation policies.
- Strengthening the participation of farmer organizations that represent marginalized populations (including smallholders and women) in policy- and decision-making processes and in the provision of services.

POTENTIAL BENEFITS

Box 1.22 presents an example of the benefits of instilling innovation capabilities in a farmer organization in Kenya. More generally, farmer organizations with stronger innovation capabilities can offer the following benefits:

- Faster development of social and economic innovations, resulting in stronger economic growth and more sustainable use of natural resources; strong farmer organizations are particularly necessary when public organizations (especially research and extension) are weak.
- Quicker development and implementation of organizational and commercial innovations.
- Better provision of services when governments are not decentralized or are weak.

Box 1.22 Benefits of Innovation Capabilities in a Farmer Organization: The Kenya Tea Development Agency

The Kenya Tea Development Agency (KTDA) was created soon after independence as a state company to regulate tea production by smallholders, but it repeatedly demonstrated its capacity to innovate over the years as it evolved into a major corporation owned by small-scale farmers. It provides production and marketing services for members, successfully manages tea nurseries and 59 factories, and represents small-scale farmers in the Kenya Tea Board. The agency's innovation capabilities include the following benefits:

- Increasing the income of its associated small-scale farmers.
- Successfully implementing collective action involving hundreds of thousands of small-scale farmers. In 1963, tea was cultivated by 19,000 smallholders on 4,700 hectares, with an annual crop of about 2.8 million kilograms of green leaf. In 2009, about 400,000 smallholders grew tea for KTDA on 86,000

hectares and produced in excess of 700 million kilograms of green leaf.
- Developing new tea products and opening new markets for them.
- Implementing new programs to support its associates (affordable credit and input supply, for example).
- Developing strong managerial capabilities.
- Developing strong competitive advantages. The agency accounts for 28 percent of Kenya's exporting earnings and is the world's second-largest exporter of black tea.
- Adapting its operations and governance in response to changes in the socioeconomic environment and market opportunities.
- Influencing strategic stakeholders to allow organizational change.
- Partnering with private actors and foreign universities to develop and diffuse sustainable production practices for small-scale farmers.

Source: Author, based on information from KTDA, http://www.ktdateas.com.

- Stronger two-way information flows, conveying technical and commercial information to a larger number of farmers as well as communicating farmers' needs and concerns to other actors in the AIS, including researchers and policy makers. ICTs can be important enablers of communication, but they should not be seen as an alternative to building social capital.
- More effective use of human, social, physical, and financial resources for innovation.
- Easier access to input and output markets and to other resources for innovation (technical advice, innovation networks, and participation in action-research projects), both for commercial farmers and marginalized populations (women and landless farmers, for example).
- Better interaction and coordination with other actors in the AIS.
- More inclusive and effective institutional innovations when marginalized groups have a stronger influence on the design and implementation of innovation policies.
- More relevant and more effective public research and extension programs when farmer organizations effectively participate on boards of research and extension organizations.
- More inclusive development, especially when affirmative action allows more effective participation of marginalized groups in decision-making bodies.

POLICY ISSUES

Farmer organizations have policy issues similar to those of innovation networks. They include considerations related to social hierarchies and inclusiveness, sustainability and dependence on external funds, and the respective roles of the public and private sectors.

Social considerations

Local social considerations inevitably come into play in many farmer organizations. For example, it is difficult to introduce organizational or institutional innovations (especially more transparent governance) in societies dominated by elders or clans. Hierarchical societies also stifle technical and commercial innovation, reducing opportunities for farmer organizations to benefit members. In such cases, efforts to strengthen farmer organizations may award more power to local groups or individuals that are already powerful. Affirmative action (reserving seats on the organization's board for marginalized groups, for example, or well-designed communications programs to reach all

subsets of farmers, such as women or poorer farmers) can ensure that weaker groups have greater influence on the farmer organization and that it meets their particular needs. These programs succeed, however, only if representatives of the marginalized groups have the appropriate capabilities. Often women are illiterate and at a disadvantage in organizations that rely on written information.

Institutional considerations

An enabling environment that allows farmer organizations to operate effectively and with as little external interference as possible is a prerequisite for innovative organizations. This environment can be achieved more easily when governments, donors, and farmer organizations themselves have a clear understanding of innovation processes. When governments or donors finance farmer organizations, the funding agencies usually try to influence them. Governments may resent having to negotiate policies and priorities with independent farmer organizations. Finally, farmer organizations operate more effectively when laws, regulations, and interventions by external stakeholders (especially governments and donors) facilitate transparency and accountability to farmers.

Sustainability

Farmers who organize spontaneously in response to a need or opportunity often demonstrate self-reliance and strong and lasting solidarity, but often these farmer organizations command limited resources. They can profit greatly from programs to build their capabilities and link them with other actors in the AIS. On the other hand, farmer organizations created by external partners usually have more resources but are less sustainable. They run the risk of not attending to members' priorities, and they may lose their autonomy and effectiveness. Bigger farmer organizations are better at influencing innovation processes—but they are more likely to be captured by governments or elites.

Public and private sector roles

Farmer organizations can substitute for weak private organizations, especially when the commercial private sector is slow to take over activities abandoned by the public sector, such as input provision. More frequently, farmer organizations are seen as substitutes for weak public research and extension organizations, as occurred in Latin America and sub-Saharan Africa when farmer organizations began to specialize in natural resource management (IAP 1). Although farmer organizations can effectively implement research and extension programs, this activity should not substitute for research and extension by public programs, because it increases the odds that nonmembers will be excluded or that their needs will not be meet. Farmer organizations should rather participate in joint research and extension efforts that complement those of public organizations. When farmer organizations take the lead in fostering innovation, research institutes may feel threatened, perceiving that farmer organizations are essentially trying to change the ways that research institutes interact with the AIS.

Appropriate policies and incentives may prevent some of these problems and at the same time strengthen farmer organizations. In addition to implementing general policies (such as research policies or policies facilitating the operation of markets), the public sector can support farmer organizations by introducing new incentives for researchers and other civil servants to work more intensively with farmer organizations; providing resources for organizational facilitators (such as NGOs) and innovation brokers to create or strengthen farmer organizations (IAP 6); and supporting the development of farmer organizations' innovation capabilities. It is vital that farmer organizations do not end up responding to government interests in lieu of members' needs.

Nonpublic actors—including private firms, international research institutes, and NGOs—can play a number of roles. They can induce the creation of farmer organizations, create new marketing channels for small-scale farmers, and work with existing organizations to build their innovation capabilities.

LESSONS LEARNED

The considerable variety of farmer organizations, experience with older forms of organization, and new organizational strategies are yielding a number of lessons about farmer organizations and innovation. The sections that follow summarize the conditions that make farmer organizations effective (especially as innovators), including lessons related to their structure, way of operating, capabilities, financing, and the environment in which they operate.

When are farmer organizations most effective at innovating?

Farmer organizations are particularly effective for developing technical and commercial innovations when they focus on specific products (the approach of the Colombian Sugarcane Research Center, CENICAÑA)[7] or subsectors and commercial farmers. Sectoral or national farmer organizations can also influence the introduction of institutional innovations, either for commercial or small-scale farmers. Finally, local farmer organizations (especially community-based organizations) are effective in developing social innovations, including facilitating poor households' access to markets, and addressing local problems that require collective action. These observations, however, are not absolute. In some cases, national associations that coordinated local groups have developed important economic or technical innovations (see, for example, IAP 1).

Innovative farmer organizations tend to have federated structures

The most innovative farmer organizations are composed of small, linked farmer groups. This structure combines the best features of large organizations and small groups and benefits the organization's capacity to innovate. Small groups have greater internal cohesion, and it is easier to monitor the members. Farmer organizations with strong community ties enable grassroots concerns to be voiced more clearly. They facilitate greater upward participation and downward accountability. The drawback is that they command so few resources. By federating into larger groups, they can achieve greater economies of scale, but they can also be more easily captured by elites or governments and find it harder to control free riding. Examples of federated farmer organizations are India's milk cooperatives (described in box 1.21), the no-till farmer associations in South America (IAP 1), and Mexico's Produce Foundations (IAP 2).

Building capabilities for innovation in farmer organizations

In helping farmer organizations develop their capacities to innovate, it is important to focus not only on technical or commercial issues such as accounting or crop management but on developing good governance, creating structures and incentives for innovation, developing external links, and building strong leadership. Capacity-building activities may include traditional training in purpose-built facilities, on-the-spot training, consulting services, extension, seminars and workshops on the dynamics of organizational innovation, exchange programs so that managers can see the capabilities of innovative farmer organizations, tutoring, and mentoring. For example, African managers at the Kenya Tea

Development Agency developed their capabilities by working over many years under senior managers from multinational companies (Ochieng 2007).

Building capabilities for innovation should be a permanent effort, because organizations tend to lose their innovativeness once they have found routines that help them reach their goals. The conservative nature of organizations is particularly strong when top management lacks a good understanding of the nature of innovation processes (Christensen, Anthony, and Roth 2004). The incentives to change in commercial firms are provided by markets; the incentives in nonprofit organizations come from their associates, their own sense of duty, and the pressure exerted by donors and governments. This pressure usually is convened through two channels: funding and policy dialogue. When funding is conditional on the implementation of capacity-building programs, funders should be careful not to impose their objectives on the farmer organizations or alienate management from the farmers. Recognizing this problem, the World Bank and other donors are providing funds for farmer organizations to build capacity based on demand (training, advisory services, and startup funds for innovative ideas), sometimes requiring matching contributions. The decision on how to allocate the funds is left to the farmer organization's management, which can contract specialists from a roster of regularly appraised service providers. This approach allows external experts to be truly independent and at the service of the farmer organization. Such independence is often harder to ensure when donors select the experts. The World Bank has funded projects with this design in Bolivia, Brazil, Colombia, Guatemala, Mexico, and Panama (IAP 6).

Development of innovations requires flexible management by farmer organizations, governments, and donors

Innovations are developed by exploring alternatives. The failed initiatives may seem to have wasted resources, yet failure is part of the learning process. Farmer organizations must be allowed to explore, but in that case, donors and governments must strengthen their own capabilities to manage innovation, because the inflexibility of public procedures runs counter to the need for flexibility and adaptability. Courses and consultancies on the management of agricultural innovation for donors and senior civil servants are necessary for flexible management practices to take root. Appropriate controls and audits are also necessary to prevent misuse of resources without hindering exploration. A committee, formed by donors and senior civil servants, should be established to review changes to the original project design at least once a year.

Financing farmer organizations for innovation

As noted, some farmer organizations have developed sustainable sources of funding by selling products or services, including lobbying, for members. These organizations do not need special support, but they should be able to access all of the programs available to private firms, including those that support innovation.

Demanding that farmer organizations be self-financing is generally not realistic, especially if they include a large share of small-scale farmers or if they manage public funds. Demanding cofinancing from farmer organizations that include marginalized groups may further alienate those groups from innovation processes. Even so, farmer organizations should ask members to contribute some funding (even in kind), because it stimulates commitment. A number of financial arrangements have been used to support the innovation activities of farmer organizations, including competitive grants, matching grants (see module 5, TN 2; IAP 2), credits, funds distributed through government channels with no mention of competitive or matching grants, revolving funds for services, beneficiary fees for services (such as advisory services, veterinary services, and maintenance of group-managed infrastructure), contracting of services, and performance-based contracts. In each case, the arrangement that is adopted should match the needs and capabilities of the organization.

Groups to generate technology have been sustainable only when formed by commercial farmers

Commercial farmers have usually solved their main organizational and commercial problems and have resources to invest in becoming more efficient. Most noncommercial small-scale farmers, on the other hand, have limited resources and a diversified livelihood strategy, in which agriculture is a dwindling source of income. They can be reluctant to invest time and money in developing new techniques, although they can occasionally benefit from innovations developed by commercial farmers.

A more supportive environment for organizational innovation

The level of innovation achieved by farmer organizations depends on the enabling environment. Governments and donors can create a more supportive environment for innovation in many ways. They can create market and social opportunities for innovation (by deregulating markets, for example), facilitate access to crucial inputs (such as affordable credit), and strengthen the flow of commercial and technical information (by strengthening universities and innovation brokers, for example). They can offer public institutions better incentives to interact with farmer organizations. Promotions in many research institutes are based mostly on the number of indexed papers published, whereas interactions with farmers are less valued. Action-research projects could be implemented to identify new instruments to build innovation capabilities in farmer organizations, such as the structure of decentralized experimentation with centralized learning described in IAP 2 for the Produce Foundations. Policy dialogues could be set up to develop consensual innovation policies; national committees for innovation are critical in implementing such dialogue and defining the policies, as discussed in the module overview and TN 1. Government could promote a legal environment that supports farmer organizations and encourages decentralization to empower local farmer organizations.

RECOMMENDATIONS FOR PRACTITIONERS

A few principles are useful to consider in designing and implementing projects that focus on developing farmer organizations that can innovate.

When selecting project participants, support organizations that can learn. Organizations tend to be conservative, and not every organization can develop innovation capabilities (Christensen, Anthony, and Roth 2004). Organizations that can learn are identified in two ways: They have already shown their innovativeness, or it can be evaluated through organizational assessments (see, for example, Ekboir et al. 2009). The assessment is critical to identify the best approaches to develop innovation capabilities. For example, if the farmer organization is particularly conservative, participatory methods may not be suitable, because management is likely to oppose change. The assessment should also look at the organization's innovation strategy, identify its mission and vision, determine the resources available for innovation and those that must be secured, and identify the most important current and potential partners. The identification of important partners can be done using multistakeholder, participatory procedures such as those described in

Devaux et al. (2009), Vermeulen et al. (2008), and Hartwich et al. (2007).

Eligibility criteria should not be too strict. Given the diversity of farmer organizations and the difficulty in knowing an organization's true potential for change beforehand, any membership organization should be eligible for support as long as it: (1) is recognized as useful by its members; (2) has an identity—in other words, a history and effective operating rules that, even if they are informal, regulate relations between members and between members and the outside world; (3) has governing bodies that function effectively; and (4) has demonstrated its willingness to develop its innovation capabilities. An organization can demonstrate its willingness explicitly (for instance, by training members or contacting external advisors to develop a plan to strengthen capabilities) or implicitly (through its formal and informal routines, the nature of its leadership, its culture, and its incentives).

Do not push farmer organizations to become formal entities. Experience shows that when a farmer organization survives a period of informality, it is more likely to succeed because its members have had time to develop a common experience and resolve the issues that emerge when implementing collective action.

Identify the networks in which the farmer organization participates. An organization's partners can indicate its own innovativeness. Consider all stakeholders that collaborate with the organization in innovation networks and identify the appropriate incentives for them. In particular, public organizations may face major hurdles to interacting effectively with farmer organizations.

Identify support and training needs. The leaders and staff of farmer organizations usually need help to identify the capabilities they lack. Even when the needs have been identified, it is difficult to build the capabilities, especially in organizations formed by marginalized groups, which need economic, managerial, and technical support. External consultants and innovation brokers are necessary to identify these gaps and design training programs for farmer organizations, given their generally limited organizational experience and resources. Support for these farmer organizations should be based on the principle of empowerment. Experience shows that building farmer groups for learning and reflection often requires continual access to external facilitation (van der Veen 2000).

Seek nonconventional instruments to strengthen farmer organizations. The effectiveness of formal courses for building innovation and entrepreneurial capabilities is quite low, because these capabilities can be developed only by doing and by observing other actors who possess the capabilities. In addition to the instruments mentioned in box 1.8 in the module overview, exchange visits have been very useful in spurring innovation in farmer organizations, but only when visitors had the capabilities to absorb the information and only when their organization had reached a certain maturity.

Develop a monitoring and evaluation system that promotes exploration and learning. Examples of indicators that can be used are given in table 1.2 in the module overview.

No-Till Networks in Developing Countries

Javier Ekboir, Institutional Learning and Change Initiative (ILAC)

SYNOPSIS

This profile summarizes lessons from the networks that arose to develop no-till technology. The most innovative aspects of the networks were their composition and the ways in which they operated. Partners and research and extension approaches varied, but in all successful cases, a flexible innovation network emerged in which farmers' participation was important. Innovation brokers and catalytic agents played critical roles. Effective networks explored organizational innovations until they found approaches appropriate to local conditions; they also explored different approaches to organize the generation and diffusion of the technical innovation. The composition of innovation networks changed as the innovation process matured and new challenges emerged. Parallel innovation efforts contributed to the effectiveness of innovation, including efforts by public organizations that gave their employees scope to explore new ideas.

CONTEXT

In the 1970s, intensification of agriculture after the Green Revolution created severe soil erosion in many temperate and tropical countries. At the same time, grain prices started to fall due to rapid production increases. Reacting to these problems, farmers and researchers in many countries started to explore more sustainable, profitable technologies.

For centuries, farmers plowed the soil to favor water infiltration and control weeds. In the 1960s, the chemical company, Imperial Chemical Industries (ICI), developed a herbicide for which there was little demand, but the company decided that it could create a market for the product if chemical weed control could effectively replace plowing.[1] At the time, ICI was not trying to create a more sustainable or profitable technological package; it was just seeking to sell a product in which it had invested substantial resources and

for which there was little use.[2] After the first successful experiments with no-till,[3] ICI realized that a new agricultural paradigm was needed. To develop it, ICI set up a research team that eventually settled in southern Brazil in 1970. The team leader recognized that for the herbicide to be of any use, new agronomic techniques, new weed management approaches, and new equipment were required. Knowing that ICI did not have all the capabilities to develop the components, the team leader looked for partners. Soon, the network included a state research institute, the Wheat Research Center of the Brazilian Agricultural Research Corporation (Empresa Brasileira de Pesquisa Agropecuária, EMBRAPA), a factory for planting equipment, progressive farmers, and the ICI team. Once the package was developed, farmer organizations diffused it.

A similar process (although with other partners) occurred in Argentina, Bolivia, and Paraguay. In all cases, the technology diffused very rapidly in the 1990s, thanks to the efforts of very effective no-till farmer associations, the technology's benefits, and favorable economic conditions. During the same period, no-till diffused to a few other countries as well. No-till programs have now been identified in more than 60 other countries, but adoption has been negligible (Ekboir 2002) because no-till methods require considerable adaptation to local conditions. No-till was used on an estimated 100 million hectares worldwide in 2005, mostly in the United States, Brazil, Argentina, Canada, Australia, Paraguay, and China. Bangladesh, Ghana, India, Nepal, and Pakistan accounted for a relatively small area because no-till is used by large numbers of small farmers (Derpsch 2010).

OBJECTIVES AND DESCRIPTION

No-till was developed and diffused by networks that included several public and private agents such as pioneer farmers, agrochemical firms, equipment manufacturers, and sometimes researchers and public extension agents. The networks had varied origins. Some were started by private companies, others by public researchers, farmers, or public programs. Massive diffusion occurred only where networks that developed strong learning capabilities and effective interactions emerged. When no-till packages were developed with traditional experimental methods or diffusion was organized through conventional approaches, adoption was minimal.

Although the partners and research and extension approaches varied, a common factor in all successful cases was the emergence of a flexible innovation network in which farmers' participation was important (Ekboir 2002). The uniqueness of no-till processes is exemplified by the Ghanaian experience. To date, Ghana is the only country where a no-till package was adapted for small-scale farmers but not for large-scale farmers (box 1.23). Table 1.4 shows the partners and defining factors of some successful networks.

THE INNOVATIVE ELEMENT

The main innovation was the emergence and evolution of new organizations—innovation networks that involved different partners, including public and private researchers, farmers, equipment manufacturers, extension agents, and agrochemical companies—that generated, adapted, and diffused no-till techniques. To reach their objectives, the networks innovated in three domains:

- *Organizational.* New modes of interaction among different actors in the AIS emerged, as well as new approaches to research and extension.
- *Equipment.* New planters, sprayers, and combines enabled the consolidation of an innovative agricultural equipment industry.
- *Agricultural practices.* New practices were devised for planting and fertilization, weed control, management of soil cover, crop rotations that span several production cycles, and harvesting.

Box 1.23   Development of No-Till for Ghana's Small-Scale Farmers

For centuries, Ghanaian farmers used traditional slash-and-burn agriculture, which is equivalent to no-till without mulch. Slash-and-burn was sustainable when abundant land allowed farmers to leave the exhausted plots fallow for many years. Demographic pressure in the second half of the twentieth century forced farmers to shorten or abandon fallows and made slash-and-burn unsustainable. In the 1980s, research to adapt no-till with mulch as a sustainable alternative to slash-and-burn was initiated by the Crops Research Institute in Kumasi in conjunction with the International Maize and Wheat Improvement Center. Farmers did not adopt the package, because it was difficult to plant with a stick through the thick cover of plant residues in the field. In the 1990s, the importer of glyphosate (a herbicide), the Crops Research Institute, and Sasakawa–Global 2000 formed a partnership. Monsanto, trying to emulate its success in Brazil, later joined the partnership. The partners' key action was to fund the doctoral dissertation of a highly motivated soil scientist. As part of his research,

he worked with innovative farmers to develop a weed and mulch management system that small-scale farmers could use and organized a successful extension program that motivated and provided resources to public extension agents. The package was disseminated to farmers in Ghana's forest, transition, and Guinea savannah zones and rapidly adopted. In 2005, no-till was used by an estimated 300,000 small-scale farmers.

Unlike other countries, in Ghana the no-till package responded to small-scale farmers' needs. Adoption was facilitated by the low-input agricultural practices that already prevailed and the fact that few farmers had animals (other than some chickens). There was no need to develop no-till planters, which were major obstacles in other countries where no-till was introduced, because crops were planted with a stick or cutlass (machete). The lack of equipment proved to be a limitation on widespread adoption among large-scale and mechanized farmers, however, because they needed suitable machinery to adopt the technology.

Source: Ekboir, Boa, and Dankyi 2002.

Argentina	Brazil	Paraguay	India
Actors that at some point participated in the network			
– Innovative and committed farmers – Argentine researchers willing to interact with farmers – Innovative equipment manufacturers – Farmer organizations – Universities – Foreign researchers – Agrochemical companies – International research centers	– Innovative and committed farmers – Brazilian researchers willing to interact with farmers – Innovative equipment manufacturers – State research institutes – Farmer organizations – Foreign researchers – Agrochemical companies – International cooperation agencies	– Innovative and committed farmers – Farmer organizations – International research centers – International cooperation agencies	– Innovative and committed farmers – Innovative equipment manufacturers – International research centers – State universities
Modes of interaction			
– Active interactions between researchers, equipment manufacturers, and farmers in the development phase – Initial mild support from agrochemical companies, which became very strong when the leading firm recognized the market potential – Effective no-till farmer association organized diffusion programs	– Strong support from agrochemical companies – Active interactions between researchers, equipment manufacturers, and farmers – Effective no-till farmer association organized diffusion programs – Alliances of several partners developed and diffused packages adapted to small-scale farmers' needs and resources	– Strong links between Paraguayan and Brazilian farmers – Support from international cooperation agencies – Support from an international research center – Effective no-till farmer association organized diffusion programs	– Strong support from an international research center and state universities – Exchange of prototypes between Bolivian and Indian artisans enabled the development of efficient planters – Active interactions between international researchers, equipment manufacturers, and farmers – Traditional extension services worked with researchers
Factors that triggered the development of no-till networks			
– Curiosity-driven research – Severe soil compaction	– Severe soil erosion – A private firm willing to develop a market for a new product	– Farmers were aware of the economic and environmental benefits of no-till	– An international research center had experience with no-till in rain-fed areas and was willing to experiment in irrigated crops – Severe weed infestation that could be controlled easily with no-till
Socioeconomic and environmental factors that facilitated the generation and diffusion of no-till			
– Well-developed commercial agriculture – No-till's economic benefits – Research institutes that allowed researchers to conduct curiosity-driven research – Widespread use of custom operators for planting and harvesting – Macroeconomic policies reduced the profitability of grain production	– Well-developed commercial agriculture – No-till's economic benefits – Research institutes that allowed researchers to conduct curiosity-driven research	– Well-developed commercial agriculture – Previous experience with no-till	– No-till's economic benefits – Affordability of planters – Existence of market for planting services

Source: Author.

BENEFITS, IMPACT, AND EXPERIENCE

In all countries listed in table 1.5, except China, no-till was adopted mainly by large- and medium-scale commercial farmers. Among countries that have adopted no-till but do not appear in table 1.5, India, Pakistan, Bangladesh, Nepal, and Ghana together account for 1.9 million hectares under the technology. As noted, this relatively small area is deceiving in the sense that no-till is used by very large numbers of small-scale farmers. No-till practices are expanding in South Africa, Venezuela, New Zealand, France, Chile, Mexico, and Colombia and are in the early stages of adoption in other countries (Derpsch 2010).

Table 1.5 Use of No-Till in Select Countries, 2007–08	
Country	**Area under no-till (hectares), 2007–08**
United States	26,593,000
Brazil	25,502,000
Argentina	19,719,000
Canada	13,481,000
Australia	12,000,000
Paraguay	2,400,000
China	1,330,000
Kazakhstan	1,200,000
Bolivia	706,000
Uruguay	672,000
Spain	650,000

Source: Derpsch 2010.

No-till had several important impacts on the AIS:

- New linkages developed among actors of the AIS and with foreign sources of technical information.
- New business models developed. For example, contracting planting and harvesting in South America and India allowed more efficient use of specialized equipment.
- New research methods were developed, accelerating the generation of the no-till package.
- New approaches were developed to disseminate the package.
- Some researchers and extension agents changed the way they interacted with other actors in the AIS.
- Farmer organizations developed capabilities for effectively exploring alternative technical and organizational solutions.
- Existing research and extension capabilities were used more effectively.
- Effective farmer organizations with strong innovation capabilities emerged.

LESSONS LEARNED AND ISSUES FOR WIDER APPLICATION

Effective networks explore different organizational innovations until they find approaches appropriate to local conditions. "Research" networks adapted the package to local conditions and "extension" networks diffused it; these networks interacted assiduously, and some actors (especially innovative famers) participated in both of them. The most innovative aspects of the process were the composition of the networks and the ways in which they operated (box 1.3 in

the module overview). Innovation brokers and catalytic agents played critical roles in the exploration (TN 2 in this module and module 3, TN 4). Appropriate resources should be provided to pay for their services and for the trial of organizational innovations.

Effective networks explore different approaches to organize the generation and diffusion of the technical innovation. In the case of no-till, the research networks used participatory, on-farm research approaches, whereas the extension networks used a larger range of instruments, including self-help groups, demonstration days and plots, conventions, publications, radio and television programs, sales forces of agrochemical and equipment companies, and farmer-to-farmer communication. In the development of complex innovations like no-till, flexible approaches for research, financing, and evaluation should be used.

The composition of innovation networks changes as the process matures and new challenges emerge. No-till networks started as small teams and incorporated new members as the innovation process evolved. Despite this commonality, in each country (or even regions within countries) different processes led to the emergence of effective innovation networks. For example, in southern Brazil the catalytic agent was a private firm (ICI), but later, farmer organizations were the key actors in diffusing no-till among commercial farmers. In Argentina and northern Brazil, a group of researchers and farmers developed the package, but a coalition of innovative farmers and agrochemical companies diffused it. In Bolivia and Paraguay, commercial farmers led the process, but private firms joined the networks after no-till was widely adopted. Innovation projects should recognize the changing needs of innovation networks and facilitate the incorporation of partners that possess assets the network needs and the withdrawal of partners that no longer contribute to the common effort. In other words, when promoting the creation of an innovation network, it is important to (1) search for willing partners, (2) empower them to decide on the best strategies to achieve the goals, and (3) encourage them to identify constantly what resources and information they lack and who can supply them. Innovation projects should contemplate resources for these search activities.

Effective innovation networks use adaptive management to facilitate collective action. When setting up a project to foster innovation, it is important to use participatory methods that help to create trust among potential partners, develop approaches adapted to local conditions, and use an adaptive management approach. In other words, prepare a project plan, implement it, periodically assess its implementation (at most yearly, but at shorter intervals in the first two years) to identify emerging problems and opportunities, and adjust the plans in response to the information collected.

Parallel efforts are necessary for effective innovation. Different approaches were used to generate the no-till packages and to create the networks that developed and diffused them. Each approach responded to local human and social resources and institutional constraints, but distant groups interacted among themselves. The effectiveness of innovation teams depends on the individuals that participate in them, their interactions, the resources they command, their learning strategies, external constraints, and unforeseen factors. It is impossible to know in advance how effective an innovation team will be; therefore, an effective innovation policy is to have more than one team working on the same innovation (Huffman and Just 2000). There are no precise methods to determine how many teams should be created. The number depends on several factors, including the quality of the individuals involved, the resources they command, the nature of the problem (for example, whether it is a major innovation or a minor adjustment), and the institutional environment in which they operate. To facilitate learning, however, the teams should interact assiduously among themselves.

Public organizations have to allow innovative employees to explore new ideas. The public research institutes in South America initially did not recognize the value of no-till, but they allowed individual researchers to explore new ideas. This approach contrasts with current trends in the management of research, which require projects to be clearly defined before they are implemented. To increase the contribution of public researchers, extension agents, and project managers in innovation processes, it is necessary to offer incentives that favor exploration, calculated risks, and participation in innovation networks (see TN 2).

Learning Organizations That Support Innovation: Mexico's Produce Foundations

Javier Ekboir, Institutional Learning and Change Initiative (ILAC)

SYNOPSIS

A number of internal and external factors influenced the evolution of Mexico's Produce Foundations (Fundaciones Produce), a federated farmer organization that funds and implements research, extension, and innovation projects. The Produce Foundations developed strong innovation capabilities that enabled them to search for new ways to support agricultural innovation. The main internal factors were the creativity of a few farmers on the boards and of some foundation managers; the development of effective collective learning routines; the creation of an organizational culture that valued exploration, creativity, innovation, and a sense of duty; and flexible governance structures. The external factors were the existence of dynamic markets that created challenges and opportunities for farmers and the presence of key policy makers who were willing to allow the foundations to experiment and change. Autonomy and independence proved important: there is a positive correlation between a foundation's autonomy from the state government and its institutional development, efficiency in achieving its mission, and innovativeness. Decentralized experimentation and centralized learning were important to success, along with willingness on the part of federal and state governments to let the foundations explore new instruments to fulfill their mandate.

CONTEXT

In the mid-1980s, Mexico began to deregulate domestic markets and trade and establish a multiparty democracy. The central government devolved power to the states and opened channels for civil society participation. These changes created new opportunities and increased competition for agricultural producers, who reacted by looking for advanced technologies. When the public research institu-

tions could not provide them, producers and other actors in the AIS imported or developed them. The federal government also saw technical change as an important instrument to boost competitiveness, but it recognized that the public research and extension system was not responding to farmers' needs. Following the prevalent model of agricultural research, the government argued that the main problem was that the research system was supply driven and had to be replaced by one that was demand driven. At the same time, the federal government restricted its support for research. Public extension organizations were closed and replaced by a program to develop markets for technical advice.

OBJECTIVES AND DESCRIPTION

Mexico's Produce Foundations (Fundaciones Produce) are an example of a federated farmer organization that funds and implements research, extension, and innovation projects. They are also examples of learning organizations. Mexico has 32 foundations, one in each state. A national coordinating body (COFUPRO, Coordinadora Nacional de las Fundaciones Produce) facilitates organizational learning and interacts with the federal government. Each Produce Foundation is governed by a board, which is dominated by progressive farmers and also includes representatives from the federal and state governments. Operations are directed by a professional manager. COFUPRO, in turn, has a board integrated by the presidents of some of the foundations, a representative from the federal government, and another from the national agricultural research organization. Day-to-day operations are delegated to a professional management team.

The Produce Foundations' main program operates an annual budget of about US$45 million, 85 percent of which is contributed by the federal government and 15 percent by the states. The foundations also receive additional funds from public and private sources for special projects. Individual farmers cannot join the Produce Foundations; instead, a few farmers are invited to participate on the boards, and each foundation sets the selection criteria.[1] Individual farmers participate in projects financed by the foundations, both as collaborators and as contributors of resources. Initially, the Produce Foundations financed projects that mainly helped commercial farmers; over the years, they have increasingly sought to meet the needs of small-scale farmers while continuing to support commercial agriculture.

The objective that motivated the creation of the Produce Foundations was to mobilize additional funds for the national agricultural research organization and transform supply-driven research and extension systems into demand-driven systems. After several iterations, the current objective of the federal government and the Produce Foundations is to explore new instruments to foster innovation, transform traditional research organizations and universities so that they can better integrate into innovation processes, explore new methods to diffuse innovations, and influence the design and implementation of research, extension, and innovation policies by participating in policy dialogues and educating policy makers on the nature of innovation.

The changes in the Produce Foundations' objectives reflect lessons learned by the government and the foundations. When the federal government created the foundations in 1995 as part of its efforts to democratize and establish a new model for agricultural research, it negotiated with each state governor to establish a foundation that would administer public and private funds for finance research and extension projects. Each governor handpicked progressive farmers to join the state foundation's board. Soon after joining the boards, a few farmers recognized that they lacked an understanding of managing public funds for research and extension, and they started to exchange information on how they operated their foundations. This process enabled innovative foundations to differentiate themselves from the others. The foundations were also influenced by the political climate in their respective states, the presence of innovative farmers on their boards, and the backgrounds of the foundation managers (especially with respect to their managerial experience in large organizations).

In the beginning, when the federal and state governments controlled the Produce Foundations' boards, farmers from about six foundations demanded independence. At first the federal and state governments opposed

their independence, but soon they recognized the benefits of civil society participation and supported the independent Produce Foundations. Eventually most foundations followed suit and requested independent status as well. The innovative farmers also realized that the individual foundations were too isolated and that a national coordinating office was needed; before long, COFUPRO was created.

Over the years, COFUPRO and the foundations have developed a better understanding of innovation processes and consolidated their structure, operational routines, and institutional culture. The process entailed a progressive transfer of authority from individual foundations to COFUPRO. The main benefits of a strong COFUPRO were (1) the consolidation of a system of decentralized experimentation with centralized learning, (2) stronger lobbying capabilities, (3) more transparent use of resources, and (4) more effective operational rules. The learning process was particularly effective. The foundations moved from financing traditional research and extension projects to supporting innovation activities and stable interactions among researchers, technical advisers, firms, and farmers. Each foundation tried new ways to support innovation and to manage its funds. Once effective operating routines were developed, they were adopted by all of the foundations.

The foundations implement a number of activities, but the most relevant for this module are their priority-setting methods, allocation of funds, and exploration of new methods to foster innovation:

- *Priority setting.* Over the years, the foundations tried several methods to set priorities. The most important was a two-year national consultation implemented in 2002, which was the basis of all agricultural policies for the following five years. The process was considered too costly and never repeated. In the following years, each foundation developed its own priority-setting method, but they are converging on a permanent dialogue among the foundation, researchers, and important stakeholders from the different agricultural clusters. In other words, they have moved from a demand-driven, linear process to a continuous dialogue that results in participatory research and innovation (there are several modes of participation, however).

- *Allocation of funds.* For many years, the foundations used a competitive fund to select the projects to be funded. In 2006, they realized that this method did not induce researchers to abandon their linear vision of science, and they started to contract research and innovation projects directly with research institutes and other service providers. Additionally, the foundations started to prioritize projects presented by groups of actors (usually farmers and researchers) that had developed stable relationships.

- *New methods.* In recent years, some foundations have played a catalytic role in the emergence of innovation networks that explore new research and diffusion methods. These activities have not been evaluated yet.

INNOVATIVE ELEMENT

The Produce Foundations' relevance resulted from their continued exploration of new mechanisms to foster agricultural innovation and to develop organizational capabilities. This exploration was made possible by the presence of a few innovative individuals operating in a socioeconomic and institutional environment that allowed the foundations to change. It was further supported by the development of an organizational culture that allowed new alternatives to be explored.

BENEFITS, IMPACT, AND EXPERIENCE

The Produce Foundations had several impacts on the Mexican AIS:

- They funded research projects that opened new export markets, boosted the profitability of agriculture, solved serious production constraints (for example, with improved pest control), and improved the sustainable use of natural resources.
- They induced the emergence of networks that explored new approaches to foster innovation.
- They implemented development projects that benefited small-scale farmers.
- They influenced the operations of public research institutes and universities.
- They opened opportunities for researchers to interact directly with farmers, helping them to replace the linear vision of science with an innovation-based model of science.
- They influenced the design and implementation of agricultural policies, especially for research and innovation programs.

- They helped other actors in the AIS, especially farmer organizations and policy makers, to understand the nature of agricultural innovation.

LESSONS LEARNED AND ISSUES FOR WIDER APPLICATION

A number of internal and external factors influenced the evolution of the foundations. The main internal factors were the creativity of a few farmers on the boards and of some foundation managers; the development of effective collective learning routines; the creation of an organizational culture that valued exploration, creativity, innovation, and a sense of duty; and flexible governance structures. The external factors were the existence of dynamic markets that created challenges and opportunities for farmers and the presence of a few policy makers in key positions who were willing to allow the foundations to experiment and change.

Autonomy and independence proved important. There is a positive correlation between a foundation's autonomy from the state government and its institutional development, efficiency in achieving its mission, and innovativeness. These features result from the presence of more innovative individuals, the greater commitment shown by the board members, and more professional management staff.

Decentralized experimentation and centralized learning were important. The foundations succeeded because they could explore more effective operating routines and new instruments to foster innovation. The emergence of COFUPRO and the foundations' ability to change resulted from the creation of variation (32 foundations, instead of one centralized organization) and an effective process of self-organization, greatly influenced by a few innovative individuals. The combination of independent foundations with strong interactions also became a mechanism to explore alternative organizational routines and diffuse the most effective ones (box 1.24).

Finally, it was crucial that the federal and state governments allowed the foundations to explore new instruments to fulfill their mandate (box 1.25). The effectiveness of the mechanism was hampered, however, by the informality of the information flows, the lack of methods to guide the exploration, weak incentives for unwilling foundations to adopt best practices, and the loosely structured activities, which made the performance of individual foundations dependent on the personalities of the people in command.

Box 1.24 Organizational Learning in Mexico's Produce Foundations: Evolution of Priority-Setting Procedures

At the beginning, the Produce Foundations had no pre-defined criteria to allocate funds. Farmers on the board would select projects to finance from among the proposals freely submitted by researchers. The foundations soon realized that the projects did not respond to farmers' needs, and they started to search for priority-setting methods. They learned of a methodology developed by the national science and technology council, which was based on the construction of restriction trees and an ex ante cost-benefit analysis of potential projects. This methodology was used for a couple of years.

In 1997, COFUPRO authorities met the research director of the International Service for National Agricultural Research (ISNAR), who provided guidelines for a methodology based on accepted practices for project design. The method involved organizing forums where the various actors in agricultural chains could diagnose their most important problems. Over the next two years, the foundations adapted this methodology to the Mexican environment. They recognized that ISNAR's methodology was better than the council's approach because it was based on wider criteria than the expected costs and benefits, and it enabled different actors to help define priorities. The foundations implemented the methodology in 2002 and 2003 to prioritize agrifood chains and identify research demands, first in each state and then at the national level—an exercise that was not repeated because it was deemed too expensive.

After a few years, the foundations realized that a demand-driven system and clearly defined priorities did not guarantee that researchers would provide solutions that farmers could use, because it still allowed researchers great latitude in defining the approaches to solve the problems. The next step was to modify the call for proposals. The calls were very narrowly defined (almost down to the title of the project desired), but after a few years, the foundations realized that this new method still did not solve their problem because it was based on a linear vision of science. More recently, the foundations experimented with different approaches to define priorities and transform how researchers defined their methodologies. Researchers have been induced to interact more closely with farmers and move from their traditional research domains to participate in innovation networks.

The foundations continuously analyzed the limitations of the prioritization procedures in use and actively sought alternatives. As they collected information, they absorbed it to develop their own methods, which in turn induced changes in resource allocations and the monitoring of projects.

Source: Ekboir et al. 2009.
Note: ISNAR was a CGIAR center that closed in 2004. IFPRI absorbed some of its work.

The Produce Foundations initially financed traditional research and extension projects, but realized after a few years that the projects had little impact on agriculture. In 2003, an external evaluation introduced the foundations to the notion of innovation systems. A second evaluation in 2004 explained how to set up innovation projects. Since then, several foundations have implemented innovation projects, while continuing to support traditional research and extension activities. The innovation projects include the following:

■ Development of an organization of small-scale farmers and their families that markets dried and processed hibiscus flowers and uses the by-products to feed chickens for egg production.

■ Development of a farmer-to-farmer system to exchange technical and commercial information; the system is based on Social Network Analysis techniques to identify the most effective communicators.

■ Creation of a company of small-scale farmers to sell processed sheep meat to domestic and foreign markets.

■ Establishment of a consortium of researchers who interact closely with farmers to validate and disseminate innovations for livestock production.

■ A joint venture with large-scale farmers to finance a stable research program (which included paying a researcher's doctoral studies) to develop innovations for pecan production.

■ With financing from foreign foundations, importing a small sewage treatment plant for two remote, impoverished villages and using the treated water to irrigate greenhouses to produce fresh vegetables.

■ Teaming up with Wal-Mart to develop a supply chain for fresh vegetables produced by small-scale farmers.

Source: Author.

Chile's Foundation for Agricultural Innovation

Rodrigo Vega Alarcón, Consultant, formerly with the Foundation for Agricultural Innovation (FIA)

SYNOPSIS

Chile's Foundation for Agricultural Innovation has facilitated partnerships, especially between public and private agencies, that promoted technical, organizational, and commercial innovations. The foundation coordinates actors in the public sector, academic institutions, private firms, and other entities to understand and meet the innovation needs of different industries and agricultural activities. The agency operates as an honest broker, free of special interests, and implements programs to generate trust among farmers and other partners. Innovation initiatives are managed responsively, quickly, simply, and not bureaucratically, and as innovation processes move forward, the agency retains organizational flexibility and freedom (for example, to form ad hoc, specialized groups and instruments for solving particular problems).

BACKGROUND AND CONTEXT

In the 1990s, the Chilean economy was liberalized, the government reduced tariffs, and free trade agreements came into effect. The agricultural sector was at a crossroads; it had to modernize and become more competitive. Support for agricultural research, development, and innovation was stepped up. The government adopted policies to promote technology transfer and raise productivity.

The subsequent rapid transformation of Chilean agriculture into an engine of regional development has become a familiar story. Current exports from the national food and forest industry (fruit, wine, salmon, wood, white meats, and other products) are valued at about US$13 billion per year and are expected to reach US$20 billion in coming years.

The transition from a traditional agrarian economy to an export-based economy was not entirely smooth. Unemployment and migration increased as traditional agricultural products such as wheat, corn, milk, and meat were replaced by imports from countries with more competitive agricul-

tural sectors. The challenge was to improve Chilean producers' competitiveness through new technologies and to develop alternative crops for regions and farmers affected by imports. At the time, agricultural research and development focused on basic rather than applied science, and little support was given to business innovation.

PROGRAM DESCRIPTION

In 1994, Chile's Ministry of Agriculture created the Foundation for Agricultural Innovation (FIA, Fundación para la Innovación Agraria), a public agency to promote and financially support agricultural research, development, and innovation. Initially FIA focused strongly on technology transfer to improve competitiveness. It identified products with high economic or commercial value and significant labor requirements that could be adapted to Chilean agriculture. In 2009, FIA became part of Chile's National System of Innovation for Competitiveness (SNIC, Sistema Nacional de Innovación para la Competitivad). FIA's strategic objective is to promote innovation processes for the agriculture sector and improve the conditions that favor those processes, by cofinancing innovation initiatives, generating strategies, and transferring information and results of innovative programs and projects carried out with Chile's private sector.

The foundation is guided by a seven-member board of directors chaired by the minister of agriculture. It is headed by a chief executive appointed by the minister of agriculture. FIA's annual budget is about US$18 million, mostly from the Ministry of Agriculture (US$11 million); other sources of finance include the national Innovation Fund for Competitiveness (FIC, Fondo de Innovación para la Competitivad), which is supported by mining royalties (US$6 million). The national innovation policy has three pillars: *Science* and *Human Capital* are supported by the National Commission for Scientific and Technological Research,[1] and

Enterprise Innovation is supported by Innova-Chile, the multisectoral innovation agency of the Corporation to Promote Production (CORFO).[2] FIA and other small funds focus on specific sectors and complement Innova-Chile. While FIA works mainly with small- and medium-scale enterprises, Innova-Chile works with medium- and large-scale agroindustrial entities and entrepreneurs.

INNOVATIVE ELEMENT

The efforts of FIA changed the paradigm for agricultural research and development, which had been confined to technology institutions and academia and yielded results that often lacked commercial application. Aside from facilitating partnerships across the public and private sector, including producers and industry, FIA supports innovation in the following ways:

- *Strategic development of information and knowledge* to anticipate future trends and technological developments in global agriculture, through observation, exploration, and analysis of such emerging issues as climate change, agriculture's carbon footprint, water, and bioenergy, among others.

- *Development of mechanisms and instruments* for disseminating projects and programs with commercial potential to agrarian enterprises.
- *Management intervention mechanisms* that support collaborative innovation initiatives submitted by clients (box 1.26).
- *Evaluation of technical, economic, and social results* of projects cofounded by FIA, prior to their transfer and implementation.

Currently FIA works with the World Bank on redesigning the system for agrifood and forestry research, development, and innovation through scenario planning. In this context, FIA plays the role of coordinator and broker, interacting closely with all parts of the system—agricultural producers, financial agencies, companies, technological institutes, or universities.

BENEFITS, IMPACT, AND EXPERIENCE

Since its creation, FIA has promoted the development of new, high-value agricultural products for the domestic and international markets. Many of these products have improved farmers' incomes and living conditions

---

Box 1.26   Mechanisms Used by Chile's Foundation for Agricultural Innovation to Support Innovation

The Foundation for Agricultural Innovation uses several mechanisms to support initiatives and innovation projects, technology transfer, and human resource development. *Territorial innovation programs* are regional initiatives led by stakeholders and representatives of an agrifood chain in a particular region or territory. *Technology consortiums* are joint ventures between technology agencies (including public research institutes and universities) and private enterprises to create a new technology-based company or pursue innovative research on new commercial technologies (for the grape, potato, and dairy industries, for example). On a smaller scale, *technology development programs* promote specific technological development and innovation led by stakeholders and representatives of an agrifood chain. *Innovation in agri-food marketing* supports market integration and market development tools for micro, small-scale, and medium-scale agrifood companies. FIA also conducts numerous *studies* to assemble and synthesize technical, economic, and/or commercial information to aid decisions on future innovation initiatives. *Projects* are funded on innovations that improve a company's competitiveness and that of the sector to which it belongs. Projects must show measurable market effects. *Visits to centers of excellence* are sponsored in Chile and abroad to observe and evaluate technological, organizational, and managerial innovations in production systems. Individuals are also sponsored to attend *national and international technical events*, such as seminars, symposiums, congresses, conferences, and technology fairs. FIA also funds *specific events* (seminars and conferences) to disseminate and transfer national and international experiences related to products, processes, and innovative tools for marketing, organization, and management to different actors in the AIS. *Consultants* provide specific skills to enhance competitiveness of specific products, processes, or organizations and their management.

Source: Author.

---

throughout Chile. An impact evaluation in 2005 determined that each million Chilean pesos (Ch$) spent by the foundation increased sales in the agricultural sector by Ch$5.14 million, leveraged Ch$1.89 million in private contributions, and created two permanent jobs in the agricultural sector (FIA 2005). For example, FIA helped to support development of a successful olive oil–processing and export industry (box 1.27), increased sheep meat production by introducing better breeding stocks, extended cranberry cultivation to new areas, introduced new varieties of flowers for the export and domestic market, and expanded peony production area. FIA has also sponsored investments in ICT technologies for rural areas. In organic agriculture, FIA supported the first projects in what has become a growing business, and it advanced the study and commercialization of biological control, using a variety of beneficial insects.

LESSONS LEARNED AND RECOMMENDATIONS FOR PRACTITIONERS

Lessons and recommendations from FIA's experience may be useful to other agencies that seek to move from a mission based on technology transfer to one that involves building partnerships (locally, nationally, and internationally) that foster agricultural innovation and access to highly competitive global markets.

A vital role of an innovation agency is to coordinate actors in the public sector, academic institutions, private firms, and other entities to understand and meet the innovation needs of different industries and agricultural activities. The key is to be participatory and create avenues for continuing analysis and discussion in which the experience and opinions of all actors can be considered.

Organizations such as FIA that specialize in promoting innovation should act as facilitators, linking the demand for research and development with the suppliers and strengthening the capacity for research and innovation throughout the AIS. There should be a distinction, understood by all actors in the AIS, between the roles of innovation-promoting agencies and centers for developing or diffusing innovations (whether they are public, private, nongovernmental, or civil society organizations). As facilitators, innovation agencies should not operate research programs. They must remain independent of research and technology institutes,

---

Box 1.27 Purposeful Innovation to Expand Chile's Olive Oil Industry

In the mid-1990s, Chile produced olives on a small scale—approximately 3,000 hectares, of which 350 hectares were for oil production. Yields were relatively low, averaging 3–4 tons per hectare and 16–18 percent edible oil. Production relied on outdated practices. The National Olive Development Program, launched in 1995 by the Ministry of Agriculture and coordinated by FIA, addressed problems (agronomic, processing technologies, zoning) that limited the industry's development. It involved all agencies under the ministry, plus producers, private companies, and other entities, including Fundación Chile. The ultimate goal was to make olive production more competitive and stable and improve access to domestic and international markets (mainly the United States, Spain, and Canada). To that end, FIA supported the following:

■ Deepening market research and identifying business opportunities.
■ Identifying and multiplying new, more appropriate genetic stocks.

■ Expanding area planted to high-yielding cultivars.
■ Acquiring international technical expertise for agronomic and processing problems and marketing.
■ Training specialists in olive cultivation and industrialization.
■ Establishing modern oil-processing plants.
■ Developing a brand and marketing Chilean olive oil.
■ Conducting international seminars to promote Chilean olives and olive products.

By 2009, planted area reached approximately 21,500 hectares (65 percent of production was destined for oil and the rest for table olives). Investments in the sector remain strong and have reached US$50 million annually. Exports of extra virgin olive oil from Chile have increased tremendously in recent years. In 2003, Chile exported 53.7 tons, with a free-on-board (FOB) value of US$158,200; in 2007, Chile exported 562 tons, with an FOB value of US$3.1 million; and by 2009, it exported 1,933 tons with an FOB value of US$12.5 million.

Sources: FIA 2009; CHILEOLIVA 2009.

---

universities, the private sector, and agricultural organizations. They must be honest brokers, trusted by all, and not captured by special interests.

Although the private sector is an important force for innovation in agriculture, a public agency that specializes in promoting agricultural innovation, with sufficient human and financial resources, is a vital complement to private investments. In many cases, agricultural innovations may not be easy to patent or otherwise commercialize, but they may be critical to the development of the sector (biological control programs and information and communications technology infrastructure are two examples mentioned in this IAP). It is also likely that the partners involved and the relative levels of public and private investment will vary when innovations are in the developmental stage compared to when they are being commercialized. In other words, as innovation processes move forward, a certain amount of organizational flexibility and freedom are needed (for example, to form ad hoc, specialized groups and instruments for solving particular problems that emerge at a particular stage in the process).

The processes of managing innovation initiatives must be responsive, quick, simple, and not bureaucratic. Otherwise farmers and businesses have little incentive to participate. Innovation programs and the innovation agency itself should be evaluated regularly to verify their effectiveness and impact and make corrections in a timely way, if necessary. The professionals within the agency must receive training in management techniques for innovation and technology management, if they are to be of real support to farmers and entrepreneurs who seek to foster and manage innovation processes.

Needs for innovation (and partners committed to developing them) can be quite location-specific, especially in a highly ecologically diverse country such as Chile. FIA now develops what it calls "territorial innovations" by working with the regions to promote innovation at the local level. The development and adoption of innovations, especially with smaller-scale and more traditional farmers, can be slow. The process benefits from complementary programs that generate trust among farmers and other partners.

A Rural Institutional Platform Mobilizes Communities to Become Effective Partners in Agricultural Innovation in Andhra Pradesh

Gunnar Larson, World Bank
Melissa Williams, World Bank

SYNOPSIS

Community organization in the form of small self-help groups and the federation of these grassroots institutions at higher levels have created support structure that enables the rural poor in Andhra Pradesh to identify workable strategies to meet their needs—in other words, to innovate. Each tier in the organization of self-help groups functions as a financial intermediary and provides specialized services to members and other stakeholder groups in a variety of sectors. Through this rural institutional platform, community members have identified, adapted, used, and spread environmentally friendly agricultural practices, obtained credit, invested in productive assets, and improved their food security and health, among other benefits. Lessons from this experience highlight the importance of developing local institutions with local people (the local commitment and relevance makes them more likely to innovate successfully) and of organizing at higher levels (where farmers gain a collective voice to empower themselves). These institutions of the poor plan, manage, monitor, and scale up new initiatives and build social capital at much lower transaction costs and with much greater purposefulness than is possible through more traditional forms of organization, in which the rural poor are more often at the periphery than the center of service provision and innovation.

CONTEXT

Organizing large numbers of poor people around economic activities and mobilizing their combined assets until they collectively achieve a scale of magnitude sufficient to attract the interest of investors and service providers have a number of precedents—most immediately in community-driven development programs in Bolivia and Northeast Brazil beginning in the late 1980s. In the late 1990s, India applied these principles in a unique way, mobilizing and federating poor agricultural communities to access credit, markets, and services on better terms. The southern state of Andhra Pradesh, where costly and unsustainable agricultural production practices were creating unmanageable levels of debt (box 1.28), quickly established itself as a pioneer of this approach.

PROGRAM OBJECTIVES AND DESCRIPTION

The Society for the Elimination of Rural Poverty (SERP), an autonomous body established by the Government of Andhra Pradesh, implements the Andhra Pradesh Rural Poverty Program.[1] Under this program, SERP works in communities to mobilize self-help groups (SHGs), each with about 10–15 members (such as poor women and/or farmers), who engage in collective saving, lending, and other activities that enable them to build an asset base. As shown in figure 1.1, each tier in the organization of SHGs functions as a financial intermediary and provides specialized services to members (and other stakeholder groups) in a variety of sectors. The SHGs federate into village organizations (VOs), and each VO manages a capital fund, from which it provides loans to constituent SHGs. The VOs organize into subdistrict federations, which access commercial credit to lend to VOs, and subdistrict federations organize into very large district federations. This platform of federated institutions brings economies of scale and scope that allow community members to build assets, smooth consumption, access services and safety nets, and invest in livelihoods to raise themselves out of poverty.

Box 1.28   Rising Input Use and Agricultural Debt in Andhra Pradesh

Andhra Pradesh is one of India's major producers of rice, cotton, groundnuts, and lentils. Agriculture accounts for 25 percent of GDP and 60 percent of employment. Most farmers in the state practice conventional, input-intensive farming that relies on periodic purchases of high-yielding seed, chemical pesticides, and chemical fertilizers. As much as 35 percent of cultivation costs go to pesticides and fertilizers alone, which is five percentage points above the national average (NSSO 2003). High production costs generate tremendous pressure to borrow heavily to pay for inputs and make ends meet. The average outstanding loan for farmers with small landholdings was more than twice the national average in 2005. Among farm households, 82 percent were in debt—the highest estimated prevalence of debt among farm households in India (NSSO 2005). Many farmers used their land as collateral, eventually becoming nothing more than tenant farmers or wage laborers on their own land. Owing to rising debt, land mortgages, and uncertain profit potential, planted area in Andhra Pradesh plummeted by more than 988,000 acres between 1980 and 2005, along with yields and agricultural growth.

(Cropping intensity during 1980–81 and 1990–91 was 1.16; gross cropped area declined from 12.5 million hectares in 1980–81 to 12.1 million hectares in 2004–05, according to the Government of Andhra Pradesh and Centre for Economic and Social Studies (2008).) The same period saw a reduction in government-provided public services that left farmers increasingly dependent on moneylenders and input traders. Traders became many farmers' sole source of credit, inputs, and related information. One result of this arrangement was that Andhra Pradesh farmers, ill-informed about the dangers of incorrect input use, applied far more pesticide than their counterparts in any other state in India: 0.82 kilograms per hectare annually, compared to the national average of 0.3 kilograms per hectare (Government of Andhra Pradesh, Irrigation and CAD Department 2007). Another result was that many traders and moneylenders entered into buyback agreements with farmers at below-market prices in return for their services (Ramanjaneyulu et al. n.d.). Purchasing inputs imposed such a financial burden on smallholders that conventional agriculture no longer offered a viable livelihood.

Source: Authors.

INNOVATIVE ELEMENTS

Andhra Pradesh has created an ecosystem of support in which a bottom-up planning process beginning at the household and community level and aggregating to higher levels enables the rural poor to identify needs and define workable solutions (in other words, to innovate). For instance, farming households develop a microcredit plan with the help of their VO and are linked to commercial banks through their subdistrict and district federations. Community groups also manage enterprises such as procurement centers for agricultural commodities and milk, which provide grading, quality control, aggregation, and value addition for products. At the subdistrict level, federations invest in enterprises such as chilling centers for milk to increase shelf life. Meanwhile, the district federation manages a number of support functions, including running an insurance scheme for members through a network of call centers. Together, these activities help farmers receive higher prices and foster an environment that favors profitable agriculture. This support along the value chain of

agriculture makes farming sustainable from an economic perspective. At higher levels, the federation organizes training, engages NGOs and government agencies for additional support, and monitors progress.

Farmers have also used this institutional platform to practice community-managed sustainable agriculture (CMSA), a knowledge-intensive alternative to input-intensive agriculture. Through their groups, farmers learn about the harmful effects of chemical pesticides and fertilizers on soil, water, and health. They gain access to farmer field schools, seed banks, equipment centers, finance, and procurement centers, all organized by their respective VOs. District and subdistrict federations market the CMSA produce. Through the district federations, farmers can use funds from the National Rural Employment Guarantee Scheme, a social safety net ensuring every rural poor household has at least 100 days of paid work each year, to pay for on-farm improvements for sustainable agriculture. Examples include transplanting tank silt to farms, leveling land, or building structures to capture rainwater. In this way, a

Figure 1.1   The Institutional Model for Federated Self-Help Groups

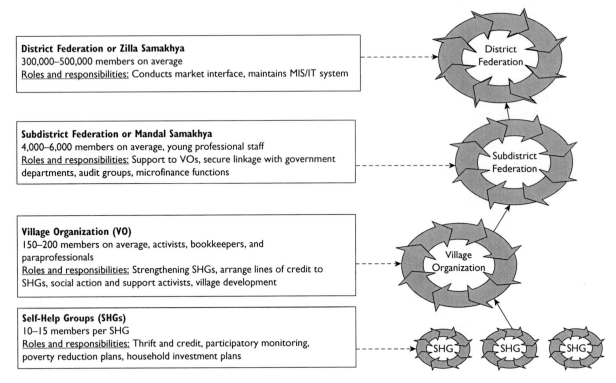

District Federation or Zilla Samakhya
300,000–500,000 members on average
<u>Roles and responsibilities:</u> Conducts market interface, maintains MIS/IT system

Subdistrict Federation or Mandal Samakhya
4,000–6,000 members on average, young professional staff
<u>Roles and responsibilities:</u> Support to VOs, secure linkage with government departments, audit groups, microfinance functions

Village Organization (VO)
150–200 members on average, activists, bookkeepers, and paraprofessionals
<u>Roles and responsibilities:</u> Strengthening SHGs, arrange lines of credit to SHGs, social action and support activists, village development

Self-Help Groups (SHGs)
10–15 members per SHG
<u>Roles and responsibilities:</u> Thrift and credit, participatory monitoring, poverty reduction plans, household investment plans

Source: SERP and World Bank 2009.

safety net entitlement is linked to productive, income-generating activities conducted by community organizations. Largely owing to the effectiveness of the institutional platform (figure 1.2), poor and marginal farmers have developed a more successful livelihood strategy.

To improve food security, households belonging to SHGs identify the amount of food they can purchase with their own funds, the amounts obtained through the public safety net, and their remaining needs. The VO aggregates the information, ensures that each family gets the safety net to which it is entitled, and then purchases the rest of the food from the market in bulk. Households belonging to the SHG can then take a low-interest, long-term loan from the VO or SHG to purchase rice on terms they can afford, thereby smoothing their consumption.

BENEFITS AND IMPACTS OF ORGANIZING RURAL COMMUNITIES TO INNOVATE

Through SERP and the rural institutional platform it provides, communities have organized to meet multiple needs.

As the discussion has indicated, the benefits of this organizational structure include improved food security as well as the accumulation of financial acumen, new agricultural knowledge and skills, and productive assets. In particular, the institutional platform's success in enabling communities to pursue alternatives to conventional agriculture has led the state government to call for the Agriculture Department's Agriculture Technology Management Agency (ATMA) to collaborate with the Rural Development Department's SERP to promote sustainable agriculture and move toward organic agriculture. SERP will train the ATMA staff to use the CMSA model; ATMA will then work with SHGs to popularize this low-cost, high-return type of agriculture. The hope is that it will yield benefits similar to those seen with CMSA (box 1.29).

LESSONS LEARNED AND ISSUES FOR WIDER APPLICATION

Experience with the rural institutional platform adopted in Andhra Pradesh indicates how the capacity to organize at

Figure 1.2   The Same Institutional Platform Provides Services to Develop Multiple Livelihood Strategies

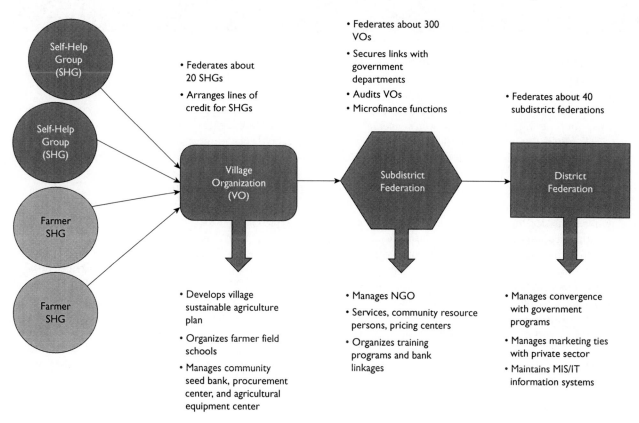

Source: Authors.

Box 1.29   Benefits of Community Organization to Pursue More Sustainable Agricultural Practices in Andhra Pradesh

Savings and incomes increased. A survey of 141 farmers found that production costs were 33 percent lower under CMSA than conventional agriculture (US$180 versus US$280 per acre). The savings translate directly to increased incomes for farmers. For crops raised without pesticides and fertilizer, farmers command a premium of 14–33 percent. Even though CMSA produce is not yet certified organic, consumers (especially in urban retail markets) increasingly recognize the benefits of pesticide- and fertilizer-free food.

Yields remained stable and diversification increased. To track changes in paddy yields after farmers switched to CMSA, 400 farmers' fields in five districts were monitored closely. Yields remained the same, ranging from 1,900 kilograms per acre to 2,200 kilograms

per acre for paddy and rice. Although CMSA brings higher labor costs, farmers are meeting this challenge by working together to manage pests and increase soil fertility. Demonstrations of multicropping and intercropping alternatives are helping more small-scale and marginal farmers in Andhra Pradesh realize the benefits of diversification. Farmers on nearly 319,000 acres now plant one or two crops in addition to the main crop.

Communities obtained debt relief. In a survey of five districts, of 467 families who had mortgaged their farmland, 386 had paid off their debt and reclaimed their land within two years through the savings from CMSA. The social empowerment associated with getting their land back from moneylenders and

(Box continued on next page)

farming on their own land is very significant for farmers, perhaps as important as the economic relief it provides.

Families increased their investments in productive assets and sustainable land and water management. Communities dug more than 10,000 composting pits and 1,200 farm ponds. Fertile tank silt has been applied to more than 13,000 acres of farmland. Lower production costs and higher net incomes have made farmers less risk averse. Primary surveys by SERP show that families are leasing additional land for cultivation, resulting in additional income for households. Farmers are also bringing fallow and government-assigned land under cultivation.

Communities saw greater business innovation and new livelihood opportunities. Villages have begun to benefit from jobs and enterprises catering to inputs for CMSA and by providing services such as quality control and procurement for CMSA produce. At least 2,000 jobs have been created in villages through shops that supply biopesticides, organic nutrients, seed, and farm implements. About 5,400 small-scale and marginal farmers are generating additional income by operating vermi-composting units.

Food security improved. Data collected by SERP from 22,000 CMSA farmers in Khammam District show that household expenditures on food grains declined by half owing to higher yields of food grain crops and the introduction of a second crop. Families purchased 44 percent less food grain from the market.

Human and environmental health benefited. Farmers reported a noticeable drop in pesticide-related health problems. Women, who traditionally sprayed the crops—and suffered the effects—are now strong advocates of the new practices. In three districts, hospitalizations from pesticide poisoning declined by 40 percent, from 242 cases per year to 146 cases. Villages that completely stopped pesticide applications are benefiting from the elimination of pesticides from groundwater and soil. Insects and birds, no longer targeted by broad-spectrum pesticides, are returning to the fields.

Sources: Authors.

---

several levels fosters innovation and adaptation in the local context, with local people, and contributes to success on a larger scale as well:

■ *Institutions (and their designers) matter.* Initiatives designed by the farmers who will participate in them carry real advantages over initiatives that outsiders might design for them. Many of these advantages relate to the fact that the people who plan and implement activities are the same people who benefit from them. Their ownership of the initiative greatly increases the likelihood that it will be sustainable. The social capital they cultivate and build upon in working together for common goals changes the rules of the game for farmers in relation to the market and the government. In organizing and mobilizing at higher levels of aggregation, farmers gain a collective voice and empower themselves, which is something that cannot realistically be done for them from the outside. Supporting their initiatives by investing in a process that over time comes to involve millions of rural people, especially women, has helped to achieve a large-scale transformation in smallholder farming. These institutions of the poor plan, manage, monitor, and scale up new initiatives at much lower transaction costs and with much greater purposefulness than is otherwise possible. Building social capital is a necessary investment in transforming human, natural, and economic capital.

■ *Small-scale and marginal farmers need a creative approach to the delivery of agricultural extension services.* Practicing farmers should be the central stakeholders contributing to an extension system. Where the system has failed to meet their needs, they should be enlisted as active extension agents. Their presence in the village makes them easily accessible and more familiar with local conditions and challenges. For example, the use of successful CMSA farmers as community resource persons was critical to building and scaling up the program. The experience of the resource persons gives their messages greater credibility among farmers. NGOs can facilitate extension-related services, but eventually the community resource persons assume this role, as they are better equipped to understand farmers' needs.

- *Farmers can experiment and develop technologies in their fields and test innovations through their local field schools.* Once technologies are proven, they can be standardized for a wider audience through training workshops. In CMSA, the menu of technology options continues to grow, and farmers no longer have to rely on the limited options available through external research and other sources in the market. Farmers should be encouraged to

look at the cost-effectiveness of options and not focus only on yields.

CMSA has already been taken to the state of Bihar with considerable success, and it will be further scaled up through India's National Rural Livelihoods Mission. The achievements in Andhra Pradesh were very much the result of tailoring solutions specifically to local contexts, and replicating those achievements elsewhere will require purposeful adaptation to local conditions.

In the Networking Age, Cassava Development Relies on Public-Private Research Partnerships

Bernardo Ospina Patiño, Latin American and Caribbean Consortium to Support Cassava Research and Development (CLAYUCA)

SYNOPSIS

CLAYUCA transcended the traditional country-based model of cassava research to develop a regional research and development network that attracted nontraditional partners and funding. All members participate in planning, financing, and implementing prioritized activities for cassava research and development while sharing costs, risks, and benefits. An emphasis on competitiveness and a value chain approach helped CLAYUCA to focus on issues limiting efficiency in production, processing, and utilization, in which the private and public sector had common research interests. CLAYUCA demonstrates the importance of allowing different interaction, coordination, and innovation mechanisms to emerge and of coordinating value chain actors to contribute to policy debates related to cassava (especially support for public-private-CSO-NGO collaboration). Another lesson is that facilitating partnerships between public and private organizations will become increasingly challenging unless stable financing is available and public organizations maintain their expertise.

CONTEXT

Cassava, the fifth most important crop in the world, can compete with coarse grains in animal feed, partially substitute for maize and other starchy crops in food products, serve as a source of industrial starch, and be used to process ethanol. The crop is especially important in Latin America, sub-Saharan Africa, and Asia. Although national and international organizations have conducted cassava research at least since the 1970s, many of these institutions experienced radical changes in the 1990s as public funding for agricultural research evaporated and the need for a less linear, more participatory model of agricultural innovation became clear. Institutions and

countries sought to establish strategic alliances to continue agricultural research and development. In 1999, the Latin American and Caribbean Consortium to Support Cassava Research and Development (CLAYUCA, Consorcio Latinoamericano y del Caribe de Apoyo a la Investigación y al Desarrollo de la Yuca) was formed as a regional mechanism to plan and coordinate research for the cassava subsector.

THE CLAYUCA MODEL

CLAYUCA is a network operating through collaborative agreements between its members—public and private entities—in which all members participate in the planning, financing, and implementation of prioritized activities to accomplish jointly established objectives while sharing all costs, risks, and benefits. An executive committee defines priorities and lines of activity, and coordinates and evaluates research undertaken by the consortium. Each member country elects one representative (not necessarily from the public sector) to serve on the committee; the participating international organizations also elect a representative. The committee approves the entry of new countries into the network and elects the executive director, who coordinates CLAYUCA's activities. Initially a regional network for Latin America and the Caribbean, CLAYUCA has attracted members from other regions, and member countries now include Colombia, Costa Rica, Ecuador, Guyana, Haiti, Mexico, Nicaragua, Panama, Peru, Trinidad and Tobago, and Venezuela, as well as Ghana, Nigeria, South Africa, China, and the United States.

A Technical Committee defines the research agenda, which is developed in detail by one representative per member country. Research focuses on competitiveness, given the tremendous challenges in identifying and strengthening market opportunities for cassava and moving from traditional to more competitive modes of production.

Aside from the collaborative agreements with its members, CLAYUCA operates through a formal agreement between its executive committee and CIAT. CIAT is a strategic partner, hosting the consortium, providing core funding, legal support, administering CLAYUCA funds, and facilitating the use of the laboratories, fields, equipment, and offices under a fee payment scheme. CLAYUCA's activities are partially financed through an annual membership fee of US$15,000 per country. In some countries, a private institution pays the fee; in others, a public agency pays (the Ministry of Agriculture, for example). In Costa Rica and Colombia, the annual quota is paid by a group of public and private entities. Additional funding is obtained through special projects and consultancy services. These resources are invested only in activities defined collectively by the members.

THE INNOVATIVE ELEMENT

The innovative aspect of CLAYUCA is its role as a regional facilitator of public-private alliances for cassava research and development, using a value chain approach and emphasizing competitiveness. Different actors in a member country's cassava subsector identify where cassava's overall competitiveness can be improved along the value chain (production, processing, or utilization). They identify organizational and technical constraints and formulate and implement technological interventions. The new emphasis on competitiveness, a prerequisite for private sector involvement in cassava-based industries, has motivated farmers, especially small-scale farmers, to adopt improved production technologies such as better varieties and improved crop and soil management practices. Increased competitiveness on the supply side is complemented by private investments and contributions to processing capacity and management. The network's regional and international character offers particular advantages for countries where cassava research has been limited by small national budgets and little external interaction.

BENEFITS AND IMPACT

CLAYUCA's structure enables members to have better control of the regional research and development agenda for cassava and participate more equitably in the distribution of benefits. The consortium also facilitates better access for public and private agencies to technologies generated by international and advanced research centers. Those centers, in turn, benefit from participating in a regional agenda for cassava research with relatively little investment.

Examples of technology and other knowledge generated and shared through CLAYUCA include the following:

- *Technology platforms for more competitive cassava production and utilization.* Technology platforms include equipment as well as the methodologies, instructions, processes, training, and other elements to use it correctly. CLAYUCA has developed a number of platforms: (1) for producing high-quality, refined cassava flour for human and industrial use; (2) for producing and using cassava leaves in animal feeding systems; (3) for producing and using sweet potato in human food and animal feed; (4) producing hydrated ethanol for local use with cassava, sweet potato, and sweet sorghum (the technology is small-scale, low-cost, and easily operated and managed by small-scale farmers); and (5) for producing cassava commercially (mechanized planting, partially mechanized harvesting, and soil fertility, pest, and disease management practices).
- *Planting and breeding materials.* CLAYUCA has facilitated access to improved, elite cassava germplasm developed by CIAT and other advanced research centers.
- *Communications.* CLAYUCA communicates relevant, current information on cassava technologies to stakeholders in the form of websites, electronic bulletins, training events, annual meetings, study tours, and technical books and bulletins.
- *Human resources development.* Through training offered at CIAT and in member countries, CLAYUCA has strengthened technical capacity in such areas as cassava processing, crop management, product and market development, tissue culture, and cassava germplasm evaluation and selection.

During its first decade, CLAYUCA has benefited various actors in the cassava subsector of each member country. It also generated the regional benefits described in the sections that follow.

Benefits to the public sector

Public institutions have taken advantage of the presence of CLAYUCA in their countries, supported by the strong research background of CIAT, to improve their capacities in areas such as managing genetic resources, training technical personnel, and improving knowledge and information about modern technologies for cassava production, processing, and utilization. At the country level, it is difficult for one single institution to possess the interdisciplinary capacity to scale up new technologies into commercial activities. Through CLAYUCA, national and regional networking ensures that experience and knowledge are shared. In some cases, the presence of CLAYUCA has helped to reconfigure relationships between the public and private sector and farmer organizations.

At a relatively low cost, CIAT benefited from CLAYUCA's role as a regional forum for planning, financing, and implementing cassava-based research and development to reestablish itself as a stronger actor in the regional innovation system for cassava. Public and private institutions that require technologies generated by CIAT now have access facilitated through CLAYUCA. At the same time, CIAT receives stronger feedback from CLAYUCA stakeholders on the performance of its technologies and emerging problems and priorities in the cassava subsector.

Benefits to the private sector (producers and processors)

CLAYUCA is a technology clearinghouse, constantly searching worldwide for competitive technologies and transferring this information to its stakeholders. Many companies lack the budget, time, and capacity to seek this information on their own. A processing technology for cassava starch would be more expensive if obtained in Europe than in Brazil or in China. The difference in the initial investment cost could affect a private entrepreneur's decision to establish a cassava processing plant in a given country. CLAYUCA organizes study tours to other countries and regions (Brazil,

China, Europe), allowing members to see technologies in operation and make informed investment decisions.

The wealth of cassava genetic resources in germplasm banks at international and advanced research centers will enable cassava to cope with the effects of climate change, among other needs. CLAYUCA has helped member countries and farmer groups gain better access to this genetic diversity for their own in-situ evaluation and selection programs. CLAYUCA has also facilitated farmer groups' access to improved varieties with higher yield potential and greater adaptation to biotic and abiotic stresses. Farmers have benefited from the new markets, additional income, and employment opportunities represented by the cassava-based agroindustries established in some CLAYUCA countries. Through CLAYUCA, some farmers have gained access to special services such as production credits under very favorable terms (box 1.30).

LESSONS LEARNED AND ISSUES FOR WIDER APPLICATION

The lessons and issues emerging from CLAYUCA's work over more than a decade reflect its experiences in inducing collaboration throughout a large network of diverse participants.

Public-private partnerships are built on trust and history

The establishment of partnerships among public and private actors, farmers groups, NGOs, and other entities is

---

Box 1.30   Raising Cassava's Profile among Policy Makers in Panama

The Panayuca Project, Panama's stakeholder in CLAYUCA, is a strategic alliance between Panayuca, a private company, and the Association of Small and Medium Agro-producers of Panama (APEMEP). Panayuca's main goal is to raise living standards in Panama's poorer rural areas by producing cassava and its derivatives. APEMEP is an association of more than 300 organizations, including farmer unions, cooperatives, women's groups, and indigenous groups, with more than 60,000 individual members. APEMEP members produce the cassava; Panayuca develops the industrial

infrastructure and handles logistics and marketing. Before Panama's cassava subsector was affiliated with CLAYUCA, credit lines for cassava at premium rates were not available to small-scale producers. The government did not regard cassava as a priority crop. After intense lobbying led by Panayuca, the government included cassava as one of the crops eligible for credit at very low, almost subsidized rates. This policy decision benefits a large group of small-scale farmers, facilitates their partnership with the private sector, and enables farmers and industry to operate at a competitive commercial level.

Source: Author.
Note: APEMEP = Asociación de Pequeños y Medianos Productores de Panamá.

facilitated when they already have a shared history and significant knowledge and trust of one another. CLAYUCA's development was strengthened by the effects of previous partnerships and activities between CIAT, civil society, and farmer groups in many member countries.

Different interactions and innovation mechanisms are important

Some countries coordinate their CLAYUCA agenda through a public agency, but in others the private sector or a combination of both sectors is more active. Other member countries have strong traditions of cooperatives and industry associations. Still others have policies that emphasize the social and economic feasibility of a dualistic agricultural sector, in which small- and large-scale producers coexist and develop. Others have policies that emphasize the importance of science and technology in agricultural development. These different traditions, practices, and attitudes have allowed different forms of interaction and coordination to emerge in work funded by CLAYUCA and its partners:

- *Partnerships* between CIAT, cooperative processing plants, and the national agricultural research organization in each member country.
- *Creation of an apex association* to link cooperatives in processing and marketing innovations.
- *Creation of a research-focused network* comprising a regional consortium, the industry (with its small-scale farmer base), national and international research organizations, government, and financial organizations, all linked to domestic, regional, and international markets.

Enabling environment for public-private partnerships

The success of public-private partnerships depends greatly on a supportive policy environment. Most Latin American and Caribbean countries (excluding the Southern Cone countries, except for Chile) import large quantities of cereals to manufacture animal feed. Most governments sought to meet growing demand for feed through policy instruments that caused producers of traditional starchy staples such as cassava to compete with imported cereals at a substantial price disadvantage. CLAYUCA has sought to counter this problem in a number of ways: through its efforts to develop more efficient production and processing methods, new cassava-based products and markets, and a

greater voice for the cassava subsector in the policy debate. The case of Panayuca (described in box 1.30) is one example of the relationship between policy and the success of a public-private partnership facilitated through CLAYUCA. Another example comes from Costa Rica (box 1.31).

After more than a decade of functioning through public-private partnerships, CLAYUCA is well aware that such partnerships have a greater impact when they are sustained by a group of well-funded technical experts with sufficient time to dedicate to the needs of the partnership's stakeholders. For example, CIAT provided strategic core support for CLAYUCA staff, operations, logistics, and management. This funding enabled successful institutional learning, the formation of a long-term network of partners, capacity building, and organizational innovation. The availability of such funding cannot be taken for granted, considering the dynamic environment in which institutions such as CIAT operate. Leaving the financing of public-private partnerships to stakeholders alone may not be sufficient for those partnerships to have an impact. CLAYUCA participants in each country must be creative to find complementary strategies for supporting the financial requirements of the partnership.

Improve coordination in value chains

Another lesson emerging from CLAYUCA's experience is that more attention should be given to coordination across value chains. Because public-private partnerships operate in an environment heavily influenced by policy, stakeholders benefit from operating in a coordinated manner, both inside and outside the partnership boundaries. For example, in Colombia, the Ministry of Agriculture's official policy is to support agricultural development and technology generation through "agro-productive value chains." The cassava subsector has not organized itself into a centrally coordinated value chain, so cassava projects do not meet government requirements for funding. They are forced to seek funding indirectly through more organized value chains (livestock, poultry, animal feed, bioenergy, human food, and so forth).

Public agencies require sufficient capacity to form productive partnerships

In each CLAYUCA country, public agencies play a central role in research and advisory services to improve the competitiveness of the cassava subsector. These agencies are often seriously affected by frequent changes in their

Box 1.31   Policy Action to Diversify the Market for Cassava in Costa Rica

Costa Rica's cassava area is not very large, but intensive cassava production and processing operations have converted Costa Rica into the world's leading exporter of frozen cassava and paraffin-coated cassava, principally to markets in the United States and Europe. In 2008, Costa Rica exported 75,000 tons of frozen and paraffin-coated cassava, with a market value of US$60 million, but a significant share of cassava is not harvested because it does not meet export standards. In 2010, the

Ministry of Agriculture created a country-level Cassava Committee to promote the use of cassava in animal feed and thus diversify the market for the cassava crop. This policy decision means that a large group of institutes and entities that were working independently will begin working in a coordinated fashion. CLAYUCA's Costa Rica group is a member of the new committee and will play an important role in transferring CLAYUCA technologies for growing and processing cassava for feed.

Source: Author.

institutional, political, and financial environment, however. Rarely can they implement long-term strategies to support cassava farmers and enhance the subsector. The skills and scientific capacity of technical personnel in some CLAYUCA countries must be strengthened. In some cases the private sector is willing to finance such training, but some of the burden must be shared by public agencies, or they will not be able to form productive partnerships with private organizations. Successful partnership will be facilitated if public agencies allocate specific funding for training through specific projects, competitive grants, donor support, and other means.

Instability and frequent changes in government support for research

In many countries, every change in the central government (such as a new president or minister of agriculture) brings a wave of new policies that alter support for agricultural research and development. Public support for cassava research in Colombia, for example, has run the gamut from full to negligible support and funding for cassava technology development projects. It is vital for public-private partnerships to seek independent, stable financing sources to avoid the vagaries of public funding and successfully pursue partners' research priorities.

Rural Productive Alliances: A Model for Overcoming Market Barriers

Marie-Hélène Collion, World Bank
Michelle Friedman, World Bank

SYNOPSIS

Rural productive alliances are formal agreements that bring commercial buyers together with producer organizations. The alliances aim to increase incomes and employment for rural producers through their participation in modern supply chains, sometimes with a particular emphasis on lagging regions or indigenous populations. Producers overcome market barriers and gain stability through consistent, higher prices while buyers receive a consistent, reliable supply of goods meeting their quality standards. Alliances are initially funded through grants for technical assistance (in production, management, and marketing) for the producer organization, along with infrastructure and equipment. Grant recipients in some organizations repay a share of the grant to the organization to create revolving funds that will provide credit to members when external funding ends. Projects to support rural productive alliances can build upon lessons from earlier projects by, for example, involving financial institutions such as commercial banks from the beginning; working with buyers to sustain and scale up activities when project funding ends; analyzing producer organizations' ability to use a grant productively; and assessing the risk that a buyer or producer organization may leave the alliance. Producer organizations need to build marketing skills and may benefit from a third-party agent or broker to enter particular high-value markets. Buyers can improve the alliance through sensitization to the benefits and transactions costs of working with small-scale producers and through support to optimize the marketability of niche products. Projects require a handover strategy so that domestic actors can fund, implement, and scale up activities when project support ends.

CONTEXT

Whether they are selling to domestic or export markets, smallholders worldwide find it increasingly challenging to enter into and benefit sustainably from modern agricultural value chains. The "supermarket revolution" has changed the parameters of market demand: Exporters, agribusinesses, and supermarkets require large quantities of consistently high-quality goods that meet sanitary and phytosanitary standards and arrive on time. Owing to the scale of their production, high transaction costs, and inability to provide goods of consistent quality, small-scale producers often are consigned to selling in less demanding but less rewarding markets, such as open-air markets, or through intermediaries. Smallholders' lack of information regarding markets, especially their poor knowledge of distribution channels and prices (in relation to product characteristics and timing of delivery), undermines their ability to negotiate with buyers.

PROJECT OBJECTIVES AND DESCRIPTION

A rural productive alliance is an economic agreement between formally organized producers and at least one buyer. The agreement specifies: product characteristics, such as size and varieties to be produced; quantity to be produced/bought; production modalities, such as how a product will be delivered, by whom, and when, as well as grading and packing requirements; payment modalities and price determination criteria; and the buyer's contribution, such as technical assistance, specific inputs, and arrangements for input reimbursement (for example, at the time of sale).

The project cycle

The project cycle begins with a call for proposals, often from the agriculture ministry to producer organizations and their commercial partners. The producer organization starts the process by preparing a basic profile of a potential business plan, which if selected is developed into a full-fledged business plan with the help of a private service provider. The private service provider also produces feasibility studies, which are reviewed by a multistakeholder committee to determine the business plan's feasibility. Plans with satisfactory technical, financial, and market feasibility receive funding.

Who funds rural productive alliances?

World Bank project funds are transferred to producer organization accounts in installments, based on evidence that the organization has used the previous installment according to the business plan and that expected outputs have been achieved. Grants from the project are matched with contributions from the producer organization and the buyer (in the form of technical assistance and inputs) and possibly funding from public and/or private institutions, such as municipal governments or commercial banks.

What do the grants finance?

The grants finance technical assistance in production, management, and marketing for members of the producer organization. The technical assistance mitigates risks for the buyer and builds trust between partners, which is essential to maintaining and sustaining the relationships. The grants also cofinance infrastructure or equipment such as irrigation equipment for individuals or collective storage and packing facilities. In certain instances, project grants fund seed or startup capital for inputs to help smallholders overcome initial financial barriers when dealing with commercial banks.

Creating savings and sustainable funding

In several countries, members of producer organizations agree to repay to their organization a share of the grant they receive from the project. (Technical assistance is typically not reimbursed.) This repayment creates a "revolving fund" that the producer organization will use to provide credit to its members when project support is over.

Implementing the rural productive alliances

For each alliance, a business agreement is signed between the agency in charge of project implementation, the commercial partner, the technical service provider, and the producer organization. An Alliance Management Committee is formed, which includes representatives from each actor,

with the objective of monitoring implementation of the business plan. The committee facilitates communication between buyer and seller.

Types of alliances

Alliances can be balanced fairly evenly between producers and buyers. They may also be dominated by either producers or buyers.

In well-balanced alliances, buyers compete to source from organized producers. Producers can meet the buyer's demands and accrue individual benefits from collective efforts. Producers in this situation improve their bargaining power with the buyer. Success comes from the productive use of technical assistance and the buyer's ability to market the product based on its particular characteristics, such as whether it is organic or has been produced for a specific niche market.

In Quindio, Colombia, plantain producers have a strong foothold in determining the prices of their products. The buyer provides technical assistance as needed, and both the buyer and producer organization are more competitive at their respective stages in the plantain value chain than before.

In alliances dominated by a single buyer, the producer organization has limited room to negotiate, even if both parties benefit from being in the alliance. In instances where the buyer is the dominant actor, the buyer helps the producers access the market. The added value of collectively organizing and creating the alliance will probably go to the buyer, however, unless special efforts are made to help producers develop negotiating skills to increase their leverage. In other situations, with a diversity of marketing possibilities for the producers, there is a risk that producers will circumvent the buyer and sell directly to alternate markets.

One alliance of this type is Agrìcola Cafetelera Buena Vista, a coffee alliance in Bolivia. The buyer provides producers with technical assistance to ensure that the coffee is certified organic. In this instance, the buyer is very involved with the producer organization and works to ensure good quality conditions for the producers. Given the high costs of organic coffee production and lack of marketing capability, the single buyer corners the market, however, leaving producers with little room to negotiate.

THE INNOVATIVE ELEMENT

In summary, rural productive alliance projects enable producer organizations to overcome the problems faced by individual, small-scale producers in accessing markets (buyers) in a *sustainable* way. A well-functioning producer organization is key to the success of the model. Working through their own organization, producers achieve economies of scale and can ensure product quality and traceability as required by the market. The revolving fund managed by the producer organization enables the organization to develop financial management skills as well as seed capital to secure future credit for members, thereby providing a means to maintain competitiveness after the project has ended.

BENEFITS, IMPACT, AND EXPERIENCE

Benefits to producers include the following:

- Around 600 rural productive alliances have been established in the four countries that have already begun implementation (Colombia, Bolivia, Panama, and Guatemala), benefitting around 32,600 rural families.
- Members of producer organizations and the organizations themselves have increased their technical and management skills and improved their market intelligence, even if the alliance between the commercial buyer and producer organization is not sustained.
- In some countries, producer organizations and their members have improved their access to private financial markets.
- Women producers benefit. In Colombia, for example, 22 percent of the beneficiaries are women producers.
- The overall result is higher agricultural incomes and increased rural employment, especially for agricultural workers and women working in postharvest activities.

Benefits to buyers include the following:

- Buyers secure access to products of consistently high quality. They meet the sanitary and phytosanitary standards that are applicable to international markets and increasingly important for domestic markets. In addition, by providing improved inputs (seed, in particular) and training, buyers can obtain raw materials of the quality they require.
- Through the alliance, private companies invest in a community, which is a time-tested way for companies to secure producer loyalty.
- Companies that buy locally, from local small-scale producers, enhance their image of being socially and environmentally responsible. This image helps them to

differentiate their products from mainstream products while meeting new consumer demands.

LESSONS LEARNED AND ISSUES FOR WIDER APPLICATION

The lessons from unsustainable alliances have been useful in developing recommendations to increase the likelihood that other rural productive alliances will be implemented successfully and become sustainable. The recommendations focus particularly on actions to ensure that alliances remain strong and can continue even after external support ends.

Lessons

A main risk of *any alliance* is that either the producer organization or commercial buyer will default from the partnership. Of the 170 alliances implemented through the first phase of the Colombia project—the oldest of the rural productive alliance projects in Latin American and the Caribbean—39 have not received World Bank financing for at least two years. Of these, 26 have maintained commercial agreements with their buyers or have identified new ones, representing a success rate of 67 percent.

Other alliances have not endured, for several reasons. Often more than one factor contributes to the demise of an alliance.

- *Producers revert to previous practices.* Producers have defaulted because there is a net benefit for them in returning to their traditional markets, despite having secured the buyer's partnership. The default arises when producers, owing to technical or managerial problems, cannot sustain the stringent requirements of high-value markets (see below).
- *Producer organizations lack social cohesion.* Producer organizations can be dysfunctional in various ways. The failure of representatives and members to communicate, poor management, lack of capacity to manage conflicts, and lack of social cohesion will all negatively affect the functioning of an alliance. In Colombia, lack of social cohesion and inability to manage conflicts were perhaps the main reasons that alliances failed. This situation often occurred when the producer organization was encouraged to include more smallholders to make the proposed alliance more socially or economically justifiable. Because the strength of the producer organization is a prerequisite for a successful alliance, it is important to identify such weaknesses early on and provide support to

improve organizational and management skills. In some cases, there is a need for additional support after subproject implementation.

■ *The producer organization cannot provide services to its members.* Producer organizations are often under social pressure from the rural community at large to make services accessible to nonmembers. Producer organizations should provide services, but only to members. This strategy keeps membership attractive and encourages producers to market through the organization so that it can continue complying with contractual arrangements made with the buyer. Otherwise the organization is likely to fail.

■ *The producer organization lacks adequate commercial or professional skills.* Producer organizations also require management, organizational, and marketing skills to provide services of good quality, such as the capacity to manage a revolving fund. In Colombia, alliances that continued for at least two years after the project ended often involved two-tier producer organizations: grassroots organizations and their union. The first-tier grassroots organizations are involved in managing production. At the union level, the organization deals with procuring inputs, marketing, and financing, with paid professional staff.

Recommendations

After nearly a decade of implementing rural productive alliances in Latin America and the Caribbean, it is possible to identify several recommendations to ensure successful implementation and sustainability of the alliances:

■ Projects should emphasize cofinancing from commercial credit sources in addition to matching grants to fund business plans. Involving commercial banks means that the issue of collateral and guarantees must be resolved. Some projects establish guarantee funds to spread the risks to commercial banks and encourage them to partner in funding rural alliances. Involving financial institutions at the beginning of the project can also build their trust in producer organizations and help producer organizations learn to deal with commercial banks. This learning on both sides is important to ensure that smallholders can access credit and partnerships can be sustained.

■ More thorough and realistic feasibility studies of business plans must analyze the ability of the producer organization to use the matching grant productively and the risk that a buyer or producer organization may default from the alliance. Feasibility studies look into the market and technical aspects of the alliance, but often they fail to analyze how the organization functions and its capacity to manage a partnership with a buyer.

■ To sustain participation in high-value markets, the producer organization needs to build its marketing skills. For example, the organization could benefit from a third-party market agent or broker to assist in breaking into particular markets. Productive alliance projects should consider establishing such brokers, whose role would be to scout the market for opportunities and identify the producer groups that can take advantage of them. These brokers should be private sector agents.

■ Three key areas of support could help buyers improve the sustainability and productivity of the alliance:[1] (1) sensitization to the benefits of working with small-scale producers; (2) support to optimize marketability of niche products; and (3) sensitization to the transaction costs associated with working with small-scale producers. Initially buyers need support to manage their relationship better with small-scale producers. An example is being aware of smallholders' cash constraints and the difficulties they face in managing deferred payments, especially with supermarkets.

■ During implementation, projects need to address the constraints to sustainability that alliances may face after project support ends. As discussed, a well-functioning revolving fund is fundamental for ensuring that alliances can be sustained. It enables producers to access credit within their own organization and to demonstrate to financial institutions their ability to manage savings and credit. It needs to be promoted more forcefully during project implementation.

■ Projects need to include a handover strategy so that domestic actors can fund, implement, and scale up activities when project support ends. At the moment, only the Colombia project is concerned with these issues. By focusing on scaling up its activities through the public sphere, however, the project is likely to face serious obstacles, such as the lack of technical capacity, budgetary pressures on public officials, and political problems (changes in government easily lead to changes in people, priorities, and policies). An alternative would be to work on the side of the buyers. Producer organizations constitute one source of procurement for an agricultural good that buyers need for their business, and they may be the main source if it is produced mostly by smallholders. Once the project has demonstrated the potential that producer organizations represent, buyers should be interested in continuing to invest in producer organizations, providing technical support, and prefinancing inputs.

ROOM TO INNOVATE IN RURAL PRODUCTIVE ALLIANCES BY SUPPORTING VALUE CHAINS

Future rural productive alliance projects should build upon the lessons of earlier projects, especially lessons about involving commercial banks from the start and working on the side of buyers and agribusinesses to sustain and scale up activities when external funding ends. Scaling up is an issue, as rural productive alliances are still relatively small-scale interventions; another issue is the need to improve the competitiveness of supply chains that benefit smallholders. In

the future, a good approach may be to combine the focus on producer/buyer partnerships with a value chain approach. The objective would be to create better productive conditions—for example, by improving the quality of services provided to actors in the chain; improving the capacity of agencies that control compliance with sanitary and phytosanitary standards; supporting research, development, and innovation; addressing the administrative and institutional aspects of certification; promoting organic production or access to other high-value niches; and improving market intelligence. Productive infrastructure, trade facilitation services, and the business environment are additional areas that a value chain approach could address. This emphasis could promote opportunities for promising subsectors and can help roll out the alliance model on a national scale.

NOTES

Module 1 Overview

1. For example, a culture of collaboration among stakeholders was a major factor facilitating collaboration and innovation in Finland and Korea (see module 6, TN 2).

2. Not all interactions result in collaboration. Interactions can also be antagonistic and result in conflict.

3. Module 4 discusses the role of, functions of, and investments that research organizations require to perform well in an AIS. TN 2 discusses public-private research partnerships and IAPs 2 and 3 describe the formation of research consortia supported by competitive grants. Module 5 provides further detail on innovation funds, including competitive research grants and matching grants.

4. Despite this weakness, a few researchers with strong research capabilities are often found in these organizations.

5. The analysis of how public research organizations can be transformed to better integrate into the AIS exceeds the scope of this module. This issue is discussed in module 4 and in Davis, Ekboir, and Spielman (2008).

6. Every individual in an organization has at the very least the power to boycott the organization's activities.

7. As noted, a "value chain" is the set of linked activities that a firm organizes to produce and market a product (Porter 1985). The value chain is a network with a commercial focus, one actor that "organizes" and commands the chain (Christensen, Anthony, and Roth 2004), and a relatively narrow and stable membership.

8. In terms of coordination, some organizations coordinate other organizations (for example, a commodity board coordinates producer organizations, traders, and manufacturers); some coordinate individuals (for example, a farmer organization); and others coordinate both organizations and individuals (as in a value chain).

9. For an example from Chile, see http://www.cnic.cl/content/view/469646/Un-camino-de-desarrollo-para-Chile.html; for one from the United Kingdom, see http://www.innovateuk.org/; for one from the Netherlands, see http://www.innovatienetwerk.org/en/organisatie/toon/11/.

10. Rural households can also form community associations to solve local problems, such as problems with water supply or education, but these organizations are not discussed in this module.

11. See module 6, TN 2 on innovation system governance.

12. Module 4, TN 4 provides further details on innovation brokers.

13. Despite their institutional weakness, most universities and research and extension organizations have some very good professionals.

Thematic Note 1

1. Empresa Brasileira de Pesquisa Agropecuária (Brazilian Enterprise for Agricultural Research).

2. Sistema Nacional de Investigación y Transferencia de Tecnología (SNITT, National System of Research and Technology Transfer).

3. Government as well as leaders of the various innovation-promoting participants in the AIS need a broad, long-term perspective on agricultural development and change, along with a sense of what is needed for such development. They will need not only to identify innovation and development opportunities but also to understand the historical, cultural, and social complexities in rural areas and among consumers. A strategic vision usually describes a set of ideals and priorities, a picture of the future—but the strategic vision is also a bridge between the present and the future, and it should be shared by the actors involved.

4. However, the actual financing and allocation of funds should belong to another entity, such as ministry or other special agency.

5. For comparison, see module 6, TN 2 on innovation system governance.

Thematic Note 2

1. Fodder innovation project, http://www.fodderinnovation.org/.

2. Golden Rice Project, http://www.goldenrice.org/.

3. For an example of this type of research and its problems, see Hall et al. 2001.

4. Networks with this combination of actors are said to exhibit a "small-world structure."

Thematic Note 3

1. See http://www.federaciondecafeteros.org/particulares/en/.

2. According to Fairtrade Labelling Organizations International (2011), Fairtrade-certified sales amounted to approximately €3.4 billion worldwide in 2008. Sales of Fairtrade-certified products grew 15 percent between 2008 and 2009.

3. A research institute (CIP) organized Papa Andina, the Andean potato network; the NGO Africare supported smallholders' access to markets, as have civil society organizations (such as the Mexican Produce Foundations) and farmer organizations (IAP 2).

4. Providing public support for extension and advisory services does not mean that they are provided by traditional public organizations. In the past two decades, many

institutional arrangements that include public organizations, private partners, and civil society have been tried (module 3).

5. Since it is not possible to produce high-value products of uniform quality, wealthier farmers who produce large volumes sell their produce through more than one channel. The poorest farmers, on the other hand, have to sell at the farm gate or in local markets.

6. For a description, see Nandakumar et al. (2010).

Thematic Note 4

1. See http://www.prolinnova.net/.

2. ROPPA, Réseau des organisations paysannes et des producteurs agricoles de l'Afrique de l'Ouest (http://www.roppa.info/?lang=en).

3. The module overview lists the capabilities required (box 1.7); IAP 3 presents an example of developing them over the long term.

4. The specialized literature refers to the creation of knowledge as "invention." An invention becomes an "innovation" only when it is first used in a product that reaches the market or produces a change in a social process.

5. See Ekboir et al. (2009) for an example.

6. AACREA is the Asociación Argentina de Consorcios Regionales de Experimentación Agrícola (Argentine Association of Regional Consortiums for Agricultural Experimentation).

7. Centro de Investigación de la Caña de Azúcar de Colombia (http://www.cenicana.org/index.php).

Innovative Activity Profile 1

1. ICI's role is described in detail because it provides important insights into the dynamics of innovation. In particular, it shows that (1) demand-driven approaches often miss important opportunities—the most important innovations start as curiosity-driven projects that eventually result in something valuable; (2) innovation processes are essentially uncertain, and it is difficult to set clearly defined objectives; (3) motivated leaders are critical for success; and (4) building an innovation network is also an uncertain process that requires a lot of experimentation.

2. Innovators have often found themselves in a similar situation—that is, they have a product that provides a new service for which there is limited demand. Subsequent innovations are necessary to create a market for the original innovation. Examples include the telephone, Internet commerce, computer hard drives, and mobile telephones (Christensen 2003).

3. No-till is defined as planting crops in previously unprepared soil by opening a narrow slot or trench of the small-est width and depth needed to obtain proper coverage of the seed. Conventional tillage practices involve multiple tractor passes to accomplish plowing, harrowing, planking, and seeding operations; no-till requires only one or two passes for spraying herbicide and seeding. In addition to reducing the number of operations, no-till requires less powerful tractors and reduces equipment depreciation. While no-till principles are the same everywhere—minimum soil disturbance, keeping soil covered, and using crop rotations—the actual packages differ greatly by location.

Innovative Activity Profile 2

1. Boards for many foundations, research institutes, and firms operate in this way. The Produce Foundations are legitimate representatives of farmers because other actors in the AIS recognize them as such, not because farmers elect their authorities (Ekboir et al. 2009).

Innovative Activity Profile 3

1. CONICYT (Comisión Nacional de Investigación Científica y Tecnológica) (www.conicyt.cl).

2. Corporación de Fomento de la Producción (http://www.corfo.cl/acerca_de_corfo/emprendimiento_e_innovacion/que_es_innovachile).

Innovative Activity Profile 4

1. Financed through community savings and thrift, the Government of Andhra Pradesh, commercial banks, and the World Bank.

Innovative Activity Profile 6

1. Currently no support is offered to buyers in the rural productive alliance projects in Latin America and the Caribbean.

REFERENCES AND FURTHER READING

Module 1 Overview

Axelrod, R., and M. D. Cohen. 1999. *Harnessing Complexity: Organizational Implications of a Scientific Frontier.* New York: Free Press.

Brimble, P., and R. F. Doner. 2007. "University-Industry Linkages and Economic Development: The Case of Thailand." *World Development* 35 (6): 1021–36.

Christensen, C. M., S. D. Anthony, and E. A. Roth. 2004. *Seeing What's Next: Using the Theories of Innovation to Predict Industry Change.* Boston: Harvard Business School Press.

CIAT (International Center for Tropical Agriculture). 2010. "Diversified Livelihoods through Effective Agro-enterprise Interventions: Creating a Cumulative Learning Framework between CIAT and Development NGOs in Central America." Alianza de Aprendizaje, http://www.alianzasdeaprendizaje.org/salon-del-conocimiento/item/60-diversified-livelihoods-through-effective-agro-enterprise-interventions-creating-a-cumulative-learning- framework-between-ciat-and-development-ngo-in-central-america.

Davis, K., J. M. Ekboir, and D. Spielman. 2008. "Strengthening Agricultural Education and Training in Sub-Saharan Africa from an Innovation Systems Perspective: A Case Study of Mozambique." *Journal of Agricultural Education and Extension* 14 (1): 35–51.

Dosi, G., R. R. Nelson, and S. G. Winter. 2000. "Introduction: The Nature and Dynamics of Organizational Capabilities." In *The Nature and Dynamics of Organizational Capabilities*, ed. G. Dosi, R. R. Nelson, and S. G. Winter. New York: Oxford University Press.

Ekboir, J., K. Boa, and A.A. Dankyi. 2002. "Impacts of No-Till Technologies in Ghana." CIMMYT Economics Program Paper 02-01.

Ekboir, J. M., G. Dutrénit, G. Martínez V., A. Torres Vargas, and A. Vera-Cruz. 2009. *Successful Organizational Learning in the Management of Agricultural Research and Innovation: The Mexican Produce Foundations.* IFPRI Research Report 162. Washington, DC: International Food Policy Research Institute.

Fagerberg, J. 2005. "Innovation: A Guide to the Literature." In *The Oxford Handbook of Innovation*, ed. J. Fagerberg, D. C. Mowery, and R. R. Nelson. New York: Oxford University Press.

Fountain, J. E. 1999. "Social Capital: A Key Enabler of Innovation." In *Investing in Innovation: Creating Research and Innovation Policy That Works,* ed. L. M. Branscomb and J. H. Keller. Cambridge, MA: MIT Press.

Freeman, C. 1987. *Technology Policy and Economic Performance: Lessons from Japan.* New York: Frances Pinter.

Hakansson, H., and D. Ford. 2002. "How Should Companies Interact in Business Networks?" *Journal of Business Research* 55 (2): 133–39.

Hartwich, F., A. Alexaki, and R. Baptista. 2007. *Innovation Systems Governance in Bolivia: Lessons for Agricultural Innovation Policies.* IFPRI Discussion Paper 00732. Washington, DC: International Food Policy Research Institute.

Hellin, J., M. Lundy, and M. Meijer. 2009. "Farmer Organization, Collective Action, and Market Access in Meso-America." *Food Policy* 34 (1): 16–22.

Intarakumnerd, P., P. Chairatana, and T. Tangchitpiboon. 2002. "National Innovation System in Less Successful Developing Countries: The Case of Thailand." *Research Policy* 31 (8–9): 1445–57.

Klerkx, L., N. Aarts, and C, Leeuwis. 2010. "Adaptive Management in Agricultural Innovation Systems: The Interactions between Innovation Networks and Their Environment." *Agricultural Systems* 103 (6): 390–400.

Klerkx, L., and C. Leeuwis. 2008. "Balancing Multiple Interests: Embedding Innovation Intermediation in the Agricultural Knowledge Infrastructure." *Technovation* 28 (6): 364–78.

———. 2009. "The Emergence and Embedding of Innovation Brokers at Different Innovation System Levels: Insights from the Dutch Agricultural Sector." *Technological Forecasting and Social Change* 76 (6): 849–60.

Lundvall, B.-Å., ed. 1992. *National Systems of Innovation: Towards a Theory of Innovation and Interactive Learning.* London: Frances Pinter.

Nelson, R., ed. 1993. *National Innovation Systems: A Comparative Analysis.* Oxford: Oxford University Press.

Neven, D., M. M. Odera, T. Reardon, and H. Wang. 2009. "Kenyan Supermarkets, Emerging Middle-Class Horticultural Farmers, and Employment Impacts on the Rural Poor." *World Development* 37 (11): 1802–11.

OECD (Organisation for Economic Co-operation and Development). 2005. *Governance of Innovation Systems.* Vol. 1: *Synthesis Report.* Paris: OECD.

Porter, M. E. 1985. *Competitive Advantage: Creating and Sustaining Superior Performance.* New York: Free Press.

Powell, W. W., and S. Grodal. 2005. "Networks of Innovators." In *The Oxford Handbook of Innovation*, ed. J. Fagerberg, D. C. Mowery, and R. R. Nelson. New York: Oxford University Press.

Robinson, L. J., and J. L. Flora. 2003. "The Social Capital Paradigm: Bridging across Disciplines." *American Journal of Agricultural Economics* 85 (5): 1187–93.

Simon, H. A. 1981. *The Sciences of the Artificial.* 2d ed. Cambridge, MA: MIT Press.

Skarzynsky, P., and R. Gibson. 2008. *Innovation to the Core: A Blueprint for Transforming the Way Your Company Innovates.* Boston: Harvard Business School Press.

Spitzer, D. R. 2007. *Transforming Performance Measurement: Rethinking the Way We Measure and Drive Organizational Success.* New York: AMACOM.

Vera-Cruz, A., G. Dutrénit, J. M. Ekboir, G. Martínez V., and A. Torres Vargas. 2008. "Virtues and Limits of Competitive Funds to Finance Research and Innovation in the Agricultural Sector: The Case of the Mexican Produce Foundations." *Science and Public Policy* 35 (7): 501–13.

Vuylsteke, A., and G. Van Huylenbroeck. 2008. "Coordination of Collective Action in the Agro-food Sector." Presented at the Twelfth Congress of the European Association of Agricultural Economists, August 26–29, Ghent.

World Bank. 2006. "Enhancing Agricultural Innovation: How to Go Beyond the Strengthening of Research Systems." World Bank, Washington, DC.

———. 2009. "Gender in Agriculture Sourcebook." World Bank, Washington, DC.

Thematic Note 1

Allegri, M. 2002. "Partnership of Producer and Government Financing to Reform Agricultural Research in Uruguay." In *Agricultural Research Policy in an Era of Privatization*, ed. D. Byerlee and R. G. Echeverria, 150–21. Oxon, U.K.: CABI.

Byerlee, D., and G. Alex. 1998. *Strengthening National Agricultural Research Systems: Selected Issues and Good Practice.* Washington, DC: World Bank.

Hartwich, F., A. Alexaki, and R. Baptista. 2007. "Innovation Systems Governance in Bolivia: Lessons for Agricultural Innovation Policies." Washington, DC: International Food Policy Research Institute (IFPRI).

Klerkx. L., A. Hall, and C. Leeuwis. 2009. "Strengthening Agricultural Innovation Capacity: Are Innovation Brokers the Answer?" *International Journal of Agricultural Resources, Governance, and Ecology* 8 (5-6): 409–38.

Klerkx, L., and C. Leeuwis. 2008. "Delegation of Authority in Research Funding to Networks: Experiences with a Multiple Goal Boundary Organization." *Science and Public Policy* 35 (3): 183–96.

Ohler, F., W. Polt, A. Rammer, and J. Schindler. 2005. "Governance in Austrian Information Society Policy." In *Governance of Innovation Systems*, Vol. 2. Paris: Organisation for Economic Co-operation and Development (OECD).

World Bank. 2006a. *Agriculture Investment Sourcebook.* Washington, DC.

———. 2006b. "Institutional Innovation in Agricultural Research and Extension Systems in Latin America and the Caribbean." Washington, DC.

———. 2006c. National Agriculture Innovation Project. World Bank Project Appraisal Document, Washington, DC.

Thematic Note 2

Devaux, A., D. Horton, C. Velasco, G. Thiele, G López, T. Bernet, I. Reinoso, and M. Ordinola. 2009. "Collective Action for Market Chain Innovation in the Andes." *Food Policy* 34 (1): 31–38.

Ekboir, J. M. 2002. "Developing No-Till Packages for Small-Scale Farmers." In *World Wheat Overview and Outlook*, ed. J. M. Ekboir. Mexico, D.F.: International Maize and Wheat Improvement Center.

———. 2003. "Innovation Systems and Technology Policy: Zero Tillage in Brazil." *Research Policy* 32 (4): 573–86.

Ekboir, J. M., G. Dutrénit, G. Martínez V., A. Torres Vargas, and A. Vera-Cruz. 2009. *Successful Organizational Learning in the Management of Agricultural Research and Innovation: The Mexican Produce Foundations.* IFPRI Research Report 162. Washington, DC: International Food Policy Research Institute.

Hall, A., G. Bockett, S. Taylor, M. V. K. Sivamohan, and N. Clark. 2001. "Why Research Partnerships Really Matter: Innovation Theory, Institutional Arrangements, and Implications for Developing New Technology for the Poor." *World Development* 29 (5): 783–97.

Hartwich, F., J. Tola, A. Engler, C. Gonzalez, G. Ghezan, J. M. P. Vazques-Alvarado, J. A. Silva, J. J. Espinoza, and M. V. Gottret. 2007. *Building Public–Private Partnerships for Agricultural Innovation.* Washington, DC: International Food Policy Research Institute.

Innovation Network. 2011. http://www.innovatienetwerk.org/en/.

Laxmi, V., O. Erenstein, and R. K. Gupta. 2007. *Impact of Zero Tillage in India's Rice-Wheat Systems.* Mexico, D.F.: International Maize and Wheat Improvement Center.

Powell, W. W., and S. Grodal. 2005. "Networks of Innovators." In *The Oxford Handbook of Innovation*, ed. J. Fagerberg, D. C. Mowery, and R. R. Nelson. New York: Oxford University Press.

Rycroft, R. W., and D. E. Kash. 1999. *The Complexity Challenge: Technological Innovation for the 21st Century.* Science, Technology, and the International Political Economy Series. New York: Cassell.

Spielman, D. J., K. E. Davis, M. Negash, and G. Ayele. 2008. *Rural Innovation Systems and Networks: Findings from a Study of Ethiopian Smallholders.* Discussion Paper 00759. Washington, DC: International Food Policy Research Institute.

USAID (U.S. Agency for International Development). 2004. "Toolkit for Alliance Builders." Washington, DC: USAID. http://pdf.usaid.gov/pdf_docs/Pnada827.pdf.

World Bank. 2006. "Enhancing Agricultural Innovation: How to Go Beyond the Strengthening of Research Systems." World Bank, Washington, DC.

Thematic Note 3

Devaux, A., D. Horton, C. Velasco, G. Thiele, G. López, T. Bernet, I. Reinoso, and M. Ordinola. 2009. "Collective Action for Market Chain Innovation in the Andes." *Food Policy* 34 (1): 31–38.

Fairtrade Labelling Organizations International. 2011. "Facts and Figures. FairTrade International. http://www.fairtrade.net/facts_and_figures.html.

Hellin, J., M. Lundy, and M. Meijer. 2009. "Farmer Organization, Collective Action, and Market Access in Meso-America." *Food Policy* 34 (1): 16–22.

Nandakumar, T., K. Ganguly, P. Sharma, and A. Gulati. 2010. "Food and Nutrition Security Status in India: Opportunities for Investment Partnerships." ADB Sustainable Development Working Paper 16. Asian Development Bank, Manila.

Neven, D., M. M. Odera, T. Reardon, and H. Wang. 2009. "Kenyan Supermarkets, Emerging Middle-Class Horticultural Farmers, and Employment Impacts on the Rural Poor." *World Development* 37 (11): 1802–11.

Poitevin, B., and S. Hassan. 2006. *Marketing Extension: A Powerful Process in 6 Steps.* Dhaka: Livelihoods, Empowerment, and Agroforestry Project.

Reardon, T., C. B. Barret, J. A. Berdegué, and J. F. M. Swinnen. 2009. "Agrifood Industry Transformation and Small Farmers in Developing Countries." *World Development* 37 (11): 1717–27.

Reardon, T., J. A. Berdegué, F. Echánove, R. Cook, N. Tucker, A. Martínez, R. Medina, M. Aguirre, R. Hernández, and F. Balsevich. 2007. "Supermarkets and Horticultural Development in Mexico: Synthesis of Findings and Recommendations to USAID and GOM." Report submitted to USAID/Mexico and USDA/Washington, August 10. N.p: N.p.

Shapira, P. 1999. "Manufacturing Extension: Performance, Challenges, and Policy Issues." In *Investing in Innovation: Creating Research and Innovation Policy that Works,* ed. L. M. Branscomb and J. H. Keller. Cambridge, MA: MIT Press.

Skarzynsky, P., and R. Gibson. 2008. *Innovation to the Core: A Blueprint for Transforming the Way Your Company Innovates.* Boston: Harvard Business School Press.

Vermeulen, S., J. Woodhill, F. J. Proctor, and R. Delnoye. 2008. "Chain-wide Learning for Inclusive Agrifood Market Development: A Guide to Multi-stakeholder Processes for Linking Small-scale Producers with Modern Markets." International Institute for Environment and Development, London; Wageningen University and Research Centre, Wageningen.

Wang, H., X. Dong, S. Rozelle, J. Huang, and T. Reardon. 2009. "Producing and Procuring Horticultural Crops with Chinese Characteristics: The Case of Northern China." *World Development* 37 (11): 1791–801.

World Bank. 2006. "Enhancing Agricultural Innovation: How to Go Beyond the Strengthening of Research Systems." World Bank, Washington, DC.

Thematic Note 4

Christensen, C. M., S. D. Anthony, and E. A. Roth. 2004. *Seeing What's Next: Using the Theories of Innovation to Predict Industry Change.* Boston: Harvard Business School Press.

Davila, T., M. J. Epstein, and R. Shelton. 2006. *Making Innovation Work: How to Manage It, Measure It, and Profit from It.* Upper Saddle River, NJ: Wharton School Publishing.

Devaux, A., D. Horton, C. Velasco, G. Thiele, G. López, T. Bernet, I. Reinoso, and M. Ordinola. 2009. "Collective Action for Market Chain Innovation in the Andes." *Food Policy* 34 (1): 31–38.

Ekboir, J. M., G. Dutrénit, G. Martínez V., A. Torres Vargas, and A. Vera-Cruz. 2009. *Successful Organizational Learning in the Management of Agricultural Research and Innovation: The Mexican Produce Foundations.* IFPRI Research Report 162. Washington, DC: International Food Policy Research Institute.

Hartwich, F., J. Tola, A. Engler, C. Gonzalez, G. Ghezan, J. M. P. Vazques-Alvarado, J. A. Silva, J. J. Espinoza, and M. V. Gottret. 2007. *Building Public-Private Partnerships for Agricultural Innovation.* Washington, DC: International Food Policy Research Institute.

Ochieng, C. M. O. 2007. "Development through Positive Deviance and Its Implications for Economic Policy Making and Public Administration in Africa: The Case of Kenyan Agricultural Development, 1930–2005." *World Development* 35 (3): 454–79.

Rondot, P., and M.-H. Collion, eds. 2001. *Agricultural Producer Organizations: Their Contribution to Rural Capacity Building and Poverty Reduction.* Washington, DC: RDV and World Bank.

Skarzynsky, P., and R. Gibson. 2008. *Innovation to the Core: A Blueprint for Transforming the Way Your Company Innovates.* Boston: Harvard Business School Press.

van der Veen, R. 2000. "Learning Natural Resource Management." In *Deepening the Basis of Rural Resource Management: Proceedings of a Workshop,* ed. I. Guijt, J. Berdegue, and M. Loevinsohn. The Hague: International Service for National Agricultural Research.

Vermeulen, S., J. Woodhill, F. J. Proctor, and R. Delnoye. 2008. "Chain Wide Learning for Inclusive Agrifood Market Development: A Guide to Multi-stakeholder Processes for Linking Small-scale Producers with Modern Markets." International Institute for Environment and Development, London; and Wageningen University and Research Centre, Wageningen.

World Bank. 2004. "Module 1: Strengthening the Capacity of Farmer Organizations to Influence Agricultural Policy." In *Agriculture Investment Sourcebook*. Washington, DC: World Bank.

———. 2011. Module 6 in *Information and Communication Technologies for Agriculture e-Sourcebook*. Washington, DC. http://bit.ly/ICTinAG.

Innovative Activity Profile 1

Christensen, C. M. 2003. *The Innovator's Dilemma*. New York: HarperCollins.

Derpsch, R. 2010. "No-tillage, Sustainable Agriculture in the New Millennium." www.rolf-derpsch.com.

Ekboir, J. M. 2002. "Developing No-till Packages for Small-scale Farmers." In *World Wheat Overview and Outlook*, ed. J. M. Ekboir. Mexico, DF: International Maize and Wheat Improvement Center.

Ekboir, J. M., K. Boa, and A. A. Dankyi. 2002. "Impact of No-Till Technologies in Ghana." CIMMYT Economics Paper 02-01, International Maize and Wheat Improvement Center, Mexico, D.F.

Huffman, W., and R. E. Just. 2000. "Setting Efficient Incentives for Agricultural Research: Lessons from Principal-agent Theory." *American Journal of Agricultural Economics* 82 (4): 828–41.

Innovative Activity Profile 2

Ekboir, J. M., G. Dutrénit, G. Martínez V., A. Torres Vargas, and A. Vera-Cruz. 2009. "Successful Organizational Learning in the Management of Agricultural Research and Innovation: The Mexican Produce Foundations." IFPRI Research Report 162. Washington, DC: International Food Policy Research Institute.

Innovative Activity Profile 3

CHILEOLIVA. 2009. *Informe anual de mercado nacional de aceite de olive: 2009*. http://www.chileoliva.cl/files/INFORME%20ANUAL%20DEL%20MERCADO2%20 2009.pdf.

CNIC (Consejo Nacional de Innovación para la Competitividad). 2007. *Estrategia nacional de innovación,* *2007 y 2008*. http://www.cnic.cl/content/view/468181/Hacia-una-Estrategia.html.

de Ferranti, D., G. E. Perry, W. Foster, D. Lederman, and A. Valdés. 2005. *Beyond the City: The Rural Contribution to Development*. World Bank Latin American and Caribbean Studies. Washington, DC: World Bank.

FIA (Fundación para la Innovación Agraria). Various years. *Memorias anuales* [annual reports]. Santiago: FIA. http://www.fia.cl/QueacuteesFIA/tabid/55/Default.aspx.

———. 2005. "Evaluación de impacto de programas y proyectos FIA." Informe Consultoria. GPI, Santiago. http://bibliotecadigital.innovacionagraria.cl/gsdl/cgi-bin/library.exe?l=es.

———. 2009. *Impacto en innovación agraria: Diez rubros exitosos*. Santiago: FIA. http://www.innovacionagraria.cl/LinkClick.aspx?fileticket=AkVeFwcosUQ%3D&tabid=64&mid=590.

Ministerio de Economía Chile. 2009. "Politica nacional para la innovación y la competitividad." Santiago: Ministerio de Economía. http://www.economia.cl/1540/articles-188772_recurso_1.pdf.

Innovative Activity Profile 4

Government of Andhra Pradesh, Irrigation and CAD Department. 2007. "Andhra Pradesh Community Based Tank Project Implementation Plan." Hyderabad. Andhra Pradesh Community Based Tank Project. http://www.apmitanks.in/Templates/files/OperationalManuals/PIP_20070228.pdf.

Government of Andhra Pradesh and Centre for Economic and Social Studies. 2008. *Human Development Report Andhra Pradesh 2007*. Hyderabad: Government of Andhra Pradesh.

NSSO (National Sample Survey Organization). 2003. *Income, Expenditure and Productive Assets of Farmer Households,* 2003. NSS Report 497. Delhi: NSSO.

———. 2005. *Situation Assessment Survey of Farmers, 2005*. NSSO Report 498. Delhi: NSSO.

Ramanjaneyulu, G. V., M. S. Chari, T. A. V. S. Raghunath, Z. Hussain, and K. Kuruganti. 2009. "Non-Pesticidal Management: Learning from Experiences." In *Integrated Pest Management*, Vol. 1: Innovation-Development Process, ed. R. Peshin and A. K. Dhawan, 534–74. New York: Springer.

SERP (Society for Elimination of Rural Poverty) and World Bank. 2009. "Ecologically Sound, Economically Viable: Community Managed Sustainable Agriculture in Andhra Pradesh, India." Washington, DC: World Bank.

Innovative Activity Profile 5

Cox, T. P. 2010. "Cassava Development in the Networking Age." *New Agriculturalist,* May. http://www.new-ag.info/en/focus/focusItem.php?a=1586.

Gottret, M. V., and B. Ospina. 2004. "Twenty Years of Cassava Innovation in Colombia: Scaling up under Different Political, Economic, and Social Environments." In *Scaling Up and Out: Achieving Widespread Impact through Agricultural Research,* ed. D. Pachico and S. Fujisaka, 105–26. Cali: International Center for Tropical Agriculture (CIAT).

Henry, G., B. Ospina, and R. Best. 1999. "Development by Linking Small Farmers to Growth Markets: Cassava in Latin America." In *Sustainable Agriculture and Environment: Globalisation and the Impact of Trade Liberalisation,* ed. A. K. Dragun and C. Tisdell, 271–86. Cheltenham, U.K.: Edward Elgar.

Mytelka, L., and I. Bortagaray. 2005. "Strengthening the Agricultural Innovation System in Colombia: An Analysis of the Cassava and Flower Sectors." Background paper for World Bank project.

Ospina, B. 2001. "CLAYUCA: Latin American and Caribbean Consortium to Support Cassava Research and Development." In *Cassava's Potential in Asia in the 21st Century: Present Situation and Future Research and Development Needs.* Proceedings of the 6th Regional Workshop, February 21–25, 2000, Ho Chi Minh City, 632–40. Cali: International Center for Tropical Agriculture (CIAT), Institute of Agricultural Sciences of South Vietnam (IAS), and Nippon Foundation.

Ospina, B., S. Poats, and G. Henry. 1994. "Integrated Cassava Research and Development Projects in Colombia, Ecuador, and Brazil: An Overview of CIAT's Experiences." In *Cassava Flour and Starch: Progress in Research and Development,* ed. D. Dufour, G. M. O'Brien, and R. Best. Cali: International Center for Tropical Agriculture (CIAT).

World Bank. 2006. "Enhancing Agricultural Innovation: How to Go Beyond the Strengthening of Research System." Washington, DC.

Innovative Activity Profile 6

Berdegué, J. A. S. 2001. *Cooperating to Compete: Associative Peasant Business Firms in Chile.* PhD thesis, Wageningen University.

Collion, M.–H., and M. Friedman. 2010. "Rural Productive Alliance Tour of Bolivia and Colombia: Lessons Learned and Scope for Improvement." Washington, DC: World Bank.

Fearne, A., D. Ray, and B. Vorley. 2007. *Regoverning Markets: A Place for Small-Scale Producers in Modern Agrifood Chains?* Burlington, VT: Gower Publishing Company. See also http://www.regoverningmarkets.org/.

KIT (Royal Tropical Institute), Faida MaLi (Faida Market Link), and IIRR (International Institute of Rural Reconstruction). 2006. *Chain Empowerment: Supporting African Farmers to Develop Markets.* Amsterdam, Arusha, and Nairobi: KIT, http://www.kit.nl/smartsite.shtml?id=SINGLEPUBLICATION&ItemID=1952.

Labaste, P., and C. M. Webber. 2010. *Building Competitiveness in Africa's Agriculture: A Guide to Value Chain Concepts and Applications.* Washington, DC: World Bank.

Quintero, J. F. 2010. "Evaluación de la sostenibilidad comercial de las alianzas productivas en Colombia: 12 estudios de caso." World Bank, Washington, DC.

World Bank. 2007. *World Development Report 2008: Agriculture for Development.* Washington, DC.

Agricultural Education and Training to Support Agricultural Innovation Systems

OVERVIEW

Charles J. Maguire, Consultant

EXECUTIVE SUMMARY

Agricultural Education and Training (AET) has a major role as a creator of capacity and supplier of the human resources that populate key segments of the AIS and enable that system to function more effectively. Past neglect and low levels of investment have prevented many national AET systems from equipping graduates to meet the needs of modern agriculture and contribute to the AIS. Aside from the technical knowledge that is the traditional focus of AET, graduates require the knowledge and tools to recognize innovative ideas and technology, catalyze communication between other AIS actors, and provide feedback to researchers and investors. Graduates particularly require new, "soft" skills, such as leadership, communication, negotiation, facilitation, and organizational capabilities. Employers increasingly demand these skills, which foster active participation in the AIS.

Serious constraints to quality education and training include weaknesses in policies that guide AET, the divided responsibilities for parts of the AET system, poor governance of AET institutions, continuing isolation of AET systems from key stakeholders, and serious underinvestment in AET systems. The major priority for reform is to develop a policy framework and (innovation) policy management capacity to guide AET. This reform underpins all others; it has wide implications for AET, interministerial cooperation,

financing, and stakeholder involvement. Another investment priority—wide-ranging, systemic reform—requires internal and external consultations with stakeholders and an analysis of gaps between stakeholders' expectations and current academic programs. Other priorities for investment include reforming curricula and teaching methods; building capacity and stakeholder partnerships for technical education and training; and developing effective in-service and life-long learning capacity among public workers who interact frequently in the AIS. Such reforms can be supported by investments in capacity building and infrastructure for ICTs to facilitate learning, research, and global and local networking and communicating. Investments in accreditation or in a regional resource for advanced degrees may also improve the likelihood that AET delivers content that meets stakeholders' needs.

Regardless of the chosen reform target, any change initiative will be subject to resistance, and leadership and commitment will be needed to see reforms through to the end. Depending on the location, capacity, commitment, and leadership for change, the time focus may shift to require longer-than-anticipated support; in other cases, reforms and changes may proceed faster than expected. A broad lesson for practitioners in planning reform programs of any length is to pay close attention to building constituencies of stakeholders at all levels to help ensure the program's sustainability.

WHY INVEST IN AET TO SUPPORT AGRICULTURAL INNOVATION?

Complementary investments in agricultural education have been neglected but are essential to ensure a new generation of agricultural scientists and leaders (World Bank 2004).

The growing focus on innovation systems in agriculture presents agricultural education and training (AET) with a challenge and an opportunity. The AIS creates demand for skills not traditionally developed in agricultural education—especially the "soft skills" that enable people to communicate better, listen more carefully and efficiently, nurture leadership, work cooperatively, and generally contribute more effectively to the AIS. The emphasis on the innovation system as a dynamic, highly interactive marketplace for ideas challenges AET to strengthen its role as one of the critical actors in agricultural innovation. All too often, agricultural education is failing to impart the knowledge, skills, and attitudes that can enable countries to feed growing populations, participate in international agricultural value chains, and cope with climate change, especially in the midst of grinding rural poverty.

The importance of enhanced skills for graduates who will interact with a wide spectrum of actors in the AIS is not in question; in fact, such skills assume even greater importance, given that they will have to compensate for a considerable educational deficit among the population in general. The majority of people with whom skilled AIS actors will interact in most developing countries have not had access to much more than basic education. The World Bank (2007b, 9) notes that education levels in rural areas worldwide tend to be dismally low—an average of four years for rural adult males and less than three years for rural adult females in sub-Saharan Africa, South Asia, and the Middle East and North Africa. Research in the 1980s established the relationship between primary education and annual farm output (Lockheed, Jamison, and Lau 1980; Jamison and Lau 1982; Jamison and Moock 1984). Basic education is a critical element for communication, understanding, and assessing innovations in the interactive process that prevails in the AIS.

For a very long time, governments and donors have invested very little, or only very intermittently, in AET (Willett 1998; Eicher 1999; Rygnestad, Rajalahti, and Pehu 2005; World Bank 2007b). The results are deteriorating physical infrastructure for education, overcrowded classrooms and residential accommodations, the exodus of teaching staff, outdated curricula, inadequate teaching and learning materials, and graduates' limited skills and employment options.

At this critical juncture, AET remains the main supplier of human resources for many of the public, private, and civil-society constituents of the AIS, through its network of agricultural universities, faculties of agriculture, vocational and technical colleges, and farmer training centers which, together, constitute the AET system. The central question in this module is whether corrective investments in particular AET models, programs, and activities will enable the AET system to take its place as a forceful and valued agent of innovation in agriculture, keeping in mind that major investments in AET systems occurred decades before the concept of "innovation systems" could influence their design and that much work remains to be done.

The general outlines of an AET system that is capable of operating successfully within an innovation system must recognize that the innovation system in which it operates is dynamic. The AET system itself will need to be agile, flexible, attuned to the needs of stakeholders in the innovation system, and acutely aware of developments in technology, communications, and markets as well as challenges to production stemming from high energy costs, declining water resources, and climate change. It will also need to channel advice to decision makers on policies to guide AET at all levels.

These generalizations aside, not all AET systems are equal. Some need deep, fundamental reform and strengthening, whereas others may require only minor adjustments to become more effective within the AIS. Before discussing specific investment needs and strategies over the short, medium, and long term, this module presents a broad review of AET—its structure, weaknesses, and strengths. The module then describes investments in education and training that will equip actors in the agricultural sector to negotiate the rapidly changing agricultural landscape with greater skill, resilience, and innovation. The need for these investments to foster gender inclusiveness in AET systems is incontestable (for one example, see box 2.1). Above all, this module will emphasize that fostering a capacity for innovation on this scale will require equally large measures of persistence and collaboration—from the agricultural and education sectors and also from government, civil society, and rural people.

THE STRUCTURE OF AET SYSTEMS

"Agricultural education and training" covers a range of organized programs and activities that serve the need for information, knowledge, and skills among those who work in various parts of the agricultural sector and the broader

Box 2.1   Gender-Inclusive AET: The Example of African Women in Agricultural Research
            and Development

In agriculture as in other domains, innovation requires communication of many kinds at many levels. The dissemination of knowledge, information, and innovations poses a special problem among women. Few women graduate from agricultural education programs—too few to work with women in societies where women are excluded from rights to land and other natural resources. Women are also marginalized from agricultural events, activities, and programs led by men or not permitted to communicate with men outside the family. Essentially, "women have been . . . underrepresented at all levels of AET institutions, from postsecondary to tertiary and higher education, although detailed gender-disaggregated data are available only very sporadically or not reported at all" (World Bank 2009, 181). An innovative program that aims at increasing the numbers of females with higher degrees in Africa may be the beginning of a change in the gender balance in academic and research institutions.

African Women in Agricultural Research and Development (AWARD, http://awardfellowships.org) is a project of the Gender and Diversity Program of the Consultative Group on International Agricultural Research (CGIAR). This professional development program was launched in 2008 after a successful pilot in East Africa supported by the Rockefeller Foundation. It seeks to strengthen the research and leadership skills of African women in agricultural science, empowering them to contribute more effectively to poverty alleviation and food security in sub-Saharan Africa. The two-year career development package builds on four cornerstones: establishing mentoring partnerships; building science skills; developing leadership capacity, and tracking, learning, monitoring, and evaluating fellowship-holders' progress. The program does not provide funds for the fellows' academic studies or offer research grants, although fellows can apply for research attachment opportunities.

Sixty outstanding women agricultural scientists received AWARD fellowships in July 2010, and the project currently supports 180 African women working in agricultural research and development who have completed bachelor's, master's, or doctoral degrees in selected disciplines. The fellows come from Ethiopia, Ghana, Kenya, Malawi, Mozambique, Nigeria, Rwanda, Tanzania, Uganda, and Zambia.

AWARD is a US$15 million, five-year project with plans to expand to a second phase starting in 2013. It is supported by the Bill and Melinda Gates Foundation, the United States Agency for International Development, and the CGIAR. AWARD partners with more than 75 national agricultural research institutions.

Sources: Author.

---

rural space. At the apex of the system for AET are the tertiary educational institutions such as agricultural universities or faculties and colleges of agriculture within comprehensive universities. Traditionally, higher agricultural education produced graduates who found employment in public agricultural research (see module 4) and extension programs (see module 3) and other technical services offered by ministries of agriculture. Over the years, as these public agencies greatly curtailed hiring, holders of agricultural degrees, diplomas, and certificates have been more likely to seek employment with agribusinesses or with NGOs operating agricultural programs.

Other institutions in the AET system include the polytechnics, institutes, or colleges that prepare technicians at the diploma level (the postsecondary, subdegree level). This category of education, often termed "agricultural technical-vocational education and training" (ATVET) or "vocational education and training" (VET), prepares technicians in a variety of specializations in agricultural subsectors. Some secondary schools offer agriculture as an elective (Tajima 1985), but in most developing countries these programs have a checkered history, influenced by the qualifications and experience of the teachers assigned to the subject and the motivation of the students who enroll. Probably the most successful secondary agricultural education model is the vocational agriculture program offered in largely rural districts in the United States. The program offers academic and practical subjects in school and, through a supervised youth organization (Future Farmers of America), helps students develop leadership skills and technical prowess by participating in contests and undertaking a supervised project (box 2.2).

Box 2.2   Future Farmers of America: A Unique Young Farmer Organization

Future Farmers of America (FFA) was founded in 1928 and brought together students, teachers, and agribusiness to solidify support for secondary agricultural education. Today (circa 2011), the FFA has some 523,000 members (38 percent are female) aged 12–21 in all 50 states. Of these, 27 percent live in rural farm areas, 39 percent in rural nonfarm areas, and the remaining 34 percent in urban and suburban areas. Now known as the National FFA Organization, FFA—with its motto of "Learning to Do, Doing to Learn, Living to Serve"—is dedicated to making a positive difference in the lives of students by developing their potential for premier leadership, personal growth, and career success through agricultural education.

The FFA is an integral part of the secondary-level Vocational Agriculture program, which has three parts: classroom instruction, the FFA, and Supervised Agricultural Experience (SAE). Students develop their SAE projects in one of four categories: entrepreneurship (the student owns and operates an agricultural busi-ness); placement (the student gets a job on a farm or at an agricultural business, school, or factory laboratory); research and experimentation (the student plans and conducts a scientific experiment, usually related to agriculture); or exploratory (the student attends an agricultural career fair or creates a report or a documentary on the work of a veterinarian or extension worker).

Through this program, more than 11,000 FFA advisers and teachers deliver an integrated model of agricultural education that provides students with innovative and leading-edge education, enabling them to grow into competent leaders. The FFA provides 23 national career development events through which members are challenged to real-life, hands-on tests of skills to prepare them for more than 300 agricultural careers. Agriculture is the largest employer in the United States. More than 24 million people work in some sector-related activity. Industry values the FFA program and its graduates, contributes to the National FFA Foundation, sponsors programs, and provides individual scholarships.

Source: Adapted from www.ffa.org, accessed December 2010.

Agricultural training, frequently delivered in training centers or training institutes, is offered to public employees as in-service training and/or to farmers as farmer training. Ministries of agriculture are usually responsible for agricultural training programs. Public extension services offer training (largely to farmers) through formal presentations, lecture-demonstrations, field days, crop and animal field trials, farm tours, and various other media. Public agricultural research systems provide educational opportunities for farmers and extension staff, usually in the form of field days combined with lectures. Public sector researchers also act as resource persons in formal, higher-level education programs, work with extension staff to train farmers, or provide in-service training for extension staff.

Formal AET

Traditionally, agricultural education has been supplied and supported largely by the public sector. Although the various elements in the AET delivery chain are often referred to collectively as a "system" (Bawden 1998, 1999; Rivera 2008), in many developing countries it is questionable whether these elements form a robust system in which communication and feedback flow between institutions and allow for adjustments and improvements on a continuous basis. Agricultural education and learning (AEL) is a variation on AET that reflects a more student-centered approach to formal programs (Ochola and Ekwamu 2008).

In many countries, responsibility for education and training for agriculture and rural development has been, and continues to be, divided between ministries of agriculture and education. Typically, higher agricultural education has been the responsibility of the education ministry, whereas training for agriculture and its subsectors has been the responsibility of the ministry of agriculture. In some countries, the ministry of cooperatives is responsible for providing training for a variety of cooperatives that deal with a range of topics, including agriculture. Given the increasing interest in farmer organizations (see module 1, TN 4 and IAP 2) (especially as precursors to large, organized agricultural cooperatives), cooperative colleges are becoming an important aspect of formal AET.

Whether it is part of a robust, well-integrated system or not, agricultural education is weakened by the division of responsibilities among ministries, the isolation of individual ministries, and their failure to collaborate in designing and delivering education and training in a manner that meets the needs of all AET stakeholders. Under these circumstances, a broad vision for AET is rarely in place. As a result, policies and strategies for modernizing agricultural education are seldom developed.

As a small but specialized element in the bigger education picture—which includes primary, secondary, and higher education together with vocational and technical education and training, teacher training, and a number of special adult education programs—agricultural education tends to lack bargaining power when investment decisions are made. This relative lack of visibility and clout is all the more critical because the mounting pressure for change in AET chiefly comes from developments in agriculture that are outside the control of AET institutions.

Informal AET

Alongside the formal AET system, a dynamic, informal learning system greatly influences how information, knowledge, and skills are channeled and used in agriculture. At the heart of this system are farmers, farming families, the services they receive, and the contacts they make on a regular basis. Informal agricultural education involves awareness-raising and training provided to farmers by public extension and research services, by traders who purchase farm products and supply farm inputs, and by the media, which convey a variety of information to rural communities.

Farmer-to-farmer communication (see also module 3, especially IAP 2) is one of the most powerful forces for education within the informal system. Farmers communicate easily with their peers, observe the techniques and skills used by others, and quickly adopt what they perceive as successful practices. The actors in this AET system are linked with one another and within the broader AIS. The many roles played by graduates of AET, regardless of how structured or organized it may be in a given setting, are detailed in modules 3 and 4.

In-service training and development

Public and private in-service training and development, which can be categorized as formal AET, serves employers, employees, extension workers, NGO technical staff, and vocationally oriented, self-directed learners (Rivera 2008). Similarly, the innovation system benefits from the communications and technical skills obtained by rural youth, either in or out of school, through participation in a variety of young farmers' organizations. The Future Farmers of America, described in box 2.2, is a very successful but unique example of this kind of skill development, which makes important contributions to human resource capacity in agriculture.

PAST INVESTMENTS IN AET

The 1950s, 1960s, and 1970s saw substantial, dedicated investments in AET. One of the largest investments, launched by USAID in the mid-1950s, was a long-term program that established universities similar to the United States land-grant universities in Latin America, Asia, and Africa. This ambitious program included technical assistance for administrative and academic activities and curriculum development, provided links to overseas advanced degree programs, modernized libraries, and paired the new universities with counterpart land-grant universities. The USAID program modernized the way that agriculture was taught and learned in many developing countries; enhanced the quality of education, research, and extension; provided current teaching materials; and created an international network of agricultural education professionals.

The impact of the investment was impressive but not always sustained. The quality of teaching and learning deteriorated. Changes in leadership, reduced funding, and the winding down of collaboration with individual overseas universities all reduced performance. On the other hand, a number of universities established under the program thrived and have continued to provide education leadership long after the investment program closed (the Brazilian system is a case in point; see box 2.3).

During the same period, multilateral organizations such as the World Bank, Food and Agriculture Organization (FAO), International Labour Organization (ILO), and the United Nations Educational, Scientific, and Cultural Organization (UNESCO) supported AET through freestanding agricultural education projects, training components in agricultural projects, and seminars, workshops, conferences, and in-country and international courses. With the exception of free-standing agricultural education projects, most of the other activities were of short duration. The impact of the free-standing projects depended heavily on the recipient ministry's or country's commitment to sustaining the new investment. The choice of participants for seminars, workshops, and training courses proved decisive in terms of the usefulness of these activities and the effectiveness with which the participant transferred knowledge and/or technology to the parent organization—a lesson that should not be forgotten in designing AET projects (see TN 4). By the end of the

Box 2.3   Brazil's National Agricultural Research Program Benefits from Long-Term Investments in Human Resources

In 1963, the Brazilian government took a high-level decision to build a human capital base for a modern agricultural sector. With financing from the United States Agency for International Development, four American land-grant universities assisted four Brazilian universities in strengthening BSc level training for a decade followed by another four years of support for postgraduate education. In 1971–72 more than 900 Brazilian graduate students were studying agricultural sciences in United States universities. This experience with building human capital in programs in agriculture is directly linked to political decisions by the Federal Government and the Ministry of Education to pass the University Reform Act of 1968, which linked promotions to higher graduate degrees and required academic staff to work full time. In 1972, when the government established the Brazilian Agricultural Research Corporation (EMBRAPA) to coordinate its national agricultural research program, EMBRAPA continued to invest in human resources. It launched a massive human capital improvement program that sent 500 agricultural researchers for PhD programs and spent 20 percent of its budget from 1974 to 1982 on training in Brazil and abroad (World Bank 2007a, 39). Today, one-third of EMBRAPA scientists have a PhD, half have an MSc, and the balance have a BSc or equivalent. The most important lesson from this experience is that Brazil did not reduce public expenditure on its core agricultural institutions some 40 years ago when foreign investment waned. Instead, by mobilizing high-level political support, Brazil built a strong human capital base to sustain a globally competitive agricultural research and extension base.

Source: Author.
Note: EMBRAPA = Empresa Brasileira de Pesquisa Agropecuária.

1970s, funding for AET began to decline dramatically (Willett 1998), overtaken by other development priorities. Numbers of AET specialists in many international organizations and bilateral donor agencies decreased. Despite pleas by numerous observers and organizations to governments, donors, and universities to rehabilitate and reform deteriorating agricultural education programs and facilities, AET continued to drop even lower on the development agenda.

A number of generic weaknesses in the planning and delivery of AET in developing countries have persisted.[1] Briefly, these weaknesses include a lack of university autonomy, weak links to stakeholders, lack of accountability for quality or employability of graduates, outdated curricula and teaching approaches, weak training in practical skills, the variable quality of programs, weak adoption of ICTs, and low remuneration of faculty and staff.

Diploma-level AET also exhibits weaknesses, including the absence of supporting policies, weak links to stakeholders, programs that fail to reflect labor-market needs, inadequate and inconsistent funding, and a shortage of skilled teachers/instructors. Agricultural training at the secondary level, which is not universally offered, is often chosen as an "easy pass" by students. It also suffers from a lack of qualified teachers and is weakened by poor cooperation between v of education and agriculture.

A 2005 review of investment in AET in projects supported in Africa by the World Bank found that the same weak level of investment had persisted since the end of the 1970s (Rygnestad, Rajalahti, and Pehu 2005). One outcome of the weaknesses and low investment in AET is the reluctance of students to choose agriculture as their preferred academic pursuit (Pratley 2008; Rivera 2008; Mulder 2010). In countries where higher education is at a premium, this reluctance inevitably creates a situation where many of the students who enroll in agricultural programs have a greater interest in possessing an academic degree or certificate than in making a career in agriculture.

KEY POLICY ISSUES

Sound policies are essential to address the recurring weaknesses of AET systems, yet weak and fragmented AET systems cannot present a united front to government or policy makers and gain support for increasing the effectiveness of AET. Policy guidance and support are needed above all to:

Clarify the role of AET. Divided ministerial responsibility for agricultural education, especially for public agricultural universities and faculties of agriculture, prevents AET from realizing its potential. Without a sense of urgency and collaboration between ministries of agriculture and education, higher agricultural education will not be reoriented and modernized to reflect the needs of a changing agricultural and rural environment. Clarifying roles and responsibilities of agricultural education in the development agenda (with clear links to the agendas of the respective ministries) would open the door to reforms that enable these institutions to fulfill stakeholders' expectations.

Ensure sustainable, regular funding. The allocation and disbursement of funds for AET on a sustainable and regular basis is essential for maintaining the quality and relevance of education programs. Policies that underpin AET need to recognize agriculture's broad contributions to the economy: as a producer of basic and export commodities, source of employment, provider of nutritional health, and engine of poverty reduction. It is essential that these policies are in place and implemented and updated as changes occur in the sector.

End political interference in university administration. Because most agricultural universities, TVET colleges, and training institutes are public entities, they are subject to political influence of one kind or another. Demand for education, especially at the tertiary level, can cause politicians to decree increases in enrollment without providing the physical, human, and financial resources to accompany higher enrollments. The lack of resources guarantees that educational standards will fall as facilities become overcrowded, teacher–student ratios become unmanageable, students struggle to study and learn in unfavorable surroundings, experienced teachers and researchers leave, and large numbers of underqualified graduates flood the job market. Appropriate policy instruments, strictly enforced, can help prevent interference of this kind.

Improve governance in higher agricultural education. A related policy issue is that agricultural universities and faculties of agriculture lack autonomy. Typically, universities have little or no control over budgets, hiring or firing of staff, links with stakeholders, or accountability for the fate of graduates once they leave. Policies are required to correct these weaknesses and improve the credibility and usefulness of higher agricultural education. Universities and other third-level entities need control over their programs, budgets, and staff. They also need to interact with stakeholders to better understand the changing agricultural sector and amend curricula to reflect these changes. (See also module 1, TN 1.)

Address the gender imbalance. An appropriate policy that stipulates active recruitment and quotas for females should be in place. The policy would also support gender balance in the faculty and teaching staff of universities, TVET institutions, and training centers. The presence of female teachers, instructors, or trainers can make a difference when female students make decisions about pursuing a qualification in agriculture.

Create a favorable environment for investing in AET and improve the balance of investments in agricultural research, extension, and education. Investment is uneven in research, extension, and education—the three pillars of agricultural knowledge and information systems. Investors find it easier to deal with research or extension agencies, with their clear lines of authority and organized networks, than with multitiered educational systems, which answer to more than one ministry and are often isolated from research and extension. Policies should be in place to ensure that research, extension, and education work together to capitalize on their respective strengths and present a holistic picture of their interdependence in bringing knowledge and services to the sector. See module 6 for a broader discussion on an enabling environment.

INVESTING IN CHANGE: PRIORITIES AND NEW DIRECTIONS FOR AET

Given this background—the declining quality of many AET systems, the general failure to articulate a strong policy framework for agricultural education, and the corresponding failure of educational institutions to build a strong and vocal constituency within the agricultural sector—it is clear that many AET systems must change. This section discusses some of the priorities and new directions that could reposition AET systems as catalysts and beneficiaries of innovation. These changes—carefully managed, with sustained commitment—should enable AET to attain the autonomy, agility, and human and financial capacity to produce graduates who meet the needs of the agricultural sector and its diverse stakeholders.

Priorities and new directions are discussed first to provide a broad sense of the kinds of reform that AET systems might undertake. Because the success of any reform depends to a great extent on how it is managed, the next section briefly presents guidelines, based on experience, for managing change in AET. To support this kind of change in AET systems, decision makers, AET entities, and donors can choose among a range of specific investments, which are discussed next. These investments can be of long, medium, or short duration, depending on the agreed objectives, identified needs, support time horizon, and funding availability.

Table 2.1 provides a useful checklist of future directions for AET in relation to current conditions.

Investment priorities

What are the most important investments to make AET a forceful contributor to the innovation system? Although every case in every country will differ, and although it is hoped that new ways will always be found to make AET more effective, this list gives an indication of the range of options for reform and the issues that often prove to be the most important.

DEVELOP A POLICY FRAMEWORK AND (INNOVATION) POLICY MANAGEMENT CAPACITY. A major investment priority is to seek policies from government to guide AET, enable it to develop effective education and training strategies, and provide it with high visibility in sector and national

Table 2.1   Current and Potential Future Directions of AET Systems

Higher agricultural education now	Future directions
Weak, unenforced, or absent policies	Clear AET policies with responsibilities defined and enforced
Weak governance	Strong governance inclusive of stakeholders
Little autonomy	Autonomy that enables staff decision making, financial control, and standards setting
Uncertain funding	Steady and regular funding guaranteed
Isolation (academically and from stakeholders)	Academic, rural community, and stakeholder connections established and maintained
Programs not accredited	Accreditation the norm

Curriculum now	Future directions
Outdated	Updated and current with stakeholders' needs
No stakeholder input	Stakeholder consultations; input solicited and incorporated in changes; stakeholders participate in governance
Teaching	Learning
Theory	Theory and practical application
No student attachments	Regular, organized, and supervised attachments
Inappropriate pedagogy	Effective pedagogy tailored to subject matter and learner needs
Little use of external teaching resources	External resources used in team teaching to expand knowledge and skill pool
Little use of ICT	Use of appropriate ICT the norm

Technical training now	Future directions
Heavily supply driven	Mostly demand driven
Managed by the public sector	Managed through public-private partnerships
Poorly qualified and remunerated instructors	Qualified and fairly remunerated instructors
Qualifications not certified by professional bodies	Certification ensured
Equipment in short supply and outdated	Equipment/practice areas obtained through public-private partnerships

Management of human resources now	Future directions
Weak human resource management leadership	Qualified human resource managers and trainers
Selection of trainees not based on need	All selection based on need and future tasks
Training needs assessments are not undertaken	Needs assessments are standard procedure
Little supervisor/manager involvement	Supervisors/managers consulted and involved
No evaluation of trainee performance on the job	On-the-job performance measured
Trainers not trained to instruct/teach	Qualified trainers standard

Source: Author.

development planning and implementation. This reform alone has wide implications for AET, as well as major implications for interministerial cooperation, financing, and stakeholder involvement. Specific investments to develop a policy framework might include the facilitation of interagency dialogue, technical assistance to help draft policies, consultations with affected parties at draft policy stage, and policy launching where all key stakeholders are represented. *This investment priority underpins follow-up investments that would fund needed organizational and academic reforms—in other words, all of the other investments listed here, whether they are pursued together or separately.* This reform effort also requires capacity-building efforts for policy management and the design of innovation policies. Most countries experience a chronic lack of capacity to design and implement public policies, and the capacity to design and implement policies that foster agricultural innovation is even more limited (OECD 2005; Rajalahti, Janssen, and Pehu 2008).[2] See also module 6, TN 1 and TN 2 for education policy and governance processes.

Support wide-ranging, systemic reform. Once the policy framework is in place, AET system reforms are the next major priority. Such reforms enable universities and other institutions of higher education to clarify their roles in relation to the educational system, to their stakeholders, and to the wider needs of the agricultural sector. Then they can make the necessary adjustments to governance, administration, curricula, pedagogical methods, the provision of in-service and life-long education for graduates and civil society, and partnerships and links with stakeholders, including the private sector.

In this "inside-out" approach to reform, the system itself (and especially the tertiary institutions) undertakes the reform process from within and reaches out to stakeholders to measure expectations and satisfaction with existing programs. Gaps between the supply, content, and quality of education and training and demand in the labor market point to reform measures that need to be taken. Investments in system reform, especially at the university level, must support internal dialogue, interaction between the educational institution and those who make policy and decisions, stakeholder consultations, analysis of gaps between stakeholders' expectations and current academic programs, formulation of the reform strategy, a schedule for its implementation, and the related costs. External facilitators (see also module 3, TN 4) are required to manage the reflection process, carry out the needs assessment, formulate the reform strategy, and determine the final reform agenda.

Change what is taught and how it is taught. Very often it is a priority to invest in reforming curricula and teaching methods—in many instances, such changes are long overdue. Coursework will include practical as well as academic knowledge and skills. Traditional teaching methods will be replaced by a learning approach that enables students or trainees to discover and internalize knowledge and skills and thus equip the AIS with people who know how to communicate and share their knowledge with others. Potential employers of graduates from the AET system consistently report that they seek (and often do not find) problem-solving skills, the ability to listen, the capacity to analyze situations, and skill with information technology (including computer applications), among other skills (box 2.4).

In-service training: continue investing in human resources. In-service training is an investment priority for continuing enrichment of the innovation system. Continuing training and learning maximizes use of previously acquired knowledge and skills, adds the lessons of practical peer experience, and expands human capacity by introducing new and updated technical and social information and knowledge. Well-managed, high-quality, flexible in-service training and learning for public, private, and civil-society clients expands the network of people in the innovation system with relevant information and knowledge. In many cases, they gain specific skills in communicating more effectively and supporting others in assessing the suitability and viability of agricultural innovations.

Tap the power of ICTs. Investments in capacity building and infrastructure that enable ICTs to facilitate learning and research, improve the delivery of subject matter, and support global and local networking and communicating are critical for effective interaction within an education system and AIS. The ICT revolution has vastly increased the flow of information and ideas throughout the rural space. It has also increased stakeholders' expectations of AET graduates, who should be familiar with the range of ICTs and use them to communicate technical and managerial information to others in the AIS. The *Information and Communication Technologies for Agriculture e-Sourcebook* (World Bank 2011) shows how ICTs are used increasingly to provide advisory services, capture and disseminate market information, and facilitate research, learning, and communication of all kinds (box 2.5).

Box 2.4 The Need for a Broader Skill Set to Foster Innovation

Effective innovation systems need technical specialists to investigate and elucidate the complex technical aspects of innovations. Effective innovation systems also require a cadre of professionals whose mindset and skill set extend beyond a particular specialization to encompass (for example) markets, agribusiness, intellectual property law, rural institutions, rural finance and credit facilitation, systems analysis, and conflict management. To foster these capacities, universities must reform their curricula to include innovation systems approaches and shift toward more client-oriented, vocational courses.

The demands placed on agricultural education programs are shifting so rapidly that programs in many locations cannot produce human resources capable of the sorts of innovation that the world's agricultural economies require. To respond to these requirements, AET programs will need to balance a highly technical curriculum with training in a wide range of skills and competencies. Researchers, extension agents, and other service providers will need to bolster their professional training with skills that support interaction with diverse actors to collaboratively address new challenges and opportunities and to share information and knowledge. For example, extension agents need to know how to build social capital by organizing rural actors, provide diverse services from technology transfer to marketing, and serve as facilitators or intermediaries among actors (see module 3). Researchers, rather than sequestering themselves in labs and field stations,

require managerial, entrepreneurial, leadership, negotiation, intellectual property law, facilitation, and partnering skills that educational institutions rarely cover.

To be capable of fostering agricultural innovation, graduates of the various levels of the AET system should possess a wide variety of skills, which may include:

Basic skills and digital literacy: Reading, writing, numeracy, and the ability to use digital technology and access and interpret information.

Academic skills to pursue disciplines in advanced educational institutions, such as languages, mathematics, history, law, and science.

Technical skills: Academic and vocational skills required by specific occupations and knowledge of certain tools or processes.

Generic skills such as problem solving, critical and creative thinking, the ability to learn, and the capacity to manage complexity.

"Soft" skills such as the ability to work and interact in teams, in heterogeneous groups, and across cultures; communications skills; motivation, volition, and initiative; and receptiveness to innovation.

Leadership skills: Building and leading teams, coaching and mentoring, lobbying and negotiating, and coordinating, with a clear understanding of ethical behavior;

Managerial and entrepreneurial skills to put innovations into practice and enable organizations to adapt and respond in competitive environments.

Source: Adapted from Snapp and Pound 2008, OECD 2010.

Despite the evolution from traditional lecturing to interactive, student-centered learning, many AET systems still operate in the "chalk-and-talk" era. Some systems of higher agricultural education rely on dated lecturer's notes as the only teaching aid, and some vocational and technical programs cannot provide students with an opportunity to practice on equipment or use laboratories as they learn. Even when teachers and instructors are willing to use ICTs, budgets are insufficient to purchase them, or unreliable supplies of electricity make their use unpredictable. Investments in institutional or curriculum reform for tertiary agricultural education, agricultural TVET, or in-service training should include provisions for introducing and/or updating ICTs and training staff to use them.

MEET RECOGNIZED STANDARDS FOR QUALITY. Investing in accreditation or certification provides universities and training institutes an incentive to raise their academic quality to recognized standards. Investments that develop close working relationships with preeminent educational institutions are another means of raising academic standards. For institutes of higher education especially, these alliances lead to collaborative learning programs involving information sharing, staff and student exchanges, and joint research. The local institution, its staff, and its graduates gradually become much stronger contributors to the AIS.

PROVIDE A REGIONAL RESOURCE FOR ADVANCED DEGREES. Given the limited resources in many AET systems, many

Many AET systems in developing countries do not employ even low-level ICT in the teaching/learning process. Reasons for this include poorly trained teaching staff who have not been exposed to ICT in their training, lack of funds to purchase ICT, unreliable power access, and no supervisory pressure to adopt and use ICT. ICT can improve the quality of teaching and learning and raise the quality and relevance of AET for greater impact within the AIS through a number of means:

- *Improved competencies among teachers.* Teachers and trainers are trained in basic ICT skills and ICT-based teaching methods.
- *Improved competencies among graduates.* The effectiveness and employability of AET graduates improves because they possess ICT skills.
- *Better educational materials.* ICT enables teachers to access information sources and create, update, and share learning materials.
- *Distance education and e-learning.* ICT is integral to creating opportunities for distance and electronic teaching and learning.
- *Improved education administration and management.* Throughout the AET system, ICT enables more effective and efficient management of human and financial resources and monitoring of student performance.

Source: Author and www.IICD.org.

The International Institute for Communication and Development (IICD), a nonprofit foundation that specializes in ICT as a development tool, has helped to introduce a number of ICT solutions in the education sector, which indicate the potential for ICT within AET. For example, teachers in Burkina Faso learned to build websites, find materials on the web, and use video, web publishing, and other applications to improve their lessons. Social media training helped these teachers start an active online community for sharing teaching materials with schools across the country. In a similar project, Bolivian teachers learned to create videos and CD-ROMs to support lessons; its success inspired the Bolivian government to launch a national program to put computer labs in 1,000 schools. At the Copperbelt College of Education in Zambia, one initiative requires that all graduating teachers be able to prepare lessons digitally. ICT skills also benefit vocational training and help make youth more employable. In Zambia, IICD helped set up a computer lab in a youth center, where young people learn basic ICT skills and access the web. As users learned more about the potential of ICT, the computer lab began to offer additional services, including secretarial and marketing support. Now ICT is also used to support training for tailors, carpenters, and mechanics.

---

universities cannot provide teaching and research opportunities at the level of excellence needed to produce graduates who can assume leadership roles in the AIS. Investments to create a center of excellence within a region can provide opportunities for qualified candidates from smaller or less-well-endowed educational systems to pursue studies at a higher level. These candidates, on returning to their academic bases, can contribute to the development of new knowledge and enhance the local AIS. See also module 4, TN 5 for lessons on organizational change.

Long-, medium-, and short-term opportunities to revitalize the AET system

The discussion here indicatively classifies investment opportunities as long, medium, and short term to emphasize the commitment required to achieve various kinds of

objectives. Changes in AET systems, whether sweeping or piecemeal, do not always conform to the expected process or timeframe, and practitioners should plan for adjustments. *Depending on the location, capacity, commitment, and leadership for change, the time focus may shift to require longer-than-anticipated support; in other cases, reforms and changes may proceed faster than expected.*

LONG-TERM INVESTMENTS (FIVE OR MORE YEARS; SOMETIMES MUCH LONGER). Regardless of whether the pressure for change is internal or external, long-term investments to support change in AET systems largely involve wide reforms. Such investments require considerable preparation in the form of discussion, creating a vision for AET (see also module 7, IAP 4 for lessons on a vision-building process), stakeholder agreements, catalyzing inputs from facilitators, and field visits to successful reform initiatives. Important investment opportunities include:

Reform of agricultural universities or faculties of agriculture in comprehensive universities. This type of reform can be classified as internally led change even though the impetus for change may originate outside the university.

Long-term "twinning" or collaborative links to universities recognized as leaders in AET. These links can reinforce reforms and lead to productive growth opportunities in research, teaching, and learning through staff and student exchanges.

Investments in the creation, capacity building, and continuing support for regional institutions that offer specialized degrees, especially at the master's and doctoral levels. These entities could become centers of excellence in particular fields of research and education.

The availability of competitive funds that are tied to an agreement to bring about reforms.

The establishment of new, privately funded and managed agricultural universities—new in approach, vision, and autonomy, not merely newly built—also fits into the long-term investment category.

Examples of long-term investments to support reform in agricultural education are included in one TN and several IAPs:

TN 1: Reforming Public Agricultural Education at the Tertiary Level. The process, benefits, risks, and lessons from comprehensive reform in higher agricultural education are described, and an example from China is presented in detail. Efforts in other settings (Australia, Ireland) are discussed in passing.

IAP 1: Reforming India's State Agricultural Universities. This IAP encapsulates lessons from a major effort to modernize university administration and management; update the curriculum; make related changes in pedagogy, learning materials, and infrastructure; set new standards for higher agricultural education; and improve human resource management in state line departments that worked closely with the agricultural universities.

IAP 4: Innovative Training Program for Midcareer Agricultural Extension Staff: The Sasakawa Africa Fund Education Program. This long-term investment focused on expanding the skills of midcareer professionals through degree programs featuring fieldwork as well as academic training. The program developed strong links between universities, the public sector, and rural communities that benefited all concerned.

IAP 5: Chiang Mai University Links with Rural Communities to Focus Research on Farming Problems and Foster Curriculum Change. Some twenty years ago, the University of Chiang Mai developed a highly influential learning and research model that integrated faculty, students, and rural communities. Feedback from the community experience continues to influence the university's research focus, its curriculum, its role in the AIS, and its international standing.

IAP 6: EARTH University, Costa Rica: A New Kind of Agricultural University. Founded in 1990, the private, autonomous EARTH University is an example of an institution that was newly developed to address the need to educate and train young people to deal with the region's numerous agricultural, social, and political problems in rural areas. Its model blends academic work with practical experience and collaboration in agrarian communities and agribusiness.

IAP 8: Agribusiness Training for Secondary School Graduates in Timor-Leste is a relatively short-term investment that requires longer-term support to be sustained.

Medium-term investments (three to five years). Medium-term investments are tied to reforms and improvements in education and training organizations, in the quality of education and pedagogical skills, in shifting from teaching to learning, boosting the acquisition of practical skills, and providing life-long education and training. More specifically, they can include needs assessments with stakeholders at all levels in the AET system; reforming the curriculum; improving teaching and learning methods; building staff capacity; enhancing communications skills; using, applying, and facilitating access to ICTs; introducing or strengthening programs for technicians; creating policies to underpin higher agricultural education and TVET; modernizing governance structures at universities, colleges, and training entities; gaining accreditation for higher agricultural degree programs, technical diploma programs, and certificate qualifications; strengthening the management of in-service training; evaluating the impact of training programs on work performance; and establishing or strengthening university capacity to offer outreach education and training to external stakeholders, including the private sector and communities. A medium-term investment could also fund competitive grants to catalyze such changes.

Several TNs and IAPs in this module discuss medium-term investments:

TN 2: Curriculum Change in Higher Agricultural Education. Practical steps involved in designing programs to foster curriculum change are discussed, along with useful supporting investments and policies.

TN 3: Education and Training for Technician Development. TVET colleges or institutes produce technicians with the practical skills to manage farm and agribusiness enterprises, provide practical leadership to farm organizations and agricultural banks, and become entrepreneurs in their own right. This TN reviews newer approaches to technician education and training, including public-private and agribusiness approaches.

IAP 2: Transforming Wageningen Agricultural University. In 1997, the Netherlands Ministry of Agriculture initiated a major investment in the knowledge infrastructure for agriculture that led Wageningen University to change its focus, structure, programs, and staffing and cooperate with a wider research, social science, and stakeholder network.

IAP 3: Curriculum Change in Agricultural Universities Strengthens Links in Egypt's Innovation System. Curriculum change in five universities has enabled course content to respond to the needs of potential employers and proven to be a good entry point for wider institutional change.

IAP 7: Technical Skills for Export Crop Industries in Uganda and Ethiopia. Through cooperation between a donor, training institution, and commodity group, workers in high-value export crop industries in Uganda and Ethiopia acquired better technical skills. Demand from employers triggered the training.

IAP 8: Agribusiness Training for Secondary School Graduates in Timor-Leste. A one-year certificate course on agribusiness, piloted for three secondary schools of agriculture in Timor-Leste, highlights the value of such programs as well as the considerable challenges, especially in rural areas of fragile states.

SHORT-TERM INVESTMENTS (ONE TO FIVE YEARS). Even short-term investments can greatly enhance the capacity of those involved in AET to be more effective communicators and agents of innovation. For example, in-service training programs or programs of farmer training can be improved by offering short courses to teachers, facilitators, and demonstrators to enhance quality, make the content more effective, and improve the delivery of the programs. Specialized study programs for key sector staff also fit under this category of investment, provided that they are based on clearly defined needs and that recipients return to undertake a work program that is more effective as a result. In-country, regional, or international sites may be chosen for specialized learning depending on need and the quality of programs offered. Support for attachments, work experience assignments, and student community interaction also falls into this category of investment. For rural youth organizations, investments could support organizer training, youth programs and competitions, study visits, and the production of educational and media materials.

While short-term activities do contribute to the strength and capacity of AET, it should be recognized that interventions designed to strengthen AET systems are, for the most part, a long-term undertaking: "Only through a long-term outlook on change can AET systems contribute to the development of more dynamic and competitive agricultural economies that engage farmers, entrepreneurs, extension agents, researchers, and many other actors in a wider system of innovation" (Davis et al. 2007). This observation is echoed by Eicher (2006), who reported that a number of successful AET reforms took sixty years or more and required a continuing commitment on the part of donors or other partners.

Short-term investments are featured in:

TN 4: Reforming the Management of In-Service Training/Learning. This note describes the rationale and steps involved in improving the management of in-service training and the capacity of trainers to assess needs, design programs, deliver training in a variety of modes, and evaluate its impact.

IAP 9: Vocational Training in Egypt Combines Technical and Innovation Skills for Agriculture. Vocational agricultural education programs in 25 secondary schools in Egypt were transformed to introduce students not only to practical training but to higher-level skills such as problem solving, critical thinking, and decision making. This effort involved additional changes in pedagogy and school management.

Reforms require reformers

It is worth emphasizing again that the duration of reforms and the examples provided here are indicative. Short-term programs have become longer-term programs when stakeholders have championed their goals; many long-term programs have foundered when they are not institutionalized in a sustainable way. *A broad lesson for practitioners in planning reform programs of any length is to pay close attention to building constituencies of stakeholders at all levels to help ensure the program's sustainability.*

MONITORING AND EVALUATING INVESTMENTS TO REFORM AET

Monitoring and evaluation help to determine whether investments are contributing to desired outcomes, enable institutions to rally support for continued reforms, and enable institutions to adjust programs and administrative processes. Although each particular investment in AET will require a specific set of progress and impact indicators, most investments to reform agricultural education have common elements, regardless of whether the reform encompasses entire systems or institutions or whether it addresses a specific aspect of AET, such as curriculum reform.

Table 2.2 shows corresponding indicators for investments in each major type of reform described in this module:

Table 2.2   Measuring the Progress of AET Reforms

Comprehensive reform in entities offering higher agricultural education

Intended outcome	Selected indicator	Needed action/step
– University an active member of the AIS. – Improved sector productivity through a more effective role in the AIS. – University offers life-long learning opportunities to stakeholders. – University seen as reliable and competent source of advice and in-service learning/training. – Improved, higher-level governance of AET, underpinned by clear policy and financial resources. – Reforms institutionalized. – Quality of university degree programs assured through accreditation. – Stakeholders participate in university governance, and university has closer ties with communities and employers of graduates. – University actively participates in and contributes to national, regional, and international AET networks.	– Tripartite body (ministries in charge of education, agriculture, and finance) established to coordinate higher-level AET. – Representative stakeholder membership of university board. – Annual university budget prepared and submitted to tripartite body. – Annual budget approved and funds allocated. – Legal status of reforms finalized and available to the public. – Reforms presented to tripartite body for approval and endorsement. – University senate approves reforms. – Staff pedagogical skills updated. – Use of ICT for learning becomes the norm. – Curriculum reform approved by university senate and tripartite body. – Community outreach program in place. – Student attachment mechanism working well. – Increase in master's and doctoral program enrollment. – Staff reflect improved gender balance.	– Initiate dialogue between ministries in charge of education, agriculture, and finance. – Prepare budgets on annual basis. – Present reforms to key stakeholder groups. – Campaign to institutionalize reforms undertaken with decision makers. – Carry out and analyze needs assessment to identify gaps between current program and labor-market expectations. – Describe and undertake curriculum adjustment. – Update pedagogy to focus more on learning than teaching. – Improve staff remuneration and incentives to attract and retain good personnel. – Improve student intake selection and gender and minority group balance. – Hold student population to manageable numbers to assure quality education. – Increase use of ICT to bring higher AET into contact with new sources of knowledge and support. – Forge links with universities at home and abroad.

Reform of ATVET and VET

Intended outcome	Selected indicator	Needed action/step
– Role of agricultural technical education and training is defined. – Institution offering life-long education and training opportunities. – Instructors' and teachers' skills enhanced, including student-centered pedagogy. – Policies in place to underpin ATVET. – Active stakeholder role in governance of ATVET. – Functioning network of ATVET/VET institutions and stakeholder groups in place. – Status of ATVET/VET qualifications improved because of certification and support from employers. – Facilities and equipment adequate for supporting quality ATVET programs in place. – Remuneration levels in place attract qualified instructors.	– Role and management of ATVET officially approved. – Policies officially approved. – Public-private cost-sharing formula for ATVET publicized. – Public-private partnership agreements in place for funding ATVET. – Programs certified to agreed standards. – Employment rates for graduates satisfactory. – Employer satisfaction with quality and skills of graduates high. – Courses certified by recognized authorities. – Employer and student/trainee satisfaction with training/learning impacts.	– Clarify responsibility for management of ATVET. – Establish links to the AET system. – Establish attachment and work experience programs, and agree on the rules and evaluation criteria. – Undertake training needs assessments. – Train instructors/teachers in student-centered pedagogy. – Undertake survey of employers and graduates. – Carry out facilities and equipment needs assessments. – Review and adapt pay scale for instructors/teachers. – Funding arrangements between the public and private sector and trainees and their sponsors agreed. – Courses/programs designed to meet needs of labor market and stakeholder employers.

(Table continues on the following page)

Table 2.2 Measuring the Progress of AET Reforms (continued)

Curriculum change (reform)

Intended outcome	Selected indicator	Needed action/step
– Curriculum reflects knowledge and skills that graduates need to meet labor-market expectations. – Strategy for implementing reform/change developed and implemented. – Functioning mechanism for continuous monitoring of sector changes in place. – Staff adopt appropriate pedagogy. – Mechanism in place for continuously monitoring the effectiveness of the curriculum.	– Curriculum change team appointed and terms of reference and work program defined. – New curriculum shared with key stakeholders. – Reformed curriculum approved by education entity and responsible ministry/ministries. – Feedback from employers positive. – Teaching staff use improved teaching materials. – Widespread use of ICT in teaching and learning. – Student satisfaction with curriculum.	– Identify leadership group to spearhead the reform process. – Explain curriculum change to key stakeholders. – Conduct needs assessment involving key stakeholders. – Define clear strategy and reform process steps and assign responsibilities for all steps in the process. – Identify curriculum gaps and prepare new curriculum to bridge them. – Plan and implement training program for teachers and instructors. – Field-test curriculum and amend as necessary. – Design, test, and produce materials to support curriculum. – Train teachers/instructors in using new curriculum. – Lessons from university-community interactions conveyed to curriculum change managers. – Design monitoring and evaluation plan. – Survey employers for satisfaction with graduate performance.

Training management and quality enhancement

Intended outcome	Selected indicator	Needed action/step
– Training replaced by learning. – Clear role defined for the management of agricultural sector human resources. – Capacity for and regulations governing HRM developed and in place. – Training program design and trainee selection improved and based on needs assessment and analysis. – Training/learning records maintained and up-to-date. – Merit-based hiring and promotion.	– Commitment to improved HRM agreed and announced. – HRM strategy in place. – Responsibility for HRM assigned. – Qualified persons appointed to manage HRM. – Budgetary support for HRM made available. – Training/facilitating staff skills updated regularly. – Training/learning materials updated and appropriate. – Higher staff morale and client satisfaction reflected in surveys. – HRM program results reviewed regularly at ministry level. – Circulate detailed annual training/learning reports.	– Managers and supervisors identify training needs. – Prepare academic and training achievement profiles of all staff. – Create HRM management capacity within units. – Conduct HRM orientation for manager-level staff. – Impart needs assessment skills to training/learning staff. – Hire designers and create HRM database. – Revisit job descriptions to indicate skill packages required. – Upgrade trainer/facilitator pedagogy. – Acquire appropriate ICT equipment and tools. – Develop and allocate budgets. – Update teaching/learning materials. – Prepare regular and detailed documentation on HRM activities. – Survey impact of training/learning programs on workplace performance.

Source: Author.

Note: ATVET = agricultural technical–vocational education and training; HRM = human resource management.

Reforming Public Agricultural Education at the Tertiary Level

Charles J. Maguire, Consultant

SYNOPSIS

Approaches for reforming higher agricultural education vary but may include comprehensive, systemwide reform, curriculum reform, or programmatic reform to increase the quality and number of degree-level graduates. Factors determining the scope of investments to reform higher education include the size and scope of the AET system. In large systems, decisions about whether the reform will include all institutions or pilot institutions are particularly important; reforms of smaller systems can be more focused. The level and type of investment are also influenced by the time horizon to which government, the university, and donors are committed. Classic examples of reform in higher education have spanned ten to sixty years, but shorter investments can yield positive results and sometimes catalyze more thoroughgoing reform. Stakeholders must agree on the need for reform, the impediments, and the steps to create an environment for successful reform within the AET system. Investment should not proceed until key stakeholders recognize that reforms are difficult to implement and understand that reforms will lead to improved university autonomy, funding, policy guidelines, and governance. Proposed reforms should either fit the existing rules governing administrative and structural change in public entities, or changes should be negotiated to ensure that essential reforms can be accommodated. Prerequisites for successful reform include support for change at the highest levels and leadership that is prepared to sustain reform over a considerable period.

BACKGROUND AND CONTEXT FOR INVESTMENT

Institutions that provide higher agricultural education are often aware that their graduates are not meeting the needs of the agricultural labor market. Perhaps graduates are not finding employment, or stakeholders report dissatisfaction with the quality of graduates. Yet policies are rarely in place to help universities identify and correct structural and programmatic weaknesses before they become serious. A strategy to correct those weaknesses would help universities organize themselves to develop and deliver programs that build the technical capacity and other skills required to foster effective innovation, interaction, and communication with others in the AIS.

Reforms that bring about meaningful, lasting change in higher education are hard to implement. They demand leadership, time, financial resources, and persistence. Reforms on this scale are not one-dimensional. They must often encompass policy and strategy formulation, stakeholder cooperation and involvement, governance changes, increased university autonomy, curriculum updating, pedagogical capacity building, increased practical work by students, and stronger links to rural communities. The weaknesses of agricultural education at the tertiary level in most developing countries are sufficient in themselves to make a convincing case for reform, but amid the challenges of persistent rural poverty, globalization of markets, climate change, and continuing demand to feed a large population from a stressed natural resource base, the case for reform is more urgent than ever.

Enabling graduates to operate effectively in the innovation system can lead to increased agricultural productivity, better livelihoods, and less poverty. These outcomes are more assured when agricultural education plays an effective role in the innovation system for agriculture. Although the particular approach to reform may vary, the priorities for reform are similar in many cases. The approaches to reform and elements of investment discussed in the remainder of this note reflect these priorities:

- *Develop or update policies* that describe and guide higher agricultural education in the broader educational system and in the agricultural sector. The policies that guide AET must be aligned with agreed recommendations for reform.

■ *Pursue an agenda of agreed, specific reforms within agricultural universities, faculties of agriculture, and agricultural colleges* to revitalize management and governance, increase autonomy, improve teaching methods, update curricula, ensure financing, strengthen relationships with clients, encourage the use of ICTs to enrich learning, and focus on stakeholders' needs as well as external influences, such as climate change and global trade in agricultural commodities.

■ *Institutionalize reforms* to ensure that reforms are agreed by decision makers and university administrators, clearly documented, approved by the university governing body, incorporated in guidelines and policy instruments, and available to all stakeholder groups.

■ *Attain accreditation* of reformed universities and their programs to give them, their staff, and their degree programs national, regional, and international credibility.

■ *Base curriculum reform on consultation with stakeholders* to ensure that the resulting degree and diploma programs are relevant.

■ *Link curriculum reform closely to pedagogical reform,* especially to adopting a student-centered, practical approach to learning.

INVESTMENT NEEDED

The heterogeneity of AET systems across the world suggests that the kind of investments required will be similarly diverse. Reforms will need to be tailored to the prevailing situation, including the needs of specific groups of stakeholders (farmers, processors, marketers, and consumers of agricultural commodities, for example). Innovations in the management, relevance, and quality of higher agricultural education will depend on factors such as the quality of leadership at a given university, the incentives to undertake reform, the university's capacity to accept that gaps may exist between current programs and standards and stakeholders' expectations and needs, and the level of support from decision makers who finance and guide education and agricultural development and management. Reforms may encompass single or multiple universities; they may be long or short term; and they may be funded by government and/or external donors.

The impetus for reform in AET can come from different sources, and organizations can pursue many different paths to reform. Regardless of the source of pressure for reform or the path (direct, indirect) taken, all efforts to foster change require consensus among stakeholders, support for change at the highest levels, and leadership to sustain change over time.

It is critical to remember that investments to reform agricultural education, especially at the higher levels, are also long-term investments in a more robust AIS and in greater productivity throughout the agricultural sector. With this ultimate goal in mind, this section introduces alternative approaches to reforming and modernizing higher agricultural education, lists the main investment elements in each, and presents examples of good practice.

Comprehensive ("big bang") reform

The most direct path to reform, often catalyzed by external pressure, leads the agricultural university or faculty of agriculture to compare stakeholders' expectations with program offerings and use the gaps between the two to create an agenda for change. The change agenda needs buy-in from university management and, when implemented, needs to be institutionalized through policy, regulations, and formal recognition by the ministries in charge of education and/or agriculture. This "big bang" reform requires leadership and prior agreement among a number of actors, including university management, faculty, and staff; decision makers at the resource allocation and policy levels; stakeholders who employ graduates; and students and their families. The main elements of investments in this type of reform include:

Facilitating dialogue between the main stakeholders to agree on the need for and scope of the intended reforms and to assign responsibility for each reform step and activity.

Undertaking a needs assessment (skills gap assessment) that reflects differences between the capacity of present graduates and the expectations of those that hire them.

Analyzing the outcome of the needs assessment and designing a change agenda.

Formulating a strategy for implementing the change agenda and supplying the facilitation capacity to move the strategy forward.

Convening stakeholder meetings to apprise all involved with progress and resolve difficulties in implementing the change agenda.

Finalizing the list of change-promoting activities and clearly specifying their implications for governance, organizations, personnel, and budgets.

Presenting the detailed change strategy to university management and to policy and decision makers at high levels in the government.

Undertaking preparatory activities to develop a refocused academic program, such as leadership strengthening, curriculum adjustment, improved pedagogy, administrative capacity enhancement, information technology upgrading, strengthened links to stakeholder/employers, and community outreach.

A concrete, stepwise example of a complex, long-term, national effort to reform higher agricultural education comes from China (box 2.6). In addition to China, a diverse group of countries has undertaken reforms in higher agricultural education. India implemented a large, externally funded project to reform its State Agricultural Universities

---

Box 2.6   Reforming Higher Agricultural Education in China, 1990–2000

By the late 1980s, as China gained momentum in moving toward a market economy, the higher education system for agriculture (created in the 1950s in the image of the Soviet system) recognized that its graduates would not have the skills to perform well in the changing labor market. Developed for a planned economy and operating under a centralized administrative system, China's institutions of higher agricultural education had little autonomy. They had no control over staff recruitment and finances. They struggled to accommodate to the changes brought about by the market economy.

Catalysts of reform. External and internal factors catalyzed the decision to reform China's higher agricultural education system. The three principal external factors were: (1) better-qualified, better-skilled graduates were needed to solve emerging technical and managerial problems in the agricultural sector; (2) government pressure to improve the efficiency of investments in education and reduce costs; and (3) lessons from reforms in other countries. Internal pressure for reform included: (1) demand from higher education institutions to gain greater authority to plan agricultural education, develop curricula, and manage personnel; (2) the need to rationalize the use of academic staff and bring about efficiencies in teaching and research; (3) the realization that teaching staff and institutions had to merge if they were to deliver graduates with the desired skills; (4) the need to use agricultural education infrastructure more efficiently; and (5) the need to reduce staff and personnel costs.

Steps in the reform process. Reform began toward the end of the 1980s with an analysis of internal and external problems and constraints that emerged from implementing reforms in the national economic system. This process—initiated mainly internally by the agricultural education institutions and the Ministry of Agriculture (MOA)—was informed by the outcomes of a conference and several meetings. Reforms were implemented in four main steps. The first was to formulate a reform strategy, concept, and guidelines to address the problems and

constraints identified and integrate them with national institutional reform. This step was taken jointly in 1993 by MOA and the Ministry of Education (MOE), with support from central government. The second step, undertaken from 1993 to 1995, was to initiate pilot reforms at some universities. These pilots, which focused on internal structural reform and merging institutions, tested the reform concept and contributed to an action plan for large-scale reform in higher agricultural education. In the third step, the agricultural education institutions, MOA, and MOE synthesized their experiences with the pilots. Based on the outcome of this learning exercise, MOA formulated an action plan for launching the national reform in higher agricultural education in 1996. The fourth step was to implement the reform throughout China from 1996 to 2000. That process was guided by MOA and MOE in cooperation with provincial governments. In some cases, the reform process continues.

Stakeholders and their contribution or involvement. Many institutions collaborated and cooperated in the reform. The MOA helped formulate the reform strategy and action plan. The MOE was involved in designing and implementing the curriculum reform and merging institutions. Central government participated in the earlier stages, and provincial governments participated in merging institutions. Employers supported curriculum and employment reform, and students and their parents pressured the higher agricultural education institutions to improve internal education management and bring about financial reform. Staff of agricultural education institutions participated in planning and implementing internal structural reforms, and education research institutions provided concepts and guidance with regard to the reform.

Changes emerging from the reform. The major changes emerging from the reform included:

- *decentralizing the administrative structure* to the provincial government level and giving *more*

(Box continues on the following page)

---

decision-making responsibility to higher agricultural education institutions;

- *changing curricula* and *pedagogical approaches; changing student enrollment; aligning employment patterns* more closely with labor-market expectations;
- *changing internal administrative structures* to enable better recruitment and motivate staff; and *reforming logistical systems* (leading to efficiency and financial savings).

Note that *no gender reforms were needed.* The institutions involved did not have gender discrimination in recruitment or promotion of professional staff.

Other results of reform. Institutions involved in higher agricultural education gained more autonomy in implementing and managing education and research and in balancing agricultural market needs with national education guidelines. A performance-linked staff recruitment and remuneration system was integrated into the management system for these institutions. Education efficiency improved significantly. Funding also improved, because student fees were raised and enrollment rose by 10–15 percent. Curricula and teaching methods are better, more flexible, and more student-friendly since the reforms took place. The establishment of a Rural Development section (faculty) has contributed to meeting the challenges of rural development.

New teaching methods introduced by the College of Rural Development set a good example for further reforms in teaching methods for other faculty.

Lessons learned. Some aspects of the reforms were very specific to the national context. For example, once reform was agreed upon, it proceeded according to plan. All key actors cooperated in meeting a major economic challenge that demanded better-educated and trained graduates from higher agricultural education institutions.

Even amid strong collaboration and support, the reforms took a decade to achieve their main objectives, and the process continues to evolve. The lesson is that if higher agricultural education is to be reformed in a comprehensive way, the investment in time and continued support will be substantial. In China, even with strong agreement over the reforms, the reforms did not always work smoothly, especially when institutions were merged and the rapid rise in enrollment placed high pressure on teachers and on space.

Reforms had winners and losers among educational institutions and staff as the entire system was made more cost-effective. Notably, reforms in student enrollment and placement systems did not seem to improve employment among graduates, who found it more difficult to gain employment owing to greater competition for jobs. Government funding is still less than required, especially for higher agricultural education institutions in poorer areas.

Source: Liu and Zhang 2004.

(IAP 1); the established, respected Wageningen Agricultural University in the Netherlands made major adjustments to management and curricula when faced with declining student numbers and imminent restructuring (IAP 2).

Curriculum reform

An often effective though less direct approach to institutional reform can begin with curriculum change across degree programs. TN 2 discusses curriculum reform in detail, but here the point is that while curriculum change is being implemented, often it exposes other organizational and academic problems and leads to further and deeper reforms. Egypt provides an example of this type of investment. Five Egyptian universities undertook radical curriculum and pedagogical change and forged strong links with

stakeholders (IAP 3). Curriculum reform can be undertaken directly as a project within the university or faculty of agriculture (see TN 2) or less directly by strengthening links with rural communities as in the cases of Indonesia (box 2.8), Thailand (IAP 5), and Africa (IAP 4). In the less direct approach, when university research and teaching staff and students become involved with farming families and communities, their understanding of agricultural and rural problems improves, giving rise to innovative solutions and to curricula that are relevant and that reflect the real circumstances of rural communities. Once the value of reforms in the curriculum or graduate-level courses is recognized and appreciated, a movement for deeper reform arises from within the organization.

Another approach to curriculum change is to repackage the traditional agricultural degree as a number of better-focused programs that attract student interest and support. Ireland offers a recent example of such an approach. In response to declining student numbers, Ireland's public universities changed their enrollment procedures during academic year 2000/01 so that new students could apply to nine agricultural degree programs (each with a specific focus) rather than to one generically defined "agriculture" program. This change greatly increased the number of students choosing agriculture, an increase that has been sustained (Phelan 2010; MacConnell 2010).

Curriculum enhancement to better prepare graduates for the demands of the modern agriculture sector is another method of supporting curriculum change. Ethiopia launched a Rural Capacity Building Project (2006) that focused on new, university-level courses that would respond to labor-market needs. The courses added would include civics, ethics, communications skills, and entrepreneurship.

Box 2.7 describes the main investment elements of direct and indirect curriculum reform. The elements of direct reform echo the experience in China (described in box 2.6) and India (IAP 1). The elements of indirect reform echo the experience in universities such as Chiang Mai University (IAP 5) and even (on a smaller scale) in postsecondary education (in Timor-Leste, IAP 8) and in-service training (in Ghana and beyond, IAP 4).

Reforms catalyzed by increasing the number and quality of graduates with higher degrees

Yet another indirect approach to reform is to invest in new or upgraded doctoral or master's degree programs that

---

Box 2.7   Main Elements of Investment in Direct and Indirect Curriculum Reform

The elements of investment in direct curriculum reform include:

- Facilitating dialogue between the main stakeholders to reach agreement on the need for and scope of the intended reforms and assign responsibility for each reform step and activity.
- Organizing and facilitating a series of meetings between university managers, academic staff, and representatives of key stakeholder groups to identify gaps between stakeholders' expectations and the quality of graduates from the academic entity.
- Undertaking a skill gap analysis to better understand what kind of curriculum change must be considered.
- Reviewing and updating the curriculum, using external expertise if needed.
- Packaging revised curricular materials to suit a variety of learning styles—for example, offer material in several media such as print, CD-ROMs, DVDs, video, learning management systems (Moodle is open-source software to create online learning sites) and formats (textbooks, teachers' aids, case studies, and so forth).
- Pedagogical upgrading for teaching/facilitating staff.
- Designing a monitoring and evaluation (M&E) instrument and implementing M&E.

- Adjusting or revising curricula, based on the results of M&E and user feedback.

The main investment elements in indirect curriculum reform through community outreach include:

- Facilitating dialogue between university management, faculty, and community leaders to agree on the need for university-community cooperation and on the operational approach.
- Agreeing on the details of the university-community program, including its goals and the responsibilities of both sides.
- Arranging for university staff and students to visit and reside in communities.
- Adjusting curricula to incorporate community involvement in the academic program.
- Facilitating staff and student experiences and observations in communities and translating these activities into processes for developing technical and social solutions.
- Organizing meetings with communities to share results of university-community interaction.
- Accommodating feedback from university-community experiences in an adjusted curriculum.
- Funding the logistical arrangements for student and faculty involvement in community links.

Source: Author.

---

strive for excellence. These programs will produce human resources with a better capacity to generate knowledge and greater leadership ability (among other skills). Graduates will staff agricultural universities and faculties of agriculture and, through their work, catalyze reform. The key investment elements in achieving curriculum and program reform indirectly, by improving the quality and quantity of higher agricultural degrees, include:

- Identifying academic institutions that could become centers of excellence or hubs for obtaining higher degrees.
- Facilitating dialogue between the centers of excellence and universities that could potentially supply them with degree candidates. Those universities may be unable to afford the staff or infrastructure to support the volume or quality of higher degrees needed.
- Upgrading facilities at the centers of excellence.

- Supporting academic staff capacity with training and pedagogical skills and tools.
- Underwriting some or all of the cost (fees, travel, subsistence) of students accepted from smaller universities to attend the center of excellence.
- Subsidizing research costs in candidates' home countries while they fulfill their degree requirements. This item would include students' costs in undertaking research and supervisors' costs in undertaking visits to the research sites.

ANTICIPATING FUTURE KNOWLEDGE AND SKILL NEEDS FOR THE SECTOR. Investment in brainstorming about future knowledge and skill needs for the changing agricultural sector can lead to better education planning and resource allocation (box 2.8 presents an example from Indonesia). In the United States, the Association of Public and Land-Grant Universities launched the Kellogg Commission on the Future of State and Land-Grant Universities in 1999,[1] and

---

Box 2.8 The Pursuit of Relevance Spurs Reform in Bogor Agricultural University, Indonesia

Bogor Agricultural University—Institut Pertanian Bogor (IPB)—was a pioneer in higher agricultural education in Indonesia. It developed the first four-year undergraduate degree program in 1974 and Indonesia's first graduate school of agriculture in 1975, actively recruited students from high schools throughout the country, introduced community extension programs, and established a regional planning board. More recently, IPB envisioned additional changes in management, organization, academic programming, community participation, and international outreach to continue developing and remain relevant.

These changes were reflected in the university's decisions to: reengineer the academic and administrative functions in education, research, and community services; develop human resources; undertake activities to generate revenue; revitalize the financial and funding systems, infrastructure, and infrastructure management; and strengthen the management of information systems and IT facilities. Improved governance was a prerequisite for these changes, including the creation of a board of trustees and an academic senate, the use of auditors, the development of university organizations, and portfolio analyses by all university units.

Source: Wirakartakusumah 2007; www.IPB.org.

The curriculum was transformed by introducing more programs that involved additional stakeholders, most importantly the community. To reinforce the involvement with stakeholders, IPB emphasized the formation of a network and good working relationship among the various stakeholders in the Bogor area. Aside from IPB, these stakeholders consist of research institutions, the private sector, NGOs, farmers, and rural communities. Examples of collaboration that are already in place include the Integrated Pest Management and Biological Control project in northern West Java and southern Sumatra, a shrimp restocking project in Sukabumi, an animal husbandry project, and reforestation and community participation projects. An entrepreneurial spirit was developed by involving students, communities, and the private sector in joint projects (such as a fish-processing project currently underway), by refocusing research at university centers, encouraging the development and submission of patents by the university community, encouraging agribusiness incubator programs, renewing the focus on community outreach and rural mediation programs, emphasizing programs featuring student-centered learning, and developing effective international networks and linkages.

the Secondary Vocational Agricultural Education Program undertook a three-year (1996–99) program, Reinventing Agricultural Education for the Year 2020, with funding from the W.K. Kellogg Foundation (National Council for Agricultural Education 2000). See also module 7, TN 3 and IAPs 4 and 5 for foresighting and visioning.

Designing new agricultural education programs

Occasionally opportunities arise to create a new educational institution, giving designers the freedom to develop a curriculum and adopt pedagogical approaches that best suit the new institution's mission and goals. New, privately funded universities were established in Costa Rica (IAP 6) and Honduras (in Zamorano, to produce graduates with the technical and entrepreneurial capacity to work as self-directed entrepreneurs or as employees in agribusiness, the public sector, or NGOs).

POTENTIAL BENEFITS OF INVESTMENTS TO REFORM HIGHER AGRICULTURAL EDUCATION

Reform that is supported by the academic entity and taken to a satisfactory conclusion can have a number of important impacts on the effectiveness of the innovation system. As mentioned, higher agricultural education can contribute more effectively to agricultural and rural development by producing human resources who can solve problems related to technical agriculture, social issues, and external factors such as the changing climate and globalizing markets. The AIS functions better, because graduates are better at listening, analyzing technical and social situations, proposing solutions, and interacting well with all actors in the AIS. Investments in higher education foster greater scope in agricultural research systems to identify and internalize community issues and problems and make them part of the research agenda. Finally, investments in higher education can endow extensionists with the technical and soft skills to build a better two-way bridge between researchers and their rural clients, because extensionists become more proficient in identifying, analyzing, and communicating issues and technical responses.

POLICY ISSUES

The policy issues that apply to AET in general (see the module overview) are relevant for reforming public agricultural education at the tertiary level. Such policies would clarify the role of higher education in agricultural development, give educational institutions the autonomy to provide high-quality education and cooperate with stakeholders, and guarantee the financial resources to underpin comprehensive reforms.

LESSONS LEARNED

An important general lesson is that the reform of tertiary agricultural education is a complex undertaking, involving numerous actors with varied interests in the outcomes and requiring considerable commitment on the part of reformers, the university or other institutions involved, stakeholders, and decision makers. In many respects, reform needs to be continuous if higher agricultural education is to respond to needs that will always be changing. The following sections discuss the conditions needed for reform to succeed, factors that influence the level of investment in reform, and more specific lessons from reforms undertaken over the years.

Conditions for reform

Investment in AET reform is appropriate when the reform/change, regardless of its source, is supported from within the system. Even then, high-level decision makers, university administrators, and other key stakeholders must clearly understand that reforming higher agricultural education is a long-term process that must be seen to its conclusion. Experience indicates that the length of the process depends on where the reforms are undertaken, but periods of 10 to 20 years are not unusual. In fact, if the reform is truly successful, the change process continues as the higher education system monitors changing needs for knowledge and skills and encourages organizational responses on an ongoing basis. In other words, a short-term project approach to reforming major educational entities or an entire system is unlikely to have a lasting impact.

Another important condition for reform (recall the example from China) is that the ministries in charge of education and agriculture must agree on the responsibilities of each ministry for the reformed entity and must be willing to modify those responsibilities if the assured future of the university is at stake.

Decision makers and the higher education entity must also understand that change will almost certainly include greater autonomy for the university. The lack of autonomy is a key generic weakness of higher agricultural education. It curtails institutional agility in responding to developments in the agricultural sector and stakeholders' concerns. The lack of autonomy prevents university administrators from managing a range of human resource issues, from the hiring of staff with new skills to the termination of nonperforming personnel, and the lack of autonomy prohibits financial decisions from being made at the entity level.

Stakeholders must be involved in formulating as well as implementing the reform program. In this process, the needs of women and minority groups must be taken into account. Equity and gender balance concerns should be debated in formulating the reform investment program and set of achievable indicators.

Deciding on the level of investment

The level of investment is influenced by the size and scope of the AET system. In some countries, a network of universities and colleges offers higher agricultural education (China, Brazil, India, Indonesia, and South Africa are examples). In such large systems, decisions about the scope of reform are particularly important: Will the investment cover all institutions in the AET system or only selected institutions to begin with? Other countries typically have only one or two higher education institutions in their AET systems, so reform initiative can be more focused.

Another important decision that influences the level and type of investment is the time horizon to which government, the university, and donors are committed. Although educational institutions are conservative and slow to change, slowness can provide a sense of continuity. Paarlberg (1992) has observed that "while the pace of change in our institutions and rhetoric has been too slow, some lag does permit accommodation without inducing chaos." Reforms take time and require adjustments as the reform process unfolds.

The classic examples of reform in higher education have spanned periods from ten to sixty years, but shorter investments can yield positive results and sometimes catalyze more thoroughgoing reform. Stakeholders' roles in the process also have an important influence on the type and scope of reforms. External stakeholders can often articulate demands to policy makers more freely than those who work in the AET system, and their influence can convince decision makers to support organizational and curriculum change. Another factor shaping the investment is the ability of the AET system and its supporters to persuade the ministry of education that AET merits an adequate share of education funding to support the outcomes of reform. Because AET is a small element in the larger national education picture, a very convincing case must be made.

RECOMMENDATIONS FOR PRACTITIONERS

Donors, in addition to committing to the reform process, should be prepared to create a suitable setting in which dialogue between stakeholders can bring about a common understanding of the need for reform, the impediments to reform, and the steps that must be taken to create an environment within the university or university system where reform can succeed. The amount of dialogue and interaction that should occur among stakeholders before reforms can be initiated must not be underestimated. Sufficient time must be allocated to enable that phase of interaction.

Once decision makers—ministers of education and/or agriculture—and managers in the university/faculty of agriculture agree on the approach to reform, the details should be developed by the university or faculty of agriculture or a joint working group comprising high-level ministry personnel and a representative reform team from the university. When systemwide reform is the objective, considerable investment must be made in ensuring that all key actors in the reform process participate in the conceptualization and design of the reform. (See also module 6, TN 2.)

Investment should not proceed until all key stakeholders recognize that reforms are difficult to implement and can be affected by political, economic, and social factors. Nor should investment proceed until it is evident that well-articulated reforms are supported by decision makers who are convinced that change is needed and will persist until reforms are institutionalized.

Investments in major reforms should proceed based on the understanding that reforms will lead, among other outcomes, to improved university autonomy, funding, policy guidelines, and governance (with clear lines of responsibility for university oversight and stakeholder involvement). Reforms should also lead to greater staff support for meeting standards for higher agricultural education and better faculty and staff remuneration and incentives. See modules 6 and 7 for a broader discussion.

Leadership is vital to remain focused and maintain momentum. Reform leaders need to be identified and exposed to examples of effective reform programs in other entities of higher agricultural education so that they may provide the leadership and articulate the vision for the process to succeed.

> **Box 2.9   Key Steps in the Reform of Higher Agricultural Education**
>
> The key steps in the reform of higher education include:
>
> - Decision makers agree on the reform strategy and its content, which is publicized to university/faculty administrators, staff, and other key stakeholders.
> - A timetable for the reform is drawn up. It identifies tasks and responsibilities in detail for all stakeholder groups.
> - Academic programs are analyzed with the help of facilitators.
> - Consultations with key stakeholders are conducted.
> - Gaps are identified between current program content and stakeholders' needs and expectations.
> - A change agenda is drawn up to bridge the gaps.
>
> - A timeframe for implementing the change agenda, the human and other resources requirements, and cost estimates are produced.
> - The reform proposal is presented to decision makers for approval and funding.
> - The change agenda is implemented. In some instances, the strategy is piloted in a limited number of locations in the AET system. Later, reforms are scaled up across the system.
> - As reforms proceed, the process is adjusted, based on lessons learned.
> - Decision makers formulate appropriate policies to support the reforms.
> - Reforms are institutionalized within the higher educational institution.
>
> *Source:* Author; Liu and Zhang 2004.

The capacity of organizational units to implement tasks in the reform process must be assessed (See also module 7, TN 2 in particular) before it starts. Investors need to understand the rules that govern administrative and structural change in public entities. Proposed reforms in the AET system should either fit the existing rules, or changes should be negotiated to ensure that essential reforms can be accommodated. This critical issue is discussed in detail in TN 4. Once stakeholders agree on the scope of the reform and understand the commitments needed, a series of logical steps must be taken to develop and implement a practical reform agenda (box 2.9).

Curriculum Change in Higher Agricultural Education

Charles J. Maguire, Consultant

SYNOPSIS

Agricultural curricula require frequent updating to remain relevant to the agricultural sector, its many stakeholders, and the AIS. Updated curricula must reflect technological and social change, consumers' preferences and concerns, external phenomena (climate change, globalization), and governance issues related to natural resources. Modern agricultural curricula feature a learning approach centered on students, use a range of technology for instruction, and emphasize practical experience gained at university laboratories, farms, other facilities or external sites. Curriculum change can be undertaken directly through the academic institution, key stakeholders, and faculty or indirectly through student and faculty links to rural communities and other stakeholders. Key recommendations are to take the time needed to communicate with all key stakeholders, base changes on a thorough analysis of needs as perceived by all stakeholders, institutionalize changes to the curriculum and in the roles of stakeholders within and outside the university, and press for adequate, sustained financial resources. If new skills (beyond the technical and production agriculture skills traditionally taught in AET systems) are identified in the needs analysis, the question of how and where they will be taught must be answered. Establish a mechanism for regularly monitoring and evaluating the impact of curriculum change; this information is important for decision makers who make policies supporting higher agricultural education. Ensure that curriculum change is sustained through organizational mechanisms that involve all key stakeholders on a regular, ongoing basis. Encourage links with universities that have completed or are in the process of implementing curriculum change. Finally, changing the content of the curriculum may be the easiest part of investing in curriculum change. Additional investments will be required in organizational arrangements, staff and faculty capacity, and monitoring and evaluation. This expanded agenda, its cost implications, and the time needed to complete the change process, must be considered when making investments in curriculum change.

BACKGROUND AND CONTEXT FOR INVESTMENT

For the agricultural sector, the AIS, and the rural space more widely, agricultural curricula require frequent updating to reflect technological and social change as well as consumers' preferences and concerns, external phenomena such as climate change and globalization, longstanding agricultural and social issues affecting the rural poor, and governance issues related to natural resources, including legal rights to those resources (Villarreal 2002). Aside from technical knowledge, awareness is growing that students also require skills in listening, analyzing problems, seeking innovative responses, and communicating if they are to work effectively in any setting. Although specific reforms will be driven by clearly defined needs, curricula that fit the human resource requirements of modern agriculture have been described (Ruffio and Barloy 1995; Wallace 1997; McCalla 1998).[1]

Universities typically are alerted to the need to revise, update, and reform curricula by concerned stakeholders, international exposure of staff to other university curricula, and the internal realization that students are not well prepared for the world of work. Curriculum change, once the need is recognized internally, is perceived as less threatening than major organizational reform and can usually be completed in a shorter time. The danger of opting for curriculum reform, however, is that the institution may be avoiding true organizational reform (Hansen 1990). Given that a relevant curriculum must reflect the labor market's needs for graduates, however, sooner or later the university may be forced to confront and correct the governance, administrative, and academic weaknesses that prevent it from meeting the needs of stakeholders in the agricultural sector.

INVESTMENT NEEDED

The specific investments that make agricultural curricula more relevant can be divided into those that catalyze curriculum reform and those that support the individual activities through which a curriculum is revised and integrated into academic programs. Although external actors—whether they are government, donor, or other stakeholders—can catalyze the decision to revise the curriculum, the AET institution itself must assume the role of champion. Internal leadership that mobilizes faculty, staff, and key stakeholders in favor of curriculum change is critical to success.

Investments for catalyzing reform include:

- Support for stakeholder interactions that examine the relevance of the present curriculum to labor-market needs.
- Resources to ensure that a strategy for undertaking curriculum change is developed.
- Investments that support the curriculum change strategy may include:
- Support for staff to design, develop, and produce new content for the curriculum and the corresponding learning methods and materials (IAP 3 provides an example from the Arab Republic of Egypt).
- Training for teachers, instructors, and facilitators to ensure that they are comfortable with the new material and have the confidence to guide students in a learning rather than a teaching mode.
- Support for teaching and facilitating staff to learn to use ICT effectively in teaching, learning, and more effective communication in the wider AIS.
- Training to ensure that practical sessions are planned and implemented effectively (see the discussion that follows). Staff and students who are unaccustomed to practical skills programs will require orientation. The programs themselves will require added inputs to ensure that they meet high standards and are delivered well.
- Faculty and staff may need to be introduced to the successful use of external resource persons for enriching academic programs (see the discussion below).
- Support for university-community interaction to foster curriculum change indirectly (see the section on "Potential Benefits").

A critical area for support is to ensure that the new curriculum reflects and improves links to stakeholders in the agricultural sector. Such links have been weak in AET in developing countries. Curriculum change places new pressures on administrators and faculty, and investments in capacity building are needed for these groups to cope with new and more intensive demands from stakeholders and students. Additional investment will be required to establish a mechanism that ensures continuous contact with stakeholders and reviews and analyzes feedback to adjust the curriculum.

One of the most important opportunities that arise during curriculum reform is the opportunity to use stakeholders as external learning resources. The wealth of knowledge, skills, and goodwill for agricultural education among most stakeholder groups can be captured and presented to students in the classroom or at practical work sites. A significant means of improving links between the university and external learning resources in the agricultural sector is to develop practical skills programs in which students are attached to stakeholders. To initiate and manage these programs, support is needed to visit field sites where such schemes have been implemented successfully and, based on these visits, to design a local attachment program. Next, the program concept has to be shared with potential participants—such as farm owners and agribusinesses (see also module 5 and module 3, TN 2)—to further define and establish the program. Once agreement to initiate the program is reached, it is vital to clearly state the terms under which it will be implemented, including statements of the responsibilities of trainees, hosts, and the university or educational institution. The issue of cost sharing must be part of the attachment program design and must be agreed in negotiation with attachment site hosts.

As a final note, it is important to recognize that all curriculum change initiatives will lead to the realization that some aspects of implementation require investments beyond amending or updating content and packaging.

POTENTIAL BENEFITS

Curriculum change can be undertaken directly, with the academic institution and the involvement of key stakeholders and faculty, or indirectly through student and faculty links to rural communities, which channel real issues in rural livelihoods back to the university, where they influence the research program and curriculum. Either way, curriculum change can have a number of impacts.

Decision makers as well as prospective students and employers come to see higher education institutions as a prime source for human resources for the agricultural sector. In Costa Rica, where the EARTH University curriculum was designed to meet a range of stakeholders' needs, the evidence that considerable value is placed on the university is reflected in its admission applications; requests for places outnumber slots by 40 to 1 (IAP 6).

Graduates of the institution are well prepared to meet employers' needs (the labor market) and the challenges of the agricultural sector and rural development. Graduates from EARTH University and Zamorano University (the Pan-American agricultural school in Honduras) are in high demand because their mix of skills prepares them for a wide variety of employment and entrepreneurship.

Graduates are confident and involved in the AIS as researchers, service providers, observers, listeners, communicators, and contributors of innovative ideas and solutions to agricultural and rural development problems.

The higher educational institution is aware of researchable issues and problems through its interactions with stakeholders. Faculty and students from Bogor University in Indonesia (box 2.8, TN 1) and Chiang Mai in Thailand (IAP

5) work with rural communities to better understand real-life challenges and identify researchable topics.

The morale of faculty and staff is boosted by interaction with stakeholders and new approaches to teaching and learning (the Egypt Curriculum Change project described in IAP 3 reflects such an interaction).

The institution gains a heightened awareness of the importance of organizational reforms beyond curriculum change, perhaps including reforms to improve governance, increase autonomy, and improve remuneration and benefits for faculty, staff, and administrators.

The higher educational institution attracts additional resources because it prepares human resources effectively for the agricultural sector. For example, through the Sasakawa Africa Fund for Extension Education (SAFE), students in Africa undertake a supervised enterprise project in communities that brings in support from the private sector (IAP 4). The Regional Universities Forum for Capacity Building in Agriculture (RUFORUM) includes a community-based master's research program that has attracted considerable donor funding (box 2.10).

Box 2.10   Regional Universities Forum for Capacity Building in Agriculture: Fostering Capacity for Innovation and Adaptation among Students

The Regional Universities Forum for Capacity Building in Agriculture (RUFORUM) is a consortium of 25 agricultural universities and faculties in Eastern, Central, and Southern Africa. To date, its main activity is to offer a small grants program for MSc studies in agriculture, although the program is being expanded to include doctoral studies. Through its community action research program, RUFORUM also provides grants for strengthening the links between rural communities and member universities. RUFORUM's mission is to "foster innovativeness and adaptive capacity of universities engaged in agricultural and rural development to develop and sustain high quality in training, innovative and impact-oriented research, and collaboration."

RUFORUM was originally sponsored by the Rockefeller Foundation and covered 12 universities in Eastern and Southern Africa. It was established as a consortium under African ownership and management in 2004 and is registered as a nongovernmental organization in Uganda with a secretariat in Kampala. Each member university pays an annual fee and several international donors, including the Bill and Melinda Gates Foundation, have made grants to the organization. Fundraising will be key to the consortium's future, as donor grants will eventually end.

Under its competitive grant program for MSc students, RUFORUM awards around US$60,000 to enable two students to work under faculty supervision for two years. Relevance of the studies to African agriculture is a key criterion in making the awards. Each MSc thesis must focus on a topic of local importance for agricultural development.

RUFORUM provides MSc students with field-tested, problem-solving skills that many African universities are still unable to provide. During the first semester of year two of the study program, students are based in the field and their work is supervised in several visits by the faculty supervisor. After five years of experience, RUFORUM has succeeded in connecting graduate students directly with farmers, rural communities, and the reality of African agriculture.

Source: RUFORUM 2010.

POLICY ISSUES

Demand for curriculum change articulated by the AET system and supported by key stakeholders is an important precondition for gaining support from the ministries in charge of education and agriculture as well as from planning and finance decision makers. In turn, those stakeholders play a crucial role by clarifying the policies and channeling the resources that will support change.

Curriculum change has implications for the administrators, staff, students, and external stakeholders of higher educational institutions. These institutions, as bureaucracies, are often slow to agree to change. Change often proceeds slowly, caused by the perception that it is not approved at higher levels of government, that the old curriculum is good enough, that incentives to change are unattractive, and that, given time, the pressure to change will decrease. Policies are required to support curriculum change, to make it clear that change is a priority, that administrators and staff of educational institutions are required to implement the changed curriculum as part of their terms of reference, that stakeholders' input will continue to be integral to evaluating the impact of the new curriculum on graduates' performance, and that adequate funding will be made available to make the changed curriculum work.

Some of the most important policies encourage educational institutions to monitor the quality and relevance of their programs in relation to the evolving needs of agriculture and rural development. Such policies provide guidance on governance of the educational institution, especially the role of stakeholders in providing feedback on graduates' readiness for the modern agricultural workplace and in alerting the institution to changing requirements for knowledge and skills.

High-level decision makers outside higher educational institutions must be aware that they require a continuing flow of information to update or create policies that give the institutions the authority to act. Managers of educational institutions must establish systems that continuously and critically evaluate their programs (especially needs assessments that capture changes in the agricultural sector). This information, regularly provided to policy makers, should help to prevent educational policy from falling out of step with the needs of agricultural education institutions and their stakeholders throughout the agricultural sector. An essential policy, of course, is one that allocates funds to support change and ensures that such funding is sustained.

LESSONS LEARNED

Important lessons from previous attempts at curriculum change concern the impetus for change (who demands change, and who supports it); the steps that must be followed in the change process; and the possibility that curriculum change will reveal the need for further reform. Curriculum reform is not a one-off activity. To be meaningful, curriculum changes must continue to be updated as technological innovation develops new approaches to producing, storing, processing, and marketing agricultural products. Curriculum change must be institutionalized in AET systems and given the approval and recognition that ensures the sustainable, continuous evolution of the curriculum.

Impetus for change

As noted, the need for change may be articulated by employers who are dissatisfied with the graduates of agricultural degree programs, signaled by a serious drop in applicants to the degree program, or emerge from an internal awareness that the degree program is obsolete. Even if the pressure for change comes from one or more external groups of stakeholders, support for change must come from inside the higher educational institution as well.

Steps to bring about curriculum change

The previous section refers to some of the steps in curriculum change, but it is important to list them in their entirety. Based on experience with curriculum change in a variety of settings, these steps appear to work well, although processes will differ from one place or time to another:

- All key stakeholders participate in describing shortcomings in the current graduate knowledge and skill mix and/or needs not identified by educational institution.
- Draw up a clear "roadmap" for implementing curriculum change and make it available to all.
- Clearly spell out roles and responsibilities of the educational institution and other stakeholders in the change process.
- Describe incentives for participation in the change process (for example, capacity building for faculty and administrators or out-of-pocket expenses for stakeholders who have to travel to participate in meetings).
- Select milestones for reviewing progress in curriculum change. Such milestones could include: presentation of the analysis of the needs assessment; drafts of new

content for specific subject matter; proposed pedagogical changes to support the revised curriculum; infrastructure and teaching/learning tools required; and curriculum field test results and indicative amendments. Share these milestones with all stakeholders in workshop settings. Note and act upon stakeholders' comments.

- Submit the final draft of the proposed new curriculum for peer review by regional and international entities in higher agricultural education. Act upon the relevant comments.
- Present the new curriculum to policy makers and key stakeholder groups, highlighting the changes and seeking approval.
- Develop or purchase materials to support the implementation of the updated/new curriculum.
- Train faculty and staff to use the new curriculum, especially in the area of pedagogy, student-centered learning, field visits, and attachments to employer sites.

The "iceberg" phenomenon

Achieving curriculum change appears to be a relatively simple objective, but it can be like the tip of an iceberg: The immediate task (changing the curriculum) is the visible part of the iceberg, although in fact a number of less-visible, interconnected tasks must be undertaken for curriculum change to be effective. In addition to revising or developing a curriculum, it may also be necessary to improve capacity for pedagogy, update teaching materials, involve teachers in managing practical agricultural activities, and create and maintain active networks with stakeholder groups that include agribusiness, NGOs, and communities. This potentially expanded agenda, its cost implications, and the time needed to complete the change process, must be considered when making investments in curriculum change projects or programs.

RECOMMENDATIONS FOR PRACTITIONERS

The technical process of curriculum change needs to be preceded by dialogue with key stakeholders. It should conclude by putting mechanisms in place to monitor the impact of curriculum change and support its sustainability. Key recommendations include:

- Take the time needed to communicate with all key stakeholders. While this may seem to delay the commencement of curriculum change, it is critical that all parties understand why the change exercise is being undertaken, support the objectives of the changes to which they have agreed, and appreciate their roles in ensuring a successful outcome.
- Changes to curricula must be based on a thorough analysis of knowledge and skill needs as perceived by all stakeholders
- Press for institutionalizing changes to the curriculum, roles of university faculty as well as stakeholders from outside the university, and adequate and uninterrupted financial resources.
- Underline the need for a mechanism for regular monitoring and periodic evaluation of the impact of curriculum change. This information is important for decision makers, who will want to be convinced that there is a positive return from investments in curriculum change and who will make the policies that support higher agricultural education.
- Ensure that curriculum change is sustained through organizational mechanisms that involve all key stakeholders on a regular, ongoing basis.
- Encourage links between universities (in country, regionally, or internationally) that have completed or are in the process of implementing curriculum change.
- Modern and modernizing employers in the agricultural sector seek graduates with skill sets that go beyond technical and production agriculture. This requirement presents a major challenge to curriculum reform. If new skills are identified in the needs analysis, the question of how and where they will be taught must be answered.
- Be conscious of the fact that changing the content of the curriculum may be the easiest part of investing in curriculum change. Additional investments will be required in organizational arrangements, staff and faculty capacity, and monitoring and evaluation. The need for these investments may not be obvious, but failure to make them will jeopardize the success of the investment in curriculum change.

Education and Training for Technician Development

Charles J. Maguire, Consultant

SYNOPSIS

The modern agricultural sector demands that technical vocational colleges or institutes produce highly skilled personnel who can manage a variety of farms and production units, run processing enterprises, service market chains, manage and repair farm and processing machinery, monitor food quality and safety issues, and support AIS actors in assessing the suitability of a particular innovation for specific persons, communities, farming systems, and value chains. In addition to technical knowledge, technical personnel need critical behavioral skills such as teamwork, diligence, creativity, and entrepreneurship. Reforms in the governance and management of training colleges and institutes are needed for traditional ATVET and VET programs to deliver this array of skills in an equitable, gender-sensitive way. Demand for and content of vocational education and training must be clarified through dialogue with key sector actors (ministries of education and agriculture, employers of TVET/VET graduates, and staff and administrators of entities that train them for the public and private sectors). If the system produces graduates valued by stakeholders in the labor market, those stakeholders will help to sustain the system financially. The VET system must be underpinned by clear and supportive policy that defines the role of VET; guarantees public financial contributions to VET; clarifies roles and contributions of stakeholders who employ graduates; and spells out governance arrangements for the system. Based on documented needs and stakeholders' expectations, invest in the capacity of teachers and instructors; develop certified programs that meet stakeholders' needs (begin with a small number of good programs); and encourage cross-sector dialogue, because agricultural TVET/VET is often linked to other sectors' programs and development plans (the environmental sector is a good example). IAPs 7, 8, and 9 examine specific investments in TVET and lessons learned.

BACKGROUND AND CONTEXT FOR INVESTMENT

Technicians interact on a regular basis with other actors in the AIS and are a unique repository of knowledge and—more important—skills that enable farmers and others to decide whether to adopt innovations. Public and private entities in agriculture have always relied on educated and trained technicians to undertake a range of tasks and activities that support production agriculture and form links in the value chain that leads to the local or international consumer. This reliance has been heightened by growth in agribusiness and the adaptation of advanced technology for producing, processing, and distributing agricultural goods and services (see also module 5).

Despite the need for skilled technical personnel generated by modern, knowledge-based commercial agriculture, with its emphasis on value-added and marketing, the demand for appropriately educated and trained technicians exceeds supply, and agribusiness entrepreneurs have difficulty hiring suitably qualified people. The shortage of qualified technicians persists amid significant reductions in public employment opportunities for technicians.

The shortage of skilled technicians is not new (it was noted during the 1970 FAO, UNESCO, and ILO Conference on Higher Agricultural Education). This shortage has been exacerbated in some countries by filling technician-level jobs with graduates who do not necessarily possess technicians' practical skills and knowledge, to work directly with farmers (Muir-Leresche, pers. comm., 2010). A 2009 study (Blackie, Mutemba, and Ward 2009) showed that African university graduates typically lacked the hands-on skills and capabilities that farmers value most.

Government, the private sector, and civil society have a stake in ensuring that enough qualified, skilled agricultural technicians are available for the labor market. The preparation of technicians, traditionally sponsored largely by the public sector, can be expensive. It involves equipment, land, qualified instructors, and a low student-to-instructor ratio. While rapidly industrializing countries are investing heavily in technical education and training to meet high demand for engineering and IT workers, public investments in formal agricultural diploma education (usually classified as ATVET) or certificate training (described as VET) vary and are unpredictable. The unevenness of investment gives rise to fluctuating student enrollment, poor staffing, weak program content, and a variable supply and quality of graduates.

As mentioned, modern agriculture emphasizes greater productivity, value added activities, and agribusiness. Agricultural products increasingly are consumed by growing urban populations or, in the case of high-value crops, sold in international markets. The focus on intensive farming and sustainability has created management complexity, altered energy requirements compared to traditional agriculture, and led to unforeseen consequences for human and animal health. The modern agricultural sector is also quite knowledge intensive. It demands highly skilled technical personnel who can manage a variety of farms and production units, run processing enterprises, service market chains, manage and repair farm and processing machinery, and monitor food quality and safety issues (see also module 6, TN 5).

In addition, employers in many economies seek workers who possess behavioral skills such as teamwork, diligence, creativity, and entrepreneurship, which are essential to thrive in rapidly evolving, technologically driven, and globalized economies (Blom and Cheong 2010). For this reason, improvements only in workers' technical and vocational skills will not always meet employers' needs. Systems that build skills will also have to ensure that these added behavioral attributes are in place (Blom and Cheong 2010). Any realistic attempt to train technical personnel who can meet these various needs for 21st century agriculture must revitalize the knowledge and skills provided by traditional ATVET and VET programs (box 2.11) and reform the governance and management of training colleges and institutes.

INVESTMENT NEEDED

Investment in technical education and training for agriculture has, like investment in all AET, been low. Much recent World Bank funding for formal TVET is industry-related,

---

Box 2.11   Producing Technical Human Resources for the Agriculture Sector in Australia

In Australia, technicians for the agricultural sector are trained through public and private institutions ranging from comprehensive technical and further education institutes or colleges, specialist agricultural colleges, private companies, and not-for-profit organizations operating as registered training organizations. Australia's six states and two territories have legislative responsibility for education. The national government has become more active in vocational education over the past twenty years, developing a national system for vocational training, providing significant funding to deliver training, and organizing national curricula around the principles of competency-based training.

Each sector of the economy, including agriculture, is covered by a comprehensive suite of competencies known as a "training package." The content of these training packages is based on the needs of the economy and industry, which are communicated through Industry Skills Training Councils.

The council covering vocational training in the agricultural sector, Agrifood Skills Australia, was established in 2004 as one of 11 Industry Skills Councils to provide accurate industry intelligence on current and future skill needs and training requirements for the agrifood industry. Agrifood is a public company with an industry-led board of directors and industry advisory committees. Funding is provided under contract by the Department of Education, Employment, and Workplace Relations. In New South Wales (NSW), Tocal College is a Registered Training Organization within the NSW Department of Primary Industries (DPI). A trademarked brand name—PROfarm—is used for all short courses for farmers run by DPI technical specialists through Tocal College. Attendance at PROfarm courses from the third quarter of 2009 to the second quarter of 2010 was 4,309, and the projected income was 862,377 Australian dollars. Full-time courses at this college have a 90 percent completion rate.

Source: Agrifood Skills (http://www.agrifoodskills.net.au) and Cameron Archer (Principal, Tocal College, Paterson, NSW, Australia; see cameron.archer@industry.nsw.gov.au and www.tocal.com), personal communication, 2010.

and recent projects supported in China and India reflect this trend.[1] In both countries, projects were designed to meet growing deficits in the availability of skilled workers, particularly in engineering and IT. These investments reflect the economic importance of engineering, manufacturing, and IT in both economies.

Innovative aspects of more recent TVET projects outside agriculture include:[2]

- Greater industry involvement in defining training standards.
- Development of public-private partnerships that enable costs to be shared.
- Creation of attachments and apprenticeships (see module 4, TN 2; and module 5, TN 1).
- A change in focus from training inputs to training outcomes (that is, competency-based training).
- Encouraging life-long learning by offering modular curricula that enable learners to enter and exit and continue to upgrade skills.
- Making the shift from teacher-centered to learner-centered pedagogy.
- Seeking accreditation for programs.

In Africa, these innovative aspects can be seen in Ethiopia, where TVET is provided to the agricultural sector under the management of the Ministry of Agriculture and to other sectors through the Ministry of Education and the Regional Education Bureaus. A 2004 World Bank Post-Secondary Education Project included a small component for TVET innovation to expand and deepen system reforms spearheaded by the Ministry of Education. Agriculture was not included in the project, but the project's five pillars of TVET transformation would be a good fit for vocational training in agriculture: (1) decentralization of service design and delivery; (2) strengthening partnerships among stakeholders, especially between training providers and employers; (3) development and implementation of a trades testing and certification system; (4) cost sharing by beneficiaries; and (5) an orientation to market-based demand to shape training.

A 2006 World Bank project in Ethiopia, the Rural Capacity Building Project, included an agricultural TVET component. The Project Appraisal Document notes that "capacity-building of middle-level technical workers is an important factor in the drive to enhance productivity, stimulate economic competitiveness, and raise people out of poverty." Implementation of the TVET component is encountering some difficulties, in part due to the lack of cooperation between the Ministries of Agriculture and Education and between central and regional governments.

POTENTIAL BENEFITS

The presence of skilled agricultural technicians has the potential to strengthen all links in the agricultural value chain and lead to still other benefits:

- Greater productivity and efficiency in public and private entities employing technically educated and trained graduates of the AET system.
- A VET system that is networked with stakeholders in the public and private sectors and with civil society organizations (CSOs).
- Well-crafted and successfully implemented policies to guide VET for the agricultural sector.
- Financing of TVET/VET on a sound footing that includes public and private contributions and fees levied on trainees or their sponsors.
- Assured quality of TVET/VET through certification of courses and programs by internal and external accreditation bodies.
- TVET/VET pedagogy that reflects a student-centered approach to learning, coupled with attachments and work experience opportunities with stakeholder enterprises, farms, and CSOs.
- TVET/VET networked with other parts of the AET system.
- Well-qualified and high-performing graduates of TVET/VET enrich the AIS through their interactions with a variety of actors in the system.

Investments in the production and upgrading of technical sector specialists ensure that the continued modernization and growth of agriculture and its numerous areas of focus meet the needs of a public sector that requires technicians to disseminate technology and undertake regulatory functions; of a private sector that invests in agribusiness, input supplies, and domestic and international marketing; of farmers' associations and cooperatives that represent producers; and of consumer organizations that rely on food supplies that are fresh and healthful.

Skilled technicians with the potential to be self-employed entrepreneurs who, in turn, create rural employment and serve as role models for farmers and others with whom they interact in the AIS.

POLICY ISSUES

A major policy challenge is to arrive at an agreed formula for financing TVET/VET. Should the public sector fully underwrite such education and training, or should employers and students share the costs? Given that stakeholders' involvement in describing the real needs of the labor market has been minimal and that the standards set by VET entities have been poor, the private sector has had little incentive to share the cost. Incentives for cost sharing emerge only when programs and courses meet labor-market needs, standards of education and training are high, and education and training entities are flexible and responsive to change.

Aside from policies that improve TVET financing, policies must ensure fair and equitable recruitment of students/trainees from all segments of society and seek a gender balance that reflects societal structure and labor-market needs.

Policies are also needed to support more effective governance of TVET. Such policies would formalize stakeholders' contribution to the content and focus of curricula and programs. They would also clarify the roles of ministries and regional bureaus in a decentralized education system and end the administrative paralysis that stems from poorly defined roles.

Policy guidance is also needed to ensure that apprenticeship and attachment programs are well designed, that responsibilities of TVET/VET entities and hosts are clearly defined, and that evaluation of student/trainee performance is fair and open. Finally, policies must guide TVET/VET entities to seek accreditation or certification for their education and training programs and courses.

LESSONS LEARNED

Supply-driven VET for agriculture has not been very successful for a number of reasons, including: weak or absent links to employer stakeholders; the poor quality of graduates resulting from inappropriate curricula and the poor availability and quality of teachers and instructors; intermittent funding; poor governance of the VET system; and the related absence of good policies to guide VET. Courses are rarely certified and competencies rarely tested, with the result that only a low value is placed on the VET system and its products.

Another lesson is that divided ministerial responsibility is a primary source of poor support for and performance of VET entities (just as it is a fundamental problem for all public agricultural education). For example, the ATVET component of the rural capacity project in Ethiopia proved difficult to implement because of poor coordination and cooperation between the Ministries of Education and Agriculture at the national and regional levels.

Policy to remedy this and other weaknesses of agricultural vocational education cannot be formulated in a vacuum. A precondition for sound policies is the development of a clearly expressed and understood needs statement from the concerned stakeholders. Policymakers must be convinced that technician education and training, if performed to a high standard, will produce graduates who will contribute to the productivity of the agricultural sector. To clarify the demand for and content of vocational education and training, key sector actors need to be involved in a series of dialogues. The key actors include ministries of education and agriculture, employers of graduates from TVET/VET, and staff and administrators of education entities that offer programs and courses that educate and train technical personnel for the public and private sectors.

Clearly, the sustainability of a quality VET system is based on its utility to the labor market. If the system produces graduates that are needed and valued by stakeholders, resources will be made available by those stakeholders. The VET system must be underpinned by clear and supportive policy that defines the role of VET; guarantees public sector financial contributions to VET; clarifies the roles and contributions of stakeholders who employ graduates of the system; and spells out the structure of the governance arrangements for the VET system.

The diverse and fragmented nature of the agricultural sector robs it of the high-profile, high-energy features of modern high-technology industries that generate jobs, export earnings, and strongly pressure decision makers to support education and training for those industries. Evidence of this phenomenon can be seen quite clearly in the lopsided investment profile in technical education and training in recent times.

RECOMMENDATIONS FOR PRACTITIONERS

By and large, developing countries have neglected TVET for the agricultural sector, but the time is right for new instruments to support the human resource needs of modern agriculture. Successful investments in ATVET require innovative governance and technical approaches, but they have high potential for making positive contributions to sector productivity. Recommended steps and considerations include:

■ Initiate dialog with key stakeholders to clarify needs and expectations and share supply capacity and constraint information and data.

- Invest in dialogue with major stakeholders to arrive at a widely understood and supported view of the importance of TVET/VET.
- Involve stakeholders in high-level governance of ATVET.
- Create a shared and well-documented message for decision makers to get policy and material support for VET.
- Encourage decision makers to clarify responsibility for TVET. Is TVET solely a public responsibility, or is responsibility shared with the private sector?
- Undertake a collaborative assessment of labor-market needs and expectations from the TVET/VET system.
- Obtain agreement on funding for TVET, based on documented needs and stakeholders' expectations. For example, TVET could be funded through public-private partnerships that provide private support in cash or kind (equipment, facilities).
- Invest in the capacity of teachers and instructors; it is important for the viability of TVET.
- Begin with a small number of good courses or programs that meet stakeholders' most important human-resource needs.
- Form an active network with other TVET and VET systems and, when resources allow, undertake study visits.
- Ensure that TVET/VET is linked to AET.
- Encourage cross-sector dialogue, because agricultural TVET/VET is often linked to other sectors' programs and development plans.
- Seek certification of all TVET programs to ensure quality.
- Include environmental considerations in the preparation of technicians for agriculture, because the sector is recognized as a major contributor to pollution and environmental stress. Programs and courses for technicians should identify the issues of concern and underline the need for environmental sensitivity on the part of their graduates. The environmental focus of the curriculum would be examined in certification and accreditation exercises undertaken by internal and external bodies.

Reforming the Management of In-Service Training/Learning

Charles J. Maguire, Consultant

SYNOPSIS

In public agencies or projects, the benefits of in-service training include enhanced capability to contribute to growth in the agricultural sector and operate in the AIS; in private enterprises, benefits of in-service training include more productive workers and higher profits. Opportunities for improving human resource development (HRD) and human resource management (HRM) through training arise when countries develop national, agricultural, or rural development strategies or their poverty reduction strategies; when new projects are planned; or in discussions about adjustments or extensions of current projects. Similar opportunities arise when the private sector is creating demand for skilled human resources. The three main investments in HRD and HRM include building capacity in public agencies, developing specific capacities in development project personnel, and building capacity in private enterprises. Elements in improving in-service training in the public sector include a clear policy articulating responsibilities for HRD; training personnel to implement the HRD approach; linking with learning providers; and supporting specific, short-term programs as well as life-long learning programs. Investments in project-related training begin when projects are being developed and include (1) assessing capacity needs, (2) evaluating capacity-building activities in every component of project-related training, and (3) supporting training in communications and problem-solving skills in every component for effective participation in the AIS. Large private enterprises with a long-term need for personnel will build capacity in human resources through joint ventures with the public sector; smaller entities may be more comfortable purchasing training packages from the AET system. Major recommendations include: provide training for faculty and staff managing in-service training; include training/learning specialists in project design and implementation; keep capacity-building objectives realistic in projects; consider investing in higher degrees for qualified HRM candidates; and identify potential partnerships between public and private entities to add to in-service training capacity.

BACKGROUND AND RATIONALE FOR INVESTMENT

Investments to improve how capacity building is designed, implemented, and managed—especially in public agricultural organizations and development projects undertaken in the public sector—add knowledge and skills that make public organizations and projects more effective and enable the public sector to participate more effectively and actively in the AIS. People who interact with the rural population on behalf of public or private agencies and projects are working exactly where some of the most important possibilities for knowledge sharing, innovation dialogue, observation, assessment, and adoption occur. They are a critical vector in the innovation system. Investing in, adding to, and sharpening their skills should be a priority. Investments can also be considered in building human and institutional capacity in private agricultural enterprises, especially where there is scope for joint activities through public-private partnerships. It is also important to distinguish between building capacity to implement projects and broader institutional capacity-building. The former is undertaken to ensure that each individual has the knowledge and skills to perform project-related responsibilities effectively, whereas the latter is concerned with the capability of the organization (ministry, department, unit, private business) to implement its terms of reference. While individuals collectively contribute to the capacity of an organization, factors such as policy frameworks, organizational structures, links, management, financing, and interface with clients or stakeholders also influence this capacity.

INVESTMENT NEEDED

Three main types of investments can be considered to build capacity in the staff of public agricultural agencies and development projects and those in the private sector who provide a variety of agriculture-related services and supplies to farmers, processors, marketers, and civil society. The first type builds capacity in public agencies; the second builds private sector capacity; and the third develops specific capacities in development project personnel.

Managing and improving in-service training in ministries of agriculture

In-service training in ministries of agriculture is generally poorly managed. Managers exert weak oversight of training programs and impacts. Ordinarily, personnel appointed to plan and implement training/learning programs are not selected from a pool of professionally qualified or practicing human resources development specialists. Nor are ministry and other public sector staff systematically identified for specific training and learning experiences that prepare them for future responsibilities. Seniority in the civil service is often the criterion that decides who is selected and promoted in the system; competence is not always a deciding factor. Shrinking budgets in the public sector have left smaller cadres of professional research, technical, and extension staff to deal with additional responsibilities that require enhanced knowledge and skills. Under these circumstances, government agencies must make their capacity-building efforts a priority and ensure that they are as effective as possible. Key elements in bringing about change include:

- *Policies.* Following high-level management commitment to adopting change, a policy directive is issued describing responsibilities for human resource development (HRD).
- *Capacity to implement the policy.* Personnel to implement the new HRD approach are selected and enrolled in appropriate training programs.
- *Review of training/learning programs.* HRD and human resource management (HRM) rules, needs assessments, training/learning responses, selection criteria, and monitoring and evaluation of training/learning impacts are reviewed in detail and adjusted, based on the new approach to HRM.
- *Links to learning resources:* Establish links with learning providers such as the research system, universities, TVET entities, private sector leaders in agriculture and rural

development, and regional and international training agencies.
- *Build AIS-ready skills capacity.* Provide selective support for specific, short-term programs for in-service training using public funding or with public-private-partnership resources. Specific responses may include: study visits; short-term overseas training; attachments to research, community, or private enterprises; or acquiring urgent adaptation strategies for dealing with the impact of natural phenomena, such as insect or disease outbreaks or climate change.
- *Create capacity for life-long learning.* Support may be given for the development of seminars and short learning activities at universities. These options would be available to public sector decision makers, senior technical managers, private sector managerial and technical personnel, and CSOs. This investment category serves to increase knowledge and skills as well as to exchange experiences and ideas among influential actors in the AIS.

Investments in building private sector capacity

These investments include support for mutually beneficial joint ventures in which, for example, the public sector supplies in-service or life-long-learning opportunities for private workers and the private sector supplies skilled operatives as teachers, facilitators, demonstrators. The private sector may also offer practical training internship spaces for public employees or students from the AET system (IAPs 3, 8, and 9). See also module 5 on private sector development.

Investing in project management and implementation personnel

Almost all development projects for the agricultural sector seek to strengthen the capacity of the people who implement them.[1] A variety of training and learning activities are often funded for these staff through the project's larger capacity-building components. The range of capacity-building activities for project personnel and beneficiaries is wide. Although there are generic lists of activities that can meet basic needs, each situation and its particular needs will determine the shape, content, duration, and participants in the capacity-building intervention.

These activities may be labeled capacity-building components or training components and typically are funded at about two or three percent of project costs. Such components are project specific. They operate alongside in-service training programs provided by public agencies for agricultural sector staff. A project's capacity-building components are usually managed by a unit subsidiary to the project management unit. Not surprisingly, the quality of management varies from project to project.

As noted, investments in project-related training need to begin when projects are being developed. The team preparing the project should include expertise in capacity building at every stage. Managers from the beneficiary entity should be briefed on the importance of the project's capacity-building component and advised to appoint suitably qualified staff to manage the component. Managers from the beneficiary entity should provide a profile of each appointee (qualifications, experience, and training history). They should receive technical assistance to be able to undertake needs assessments, propose corresponding learning programs or activities, create (with the help of resource persons in the network) training/learning content, and plan and implement training/learning activities for carefully selected project staff. They should learn techniques to measure the impact of training and learning, to inform managers of the results, and to maintain clear, up-to-date records of all capacity-building activities.

Capacity-building specialists should be included in project preparation teams (in the case of World Bank projects, during the time from identification to appraisal) to ensure that training for project personnel is relevant, clearly defined, and can be implemented in time for the project to operate smoothly and achieve its objectives. Project funds will need to be allocated specifically to train staff assigned to project management to ensure that they understand the project's overall objectives and can manage its components. Resources will also be needed to (1) assess capacity needs, (2) evaluate capacity-building activities in every component of project-related training, and (3) provide support in each capacity-building component of the project for training in communications and problem-solving skills (for effective participation in the AIS).

FACTORS DETERMINING THE TYPE AND DURATION OF INVESTMENT IN ENHANCED IN-SERVICE TRAINING

The most powerful factor guiding in-service training investments (as in higher and technical education investments) is the realization among stakeholders and within the organization that change is needed. Investors should initiate a dialogue that highlights the benefits of changing the way that human resources are managed and details the steps that need to be taken. The short implementation span of projects and programs enables decision makers and project management to observe relatively quickly how a capacity-building component benefits implementation and sustainability. Success at the project level can convince management of the benefits of adopting a more comprehensive HRD approach at the ministry or department level. With respect to the duration of the investment, it is important to note that installing quality management of human resources in a ministry or department is a process dictated by organizational and civil service rules, the pace of learning, satisfaction with progress and results on the part of senior management, and on the commitment of donors to see the process through to the end. The source of demand for enhanced in-service training influences the most appropriate investment response. In the case of short-term investments in specific courses, study visits, or attachments, requests may originate from government or from project implementers through government channels. Responses from investors may be made bilaterally or through existing project channels. The duration of such investments can range from weeks to one year. Support for the development and implementation of seminars or short learning activities at universities would most likely cover an agreed time period measured in years (or in numbers of events without time constraints). Underwriting the strengthening of advanced degree programs at regional or hub universities would fit in a medium- to long-term framework of support. Scholarship support for candidates would take into account the normal cycle for completing advanced degrees, so an initial commitment might be for ten years, dictated by successful completion of programs.

In the case of public-private partnerships, much depends on the type and size of the private enterprise. When the private enterprise is, for example, involved in production, processing, and marketing nationally and internationally, it will have a long-term need for personnel and will be keen to invest in building capacity in human resources. It can benefit from technology training from academic and research institutions and from offering internships to students that it can assess and possibly hire. Smaller private entities may be more comfortable purchasing training packages from the AET system and offering a small number of internship or attachment places for which the AET system would pay. A minimum commitment of five years would give both partners time to adjust the program in light of experience and to evaluate the impact of the partnership.

POTENTIAL BENEFITS

Benefits of interventions to improve how HRD is managed in public agencies or projects can include: enhanced capability of a ministry, department, project management unit, and individuals to contribute to growth in the agricultural sector; heightened capability of individuals and service units to operate effectively in the AIS; greater clarity of purpose in capacity-building goals and activities; heightened morale of better-educated, better-trained, and more confident public sector staff; greater respect for public service providers among farming and rural communities; and more successfully implemented projects. Private enterprises can benefit from more productive workers and increased margins of profit.

IMPLEMENTATION ISSUES IN MINISTRIES AND DEPARTMENTS

In a practically oriented government ministry such as agriculture, it is often difficult to gain support for in-service training and learning, especially for reviewing and reforming the content and management of that training. The application of technology developed through agricultural research dominates the service activities that support growth in agriculture. Compared to these practical pursuits, capacity building and HRM can be perceived in the public sector as less important, "soft" areas that do not justify funding when other ministry tasks are so urgent. (In the private sector, capacity building and HRM are connected clearly to profit and loss and get much more attention from managers.) Yet capacity building in public entities is usually a small fraction of total project costs or ministry operations. Few truly qualified persons are assigned to capacity building. Even at higher ministry levels, this phenomenon holds true.

LESSONS FOR PRACTITIONERS

Opportunities for discussing the effectiveness of in-service training and project-related training can arise when countries are developing their national, agricultural, or rural development strategies or their poverty reduction strategies; when new projects are planned; or in discussions about adjustments or extensions of current projects. Similar opportunities arise when the agricultural private sector is creating demand for skilled human resources.

Several points are critical in taking action to improve in-service training and learning. Perhaps the most important point is to be certain that decision makers are committed to reforming in-service and project-related training and that civil service rules and regulations allow organizational change in HRM. Keep decision makers informed of design and implementation progress and problems. Consider investing in scholarships for qualified HRM candidates to earn higher degrees, and provide training for faculty and staff with responsibility for managing in-service training. Include training/learning specialists in project design and implementation. Given project timeframes and the capabilities of implementation staff, keep capacity-building objectives realistic. Be prepared to invest in short-term support for in-service training and capacity building. Identify potential partnerships between public and private entities to add to in-service training capacity, and discuss these partnerships at decision maker and AET management levels.

Reforming India's State Agricultural Universities

Charles J. Maguire, Consultant

SYNOPSIS

A major, systemwide reform of India's state agricultural universities (SAUs) aimed to modernize administration and management, update curricula and pedagogical approaches, upgrade teaching materials and laboratories, set new norms and standards for higher agricultural education, and improve human resource management in state line departments working closely with the agricultural universities. The process, initiated with four universities, was viewed as a ten-year effort but confined to a project of six years (1995–2001). The project improved the quality and relevance of India's SAUs by establishing an Accreditation Board, demand-oriented curriculum reforms, and complementary investments in staff training and educational infrastructure. The quality and diversity of student intake and the quality of faculty improved. In hindsight, however, it is clear that the process could have been introduced more slowly, with fewer objectives and a longer time frame to institutionalize the reforms. The capacity, readiness, and commitment of project actors could have been better assessed. The project might have done better to focus on higher education alone, reserving human resource management in line agencies for a separate project. Donor coordination should have been emphasized to prevent universities from attaining funds that made it possible to delay reforms.

CONTEXT

By working with a small number of universities, a project sought to create a compelling demonstration of the potential benefits of systemwide reform in India's state agricultural universities (SAUs). As state institutions, SAUs receive funding from their respective states but also from central government and other sources. At the national level, the agricultural university system is coordinated by the Indian Council of Agricultural Research (ICAR), which is also a source of research funding for the universities. A long-term goal of the reforms was to establish an Agricultural Education Council with statutory power to set norms and standards in agricultural education. For more on ICAR and India, see also module 1, TN 1; and module 4, IAP 2.

In the 1960s and 1970s, the SAU system was strengthened and expanded under a large USAID project (Busch 1988). In the 1990s, concerned about declining standards in the system, ICAR approached the World Bank about the need to reestablish SAUs as centers of high-quality agricultural education. The resulting project concentrated initially on four SAUs.

PROJECT OBJECTIVES AND DESCRIPTION

The five-year project was the first phase in a long-term program to improve agricultural HRD. Estimated project cost at appraisal was US$74.2 million (World Bank US$37.1; India US$37.1). The project had four components: (1) strengthening ICAR; (2) university programs; (3) in-service HRD and HRM; and (4) manpower needs assessment.

Through these components, the project initiated several broad reforms:

- It assisted India's drive to modernize its agricultural sector by promoting changes in the way the center and states developed and employed human resources.
- It supported these changes with policy and institutional reforms.
- It began related processes of (1) improving the quality and relevance of higher agricultural education and in-service training programs and (2) strengthening the capacity of participating states to develop and manage agricultural human resources.

To foster academic improvement, the project strengthened capability within ICAR to establish norms and standards in agricultural education and monitor compliance with these standards. While the proposed Agricultural Education Council was being established, ICAR's Education Division was strengthened, and the Norms and Accreditation Committee was restructured.

The project encompassed four subprojects from three participating states—one each from Haryana and Andhra Pradesh and two from Tamil Nadu—selected as good candidates for demonstrating the effects of reforms to other states. The subprojects sought to reform the curriculum and syllabus, improve faculty quality, revitalize teaching methods, organize faculty exchanges within India and with foreign universities, modernize university administration and management systems, upgrade infrastructure (teaching laboratory equipment, computer systems, communications, farms, libraries, and hostels), and establish placement centers and programs for student attachments to agroindustries. The project also promoted initiatives to involve university clientele more in university management and programs and improve education-related financial management.

To upgrade human resource management, at the state level the project supported:

■ *In-service HRD and HRM programs in 14 line departments that worked closely with SAUs.* This support included training focused on job-oriented needs; systematic training needs assessments; training of trainers; evaluation of training effectiveness; better instructional facilities; and improved management of state agricultural employees.

■ *Manpower needs assessment,* involving the establishment of broad-based Manpower Advisory Councils to sponsor rigorous studies of labor-market requirements and trends (that is, to begin developing labor-market intelligence) within each state. Data from the studies were expected to provide state authorities and university officials with technically sound information for crafting public policy, academic programs, budgets, and adjustments to university intake numbers.

INNOVATIVE ELEMENTS

In the context of India's SAU system at the time, the project delivered some innovative interventions.

It raised academic and administrative standards in the SAUs through updated curricula, improved pedagogy and teaching/learning materials, an emphasis on practical exposure for students, and accreditation of academic programs.

It upgraded the relevance and quality of in-service training, increased client involvement in identifying training needs, modernized training facilities, and enhanced staff knowledge and skills.

It emphasized the management of human resources in line departments to ensure that staff knowledge and skills were of the highest quality and that human resources were employed effectively.

It initiated the reform process with four universities and used the experience to scale up reforms to other universities in the SAU system.

IMPACT

By and large, the project achieved its development objectives, although the Project Completion Report (June 2002) identified some design flaws and less-than-satisfactory outcomes.

The quality and relevance of higher agricultural education was improved by establishing an Accreditation Board, demand-oriented curriculum reforms, and complementary investments in staff training and educational infrastructure.

A participatory system of institutional accreditation was developed, and ICAR was implementing it throughout the SAU system. Academic norms for all undergraduate and postgraduate programs were revised and implemented.

Education programs were more relevant. Curricula were updated for 11 undergraduate and 32 postgraduate programs. Courses were introduced in new areas such as biotechnology, computer applications, agribusiness management, and sustainable agriculture. Coursework was broadened to include skills-oriented, hands-on training programs developed through wide consultation with stakeholders.

These changes were reflected in new and improved teaching materials (laboratory manuals, course modules, textbooks, and so forth) and methods, along with substantial investments to train research and teaching faculty and upgrade classrooms, laboratories, libraries, and IT facilities. These efforts improved the quality and relevance of the education programs and the teaching/learning environment.

In-service training improved in quality and relevance through the establishment of needs-based training programs, greater client involvement (farmers, agroindustry, input suppliers, and others), modernized training facilities, and investments in staff training. Improved training programs and the adoption of more effective practices to disseminate agricultural technology appear to have improved extension performance.

The capacity of participating states to develop and manage agricultural human resources was enhanced by the creation of skills, institutional capacity, and infrastructure. These new resources enabled line departments to assess their HRD needs, formulate and implement human resource management plans, provide in-service training, and liaise with other relevant institutions.

Additional positive results

Policy changes improved the quality and diversity of student intake through a nationwide testing system and the introduction of national talent scholarships and research fellowships. A state-level Common Entrance Test was in place, and examinations were revitalized through a system of internal and external assessors. Aside from the training mentioned previously, a National Eligibility Test helped to improve the quality of faculty.

To enforce national norms and standards in agricultural education, ICAR set up a monitoring unit. A system of rewards and incentives was put in place to speed the adoption of norms and encourage self-improvement in educational standards in SAUs and among students. A manual for accreditation was issued. By the end of the project, 32 of 35 colleges in the project SAUs had been accredited by the newly established Accreditation Board. All project SAUs were accredited, and the process was continuing with other SAUs throughout the country.

University governance had improved with the establishment of broad-based Advisory Groups and an expansion of the University Board of Management to include representation from the private sector. Financial powers were delegated to deans and heads of departments, among other financial reforms. Students' records and evaluations of teachers were digitized.

Although Educational Technology Development Cells and the Student Counseling Centers were set up, by the time the project ended, their impact on graduate employment had not been evaluated.

Effects on institutional development

The new accreditation system enabled SAUs to analyze their strengths and weaknesses, develop strategic plans for academic excellence, and formulate action plans to improve the quality of agricultural education by involving all stakeholders. Administrative, financial, and governance changes, especially closer links with clients in the agricultural sector, are likely to continue and spread to other universities through positive demonstration effects. Line departments also experienced changes in their attitude to, and organization of, personnel and training matters.

Sustainability

When the project ended, changes in institutions and procedures, including managerial and administrative changes, were partly internalized, and the relevant stakeholders favored continuing the reform program (for example, by addressing governance reform and individual performance incentives). Staff from SAUs and line departments reported a greater sense of achievement and job satisfaction. Overall, the sustainability of the project was rated as "likely." The likelihood of a follow-up project provided incentives to continue project activities. Yet as the next section will show, these expectations were not fulfilled.

LESSONS LEARNED

The process could have been introduced more slowly, with fewer objectives. The capacity, readiness, and commitment of project actors could have been better assessed. The project might have done better to focus on higher education alone, reserving human resource management in line agencies for a separate project. Donor coordination should have been emphasized to prevent universities from attaining funds that made it possible to delay reforms.

Some lessons are particular to the project:

In hindsight, the project overestimated the capacity and readiness for reform in the government and implementing agencies. Many assurances obtained when the project was negotiated did not materialize within the agreed timeframe. Implementing agencies' capacity to identify priorities and needs and develop workable annual plans for procurement, financial management, and training were overestimated.

The project was too complex. It should not have attempted to initiate and manage change in institutions (universities and line departments) with widely differing organizational structures, cultures, and objectives.

The project did not give enough attention to monitoring and evaluation. It used an undifferentiated list of indicators that were not prioritized, quantifiable, or possible to substantiate. No benchmarks were in place for progress to be assessed more objectively. The project lacked a framework to link the project's various activities clearly with its objectives.

The manpower needs assessment was not satisfactory. Data were delivered late, were not used, and the substance and style of teaching did not change in any substantial way, even though trainers were using more instructional aids.

Other lessons arising from the project have broader application:

As emphasized in the module overview, the case for reform needs to come from within the system—in this instance, from the coordinating body for higher agricultural education. The weak impetus for reform underlies the other lessons presented here.

Reform takes time. Although this project recorded many successes, the difficult issues were not resolved. A second phase of the project, which would have institutionalized the reforms, was never funded.

The number of universities in the project was limited to four to achieve a convincing demonstration effect. This decision, in a system as large as India's, was perhaps good—but it remains an open question.

Human resource management—intended to make training more meaningful in the state line departments that worked closely with the SAUs—proved more difficult to manage than expected, because the universities and line departments were administratively separate. (As noted, it may have been preferable to focus on line agencies in a separate project.)

Unless a project can tap into or build durable constituencies for reform, the "champions for change" will turn their attention elsewhere, reforms in governance will not continue, and the initiatives started by the project will not be sustained. Creating or strengthening these constituencies requires delegation, decentralization, and empowerment of different functionaries as well as the institutionalization of key reforms, such as changes in university statutes that give a genuine voice to the private sector or "teeth" to student evaluations. Projects need to be designed in ways that will initiate and elicit such governance and institutional reforms, perhaps by phasing in project investments that are explicitly linked to realizing agreed elements of reform.

Although the project devoted attention to building up management and implementation capacity at various levels, in retrospect the ability of the implementing agencies to manage procurement, financial management, and selection and placement of staff for overseas training was overestimated. It is critical to make a thorough assessment of capacity and readiness for reform before designing the project in detail.

Finally, when supporting university reform in entities that focus on teaching, research, and extension, it is important for donor organizations to ensure that they do not offer those entities competing funds that would allow administrators or faculty to ignore the more difficult elements of the reform agenda and delay or derail the process.

Transforming Wageningen Agricultural University

Charles J. Maguire, Consultant

SYNOPSIS

A Ministry of Agriculture decision to make a major investment in the knowledge infrastructure for agriculture, strengthen research, and make research more accountable to its clients induced Wageningen Agricultural University to change its focus and work with a wider research network and with stakeholders. The change process began in 1997 when the university merged with the Ministry of Agriculture's research department to create the Wageningen University and Research Center. The vision for education at Wageningen University and Research Center is to create a content-inspired international learning environment based on (1) an orientation to competencies (knowledge, skills, and attitude), (2) professional and academic education, and (3) diverse, dynamic, and flexible learning tracks. To achieve this vision, the university has undertaken major administrative, staffing, and program changes, in the course of which it has merged with other education and research entities. A better balance was attained between discipline-oriented and integrated courses and between a focus on the development of knowledge and skills and on the competencies needed to use knowledge in society. Students were expected to understand the synergy between natural and social sciences to increase the societal relevance of the university's programs. As it continues to evolve, the new organization has attracted more students and funding, become more client oriented, and expanded its international cooperation network with academic institutions.

CONTEXT

The transformation of Wageningen Agricultural University into the Wageningen University and Research Center (www.wageningenuniversity.nl/uk/) reflected the Ministry of Agriculture's desire to regain its high profile in the economy and improve agriculture's image and political clout.

The agricultural sector's poor image was reflected in low student enrollment in agricultural education programs. Budget negotiations in 1994 suggested shifting agricultural education and research (one-third of the ministry's budget and staff) to the Ministry of Education. The response was to create a strong education and research center—the Wageningen University and Research Center—to strengthen the Ministry of Agriculture's position and give it a guiding/steering role. For more information on research networks in the Netherlands, see module 1, TN 1.

REFORM OBJECTIVES AND DESCRIPTION

The reform was a total reorganization that involved budget cutting, staff reductions, rationalization of course offerings, and a public relations campaign to inform future students and the public of new and revised academic programs. The university also reached out to partner with other education institutions, private sector clients, and the European and world network of academic agricultural education and research institutions.

More specifically, 24 academic chairs were eliminated, several education programs closed, and 280 staff positions cut. A strategic plan developed and introduced in 1999 helped bring about desired improvements in organizational relevance and enrollment. A key element in the strategic plan was that the university abandoned the generalized "agricultural" track to focus on "life and social sciences," given that health, food, nutrition, lifestyles, and livelihoods had gained importance.

Research institutes that were under the Ministry of Agriculture, Nature, and Food Quality merged with the university and now operate in so-called Science Groups on animal, environmental, food, plant, and social sciences. Van Hall Larenstein, a college of higher agricultural and vocational education, joined Wageningen, and the former tropical school for higher vocational education relocated to Wageningen from Deventer.

INNOVATIVE ELEMENT

An innovative aspect of the reform was the leadership from the Ministry of Agriculture, which saw the importance of creating a strong, competitive research and education center that would give the ministry a prominent guiding/steering role and strengthen its political position. Equally innovative was the university's capacity to see change as an opportunity to increase its competitive power by combining different levels of research and education that included the university, research centers, experiment stations, and professional education, complemented by special centers for knowledge valorization, business schools, and professional midcareer training and capacity building.

EXPERIENCE

The Wageningen University and Research Center's vision for education is to create a content-inspired international learning environment based on (1) an orientation to competencies (knowledge, skills, and attitude), (2) professional and academic education, and (3) diverse, dynamic, and flexible learning tracks. The university's reformulated mission—"to explore the potential of nature to improve the quality of life"—has given rise to programs that attract more students to pursue a career in agriculture. Student numbers have risen since the reorganization, and the number of students entering the university from secondary school doubled. This trend was reinforced by many initiatives to inform the general public and students in secondary education about the university's new mission and emphasis on exploring nature to improve the quality of life.

The university has more than 11,000 students. Approximately 7,000 participate in the bachelor-level program, and of the university's 2,500 MSc and 1,600 PhD students, more than half are from abroad. The relatively large number of doctoral students signifies the university's true research character and international scope (it draws students from more than 100 countries). A new campus with cutting-edge teaching and research facilities is under construction. The university has developed a flexible funding structure that attracts financing for research fellowships; its scientific excellence helps to secure operational funding. It participates in international research programs oriented to development and plays a leading role in large, privately funded programs set up by foundations. The university uses core funding to support development-oriented research to help solve important societal problems and at the same time to build skills and competencies of partners and partner institutions in developing countries.

Curriculum Change in Agricultural Universities Strengthens Links in the Arab Republic of Egypt's Innovation System

Charles J. Maguire, Consultant

SYNOPSIS

Falling enrollments and mismatch between graduates' skills and labor-market requirements inspired a curriculum change project in five agricultural universities in Egypt. The project's strategy was to strengthen connections between important institutions in the innovation system (universities, private firms, and commercial farms) while transforming academic programs. Leaders from the academy and the private sector participated in a steering committee that guided the project's implementation. Based on a skill gap analysis that revealed the human resource needs of private employers and the corresponding weaknesses in academic programs, faculty updated core courses and made them more consistent in content as well as academic standards. The project also trained faculty, instituted active learning and recognition of good teaching, and improved the use of teaching aids. An overseas study tour formed the basis for significant institutional changes in the participating universities; for example, the universities organized external advisory committees to provide feedback on sector development and labor-market needs to university management. The universities also established internship programs and career resource centers. Additional links were forged between the university and others in the AIS through the establishment of extension-outreach centers, which enabled universities to provide direct assistance to communities and, in turn, learn about real community needs. The main lesson from the design and implementation of this project is that curriculum reform is complex, involving many aspects of the academic program, the university administration, and stakeholders. One approach to ensure institutionalization of the reforms would be for the project to include a mechanism for continuing high-level dialogue with stakeholders.

CONTEXT

Egypt's agricultural universities were losing students. Outdated courses and limited contact with prospective employers produced a significant number of unemployed graduates whose skills did not match labor-market requirements. The quality of education had deteriorated because faculty lacked the resources to conduct research or pursue professional development. The Institutional Linkage Project, a component of the USAID-funded Agricultural Exports and Rural Income (AERI) Project, was designed to help five agricultural universities in Upper Egypt reassess their course content and create a model curriculum that other universities could adopt. (See also IAP 9.)

PROJECT OBJECTIVES AND DESCRIPTION

The strategy of the AERI Linkage Project was to strengthen connections between important institutions in the innovation system (universities, private firms, and commercial farms) while transforming academic programs. The process began with a skill gap analysis that revealed the human resource needs of private employers and the corresponding weaknesses in academic programs. Based on this analysis, three broad steps emerged to support the project's overall goal:

- *Step 1.* Active teaching and learning methods were introduced to faculty to develop the higher cognitive skills (such as critical thinking, problem solving, and decision making) that private employers were seeking in graduates.
- *Step 2.* Content of the "core courses" (taken by all agricultural students) was updated.
- *Step 3.* Based on a strategic planning exercise by academic and private sector leaders, the universities organized external advisory committees and internship programs as well as career resource and extension outreach centers.

INNOVATIVE ELEMENTS

Innovations in project design and implementation included:

- Leaders from the academy and the private sector participated in a Steering Committee that guided the project's implementation.
- A skill gap analysis identified knowledge and skill deficits in recent graduates.
- Academic staff participated in redesigning and improving courses and learning materials.
- University deans and private sector leaders gained first-hand views of overseas university systems.
- External Advisory Committees were created and provided feedback on sector development and labor-market needs to university management.
- Student internship programs were developed.

Details on the project's experience with these innovations follow.

Role of the steering committee

A Steering Committee of Egyptian academic and private sector leaders guided planning and implementation of the project's capacity-building component. Their participation helped bridge the gap in understanding and cooperation between the private sector and the participating institutions. The committee met periodically to review and approve all major project activities and to select and/or approve nominees for major activities such as overseas study tours.

Skill gap analysis

The skill gap analysis involved 254 private employers and 1,000 university graduates who had been out of school and/or working for at least one year. Private firms and recent graduates agreed that university students needed to develop critical thinking, problem solving, and decision-making skills, as well as effective communication and teamwork skills. The common method of teaching through lectures did not actively involve students in learning. It emphasized knowledge recall at the expense of higher cognitive skills such as analysis, synthesis, and evaluation. The first step in providing students with the skills they lacked was to transform the teaching/learning methods at the universities.

Active teaching and learning

Altogether, 239 university faculty members participated in seven in-service workshops on active teaching/learning

methods. These two-day events focused on the science, mechanics, and art of teaching as well as skills to elicit learning. The workshop emphasized active, problem-based learning strategies. Educators from the United States conducted three follow-up workshops with 139 faculty members who had completed the first round of workshops. Outstanding Egyptian teachers who emerged from the first workshops organized additional in-service workshops for faculty who did not speak English. Nine such workshops were conducted for 234 faculty members from all five universities.

Leadership study tour generates ideas for change

Nineteen deans and department heads traveled with private sector leaders interested in strengthening the universities to visit four United States land-grant universities with important faculties of agriculture. The tour emphasized how each university worked closely with the private sector to modify the curriculum, established internship programs, operated career centers, and incorporated practical skill training into their respective academic programs. During a two-day strategic planning workshop following the tour, participants discussed and developed the elements of a plan to incorporate similar ideas in their universities. The tour enabled participants to see the value of advisory committees involving the private sector, including their potential to promote partnerships between the private sector and the university system. The conclusion was that the study tour "formed the basis for significant institutional changes in the participating universities" (Swanson, Barrick, and Samy 2007).

Course development workshops

The skill gap analysis demonstrated the need for the universities to change their curricula, but that would be difficult in view of the time and effort needed for the Supreme Council for Higher Education to approve the modifications. As an alternative strategy, the basic structure of the curriculum was retained and individual courses were modified to reflect the current knowledge base in each field of study. The focus shifted to updating basic course content and teaching methods and developing common academic standards across all five universities, especially for the common core courses.

Course development workshops, attended by faculty members and teaching assistants responsible for the core courses, adapted courses to become more consistent in content as well as academic standards. The Midwest Universities Consortium for International Activities (MUCIA) selected outstanding teachers from partner universities in the United States to conduct the workshops. They brought their recommended book for the particular course plus an array of teaching materials (such as the course outline or syllabus, PowerPoint presentations, videos, and classroom exercises, quizzes, and examinations).

Implementing faculty development programs at United States universities

Faculty development programs consisted of a six-week program split between two United States universities in MUCIA or a four-week program at one university. The vast majority of participants (93 percent) intended to modify their teaching methods in various ways, by promoting greater student-teacher interaction, encouraging more use of the Internet, making courses more market driven, bringing in more guest lecturers, stressing practical applications, increasing field visits, and promoting more team-based learning.

Developing external advisory committees

Workshops led by MUCIA specialists helped to organize this institutional innovation. Following the first workshop, three follow-up workshops were organized to discuss progress. At a separate, one-day workshop held later (April 2006), academic leaders from each university and the private sector members of their respective committees met to compare experiences and outline plans for building public-private partnerships.

Forming internship programs and other links in the innovation system

The external advisory committees promoted the development of student internship programs in collaboration with private firms. Additional links were forged between the university and others in the AIS through the establishment of Extension-Outreach Centers. The centers enabled universities to provide direct assistance to communities and, in turn, learn about real community needs. The establishment of Career Resource Centers helped students plan coursework and prepare themselves for seeking employment.

PROJECT IMPACT AT MID-TERM

The project created greater awareness of the importance of supporting major change in the knowledge and skill profiles of graduates from agricultural universities, and it successfully engaged university administrators and faculty in bringing about those changes. Among the achievements noted in mid-term project reviews were:

- *Active teaching/learning methods.* Sixty-seven percent of professors had adopted and thirty-three percent had partially adopted the new methods. The methods seem to have improved interactions between professors and students.
- *Course development workshops.* All participants found the workshops conducted to review courses and introduce new material relevant to their particular courses; 53 percent fully adopted the recommended modifications to their course. The remaining 47 percent adopted some of the materials or recommendations. (Note that the project also supplied computers and LCD projectors to use with the updated teaching materials.)
- *Links to private employers.* The study tour was effective in convincing university administrators that public-private cooperation benefits future employers of graduates and the university.

LESSONS LEARNED AND ISSUES FOR WIDER APPLICATION

The main lesson from the design and implementation of this project is that curriculum reform is not as straightforward as it might seem.[1] It is a complex process involving many aspects of the academic program, the university administration, and stakeholders. For example, a revised or updated curriculum without improved teaching materials and appropriate pedagogical skills is unlikely to have much impact. The benefits of a revised curriculum will not be sustained unless the curriculum keeps pace with stakeholders' evolving needs. Key stakeholders inside and outside the university must contribute their perspectives on the knowledge and skills needed in a developing agricultural sector. To ensure support for curriculum change, teaching staff, administrators, and stakeholders must be consulted and engaged as partners in making the desired changes. Despite impressive reforms in the curriculum, capacity building for academic staff, and improved links to agribusiness, the question of whether these five universities can sustain their efforts appears unanswered. One approach to ensure institutionalization of the reforms would be for the project to include a mechanism for continuing high-level dialogue.

Innovative Training Program for Midcareer Agricultural Extension Staff: The Sasakawa Africa Fund Education Program

Charles J. Maguire, Consultant

SYNOPSIS

The Sasakawa Africa Fund Education Program builds capacity of midcareer practicing extension workers by enabling them to participate in degree programs. This opportunity enriches university understanding of field conditions and problems (leading to more focused curriculum content), benefits participating communities, enhances the status of the university in the agricultural sector and rural space, gives ministry of agriculture services greater credibility and effectiveness, and confers greater skills, status, and influence on midcareer extension workers—all of which is beneficial to the AIS. The process began in 1993 with cooperation between Ghana's Ministry of Food and Agriculture, the Sasakawa Africa Fund for Extension Education (SAFE), and Winrock International Institute for Agricultural Development. By January 2008, 2,292 students had benefited from the program. One key to the program's success was the Supervised Enterprise Projects (SEP) undertaken by each midlevel extension worker accepted to the program, which forged the links between extension, the university, and the community. The main lesson is that initiating a program of this type requires considerable discussion with higher-level decision makers and university managers so that the potential value of the program is understood. Agreement on program design is essential to ensure that the community attachment portion of the program (SEP) is implemented effectively. Careful selection of candidates for the degree program is vital. The financial lesson is that SEP is expensive to implement, as it requires supervision and travel for faculty and students.

CONTEXT

The persistence of formal teaching methods characterized by the lecture model is a generic weakness of agricultural education, especially in sub-Saharan Africa, where students lack practical experience gained through demonstrations, field visits, and interactions with rural communities. University graduates who work in agriculture ministries and their subsidiary agencies often lack the skills and confidence to understand, respond to, and communicate effectively with sector stakeholders. To respond to this need and boost field experience in the leadership ranks of the extension service, Sasakawa Africa Association (SAA), an NGO funded by the Nippon Foundation, launched the Sasakawa Africa Fund for Extension Education (SAFE) (http://www.safe-africa.net), in 1993. A pilot program was implemented in Ghana in collaboration with Winrock International Institute for Agricultural Development, a nonprofit NGO based in the United States.

PROGRAM GOALS

The SAFE program has four objectives:

1. *Create midcareer training opportunities* to improve the technical and human relations skills of outstanding male and female field-based extension staff with certificates and diplomas in agriculture and related fields.
2. *Help reform agricultural extension curricula* in selected African universities. Coupled with this reform, participating universities are assisted to acquire relevant instructional materials and to network with other universities in the SAFE program to build strong pan-African academic partnerships. See also module 3.
3. *Develop agricultural leaders for extension organizations in sub-Saharan Africa.* This objective does not necessarily imply helping extension staff to occupy high positions within the extension organization. Rather, it means helping them achieve the more important goal of developing new, positive attitudes towards their work

and responsibilities and to become systems thinkers, catalysts, facilitators, and effective managers of change within their extension organizations.

4. ***In the long term, bring about institutional reform within African universities,*** not only in the development of responsive agricultural extension curricula, but also in the wider institutions themselves.

In summary, SAFE strengthens the in-country capacity of African universities to be adaptable organizations that can develop client-focused training programs, acquire relevant core instructional materials in agricultural extension and related fields, mobilize internal and external resources to sustain the programs, and forge partnerships with other local and international institutions and agencies. Specific problems that SAFE addresses are (Zinnah 2003):

- Outdated extension curricula that do not reflect changing needs of providers and users of extension services, including the private sector.
- Inadequate off-campus, farmer- and client-focused practical training activities for students in agricultural colleges and universities.
- Lack of appreciation of the experience of midcareer extension professionals in the agricultural and rural development process.
- Lack of appreciation for midcareer professionals, especially certificate and diploma holders, as a prospective group of learners.
- Lack of appropriate training opportunities for midcareer extension staff.
- Lack of partnerships among agricultural colleges and universities and employers/clients of their graduates.

SAFE'S DEVELOPMENT AT THE UNIVERSITY OF CAPE COAST, GHANA

To develop the SAFE pilot, in 1992 Ghana's Ministry of Food and Agriculture (MOFA) and its NGO partners requested the University of Cape Coast (UCC) to create a needs-based BSc degree program in agricultural extension for midcareer extension staff with diplomas and certificates (qualifications possessed by some 85 percent of Ghana's extension staff). These men and women would undertake the program on a leave of absence from their jobs and would return to work when they completed their studies.

The resulting program has two options. The basic option is a four-year program offered to extension workers who have passed a certificate course (two years of postsecondary

education). The second option is a two-year postdiploma program. Both programs lead to a BSc in Agricultural Extension.

Innovative elements

The program's innovative feature—aside from its focus on the untapped potential of midcareer professionals—was its off-campus Supervised Enterprise Project (SEP). After an initial period on campus, students return to their work environment to undertake a SEP for 4–6 months. These community-based experiences facilitate experiential learning as well as linkages between the major subsystems of the agricultural knowledge system: the farmer subsystem, which includes agribusiness, the extension and education subsystem, and the research subsystem. The SEP embodies the innovative link that connects the university and the community, thereby enhancing AIS interaction. Through the student extension workers, the SEP conveys community problems and successful practices to university researchers and enriches the curriculum.

Impacts of the Cape Coast program

The main beneficiaries of the SAFE Program are:

- ***Frontline extension staff of MOFA,*** because without the program, they would have little chance to get a BSc, given the perceptions of diploma and certificate holders among university admission officers.
- ***Farmers,*** because the SEPs concentrate on farmers' problems and have generated income-producing projects.
- ***Academic programs at the university,*** which are exposed to real farming community problems and concerns.
- ***UCC,*** through heightened visibility and links to MOFA, farmers, extension staff, NGOs, and District Assemblies. The university received international recognition for the innovative nature of the program.
- ***MOFA,*** which gains an infusion of newly motivated staff with practical skills and experience who serve in key positions and supervise the work of other extension staff.

By 1999, 51 students (22 percent female) had completed BSc degrees; of these, 22 percent gained first-class honors. Seventy-six students, including three from Nigeria and one from Mozambique, were enrolled (29 percent were female); 51 were certificate and 25 were diploma holders. The participation of the women in particular should increase the gender balance in the upper levels of the extension service.

A seven-member Consultative Committee was formed with representatives from MOFA, UCC, SAA/Winrock, and farmers/agribusiness to provide recommendations, including recommendations for fund raising. Greater interaction was fostered between field staff of some bilateral agencies and NGOs, UCC, and students because of the SEP.

By January 2008, 246 enrollees at UCC had obtained BSc degrees and 231, a diploma. At the same time, 81 persons were enrolled in the degree program and 112 in the diploma program.

WHAT IS THE FUTURE OF SAFE IN AFRICA?

It is becoming evident that partnerships are the only solution to the multifaceted problem of training Africa's agricultural extension staff. In extension education (in fact, throughout the AIS), the tradition of working alone in a competitive rather than a collaborative mode is no longer valid. No single player has the capacity to cover all education and training needs, and partnerships with donor agencies will continue to be critical in this poorly resourced area.

The motives and methods of the SAFE program resonated with African universities outside Ghana. Following the UCC pilot, other universities in Ghana and elsewhere joined, including: Kwadaso Agricultural College (Ghana); the University of Abomey-Calavi (Benin); the Polytechnic University of Bobo-Dioulasso (Burkina Faso), Alemaya University of Agriculture and Hawassa University (Ethiopia); the Rural Polytechnic Institute of Training and Applied Research and Samako Agricultural Institute (Mali); Bunda College of Agriculture (Malawi); Ahmadu Bello University and Bayero University (Nigeria); Sokoine University of Agriculture (Tanzania); and Makerere University (Uganda).[1] Donor support for the SAFE program was lengthy (UCC Ghana, the pioneer university, was supported from 1992 to 2002), but once universities and diploma-granting institutions accumulated sufficient experience with running the new curriculum, direct support ended.

A number of lessons were learned as SAFE expanded to new settings. Perhaps the most important lesson was that the seemingly impossible task of bridging the technical-professional gap in civil service employment could be solved for midcareer, technically qualified extension staff.

A relevant program that shows positive results creates further demand among academic entities and potential enrollees.

Including a practical stage (the SEP) in the program was central to success, because it fostered the alliances that spread the benefits to all participants. Communities gained from the external contacts. Ministries of agriculture gained better-trained staff with more field experience, which made their contribution to sector development more effective. Universities gained greater community visibility and access to real rural training settings and challenges, and university research programs and curricula were enriched to reflect changes in agriculture on the ground.

Aspects of the program can be improved. The four-year certificate program is too long for some midcareer workers who would be content with a diploma rather than a degree. The program also needs more diversity, especially with respect to female recruits, but the pool of midcareer women extension workers for the program remains small.

One lesson for implementing this kind of program is that the need for qualified and committed core staff is paramount. The lack of such staff has been a big constraint, affecting not only the implementation but the long-term sustainability of the program. Because the SAFE concept is new to the university and the extension service, administrative and academic staff members must be oriented to it. University staff members must acquire the improved pedagogical methods and skills to better meet midcareer students' learning needs. The acquisition of such skills is facilitated by staff involvement in the SAFE program through team teaching and joint supervision of SEPs.

Financing is crucial. SAFE, with its strong off-campus focus, is expensive to run, even if the returns on investment are large. A clear exit strategy for the funding partners is vital to enable the university to develop its own means of continuing the program. In Ghana, MOFA is considering including the SAFE program in its budget. Incentives also help universities adopt the program. At UCC, for example, a multipurpose building (the Sasakawa Center) was completed and is used to generate income for the SAFE program at the university. It remains vital to build constituencies that can pressure decision makers on behalf of the university; some administrators continue to regard SAFE as extra work rather than a strategic necessity.

Chiang Mai University Links with Rural Communities to Focus Research on Farming Problems and Foster Curriculum Change

Charles J. Maguire, Consultant

SYNOPSIS

This reform fostered interaction between the University of Chiang Mai and communities to identify problems and find solutions jointly through Community-Based Research (CBR). In the ten years since the program began, students and researchers have gained hands-on experience in communities; through this interaction, they identified researchable topics that enriched the university's research program. The curriculum also benefited from the enhanced understanding of rural problems and challenges. The innovative aspects of the university-community link were that it used an ongoing project (the CBR project) as a vehicle for introducing students to the realities of rural life; it created a university Center for Community-Based Research; and it used CBR projects created from this center as a learning resource for undergraduate and graduate students. Apart from the core faculty in the Center for Community-Based Research, other faculty members began to adopt this approach to building student capacity to deal with real problems in rural communities. The university became a visible and active actor in the farming community and the AIS. The integration of learning/teaching, research, community service, and faculty required considerable collaboration among the CBR projects and communities, the Practical Training Unit, the Student Affairs Unit, and the Department of Agricultural Extension. At the same time, the program required support from the national research funding agency as well as financial support and collaborative organization inside the university and with communities.

CONTEXT

A major policy of Thailand's Chiang Mai University is that research should be of world-class quality and serve local communities. The university has established a center to strengthen research by systematically accessing new research funding and strengthening faculty research capacity. A faculty research manager encourages faculty, staff, and students to engage more actively in research. Research performance is a major criterion for promoting faculty members.

Yet the Department of Agricultural Extension in the Faculty of Agriculture at Chiang Mai University realized in the 1990s that students and faculty were poorly prepared for work in rural areas. Students and faculty were unfamiliar with rural living conditions, technical agricultural challenges, and social problems. The university placed heavy emphasis on classroom learning, literature reviews, and laboratory experimentation but offered little contact with communities. Many undergraduates had little capacity to analyze and synthesize information on social situations or conduct community-based research, and their facilitation, communication, and writing skills were poor. In 1996, the Thailand Research Fund (TRF) had resources to support a Community-Based Research (CBR) program to answer the question: "How can research findings be used by local people—the users of research results?"

THE COMMUNITY-BASED RESEARCH PROGRAM

The CBR program employed an empowerment and people-centered approach that:

- Identified problems and research questions in a participatory manner with local residents.
- Involved local people as community research in the CBR program (farmers, women leaders, village leaders, school teachers, administrative officers, community development workers, community extension workers, doctors, monks, and youth groups).

- Planned projects with care and implemented agreed actions to solve problems.

Each CBR project would be developed and implemented in 12–18 months and would receive a grant of 200,000–300,000 baht (B). To facilitate CBR projects, CBR centers were set up. Through counselors and staff, they would identify community researchers, identify local problems, develop research questions, develop a CBR proposal and action plan, sign research contracts, and monitor the CBR from the beginning to final report phase.

In 1998 the Extension Department saw that CBR projects could provide an opportunity for students to become involved in communities. Student learning was integrated with community research through a number of initiatives. Students enrolled in Extension Communications visited active CBR projects and, as an exercise, could write an article on their observations and village issues; produce a script to be broadcast over community radio; or submit an article for community newspapers. Students specializing in Media Production for Extension visited CBR projects and developed media products that reflected the needs of community researchers, such as posters, newsletters, photographs, DVDs, and radio programs. Graduate students enrolled in Agricultural Communities Studies undertook study visits to CBR communities; participated in discussion and dialogue; listened carefully when interacting with CBR researchers, counselors, and staff; took detailed notes; and produced a review of their visits. The materials produced, together with their experiences in the communities, formed the basis of their thesis topics.

Efforts to widen the interaction between the broader university community, the research funding agency, and CBR communities proved difficult to achieve. For this reason, a project specific to the university—the Research Management Fund—was established in 2002 with support from TRF to foster wider collaboration among the faculty of agriculture, TRF, and CBR communities for integrating teaching/learning, research, and community service. The ultimate objective was to create a learning community of undergraduate students based on CBR projects. To reach this objective, a Center for Community-Based Research was established in the Faculty of Agriculture to develop CBR projects as a means of empowering community researchers. In 2003, 11 CBR projects were developed.

The next challenge was to get the university's Student Affairs Unit and Practical Training Unit involved in using CBR projects as a learning community base for undergraduates. A start was made during the practical training period in 2003, when 120 second-year students undertook practical training in five CBR communities supported by staff from the Center for Community-Based Research in the Faculty of Agriculture. The university's Practical Training Unit staff learned, for the first time, how to organize such training in rural communities and how to communicate with community members and undergraduate students. A practical training manual was prepared for the 2004 program. As a result, the Faculty of Agriculture has allotted an annual budget for this training activity. It was also the first time that communities had hosted 30 university students for a five-day visit, helped organize the practical training program, and interacted closely with such a group of visitors. For the students, the visit was a true learning experience. For the first time, students lived with rural families, communicated, understood the realities of rural life, and appreciated the value of local wisdom in dealing with livelihood issues.

INNOVATIVE ELEMENT

The innovative aspects of the university-community link were that it used an ongoing project (the CBR project) as a vehicle for introducing students to the realities of rural community life; it created a university Center for Community-Based Research; and it used CBR projects created from this center as a learning resource for undergraduate and graduate students. Apart from the core faculty in the Center for Community-Based Research, other faculty members began to adopt this approach to building student capacity to deal with real problems in rural communities.

IMPACT

In a relatively short period, the program established the CBR link with communities and dramatically raised awareness among students and faculty of how isolated the university had been from life in rural communities. Research has become more focused, and the curriculum reflects the knowledge and skills needed by graduates who will meet technical and social needs in rural areas. The university, through its faculty and students, has gained visibility and stature among its stakeholders and has become an active AIS actor.

By 2007, 650 CBR projects had been funded with grants from TRF, and 264 had been implemented, facilitated by the core research team that runs the Center for Community-Based Research. The center collaborated with the Practical Training Unit in the faculty to enable second- and third-year undergraduate students to undertake "practical training" in communities.

Additionally, these students were encouraged to practice further "student activity" in the communities through various student clubs. Through this learning process, students, the Faculty of Agriculture, the Center for Community-Based Research, the TRF, and local communities have gradually developed collaborative interaction. For example, research problems identified by community members have increased the effectiveness of the students' clubs. Local communities have increased their ability to manage "student practices" as a vehicle for identifying community problems, to analyze causes of problems, and to develop solutions through a participatory research process in which students and faculty become their co-researchers.

LESSONS LEARNED

To summarize, the CBR program—initiated through one university department in the faculty of agriculture—created and sustained links between students, teachers, researchers, and rural communities, all of whom benefit. The university can focus on real-life problems through research; supervising teachers gain from the field experience; students learn how to communicate with rural people; and communities gain solutions to problems based on academic and local knowledge and see the university as a development partner. Box 2.12 lists specific lessons from the program.

Box 2.12 Specific Lessons from the Community-Based Research Approach Adopted by Chiang Mai University's Faculty of Agriculture

- The need for change in academic programs can be identified from within the educational institution.
- Available resources (in this case, the Thailand Research Fund) can be tapped to begin the change process.
- Involving students and researchers in the change process is likely to ensure that the change model is sustainable.
- It was not easy to convince all faculty members to adopt the CBR model, especially because projects run from 12 to 18 months, whereas mainstream student research projects run only from 6 to 8 months.
- Collaboration within the university and with communities is essential.
- The participation of academic staff as co-researchers allows them to increase their knowledge and facilitating skills and to change their world view. For example, teachers adopted a participatory learning process, respected local wisdom, and recognized villagers' tacit knowledge. These practices inspired changes in their teaching style, which has become more participatory and facilitates learning more effectively.

- Through the CBR projects, academic knowledge was applied to problems that community members could not solve by themselves.
- The involvement of faculty and students with communities brought the university to the field and strengthened AIS linkages between the university and a major stakeholder group.
- The collaboration of the CBR projects and communities, the Practical Training Unit, the Student Affairs Unit, and the Department of Agricultural Extension contributed to the integration of learning/teaching, research, community service, and faculty.
- Information and knowledge generated by CBR projects can be used to develop research projects for undergraduate and master's students. At the same time, community service is built in throughout the research process and culminates in the development and use of new knowledge to improve life in the communities.
- As noted, the CBR program was successful but required financial support and collaborative organization inside the university and with communities, and the support of the research funding agency.

Source: Author.

EARTH University, Costa Rica: A New Kind of Agricultural University

Charles J. Maguire, Consultant

SYNOPSIS

EARTH University, an autonomous educational entity, was newly designed and established in Costa Rica in 1990. Although a significant part of the plan of study focuses on agriculture and natural resource management, the university's core purpose is to prepare leaders capable of promoting sustainable development in the tropics. EARTH's unique educational model is based on technical and scientific knowledge, entrepreneurial mentality, values and ethics, and social and environmental commitment. In addition to inculcating practical skills that benefit interactions at many levels throughout the AIS, EARTH's experiential curriculum balances theory and practice through work experience, community development, the academic program, entrepreneurial projects, and internships. The emphasis on entrepreneurial skills equips graduates with a greater range of career choices, and exposure to rural communities is vital for producing graduates dedicated to serving the sector and rural people. A major lesson is that the development of robust rural economies requires individuals capable of applying knowledge and skills from across disciplines, often in very practical and applied ways. Faculty must engage directly with their students in production, processing, and marketing as well as in research and extension. Reward and promotion criteria must be designed to encourage innovative teaching and engagement as well as research. Assured funding combined with continuous fund raising enable the university to maintain high academic and infrastructure standards.

CONTEXT

EARTH—an acronym for the university's name in Spanish, Escuela de Agricultura de la Región Tropical Húmeda—is a private, international, four-year undergraduate institution located in the Caribbean lowlands of Costa Rica.[1] An initiative of a group of largely Costa Rican business, academic, and government leaders, EARTH was founded in 1990 with significant investments from USAID and the W.K. Kellogg Foundation, with strong support from the government of Costa Rica. The support provided by USAID allowed for the establishment of an endowment that covers between one-third and one-half of the university's annual expenses.

EARTH University was established in response to urgent problems in Central and South America, including rural poverty, high population growth, low productivity, migration to cities, destruction of fragile ecosystems, and political instability and war throughout the region. EARTH is a private, nonprofit, international university, autonomous and independent of political pressure. It has an international faculty, a student body originating from 25 Latin American and Caribbean countries, a small number of students from Africa, and is small, with 400 students and 40 faculty. The university's 3,300-hectare farm is used for training as well as commercial, income-generating crop production.

PROJECT OBJECTIVES AND DESCRIPTION

Although a significant part of the plan of study at EARTH focuses on agriculture and natural resource management, the university has defined its core purpose as preparing leaders capable of promoting sustainable development in the tropics. The curriculum is highly integrated and transdisciplinary. While the acquisition of technical and scientific skills and expertise is obviously important, the development of values, leadership, commitment, and a diverse set of skills, abilities, and attitudes are considered equally important.

The development of EARTH in the 1980s coincided with a significant change in the agricultural sector and a marked shift in the structure of employment opportunities in agriculture. Whereas previous generations of agronomists and other professionals had found relatively abundant employment in Ministries of Agriculture, development banks, and other public agencies, by the 1980s structural adjustment programs and other changes had largely eliminated these possibilities. Consequently, EARTH's academic program was focused largely on preparing professionals for the private sector. Providing graduates with entrepreneurial skills and abilities became fundamental to the program. The centerpiece of entrepreneurial training at EARTH is the Entrepreneurial Project course (see the next section).

INNOVATIVE ELEMENTS

To create the type of leader capable of responding to the social and environmental problems facing rural communities of the humid tropics, EARTH developed a unique educational model based on four pillars: technical and scientific knowledge, entrepreneurial mentality, values and ethics, and social and environmental commitment. As part of this model, EARTH created an experiential curriculum that balances theory and practice. Five keystone programs within EARTH's curriculum are based on experiential learning:

- *Work Experience.* This course is taken by all first-, second-, and third-year students and continues in the fourth year as the Professional Experience course. In the first and second years, students work in crop, animal, and forestry production modules on the EARTH farm. In the fourth year, students identify work sites or activities on campus or in surrounding communities that correspond with their career goals and develop and implement a work plan, dedicating a minimum of 10 hours per week to the "job."
- *Community Development.* In their second year, in an extension of the Work Experience course, students work on an individual basis with small-scale, local producers on their farms and with organized groups in sustainable community development. During this experience in the community, students try to resolve problems facing the region's inhabitants. The community transmits real-world experience and provides learning opportunities for students.
- *Academic Program at EARTH-La Flor.* In their third year, students spend seven weeks living with a host family near EARTH's education and research center in Guanacaste, a province in the dry tropics of Costa Rica, where they have the opportunity to become actively involved in the region's development process. Students contribute to improvements in the communities and also gain experience by

working with companies in the region in such areas as: crop management techniques and practices for melons, sugarcane, rice, and citrus; livestock management practices; small-scale and industrial aquaculture; and regional efforts in alternative energy generation from wind, geothermal, solar, and sugarcane-derived sources. Students also learn about water management and biodiversity in the dry tropics in relation to each of the above-mentioned activities.

- *Entrepreneurial Projects.* Students develop a business venture from beginning to end during their first three years at EARTH. Small groups of 4–6 students, of different nationalities, decide upon a business activity related to agriculture and natural resources and conduct a feasibility study (including financial, social, and environmental criteria). If the study is approved by a panel of professors, other students, and external experts, the university loans money to the company, and the team implements the project, including the marketing and sale of the final product. After repaying their loan, with interest, the group shares the profits. The Entrepreneurial Project is accompanied by a series of classroom modules related to business organization, accounting, marketing, and similar themes.
- *Internships.* In their third trimester of their third year, students leave campus and take part in an internship program with a host organization such as a business, NGO, or farm. This internship program lasts 15 weeks and is a crucial component of the student's experiential education. Using knowledge and skills acquired in their first three years at EARTH University, students obtain real-world practical experience upon which they can reflect during their fourth and final academic year.

These programs give students opportunities to develop planning and leadership skills, foster responsibility, encourage them to become decision makers and critical and creative thinkers, improve their ability for analysis, synthesis, and evaluation, and apply technical and scientific knowledge in real situations.

IMPACT

In its twenty years of operation, EARTH University has graduated some 2,000 students, the majority of whom have returned to their home countries to work in agriculture and rural development. The EARTH educational model is widely recognized as suitable for developing graduates who have the academic and practical knowledge, skills, and confidence to take leadership positions in the sector. Many graduates create their own businesses and become employers.

LESSONS LEARNED

The key to graduating professionals capable of successfully promoting change and sustainable rural development is to begin with young people who have a vocational interest in agriculture, natural resource management, rural development, and related areas. Such graduates are likely to return to rural areas and engage in the complex and difficult work that drives development. Investment in choosing and recruiting new students is a major contributor to the production of graduates dedicated to serving the sector and rural people.

Universities have to be engaged with rural communities. Too often, universities are located in the capital city and lack the resources (or the will) to get faculty and students into rural areas to engage directly with farmers and their families. Frequent and direct contact with the realities of rural life and the challenges of agricultural production are essential in the formation of future change agents.

Higher education in agriculture has become increasingly specialized, with the result that many graduates have great difficulty integrating knowledge across disciplinary boundaries. Yet the development of robust rural economies requires individuals capable of applying knowledge and skills from across disciplines, often in very practical and applied ways. Particularly at the undergraduate level, a generalist formation would seem to be more relevant to the needs of most developing countries. An emphasis on entrepreneurial skills equips graduates with a greater range of career choices. Exposure of students and faculty to agricultural communities leads to better communication, greater understanding of rural living conditions and livelihood challenges, and the formulation of technical and social solutions.

A favorable student-to-faculty ratio (10–1 at EARTH) allows for quality interaction and instruction. Faculty have to be willing to get their hands dirty and to engage directly with their students in production, processing, and marketing as well as in research and extension. Reward and promotion criteria must be designed to encourage innovative teaching and engagement as well as research. At the same time, teachers and professors must be provided decent compensation for their service. Substandard salaries and working conditions only serve to drive the best out of education, leaving the mediocre in charge.

Agricultural institutions, faculties, and schools require investment and must be equipped with the latest technological advances to make a meaningful contribution to building human resources for agriculture. Although assured funding enables the university to maintain high academic and infrastructure standards, fund-raising remains a constant task for a private autonomous university, since its future depends on such income.

Technical Skills for Export Crop Industries in Uganda and Ethiopia

Charles J. Maguire, Consultant

SYNOPSIS

A shortage of skilled technical personnel and high demand from commercial producers led to the cooperation between a donor, training institution, and commodity group to provide technical skills for workers in high-value export crop industries (floriculture and horticulture) in Uganda and Ethiopia. The innovative element of the project was the introduction of competence-based training, which ensured that trained technicians were equipped with work-ready skills; in fact, trainees completing the program had a high level of employment. A lesson learned from operating the program over three years is that demand from employers can trigger technical/vocational training and that close cooperation between a qualified training supplier and clients can lead to a successful outcome.

CONTEXT

Floriculture is a large market in East Africa. Local and foreign investors have developed flower farms that produce large quantities of roses and chrysanthemums. Dutch flower growers are also active in this market, which is understandable, since flower auctions in the Netherlands trade one-third of the global market for flowers and plants. The agronomic and economic conditions for raising high-value export crops are favorable in many parts of East Africa: Cheap land, labor, and energy combine with good soil, water, and climatic conditions. Despite this potential, the associated growers in Uganda and Ethiopia (the Ugandan Flower Exporters Association and Ethiopian Horticulture Producers and Exporters Association) lacked well-trained middle management workers. Various farmers were hiring managers from abroad, mostly from India or Kenya. Producers and growers felt the need to train local workers and students and proposed a project to the Netherlands Foundation for International Cooperation to develop education

and training programs at various levels, ranging from short courses for farm workers to certificate and diploma courses and bachelor's and master's degree programs.

PROJECT OBJECTIVES AND COMPONENTS

The overall objective of the project was to expand the numbers of qualified technical workers for the export flower industry in Uganda and Ethiopia. The project had two components. The first identified the knowledge and skills required and designed corresponding training activities. The second developed a new pedagogical approach.

To develop training programs, two consortia, with experts largely drawn from the same institutions (the Department of Education and Competence Studies of Wageningen University, PTC+, the Agricultural Economics Institute, and a practical research center in horticulture), worked with local stakeholders. The consortia are working with the respective grower associations and academic institutions (in Uganda with the Mountains of the Moon University in Fort Portal and Bukalasa Agricultural College in Wobulenzi and in Ethiopia with Jimma University College of Agriculture and Veterinary Medicine). The short courses have been effective at the thematic level. Farm supervisors and assistant managers of various departments, such as the greenhouse, "fertigation" (fertilization and irrigation), postharvest and handling, and pest management, completed the training and were able to apply their new skills immediately. In Ethiopia, the development and implementation of the curriculum is still underway, but the first reactions are positive. An impact study will be undertaken to show the real effectiveness of the approach.

A competence development philosophy was employed in developing the curriculum. A group of staff members went to the Netherlands for a Training of Teachers (ToT) program. Apart from floriculture and horticulture training, they also learned principles of competence-based education

(Mulder 2007; Mulder et al. 2009). They were expected to disseminate the knowledge obtained in the ToT program to lecturers in their colleges. The core of the competence-based education philosophy is that the flower (or other) industry articulates the need for training. Through a labor-market analysis and needs assessment, all stakeholders obtain a picture of how the educational program should be structured. Occupational profiles and competency profiles are developed. The competency profiles contain knowledge, skills, and attitudes.

A major premise in competence-based education is that knowledge alone is not sufficient to bring about improved practice. In many cases, unbalanced concentration on knowledge develops graduates who may know a lot but cannot apply their knowledge in practice. Applying knowledge in practice, however, is exactly what is needed in many developing countries. A matrix comprising eight principles and four competence-based implementation levels was used to decide the extent to which the horticulture training would be competence-based. The matrix structure resembles that of models used by the European Foundation for Quality Management.

INNOVATIVE ELEMENT

The innovative element of the project was the introduction of competence-based training, which ensured that trained technicians were equipped with work-ready skills.

BENEFITS, RESULTS, AND LESSONS LEARNED

As noted, the farm supervisors and assistant managers of various floriculture departments were able to apply their new skills immediately after training. In Uganda, the first batch of sixteen diploma students graduated, and 14 are working on flower farms. Because of the training program's competence focus, employers appreciate the trained technicians. Placing students in internships was not difficult. Some farms asked for as many students as were available. Various farms retained the interns and offered them labor contracts (Mulder, pers. comm., 2010).

The lessons from this experience to date are that demand from employers can trigger technical/vocational training and that close cooperation between qualified training supplier and clients can lead to a successful outcome.

Agribusiness Training for Secondary School Graduates in Timor-Leste

Charles J. Maguire, Consultant

SYNOPSIS

In Timor-Leste, capacity improvements are needed at all levels and across the spectrum of participants in the agricultural sector (public agencies, agribusinesses, individual farmers, cooperatives, and others). Building Agribusiness Capacity in East Timor (the BACET project) offers supplementary training in agribusiness skills through a one-year, postsecondary certificate program. The project—implemented by the National Directorate of Agricultural Education and Training in the Ministry of Agriculture and Fisheries (MAP) and Land O'Lakes (an international agribusiness)—graduated the first set of students in 2008 (121) and the second set in 2009 (111). For BACET students, the overarching goal is to learn the problem-solving and other skills required to begin careers as agribusiness entrepreneurs, agricultural extension workers, and middle managers to benefit communities in Timor-Leste. The graduate of this program is envisioned to be a learner rather than a passive absorber of information, capable of contributing to practical solutions in agribusiness. The curriculum's focus on agribusiness and its emphasis on learning by doing and practical application of theoretical classroom materials were innovative in Timor-Leste. Teacher skills that emphasized learning versus teaching were another novel aspect of the project and considered fundamental to success. A major lesson from the experience is that the capacity of staff and facilities needs to be carefully assessed prior to program design, especially in countries classified as fragile states. Sites for the project must be chosen carefully to take into account logistics and the availability of services. Program design needs to incorporate lessons from implementation in redesigned curricula. To sustain the program over the long term, one strategy would be for MAP to shift the BACET program from the certificate level to a formal, two-year diploma course, similar to those offered by other countries in the region. Flexibility is needed to shape the future of the program in light of lessons from the first phase of implementation and the fragile, postconflict setting.

CONTEXT

Timor-Leste is a postconflict, fragile state that depends heavily on agriculture, which accounted for 31.5 percent of GDP in 2007 and employs about 84 percent of the population. Food security is critical for most people in Timor-Leste, where agricultural productivity is low in comparison to neighboring countries. Agricultural development is essential for improving food security, promoting stability, and ensuring conservation of environmental and natural resources.

The human and institutional capacity to bring about agricultural innovation and development suffered greatly during the conflict that ravaged the country for a quarter of a century. Education programs were disrupted; experienced and educated individuals fled; there was much loss of life; institutions were destroyed. By one estimate, 53 percent of rural people have never attended school. Capacity improvements are needed at all levels and across the spectrum of participants in the agricultural sector—public agencies, agribusinesses, individual farmers, cooperatives, and many others. In this context, only long-term investment can build the foundation for the country to recover.

PROJECT OBJECTIVES

Building Agribusiness Capacity in East Timor—the BACET project—was designed to supplement agricultural training at the secondary level with training in agribusiness skills. The one-year, postsecondary certificate would be offered to graduates of the three-year program at three of Timor-Leste's agricultural secondary schools. A five-year project costing US$6 million, BACET was funded by USAID. It was implemented by the National Directorate of Agricultural

Education and Training in the Ministry of Agriculture and Fisheries (MAP)[1] and Land O'Lakes, an international agribusiness. The project, which started in 2006, was to run for two years, but that timeframe proved insufficient to evaluate and refine the curriculum, and the project was extended to 2011. The target for each secondary school was to produce 50 graduates from the BACET program each year, beginning in 2007. The first set of BACET students graduated in 2008 (121) and the second set in 2009 (111). For more information on agribusiness skills, see also module 3, TN 2; and module 5.

Overarching objectives

The project's initial objectives (2006–08) were to:

- Develop and deliver an agribusiness curriculum that provides practical skills in crop and livestock production, agricultural mechanics, English language, IT, and business management. Courses would include instruction in basic bookkeeping, organizational management, marketing, financial analysis, cooperative business, and farming systems in Timor-Leste.
- Develop an ongoing agribusiness at each school and support agricultural enterprises and improved production in nearby communities. These efforts might include the development of a functioning poultry business (eggs) at the Fuiloro School and horticultural crops (a nursery, orchard, and vegetables) and livestock (goats) at the Natarbora and Maliana schools. Train students and faculty in the respective technical disciplines at each school with a view to taking a lead role in operating the business.
- Develop local and regional agribusiness case studies and identify published case studies appropriate for use in Timor-Leste. Train teachers in the use of case studies as a teaching tool.
- Develop career advancement services for BACET graduates.

In Phase 2 (2009–11), Land O' Lakes, using feedback from the Ministry of Agriculture, USAID, teachers, and students, enhanced the initial program to reinforce practical and market-oriented skills. Ideally, in this phase, the project will finalize an agribusiness course that can supply graduates to be problem solvers and meet labor-market needs for agribusiness entrepreneurs, agricultural extensionists, and agricultural middle managers. The project will continue to facilitate enterprise creation at schools and in communities and expand school outreach to farming communities. This phase of the project will also develop a placement system to help students understand the professional recruitment process and secure jobs in agribusiness. An agribusiness education program will be coordinated to enroll young people in the one-year agribusiness course, which will be handed over to the Ministry of Agriculture and Fisheries (MAP) in June 2011.

Program content: skills and coursework

For BACET students, the overarching goal was to learn the problem-solving and other entrepreneurial skills required to begin careers as agribusiness entrepreneurs, agricultural extension workers, and middle managers to benefit communities in Timor-Leste. During the one-year certificate program, students build their knowledge of agricultural systems, with a particular emphasis on markets and practical training. They learn to apply their agricultural knowledge practically and effectively in agribusiness, demonstrate practical agricultural skills and transfer them to rural communities, analyze issues and problems and find appropriate solutions, and adapt to changes and needs of the agricultural systems in Timor-Leste. Students also develop their capacity to be independent and reflective in their practices and to perform their duties in a moral and positive manner.

The core of the BACET curriculum is agribusiness, with an emphasis on agribusiness management and planning. In addition, students learn the principles of cooperative business models, farming as a business (see module 3, TN 2), and providing business services to agricultural producers. Additional courses required by agricultural professionals included production agriculture (plant science, crop production, animal science, and livestock production); agricultural mechanics; computers and IT; and English. All of these courses are linked to the core agribusiness focus. Other themes that students gained exposure to included sustainable agriculture, gender awareness, healthy lifestyles, and ethics.

INNOVATIVE ELEMENTS

The curriculum's focus on agribusiness was an innovation in Timor-Leste. The emphasis on learning by doing and practical application of theoretical classroom materials was also innovative. The graduate of this program is envisioned to be a learner rather than a passive absorber of information, capable of contributing to problem solving and practical solutions to agribusiness. Teacher skills that emphasized learning versus teaching are an innovative aspect of the project and considered a fundamental principle for success.

IMPLEMENTATION HIGHLIGHTS: BENEFITS AND IMPACT

With project assistance, infrastructure for water and sanitation, dormitories, and school buildings was improved. Each school has enough land for practical work and demonstrations, but additional investment is needed in fencing, irrigation, livestock facilities, and storage. The computer centers and Internet connections at each school are important project contributions. Support from MAP is limited, and Internet connection costs are high. Priority should be given to sustaining the existing computer and Internet arrangements and planning the expansion of the facilities.

The variation in capacity at the three schools affected implementation of the BACET project. Project designers overestimated faculty capacity: The new program required skills that are not found in traditional, technology-oriented programs. The curriculum was adjusted quickly to take faculty capacity into account, but clearly faculty capacity building must continue.

Texts for the various courses were prepared in Bahasa Indonesia by consultants and with inputs from staff and BACET coordinators, but they proved too difficult for teachers to use as testing materials and for students to understand. In response, the curriculum was streamlined and revised courses are being used. The curriculum is designed to be 80 percent practical and 20 percent theoretical, which proved challenging for teachers more familiar with rote theoretical learning in a classroom setting. During the remaining project implementation period, a detailed syllabus for the BACET program will be completed.

Recruiting students is another challenge, perhaps partly owing to the fluidity of society in a postconflict country. The BACET program competes with university recruitment. Efforts are being made to publicize the program, and the midterm evaluation notes that "enrollment is likely to fluctuate until the BACET program is well established, with adequate school facilities, a clearly defined curriculum, and recognition of the enhanced skills and knowledge of the graduates."

The program established successful internships with about 18 organizations, even if some placements were less than satisfactory. Students appreciated the widening of their horizons, and some employers offered jobs to interns after graduation. A big gain was the increase in student confidence after completing the internships.

Employment of graduates

Timor-Leste is a country of high unemployment. Finding a job is difficult for all graduates, including those from the BACET program. Graduate student placement services were part of the project's design. The service is labor intensive and difficult to institutionalize, but the benefit to the schools is that they can learn where their graduates go to work and gain information on where job openings can be found. As of September 2009, 42 percent of the initial group of 121 graduates (2008) was employed, and 26 percent were continuing their education. A slightly higher percentage of women graduates were employed. Groups of graduates have established their own farming businesses and many others have gained employment as community agriculture extension workers or with agriculture livelihood NGOs.

Government commitment

The government was strongly committed to the BACET program. Memoranda of Understanding were signed with GIZ[2] (for teacher development in Indonesia); with Land O'Lakes (for sharing responsibility for the program, including curriculum development, agribusiness support, infrastructure, technology, operational support, and phased-in support for teacher's salaries and student scholarships from 2009); and with Udayana University in Indonesia (to enroll agricultural school teachers to pursue bachelor's degrees). Financial commitment from government proved difficult because of the overall national economic situation, but in-kind support was provided. As noted, MAP will assume full responsibility for the agribusiness course in mid-2011.

LESSONS LEARNED

Lessons include implications for the program's long-term operation as well as the initial lessons from the attempt to augment secondary education with subject matter and teaching methods that were innovative in the context. Flexibility is needed to shape the future of the program in light of lessons from the first phase of implementation and the fragile, postconflict setting.

Long-term strategy for a diploma course

To sustain the program over the long term, one strategy would be for MAP to shift the BACET program from the certificate level to a formal, two-year diploma course, similar to those offered by other countries in the region. Presently, graduates of BACET and the agricultural secondary schools are expected to work as technicians with agricultural skills, field managers, and extension workers, but in the future the agricultural sector would benefit from a pool of graduates trained to the diploma level. There is logic in focusing on the practical planning and agribusiness aspects of commercial agriculture in the first year, followed by a second year of further training and skills development in technical specialties. A strategy to upgrade to a diploma course could have the following elements: continue to seek donor support to increase teachers' level of education (this support is needed because of government budget shortages); send teachers to Udayana University in Indonesia to improve skills and knowledge; review the curriculum periodically; and enhance and upgrade facilities (such as laboratories, libraries, and agricultural equipment) to facilitate learning.

Lessons from the midterm evaluation

At midterm, it is clear that the difficulties of establishing the new agribusiness curriculum were underestimated. As noted, the greatest challenge was that the initial curriculum was too academic and difficult for students, and the faculty lacked the knowledge and pedagogical skills to teach it. Owing to the new subject matter, developing new courses and teaching the new material was a challenge. An additional complication was that 80 percent of the curriculum involved practical work, which required considerable innovation and flexibility to plan production and marketing activities, manage machinery workshops, and guide computer work that maximizes practical learning.

A second challenge was the poor infrastructure (from sanitation and housing to the kinds of equipment needed to support training). Logistics were another problem, because of the geographic location of the schools and the condition of roads and communications. The midterm evaluation observed that it would have been better to start with one school and, based on that experience, scale up. Finally, few students or staff members were familiar with Portuguese (the official curriculum language). Although Tetum is the national language, Bahasa Indonesia is widely used. Teaching materials were not available in Bahasa and had to be developed.

The midterm evaluation contained a number of other observations. Students should be involved in two types of enterprises while in school. One is a school-sponsored enterprise in which students interact with the public by marketing school produce and work in production projects on school farms. The second enterprise, based on the school farms, should have activities structured as agribusinesses, with annual budgets and financial records to improve management and to use in teaching. Budgeting by farm enterprise can help with planning and management. Some farm enterprises will be profitable and some will not, but this is a good teaching/learning situation for students. Graduate enterprises established after students leave the program should receive continuing support from BACET and from the agricultural secondary schools. GIZ will help graduates who wish to start their own enterprises.

Faculty training should also be emphasized. School management should be strengthened so that all responsibilities for administering BACET will reside with BACET when external support ends. Finally, semiannual or annual tripartite reviews should take place to plan for sustainability beyond project completion.

Vocational Training in the Arab Republic of Egypt Combines Technical and Innovation Skills for Agriculture

Charles J. Maguire, Consultant

SYNOPSIS

A pilot intervention in 25 agricultural technical (secondary) schools in Upper Egypt introduced students to higher-level cognitive skills including problem-solving, critical thinking, and decision making, as well as practical training, and grew into a four year project (2008–12) involving 54 such schools. The project was funded by USAID with US$4.9 million and implemented by the Midwest Universities Consortium for International Activities (MUCIA) from 2003 to 2007. The reforms involved pedagogy, practical training skills, and school management. Using technical school farms to train students in the production of export crops proved difficult to implement. On the other hand, the Supervised Internship Program, which enabled students to participate in practical activities on commercial farms and with fruit and vegetable producers and exporters, proved very successful.

CONTEXT

Egypt has 130 secondary-level agricultural technical schools (ATS), each with an average enrollment of 2,750 students. These vocational agricultural high schools have about 154 teachers each, of which 42 percent teach agricultural courses. The teachers are organized into technical departments that include field-crop production, livestock production and animal health, horticulture, agricultural economics, and agricultural mechanics.

All ATSs follow the same basic curriculum, and teachers prepare students for standardized tests at the end of each school year. Most technical agricultural teachers have had no training in pedagogy or preparing lesson plans. Teaching features lectures and rote learning, so little attention is paid to higher-level cognitive skills. Because graduates of ATSs are poorly prepared to use the knowledge and skills obtained in the courses, their employment level is very low. Teaching

equipment is in short supply, and audio-visuals are rarely used. Each school has a farm of about 25 acres, but it is not used for practical demonstration or practice by students.

PROJECT OBJECTIVES AND DETAILED PROJECT COMPONENTS

The pilot project aimed to improve teaching methods used by over 1,600 agricultural teachers in the 25 selected ATSs. It would provide lesson plans, teaching aids, and overhead projectors. It would also introduce active learning, which would promote instructional activities involving students in doing things and thinking about what they were doing. Educators moved from knowledge recall and comprehension toward more advanced cognitive skills, including analysis, synthesis, and evaluation of information and knowledge to solve problems. The school farms enabled students to produce export crops and provided them with the skills to become mid-level technicians on agribusiness farms. The project approach included seven steps, described in box 2.13. See also IAP 3.

INNOVATIVE ELEMENTS

In summary, the innovative elements of the project were its revitalization of the agricultural technical curriculum, especially through teachers' improved capacity to use new pedagogy centered on student learning and equip students with the skills and confidence to become problem solvers. Another innovation was the repurposing of the ATS farms for practical work and high-value agriculture.

BENEFITS AND IMPACT

The project's initial phase was successful in introducing active learning and raising the quality of the curriculum. The Ministry of Education supported an expansion and

Box 2.13 Seven Steps to Improve Teaching Methods and Introduce Active Learning in Egypt's Agricultural Technical Schools

Step 1: Training the ATS teacher in active teaching-learning methods. Forty-five Egyptian university faculty members from different subject matter areas were trained by two highly experienced teacher-educators from partner universities in the Midwest Universities Consortium for International Activities (MUCIA). The teacher-educators conducted a 28-hour practical workshop on active learning strategies for the faculty members, who would serve as future trainers. During the workshop, 15 active learning strategies were taught and practiced. The most effective faculty members emerging from the workshop were selected to conduct similar workshops for teachers from the agricultural technical schools (ATSs). The MUCIA specialists handed over full implementation of the process to the Egyptian trainers over a series of three workshops, and these trainers were given the task of conducting 20 two-day active learning techniques workshops for over 1,000 ATS teachers.

Step 2: Developing instructional materials for ATS teachers. A four-person MUCIA team worked in Egypt for two weeks to initiate this second step. It became clear that ATS teachers did not have access to audiovisual equipment, so the team emphasized developing transparencies that could be produced cheaply and easily and distributed to ATS teachers. Action plans were developed for the procurement of audiovisual materials suitable for the ATSs. Some 4,000 new transparencies were produced in Arabic and distributed to ATSs. In addition, 1,100 overhead projectors and screens have been purchased and installed in each ATS classroom.

Step 3: Developing lesson plans for each ATS course. Lesson plans were prepared, following the basic content of each course derived from the textbook used in the ATSs. A workshop was planned to help teachers use the lesson plans (Step 6).

Step 4: Headmaster study tour to the Netherlands. A one-week study tour for headmasters was implemented in the Netherlands. The objective was to introduce the headmasters to the Dutch vocational education system and to investigate innovative ideas that could be used in Egypt's ATS system.

Step 5: Refocusing ATS school farms and using them for practical skill training. MUCIA sent a university farm manager to develop a work plan that would change the focus of the ATS school farms toward the production of high-income, labor-intensive export crops and give more emphasis to hands-on practical training for students. The project provided a grant to each school to purchase inputs, including seed, equipment, facilities, and tools. The goal was to have these innovations become operational by the beginning of the 2007/08 school year.

Step 6: Training ATS teachers in the use of lesson plans and instructional materials. A series of two-day workshops enabled teachers to effectively use the lesson plans developed in Step 3. The methodology for running the workshops was the same as used in Step 1 (training teachers in active teaching-learning methods).

Step 7: Assessing progress and refining the lesson plans and instructional materials. Project management would meet with ATS teachers and their students to assess the value and impact of the different innovations in improving the teaching-learning process at the 25 ATSs. Adjustments would be made to selected lesson plans and transparencies. The Ministry of Education expressed interest in having lesson plans and transparency sets for each of the 33 courses reproduced and made available to teachers in all 104 ATSs.

Source: Author.

---

continuation of the project to cover 54 ATSs for 2008–12. To date, results from the expanded project include:

- Administrators and headmasters of the 54 ATSs have been introduced to new approaches to managing technical schools, providing a more effective teaching-learning environment, and identifying skills needed by the private firms through observation study tours in the Netherlands and Greece and through in-country training.

- Over 3,700 ATS teachers have been trained to use active teaching-learning methods, and they have been using these new methods in their classrooms.

- Classrooms and laboratories at each of the 54 ATSs have been equipped with 386 computers, 910 overhead projectors, and 54 LCD projectors and screens. Overhead transparencies for 51 technical agricultural courses (approximately 120 transparencies per course) were developed and are in use.

Box 2.14 Views on the Impact of the Supervised Student Internship Program in Egypt

"These are absolutely the best employees who have ever worked on my farm and in my packing station," said Mr. Samy Ibrahim, Managing Director and owner of CELF, one of Egypt's leading exporters of horticultural products to Europe. He was referring to the 122 agricultural technical school interns his company had hired since 2008. Mr. Hussein Mohamed, a teacher who participated in the internship training as supervisor, noted that "The greatest impact of internship on students is that they gain more than just skills. They learn about their strengths, interests, problem-solving skills, and abilities to deal with clients. They also gain greater understanding of science-based agriculture and develop a positive attitude about working in agriculture."

Source: Adapted from USAID Egypt, n.d.

- More school farms at selected ATSs are being transformed to provide more practical and hands-on training for students, especially related to export crops.
- Supervised Student Internship Programs paid by commercial farms and Ownership Programs helped more than 8,000 students to improve their technical and managerial skills in horticulture and livestock farms and agribusiness firms. The internship program, which is part of the USAID-funded Value Chain Training initiative, has been highly successful in increasing capacity and confidence in both students and the teachers who supervise their practical experience (box 2.14). Employers have been impressed at the capability and dedication of the student trainees and have hired many of them in full-time or part-time positions.
- Career development activities provided communication, leadership, and personal growth skills to 7,500 students to prepare them for successful careers in agriculture.
- Student competence and confidence improved as a result of the new pedagogical approach.
- Demand for the newly graduated students from the project has increased.

LESSON LEARNED

This relatively low-cost approach to transforming the teaching-learning process is directly applicable to the other ATSs in Egypt and to vocational agricultural programs throughout the developing world.

The initial phase that focused on creating teacher and administrator capacity to manage the new approach to learning and on creating links to potential employers of ATS graduates enabled all parties to evaluate progress and impact. The result was very favorable and led to an extension of the project for four more years and involving 54 additional schools.

NOTES

Module 2 Overview

1. These generic weaknesses have been identified and documented by individual researchers and by gatherings of specialists around the world. See, for example, FAO, UNESCO, and ILO (1970); Busch (1988); Hansen (1990); GCHERA (1999); Magrath (1999); Foster (1999); Csaki (1999); Maguire (2000, 2007); Hazelman (2002); Muir-Leresche (2003); FAO and UNESCO (2003); Eicher (2006); Ochola and Ekwamu (2008).

2. Globally, few educational institutions consistently provide learning opportunities in public policy management, a need that requires further attention from governments, educational institutions, and donors alike.

Thematic Note 1

1. Association of Public and Land-Grant Universities, "Kellogg Commission on the Future of State and Land-Grant Universities," http://www.aplu.org/page.aspx?pid=305, accessed July 2011.

Thematic Note 2

1. See also "Kellogg Commission on the Future of State and Land-Grant Universities," APLU, http://www.aplu.org/page.aspx?pid=305.

Thematic Note 3

1. In China, the World Bank has supported the Guangdong Technical and TVET Project (2009), Vocational Education Reform Project (1996), Vocational and Technical Project (1997), Vocational Education Reform Project (2003), and Liaoning and Shandong Technical and Vocational Education and Training Project (2010). In India it has supported the Technical/Engineering Education Quality Improvement Project (2003) and the Third Technician Education Project (2001).

2. Based on Liang (2010) and World Bank (2010).

Thematic Note 4

1. These tasks include implementing the project's larger capacity-building components for beneficiaries, who are a significant constituency within the AIS.

Innovative Activity Profile 3

1. This section draws on Swanson, Barrick, and Samy (2007).

Innovative Activity Profile 4

1. Adapted from Adjepong (1999) and Akeredolu and Mutimba (2010).

Innovative Activity Profile 6

1. This activity profile has been adapted from Sherrard (2009) and Zaglul (2011).

Innovative Activity Profile 8

1. Ministério da Agricultura e Pescas.

2. Now GIZ (Gesellschaft für Internationale Zusammenarbeit, http://www.giz.de/).

REFERENCES AND FURTHER READING

Module 2 Overview

Bawden, R. 1998. "Agricultural Education Review. Part II: Future Perspectives." Washington, DC: AKIS Rural Development Department, World Bank.

———. 1999. "Education Needs of the Rural Sector: Looking to the Future." AKIS Discussion Paper, based on a workshop on Education for Agriculture and Rural Development: Identifying Strategies for Meeting Future Needs, December 1–3, World Bank, Washington, DC.

Busch, L. 1988. "Universities for Development: Report of the Joint INDO-U.S. Impact Evaluation of the Indian Agricultural Universities." AID Project Impact Evaluation Report No. 68. Washington, DC: United States Agency for International Development (USAID).

Chacharee, N., and S. Chinnawong, 2009. "Farmer Participatory Approach for Sustainable Agricultural Development: A Case Study on GAP in Vegetable Production Extension in Nakhon Pathon Province, Thailand." Paper presented at the 4th International Conference on Agricultural Education and Environment of the Asia-Pacific Association of Educators in Agriculture and Environment (APEAEN), 3–6 August 2009, Obihiro, Japan.

Csaki, C. 1999. "Change in Higher Agricultural Education." In *Leadership for Higher Education in Agriculture: Proceedings of a Conference Held in Amsterdam, the Netherlands, July 22–24, 1999*, edited by D.G. Acker. Ames: Iowa State University, Global Consortium of Higher Education and Research for Agriculture (GCHERA). Pp. 67–70.

Davis, K., J. Ekboir, W. Mekasha, C.M.O. Ochieng, D.J. Spielman, and E. Zerfu, 2007. "Strengthening Agricultural Education and Training in Sub-Saharan Africa from an Innovation Systems Perspective: Case studies of Ethiopia and Mozambique." IFPRI Discussion Paper No. 736. Washington, DC: International Food Policy Research Institute (IFPRI).

Eicher, C.K., 1999. "Institutions and the African Farmer." Issues in Agriculture 14. Washington, DC: Consultative Group on International Agricultural Research (CGIAR).

———. 2006. "The Evolution of Agricultural Education and Training: Global Insights of Relevance for Africa." Staff Paper No. 2006-26. East Lansing: Department of Agricultural Economics, Michigan State University.

Falvey, L. 1996. *Food Environment Education: Agricultural Education in Natural Resource Management.* Melbourne: Crawford Fund for International Agricultural Research and Institute for International Development Limited.

FAO and UNESCO (Food and Agriculture Organization and United Nations Educational, Scientific, and Cultural Organization). 2003. "Higher Education and Rural Development: A New Perspective." Chapter 5 in *Education for Rural Development: Towards New Policy Responses,* edited by D. Atchorena and L. Gasperini. Rome and Paris.

FAO, UNESCO, and ILO (Food and Agriculture Organization, United Nations Educational, Scientific, and Cultural Organization, and International Labour Organization). 1970. "Report of the World Conference on Agricultural Education and Training, July 28–August 8, Copenhagen." Rome.

Foster, R.M. 1999. "From Local to Global: The Challenge of Change in Agriculture and the Food System." Chapter 11 in *Leadership in Higher Education for Agriculture: Proceedings of the Inaugural Conference of the Global Consortium of Higher Education and Research for Agriculture, July 22–24, Amsterdam,* edited by D.G. Acker. Ames: Iowa State University. Pp. 71–76.

GCHERA (Global Consortium of Higher Education and Research for Agriculture). 1999. *Leadership for Higher Education in Agriculture: Proceedings of a Conference Held in Amsterdam, the Netherlands, July 22–24, 1999,* edited by D.G. Acker. Ames: Iowa State University.

Hansen, G.E. 1990. "Beyond the Neoclassical University: Agricultural Higher Education in the Developing World—An Interpretive Essay." USAID Program Evaluation Report No. 20. Washington, DC: United States Agency for International Development (USAID).

Hazelman, M. 2002. "The Need for Change: From Agriculture Education to Education for Rural Development." In *Searching for New Models of Agriculture Education in a Disturbed Environment: Proceedings of the 1st International Conference on Agriculture and the Environment, Los Baños, Philippines.* Bangkok: Regional Office for Asia and the Pacific, Food and Agriculture Organization (FAO).

Jamison, D.T., and L.J. Lau. 1982. *Farmer Education and Farmer Efficiency.* Baltimore: Johns Hopkins.

Jamison, D.T., and P.R. Moock. 1984. "Farmer Education and Farmer Efficiency in Nepal: The Role of Schooling, Extension Service, and Cognitive Skills." *World Development* 12:67–86.

Lockheed, M.E., D.T. Jamison, and L.J. Lau. 1980. "Farmer Education and Farm Efficiency: A Survey." In *Education and Income,* edited by T. King. World Bank Staff Working Paper No. 402. Washington, DC: World Bank.

Magrath, C.P. 1999. "Reforming U.S. Higher Education." Chapter 14 in *Leadership in Higher Education for Agriculture: Proceedings of the Inaugural Conference of the Global Consortium of Higher Education and Research for Agriculture, July 22–24,* Amsterdam, edited by D.G. Acker. Ames: Iowa State University. Pp. 23–30.

Maguire, C.J. 2000. "Agricultural Education in Africa: Managing Change." Paper presented at Workshop 2000, sponsored by the Sasakawa Africa Association in Accra and at the University of Cape Coast, Ghana.

———. 2004. "Effective Educational Strategies for Poverty Alleviation." In *Proceedings of the 2nd International Conference on Agricultural Education and Environment, October 2004, Souwon, Korea.* Bangkok: Regional Office for Asia and the Pacific, Food and Agriculture Organization (FAO).

———. 2007. "Preparing for the Future: Revisiting Agriculture and Environment Education in Asia." In *Proceedings of the 3rd APEAEN Conference on Agriculture and Environment, held at Munoz City, Nueva Ecija, Philippines.* Bangkok: Regional Office for Asia and the Pacific, Food and Agriculture Organization (FAO).

Moulton, J. 2001. Improving education in the rural areas: Guidance for rural development specialists. Unpublished paper for AKIS, World Bank, Washington, DC.

Muir-Leresche, K. 2003. "Transforming African Agricultural Universities and Faculties: Examples of Good Practice." Paper prepared for the International Seminar series, Sustainability, Education, and the Management of Change in the Tropics (SEMCIT). N.p.: n.p.

Mulder, M. 2010. Personal communication.

Mulder, M., and N. Ernstman. 2006. "The Public's Expectations Regarding the Green Sector and Responsive Practices in Higher Agricultural Education." In *Proceedings of the 8th European Conference on Higher Agricultural Education, The Public and the Agriculture and Forestry Industries, the Role of Higher Education in Questioning Assumptions and Matching Expectations,* edited by M. Slavík and P. Žáková. Prague: Czech University of Agriculture. Pp. 35–53.

Munshi, K. 2004. "Social Learning in a Heterogeneous Population: Technology Diffusion in the Indian Green

Revolution." *Journal of Development Economics* 72(1):185–213.

Ochola, W.O., and A. Ekwamu, 2008. "Agricultural Education and Learning: Challenges, Opportunities, and Experiences from Eastern and Southern Africa." Paper presented to the Conference on Improving Agricultural Education and Learning through Collaboration and Partnerships, December 1–5, Maputo.

OECD (Organisation for Economic Co-operation and Development). 2005. *Synthesis Report.* Vol. 1 of *Governance of Innovation Systems.* Paris.

———. 2010. *The OECD Innovation Strategy: Getting a Head Start on Tomorrow.* Paris.

Pratley, J. 2008. "Workforce Planning in Agriculture: Agricultural Education and Capacity Building at the Crossroads." *Farm Policy Journal* 5(3): 27–41.

Psacharopoulos, G., and M. Woodhall. 1985. *Education for Development: An Analysis of Investment Choices.* New York: Oxford University Press.

Rajalahti, R., W. Janssen, and E. Pehu. 2008. "Agricultural Innovation System: From Diagnostics to Operational Practices." Agriculture and Rural Development Discussion Paper 38. Washington, DC: World Bank.

Rivera, W.M.2008. Transforming Post-secondary Agricultural Education and Training by Design: Solutions for Sub-Saharan Africa. Unpublished report for the World Bank (AFTHD), Washington, DC.

Rygnestad, H., R. Rajalahti, and E. Pehu. 2005. Agricultural Education and Training in Sub-Saharan Africa FY'98–'04. World Bank, Washington, DC.

Snapp, S., and B. Pound 2008. *Agricultural Systems, Agroecology, and Rural Innovation for Development.* Burlington, MA: Elsevier.

Tajima, S. 1985. *Types, Development, Processes, Characteristics, and Trends of World Agricultural Education.* International Encyclopedia of Education. Oxford: Pergamon.

Willett, A. 1998. Investment in Agricultural Education and Training 1987–1997. AKIS, World Bank, Washington, DC.

World Bank. 2004. *Agriculture Investment Sourcebook.* Washington, DC. http://web.worldbank.org/WBSITE/ EXTERNAL/TOPICS/EXTARD/EXTAGISOU/0,,menuP K:2502803~pagePK:64168427~piPK:64168435~the-SitePK:2502781,00.html, accessed July 2011.

———. 2007a. "Cultivating Knowledge and Skills to Grow African Agriculture: A Synthesis of an Institutional, Regional, and International Review, 2007." Agriculture and Rural Development Department and Africa Region Human Development Department. Report No. 40997-AFR. Washington, DC.

———. 2007b. *World Development Report 2008: Agriculture for Development.* Washington, DC.

———. 2009. *Gender in Agriculture Sourcebook.* Washington, DC. http://web.worldbank.org/WBSITE/EXTERNAL/ TOPICS/EXTARD/EXTGENAGRLIVSOUBOOK/0,, contentMDK:21348334~pagePK:64168427~piPK:641684 35~theSitePK:3817359,00.html, accessed July 2011.

———. 2011. *Information and Communication Technologies for Agriculture e-Sourcebook.* www.ICTinagriculture.org, http://bit.ly/ICTinAG. Washington, DC.

Thematic Note 1

Liu, Y., and J. Zhang. 2004. "The Reform of Higher Agricultural Education Institutions in China: A Case Study." Rome and Paris: Food and Agriculture Organization (FAO) and International Institute for Education Planning (IIEP).

MacConnell, S. 2010. "Greater Demand for Agri-food Courses Drives up CAO Points." *The Irish Times,* Aug. 28.

National Council for Agricultural Education. 2000. "The National Strategic Plan and Action Agenda for Agricultural Education. Alexandria, VA. http://www.teamaged.org/ council/images/stories/pdf/plan2020.pdf, accessed July 2011.

Paarlberg, D. 1992. "The Land Grant College System in Transition." *Choices Magazine,* third quarter.

Phelan, James, University College Dublin 2010, personal communication.

Wirakartakusumah, M.A. 2007. "Government-University Relations and the Role of International Cooperation: An Experience of Bogor Agricultural University." Presentation at the 5th GCHERA Conference, March, San José, Costa Rica.

World Bank. 2007a. "Cultivating Knowledge and Skills to Grow African Agriculture: A Synthesis of an Institutional, Regional, and International Review, 2007." Report No. 40997-AFR. Agriculture and Rural Development Department and Africa Region Human Development Department. Washington, DC.

———. 2007b. *World Development Report 2008: Agriculture for Development.* Washington, DC.

Thematic Note 2

Hansen, G.E. 1990. "Beyond the Neoclassical University: Agricultural Higher Education in the Developing World—An Interpretive Essay." A.I.D. Program Evaluation Report No. 20. Washington, DC: United States Agency for International Development (USAID).

McCalla, A.F. 1998. "Agricultural Education, Science, and Modern Technology's Role in Solving the Problems of Global Food Resources in the 21st Century." Paper prepared for the conference "Globalizing Agricultural Higher Education and Science: Meeting the Needs of the 21st Century," National Agricultural University of Ukraine, Kiev, September 28–30.

Regional Universities Forum for Capacity Building in Agriculture (RUFORUM). 2010. "Report of the Strategic Reflection Meeting of the Regional Universities Forum for Capacity Building in Agriculture (RUFORUM), April 12–16, 2010, Bellagio, Italy."

Ruffio, P., and J. Barloy. 1995. "Transformations in Higher Education in Agricultural and Food Sciences in Central and Eastern Europe." *European Journal of Agricultural Education and Extension* 2(2):11–19.

Swanson, B.E, R.K. Barrick, and M.M. Samy. 2007. "Transforming Higher Agricultural Education in Egypt: Strategy, Approach, and Results." In *Proceedings of the 23rd Annual Meeting of the Association for International Agricultural and Extension Education: "Internationalizing with Cultural Leadership."* Pp. 332–45. AIAEE, http://www.aiaee.org/attachments/484_Compiled.pdf, accessed February 2011.

Villareal, R. 2002. "Status of Agriculture and Natural Resources Management Education in Southeast Asia." Paper presented at SEMCIT, Session IV, Maejo University, Chiang Mai, September 29–October 5.

Wallace, I. 1997. "Agricultural Education at the Crossroads: Present Dilemmas and Possible Options for the Future in Sub-Saharan Africa." *International Journal of Educational Development* 17(1):27–39.

World Bank. 2007a. "Cultivating Knowledge and Skills to Grow African Agriculture: A Synthesis of an Institutional, Regional, and International Review, 2007." Report No. 40997-AFR. Agriculture and Rural Development Department and Africa Region Human Development Department. Washington, DC. P. 58.

Thematic Note 3

Asian Development Bank (ADB). 2008. *Education and Skills: Strategies for Accelerated Development in Asia and the Pacific.* Manila.

Blackie, M., M. Mutema, and A. Ward. 2009. A Study of Agricultural Graduates in Eastern and Central Africa: Demand, Quality, and Job Performance Issues. Report prepared for the Association for Strengthening Agricultural Research in Eastern and Central Africa (ASARECA), Entebbe.

Blom, A., and J. Cheong (eds.). 2010. "Governance of Technical Education in India: Key Issues, Principles, and Case Studies." Working Paper No. 190. World Bank: Washington, DC.

Liang, X. 2010. "Global Trends in Technical and Vocational Education and World Bank Support in China." Presentation at the Innovation Workshop, Quingdao, June 11–12.

UNEVOC (International Project on Technical and Vocational Education). 1996. "Current Issues and Trends in Technical and Vocational Education." Study No. 8. Paris: Section for Technical and Vocational Education, United Nations Economic, Scientific, and Cultural Organization (UNESCO).

World Bank. 1991. "Vocational and Technical Education and Training: A World Bank Policy Paper." Washington, DC.

———. 2004. Ethiopia Post Secondary Education Project. Report No. 28169-ET. Internal document. Washington, DC.

———. 2006. Ethiopia Rural Capacity Building Project (ID number PO79725). Internal document. Washington, DC.

———. 2010. "Governance of Technical Education in India: Key Issues, Principles, and Case Studies." A. Blom and J. Cheong (eds). Working Paper No. 190. World Bank, Washington, DC.

Thematic Note 4

World Bank. n.d. "Indonesia: Tree Crops Human Resources Development Project." http://web.worldbank.org/external/projects/main?pagePK=64283627&piPK=73230&theSitePK=40941&menuPK=228424&Projectid=P003899, accessed July 2011.

Innovative Activity Profile 1

Busch, L. 1988. "Universities for Development: Report of the Joint INDO-U.S. Impact Evaluation of the Indian Agricultural Universities." AID Project Impact Evaluation Report No. 68. Washington, DC: United States Agency for International Development (USAID).

World Bank. 1995. "India: Agricultural Human Resources Development Project." World Bank Report 13517. Washington, DC.

———. 2002. "Implementation Completion and Results." Report 24287. Washington, DC.

Innovative Activity Profile 2

Kropff, M. 2010. "Connecting Knowledge and Society: Scientific Capacity Building for Impact." Presentation to the RUFORUM Ministerial Meeting on Education in Agriculture in Africa (CHEA), November 15–19, Kampala.

Mulder, M. 2010. Personal communication.

Rabbinge, R., and M.A. Slingerland. 2009. "Change in Knowledge Infrastructure: The Third-generation University." In *Transitions towards Sustainable Agriculture and Food Chains in Peri-urban Areas,* edited by K.J. Poppe, C. Temeer, and M. Slingerland. Wageningen: Wageningen Academic Publishers. Pp. 51–62. http://www.onderzoek enadvies.org/wp-content/uploads/2010/01/transitions.pdf, accessed July 2011.

Innovative Activity Profile 3

Swanson, B.R., K. Barrick, and M.M. Samy. 2007. "Transforming Higher Agricultural Education in Egypt: Strategy, Approach, and Results." In *Internationalizing with Cultural Leadership, Proceedings of the 23rd Annual Conference of the Association for International Agricultural and Extension Education (AIAEE), 20–24 May, Polson, Montana,* edited by M. Navarro. Pp. 332–42. AIAEE, http://www.aiaee.org/attachments/484_Compiled.pdf, accessed February 2011.

Vreyens, J.R., and M.H. Shaker. 2005. "Preparing Market-Ready Graduates: Adapting Curriculum to Meet the Agriculture Employment Market in Egypt." In Proceedings of the 21st *Annual Conference of the Association for International Agricultural and Extension Education (AIAEE), San Antonio, Texas.* Pp. 227–235. http://www.aiaee.org/proceedings/119-2005-san-antonio-texas.html, accessed July 2011.

Innovative Activity Profile 4

Adjepong, S.K. 1999. "The University and National Development: An Overview of an Innovative Programme at the University of Cape Coast, Ghana for Mid-career Agricultural Extension Staff." Presentation to World Bank staff at Rural Week, Washington, DC.

Akeredolu, M., and J.K. Mutimba. 2010. Winrock International, Kano presentation at the First Intercontinental Meeting of the Global Forum for Rural Advisory Services and the 16th Annual Meeting of the Neuchatel Initiative Viña del Mar, Chile, 2–5 November 2010.

Sasakawa Africa Association. 2008. "SAFE." *Feeding the Future: Newsletter of the Sasakawa Africa Association* 24:10–11. http://www.saa-tokyo.org/english/newsletter/pdf/issue24.pdf, accessed July 2011.

Zinnah, M.M. 2003. "Innovations in Higher Agricultural Education in Africa: The Case of Sasakawa Fund for Extension Education (SAFE) Programme in Selected Universities and Colleges in Sub-Saharan Africa." Paper prepared for SEMCIT Seminar Session V, Oslo, Norway. ILN-Africa, http://www.iln-africa.net/uploads/documents/case_studies/Case_Study_Sasakawa.doc, accessed February 2011.

Innovative Activity Profile 5

Opatpatanakit, A. 2007. "Community-Based Research (CBR) Approach as a Research Methodology to Enable Students to Empower Rural Community." Presented at the 3rd Asia Pacific Association of Educators in Agriculture and Environment (APEAEN) International Conference on Agriculture, Education and Environment, 4–7 November 2007, Philippine Carabao Center, Munoz, Nueva Ecija.

Innovative Activity Profile 6

Sherrard, D. 2009. "A Private University's Model and Vision for Agricultural Education – EARTH University, Costa Rica." Seminar at the World Bank, October 14, Washington, DC.

Zaglul, J. 2011. Paper in Proceedings of the 4th International Conference on Agriculture Education and Environment, August 3–6, 2009, Obihiro University of Agriculture and Veterinary Medicine, Obihiro, Hokkaido, Japan. Bangkok: Food and Agriculture Organization (FAO), Regional Office for Asia and the Pacific.

Innovative Activity Profile 7

Mulder, M. 2007. "Competence: The Essence and Use of the Concept in ICVT." *European Journal of Vocational Training* 40:5–21. [Available in FR, DE, ES, and PT.]

———. 2010. Personal communication.

Mulder, M., J. Gulikers, H. Biemans, and R. Wesselink. 2009. "The New Competence Concept in Higher Education: Error or Enrichment?" *Journal of European Industrial Training* 33 (8/9): 755–70.

Innovative Activity Profile 8

Land O'Lakes Inc. n.d. "Building Agribusiness Capacity in Timor Leste. http://www.idd.landolakes.com/PROJECTS/Asia/ECMP087511.aspx, accessed July 2011.

Land O'Lakes Inc. n.d. "Young Women in Timor-Leste Find Opportunity and Income from Agribusiness Training." http://idd.landolakes.com/PROJECTS/Asia/ECMP2-0113018.aspx, accessed July 2011.

Land O'Lakes Inc. 2009. Cooperative Agreement # 486-A-00-06-00011-00 Program Update. USAID, http://pdf.usaid.gov/pdf_docs/PDACO315.pdf, accessed July 2011.

USAID (United States Agency for International Development) and Land O'Lakes Inc. 2009." Building Agribusiness Capacity in East Timor (BACET)." http://www.idd.landolakes.com/stellent/groups/public/documents/web_content/ecmp0095377.pdf, accessed July 2011.

Innovative Activity Profile 9

Swanson, B. E., R. K. Barrick, and M. M. Samy. 2007. "Transforming Higher Agricultural Education in Egypt: Strategy, Approach, and Results." In *Proceedings of the 23rd Annual Meeting of the Association for International Agricultural and Extension Education: "Internationalizing with Cultural Leadership."* Pp. 332–45. AIAEE, http://www.aiaee.org/attachments/484_Compiled.pdf, accessed February 2011.

Swanson, B., J. Cano, M. M. Samy, J. W. Hynes, and B. Swan. 2005. "Introducing Active Teaching-Learning Methods and Materials into Egyptian Agricultural Technical Secondary Schools." In Proceedings of the 23rd Annual Meeting of the Association for International Agricultural and Extension Education (AIAEE), Polson, Montana. Pp. 343–51. AIAEE, http://aiaee.tamu.edu/2007/Accepted/343.pdf, accessed September 2011.

USAID (United States Agency for International Development) Egypt. n.d. "Success Story: Turning Internships into Jobs: Major Agricultural Exporters Hiring Upper Egypt ATS Students," MUCIA, http://www.mucia-vct.org/successstories/Internship-%20En.pdf, accessed July 2011.

Investment in Extension and Advisory Services as Part of Agricultural Innovation Systems

OVERVIEW

Kristin Davis, Global Forum for Rural Advisory Services (GFRAS)
Willem Heemskerk, Royal Tropical Institute (KIT)

EXECUTIVE SUMMARY

Extension and advisory services are integral to the AIS, where now more than ever they play a brokering role, linking key actors such as producer organizations, research services, and higher education. This module looks at the history and current status of extension and advisory services and examines important topics such as pluralism, new roles for extension, new kinds of service providers, ICTs, and agribusiness.

For strong extension and advisory services, it is important to have coordination and linkage within pluralistic, multistakeholder AIS. Less traditional actors such as farmer organizations and agrodealers are important extension and advisory service providers who are vital to include in the design of investments and programs. Extension and advisory services must be ever-adapting to the needs of clients, and they must monitor and evaluate their services.

Perhaps the broadest challenge is the tremendous need for new capacities within extension. Throughout the developing world, evolving demands and new roles for advisory services in the wider innovation system will require investments in the capacity of individual extension workers and organizations for value chain approaches, in market-oriented extension, in group and organizational development, in agribusiness, and in mechanisms to share information (networks, platforms, and the like). Recent global developments require advisory services to focus on climate change, food security, and equipping rural people to deal with risk in general.

To better serve their constituencies and influence policies, advisory services need a stronger voice at the global and regional level. There is a need for evidence-based direction regarding investment priorities and programming options for agricultural advisory services within innovation systems. Policy issues related to pluralistic advisory services and extension include the changing roles of various extension providers, the comparative advantage for different providers in carrying out specific extension functions and advisory services, sustainability, and equity. Paradigm shifts—from the perception that research knowledge can drive innovation to the notion that change in the whole system is needed for innovation—must take place not only in the programs and the thinking of field staff but in the thinking of extension administrators and policy makers.

DEFINING AGRICULTURAL EXTENSION AND ADVISORY SERVICES FROM AN INNOVATION SYSTEMS PERSPECTIVE

Many definitions, philosophies, and approaches to agricultural extension and advisory services exist, and views of what extension is all about have changed over time. When agricultural extension services were implemented widely in developing countries in the 1970s, the needs, expectations,

> **Box 3.1** **Extension and Advisory Services, Defined**
>
> In this module, extension and advisory services are defined as systems that facilitate the access of farmers, their organizations, and other value chain and market actors to knowledge, information, and technologies; facilitate their interaction with partners in research, education, agribusiness, and other relevant institutions; and assist them to develop their own technical, organizational, and management skills and practices as well as to improve the management of their agricultural activities.
>
> *Sources:* Birner et al. 2009; Christoplos 2010.

perceptions, and tools that defined extension differed from those we have today. At that time, extension focused very much on increasing production, improving yields, training farmers, and transferring technology. Today extension is no longer viewed as an *agency* but as a *system* that is integral and central to innovation systems and that focuses on facilitating interaction and learning rather than solely on training farmers. CGIAR research on agricultural extension from an innovation systems perspective shows that it has a vital role to play in helping to strengthen capacities to innovate and broker linkages (Spielman et al. 2011). Box 3.1 explains how the term "extension and advisory services" is used in this module. Many other extension terms are included in the glossary for this sourcebook.

OTHER ROLES AND IMPACTS OF EXTENSION AND ADVISORY SERVICES

Besides being an important part of innovation systems, extension and advisory services contribute directly to economic growth, poverty reduction, and environmental well-being. Extension is an essential tool for dealing with the serious challenges facing agriculture—such as climate change, high food prices, and the degradation of natural resources—while helping to increase productivity and reduce poverty (Davis 2009). Other roles for advisory services therefore involve such diverse functions as providing market information, phytosanitary and epidemiological information, information on access to credit, or the facilitation of access to sources with this information (see also TN 1 and TN 4). Although it is very difficult to show the impact of extension services, and while evidence on the impact of some major extension models has been mixed, extension

has proven to be a cost-effective means of increasing economic returns for farmers and has had significant and positive effects on knowledge, adoption, and productivity (see, for example, Birkhaeuser, Evenson, and Feder 1991). A CGIAR meta-analysis of 292 research studies found median rates of return of 58 percent for investments in advisory services (Alston et al. 2000; Dercon et al. 2008).

Apart from yielding significant financial returns, advisory services have also yielded positive social returns, particularly for women, people with low literacy levels, and farmers with medium landholdings (as shown by CGIAR research on extension by Davis et al. 2010b). Some extension programs, such as Farmer Field Schools (FFSs), have shown positive impacts on the environment and health (Praneetvatakul and Waibel 2006).

Despite calls for privatization, government must play a continuing role in extension (see Rivera and Alex 2004; Swanson and Rajalahti 2010). Although a variety of public and private services are available to farmers, many tasks of extension and advisory services have a public goods nature, including tasks related to regulation, quality control in the produce supply chain, the coordination of service provision, and natural resource management, as well as the provision of services to marginal groups, which are unlikely to access or afford private advisory services. The public sector's role is to fund the provision of advisory services (directly or through outsourcing) where demand for services is not being met, to support advisory services in addressing issues of long-term social and ecological sustainability (including food security), and to manage extension and advisory services (including quality control and knowledge management). The public sector can also provide incentives for nonpublic actors to play a greater role in providing services. In pluralistic extension systems, space can be created by the public sector to shift some public investment toward the management of extension systems and strengthening of private actors' capacities, although this shift can come about only when there is ownership within the public sector for such changes (Christoplos 2010; Spielman et al. 2011). Embedded advisory services in input supply services are widespread and increasing (IAP 1), but coordination by the public sector at the local level is needed to regulate and certify those services (to prevent them from providing biased information, for example) and to facilitate interaction between public and private service providers.

EVOLUTION OF EXTENSION AND ADVISORY SERVICES

The renewed prominence of agriculture on the development agenda has renewed the focus on agricultural extension and advisory services. At the same time, strong demands for "more extension" have emerged from unexpected sources: the growing need to provide more climate information, increasing food security programming, the changing aid-for-trade agenda, value chain development programs, and comprehensive reform in global agricultural research for development. These demands imply a need to apply existing knowledge as well as a need to explore the relevance of changing extension forms within new development agendas, aid architectures, and institutional structures (Christoplos 2010, 6,9).

Despite the recognition that traditional approaches to advisory services are not always appropriate or effective, no consensus has emerged on what expanded extension services should actually include. Past mistakes will be repeated if there is not greater awareness of what has worked and what has not, what has proven sustainable and what has not, and who has accessed and benefited from different forms of extension services. Several publications discuss these issues at length (see, for example, Leeuwis and van den Ban 2004; Birner et al. 2009; Christoplos 2010; Hoffmann et al. 2009; and Swanson and Rajalahti 2010). Here we briefly examine the changing nature of extension investments over time, outline how and why advisory services have evolved, and present some of the newer approaches and their goals.

Changing investment levels

Extension investments have been made by donors, various governments, (international) NGOs, and the private sector. The type and level of investments varied considerably over the past few decades, especially as extension approaches rose and fell in popularity (box 3.2).

Many governments have over the years reduced their investment in extension and advisory services, leaving the services without operational resources and forced to

---

Box 3.2   Past and Current Investment Levels in Agricultural Advisory Services

Numerous donors, investors, private companies, and virtually all governments invest in extension, although the precise amounts of their investments are difficult to obtain. Global public investments in extension were estimated at US$6 billion in 1988, and currently two initiatives seek to update this estimate. The Food and Agriculture Organization (FAO) recently surveyed investments in nine agricultural sectors worldwide, including extension. With the International Food Policy Research Institute, FAO is also conducting a worldwide extension assessment that will provide a better idea of investments in physical and human capital, as well as other data, in the near future.

Bilateral and multilateral donors have invested in national extension systems and in extension approaches such as Farmer Field Schools and farmer research groups. World Bank lending to the agricultural sector more than doubled between 2006 and 2009, to US$5.3 billion in FY09 from US$2.9 billion in the baseline years 2006–08. Agricultural research, extension, and education services did not benefit from this increase nearly as much as other agricultural subsectors. Most of the additional lending went for productive infrastructure and policy lending. World Bank support for agricultural research, extension, and agricultural education has been around US$120 million per year during 2007 and 2008, with a significant share going to Africa. Annual lending to these subsectors has fluctuated widely, with lows of around US$100–126 million in some years (2003, 2008, and 2007) and highs of US$499 million in 2006, US$582 million in 2009, and around US$300 million in 2010.

World Bank investments in extension services often consist mainly of small investments accompanying investments in improved agricultural productivity and market linkages. Notable exceptions have included large investments in research and extension system linkages as well as sweeping reforms of extension systems. For example, with World Bank and other support, governments have invested heavily in designing and implementing new extension models such as Uganda's National Agricultural Advisory Services approach (described in box 3.7) and Ethiopia's farmer training center approach. The private sector has also invested in extension, including British American Tobacco, Nestlé, and horticultural and brewing companies. In many (particularly East African) countries, the export crop subsectors have organized the delivery of services, including extension, by sector, financed through export levies and district marketing fees and taxes.

Sources: Swanson, Farner, and Bahal 1990; Davis 2008; Davis et al. 2010b; World Bank Rural Portfolio Team.

continue providing blanket recommendations promoted through ever-repeated demonstration trials. The newly developing extension constituency, based on strengthening farmer organizations, the private sector, and NGO-supported advisory services, has evoked strong attention to extension in the Comprehensive African Agriculture Development Programme (CAADP) and the related Framework for African Agricultural Productivity. Outside Africa, increased attention to extension is expressed through the Global Forum for Rural Advisory Services (GFRAS). CAADP and the corresponding compact agreements at the country level advocate sharpening the focus and efficiency of service provision by basing it on farmers' actual demands, avoiding blanket recommendations, working with existing farmer groups, aiming for matching funds from value chain actors, and using new tools such as ICTs (box 3.3). The sustainability of service provision has become an important part of advisory service strategies. CAADP compact agreements also commit national governments to invest more in extension and not to rely on donor funding. In Uganda, for example, the percentage of the national budget allocated to extension (the National Agricultural Advisory Services—NAADS) gradually increased from 0.3 percent in 2003 to 2.6 percent in 2011, while significantly increasing as a percentage of the agricultural budget.

Changing approaches

Traditional approaches to extension changed as they encountered criticism for being top-down, unaccountable to users, biased against women, oriented to production and technology rather than to markets, and focused on blanket recommendations that did not take the diversity of farm households' circumstances into account. Such criticism generally stemmed from a combination of factors: a lack of relevant technology; failure by research and extension to understand and involve their clients in defining and solving problems; a lack of incentives for extension agents; and weak links among extension, research, farmers, and market actors (Davis 2008). In many countries, policies that favor economic liberalization have enabled farmers to become more market-oriented and entrepreneurial, creating the demand for extension services to advise farmers not only on production issues but on issues related to accessing markets. Training in marketing skills has become much more important for extension workers (Dixie 2005).

A number of approaches sought to overcome these problems and meet new demands on advisory services. The more traditional training and visit (T&V) extension model

(Benor and Baxter 1984) was superseded by approaches pioneered on a small scale by NGOs, FAO, and bilaterally funded projects. These approaches emphasized participatory learning and action models, with farmer participation and more tailor-made services, including facilitation of access to financial services and access to markets. National and international efforts to revitalize extension brought about a variety of institutional reforms (Rivera and Alex 2004), informed primarily by market-led and demand-driven perspectives. For an example from India, see box 3.4.

Particularly in open and democratizing societies, and especially through innovations in communications, farmers are drawing information from an increasing range of sources. Their knowledge and innovation system has become quite diverse (Engel and Salomon 1997). Modern advisory service systems reflect this diversity and complexity in the range of approaches they use, their content, and their interaction with public and private entities. The term "pluralistic" is often used to capture the emerging diversity of institutional forms for providing and financing agricultural extension (TN 1). New actors are offering and funding advisory services, including NGOs, farmer organizations, the private sector, and community-based organizations. This pluralism is almost certain to prevail and deepen with respect to organizational forms, methods, and institutional structures.

Emerging innovative approaches

Group-based and participatory approaches to providing advisory services are gaining ground. These methods have the potential to overcome barriers to participation, foster inclusiveness, and lead to more demand-driven services. They all aim to strengthen the voice of farmers and channel their knowledge into agricultural extension, eventually contributing to farmer empowerment in service delivery and in value chain development (Nederlof, Wennink, and Heemsekerk 2008; KIT, Faida Mali, and IIRR 2006).

Farmer groups (contact groups) were introduced in the T&V extension model, mainly because it was more efficient to transfer information to groups rather than individuals. Subsequent experience with farmer extension groups in participatory planning and field schools (FFSs) has expanded farmer organizations' involvement in providing extension services and in farmer-to-farmer ("F2F") extension, further facilitated by mobile telephony (subjects discussed in boxes 3.3 and 3.5–3.6). For example, in the district participatory planning model used in Mozambique, farmer consultative

Researchers associated with the Consultative Group on International Agricultural Research have shown that telecommunications infrastructure helps to reduce poverty and provide opportunities to people in developing countries (Torero, Chowdhury, and Bedi 2006). In the context of rural advisory services that support innovation, ICTs have three broad functions:

- ***ICTs address the need for localized and customized information***—adapted to rural users in a comprehensible format and appropriate language—to give small-scale producers as well as providers of advisory services adequate, timely access to technical and marketing information.
- ***ICTs store information for future reference.*** In many cases, information on technologies and good practices is available only in hardcopy, and data are incomplete, scarce, or useless. Local and indigenous knowledge is often transmitted orally, records are often unavailable, or the information is dispersed. A proper information system for rural users with standardized formats to compile, document, and share information renders that information more useful, secure, and accessible.
- ***ICTs facilitate the creation of networks*** locally, regionally, and globally, leading to collaborative and interdisciplinary approaches to problem-solving and research diversification through shared knowledge bases, online forums, and collaborative spaces.

Sources: World Bank 2011; Davis and Addom 2010.

Many NGOs, research organizations, and national ministries have improved access to technologies and knowledge for their rural advisory services by means of rural telecenters and online forums.

Throughout the developing world, ICTs are being integrated into rural advisory services in a variety of forms, including rural radio, television, Internet, and mobile services. The advice and information provided via ICTs is becoming more varied, ranging from information about specific technologies and practices to information that enables climate change mitigation and adaptation; disaster management; early warning of drought, floods, and diseases; price information; political empowerment; natural resource management; agricultural information; production efficiency; and market access. ICTs also open new channels for farmers to document and share experiences with each other and with experts. The *Information and Communication Technologies for Agriculture e-Sourcebook* (World Bank 2011) features many examples of these applications.

Although many extension and advisory service providers are using "e-extension" or "cyber-extension" to improve their outreach to farmers and farmers' access to information, most of these initiatives are at early pilot stages and limited empirical evidence is available on the effectiveness of ICTs in extension.

councils orient the investment of district economic development funds in local projects developed by farmer associations. The associations receive support to develop business plans for the selected projects, many of which include the provision of extension services (see TN 2). The FFS approach (see box 3.5 and IAP 2) enhances interactive learning between farmers and between farmers and service providers. More recently, the involvement of farmer groups has been emphasized in the formation of "modern" cooperatives to develop enterprises and access financial services—savings and credit cooperatives (SACCOs) are an example (Heemskerk and Wennink 2004; Wennink, Nederlof, and Heemskerk 2007).

ICTs have created more options for providing advisory services (box 3.3, table 3.1) and are increasingly used to circulate market, price, and weather information as well as to

offer specific kinds of extension advice (see World Bank 2011 and an example for animal health services in Kenya in box 3.6). At the same time, informal advisory systems, such as farmer-to-farmer dissemination of knowledge and technology, are increasingly recognized and built upon in pluralistic extension systems (see TN 1, box 3.12).

PRINCIPLES FOR DEVELOPING EFFECTIVE EXTENSION AND ADVISORY SERVICES

The specific level of investment in extension and the particular reform strategies to be followed will depend on the national context, including the current configuration of the actors in the extension and advisory service system (Birner et al. 2009). It is not sufficient to find an approach that worked in one country or district and implement it in

Box 3.4   Agricultural Technology Management Agency in India

The Agricultural Technology Management Agency (ATMA) is a market-oriented, decentralized approach to extension that many regard as a successful model of extension reform. The ATMA model attempts to increase farm income and rural employment by integrating extension programs across line departments, linking research and extension, and using bottom-up planning. Building blocks of ATMA include empowerment of farmers through farmer interest groups (FIGs), delivery of services to FIGs by diverse service providers, use of bottom-up planning relying on FIG representatives (consultation on farmers' needs and demands), and autonomy of the extension system. Coordination of extension service providers is an essential element. The impact of ATMA is well detailed (Swanson and Rajalahti 2010, 114).

Among the many lessons learned from ATMA, one of the most valuable is that extension should be more decentralized and bottom-up for the following reasons:

- Like agroecological conditions, markets for high-value crops and products are location-specific. Extension and farmers must identify and consider *which* high-value crops have the highest potential for success in each area. The most effective approach is to identify innovative farmers within similar areas

who have started producing and marketing specific products.

- Extension must formally establish steering or advisory committees to identify the specific needs and priorities of representative farmers in each district, including rural women. For example, under the ATMA model, 30 percent of the places on each Farmer Advisory Committee and Governing Board were allocated for rural women.
- Extension can better serve male and female farmers by allowing private firms to play a role in "disseminating" product innovations and focusing public extension services more on process innovations, in which extension personnel serve as facilitators or innovation brokers (see TN 4).
- Innovative farmers play a key role in identifying and then scaling up process innovations (in farmer-to-farmer extension).

Scaling up of the ATMA model has been attempted with varying success. Successful scaling up often relied on sufficient attention to capacity-building to public extension providers (bottom-up planning, group formation, new extension methodology) as well as the allocation of sufficient resources for operational costs. In the absence of these characteristics, the model was less successful.

Sources: Singh et al. 2006; Anderson 2007.

another. Even though extension reforms must be tailored to local conditions, it is valuable to begin designing and developing more effective and sustainable extension and advisory services by considering several approaches to reform. These include reforms in governance structures, reforms in capacity and management, and reforms in advisory methods (table 3.1). Investment options and examples of these principles are provided in TN 1–4.

Many countries, especially those under pressure from democratic decentralization, have embarked on reforms that bring services closer to farmers. Under these reforms, participatory planning and resource allocation occur at the district level, and district agricultural offices coordinate the

provision of services. Examples include NAADS in Uganda (box 3.7) and the National Agricultural Extension Program (PRONEA, Programa Nacional de Extensão Agrária) in Mozambique (see box 3.12 in TN 1). Ethiopia has embarked on an ambitious plan to bring advisory services to its most local administrative level. An intensive review of the extension system was led by CGIAR researchers in 2009 (box 3.8).

Decentralization and the demand for market-oriented services have heightened the need for district and provincial governments to involve private service providers in extension, either through close coordination with private agencies or by contracting them to provide services. These kinds of outsourcing models exist in Uganda,

Box 3.5	Farmer Field Schools for Participatory Group Learning

Farmer Field Schools (FFSs) consist of groups of people with a common interest, who get together on a regular basis to study the "how and why" of a particular topic. The FFS is particularly suited and specifically developed for field studies, where hands-on management skills and conceptual understanding (based on nonformal adult education principles) are required.

So what are the essential and original elements of FFSs? FFSs are a participatory method of learning, technology development, and dissemination based on adult-learning principles such as experiential learning. Groups of 20–25 farmers typically meet weekly in an informal setting on their farms with a facilitator. The defining characteristics of FFSs include discovery learning, farmer experimentation, and group action. The approach is an interactive and practical method of training that empowers farmers to be their own technical experts on major aspects of their farming systems. Farmers are facilitated to conduct their own research, diagnose and test problems, come up with solutions, and disseminate learning to others.

Source: Davis 2008.

Box 3.6	Mobile Telephony for Delivering Animal Health Services

FARM-Africa, an NGO working in Kenya in conjunction with the government and other stakeholders, developed a decentralized animal healthcare system in its Kenya Dairy Goat and Capacity Building Project (KDGCBP). To link key participants in the system, the project approached the Safaricom Corporation, the corporate social responsibility arm of the mobile phone company Safaricom. The KDGCBP system works with a community animal health worker, who purchases a veterinary drug kit and mobile phone at a subsidized price. The project also installs community phones, which have solar panels and batteries where there is no electricity, at veterinary shops. The owner of the community phone is responsible for repairs and can make a profit by charging for its use; for the private veterinarians, the phone is a means of diversifying income. Animal health assistants and vets working with the project also receive mobile phones. The phone system allows animal healthcare providers to update one another, share information, and conduct referrals. This system has reduced transaction costs and increased the efficiency of animal healthcare in the area.

Source: Kithuka, Mutemi, and Mohamed 2007.

Table 3.1	Approaches for Developing Effective Extension and Advisory Services	
Approach	**Definition**	**What is needed***
Reform of governance structures		
Decentralization and deconcentration	Based on the subsidiarity principle, the planning, financing, and administration of extension services occur at the lowest possible state administrative level.	General decentralization policies that are effectively implemented; demand-driven services for diverse farming systems; limited public goods character and nonlocal externalities of the extension messages; earmarking of funding in case of fiscal decentralization of extension to local governments; political will to build and maintain capacity for extension at the local level.
Strengthening of pluralism through outsourcing between public and private sector	Local extension systems that are based on coordination between public and private service delivery, complemented by contracting for services based on needs.	Capable service providers from private and third sector,** or sufficient resources to build this capacity; competition among service providers; recognition of the governance and procurement problems involved in outsourcing and adequate steps to overcome them, including building the extension agency's capacity to manage contracts.
Involving farmer organizations	Farmer involvement in extension service provision, from participatory planning to procurement to farmer-to-farmer extension and paying for services.	Existing social organizations (social capital); absence of strong social hierarchies; availability of sufficient resources to invest in social mobilization and group formation, especially if previous conditions are not met.

(Table continues on the following page)

Table 3.1 Approaches for Developing Effective Extension and Advisory Services (continued)

Approach	Definition	What is needed*
Privatization and public-private partnerships	Services (partially) paid by farmers themselves, directly or indirectly.	Commercialized farming systems with adequate market infrastructure; suitable business climate for the agribusiness sector; required market-oriented extension services. An example is the marketing extension approach, based on farmer training and market information.
Cost-recovery	Part of the operating costs of services paid by farmers in cash or kind to ensure that they get what they want and that the system is more financially sustainable.	Commercialized systems; possibility to embed in contract farming or link to the sale of inputs; possibilities to raise levies on commodities (such as export crops).
Reform of capacity and management		
New public management	Use of private sector principles such as those for human and financial resource management (performance contracts, costing, and financial transparency).	Fit with general public sector reform approaches and relatively autonomous extension organizations.
Business process reengineering	The analysis and design of workflows and processes within an organization.	In reviewing hierarchical structures and reporting systems.
Reform of advisory methods		
Farmer Field Schools	Farmer-centered learning groups, eventually facilitated by farmers (farmer-to-farmer extension).	Complex technologies that require substantial learning (for example, technologies that must be adapted to diverse agroecological conditions) and/or behavioral changes.
Use of information and communication technologies (ICTs)	ICTs as a means for wider access to information.	Adequate countrywide ICT infrastructure. Capacity of users (e.g., literacy) required in many cases. Appropriate language needed.

Source: Birner et al. 2009.

* See "New Directions, Priorities, and Requirements for Investment" (in this module) and TN 1 for ideas on how to implement advisory services of this kind.

** Consisting of NGOs and organizations based on collective action.

Box 3.7 National Agricultural Advisory Services in Uganda

The Government of Uganda created the National Agricultural Advisory Services (NAADS) through the 2001 NAADS Act to provide a decentralized, pluralistic, contract-based agricultural advisory system that would improve farmers' productivity and livelihoods. Local governments contract for NAADS advisory services based on needs identified by local farmer groups, organizations, and farmer forums. District governments provide some additional funding for those extension activities and help set priorities.

Creating a totally new organizational and management structure for a national extension system takes considerable time, both for hiring new staff and for organizing farmers to help set extension priorities, monitor extension programs, and track expenditures. Under NAADS, public extension workers were phased out progressively across regions of the country. Most of these workers were rehired by the private firms and NGOs that participate in NAADS and were assigned to new positions and service areas. This transformation

has had its challenges, such as public extension workers' dissatisfaction with short-term, performance-based contracts and the lack of a civil service job guarantee. Another challenge was the limited availability of resources to train and improve the skills and knowledge of the "new" privately employed advisors, who needed to know how to organize farmer groups and train different types of farmers, including women, to diversify their crop/livestock farming systems. Along with creating a new management structure and hiring new employees, the decentralized, private NAADS system had to arrange for new facilities (offices), equipment, transportation, and a communications system. Because the advisory services were to be managed by new farmer-based organizations, about 80 percent of the organizational and operational costs were still financed by donors as of 2008. In addition, the central government covered 8 percent of the recurrent costs, local governments financed about 10 percent, and 2 percent were financed by the farmers themselves.

(Box continues on the following page)

Box 3.7   National Agricultural Advisory Services in Uganda (continued)

In 2007/08, NAADS reached 760,000 households in 712 subcounties in 79 of the 80 districts, which is still less than 20 percent of all farming households that accessed agricultural extension advice. Apart from NAADS, Uganda had 1,600 public extension agents (due to be fully integrated in NAADS in 2010) and parallel extension programs operated by NGOs and private service providers.

The investment in Phase I of NAADS (2001–09) was estimated at about US$110–150 million. For Phase II (2010–15), an investment of US$300 million is foreseen.

Sources: Authors; for more information on NAADS, see www.naads.or.ug; Benin et al. 2007; Heemskerk, Nederlof, and Wennink 2008; and Swanson and Rajalahti 2010.

Box 3.8   Ethiopia: Investing in Human Resources

Recently the government of Ethiopia invested heavily in putting farmer training centers (FTCs) in every local administrative area (there are 18,000 nationwide) and three extension agents at every training center. From 2000 to 2008, the number of extension agents increased from 15,000 to at least 45,000, with a goal of reaching about 66,000. Reaching that goal would probably give Ethiopia the world's highest ratio of extension agents to farmers (see box figure).

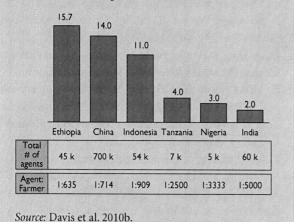

	Ethiopia	China	Indonesia	Tanzania	Nigeria	India
Total # of agents	45 k	700 k	54 k	7 k	5 k	60 k
Agent: Farmer	1:635	1:714	1:909	1:2500	1:3333	1:5000

Source: Davis et al. 2010b.

Mozambique, Mali, and Tanzania, among other countries (Heemskerk, Nederlof, and Wennink 2008).

Farmer organizations are becoming much more involved in delivering extension services. Their involvement is growing because group approaches are becoming more common (as mentioned earlier) and improve the cost-effectiveness of providing extension services.

The private sector increasingly finances extension services for specific objectives and/or value chains. Contracting public extension workers for specific tasks is a common practice among NGOs as well as specific commodity development programs, such as the program for cashew production in Mozambique. Some export commodity chains finance extension services through a government-instituted export levy, as in Mozambique and Tanzania. The private sector also finances extension services directly, as is the case with large tobacco companies in Malawi and Mozambique. Many of these arrangements are in transition to become systems of cost-sharing with farmers, first by assuring effective demand for relatively costly services and eventually by having farmers fully finance extension services, as a complement to services they already provide one another (F2F extension) (box 3.9).

To increase efficiency and performance, service provision systems financed by the public sector increasingly apply principles from the private sector, such as the development of a business plan for service provision, the costing and financial transparency of services provided for farmers, and the use of performance contracts for service providers. These reforms are generally referred to as "new public management" (Heemskerk et al. 2003).

KEY POLICY ISSUES

In conjunction with efforts to make advisory services more effective, what key policy issues must be considered? Extension and advisory service systems need to build new constituencies if they are to influence policies. Constituencies could be based on alliances of public and private service providers with farmer organizations and key value chain actors from the private sector. National networks can relate to international networks such as GFRAS

Box 3.9 Fee-for-Service Extension: Pros and Cons

Fee-for-service extension is provided by the public (or another) sector and paid for by farmers. Small groups of farmers usually contract the services. This arrangement allows clients to "vote" on the programs and the scale of the programs they want by paying for them. Most examples of this model come from developed countries, such as New Zealand, where agricultural advisory services are completely privatized.

In addition to providing feedback to public extension efforts, fee-for-service extension also can provide revenue to public extension. It is suitable for rival and excludable products. Hanson and Just argue that universal paid extension is not in the public interest but that there is an optimal mix of public, private, and paid extension. A problem with implementing this type of extension service in developing countries is that farmers who do not produce for the market may purchase fewer services. One solution to this difficulty is to stratify farmers, allowing commercial farmers to purchase services and offering public extension services to smaller-scale, poorer farmers.

Sources: Hanson and Just 2001; Anderson and Feder 2004.

(http://www.g-fras.org/en/) and the African Forum for Agricultural Advisory Services (AFAAS, http://www.afaas-africa.org/) for effectively influencing policy (both forums are discussed in box 3.10).

General policy issues for extension and advisory services

Two main opportunities for developing policies will improve the effectiveness of advisory services, based on evidence of what really works. The first opportunity is provided by the many lessons and pilot experiences emerging from structural reforms to develop pluralistic, demand-led, and market-oriented extension systems. The second opportunity lies in the new requirements for advisory services to meet the demands arising from climate change, food security programming, the new aid-for-trade agenda, and reform in the agricultural research-for-development agenda (Christoplos 2010). In realizing these opportunities, several important challenges must be addressed (Christoplos 2010):

- *Proceed with extension system reform without relying on a single grand model,* as one model cannot accommodate all situations: Extension is to be location- and even value chain-specific.[1]
- *Move toward pluralism* in extension service provision while retaining public financial commitments and coordination (see TN 1).
- *Increase downward accountability* to farmer organizations (also through decentralization and based on the subsidiary principle) (see table 3.1 and related text).
- *Create an effective, efficient market for service providers,* which will control the costs of scaling up promising experiences by different public and private actors (see IAPs 3 and 4).
- *Face the enormous need for human capacity development* in management and implementation; extension workers, to cite just one example, need a wide range of new skills.
- *Move away from projects* to programs based on long-term vision and commitments (move toward national extension systems based on public-private partnerships).
- *Balance investments in extension supply and extension demand,* because both types of investment are needed for effectiveness (introduce new public management principles).
- *Focus on institutions* rather than grand methodological or technological solutions (see IAPs 3 and 4). As noted, extension approaches must be adapted locally, and there is no single correct method (emphasize institutional innovations and organizational change).
- *Move from standard packages to tailored services* provided at the right place, at the right time, and in the right format. Critical thinking and problem solving are integral to developing tailored services. (Participatory planning as part of decentralization and deconcentration—including downward accountability—and facilitation rather than teaching are crucial, as well as an emphasis on learning and business plan development.)
- *Address equity issues.* It remains a challenge to ensure that extension adequately reaches different groups of farmers and entrepreneurs: women, youth, the landless, resource-poor farmers, minority ethnic groups and castes, and others. (Different extension strategies are needed for small-scale commercial farmers, emerging commercial farmers, and farmers producing for food security, subsistence, or part-time. Women require specific extension programs. Priority setting needs to be addressed in this context for younger and older farmers as well as male and female farmers and

Box 3.10 Global and Regional Coordination to Strengthen Agricultural Advisory Services

Many stakeholders recognize that advisory services require a more formal, dynamic, and proactive structure to gain a more credible, authoritative voice. Two forums provide advocacy and leadership for advisory services at the regional and global levels.

The Global Forum for Rural Advisory Services (GFRAS) (www.g-fras.org, established in January 2010) is designed to provide a voice within global policy dialogues and promote improved investment in rural advisory services; support the development and synthesis of evidence-based approaches and policies for improving the effectiveness of rural advisory services; and strengthen actors and forums in rural advisory services through interaction and networking. GFRAS will link closely to regional networks such as AFAAS.

Source: Authors.

The African Forum for Agricultural Advisory Services (AFAAS) (www.afaas-africa.org, established in 2004), was conceived when the leadership of the National Agricultural Advisory Services realized that extension services, unlike research services, had no mechanism to share experiences. AFAAS envisions agricultural advisory services that "effectively and efficiently contribute to sustained productivity and profitable growth of African agriculture" in ways that are oriented toward countries' individual development objectives. Through increased professional interaction and information sharing, AFAAS participants build on lessons learned in agricultural advisory initiatives and enhance the use of knowledge and technologies by actors in agricultural value chains.

---

farmers oriented more to markets or more to food security).

Other policy issues related to pluralistic advisory services and extension include the changing roles of various extension providers and the comparative advantage for different providers in carrying out specific extension functions and advisory services. For instance, publicly funded advisory services should not involve themselves directly in the provision of physical inputs (including credit). Also, many bureaucrats still regard extension in a very linear way that focuses on extension functions such as transferring technologies to ensure better food security. Paradigm shifts must take place not only in the programs and the thinking of field staff but in the thinking of extension administrators and policy makers. Finally, the sustainability of extension institutions is another major issue for policy to address (Swanson and Rajalahti 2010), as is equity. These three topics (the respective roles of public and private extension providers, sustainability, and equity) are covered in the sections that follow.

Public and private sector roles

In principle, agricultural advisory services can be provided and financed by the public sector, the private sector (individual farmers or companies), and what can be referred to as the "third sector," consisting of NGOs and organizations based on collective action. These providers can be organized on the basis of who provides and who finances the services (table 3.2). The functions of service provision and financing often are separated to ensure that services are financed by clients or the corresponding sector and reflect their demands. Combinations of implementation and financing of services are presented in each cell of table 3.2.[2]

Institutional base for sustainability

Different aspects of sustainability can be considered with respect to advisory services, but most often the concern involves the sustainability of financing. Several approaches have been criticized for their lack of financial sustainability, including the T&V system promoted in the 1980s and the more recent FFS approach (Quizon, Feder, and Murgai 2001; Anderson 2006). Current FFS programs, especially in Africa, address sustainability in various ways, including revolving FFS funds, self-financing, and FFS loan and repayment schemes. The use of farmer facilitators reduces costs dramatically.[3] More agribusiness development services and market-oriented advisory services aim for farmers, the subsector, or the commodity chain to pay at least partially for services.

Sustainability can also be addressed through innovative modalities for financing advisory services. Cost-sharing

Table 3.2 Options for Providing and Financing Pluralistic Agricultural Advisory Services

Service provider	Finance provider				
	Public sector	**Private sector: Farmers**	**Private sector: Companies**	**Third sector: NGOs**	**Third sector: Farmer-based organizations (FBOs)**
Public sector	Public advisory services (different degrees of decentralization)	Fee-based public advisory services	Private companies contract staff from public advisory services	NGOs contract staff from public advisory services	FBOs contract staff from public advisory services
Private sector: Companies	Publicly funded contracts to private service providers	Private companies provide fee-based advisory services	Embedded services: Companies provide information with input sale or marketing of products	NGOs contract staff from private service providers	FBOs contract staff from private service providers
Third sector: NGOs	Publicly funded contracts to NGO providers	Advisory service staff hired by NGO, farmers pay fees	Private companies contract NGO staff to provide advisory services	NGOs hire own advisory staff and provide services free of charge	
Third sector: FBOs	Publicly funded contracts to FBO providers	Advisory service staff hired by FBO, farmers pay fees		NGOs fund advisory service staff who are employed by FBO	FBOs hire own advisory staff and provide services free to members

Sources: Birner et al. 2009, adapted from Anderson and Feder (2004, 44).

arrangements (such as those used in Uganda's NAADS program) allow resources to be mobilized from various sources. These resources can be pooled and distributed to end-users based on demand.

Stakeholder forums consisting of farmer groups create a critical mass for services required from either public or private bodies and can reduce service costs. Forums empower farmers to identify and use selected qualified service providers (Government of Kenya 2005). Other potential methods for mobilizing and managing funds include levies on export commodities (Tanzania, Kenya), community-driven development funds (Guinea, Kenya), and contracting by the government (Mozambique) (Rivera and Alex 2004).

Financing for advisory services may also come from resources provided through decentralization programs, the involvement of farmer associations and NGOs, contracting-out of extension services, public-private partnerships, privatization, and embedding advisory services in other types of contracts (Anderson 2007). More information on these subjects is available in module 3 of the *Agriculture Investment Sourcebook* (World Bank 2006b).

Financing alone cannot guarantee the institutional sustainability of advisory services. Capacity within the advisory service is another major concern. Extension workers must be able to apply new approaches that focus more on facilitating processes than on teaching models and are more

oriented toward the development of businesses, markets, and enterprises. Apart from absorbing these individual capacities, public providers of advisory services will have to undertake major organizational changes, such as the use of performance-based contracts and incentives. Institutional development is also important (IAP 2). Public advisory services must develop the institutional capacity to coordinate and manage local extension systems. For example, they will need the capacity to facilitate interactive learning between different extension service providers.

Gender and equity considerations

Women make up 60 percent of the rural population worldwide (Hafkin and Taggart 2001), yet they receive only 2–10 percent of extension contacts and 5 percent of services (Swanson, Farner, and Bahal 1990). In sub-Saharan Africa, where women play a major role in agriculture and account for more than half of agricultural output, they continuously receive a less-than-proportional share of the total investment in agriculture (Blackden et al. 2006; Quisumbing 2003). Only 7 percent of extension resources are spent on African women (Blumberg 1994, cited in Haug 1999). African women remain especially disadvantaged in interventions relating to education, extension, capacity strengthening, empowerment, and market access (Rahmato 1993; Alawy

1998; Frank 1999; Haug 1999). This problem is especially pronounced in areas emerging from conflict (World Bank, UNDP, and UNIFEM 2010).

Despite this evidence of neglect, recent studies conclude that some programs have reached women farmers by taking practical steps to address the lack of inclusiveness in providing advisory services (Davis et al. 2010a; Gender and Governance Research Team 2009). A major first step is to develop transparency in service provision by segregating data on the participation of men and women, young and old, and different categories of farmers (subsistence, emerging, and small-scale commercial) in all activities, from planning and training to monitoring and evaluation. This information can form the basis for developing joint action plans to address any problems with inclusiveness for any of these groups. Second, extension agents and others (including policy makers and local government officials) must be equipped with the skills to respond to the needs of a diverse clientele with respect to age, gender, socioeconomic background, ethnic differences, age, livelihood source(s), and so on. For more information see Christoplos (2010).

NEW INVESTMENT DIRECTIONS, PRIORITIES, AND REQUIREMENTS

In Africa, CAADP and the compact agreements are guiding advisory services into efficient and well-focused service delivery at the center of the AIS. This is a complete paradigm shift from the perception that research knowledge can drive innovation to the notion that change in the whole system is needed for innovation.

Throughout the developing world, similar evolving demands and new roles for advisory services in the wider innovation system will require new investments—among others, investments in the capacity of individual extension workers and organizations for value chain approaches, in market-oriented extension, in group and organizational development, in agribusiness, and in mechanisms to share information (networks, platforms, and the like). Recent global developments require advisory services to focus on climate change, food security, and equipping rural people to deal with risk in general. There is a need for evidence-based direction regarding investment priorities and programming options for agricultural advisory services within innovation systems. To influence policies and better serve their constituencies, including the poor and women, advisory services need a stronger voice at the global and regional level; box 3.10 describes approaches to achieve this goal. Finally, investment in nonagricultural issues will be essential.

This topic is somewhat outside the scope of this sourcebook, yet it must be considered for investments in extension. The extent of attention to such issues as nutrition, community organization, microenterprise development, health, youth activities, women's empowerment, and rural development varies. There may well be a case for extension to facilitate off-farm employment as a means of improving opportunities for agricultural commercialization.

The thematic notes and innovative activity profiles in this module offer an array of strategies that may be adapted to meet these needs for investments, policies, and innovative approaches:

- **TN 1: Pluralistic Extension Systems.** Pluralistic extension recognizes the inherent plurality and diversity of farmers and farming systems and the need to address challenges in rural development with different services and approaches. This note describes pluralistic systems, their strengths and challenges, investment opportunities, policy issues, and emerging lessons.

- **TN 2: Farming as a Business and the Need for Local (Agri-) Business Development Services.** Farming as a small-scale business requires access to markets, financial services, and inputs, as well as a suitable mix of farmer entrepreneurial skills and attitudes and bankable business plans. This note discusses approaches to address farmers' business development needs, such as reorienting programs and staff, recruiting new staff, involving communities as agribusiness promoters, and developing multistakeholder platforms to support agribusiness development.

- **TN 3: Extension-Plus: New Roles for Extension and Advisory Services.** Extension can move beyond its traditional technology transfer role to operate as a nodal agency within the AIS, providing technological and nontechnological services to farmers. This "extension-plus" approach emphasizes locally developed strategies for participants to learn through experimentation and adaptation. It is a "best fit" rather than a "best practice" approach, requiring changes in extension and other institutions in the AIS.

- **TN 4: The Role of Innovation Brokers in AISs.** Some extension agents and other actors (such as researchers and staff of NGOs) have chosen to operate as innovation brokers. Innovation brokering expands the role of agricultural extension. Extension is no longer a simple, one-to-one intermediary between research and farmers but an intermediary that creates and facilitates many-to-many relationships (a key concern within AISs).

- **IAP 1: Agrodealer Development in Developing and Emerging Markets.** Agrodealers have an increasing presence as providers of advisory services. A holistic, market-oriented approach to agrodealer development facilitates improved efficiency in resource allocation, operations, and economic performance and helps to develop sustainable input supply systems.
- **IAP 2: Federating Farmer Field Schools in Networks for Improved Access to Services.** By federating, farmer groups increase their effectiveness in obtaining the advisory services they identify as important at the local level, often at a lower cost.
- **IAP 3: INCAGRO: Developing a Market for Agricultural Innovation Services in Peru.** This case study of INCAGRO describes how Peru developed a demand-driven market for agricultural innovation services; two competitive grant funds were important features of the program.
- **IAP 4: Combining Extension Services with Agricultural Credit: The Experience of BASIX India.** Recognizing that agricultural credit alone did not equip India's rural poor with the knowledge, skills, and support services to improve incomes, BASIX developed a triad of integrated services—financial services; agricultural, livestock, and enterprise development services; and institutional development services—to improve livelihoods.

MONITORING AND EVALUATING INVESTMENTS AND SCALING UP

To ensure the proper implementation of extension investments, M&E exercises and tools are crucial. Because attempts to monitor and evaluate extension and advisory services have been weak or nonexistent, GFRAS developed a guide for the evaluation of extension and advisory services (box 3.11). General studies find high rates of return to investments in advisory services, but the challenges and difficulties in estimating the benefits are many. Efforts to strengthen the understanding of how to improve M&E in extension continue, but it is already clear that some of the most important components of M&E are participation by all parties; the clear definition of objectives, indicators, outputs, outcomes, and desired impact; continual assessment throughout the investment period; and the collection of baseline data. Important indicators include benchmark and baseline indicators as well as input, output, outcome, and impact indicators (Swanson and Rajalahti 2010). For extensive lists of indicators, see Swanson and Rajalahti (2010) and Rajalahti, Woelcke, and Pehu (2005).

Box 3.11   Guide to Extension Evaluation

The Global Forum for Rural Advisory Services (GFRAS) publishes a guide to conducting more comprehensive, rigorous, credible, and useful extension evaluations. The guide describes different types of evaluation, explains how to select the approach that is most appropriate to the particular context, and identifies additional sources of theoretical and practical information. It includes guidance on such issues as preparing terms of reference and links to evaluation manuals in different sectors. The guide is intended to be used primarily by:

- Those commissioning and managing evaluations.
- Professional evaluators and staff responsible for monitoring systems.
- Those involved in knowledge- and results-based management within a range of organizations involved with extension.
- Staff of public extension agencies, farmer associations, and other organizations directly or indirectly engaged in providing extension services.
- Professionals involved in training and educating evaluators.
- Researchers looking for ways to synergize their efforts with evaluation initiatives.

Source: Adapted from the website for the guide at GFRAS, http://www.g-fras.org/index.php/en/knowledge/gfras-publications/file/20-guide-to-extension-evaluation, accessed July 2011.

Indicators are needed for judging the effectiveness of extension programs, including the share of farmers with regular access to services and their perceived satisfaction with the services. Such indicators should not be limited to farmers but also developed for male and female producers, other (mostly private) actors in the chain, and (local) governments (Spielman and Birner 2008). To monitor and evaluate pluralistic advisory services, *output indicators* include capacity level in terms of business development services and local certification services; management of pluralistic extension systems at the district level; and learning taking place between agencies. *Outcome indicators* include the quality of investment plans to improve associations and indicators measuring whether the voices of female farmers are heard in farmer forums.

Given the increasingly pluralistic character of extension systems, many innovations—technological as well as institutional—will develop. The system needs a mechanism for monitoring and learning from new, often very local, practices and experiences, which is a major new task for extension management. Management will have to be open to experiences from the private as well as the public sector, involving all key innovation system stakeholders. Local good practices can be identified, documented, and then considered for use on a wider scale through a specific strategy for scaling up.

The process of scaling up agricultural practices is complex and influenced by many factors. Scaling up can largely concern the more quantitative aspects of increasing the number of farmers adopting or adapting technology to their own situations, but it can also concern the policy, institutional, and organizational aspects of implementing a practice on a wider scale. Based on innovation system concepts, the factors potentially influencing success in scaling up good practices need to be analyzed ex ante.[4]

Pluralistic Extension Systems

Willem Heemskerk, Royal Tropical Institute (KIT)
Kristin Davis, Global Forum for Rural Advisory Services (GFRAS)

SYNOPSIS

Pluralistic extension recognizes the inherent diversity of farmers and farming systems and the need to address challenges in rural development with different services and approaches. It is characterized by the coexistence of multiple public, private, and mixed extension systems and approaches; multiple providers and types of services; diverse funding streams; and multiple sources of information—all of which benefit from some degree of coordination and regulation that facilitates interaction and learning. Ideally, the outcome of pluralistic extension services is that different client groups in different contexts are satisfied with their access to services that they have demanded. Although pluralism in advisory services makes it possible to capitalize on the competitive advantages of different actors, one of pluralism's greatest challenges is to coordinate organizations that have vastly different mindsets and worldviews. A key message is that the public sector's primary role is to ensure that this mix of providers achieves jointly developed objectives. Public coordination and management of pluralistic extension services should be based on a program for action developed jointly by multiple stakeholders and service providers. The action program should reflect stakeholders' agreement on the roles for the different service providers and on who is best suited to perform each function under the program. The variety in services demanded is then matched with the existing variety of service providers. The emphasis is on coordination, which can lead to regulation and performance-based contracts for additional services, all based on complementarity.

WHY PLURALISTIC SERVICE SYSTEMS?

Many types of advisory service providers and approaches exist side by side. This situation is good, as the diversity of rural life and needs should be matched by diversity in services, approaches, and providers. Differences between the poor and resource-poor farmers; crop, livestock, and fisheries systems; production and conservation objectives; and local and export value chains—to name only a few—will affect which organizations can best provide services and by which methods. These differences are a major reason for encouraging pluralistic systems.

Farmers, often impelled by market opportunities but also by environmental, labor, and land productivity challenges, look for information and knowledge to strengthen their production systems. Trends such as market liberalization and development, as well as democratization and the communications revolution, drive farmers to obtain agricultural information through a wider range of means and from a wider range of sources than ever before. Even traditional mass media such as (community) radio,[1] television, and newspapers can reach quite different audiences. For farmers, public extension services are just one source of information, often the one focusing purely on production issues (Spielman et al. 2011). Farmers procure other, more business-related services in the private sector and access facilitation services (for group processes, as well as interaction with input and market actors) through NGOs and farmer organizations. Technology and information are no longer transferred through a linear system (Wennink and Heemskerk 2006), leaving national extension and advisory systems in many parts of the developing world struggling to meet new demands from farmers and other actors in the innovation system.

A useful alternative is the coordination and management of pluralistic extension services based on a program for action developed jointly by multiple stakeholders and service providers. The action program reflects stakeholders' agreement on the roles for the different service providers and on who is best suited to perform each function required by the program. The variety in services demanded is then matched with the existing variety of service providers. The emphasis is on coordination, which can lead to regulation and performance-based contracts for additional services, all based on complementarity. Some systems are self-organized (value chains driven by the private sector) and do not require this public role in coordinating service provision.

BACKGROUND AND CONTEXT FOR INVESTMENT

Aside from the trends mentioned earlier, the provision of advisory services to smallholders in developing economies is influenced by the decentralization of governments and governance as well as by the deconcentration of public service delivery processes. In such dynamic environments national agricultural extension services are starting to play new roles, based largely on principles of demand-driven planning, management, facilitation, and learning through interaction.

All of the newly recognized actors in advisory services can equally bring about new ideas and innovations in agricultural extension, contributing to a system in which the different roles can lead to synergy. National agricultural advisory service systems are attempting to capture these institutional innovations by contracting-in different services at the district, provincial, and sometimes national levels. Advisory services are growing more varied; rather than being limited to technology services, they are offering more general information and brokering services (see TN 4). They also facilitate access to other services, such as financial and market information services, through different means, including the mass media and social media.

Advisory service providers increasingly vary as well. Traditionally, the private sector provided the more market-oriented and business development services (TN 2), and the public sector provided services focused on using technology to enhance agricultural productivity. In pluralistic extension systems, the services in demand are supplied by the right mix of providers. Certain service providers often perform specific advisory functions (as shown by the "x" in table 3.3). The matrix in table 3.3 will differ in every situation and context. It can be used to develop the best mix of services required and can ultimately lead to pluralistic extension, as described in box 3.12.

Under pluralistic systems, different types of agricultural and agribusiness advisory services or different providers work together to provide extension services. Services can be provided by:

1. **Subsectoral bodies** representing private, market-oriented farmers, such as a coffee board or national commodity association. This practice often occurs in cash crop subsectors such as coffee in Colombia, cotton in Benin, cashews in Tanzania, or the Kenya Tea Development Agency (see box 1.22 in TN 4 of module 1).

2. **Producer/farmer organizations and cooperatives**, not on the national level but at the meso level—for example, when a farmer association provides services through volunteer members, as in Mozambique's National Union of Smallholders or Mexico's Produce Foundation (see module 1, IAP 2)—and on the individual level (for example, a milk producer cooperative or a vegetable producer association).

3. **Local NGOs** usually working with farmer groups and community-based organizations, mostly in subsectors that do not involve cash commodities but increasingly in market-oriented services.

4. **International NGOs (mostly donor-funded)** usually working with farmers' groups and community organizations in subsectors for cash and noncash commodities, which may at times overlap, but also **agri-agencies of developed country farmer organizations**.

5. **Governments** that support activities under 1, 2, 3, and sometimes 4 in a sort of "joint venture" at the national, provincial/regional, or local/district level; or **public agencies** working with civil servants. In the public sector, different extension systems (for example, for crops, livestock, and forestry) can exist side by side within the same or different ministries. Many countries have taken a step forward in coordinating this multiplicity of public extension programs by adopting a unified (public) extension system.

6. **Input suppliers** and **agrodealers** supplying agrochemicals and veterinary products and **buyers of products** (such as buyers of flowers and fresh vegetables) (see IAP 1).

7. **Private business contacts** and relationships that provide informal advisory services, like playing a brokering role (TN 4). Increasingly, local business development services are also provided by **financial services** (microcredit organizations and banks), **actors in the value chain**, and **other private actors** (TN 2).

8. **Village/community extension workers**, often connected to input supply programs (such as cashew spraying services or chicken vaccination). **Lead farmers** and local

Table 3.3 Extension Service Functions and Service Provider Categories
(the number of "x's" indicates the general prevalence of specific services)

Functions versus providers	Sectoral bodies	Producer organizations	Local NGOs	International NGOs	Governments	Input suppliers	Private business sector	Community extension workers
Information	Sector-specific		Only general			Market	Market info	
Training and advice	x Quality		xx	xx	x		Quality	xxx
Technology testing	Cash crops	x		x	xx	Demos		xx
Business development	x	x	x	xx		xx	xxx	
AIS linkage facilitation/ brokerage	x	xx	xx		Only w/ research		x	
Institutional development		Bonding, bridging, linkage social capital*	xx Bonding	xx Bridging	Bonding		Market links	Research linking
Legal advice			Land rights group registration				Outgrower contracts	
Green services		Ecosystem services	xx	x			Certification	

Source: Authors.

* Bonding social capital through strengthening the group, bridging social capital through federation and unionizing, and linking social capital through developing the capacity of groups to interact with other stakeholders (Heemskerk and Wennink 2004).

Box 3.12 Pluralism in Action: Government-Funded Public, Nongovernmental, and Privately Managed
Extension Systems in Mozambique

Mozambique became independent in 1975, but civil strife prevented the government from establishing public extension services for its farmers until 1987. The government used (international) public funds to contract several local and international NGOs to organize and provide extension services to farmers in selected regions. Over the years, the size of the public and NGO extension systems has fluctuated from around 600 to 800 extension workers each. The government has also outsourced some extension services. For example, it contracted several private, large-scale farmers, companies, and NGOs and hired an additional 200 extension workers under short-term contracts to focus on specific assignments. These combined actions have resulted in an average of 10–14 extension workers in each rural district, who reach between 10 and 20 percent of farm households.

The interplay of NGOs, the private sector, farmers, and their organizations in extension has led over time to a new, pluralistic extension system in Mozambique. The public sector concentrates on strengthening and gradually expanding the size and improving the quality, accountability, and relevance of its public extension services,

because the public sector remains the cornerstone of Mozambique's pluralistic extension system. Extension activities are coordinated at the district level by local government officials. They coordinate NGOs, farmer organizations, and private service providers, whereas provinces (and also districts) may also outsource specific assignments complementary to the public extension system. Long-term public financing for extension is crucial, as it will be difficult, if not impossible, for low-income men and women farmers to pay for extension services themselves. Poor farmers will pay for specific services, such as cashew spraying and veterinary services, as these are embedded extension services, but they are generally unwilling and unable to pay for advisory services that deal with "public" knowledge and information.

The current publicly financed extension system has three main programs laid out until 2015. First, it will strengthen service provision in the public sector, the private/NGO sector, and farmer-to-farmer extension. Second, it will empower farmer associations in planning, pursuing economic activities, and providing services. Lastly, it will emphasize the coordinated management of service provision at the district level.

Sources: DNEA 2007; Swanson and Rajalahti 2010.

---

facilitators such as those working with FFSs are also providing such services.

The brokering and facilitation function can be performed by any of the entities listed above. This function in particular has become more important with the growing realization that catalyzing innovation involves more than transferring knowledge and requires strong interaction between a variety of actors (Klerkx, Hall, and Leeuwis 2010; TN 4). Such interaction responds to a key concern within AIS (TN 4).

INVESTMENT NEEDED FOR INNOVATIVE AND PLURALISTIC EXTENSION SERVICES

The principles discussed in the following list are central to pluralistic extension systems:

- *Deconcentration.* The public sector has an important but differentiated role at the local, meso (provincial,

regional), and national levels, particularly in providing coordination, technical backstopping, and knowledge management. The public sector should facilitate learning and scaling up, as well as ensure quality assurance and oversight. Advisory service systems supported by public funds are increasingly planned, financed, implemented, and coordinated at the district level. The meso level coordinates and implements crosscutting services (mostly on contracts), such as seed services, environmental management services, food security services, and other services that extend beyond district boundaries. The public sector at the national level plays a supportive and backstopping role for all service providers and provides the enabling environment—conducive policies, strategies, and regulations.

- *Decentralization.* As local governments are empowered to run their own affairs, it is becoming common (notably in Anglophone and Lusophone Africa) for district governments and administrations to operate a budget obtained from the treasury and allocated on the basis of an integrated district development plan. Agricultural planning priority is shifting from sectorwide agricultural planning to higher-quality district agricultural development plans.

- *A system for providing multiple services.* Extension managers and partners recognize that the quality of service provision can be improved through performance-based contracts and that the choice of provider must be based on the comparative advantages of the public sector, private sector, and civil society. The best mix of services can be identified for every situation, depending on the demand for and availability of services.
- *Farmer empowerment.* Farmer organizations represent the voice of their clients, but they are also partners in extension when it comes to planning, allocating resources, M&E, and providing services. Empowerment is twofold, consisting of economic empowerment as well as involvement in decision making. As farmers' economic empowerment in value chains and local economic development grows, farmers gain a more forceful role in setting priorities, planning, and providing services. In pluralistic systems, downward accountability and user involvement make quality control possible only at the local level. Downward accountability of service providers to farmers becomes more important for quality control than upward accountability to financers (see also module 1).
- *Outsourcing services.* Local governments (districts, communes, and so forth) are contracting-in the services directly demanded by farmers in district agricultural development plans, based on the comparative and competitive advantages of the various service providers. This trend should improve synergy and complementarity in service provision.
- *Partnerships.* Example of partnerships and linkages between agricultural advisory services and other actors in the innovation system and services include partnerships between advisory service providers and agricultural research agencies, agricultural chambers of commerce, microfinance organizations, and agroprocessing services.
- *Extension approaches.* A major challenge is to continue shifting extension from a top-down approach offering blanket, production-oriented recommendations toward a more interactive learning approach. The interactive approach provides room to differentiate among categories of clients, messages, and approaches. Extension officers play more of a facilitating role; based on their technical expertise, they stimulate learning among farmers (as in FFSs) and with other actors, particularly market

actors. District extension systems need to be supported by provincial and national services and knowledge centers in case demand for knowledge services extends beyond the district level, as this is part of the new extension.

Depending on the needs identified in a given situation, investments can be made to support the conditions that will enable extension to become more pluralistic and meet those needs. As indicated in table 3.4 and the discussion that follows, capacity strengthening is a major area for investment, and investment is needed at all levels.

Capacity development: A major area of investment

In general, capacity can be considered with respect to institutions, organizations, and individuals. In supporting extension services that enhance innovation dynamics, there is in general a shift from strengthening organizations to strengthening extension systems. System or organizational learning requires five core capabilities: (1) to commit and engage; (2) to carry out the mandate and deliver results; (3) to link with, attract, and mobilize resources; (4) to adapt and self-renew; and (5) to balance coherence and diversity (ECDPM 2008). Capacity development is the external facilitation of this internal learning process. System or organizational learning processes can be effective and lead to innovation only if the actors involved have adequate capacity to participate, to actively engage, and to potentially facilitate innovation processes. They also require the mindset and flexibility to allow others to participate.

In different contexts (under various governance structures, for example), investments in developing capacity will require adjustments, including a move away from agricultural sector programs and a link to more local economic development programs. Programs will need to focus not only on the public sector but also on community extension workers and private agencies. Two particular opportunities for investment, described in greater detail in the next section, are (1) to develop national capacity (independent agencies, universities, and other mediums) and higher education courses for a new type of advisory service provider in the public and private sector and (2) to develop capacity at the district level to coordinate and manage pluralistic extension systems.

Specific areas of capacity strengthening

The capacity of new extensionists is central to the success of pluralistic extension systems. They must master highly

Table 3.4 Investment Opportunities to Foster Pluralistic Extension Systems

Mechanisms and principles	Examples of investment	Cases and references*
Deconcentration	Programs to develop capacity in managing and implementing extension at the local level, including planning, monitoring, and evaluation. Differentiation of the public sector's roles at the local, meso, and central level in technical backstopping, coordination, and quality assurance.	National Agricultural Extension Program (PRONEA), Mozambique (DNEA 2005, 2007)
Decentralization	Develop integrated local government planning skills, as well as local governance skills and mechanisms, including skills to deal with downward accountability.	Agricultural Sector Development Program, Tanzania
Multiple service provision	Develop the capacity among service providers to coordinate and use learning mechanisms and skills. Support the development of local private service provision through capacity development and local matching investment funds for service providers. Develop farmer advisory service providers.	PRONEA, Mozambique (box 3.12)
Farmer empowerment	Develop associations and cooperatives to articulate clients' demands, empower them economically, and improve service delivery. Develop the triangle of (1) entrepreneurship, (2) access to (cooperative) credit, and (3) higher-level farmer lobby organizations.	Many international NGO programs and the Farmers Fighting Poverty Program (www.agricord.org); see also module 1
Outsourcing services	Develop capacity in the local government to contract for services based on principles of complementarity, synergy, and subsidiarity. Provide matching fund to contract local services based on cost-sharing and cost-recovery arrangements.	NAADS, Uganda (box 3.7 in overview) and the Agricultural Service and Producer Organization Support Project (PASAOP, Programme d'Appui aux Services Agricoles et aux Organisations Paysannes), Mali (www.maliagriculture.org)
Partnerships	Develop public-private partnerships to deliver services. Develop the capacity for coaching and facilitation.	PRONEA, Mozambique (see above)
Extension approaches	Strengthen facilitation skills and capacities to use learning approaches such as the Farmer Field Schools. Strengthen the market orientation of services at the national and district levels. Strengthen the targeting and differentiation of services for different categories of farmers and households, based on demand.	Farmer Field School program and marketing extension (http://www.farmerfieldschool.info/) African Forum for Agricultural Advisory Services (AFAAS–FARA 2009, TN 2); many (mostly international) NGO programs

Source: Adapted from DNEA 2007.

Note: See also table 3.1 in the overview for definitions and needs.

* The seven principles and/or a mix of them are applied in a number of national programs with support from organizations such as the International Fund for Agricultural Development (IFAD) and the World Bank.

technical information and skills as well as sophisticated facilitation and process skills (Blewett et al. 2008). More specifically, extension workers and their clients require:

■ *Specific skills for planning and collaboration.* Stronger capacity is needed at the local level for planning, management, and coordination. As noted, major attention must be given to enhancing facilitation skills. These skills are instrumental in the multistakeholder platforms and processes that foster capacity development across stakeholders in innovation systems. Skills for communication with male and female farmers as well as differ-

ent kinds of stakeholders in the value chain or innovation system are needed.

■ *Extension management skills.* As emphasized previously, the presence of multiple actors and approaches in pluralistic systems means that there is a strong need for coordination to avoid duplication of effort and wasted resources. Managing pluralistic extension systems at the local level requires individuals to develop new knowledge, skills, and attitudes. Management of synergetic services for local economic development, outsourcing, M&E, and quality assurance must take place in a satisfactory manner. New performance plans and indicators must be established. Management for performance and outcomes must be a focus.

- **Skills related to understanding and improving accountability.** Clients must gain the capacity to participate in, monitor, and evaluate extension. They must be equipped to express their perceptions of the performance of advisory services, both in an upward (local, provincial, and national government) and downward (famer groups, farmer forums, district councils) direction.
- **Technical knowledge** and skills are required for relevant actors in the value chain (production, processing, and marketing), including knowledge about access to all assets of the livelihood system.
- **Other skills that improve the quality of service provision.** At the local level, the variety of demand for services and the supply of services will present challenges not only for coordination but also for supervision and quality control. Service providers need to be registered and certified using established criteria and conditions in a transparent manner, mostly at the meso level. To a large extent, the actual quality of service providers' performance must be controlled by users themselves.

Other investment needs

This sourcebook presents other examples of investments to support pluralistic extension services. Examples in this module include enhancing facilitation and coaching skills (TN 4), capacity development in extension management and the development of agribusiness services (TN 2), and green services (TN 2).

POTENTIAL BENEFITS OF PLURALISTIC EXTENSION

To recapitulate, the need for pluralistic extension arises from the perception that specific services are needed for specific contexts, economic enterprises, livelihood functions, and above all different farmer categories, based on differences in entrepreneurship, poverty and gender. The development of pluralistic extension systems should enhance the competitiveness of local agricultural production within the context of local economic development, enhance local livelihoods, and ultimately reduce rural poverty, improve food security, and promote greater gender equality. Ideally, the outcome of pluralistic extension services is that different client groups in different contexts are satisfied with their access to services that they have

demanded. Pluralistic extension systems provide services on demand as identified in the joint planning process, and based on the services available for each demand.

Models and lessons of pluralistic advisory services include Mozambique's PRONEA (box 3.12), Uganda's NAADS (box 3.7 in the module overview), and the programs in Mali (PASAOP) and Tanzania (Agricultural Sector Development Program) cited in table 3.4.

POLICY ISSUES OF PLURALISTIC ADVISORY SERVICES

The policy issues pertaining to pluralistic advisory services are correspondingly diverse. As discussed below, some of the more pressing issues involve ensuring the sustainability of pluralistic advisory services and preventing them from exhausting public resources; ensuring that services are provided in a more equitable way; promoting the institutional development of advisory systems; attending to the growing demand for advice on a host of environmental issues; and clarifying the changing roles and contributions of the public sector, private sector, and civil society within a pluralistic extension system.

Sustainability

Pluralistic extension systems are in principle more effective than other kinds of extension, but the outsourcing of public services will act as a major drain on public resources if not properly implemented. Outsourcing whole systems and creating parallel structures is costly and not very effective, as shown by the experience with NAADS and pilot activities in Mozambique (Heemskerk, Nederlof, and Wennink 2008). Instead it has proven more effective and cost-efficient to outsource specific functions, such as the development of bankable business plans. Enhanced coordination between public and private services at the local level will also make the system more efficient, while quality control of service provision will make it more effective. The best mix of public and private service provision and the level of public financing of such pluralistic systems will be subject to national and local policies. These policies in turn will be determined by the broad national vision for rural development, by locally empowered smallholders, the level of focus of local development plans, and the relative strength of public and private service provision. Other services can be provided in a better, or at least a more cost-effective, way by community extension workers. In local development plans, coordination is planned and financial sustainability can be pursued.

Social considerations: Equity, gender

Agricultural production is one of the main economic and income-generating activities for rural people, yet not all rural households have the same objectives in economic development, and they can place wide-ranging demands on advisory services. Local agricultural development plans can clarify the priorities for different categories of farmers (small-scale commercial, emerging and subsistence, food-security-focused, or part-time farmers, for example) and between male and female farmers. Services for local economic development need to differentiate among many categories of clients (households based on their different objectives, for example, and clients based on gender, age, and physical abilities). The emphasis on market-oriented services, cost-sharing arrangements, and the increasing role of community extension workers will influence access to services among different categories of clients. In users' assessments of service providers' performance, as well as in the downward accountability of service providers and extension managers, the consideration of equity issues remains important (Nederlof, Wennink, and Heemskerk 2008).

Institutional considerations

Pluralistic extension systems are expected to better address the wide variety of demands and at the same time make better use of the variety of service providers available. Although eventually the right mix of services is determined by the client and through payment for services, in the foreseeable future the public sector will still finance many of the required services (also based on cost-sharing arrangements but adjusted for different categories of farmers).

Pluralistic extension systems aim to develop better service provision for all, based on the complementarity and synergy of the public and private sectors. This public sector (at the local, meso, and national levels) will need to play a strong role in managing and coordinating extension activities in such a way that demand is adequately addressed, service providers are accountable, quality is assured, and lessons are learned among service providers, who are in competition at the same time. Eventually this coordination and accountability role will gradually shift to farmers and their organizations, once they will finance these services themselves.

Environmental implications

Demand for services related to wider environmental issues is increasing, owing to such factors as increasing pressure on land, questions of access to land, market demands for sustainably produced products, and climate change. Public sector coordination of environmental and/or green services is needed to ensure that services are provided synergistically by the array of actors involved, such as:

- Farmers themselves, providing ecosystem services with and without incentives. Examples include maintaining biodiversity or soil fertility or receiving premium prices for using sustainable production methods.
- The private sector, dealing in CO_2 emission rights or the certification of sustainable and/or organic production.
- The public sector, engaging in climate change mitigation, erosion control, watershed management, and similar public good activities in environmental management.

In local development planning, an integrated approach to environmental management and the role of different service providers is needed.

Public and private sector roles

Private extension service provision, although publicly funded, contributes to the development of a new incentive system in which the quality and content of extension provision is more responsive to farmers' priorities. The transition to a system with privatized extension modalities and improved incentives takes time, public investment, and appropriate long-term plans. Private extension provision requires well-trained service providers and a certain level of capacity among farmers and local governments. Farmers' organizations must increase their capacity to contract, manage, and evaluate private extension provision. Decentralized political structures need the capacity to manage such systems.

An evolution toward private extension modalities should begin with themes that are most likely to elicit farmer demand and investment and are rarely provided by the public sector, such as the demand for agribusiness development services, particularly at the local level. Farmers in Ethiopia, Uganda, Mozambique, and Kenya have all identified the need for agribusiness development services (see TN 2). The public sector is likely to retain its responsibility for financing extension on themes such as environmental protection, although private delivery modalities may prove useful (Chapman and Tripp 2003).

LESSONS LEARNED

The lessons summarized here draw on several sources that have recently examined innovations and experiences with advisory services. The clusters of challenges identified include the management of pluralistic advisory service systems; the quality of the demand for advisory services, the quality of the supply, and the quality of the enabling environment (Nederlof et al. 2008).

Management of pluralistic advisory service systems

The decentralization of advisory service systems to the provincial or district level provides a major opportunity to improve the coordination of services at the local level. In most cases, to deliver a mix of public and private services effectively, local governments and authorities will need to improve their capacity to coordinate, manage, and direct services and service systems. Areas such as the facilitation of joint planning, the facilitation of learning among stakeholders, the regulation and certification of service providers, and quality control all need strengthening. It may be necessary to begin by building capacity in the institutions responsible for training administrators.

Capacities of the rural poor as service users

Investments are also needed for farmers and farmer organizations to strengthen their capacity to articulate their demands. To identify and address opportunities, smallholders need information about production, markets, and financial services. The more vulnerable farmers need specific services related to household food security. Farmer organizations must be able to: (1) lobby for an enabling policy and institutional environment; (2) give the rural poor a voice; (3) influence the adoption of socially inclusive research and advisory service agendas; and (4) become involved in the implementation of research, advisory, and business development services.

Provision of relevant, sustainable, and high-quality services

Extension services need to be relevant, sustainable, and of good quality. Service providers need to differentiate their offerings depending on the intended clients and their demands—for example, some services may focus specifically on vulnerable groups to enhance social inclusion, whereas others focus on value chain empowerment. Service providers also need to design, in close participation with the rural poor, services that respond effectively to poor people's needs. Offering diverse services for different groups of clients will require different financing strategies. Some services for the very poor will have more of a social nature and be supported by the public sector, community, or farmer organizations. Business development services, in contrast, will evolve gradually from cost sharing to full payment (direct or indirect) by clients. The deployment and financing of service providers and services at the local government level must be managed and coordinated to enhance the coherence and synergy of services, increase the efficient use of services by the rural poor, and stimulate interaction and learning between service providers. The public sector also has an important responsibility to control the quality of service provision (through registration and certification, for example) and prevent bias (among agrodealers providing embedded services, for example).

To provide services that are relevant and of high quality in a financially sustainable way, effective linkages are needed between (1) productive investment and technological innovation and (2) financial services, risk management, and the reduction of vulnerability. Intermediary and facilitation services (not just the dissemination of information) are needed to secure those links (Nederlof, Wennink, and Heemskerk 2008; Wennink and Heemskerk 2006) (see also TN 4).

Enabling policies and institutional arrangements for pro-poor services

For pluralistic extension systems to grow and thrive, they will need to draw on evidence from the experiences of their wide stakeholder base to influence policy. Policy changes are likely to be needed to promote innovation, decentralization, and public-private partnerships and to empower rural people. Institutional innovations are also likely to be needed to foster interaction between farmer organizations and the private sector as well as research and advisory organizations. Examples include platforms for interaction, funding mechanisms, regulations, and certification mechanisms. Access to rural services will not improve without continuous interaction and flows of information between rural service providers and the rural poor to prevent information asymmetry. Policies that support the strengthening of social capital and farmer networks will create additional institutional pathways for improving interaction among stakeholders and enhancing the performance of the AIS.

RECOMMENDATIONS FOR PRACTITIONERS

The experience with pilot and larger programs for pluralistic extension systems offers a number of recommendations for practitioners. Practical, step-by-step recommendations include:

- Sensitize and get agreement among actors at all levels on the need to: (1) strengthen interaction and learning between public and private service providers; (2) involve public and private service providers on the basis of comparative and competitive advantage; (3) make an inventory of existing service providers (public, private, and embedded services); and (4) strengthen coordination at the local level between service providers by enhancing downward accountability (for example, to farmer organizations).
- Make sure that an enabling environment is in place for a pluralistic extension service system to develop. Specifically, develop a sectoral or local government policy that supports public-private interaction in service delivery.

- Open up the public service delivery system by introducing downward accountability mechanisms and performance contracts, and involving farmer organizations in service procurement (see IAP 3 for examples).
- Make provisions for local authorities to manage the coordination of service provision, contract services locally, and handle integrated budget management. For example, local authorities (such as farmer groups) may to acquire a legal identity.
- Empower farmer groups and organizations to articulate demand (for example, in planning and M&E).
- Develop local capacity for small-scale service providers.
- Develop capacity to use new extension approaches based on participatory action learning, such as the FFSs, Farm Business Schools, and so on.
- Develop the capacity of local smallholders' private service providers.
- Develop, use, and manage local performance contract and outsourcing mechanisms.
- Develop local extension management capacity, including capacity in planning, M&E, and downward accountability and transparency.

Farming as a Business and the Need for Local (Agri-) Business Development Services

Willem Heemskerk, Royal Tropical Institute (KIT)

Kristin Davis, Global Forum for Rural Advisory Services (GFRAS)

SYNOPSIS

Small-scale farmers, local farmer organizations, and other local entrepreneurs (such as traders and processors) benefit from advisory services with a business orientation. By helping these groups to access markets, financial and input supply services, as well as knowledge oriented to their particular value chains, local (agri-) business development services (LBDSs) support innovation and entrepreneurship. Capacity to provide LBDSs must be developed at the individual, organizational, and institutional levels. National universities and business schools need to develop curricula that reflect an entrepreneurial mindset in public and private service delivery. Farmer organizations have an important role to play in articulating demands related to local economic development plans, lobbying for the right type of business services, and providing services themselves. The development of private LBDSs for smallholders will require public investments. Many programs have realized that working with smallholders to strengthen agribusiness management, business plan development, and other elements of agricultural enterprises benefits from a process approach that starts with existing business service providers. If agribusiness services are fully subsidized, they may not reflect agribusiness demands and their sustainability will remain uncertain, yet services supported fully by demand will not be inclusive. Involving other value chain actors in financing local services for enhanced quality of production increasingly appears to be the most sustainable approach. The primary element for success in establishing local business development organizations is to base the intervention on local human and financial resources, which means that practitioners must emphasize linking with product marketing and processing, creating links with savings and credit activities, developing cooperatives, and building capacity.

BACKGROUND AND CONTEXT FOR INVESTMENT

A new appreciation of small-scale entrepreneurship has emerged with the growing awareness that subsistence agriculture cannot eliminate rural food insecurity and that the commercialization of smallholder agriculture is integral to economic growth and development in many countries (Pingali and Rosegrant 1995; UNDP 2004b; DFID and SDC 2008; Jaleta, Gebremedhin, and Hoekstra 2009). Welfare gains from market-oriented production arise from specialization that builds on and creates comparative advantages, from the potential for large-scale and/or intensive production, and from the dynamic effects of technological, organizational, and institutional change that arise through the flow of ideas from exchange-based interactions (Jaleta, Gebremedhin, and Hoekstra 2009).

Small-scale farmers and their emerging enterprises require local services that help them integrate into value chains (chain empowerment) and relate to other chain actors (traders, processors) and services (value chain finance) (Webber and Labaste 2010; KIT, Faida Mali, and IIRR 2006; KIT and IIRR 2008, 2010). These "local agribusiness development services" (LBDSs) improve the performance of a small-scale enterprise oriented to agricultural production, be it individual or cooperative, in accessing markets, financial services, and enhanced agribusiness environments.[1] LBDSs encompass training and advisory services, market information services, technology, and business linkage information (UNDP 2004a). By improving the efficiency and competitiveness of agribusinesses in specialized and quality production, agroprocessing, input use, and produce marketing, LBDSs help to close a critical two-way gap between smallholders and markets: Small-scale, entrepreneurial farmers need better links to markets and value chains, while market actors (including small-scale traders, processors, manufacturers, and exporters) need sustainable sources of produce from smallholders. Figure 3.1 depicts the roles of LBDSs in relation to local product value chains. Box 3.13 summarizes the effects of successful LBDSs in Uganda.

Capacity for providing business services is generally confined to nonprimary production and/or medium-scale enterprises in the private sector in urban areas. In rural areas, the public sector and civil society provide most advisory services and concentrate on agricultural production. In their current form, these advisory services cannot cope with farmers' growing demand for services oriented to markets and value chains. Nor can they cope with growing demands for sustainable sources of produce from other actors in the value chain, including national and international exporters.[2] The limited individual, organizational, and institutional capacity to develop small-scale agribusinesses locally clearly remains a major constraint to market-oriented production.

INVESTMENT NEEDED

Different forms of investment can strengthen agribusiness development services catering to diverse groups of farmers and entrepreneurs. These strategies include reforming public extension systems (building capacity and balancing public and private service provision), developing the capacity for private LBDSs, and developing agribusiness centers for service provision and learning. In practice, a mix of strategies is usually followed, as in pluralistic extension systems (TN 1).

Reforming public service providers and offering services through public and private channels

Public extension systems can respond to demands for local business development through various combinations of reforms involving capacity building, reorganization, and complementary public, private, and nongovernmental service provision. Some public systems strengthen competencies in business development by reorienting programs and staff; others choose to recruit personnel with the required skills (usually recent graduates of universities and professional training institutes). Mozambique strengthened capacity to provide more market-oriented, demand-driven services by outsourcing some services to large-scale farmers, companies, and NGOs (see box 3.12 in TN 1) and involving more smallholders in Farm Business Schools and in developing business plans. At the district level, Mozambique organized farmer-promoters to supply advisory services and inputs (box 3.14). In Ethiopia and Uganda, business services were provided by cooperative unions to primary cooperatives (see http://apf-ethiopia.ning.com/page/business-development and http://apf-uganda.ning.com/page/farmers-organisations). Some business development services are simply unavailable from the public advisory system and need to be outsourced to the private sector or civil society, as with NAADS in Uganda (box 3.15) (Friis-Hansen and Aben 2010; van Weperen 2011).

Some countries leave the provision of business development services solely to private entities such as produce boards (for any number of commodities, such as cotton,

Figure 3.1   Roles of Local Agribusiness Development Services in Relation to Actors in the Agricultural Product Chain and to Support Services

Sources: Adapted from Wilk and Fensterseifer 2003; Roduner 2007; NAADS 2010 (unpublished); and KIT and IIRR 2010.

Box 3.13   Effects of Local Business Development Services for Farmers in Uganda

In Uganda, local business development services were identified as offering the key support required to prepare business appraisals, develop marketing plans, apply for bank credit, and obtain advice on financial and legal matters. Farmers' demand for these services widened the involvement of private agricultural service providers in helping farmers with market-oriented production.

Sources: NAADS 2010 (unpublished); Friis-Hansen and Aben 2010.

Box 3.14 Farmer Agribusiness Promoters in Mozambique

Owing to major public and international NGO investments in agribusiness at the district level in Mozambique, farmers' demand to become more market-oriented has grown rapidly. Because only a limited number of public extension workers could meet that demand, District Services for Economic Activities (SDAE, Serviço Distrital de Actividades Económicas) involved large numbers of farmer-promoters in different advisory services. Farmer promoters are involved in small-scale input supply and related advisory services, such as cashew spraying, chicken vaccination, groundnut pest control, and similar activities. Smallholders with knowledge related to market access increasingly participate in the development of smallholder business plans, as pioneered by NGOs in various Local Economic Development Projects. The Farmer Field School (FFS) program, in which farmers serve as facilitators, has been expanded to all provinces. The Field School approach is also used widely for enhancing market-oriented farm management in Farm Business Schools.

Sources: DNEA 2007; Kahan 2007, 90–92.

---

coffee, and cashews) or value chains with their own brands (box 3.16). The risk is that services geared to the demands from particular subsectors or value chains will exclude many small-scale farmers and entrepreneurs. One possibility is to provide startup capital for private providers of business development services at the local level, based on business plans and matching funds, and at the same time offer incentives for current public and private business service providers to meet the needs of small-scale entrepreneurs rather than focusing exclusively on medium-scale operations, as done in Mozambique.[3]

An important aspect of these various innovations in offering LBDSs is that public advisory systems need mechanisms to capture and share the lessons emerging from them. They must open up and develop alliances and networks for learning and interaction between different actors from different value chains and services, especially through partnerships, multistakeholder platforms, and networks (www.kit.nl; http://www.delicious.com/tag/apf_ethiopia).

Developing private capacity to deliver local business services

Service providers' skills can be strengthened through a series of related investments at different levels. Often an initial step is to develop the capacity of existing, often urban-based private organizations to work in rural areas with

---

Box 3.15 Developing Small-Scale Agribusinesses in Uganda: Strategies and Outcomes

The National Agricultural Advisory Services (NAADS) program sought to enhance small-scale entrepreneurship through efforts to develop value chains and widen access to market-oriented production services. Since 2002, NAADS have made considerable progress. About 50 enterprises have been selected for development and promotion, more than 45,000 farmer groups were engaged in market-oriented enterprise development and promotion, and 200 higher-level farmer organizations were established. Public-private partnerships were formed to support a substantial number of outgrower schemes involving "nucleus" farmers. The nucleus farmer strategy encourages small-scale farmers to increase their market orientation by providing value-adding and agroprocessing facilities as well as links to markets.

As a result of these efforts, more small-scale commercial farmers started to emerge, and their production rose to such an extent that more organized marketing and agroprocessing facilities were needed. The private enterprises and private service providers emerging to support this growth still require capacity building and enabling policies to sustain their development. Other challenges remain in scaling up the successful aspects of NAADS. Access to credit and inputs can be problematic, and farmers' empowerment in value chains is still limited by a lack of capacity, information asymmetries, poor links among key players along the value chain, and markets characterized by low activity, low volumes, and other symptoms of poor competitiveness. Additional public investment is needed to complement the efforts of the private sector in developing agribusinesses at the smallholder level.

Sources: Authors; Benin et al. 2007; Friis-Hansen and Aben 2010.

Box 3.16 KILICAFE, a Local Agribusiness Service Provider in Tanzania

KILICAFE is the name of a brand and farmer organization for specialty Kilimanjaro coffee. The smallholders who are members of KILICAFE have come to play a major role in interactions with agricultural service providers, including providers of research, advisory, financial, and input services. Aside from linking producers to markets, traders, and millers, KILICAFE provides a range of services to its members. At the Farmer Business Group level, KILICAFE provides pulping services; at the chapter level, it works to strengthen Farmer Business Groups; and at the national level, it offers marketing and financial services. Services include credit links and financial management of loans for working capital and for establishing central pulping units. KILICAFE sources financing from donors and/or financial institutes to purchase central pulping units and issues repayments from coffee sales to the Farmer Business Group on four-year term loans. Input credits are organized at the chapter level, where each chapter Annual General Meeting sets limits on how much to spend on inputs per kilogram. These credits are not cash loans but guarantees to input suppliers for future payments. Marketing is done by sending green coffee samples to the Coffee Board (for buyers at local auctions) and shipping samples directly to overseas coffee roasters (for direct exports).

KILICAFE provides technical advisory services and training, such as training farmers in production methods to improve the quality of their product, training in central pulping unit operations, and training in business management. These services are provided at the Farmer Business Group level through seminars that are open to all members. In addition, leadership training is conducted at the chapter level for all Farmer Business Group Management Committee members (chairpersons, secretaries, and treasurers). This leadership training empowers smallholder farmers to own fixed assets, which can be used as collateral for bank loans. KILICAFE also provides communication services such as a quarterly newsletter, radio broadcasts, and website (www.kilicafe.com), all containing information on coffee price trends, a farm activities calendar, association events and activities, and new developments.

Farmer Business Group members also demand other services from KILICAFE, such as supplying agricultural inputs well in advance. Some groups want KILICAFE to clearly specify coffee processing quality standards and ensure adherence by all Farmer Business Groups. To achieve uniform quality and obtain premium coffee prices, producers need to use only recommended technologies, although low prices also influence the adherence to quality-enhancing standards.

Source: Wennink, Nederlof, and Heemskerk 2007.

---

entrepreneurial smallholders (TN 4). Another important building block is to strengthen the capacity of private business development service providers in rural areas to support local development of basic farm business plans, market studies, and feasibility studies (DNEA 2007). Complementary investments involve developing multistakeholder innovation platforms or networks to assemble all actors in a particular value chain at the local level (to initiate collective action for local business development) and at the national level (primarily for advocacy). Investments in Farm Business Schools, such as those implemented through FAO in Botswana, Kenya, Malawi, Nigeria, and Zambia, will assist farmers in expressing demand for LBDSs (Kahan 2007:90–92; Malindi 2011). Finally, management capacity will be needed at the district or local government level to match the demand and supply of LBDSs (TN 1) in addition to registering, supervising, and evaluating them.

The source of human resources to provide LBDSs must not be neglected. Universities, professional education institutes, and vocational training institutes need support to update their curricula to reflect the growing demand for capacity in agribusiness, marketing, and entrepreneurial skills. A vital part of curriculum change is for these institutions to develop the capacity to train, coach, and support local providers of agribusiness development services (see http://ruforuminnovationsproject.blogspot.com). One such initiative is Business Minds Africa: Professionals for Agricultural Entrepreneurship in East-Africa (http://www.businessmindsafrica.org), a partnership between East African Universities, RUFORUM (see module 2), International Institute of Rural Reconstruction (IIRR), the Royal Tropical Institute, and Van Hall Larenstein University of Applied Sciences, Wageningen.

Business development service centers

Another potential area for investment is to provide services through a "one-stop shop" mechanism, in which any number of services (technological, business development, financial, and input supply, for example) are offered in a central location. These services centers can have additional objectives of learning and training and are mostly run through public-private partnerships. Examples include agribusiness centers (box 3.17), agribusiness incubators, and local economic development agencies, all of which could be designed to provide integrated LBDSs for small-scale farmer entrepreneurs. Services could include starting and registering a business, farm business planning, access to finance, training, and technical advice. Agribusiness incubator programs support emerging small-scale farm businesses and build capacity through learning by doing (box 3.18; see also TN 3 in module 5). Local economic development agencies, funded by the public sector, foster public-private partnerships at the local level (examples have been documented in Mozambique; see UNCDF 2009).

Investments are also needed at a higher level for mentoring and coaching these services. As part of public investment programs, national capacity should be developed for supporting business development services at the district level, with the aim of establishing and strengthening private local providers (boxes 3.14 and 3.18). In Mali (box 3.18) and Mozambique (Eduardo Mondlane University's Sustainable Trade Academy in Chibuto), an agribusiness incubator concept was also used to improve university graduates' capacity in small-scale enterprise development.

POTENTIAL BENEFITS

The potential benefits of LBDSs for smallholder farmers include increased entrepreneurial knowledge, better market linkages, enhanced access to credit, and better marketing opportunities. Smallholders have expressed this demand and given the opportunity will refocus extension on these more market- and value chain-oriented agricultural advisory services (Webber and Labaste 2010; Friis-Hansen and Aben 2010; KIT, Faida Mali, and IIRR 2006). Services to develop local farming businesses provide support to producers, traders, processors, and other actors farther along the value chain. Each group has different needs and requires different funding arrangements, which can be local, regional, or national. Service providers can also support the development of local economic development plans and strategies. Anticipated impacts among smaller-scale farmers and entrepreneurs include enhanced rural income (both directly and through employment) through enhanced small-scale entrepreneurial activity, based on the use of local resources and competitiveness (Webber and Labaste 2010).

POLICY ISSUES

LBDSs, public as well as private, need an enabling environment to make an impact. Wherever options and opportunities exist with respect to financial services, and wherever markets and market infrastructure exist and function

---

Box 3.17   Casas Agrárias in Mozambique: Lessons from One-Stop Agribusiness Centers

Around the town of Lichinga in Mozambique's Niassa Province, farmer associations established Casas Agrárias with support from OIKOS (a Portuguese NGO) and Estamos (a local NGO). Casas Agrárias are agribusiness centers for marketing crops—for example, they can offer temporary storage and processing facilities—and facilitating access to credit, inputs, and agricultural advice. The centers' processing activities include milling maize, processing rice, and extracting vegetable oil from groundnuts and sunflowers. The Casas Agrárias are special entry points for supporting public-private partnerships, developing capacity in farmers' cooperatives, and involving national and provincial farmer organizations. Farmer management committees manage these centers, which have a limited number of extension staff

(four to five) from public or nongovernmental agencies, trained in input and output marketing. Important lessons from the Casas Agrárias are that investment in these centers must focus on developing capacity among at least four to five people for each center to maintain its services, training staff and farmers in agribusiness management, and offering specialized training in storage and input supply for farmers and cooperatives. Another lesson is that Casas Agrárias need to become sustainable and autonomous. As soon as possible, they must be handed over to farmer cooperatives and unions to operate, supported with adequate financial and administrative management training, and linked with district savings schemes.

Source: Authors.

Box 3.18   The Cheetah Network Integrates Agricultural Education and Business Incubation in Mali

Business incubators are programs designed to accelerate the successful development of entrepreneurial activities through an array of business support resources and services, developed and orchestrated by incubator management and offered both in the incubator and through its network of contacts. Incubators vary in the way they deliver their services, in their organizational structure, and in the types of clients they serve (see module 5, TN 3).

Mali's national agricultural research organization (Institut d'Economie Rurale, IER), national agricultural university (Institut Polytechnique Rural de Formation et de Recherche Appliquée, IPR/IFRA), United States universities, and small-scale subsistence farmers formed an alliance to develop the Mali Agribusiness Incubator Network ("Cheetah Network"). Through its business incubators, the network identifies and assists entrepreneurs in efforts related to agriculture. The alliance led university staff and graduates to review and revise course curricula significantly to

develop more skills and change mindsets related to promoting small-scale agribusiness in Mali and the United States.

The Cheetah Network supported male and female graduate students from the agricultural university in creating a number of entrepreneurial incubators (essentially, small clusters of small-scale enterprises focused on a similar goal). One cluster involved a women's cooperative in Zantiebougou focused on producing, processing, and storing shea butter for export and on developing a high-quality market for its products in the United States and Canada. Another cluster involved the production of certified seed potatoes in Borko and Gao for regional export. The entrepreneurial incubators have strong elements of learning for farmers as well as graduate students and staff. They aim at institutional sustainability through their links with research and the university and at financial sustainability through the introduction of fee-for-service systems.

Source: USAID 2009.

properly, LBDSs can be appropriate. National policies and local government regulations (for example, those governing input supply and marketing) must foster the development of entrepreneurship by ensuring a level playing field, and public agencies must not interfere in input and output markets. Additional policy issues for LBDSs include:

■ *Social targeting.* Small-scale agribusiness entrepreneurs often constitute only 1–2 percent of rural households, yet *emerging* small-scale entrepreneurs may constitute up to 25 percent. Another large category of households has limited capacity for risk, focuses on food security, and relies on multiple income sources (remittances, local agricultural labor, petty trading, and others). Policies that support efforts by these groups to organize—in cooperative enterprises, outgrower schemes, contract farming, and farmer shareholding in marketing and processing enterprises—make it easier to address their varied entrepreneurial capacities and level the playing field. Close attention should also be given to the implications of gender in value chain development and agribusiness development services.[4]

■ *Local governance.* Local governance influences the local development context (including prevailing policies), which influences investments in developing private agribusiness services (Friis-Hansen and Aben 2010). Elements of the local development context that influence LBDSs include: (1) the emphasis on local public-private partnerships, (2) synergy between local economic development programs oriented to value chains and livelihood systems, (3) the involvement of farmer organizations; and (4) local capacity to manage multistakeholder platforms, networks, and interaction between service providers (TN 1).

■ *Public and private sector roles.* Policies influence whether and how interaction between value chain actors and private supporting services are brokered by public agencies, locally and nationally (Webber and Labaste 2010). Locally, the public sector is more prominent in empowering farmers through local economic development, whereas the private sector often predominates in value chain development. These roles need to become synergetic for value chain integration (KIT, Faida Mali, and IIRR 2006). Three additional considerations affect public and private sector roles in LBDSs. First, LBDSs facilitate integration between farmers and others in the value chain, increasing the likelihood that value chain partners will provide financial services for farmers (KIT and IIRR 2010). Second, private LBDSs tend to have greater capacity for facilitating access to financial services. Third, public support is needed to develop the capacities of service providers, coordinate local providers, empower farmers, and provide professional and vocational training in business development.[5]

■ *Local entrepreneurs and environmental services.* LBDSs can support smallholders in identifying incentives for sustainable production, coping with the effects of climate change, supplying green services (to enhance farmers' ecosystems or sustain the environment), and addressing demands from specific value chains (organic food). Incentives can include sharing in the profits from forestry concessions, hunting licenses, and carbon emission rights; premium prices for certified organic produce; or agronomic strategies promoted by the public sector to prevent soil nutrient mining (Odada et al. 2008; Pyburn, van der Lee, and ter Heegde 2011; box 3.19).

■ *Sustaining local service provision.* For LBDSs to be sustainable, they must be supported by the value chain actors or customers involved (see module 5). In the transition to more value chain- and market-oriented production, however, the public sector has a role in providing services, especially to smallholders who cannot afford them at first. Farmers indirectly finance services for traditional export commodities and are beginning to do so in emerging value chains (IAP 3 gives other examples of embedded services).[6] Some national programs have partial fee-based systems for delivering LBDSs, including Uganda (for NAADS; see box 3.7 in the module overview) and Azerbaijan (Lamers et al. 2008).

LESSONS LEARNED

As the examples in this note indicate, much of the experience in providing LBDSs to small-scale farmers, cooperatives, and other agricultural entrepreneurs has been gained through pilot projects funded by donors and NGOs. National programs for advisory services, such as those in Tanzania, Mozambique, and Uganda, have started to incorporate these

---

Box 3.19   A Successful Business Model for Mozambique's Farmers to Provide Environmental Services

With the support of a specialized service provider, Envirotrade, a prize-winning community project in Mozambique developed a successful business model for the sale of carbon offsets to support the conservation of forests and the planting of new ones. The scheme, one of three winners of an international climate grant competition, is being rolled out to other environmentally sensitive sites in Africa.[a] Sustainable farming practices introduced as part of the Nhambita Community Carbon Project increased cashew and fruit yields and improved livelihoods for about 1,300 families. Since its launch six years ago, the initiative, based in the buffer zone of the Gorongosa National Park, has traded more than 120,000 tons of CO_2, earning the community over US$1 million. Participants are paid for carbon stored by the trees they plant, the forests that they manage,

and the fires that they prevent. For example, Felicio Lucas Melo, 33, has two plots that can sequester over 55 tons of CO_2 per year, earning him US$244 in direct payments and an additional US$25 that is paid into the community carbon fund, which is used for improvements to schools, clinics, and wells.

Envirotrade is a Mauritius-based company with offices in both the United Kingdom and South Africa and project operations in Mozambique. Its business model is not a substitute for resolute international action to address the issues associated with human-induced climate change, but it offers a means for concerned businesses and individuals to link with forest farmers in developing countries to change how natural resources are used and reduce harmful environmental impacts.

Source: "Cash from Carbon," *Spore* (143) October 2009, http://spore.cta.int/index.php?option=com_content&task=view& lang=en&id=1016&catid=7, accessed July 2011.

a. Another green community program, the Kakamega Forest Again Project in Kenya, also won the top US$35,000 prize in the contest, organized by Hyundai Motor America and Carbonfund.org, in conjunction with the Climate, Community, and Biodiversity Alliance.

---

experiences in the drive toward more value chain- and market-oriented agricultural advisory programs and systems (van Weperen 2011). Demand from farmers and their organizations for high-quality business development services at the local level is burgeoning. Confirmation of this demand emerged in an analysis by the AgriProfocus country focus program with farmer organizations and their supporting agencies in Ethiopia, Kenya, Mozambique, Niger, Rwanda, Uganda, and Zambia.[7] Many countries have sought to meet the demand for local business services through an increasing emphasis on local economic development planning at the district level, with the involvement of key local stakeholders in the public and private sector (see http://go.worldbank.org/EA784ZB3F0). It is clear, however, that demand for LBDSs cannot be met by public service providers unless they receive adequate public funding.

Based on the large number of pilots to date, a number of lessons have emerged for practitioners. The lessons are grouped around key issues: (1) creating awareness; (2) building agribusiness service capacity; (3) implementation; (4) and developing dedicated agribusiness service organizations.

Creating awareness and enhancing demand articulation

At the start of a program for LBDSs, the level of commercialization of smallholder agriculture and the corresponding demand for different types of service providers must be analyzed. Demand is strongly determined by the economic, regulatory, and service context (Jaleta, Gebremedhin, and Hoekstra 2009).

An inventory of available local agribusiness service providers at different levels and by sector (public, private, NGO, and civil society) can avoid duplication and contribute to synergy based on public-private partnership. This kind of inventory is often the basis for a local farmer entrepreneur development strategy, as part of a local economic development strategy.

Farm Business Schools and cooperatives have an important learning role in promoting entrepreneurship among farmers, but initially they require external facilitation. Farm Business Schools facilitate learning about production, management, business finance, and marketing. Useful tools have been developed for this purpose by FAO (Dixie 2005) and the Swiss Agency for Development and Cooperation (SDC) (Poitevin and Hossein 2006), including modules on understanding the market; supply and demand; helping farmers decide what to do; producing for the market; producing

profitably; postharvest handling; and improving market arrangements (KIT, Faida Mali, and IIRR 2006).

Farmer organizations have an important role to play in articulating demands related to local economic development plans and in lobbying for the right type of services. They also have a central role in ensuring that services remain oriented to their demands and in providing some services themselves (boxes 3.14–3.16). Higher-level farmer organizations are needed to influence the agribusiness context and interact with actors farther along the value chain.

Building agribusiness service capacity

An analysis of the capacity development context for business services is key. Often the capacity to strengthen providers of agribusiness services to smallholders does not exist. Nor is there capacity at the district level to manage local public-private partnerships to provide such services. The capacity gaps are particularly wide among local service providers and in the capacity available to develop, mentor, and coach these providers, leading to a need for capacity development at the individual, organizational, and institutional levels.

National universities and business schools still aim to produce civil servants rather than self-employed service providers. Aside from skill development, a special challenge for these institutions is to develop an entrepreneurial mindset in public and private service delivery. Interaction between course programs and the professional sector is needed for curriculum improvement based on demand (see module 2 and Spielman et al. 2008).

The development of private providers of local agribusiness services for smallholders also requires public investments, preferably through training and matching grants/credits for starting small-scale agribusiness services based on a business plan.

Different types of agribusiness development services are needed for different types of entrepreneurs and farmers. These services will often be specific to certain value chains or even to different levels of a given value chain. Many programs developed to offer small-scale agribusiness services have realized that working with smallholders to strengthen agribusiness management, business plan development, and other elements of agricultural enterprises requires a process approach. Programs often begin by working with existing associations and individuals, some of which, through coaching and facilitation over time (often two years), develop and graduate into small-scale entrepreneurs and enterprise cooperatives. A typical process like this in

Mozambique cost US$300 to move a loosely formed association through seven steps to become a registered, market-oriented cooperative qualified to obtain financial services.

Implementing business development services at the local level

Business development is about chain development, facilitated by chain mapping and assessment (market orientation and risk assessment; local versus international markets; fostering an enabling business environment); chain engagement (developing a vision; building trust); chain development (participatory approach and ownership; addressing risks and savings; engineering an organizational development program; promoting entrepreneurial attitudes); chain monitoring and evaluation; and chain learning and innovation (KIT, Faida Mali, and IIRR 2006, Webber and Labaste 2010).

Services to develop small-scale farming businesses have a central role in facilitating access to input and financial services, but they have no role in directly supplying inputs and rural finance, which occurs in embedded services (IAP 3 and Roduner 2007). A distinction must be made between business development services and financial services. The public sector has a role in providing LBDSs to smallholders but not in directly providing financial services.

Who pays for LBDSs remains a major dilemma. If agribusiness services are fully subsidized, the services provided may not genuinely arise from agribusiness demands, private services may be crowded out, and the financial sustainability of the services will remain uncertain. If services are to be fully supported by the demand, larger enterprises may be able to pay, but others may be excluded. Involving other value chain actors in financing local services for enhanced quality of production increasingly appears to be the most sustainable approach (KIT and IIRR 2010).

To be scaled up, successful models of agribusiness development require a systems approach that pays attention to access to credit, access to high-quality inputs, the development of farmers' capacity, and the formation of public-private partnerships (box 3.15). A final lesson is that the innovation required for farmers to improve the quality of their produce cannot be sustained by farmer organizations without proper incentives or premiums (box 3.16).

Developing dedicated agribusiness service organizations

The primary element for success in establishing local business development centers is to base the intervention on local human and financial resources, which means that practitioners must emphasize linking with product marketing and processing, creating links with savings and credit activities, developing cooperatives, and building capacity. More specifically (see the discussion of incubators in module 5):

■ An agribusiness center requires a critical mass of staff trained in agribusiness management (at least four or five people) to maintain its services. Ideally it is governed with involvement from the private sector, but for smallholder producers it is often also supported by the public sector. Centers require adequate links with market actors, input suppliers, and financial services, including local credit and savings schemes.
■ Incubators can incorporate an element of capacity building for more business-minded and market-oriented service providers through interactions with universities or business schools. This interaction builds capacity in the staff of the business development center, builds capacity in the students involved in the work, and influences the content of the related academic programs.
■ Through training in financial and administrative management, centers must become autonomous as soon as possible and handed over to farmer organizations (local and national) and/or the private sector.

All programs directed at developing local agribusiness services must give considerable attention to fostering a long-term commitment to building these institutions in a stable policy environment as well to strategies that will ensure financial sustainability, based on cost sharing for the services that are delivered (World Bank 2010).

Extension-Plus: New Roles for Extension and Advisory Services

Rasheed Sulaiman V, Centre for Research on Innovation and Science Policy (CRISP)

SYNOPSIS

"Extension-plus" is a framework for investment in strengthening and reforming extension to be a strong partner and nodal agency within the AIS, providing technological and nontechnological services to farmers. The larger goal of investments in extension-plus is to strengthen the capacity of extension and advisory services to play a much wider role (a bridging role) and at the same time enhance the ability of other actors in the AIS to support producers in an integrated way. All current extension-plus arrangements have emerged from small pilot efforts that have expanded their scope and service provision based on experimentation, learning, and adaptation to local circumstances over time. Investments should focus on encouraging and enabling staff capacity to initiate small experimental projects in partnership with other organizations. Promoting the vision of extension-plus will prove challenging among public extension organizations unless reforms target the macro-institutional and policy context in which extension is practiced. Extension-plus can flourish only in organizational settings that have a culture of experimentation and learning. For cultural change in this direction to occur, it must be supported and legitimized unambiguously at the most senior levels of the extension service and allied organizations. Before designing the program and operational strategy for investment, it is advisable to undertake an institutional diagnosis to understand the range of organizations within the AIS, their expertise and activities, and their patterns of interaction. The scope of the specific extension investment and the priorities will vary in relation to the national, district, and local situations.

BACKGROUND AND CONTEXT FOR INVESTMENT

The limitations of a single model of extension and advisory services for all kinds of situations are now well recognized.

There is an increasing realization that new extension approaches need to emerge locally, based on experimentation, learning, and adaptation to prevailing circumstances. The need for extension to partner with other organizations and individuals with varied skills and competencies to provide integrated support (technical, organizational, marketing) to producers is also apparent. A number of examples emerging in the public and the private sectors illustrate how the conventional technology transfer role of extension is being expanded to improve its relevance to contemporary agricultural and rural development (Sulaiman and Hall 2004a, 2004b). Many of these examples appear to demonstrate the value of an expanded mode of extension referred to as "extension-plus" and provide important guidelines on design and implementation of new investments.

Extension-plus is a framework for investment in strengthening and reforming extension to be a strong partner in the AIS. It is especially relevant in the context of reforming public extension organizations in developing countries, where extension is struggling to find a relevant role to deal with contemporary rural and agricultural development challenges. The key elements of extension-plus are:

- A broad scope of service provision (beyond technology transfer).
- The extensive use of partnerships to fulfill an expanded mandate.
- A learning-based approach.
- Negotiations with a wide range of stakeholders for developing workable and effective service arrangements.
- An institutional mechanism to represent clients' interests at the management level, so the program remains accountable to its clients.

The larger goal of investments in extension-plus is to strengthen the capacity of extension and advisory services to play a much wider role (a bridging role) and at the same

Table 3.5 Shifting Extension to Extension-Plus

Aspect of extension	Shifts from:	Shifts to:
Form/content of extension	Technology dissemination	Supporting rural livelihoods
	Improving farm productivity	Improving farm and nonfarm income
	Forming farmer groups	Building independent, farmer-operated organizations
	Providing services	Enabling farmers to access services from other agencies
	Market information	Market development
Monitoring and evaluation	Input and output targets	Learning
Planning and implementation strategy	Doing it alone	Through partnerships
Sources of innovation in extension	Centrally generated blueprints for wider implementation	Locally evolved (through ensuring right kind of support for local experimentation), with diverse approaches and multiple partners
Role of technical research	Technology development	Source of technical expertise and supporting adaptive research
Approaches	Fixed/uniform	Evolving/diverse
Capacity development of staff	Training	Learning by doing, facilitated experimentation
Capacity development of extension system	Personnel and infrastructure	Development of linkages and networks
Policy approach	Prescriptive/blueprints	Facilitating evolution of locally relevant approaches
Introducing new working practices	Staff training	Changing organizational culture through action learning
Underpinning paradigm	Transfer of technology	Innovation systems

Source: Sulaiman and Hall 2004a.

time enhance the ability of other actors in the AIS to support producers in an integrated way. Table 3.5 describes key shifts needed to operationalize extension-plus.

INVESTMENT NEEDED

The most innovative investment element of this approach is the explicit acknowledgment that investment should be concerned with creating or enhancing the capacity of the current innovation system for interaction and coordinated action, so that the producers receive a wider range of support and services. Extension organizations traditionally have some capacity for interaction with research. In this case, however, extension has to widen its networks to connect producers with different sets of service providers. This means that extension should partner with a number of different agencies and develop specific arrangements in line with local circumstances. Investments should focus on encouraging and enabling staff capacity to initiate small experimental projects in partnership with other organizations. By facilitating small projects experimentally and assisting staff to reflect on their meaning and outcomes, these investments will build skills related to experimentation and learning. Table 3.6 summarizes the kinds of investments needed under extension-plus.

All current extension-plus arrangements have emerged from small pilot efforts that have expanded their scope and service provision based on experimentation, learning, and adaptation to local circumstances over a period of time. Box

3.20 provides three examples—two from India and one from Bangladesh—of initiatives that served as nodes linking producers to technology and nontechnology services, including marketing. Each initiative supported the development of user groups that became the basic units for implementing programs. For example, in India's Kerala State, where smallholder and marginal farmers dominate agricultural production, almost 93 percent of land holdings are marginal (less than 1 hectare), and about 5 percent are small (1–2 hectares). Kerala imports around 80 percent of its fruit and vegetable requirements, primarily from neighboring states. In view of this dependency, the larger objective of the program described in box 3.20 was to develop a replicable model for horticultural development to diversify agriculture. The model, piloted in seven districts, was scaled up to cover all districts in the state after donor funding ended.

In the second example in box 3.20, BRAC (an international NGO) shifted from community development in Bangladesh toward a more targeted approach based on village organizations in 1977. Currently BRAC's operations reach about two-thirds of the population of Bangladesh. BRAC's outreach covers all 64 districts and 78 percent of villages in Bangladesh. Eighty percent of its funds are internally generated.

In the second example from India in box 3.20, a program to foster horticultural production in South Gujarat expanded from 44 families in 1982 to more than 23,000 families in 2010. Most activities are now managed by

Table 3.6   Investments Needed under an Extension-Plus Scenario

Major investment areas	Purpose
Pre-project phase	– Analyze past and ongoing interventions by different agencies. – Institutional diagnosis to understand the patterns of interaction among the different agencies and the institutional and policy environment. – Understand demand for support. – Develop a shared vision of objectives and potential approaches and identify potential partners.
Institutional and human capacity strengthening	– Place staff with diverse expertise (networking, technical knowledge, organizational development, market/business development, credit and financial operations). – Develop a new organizational culture that focuses on experimentation, openness to new ideas, reporting and learning from mistakes, regular staff reflection, incentives for good performance, and guidelines for staff assessment. – Encourage appropriate institutional changes to enhance the organization's ability to act as a nodal agency, capable of brokering relations with other actors, by broadening its mandate and using partnership and learning as the key operational strategies.
Technical support	– Address the current weaknesses in technology use and find opportunities to bring in new technologies for production, postharvest handling, value addition, and export through contract research, recruitment or secondment of technical specialists in the program, or bringing experts on short consultancy assignments.
Credit and financial support	– Identify the current bottlenecks related to the availability of credit at reasonable rates and address them. This process might include bringing the credit issue to the right policy actors and negotiating with financial institutions.
Organizational development	– Organize producers to enhance their capacity to deal with different agencies, work collectively, and evolve new governance arrangements (see module 4, TN 5, on organizational change).
Market development	– Improve farmers' ability to negotiate and receive a fair price for their produce. This process would involve strengthening the existing value chains, developing more equitable institutions related to procurement and pricing, and, at times, creating new value chains by linking producers to new markets.

Source: Author.

Box 3.20   Extension-Plus: Examples from the Field

Kerala Horticultural Development Programme, India. Conceived in 1992, the Kerala Horticultural Development Programme (KHDP) aimed to improve the circumstances of Kerala's fruit and vegetable farmers by increasing and stabilizing their incomes, reducing production costs, and improving the marketing system. The KHDP worked with fruit and vegetable farmers to promote self-help groups. It trained three farmers from each group to become master farmers who could deal with production, credit, and marketing. It promoted the concept of credit to farmers who leased land, promoted group marketing, and established modern seed processing and fruit processing plants. To generate and access locally relevant technical knowledge, KHDP entered into contract research with the local agricultural university and strengthened the skills of farmers in participatory technology development. The total outlay for KHDP was €36.76 million, of which the European Commission contributed 78 percent and the state government contributed the remainder. Though it ended in December 2001, KHDP reinvented itself as the Vegetable and Fruit Promotion Council, Kerala (VFPCK, www.vfpck.org), a company in which 50 percent of the shares are held by producer groups. Since then, VFPCK has expanded its activities, coverage, and funding sources, obtaining some funds from government programs. All programs related to fruit and vegetable promotion are undertaken through VFPCK, whose approach was eventually extended to all districts in Kerala. The company directly reaches more than 132,000 vegetable and fruit farmers in Kerala.

BRAC's Economic Development Programme. The Economic Development Programme of BRAC (an international NGO that originated as the Bangladesh Rural Advancement Committee) is the cornerstone for all of BRAC's development work in Bangladesh. The Development Programme covers microfinance, institution building, income-generating activities, and program support enterprises (such as seed production, disease diagnostic labs, and produce processing and

(Box continues on the following page)

Box 3.20 Extension-Plus: Examples from the Field (continued)

marketing). While BRAC believes that microfinance is necessary to break the cycle of poverty, it places equal importance on microenterprise development services to maximize the return obtained by the poor. Unlike standard business development programs, which offer some mix of generic training and marketing services, BRAC has developed an integrated, sector-specific approach to enterprise development for the poor. BRAC has identified six sectors in which large numbers of low-income women can be productively engaged at or near their homes: poultry, livestock, fisheries, sericulture, agriculture, and social forestry. For each of these sectors, BRAC has developed a set of services that comprises training in improved technologies, ongoing supply of technical assistance and inputs, monitoring and problem solving as needed, and marketing of finished goods. BRAC evolved this model through continuous iteration and experimental learning. This program has so far organized 8.45 million poor and landless people into 284,825 village organizations, which are the basic units of the program.

The Wadi Programme of Dharampur Uththan Vahini, India. Dharampur Uththan Vahini (DHRUVA, "Vanguard of Awakening in Dharampur"), an associate organization of the BAIF Development and Research Foundation, works in 200 tribal villages in Valsad, Navsari, and Dangs Districts of South Gujarat. DHRUVA's Wadi Programme, which facilitated the establishment of fruit orchards (*wadis*) on land belonging to poor tribal families, started with 44 wadis in 1982. Village-level peoples' organizations have been pivotal in implementing the Wadi Programme's activities. The organization encouraged the formation of a cooperative for wadi farmers in the Vansda area to help them market their produce collectively. Produce from cashew and mango trees is sold to the cooperatives, which in turn sell them to the apex cooperative. DHRUVA helped the cooperative design appropriate systems to preserve and process horticultural produce (including cashews, mango pickles, jams, and jellies) and access local and urban markets under its Vrindavan brand name. Today, over 23,000 families from 400 villages have adapted the wadi model. Huge tracts of wasteland have been converted into orchards, which have contributed to improved livelihoods and the regeneration of natural resources. The project received funding from donors (including KfW) as well as government support for rural employment, tribal development, and funds from the National Bank for Agriculture and Rural Development.

Sources: Bhamoria 2004; VFPCK 2009; BRAC 2010; DHRUVA 2010.

producer cooperatives and village organizations, with only marginal support from the NGO that initiated the program. The initiative has been acclaimed worldwide as a sustainable and replicable model for alleviating poverty.

POTENTIAL BENEFITS

A potential benefit of investing in extension-plus is the development of a sustained capacity for innovation. For instance, the most important contribution of the investments in KHDP, BRAC, and DHRUVA has been the development of a capacity for continuous innovation even after the end of external funding. Many externally funded projects fail to deliver once funding ends because they fail to develop the capacity to keep innovating. Box 3.21 summarizes the impact of some good practices from KHDP/VFPCK. Table 3.7 lists some of the indicators that could be used to evaluate an extension-plus approach.

POLICY ISSUES

Promoting the vision of extension-plus will prove challenging among public extension organizations unless the reforms target elements of the macro-institutional and policy context in which extension is practiced. Extension-plus can flourish only in organizational settings that have a culture of experimentation and learning. For cultural change in this direction to occur, it must be supported and legitimized wholeheartedly and unambiguously at the most senior levels of the extension service and allied organizations.

Institutional issues

As indicated, some of the underlying "institutions" (norms, values, routines, and attitudes) that govern or shape extension in its current form constrain the adoption of extension-plus. For instance, many countries continue to

Box 3.21 Good Practices and Their Impacts for Kerala's Fruit and Vegetable Farmers

The good practices followed by KHDP and its successor organization, VFPCK, produced a number of impacts:

- The program promoted the concept of organizing self-help groups of vegetable and fruit farmers and training certain farmers from each group in specific skills. VFPCK currently works with about 6,800 self-help groups, of which 405 are run by women, and reaches more than 132,000 farmers.
- By working closely with 11 commercial banks, the program could help farmers obtain credit for cultivation of leased land. More than US$5.6 million in credit was distributed in 2008–09, and VFPCK also developed credit-linked insurance for farmers.
- Group marketing was promoted by establishing markets where farmers could bulk their produce for sale to traders and improve their bargaining position by obtaining information on market prices inside and outside Kerala. In 2008–09, more than 200 VFPCK Farmers' Markets operating across Kerala sold 87,000 tons of produce valued at

US$20 million. The council supports these committees with infrastructure on a limited scale.

- Through its modern seed-processing plant, VFPCK produced more than 38 tons of seed for 19 vegetable varieties in 2008–09, thereby contributing 50 percent of Kerala's internal seed production. Through participatory technology development trials with farmers, the council is promoting the cultivation of vegetables in the cool season.
- The council established a modern fruit-processing factory with farmers as stakeholders. Products from this factory are traded in domestic and international markets.
- An external evaluation and impact study of KHDP by the Xavier Labour Research Institute reported a significant increase in area under fruit and vegetables in 86 percent of the self-help groups and increased incomes in 75 percent of the groups. The same study also reported that the number of farmers receiving credit increased from 21 percent in the pre-KHDP period to 41 percent by 1999, with an increase in the efficiency of loan disbursal and an increase in the size of the loans.

Sources: XLRI 1999; VFPCK 2009.

Table 3.7 Indicators That May Be Useful for Monitoring and Evaluating an Extension-Plus Approach

Output indicators	Outcome indicators
– Farmer groups or producer associations formed; groups' sustenance, maintenance of records	– Increase in income, production, productivity; additional employment created
– Formation of new markets; marketing and price realization	– Sustenance of the arrangement; continuance, expansion, and impact
– Training organized	– Enhanced capacity for collaboration and continuance of good practices; new partnerships formed; other institutional changes generated
– New inputs and technologies distributed and/or purchased and used	– New funding generated
– Access to credit; credit use and repayment	– Ability to respond to new demands
– New value-added products developed	– Governance mechanisms: how different stakeholder views are expressed and quality of response
– Infrastructure developed; capacity utilization	
– Partnerships, new working arrangements, or new areas of collaboration; quality of interactions	
– Reforms promoted; changes in guidelines related to funding and collaboration	

Source: Author.

plan, implement, and evaluate extension centrally, which can stifle any divergence from prescribed procedures and restrict innovation and learning, particularly by mid- and lower-level staff. In many instances, extension maintains a tradition of assessing performance in terms of technology adoption and upward accountability for resource utilization rather than by examining whether outputs were achieved and whether clients are satisfied. The reluctance to change is reinforced by an extension policy dialog that continues to be couched in terms of a narrow conceptualization of extension as an agency transferring technology and improved practices from research stations to farmers (Sulaiman and Hall 2005).

Public and private sector roles

A first step in operationalizing extension-plus is to reach broad agreement that extension must be reinvented as a nodal agency that provides technological and nontechnological services to farmers. In other words, extension will need to partner with a large number of other public, private, and NGO agencies that provide many of the additional services that will be in demand. In most settings, partnership among these agencies has been the exception rather than the rule, given the great level of mistrust among them. Extension can play its wider role only after undergoing large-scale restructuring and institutional changes, which extension bureaucracies often have been reluctant to undertake. Some of these changes include a broadened mandate, partnership and learning as key operational strategies, and freedom and support for staff at district and block levels to experiment with alternative strategies.

Human resource issues

To implement this approach, extension organization would require new expertise. One way of obtaining this expertise is to create a core group of specialists with skills such as market development, organizational development, enterprise development, and agribusiness management. The extension curricula of universities and the content offered in extension training centers will also need to be reviewed to ensure that perspectives such as extension-plus are adequately covered.

Sustainability issues

To sustain the institutional changes and capacity developed through this approach, a clear exit strategy must be agreed upon by the donor and the stakeholders. Building community-based organizations (user groups, cooperatives, village organizations, self-help groups), shifting the operation and management of the program to these organizations, and enhancing the capacity of these organizations to perform their responsibilities and raise fresh resources are all important steps toward sustaining the approach, even after donor support ends.

LESSONS LEARNED

Implementing the extension-plus approach can involve a number of challenges:

- The fact that the final program details cannot be visualized in the beginning can make donors and national governments slightly uncomfortable. As the approach can be implemented only as a series of experiments, resource allocation in the initial stages can only be tentative.
- The approach requires high-quality human resources at different levels, representing more diverse kinds of expertise. Human resource costs as a percentage of the total investment can be quite high.
- Partnering with organizations with diverse types of expertise is critical. Partnering is not an easy task for organizations that have a long history of isolated or independent functioning. In such cases, implementation could be slow.
- Only when the organization has sufficient flexibility to deal with administrative and financial issues will this approach flourish. The program should have opportunities for reflection and learning and sufficient flexibility to respond to the demands and opportunities emerging from the field as the program evolves.

These operational issues are not insurmountable. The program should be fully aware of them and find ways of engaging the government and the donors to resolve some of these concerns. Box 3.22 describes how the KHDP/VFPCK program met these challenges.

Some of the lessons learned from implementing extension-plus in varied settings are:

- The goals of the investment should be broad enough to provide integrated support to producers, improve competitiveness of the sector, or upgrade the production system to improve livelihoods, and so on. Broad goals are necessary to challenge extension to broaden its agenda.
- The investment should provide for hiring a mix of global and local expertise to support program implementation. It would be useful to get human resources on a long-term basis, starting with the design and inception of the program, to provide continuity and a shared vision of the objectives and approaches for implementation.
- Partnership with other organizations having varied skills should be the basic philosophy guiding the interventions.
- Continuous experimentation, reflection, and learning should be the basic approach for identifying relevant

Box 3.22 Experience with Innovative Activity in Kerala Horticulture

The most attractive feature of the program developed for fruit and vegetable producers by KHDP and its successor organization, VFPCK, was the concept of integrating three main components of agricultural development: production (including support for research and development), credit, and markets.

To implement this concept, KHDP created a new organizational structure and management strategy. It hired the services of international and national consultants to support key areas of its operation: credit, implementing an agroprocessing program, technology, and training. The expatriate experts, who were on a long-term consulting assignment, brought new knowledge and fresh perspectives to the program as it translated its vision into action.

Another interesting feature of the program was the flexibility to change the type and nature of interventions as and when problems arose. This flexibility allowed the program to evolve over the years. In its early years, KHDP quickly found that it needed to organize farmers into groups to promote new technology, help access credit, and strengthen negotiating power through collective marketing. It entered into a contract research arrangement with the state agricultural university for technical backstopping, and when it found this arrangement was unsuccessful, it realized the importance of involving farmers directly in technology development and testing. Since

then, the program has promoted participatory technology development and testing. Initial serious setbacks in group marketing also caused the program to reflect and learn to overcome the problems. An additional problem was that traders perceived farmers' markets to be a threat, and considerable persuasion became necessary to convince them of the benefits they would receive from cooperating with the farmers' markets.

In the beginning, KHDP envisaged providing credit to farmers through cooperative credit societies. After encountering difficulty in mobilizing funds on its own, the program decided to arrange for commercial banks to supply the credit. Although the banks were reluctant to provide credit to landless farmers, KHDP's willingness to deposit its own funds with those banks encouraged them to take the risk.

The program paid explicit attention to learning along the way. It established a monthly review meeting of project managers that provided a forum for sharing knowledge and experience from implementation on the ground and served as the program's learning laboratory. Without this forum, the program probably would never have learned from its initial experiments, some of which failed. The donor (the European Commission) supported management's changes to the initial design, which to a large extent allowed the program to achieve its goals.

Source: Author.

---

strategies for implementation. The program should have sufficient flexibility to shift approaches based on this learning.

- Accountability to clients is important. Clients should be part of the governance structure.
- A well-thought-out exit strategy should be in place before funding ends. Activities in the final year of implementation should focus on how to sustain the investment's positive outcomes.

Recommendations for practitioners

Extension can and should expand its role, given its significance for the larger AIS. The principles of extension-plus provide an opportunity for expanding the role of extension

by raising questions on the nature of extension's tasks, recognizing the need for new expertise, facilitating a review of extension's current interactions, and highlighting the importance of institutional changes. These tasks are important for developing and sustaining a capacity for innovation, which should be the main focus of investing in this kind of approach.

Before designing the program and operational strategy for investment, it would be better to undertake an institutional diagnosis to understand the range of organizations within the AIS, their expertise and activities, and their patterns of interaction. The scope of the specific extension investment and the priorities will vary in relation to the national, district, and local situations. For instance, forming groups of farmers could be the starting point in one

Table 3.8 Activities to Ensure Successful Operation of an Extension-Plus Approach

Phase or aspect of operation	Activity
Pre-project phase	— Conduct individual consultations, workshops, sample surveys — Identify key partners — Develop a shared vision for the program
Institutional and human development	— Recruit experts that can bring specific skills — Negotiate to get the right kind of staff on deputation — Identify and contract consultants (short and long term) — Conduct training, exposure visits, case analysis — Conduct an organizational and management review
Technical support	— Identify best technologies and refine or adapt them to local conditions — Direct recruitment — Make available on time the best and most efficient inputs, either by producing them directly or brokering arrangements with other suppliers — Recruit qualified technical staff and train them so that they remain up to date — If necessary, fund adaptive research
Credit and financial support	— Understand the financial/credit landscape — Negotiate with financing agencies — Guarantee transactions, set up revolving funds — Organize producers for group lending — Influence policies to help mainstream credit operations
Organizational development	— Form producer organizations (self-help groups, commodity interest groups, federation of self-help groups, producer companies, and similar groups) — Enhance skills through appropriate training programs (for example, skills in group dynamics or office management, including financial management)
Market development	— Analyze and strengthen market chain — Negotiate with different actors in the value chain — Create new markets if needed — Develop new products

Source: Author.

location, whereas linking farmers to new or emerging markets could be the priority in another. It would be ideal to source expertise and other inputs by forging links with other actors rather than trying to do everything through one program. If reliable sources of expertise and inputs do not exist, however, the program will have to start its own initiatives.

Specific activities that can ensure success at different stages and for different aspects of an extension-plus approach are listed in table 3.8.

Investment should focus on potential ways for strengthening and sustaining the capacity for innovation during the project period and after its end. Developing a new office (administration/financing) manual; making arrangements for monitoring, learning, and impact assessment; creating opportunities for communication and engagement with policy; and ensuring adequate funds for addressing evolving challenges are also critical for implementing this approach.

The Role of Innovation Brokers in Agricultural Innovation Systems

Laurens Klerkx, Wageningen University
Peter Gildemacher, Royal Tropical Institute (KIT)

SYNOPSIS

"Innovation brokers" are persons or organizations that, from a relatively impartial third-party position, purposefully catalyze innovation through bringing together actors and facilitating their interaction. Innovation brokering expands the role of agricultural extension from that of a one-to-one intermediary between research and farmers to that of an intermediary that creates and facilitates many-to-many relationships. As an organization and function, innovation brokering differs from traditional extension and R&D because it represents the institutionalization of the facilitation role, with a broad systemic, multiactor, innovation systems perspective. Preliminary lessons from experience are that innovation brokers help build synergy in agricultural innovation systems, but their "behind-the-scenes" mode of operating conceals their impact and may limit financial support for their role. Their contributions to building capacity for collective innovation and preventing innovation-system failures offer a rationale for public investment in their activities, but such investments must be accompanied by improved methods for measuring the impact of innovation brokering. As "honest brokers," innovation brokers need considerable room to maneuver in building and facilitating networks from a credible position. Given that countries may have different cultures of collaboration and different stages of innovation system development (with corresponding system imperfections), a context-specific design is required for innovation brokers to attain a credible position.

WHAT IS THE RATIONALE FOR INVESTING IN INNOVATION BROKERS?

Over the past decades, the stakeholders in agricultural innovation have become more numerous and their interactions more complex (World Bank 2006). This increased complexity has made cooperation for innovation less straightforward. To function, an AIS required shared visions, well-established links and information flows among the actors, and incentives that enhance cooperation (World Bank 2006). Interaction between stakeholders that are different enough to have new knowledge but related enough to understand each other seems particularly to lead to innovation—a relationship described aptly by Granovetter (1985) as "the strength of weak ties."

Creating and fostering effective coalitions among actors is often hindered by incomplete information about what potential partners can offer, by different incentive systems for public and private actors, differences between indigenous and formal knowledge, social differences that cause exclusion of certain actors, or ideological differences (Pant and Hambly-Odame 2006). Innovation scholars (Burt 2004, Obstfeld 2005) emphasize the importance of having people who act as brokers in networks, connecting stakeholders that are not familiar to each other but may provide the "new combinations" essential to innovation. It is also recognized that a dedicated actor can fulfill this role of "innovation broker" (Smits and Kuhlmann 2004; Howells 2006).

Innovation brokers act as "systemic intermediaries" in innovation systems, forging many-to-many relationships. While the term "broker" has the connotation of a strategically acting go-between who benefits from the separation between actors and pursues objectives mainly out of self-interest, the concept of "innovation broker" derives from the notion of an "honest broker," who brings people together mainly for altruistic purposes (Obstfeld 2005). The role of the honest broker resembles a broadened notion of the role of a process facilitator (Klerkx and Leeuwis 2009). In other words, innovation brokers are facilitators of interaction and cooperation in innovation systems, and their activities extend throughout innovation processes that last several years.

In the agricultural sector, innovation is vital for sustainable economic, social, and ecological development. Efforts to overcome the many barriers to effective communication, cooperation, and ultimately innovation are thus central to the public interest and justify public investments.

WHO CAN BROKER AND HOW?

Any advisory service or related individual or organization can broker, connecting farmers to different service providers and other actors in the agricultural food chain. Examples include research organizations such as those of the CGIAR, national and international NGOs, specialized consultancy firms, temporary projects, government programs, and farmers' organizations (see Klerkx, Hall, and Leeuwis 2009 for examples). Although public organizations such as extension services and research organizations could perform innovation brokering as part of their mandates (see TN 3), many retain a linear, transfer-of-technology

mindset and lack the capacity to fulfill this role (Rivera and Sulaiman V. 2009; Devaux et al. 2009). Innovation brokers can also be independent, specialized organizations with a skill set especially tailored to innovation brokering. A broad range of specialized innovation brokers has emerged, for example, in the Netherlands (Klerkx and Leeuwis 2009). Developing countries such as Kenya (boxes 3.23 and 3.24) and India (box 3.25) have done the same in recent years (Klerkx, Hall, and Leeuwis 2009).

Innovation brokering typically comprises the following functions, to be applied in a flexible and iterative manner (Klerkx and Leeuwis 2009; Kristjanson et al. 2009):

- *Analyzing the context and articulating demand.* The participatory assessment of problems and opportunities through quick system diagnosis identifies promising entry points (in terms of prospective markets), supportive policy, and constraining factors to be overcome. The analysis provides information to stipulate a shared vision

---

Box 3.23 The Need for Innovation Brokering: Supplying Potatoes for Processing in Kenya

In Kenya, DEEPA Industries Ltd. expanded its potato crisp production capacity from 2 to 12 tons a day, but its fully automated production line required a steady supply of high-quality potatoes. The International Potato Center (CIP) and the Kenya Agricultural Research Institute (KARI) organized and facilitated a meeting in 2005 to see if an arrangement could be brokered between the processor and potato producer groups in Bomet District. During the meeting, agreements were reached on a fixed price for farmers' produce, transport arrangements, and the regular supply of produce. The parties also agreed that the local public extension office would support the producer organization's efforts to supply the processor. No stable source of funds for continued brokering beyond this one-off meeting could be identified to continue supporting development of this emerging beneficial relationship between actors.

The transporter of the first shipment sold the high-quality potatoes destined for the processor elsewhere for a higher price and replaced them with potatoes of lower quality. The processor declined to accept further deliveries from the producers because they did not meet the quality requirements, with the result that a constant supply of potatoes did not materialize. The processor had to scale down his ambition of exporting to other East African countries.

A structured and sustained innovation brokering effort could have made a big impact by building a working coalition between the different stakeholders in the innovation process. A more harmonized and effective contribution by research, extension, the private sector, and producers would have been possible through a clearly mandated broker.

Three years later, in the context of a development project funded by the Common Fund for Commodities (CFC), CIP and KARI renewed efforts to broker organizational innovation. Meetings are being organized to build trust and structure communication and economic interactions between the actors. Currently research, agricultural extension, producer groups, and DEEPA are innovating within the production chain by using high-quality, clean seed, contract farming, direct purchasing, local collection of the produce, and testing new genetic material for quality in crisp processing. These initiatives resemble types 1, 2, 3 in the typology of innovation brokering presented in table 3.9.

Sources: D. Borus and P. Gildemacher, CIP, Nairobi.

Box 3.24   The Innovation Works Unit at the International Livestock Research Institute as an Innovation Broker

The Innovation Works Unit of the International Livestock Research Institute (ILRI) sought to facilitate pro-poor innovation related to livestock husbandry through efforts with a local as well as a systemic focus. The unit created several learning platforms for public and private stakeholders in particular projects. The platforms, which were mediated by local facilitators hired by the different projects, can be characterized as hybrids of an innovation consultant and a systemic intermediary (see the typology in table 3.9). The different projects funded innovation brokering through the platforms.

The platforms often took the form of safe havens—environments outside each of the participating organizations that provided a more neutral space conducive to creativity and co-creation, bypassing dominant groups committed to maintaining the status quo. A concrete example involved facilitation of the inclusion of local pastoral Maasai communities as equal partners in drawing up a land-use master plan, in which local and scientific knowledge were combined. The Maasai gained a voice in the policy debate from which they had been excluded. A major achievement was that the facilitators tackled the huge power imbalances across their multipartner project team, such as the often unrecognized power of scientific experts. To build trust and demonstrate respect for the knowledge of all partners in the project, the facilitators pursued multiple strategies, such as hiring local community members as members of the core project research team and encouraging the joint creation of knowledge by a hybrid team of scientists and community members.

Despite these achievements, this kind of mediation often remains unrecognized and undervalued. It was difficult to make it a central function of an institute such as ILRI and get it funded.

Source: Kristjanson et al. 2009; see also www.ilri.org/innovationworks.

and articulate demands for technology, knowledge, funding, and other resources.

- *Composing networks.* Facilitate linkages among relevant actors—specifically, by scanning, scoping, filtering, and matchmaking possible partners that have complementary resources such as knowledge, technology, and funding. This also includes matching demand and supply in pluralistic advisory and research systems.

- *Facilitating interaction.* Action planning, along with the identification of and support to those taking leadership in multistakeholder activities, has the main objective of building functioning stakeholder coalitions. Considering the different backgrounds of the actors involved, coalition building requires continuous "translation" between actors, the building of trust, establishing working procedures, fostering learning, motivating, managing conflict, and intellectual property management.

Different types of innovation brokers have been observed, working at different levels of the innovation system and varying in their level of ambition and thematic scope. Table 3.9 presents a tentative typology based on the Dutch landscape of specialized innovation brokers (Klerkx and Leeuwis 2009). Although several of the types described here can be found elsewhere, the typology is subject to further research and amendment (Klerkx, Hall, and Leeuwis 2009).

Innovation brokers have been found at the supranational (across several countries), national (country), regional (province, district), and (sub)sectoral or commodity level (such as dairy or horticulture), but these levels may also mix (for example, when dealing with cross-cutting value chain innovations). With respect to their level of ambition, some innovation brokers focus mostly on incremental innovations at the farm level, in a demand-driven and bottom-up fashion. They may be reactive, responding to clients' ideas, or they can more pro-actively approach prospective clients and offer a context analysis and demand articulation session (Klerkx and Leeuwis 2008). Other innovation brokers typically focus on radical innovations that comprise complete (sub)sectors or value chains, dealing with complex problems that require a systemwide change process. In this case, innovation brokers are often proactive initiators of processes and act as change agents. With respect to thematic scope, some innovation brokers focus on one sector (dairy alone, for example), whereas others address all kinds of sectors within a region, and still others focus exclusively on a specific activity (rural tourism, for example). The optimal

Box 3.25   Agricultural Innovation Broker Initiatives in India

Several types of innovation brokers have emerged in India. They have taken different organizational forms, they operate at different levels in the innovation system, and their scope of innovation differs.

International Development Enterprises. In India and Bangladesh, an international NGO, International Development Enterprises (IDE), acted as a broker in the process of developing innovations for low-cost irrigation pumps. (In Bangladesh, aside from coordinating interaction among actors in the irrigation pump supply chain, IDE also coordinated interaction with policy makers.) Because of IDE's intervention, the focus broadened from developing a particular technology to realizing the vision of effective irrigation water provision for the poor. Institutional innovations were the key to realizing that vision and included changing the incentives for public and private actors and creating effective demand for the technology so that a self-sustaining market could emerge. IDE acted as a local innovation consultant as well as an instrument for systemic innovation.

Using ICT and social media to build awareness of innovations and other information. To truly benefit from farmers' creativity and experimentation, several initiatives use ICT and social media to identify and build awareness of little-known innovations. Participants can share experiences and scale up successful efforts. These initiatives are hybrids of an innovation consultant, a peer network broker, and a ICT-based platform that helps to articulate demands and build networks. Examples include the HoneyBee Network and Villagro Network, which scout for innovations for their databases and connect innovators to supporting agencies such as India's National Innovation Foundation. The networks also help participants to patent innovations and find investors to develop products. Sustainable inventions from the Honey Bee database comprise 34 categories, including agricultural tools and techniques, water conservation, health, education innovation, food and nutrition, traditional medicine, and industrial and household goods. (Example of specific innovations include a motorcycle-driven plow for farmers who cannot afford tractors or bullocks and matchsticks made of natural fibers sourced from agricultural waste.) Still other efforts use ICT-based brokering instruments ("infomediaries") to share operational (market and production) information (rather than strategic information) for innovation.

Sources: Authors; Gupta et al. 2003; Hall, Clark, and Naik 2007; Murthy 2010; see also www.ideorg.org, www.honeybee.org, and www.villagro.org.

innovation system level, ambition level, and thematic scope of the work can be determined only in the course of the interaction between innovation brokers and their clients. This uncertainty implies that sometimes clients will need to be referred to another type of innovation broker than the one they originally started to work with. In other instances, several complementary innovation brokers are involved within a single innovation process (Klerkx, Aarts, and Leeuwis 2010, Devaux et al. 2010).

WHAT ARE THE POTENTIAL BENEFITS OF INVOLVING AND INVESTING IN INNOVATION BROKERS?

The current imperfect interaction between the actors essential for agricultural innovation—farmers and their organizations, researchers, extension, agricultural service providers, local government, agribusiness—is often not a result of unwillingness to interact but of a lack of capacities, structures, and incentives to interact effectively. Through investments in innovation brokering, communication between the multiple actors can improve greatly. By providing fresh insights and a mirror for self-reflection, innovation brokers stimulate clients to look beyond their current situation and constraints. For example, farmers and other agrifood stakeholders can think about new possibilities to improve their businesses, or producer organizations, researchers, and extension service providers can think about innovative manners of communicating. Impartial, honest brokers, because of their less-biased position and the overview of the system that they can provide, can forge contacts between parties that would normally not cooperate. They can also mediate more easily in the case of conflict (see the first point in the section, "What Key Issues Should Be Considered?," later in this TN). Hence they can assist in promoting more perfect information.

Table 3.9 Typology of Innovation Brokers

Type of broker*	Focus
1. Innovation consultants, aimed at individual farmers and small and medium enterprises (SMEs) in the agrifood sector	Connect farmers/agrifood SMEs with relevant collaborators and service providers and also with sources of funding and policy information. Generally incremental innovation; short time horizons.
2. Innovation consultants aimed at collectives of farmers and agrifood SMEs	Similar to type 1. The main difference is that they work with collectives, first connecting farmers or agrifood SMEs with similar interests and then connecting these actors with relevant collaborators, service providers, and sources of funding and policy information. Generally incremental innovation; short time horizons.
3. Peer network brokers	Aim to bring farmers together to exchange knowledge and experience at the interpersonal and group level—in other words, to facilitate enterprise development through peer-to-peer learning resembling concepts such as Farmer Field Schools. An explicit objective is to involve actors from weak networks (surpassing regional and sectoral networks) by inviting entrepreneurs from other regions or sectors and subject matter specialists.
4. Systemic intermediaries for the support of innovation at higher system level	Catalyze radical systemwide innovation (such as an entire production chain, societal systems, or policy systems) by: (1) managing interfaces between (sub)systems in the innovation system; (2) building and organizing (innovation) systems; (3) stimulating strategy and vision development; (4) providing an infrastructure for strategic intelligence; and (5) providing a platform for learning and experimenting. Involve several societal actors, including farmers, supply and processing industry, civic advocacy organization, and policy makers, for example. Generally radical/system innovation and transition trajectories; medium to long time horizons.
5. Internet-based portals, platforms, and databases that disclose relevant knowledge and information	Portals and platforms differ with regard to their prospective audiences, which may be selective (such as farmers), all agrichain actors, or project-related audiences. Portals and platforms may have a rather passive matchmaking role. Some portals create order in a wealth of information sources and give an overview but do not serve as a selection aid. Interactive tools exist, however, to allow the provision of services adapted to users' needs. Addressing both operational or tactical problems and strategic innovation issues; short time horizons.
6. Research councils with innovation agency	Management of multiactor R&D planning networks (involving farmers, supply and processing industry, civic advocacy organization, policy makers)—e.g., facilitating a demand-driven research agenda and priority setting. Facilitation of participatory/collaborative R&D (involving end-user participation), also addressing the creation of an enabling environment for enhancing research result uptake. Incremental and radical innovations; short to medium time horizons.
7. Education brokers	Aimed at curricular innovation. Provide educational establishments with the latest insights from practice and research to enhance the fit of their education programs with business and societal needs.

Source: Adapted from Klerkx and Leeuwis 2009.

* Hybrids of different types of innovation brokers are possible within a single organization, as well as involvement of different types of innovation brokers within a project.

Broadly, innovation brokering can be expected to have immediate and long-term results. Direct results are expected through market innovations that arise when producers respond better to the needs of agribusiness and agribusiness operators develop a better understanding of production systems, as in the case of potatoes for the snack food industries in Kenya and Peru (box 3.23 and the description of Papa Andina in module 1, TN2). Brokering can facilitate technical innovation by improving how agricultural research service providers target serious bottlenecks in production or processing or by inducing required institutional change on the part of policy makers and legislators. Over the longer term, and beyond the immediate results of a single innovation brokering effort, brokering should improve how the overall innovation system functions. Once contacts have been made and working coalitions have

formed between stakeholders, the result should be more market-oriented research and advisory services, more effective agricultural value chains, and a more conducive policy environment—in other words, a better-functioning innovation system (Klerkx and Leeuwis 2008; Klerkx, Aarts, and Leeuwis 2010; see also box 3.24).

WHAT ARE THE MAIN INVESTMENTS NEEDED FOR INNOVATION BROKERING?

The main investments to mainstream the use of innovation brokers to support agricultural development are:

■ **Improving the recognition and evidence that innovation brokering is useful.** Funding the innovation broker role is problematic. Even when organizations involved in agricultural development see this role as central to their core missions, they lack the opportunity and freedom to execute the innovation broker role within their mandate (Kristjanson et al. 2009). To widen awareness of brokers' potential role in innovation and show that an investment in their role is justified, more structured documentation of successes and failures (specifically in developing countries) is required, followed by the publication and promotion of the outcomes.

■ *Improving the understanding of how to implement innovation brokering effectively as a tool for development.* Implementers should take care to not simply copy innovation brokering models from one context to the other, as best-fit solutions should be sought (Berdegué and Escobar 2002). Different approaches are needed depending on asset positions, favorable or unfavorable production environments, gender issues, and power distribution (Kristjanson et al. 2009). To increase the understanding of effective approaches of innovation brokering, action-learning cases need to be initiated and documented in different countries and agricultural systems.

■ *Improving human capacity to play the role of innovation broker.* First and foremost, innovation brokering requires skills related to process facilitation: leadership, multistakeholder facilitation, trust building, and communication; it also requires tools for managing group processes (Anandajayasekeram, Puskur, and Zerfu 2010). A system overview is required to permit stakeholders to understand and "translate" between each other. This skill set cannot be obtained through formal education alone but must be developed through a combination of formal education and practical experience. Investments are required to develop capable facilitators of innovation within organizations motivated to support agricultural innovation through brokering. A critical mass of experts and organizations in this field is still lacking, as reflected by the experience with NAADS in Uganda (Kibwika, Wals, and Nassuna-Musoke 2009) and the reorganization of agricultural service provision in Mozambique (Gêmo 2006). Traditional research and extension organizations must "retool" if they are to develop their innovation brokering capacity and abandon a mere transfer-of-technology paradigm (Devaux et al. 2009). The implication is that they must develop a service delivery philosophy and a mindset that recognizes multidisciplinarity (including topics such as agricultural economics, sociology, and gender issues), as well as facilitation skills. Capacity-building

interventions should be local and context-specific and aim to build durable and, ideally, self-sustaining systems of continuous capacity improvement.

WHAT KEY ISSUES SHOULD BE CONSIDERED IN POLICIES TO ESTABLISH INNOVATION BROKERS?

A number of criteria determine whether an organization can play a role in brokering between actors in an AIS (Klerkx, Hall, and Leeuwis 2009). The most important are:

■ *A legitimate mandate and credibility in the eyes of system stakeholders.* A key factor for the legitimacy of innovation brokers is that they must have a trusted position as a relatively neutral "honest broker." They should have a reputation that instills a degree of independence from the major stakeholders in the process and the overall innovation system. This stance is not easy to maintain, because stakeholders may exert pressure to compose and facilitate networks in a way that fits their particular objectives. An apparent connection to an organization may negatively influence credibility as a neutral, honest broker, which seems to indicate that innovation brokers might work best as independent, specialized organizations. Innovation often challenges prevailing role divisions, power relations, and profit distribution. To build productive innovation networks, sometimes parties with vested interests need to be bypassed.

■ *Both technical and methodological know-how and a clear role division.* Innovation brokers should have sufficient technical knowledge but should not become so involved with projects that they take over detailed management and take away ownership from the innovation network partners. They should also give equal attention to the goals and interests of each of the partners.

■ *Funding sustainability.* A durable source of funding is an important requirement for effective innovation brokering. Often funding is on an ad hoc, project basis, and especially in times of fiscal austerity innovation brokering services are often discontinued, despite high client satisfaction (Klerkx and Leeuwis 2008). Because the impact of innovation brokers is difficult to make visible, durable public, donor, or private stakeholder funding is hard to obtain (box 3.24). Ways need to be found to assess the impact of innovation brokers and better justify public or donor spending, starting with detailed documentation of specific cases.

LESSONS LEARNED

Well-documented experiences with innovation brokering are limited, but there appears to be a growing recognition of professionals in research, extension, and advocacy who may have the skill set and honest broker status that we are looking for in innovation brokers. The development of innovation brokering services requires continued local experimentation, adaptation, and learning (Klerkx, Hall, and Leeuwis 2009). So far several general lessons have been learned, discussed below.

Context analysis is needed prior to or as part of the innovation broker establishment

It is essential to adequately map and diagnose the strengths and weaknesses of the relevant innovation system (see Gildemacher et al. 2009) to get a clear view on missing linkages and/or deficient interaction. In doing so, it should also become clear whether some parties already fulfill an innovation brokering role and the extent to which they may complement or overlap with the envisioned task of the proposed innovation broker. Such a preparatory phase of context mapping and consultative talks with stakeholders prior to innovation broker establishment may take between one and two years.

Some innovation brokering functions are generic

To bring structure into the process of innovation brokering, several generic steps in the process can be distinguished: (1) context analysis; (2) initial network composition; (3) participatory needs and opportunity assessment, including network recomposition when necessary; (4) action planning; (5) network facilitation/coordination, problem solving, and conflict resolution; and (6) exit strategy. As progress in innovation processes is rather unpredictable, no fixed time allocations can be given for these phases.

Innovation brokers can use existing tools, methods, and approaches, but innovation brokering is learning while trying

Attention for integrated innovation brokering in agricultural development is new. The capacity to play the role of innovation broker cannot be fully obtained through formal training. However, many practitioners will recognize the role of innovation broker as a role they have played or seen being played. Although innovation brokering is thus not yet a very well-articulated and recognized role, tools from other approaches are available, such as the facilitation of multi-stakeholder interaction and value chain development. Innovation brokers can benefit from using such methods, to avoid "reinventing the wheel."

The role of AIS theory should be appropriately modest

The real proof of concept is in practice. It is important for practitioners to keep in mind that it is the experience in practice that steers the development of theory. Considering that innovation brokering has been recognized only recently as an important and deliberate function in AIS, practitioners are often pioneers. This situation implies that they should make decisions based on their own understanding, experience, and judgment rather than search for answers from AIS theory. While trial-and-error learning may incur some inefficiencies in regard to effective spending of funds (Klerkx and Leeuwis 2008), experimentation appears to be needed to create locally adapted innovation brokers, as there is no one-size-fits-all model (Klerkx, Hall, and Leeuwis 2009).

Perfect innovation brokers do not exist

When listing the skills and attitudes required in a good innovation broker, an apparently endless list of required qualities will emerge. These qualities are impossible to find combined in a single person. Still, the role of innovation broker will depend on these imperfect individuals. Each individual will have to develop a personal style as a broker that fits his or her strengths and weaknesses.

A structured exchange of experiences supports capacity building

As this field is new and capacity building is needed, peers involved in innovation brokering need to invest time and effort in exchanging experiences. As a reference point, the Netherlands took about fifteen years to develop a diverse field of innovation brokers and recognize their role (Klerkx and Leeuwis 2009). A structure of peer-to-peer exchange and support will directly improve performance as well as help to build capacity.

Innovation brokers should negotiate and defend the freedom to explore options

Once established, an innovation broker should be given considerable freedom to explore new options and establish new linkages. Brokers should not be tied to prescribed input-output schemes by either their employers or funders.

Monitoring and evaluation are needed for learning

Innovation is by definition an unsure process. It involves invention, adaptation, and changing directions as a response to the insights that are gained. It is difficult, even detrimental, to monitor progress through rigid and SMART milestones (Klerkx and Leeuwis 2008). Methods of M&E that focus on learning lessons are more suitable. Alternative M&E indicators should be identified by the stakeholders involved as relevant proof of progress, but (more important) these indicators should also serve as points of reference for learning to improve the process of innovation.

Recognize the difficulty of distinguishing and attributing outcomes

The primary work of innovation brokers is to improve the quality of interactions, which is a process that includes many intangible contributions. Innovation brokers will have to deal with the dilemma that they should sufficiently emphasize the impact of their role but not take all the credit (which may annoy stakeholders and diminish their ownership). While attribution is already a perennial challenge for extension programs, it is possibly even more problematic for innovation brokers, given their "behind-the-scenes" mode of operating. Because it is hard to distinguish and attribute the impacts of innovation brokering, it is also difficult to make the innovation broker role self-sufficient; willingness-to-pay is typically low among private actors. Long-term public investments appear to be needed in view of persistent innovation system failures such as fragmentation and lack of coordination. A focus on short-term funding may engender a vicious circle of short-term funding, leading to the disappearance of the innovation broker and renewed funding of a similar innovation broker.

Short-term results and long-term outcomes

It is important to keep in mind the two levels of results, direct and indirect, of innovation brokering. Direct and concrete activities and results are needed to keep the innovation coalition together long enough to build trust and build relationships. Without direct and concrete results and activities, it is impossible to keep actors motivated to invest in interaction and collaboration. Direct innovation results are also needed to justify investments in coalition building and brokering. The biggest potential for impact is, however, through the long-term outcome of improved collaboration between actors, transforming the innovation system in such a way that it becomes responsive and contributes to a durably competitive agriculture sector.

RECOMMENDATIONS FOR PRACTITIONERS, POLICY MAKERS, AND PROJECT LEADERS

From the lessons learned so far, several recommendations can be distilled for brokers themselves, policy makers, project leaders, and those who champion innovation brokering.

Recommendations for prospective innovation brokers

- *The problems and challenges that need to be tackled by innovation brokers may be different.* Although not exclusive to developing and emerging countries, but maybe even more severe and pressing in light of rural poverty and natural resource scarcity and degradation, problems that need to be addressed include: dealing with competing claims on natural resources, inclusion of the poor and giving them a voice in the development process, and equitably integrating smallholder farmers in global value chains. For this reason, different approaches are needed in designing the brokering role. Prior to setting up an innovation broker (which may be an individual, a unit in an existing organization, or a new organization) in a region or sector, start with an analysis of innovation system imperfections, and assess the need for an innovation broker and willingness of stakeholders to support and/or work with a broker. Gain the confidence of stakeholders, and work to gain credibility as an honest broker in the innovation system.
- *Plan for the nature of the different innovation broker functions* (particularly context analysis, demand articulation, and initial composing of networks as first steps) in the different steps of the innovation process. Do not apply them as a blueprint, however. Be flexible at the same time. Assist in reassessing the context, needs, and opportunities when needed, and help networks to adjust accordingly. The facilitation of interaction is a dynamic activity, given that changing visions and networks require constant attention to mutual understanding and trust.
- *The nature and intensity of the innovation broker's role will most likely change over time.* It should shift gradually from actively taking the initiative to handing over the

initiative and daily project management to project participants and acting in the more distant role of project monitor.

■ *A broker will have to deal with multiple accountabilities and conflicts of interest in the innovation process.* Conflict management and intermediation skills are important. Brokers must prepare to deal with contrasting demands and the opposition of incumbent actors in the innovation systems who do not favor change. As an innovation broker, try to be as transparent as possible about the "what" and "why" of certain actions or interventions, to avoid giving false impressions. Perception management is essential. In some countries with weak governance, additional challenges need to be dealt with, such as corruption and favoritism. Due to resource dependencies an innovation broker may nevertheless become a more or less "hidden messenger" for government or another party—a perception that may be detrimental to the broker's impartiality, credibility, and hence longevity.

■ *Take care in assigning credit for results.* As the innovation process evolves and results materialize, the innovation project partners, other actors in the innovation system, and funding agencies should be made aware of the broker's contribution in achieving these results. Brokers themselves should avoid taking credit from project participants.

■ *Expect the greatest reward and sign of accomplishment to be that an innovation broker may no longer be required when local innovation capacity has been built.* The broker should withdraw rather than force his or her presence between actors. Brokers should think about an exit strategy from the beginning.

Recommendations for policy makers in government, research organizations, and other organizations

■ *Before establishing an innovation broker, assess innovation system failures and current innovation broker capacity to avoid duplication of effort.* Remember that in some cases the need for context specificity may justify the coexistence of several innovation brokers. It is essential to stimulate interaction between different innovation brokers to demarcate mandates and complementarities. In the absence of coordination, overlap and even competition between innovation brokers can arise, engendering confusion among clients about who is facilitating what and reducing the synergies that innovation brokers should induce in innovation systems.

■ *When establishing innovation brokers, avoid maintaining an overly close organizational and ideological connection with the respective policy domain.* Distance will enable the innovation broker to develop a clean, "honest broker" image and sufficient operational maneuvering space. Do not try to use innovation brokers as messengers to bring about government interventions. Sometimes radical innovation goes against current government policy.

■ *Accept that innovation brokers, by counteracting imperfections in the innovation system, also change the innovation system's configurations and interaction patterns.* Although such actions may challenge certain policy lines, policy makers who champion the role of honest brokers should accept this situation and defend it with their constituencies and peers in the policy domain.

■ *Allow sufficient freedom* to forge unexpected linkages (for example, to connect agriculture with the gaming industry) and experiment (touching themes that at first sight do not have anything to do with current ways of agricultural production), but agree upon certain deliverables. Such deliverables could include the number of new concepts developed or the number of productive innovation networks forged, supported by narrative case reports of innovation dynamics. A supervisory board drawn from the different domains with which the innovation broker works should be installed to monitor the extent to which these deliverables have been realized. The board members are ideally well and widely respected but at the same time visionary and open to change.

■ *As in the case of market failure, innovation system failure justifies public investment,* even though innovation is unpredictable and difficult to plan. Investment in innovation typically is of a "best bet" nature, and stimulating innovation means that investments are made in projects with an unknown and sometimes unviable outcome. Despite innovation brokering, failure may occur, but this does not mean that innovation brokering does not merit investment.

■ *Stimulate the development of M&E indicators that capture the rather intangible activities of innovation brokers,* particularly indicators that move beyond case documentation and satisfy the need for quantitative justification of investment. Methods and indicators are especially needed to capture causal relationships between innovation network performance and the activities of innovation broker activities as well as the spillover effects of innovation brokers in innovation systems. Methodologies such as social network analysis may be promising in this regard (see Spielman, Ekboir, and Davis 2009). Stimulate policy learning and institutional memory in relation to the roles and effects of innovation brokers to avoid a vicious circle in which innovation brokers appear, only to disappear and reappear.

■ *Creating innovation brokering capacity within an existing organization, such as an extension or research organization, requires the proper institutional conditions to be shaped.* Innovation brokering cannot be judged on the basis of traditional performance criteria for research and extension, such as publications or numbers of field visits. Furthermore, management and staff need to gain an understanding of the role of innovation brokering so that it is not seen as extrinsic to the organization's core mission.

■ *Brokering is influenced by the nature of the AIS and institutional frameworks as well as cultures of collaboration.* Many countries are characterized by "immature" innovation systems that lack a functioning knowledge infrastructure (research, education, advisory services) and by inadequate institutional frameworks (in terms of well-functioning legislation, markets, and interaction patterns). Policy makers should keep in mind that different cultures of collaboration may affect the potential effectiveness of innovation brokers (for example, in building trust, achieving a collective goal) because of the cultural organization of interaction among actors at different social and economic positions and issues like clientelism, social exclusion, nepotism, and corruption.

Recommendations for project leaders, project implementers, and/or innovation champions

■ *Prepare to give up preconceived ideas.* Stimulating critical and creative "out-of-the box" thinking is a key role of innovation brokers.

■ *Establish a clear division of tasks* in innovation process management, to avoid overlap and a lack of project ownership. Depending on the innovation network's internal capacity to manage innovation processes, the intensity of the broker's involvement may vary. Generally, daily project management is a principal task of project leaders/implementers and/or innovation champions, while issues like process monitoring and conflict mitigation are a principal task of innovation brokers. The involvement of innovation brokers implies that reflection on project progress, the role of different partners, the viability of the vision, and objectives becomes an integral part of the project.

■ *Coordinate actions of the innovation network partners and the innovation broker when forming the network,* to avoid confusion among the parties approached as to whom they should regard as their main contact person.

■ *Recognize that the innovation broker cannot always take a clear stand in advocating the interests of the innovation project versus external parties,* although the broker is regarded as part of the innovation network. Advocacy is needed to some extent, but within certain limits. Innovation brokers that become too institutionalized in the project may benefit from having another innovation broker give a "second opinion."

■ *Although innovation brokers are often subsidized, if a private contribution is requested, realize that this investment is generally compensated by a lower failure rate and better access to external resources.* In general, integrate the cost of innovation brokering in the overall project sum, and do not see it as an unnecessary investment lowering the research budget.

■ *Accept that innovation brokers cannot perform miracles.* Some obstacles may be of such magnitude that they require prolonged action by innovation brokers (for example, through mediation) but nonetheless cannot be overcome.

Agrodealer Development in Developing and Emerging Markets

John Allgood, International Fertilizer Development Center (IFDC)

SYNOPSIS

A holistic, market-oriented approach to agrodealer development facilitates improved efficiency in resource allocation, operations, and economic performance and helps to develop sustainable input supply systems. The benefits of agrodealer development accrue at various levels in the value chain and reach stakeholders at the micro and macro level. Developing agrodealers' technical capacity allows them to provide high-quality advisory services to farmers, accelerates the introduction of technology, and enhances the potential economic returns for farmers who invest in yield-improving technologies. Developing their business acumen is paramount for improving dealers' operations, cost-effectiveness, and potential for long-term economic success in serving farmers. Developing business linkages is critical in enabling agrodealers to capitalize on opportunities to improve the cost and operational efficiency of value chain and credit management and to expand the scope of their operations. It is vital to tailor each agrodealer development intervention to the specific conditions and market characteristics of a given country and region. Agrodealers learn best either through direct, one-on-one assistance or through group participation with hands-on interaction. Longer-term interventions are more effective than short ones (program continuity allows for timely interaction with policy makers and donors). Broad-based stakeholder involvement and attention to cost sharing (when feasible) are essential to sustaining progress.

CONTEXT

Agrodealers play a crucial role in servicing farmers' needs related to agricultural inputs.[1] Ideally that role includes providing farmers with (1) affordable, convenient access to appropriate, high-quality technologies to enhance yields and (2) proper advisory services on the best way to use those technologies to achieve favorable economic returns. The functions performed by agrodealers in developing and emerging markets are substantially influenced by the stage of agricultural development and the prevailing macro environment (for example, the government's role in agricultural input markets, the availability of finance to buy inputs, and so on).[2]

In an early phase of development, agriculture is mostly extensive; the sector is characterized by weak, seriously underdeveloped agricultural input and output markets. The public sector typically dominates the supply of agricultural inputs to farmers. In almost all cases, public systems that perform the function of agrodealers focus on logistics management with little (if any) emphasis on stimulating demand (through farmer advisory services or technology promotion campaigns, for example). Public systems rely strongly on the agriculture ministry and public extension service to create awareness and educate farmers. In the early phase of market development, private agrodealers' role usually is limited to bridging the gap between suppliers of agricultural inputs and farmers, often in competition with the public sector. Private agrodealers perform the essential basic functions of determining the product mix, physical distribution, pricing, and sales.

In markets where agriculture is more developed, as in India and Pakistan, agrodealers may assume more complex roles. In addition to providing convenient and timely access to appropriate, high-quality products, they may provide farmers with advisory services, participate in campaigns to introduce new technologies, and provide sales on credit to their best farmer customers. In more advanced markets, agrodealers may serve as an important source of information that is useful from both a commercial and policy perspective.

As the final link in the agricultural input value chain,[3] those entities (public and private) that function as agrodealers are able both to influence farmers' demand for yield-improving technologies and to improve the transfer of knowledge related to the proper, safe use of agricultural inputs. They have a major influence on farmers' incomes. Efforts to improve food security and accelerate income growth in rural areas can be significantly affected by the presence and effectiveness of agrodealers.

INNOVATIVE ELEMENT

The International Fertilizer Development Center (IFDC) takes a holistic, market-oriented approach to agrodealer development, whether the challenge is to improve efficiency in public systems or strengthen the capacity of private agrodealers (box 3.26). Improving the performance of the members of the value chain, thereby achieving operational and cost efficiencies that directly benefit farmers, is a priority. The IFDC approach to agrodealer development incorporates the marketing concept—*an agrodealer's long-term success can best be achieved through better serving its farmer customers*—in all activities directed at capacity building and promotion.

FOCAL AREAS AND INTERVENTIONS

Focal areas in agrodealer development comprise technical knowledge transfer, business acumen development, business

---

Box 3.26   Philosophy on Agricultural Development Drives the Approach in Agrodealer Development

The philosophy on agricultural development at IFDC encompasses two premises:

- Improved use of agricultural inputs (such as fertilizer, high-quality seed, or crop protection products) is essential, along with good water management, to sustainable improvement in agricultural productivity per unit of land.
- Efficiency in resource use can be maximized by employing a market-oriented approach to development.

The IFDC goal in agrodealer development is to foster the development of agrodealers so that they may effectively serve farmers' immediate and long-term agricultural input needs.

Source: Thompson 2003, 2005.

---

linkage development, and efforts to strengthen the support systems needed for agrodealers to become successful in a competitive marketplace. The following focal areas are a priority for IFDC in the design and implementation of activities to accelerate agrodealer development.

Technical knowledge transfer

Strengthening the technical capacity of agrodealers allows them to provide high-quality advisory services to farmers, accelerates the introduction of technology, and enhances the potential economic returns for farmers who invest in yield-improving technologies. Efforts to improve dealers' knowledge and understanding of agricultural input products focus on analyzing problems in soil and crop health management and on the field performance of products (for example, in maintaining soil and plant health); their safe use, storage, and handling to minimize human and environmental damage; and proper application to achieve maximum efficiency from use and optimum economic returns. Two crucial steps in increasing farmers' demand for agricultural inputs are to create awareness and transfer knowledge related to yield-enhancing technologies. Particularly in developing countries, the "seeing is believing" concept is highly effective in educating agrodealers and farmers and stimulating farm-level demand for inputs. Some of the most effective approaches for promoting agricultural inputs include the design and implementation of collaborative technology demonstration plots as well as technology field days and crop cuttings. IFDC also provides agrodealers with point-of-purchase technical leaflets, wall hangings, and poster boards that build farmers' awareness and knowledge of agricultural input use (box 3.27).

Development of business acumen

Strengthening the business acumen of agrodealers is paramount for improving dealers' operations, cost-effectiveness, and potential for long-term economic success in serving farmers. Agrodealer development activities of IFDC, CNFA, and others emphasize improving dealers' understanding of the financial, marketing, and management functions that must be performed well for a business to survive and grow. Training sessions cover the basics of marketing and business management; strategic planning to ensure that sufficient inputs are supplied in a timely manner to farmers; record-keeping to support profitability analysis, business planning, and credit management; understanding the total cost

Box 3.27   Technical Knowledge Transfer:
            A Public-Private Approach
            in Bangladesh

In Bangladesh, IFDC conducts a range of technical knowledge transfer activities, engaging both the public sector (such as the Department of Agricultural Extension, DAE) and private fertilizer dealers. Providing education to DAE field staff improves their awareness of appropriate agricultural technologies, the best practices for using them, and the resulting benefits through a "train-the-trainer" approach to knowledge transfer. Building the capacity of dealers to provide advisory services to farmers is a well-accepted practice. Various knowledge transfer activities are used and target agrodealers as well as DAE staff, including classroom training sessions that last one to three days, collaborative field demonstrations/field days, and direct, one-on-one technical support through site visits.

When a new technology is introduced, rapid penetration promotion campaigns are effective. Specific tools to introduce the targeted technology include technical leaflets, signboards, point of purchase displays, billboards in heavily populated areas, and open sky shows.

Source: IFDC 2011.

incurred in the agrodealer business and the associated record-keeping needs for marketing and accounting; price determination and pricing strategies; the design and implementation of promotional campaigns to improve sales; procurement planning and negotiation; credit management; and extending agrodealer networks to improve geographic coverage. IFDC's experience is that short group training sessions with a blend of lectures, group exercises, and case studies are highly effective in building business acumen.

Business linkage development

Business linkage development is critical in enabling agrodealers to capitalize on opportunities to improve the cost and operational efficiency of value chain and credit management and to expand the scope of their operations. Strengthening linkages within the agricultural input value chain generates several advantages. It allows for efficiencies in logistics planning and inventory management, improves

awareness of new technologies, offers opportunities to realize economy-of-scale benefits through joint procurement, improves access to credit through banks and supplier credits, and facilitates the expansion of dealer networks. IFDC fosters business linkage development through workshops, direct technical assistance, publication of monthly market news bulletins, study tours, training programs, and the development of alliance agreements (box 3.28).

Strengthening support systems

Advantages in market development are afforded by forming groups in a manner that does not impede competition. Market efficiency requires a relatively high degree of market transparency at all levels; market information is essential to successful planning and decision-making. Various support systems facilitate agrodealer development. IFDC often works to create agrodealer associations and build their capacity to provide dealer education programs, advance technology introduction, provide policy advocacy, facilitate business linkage development, improve access to commercial finance, and enhance market transparency (box 3.29). The emphasis is on creating agrodealer associations that provide a formal structure to support long-term dealer interests.

It is important to tailor each agrodealer development intervention to the specific conditions and market characteristics of a given country and region. Human capacity building is emphasized in all activities. Educational programs and the provision of resource materials are crucial. One or more combinations of the following may be included: formal (classroom-type) training programs tailored to a particular audience on specific subject matter;[4] development/dissemination of an agrodealers' handbook; informal, one-on-one site visits with agrodealers to provide guidance on business management, product display, product storage, and safe input use and handling practices; and regional and international study tours to observe agrodealers and technology suppliers in more advanced markets. Cost-sharing (for technology demonstrations and field days, for instance) is emphasized. As an example of the type of resource material provided to agrodealers, an agrodealer handbook was developed in Uganda and Bangladesh to serve as a ready reference for agrodealers.

The scope of interventions may range from a one-time event such as a two-day training program to a more extensive, multiyear, comprehensive market development effort that includes agrodealer development. It is important to be aware of the peak agricultural input use season and avoid scheduling programs at those times.

Box 3.28   Business Linkage Development and Leveraging Resources

Global Development Alliance. In collaboration with the United States Agency for International Development and the Eurasia Group (Pioneer, John Deere, DuPont, and Monsanto), IFDC helped to create a Global Development Alliance in Kyrgyzstan. The alliance has been instrumental in introducing technology and building human capacity through cost-sharing and in linking Kyrgyzstan's agrodealers to suppliers of improved technologies.

Linking South Asian and African entrepreneurs. IFDC is organizing study visits and technical workshops to link suppliers of fertilizer briquette machines in Bangladesh and agrodealers in Kenya, Nigeria, and Rwanda. The emphasis is on creating awareness and establishing business contacts among agrodealers, entities that directly impact their businesses (including banks and microfinance institutions), and agricultural input distributors/wholesalers with local, national, regional, and international markets.

Sources: IFDC, unpublished project documents, 2009 and 2010.

Box 3.29   Agrodealer Associations Support Common Interests

The role and benefits of agrodealer associations are reflected in improvements in four key areas: access to finance, advocacy, communication, and education. IFDC implemented the Fertilizer Distribution Improvement (FDI) II project in Bangladesh during 1987–94 with funding from the United States Agency for International Development. With FDI II project support, the Bangladesh Fertilizer Association (BFA) was established in June 1994. A decade and a half later, the BFA is a 7,000-member-strong association that provides varied services to its members, including policy advocacy, knowledge transfer, and improved market transparency.

It is important to avoid dependency on donor funds for association operations. Long-term survival requires the association to create a revenue flow that comes substantially from membership dues.

Source: USAID 1996.

BENEFITS AND IMPACT

A holistic, market-oriented approach to agrodealer development facilitates improved efficiency in resource allocation, operations, and economic performance. The benefits of agrodealer development accrue at various levels in the value chain and impact stakeholders at the micro and macro levels. For instance, beginning in 2008, the Government of Bangladesh endorsed fertilizer deep placement as a technology that would help to improve rice production systems substantially, thereby contributing to food security and farmers' incomes. IFDC, with support from USAID and the Government of Bangladesh, designed and introduced a program to diffuse the technology and concurrently address demand and supply issues. The role of agrodealers in Bangladesh continues to evolve and is having a substantial impact on food security, farmers' incomes, and the national budget (box 3.30).

LESSONS LEARNED

IFDC's long experience in working with agrodealers in emerging markets throughout the world can help practitioners plan or support similar activities. Key lessons are summarized below.

- *Understand the challenge.* The agrodealer market environment and challenges faced by farmers differ from country to country. A *key lesson* is that a "one-approach-serves-all" philosophy does not work. Each intervention must be tailored to the prevailing conditions in the target area. Achieving success in agrodealer development requires a clear understanding of the overall agricultural input marketing system, the stage of development, and the influence of macroenvironmental factors at a given time.

- *Engage the public agricultural extension service to the maximum extent feasible and use its extensive networks to provide knowledge-building services to farmers.* A *key lesson* is that the extension service, other public officials, and private agrodealers must provide farmers with a consistent, clear message on the need for and appropriate and safe use of agricultural inputs.

- *Keep learning practical and interactive.* A *key lesson* is that agrodealers learn best either through direct, one-on-one assistance or through group participation with hands-on interaction. Study tours in more advanced markets often are beneficial to build business linkages and to further awareness and knowledge of technologies and the advisory role of agrodealers.

In close collaboration with the Department of Agricultural Extension (DAE) and private dealers in Bangladesh, IFDC is supporting the rapid diffusion of fertilizer deep placement (FDP) technology. Increasing farmers' demand for FDP technology and stimulating the supply and marketing system to improve farmers' access to the technology are crucial to sustainable success. At the macro level, the focus is on gaining government and donor endorsement of the technology. At the micro level, the primary focus is on:

■ *Creating farmers' awareness and demand for FDP technology:* Over four cropping seasons, IFDC completed 3,880 farmer training programs, installed 386 technology demonstration plots, completed 109 FDP technology field days, and conducted 67 train-the-trainer programs for DAE staff. Advertising activities included (among others) the installation of more than 2,000 signboards and billboards, the development/dissemination of 135,000 technical brochures and the development of 72 cinema slides.

Source: IFDC unpublished project documents, 2009–11.

■ *Stimulating supply system development:* Over 18 months, IFDC stimulated private entrepreneurs to invest (on a cost-sharing basis) in 157 FDP product briquette machines. Eighteen training programs targeting entrepreneurs were conducted. The results of concurrently addressing demand and supply issues related to diffusing FDP are impressive (see table B3.30).

Table B3.30	Impacts of Addressing Supply and Demand Issues Concurrently in Diffusing a New Fertilizer Technology
Rice area under fertilizer deep placement (FDP) technology	94,380 ha
Number of farm families adopting FDP	408,000
Incremental rice production	24,000 t
Farm family income increase	US$8 million
Urea fertilizer savings	7,000 t
GOB subsidy reduction	US$1.6 million
New urea briquette machines at dealer level	121

■ *Duration and continuity are important.* A *key lesson* is that while one-time interventions in the priority areas described earlier are beneficial, they are less effective than longer-term interventions that provide agrodealers with continued support for development. Both remedial and more advanced training are important for achieving a sustainable impact. Program continuity allows for timely interaction with policy makers and donors.

■ *Foster broad-based stakeholder involvement.* Strengthening dealers' capacity, building knowledge, fortifying support systems, and establishing business linkages are essential, but they are not enough to ensure productive agriculture and sustained economic development. A *key lesson* is that broad-based stakeholder involvement is essential to sustaining progress. It is important to engage the ministry in charge of agriculture, the public extension service, commercial bankers, donors, and other relevant development agencies to the maximum extent in agrodealer development initiatives. Establish links with other projects to achieve synergies when possible. To achieve an element of ownership, dealers' cost-sharing is emphasized when feasible. The feasibility often depends upon the stage of market development. In seriously underdeveloped input markets, where demand from farmers is weak and risks are high, cost-sharing opportunities are quite limited. Dealers lack the resources to make a significant contribution to development. In more advanced markets, dealers' cost-sharing may range from providing the inputs for technology demonstrations to covering a portion of the costs involved in field days.

Federating Farmer Field Schools in Networks for Improved Access to Services

Arnoud R. Braun, Wageningen University

Godrick Khisa, Farmer Field School Promotion Services

Deborah Duveskog, Consultant

Kristin Davis, Global Forum for Rural Advisory Services (GFRAS)

SYNOPSIS

Field school networks, which mostly developed spontaneously, consist of informal or formal groupings of FFS groups with a common interest that draw their membership from all the FFSs within a given geographical or administrative boundary. Common interests at the network level are mainly marketing, advocacy, sharing information and experience, access to finance, and representation. From an innovation systems perspective, the role of extension and advisory services here is to assist in developing the needed capacity and linking producers to markets and service providers. Network operations are supported through subscription fees from constituent FFSs and other sources of income, such as interest on revolving funds, commissions on sales, registration fees, profits from input sales, and grants. Many networks operate a revolving loan system and therefore generate more funds to support operations and activities. Although the networks have shown themselves to be sustainable, their strength varies. Management, leadership, and organizational skills are needed to federate FFS networks. Networks are also vulnerable to individual motivation and capacity, since they are mostly managed by voluntary efforts among members and committee members. The networks that have been most successful are those with very committed, dynamic, and democratic leaders. Facilitating the formation of FFS networks should be considered in large-scale agricultural projects as a means to make a larger impact and make greater use of the social capital they generate. Lessons learned from the networks are that market information, while needed, is not obtained easily, and that network capacities for financial management, standards, and use of ICTs must be developed.

CONTEXT

Networks of Farmer Field Schools (FFSs) (see box 3.5 in this module's Overview) started emerging in East Africa in 2000 as an unforeseen effect of the IFAD/FAO East African Integrated Pest and Production Management project conducted with the Government of Kenya. The FFS networks took off because farmers wished to continue with the dynamics and positive attributes generated by the first phase of the project, even without external funding. To date, at least nine FFS networks in Eastern Africa support about 2,000 FFSs with close to 50,000 direct beneficiaries. They have no support from the initial project that established the field schools but have established new partnerships and collaboration with other stakeholders.

Definition of an FFS network

FFS networks consist of informal or formal groupings of FFS groups with a common interest that draw their membership from all the FFSs within a given geographical or administrative boundary. Each FFS elects one representative to the higher network level. These networks offer a number of services to FFS member groups and individual farmers. The networks are characterized as FFSs clustered in an association or not-for-profit company. They usually have an elected core executive board and at least three working committees, such as finance and planning, loans, and market information service. They have a constitution, by-laws, are registered, and have a bank account. The operations are supported financially by member FFS through subscription fees, commission on bulk sales, shares, or profit from the sale of farm inputs.

Evolution from individual groups into networks

As the number of FFS groups in the program grew and broadened their level of operation, new challenges and issues emerged that could not be solved by individual groups. There were also increased opportunities for the FFS to take advantage of economies of scale, necessitating more interaction and coordination. Based on these developments and exchange visits and interactions between farmers, facilitators, trainers, and project staff in Western Kenya in early 2000, FFS networks emerged. The networks were formed mainly by FFS graduates. Aside from taking advantage of the opportunities just mentioned, the graduates wanted to continue the dynamics generated by the FFS process—to build local institutions to ensure the continuation of farmer-led FFS and gain a stronger voice in expressing their demand.

The inherent attributes of the FFS approach of cultivating cohesion and a willingness to learn together while solving problems that affect them as a community help to build their social capital. Common interests at the network level are mainly marketing, advocacy, sharing information and experience, access to finance, and representation. There is therefore no conflict of interest when different FFSs come together to form the network. As a precursor to transformation, the level of empowerment and organization developed in an FFS is critical and can have a significant impact on the marginal returns of a subsistence-based farming system. This strong cohesion within and among FFS groups is one of the main factors contributing to the emergence of higher-level federations[1] like the FFS networks.

BENEFITS TO FARMERS, IMPACT, AND EXPERIENCE

FFS network members state a range of benefits experienced by the networks. Important benefits appear to be increased voice and power and access to services and markets. Some of these benefits are especially important for women (box 3.31).

Despite the market barriers experienced by smallholders, the networks have been able to arrange collective marketing with its many advantages. The networks have assisted in identifying markets and collecting marketing information. A network-based monitoring and record system helps group members track the availability and quantity of their produce, making it possible to plan bulk sales and negotiate with buyers in advance of harvests. By selling in larger quantities, FFSs reduce transaction costs, gain bargaining power, and thus command better prices for their products. They have also been able to break or weaken manipulative relationships with market intermediaries and thereby gain access to more lucrative markets for their produce. Obtaining funds from government programs also appears to be easier for federated FFSs (box 3.32).

Box 3.31   Gender Issues in FFSs

FFSs have been shown to change gender relationships, mainly because they introduce a formal group structure but also because they operate under agreed group norms and rules. *All* issues brought to the group are accepted for discussion, so fewer subjects are off-limits. As a result, the position of women participating in FFSs has generally strengthened. In some East African communities, women dominate FFSs. Men are less interested in working in groups, although in some cases, they join at later stages of the FFS process.

Source: Authors.

To assist members in access to affordable inputs of reliable quality, such as seed and fertilizer, the networks have arranged bulk purchases of inputs for resale among members in smaller quantities, thus improving access and reducing costs. Many networks also operate small input kiosks at their offices.

By joining together, FFSs also gained access to technical and advisory services not normally available to individual FFSs or farmers. Government and other extension agents have been very responsive to requests for assistance by the networks, because they can reach more people.

Networking acts as a safety net and sustains the FFS process long after a given project ends. By jointly applying for/guaranteeing loans for individual members or groups and helping each other in the development of proposals, the federations have found it easier to obtain formal credit. Further, a savings fund is in place in most networks from which individual FFSs can borrow money through informal credit arrangements.

Finally, farmers appreciate the sharing of information and experience that networks facilitate. Through connections with other networks, member farmers exchange technical knowledge and new farming ideas in addition to benefitting from the social network in terms of mentoring, encouragement, and a feeling of togetherness. Farmers attribute their involvement in network activities to the social bonding and trust building taking place within the FFS. After networking and strengthening their capacity for collective action, member of FFSs have in many instances gained access to governance and policy processes, and they have also been invited by the government and other service organizations to represent farmers in official functions.

As a result of becoming federated, the Kakamega FFS Network has obtained funds from government programs such as Njaa Marufuku ("ban hunger in Kenya") and the Kenya Agricultural Productivity Project (KAAP), a World Bank-funded program. For example, the network obtained US$10,000 from the Livelihoods Diversification and Enterprise Development Fund for building marketing capacity of network leadership, developing and introducing a farming-as-a-business curriculum, establishing a marketing office that uses a blend of appropriate technology, developing links to established agricultural commodity exchanges, intro-

ducing financial diversification and the role of credit in agriculture, and teaching an agricultural planning process called "plant-to-meet-market." The network also received 2.8 million Kenya shillings (K Sh) (US$40,000) from KAPP for coordinating marketing activities and agricultural commodity marketing. The funds were used mainly to purchase computer hardware and software, provide training in agribusiness and value chain management, and support service providers. Finally, through the network three FFS Groups obtained K Sh 120,000 (US$1,715) each for vegetable production and marketing.

Source: Authors.

SUSTAINABILITY

FFS networks ensure their sustainability in a range of ways. Financially the operations of the networks are supported by the constituent FFSs through regular contributions in the form of subscription fees. Other sources of income include interest charged on revolving funds, commissions on bulk network sales, registration fees, fines or penalties, donations and grants, shares from FFS members, and profits from sales of inputs. Many networks operate a revolving loan system and therefore generate more funds to cover operations and fund activities. Some have managed to secure donor support.

Politically and institutionally the networks can be considered independent of government and development support. The networks are fully locally grown, owned, and managed. Donor support, where involved, has provided infrastructure and education. In no case have donors supported the networks' recurrent operations. Running of FFSs by FFS networks is one way to reduce the costs of running FFSs; FFS networks have lower operating costs owing to lower transport costs, lower overheads, and cheaper facilitators (most are farmer facilitators).

To date the networks have shown themselves to be sustainable; all are still active. Their strength varies, however. Management, leadership, and organizational skills are needed to federate FFS networks. Networks are also vulnerable to individual motivation and capacity, since they are mostly managed by voluntary efforts among members and

committee members. The networks that have been most successful are those with very committed, dynamic, and democratic leaders.

Environmental sustainability is supported by using environmentally sound farming methods. FFS learning revolves around principles of integrated production and pest management, in which farmers balance the ecological and economic implications of particular practices for their farms and businesses.

SCALING UP

In most locations where a considerable number of FFSs have been implemented, FFS networks have spontaneously emerged. Currently FFS networks operate at different levels in many districts of East African countries (Kenya, Uganda, and Tanzania) and elsewhere in Africa (mainly Sierra Leone). Most networks have emerged in relatively high-potential (high-rainfall) areas, although some operate in semiarid and arid areas. As noted, the replication of FFSs is stimulated by FFS graduates' wish to continue the dynamics generated by the FFS process and the recognized need to build local institutions. Through various modes of information sharing, networking is also promoted when farmers hear success stories from other places. In East Africa, the virtual network "Linking Local Learners," which connects farmer groups and networks online, contributed to the growth and development of FFS networks.

The emergence and expansion of FFS networks has also been attributed to the "foci model"[2] adopted for the establishment of FFSs in East Africa. In this model, successive FFSs are established in the immediate neighborhood of existing ones to form a cluster. This strategy has enhanced the frequency of interaction, experience sharing, and the horizontal flow of information among groups. The model also reduces the cost of implementing collective activities because the FFSs can procure inputs and market their produce in bulk. Facilitating the formation of FFS networks should be considered in large-scale agricultural projects as a means to make a larger impact and make greater use of the social capital they generate.

LESSONS LEARNED AND ISSUES FOR WIDER APPLICATION

Farmer Field Schools are considered "stepping stones" to networks, federations, and associations and are an effective platform for farmer organization and empowerment. These networks serve an important role for farmers both in terms of social and technical support. The probability that networks will form can be increased if projects give attention to the longer-term prospects, if farmer facilitators and FFS alumni receive follow-up support, and farmer-driven network development is encouraged.

In the case of the East African FFS networks, market information clearly was crucial for enhancing farmers' access to markets. Market information is not always easy for rural, often illiterate farmers to obtain. Extension advisors often are uncomfortable or incapable of changing their role from providing technical messages to acting as more of an information broker. The need to rethink the role of extension and (re-)train extension agents accordingly, discussed throughout this module, is clear.

Based on needs realized and expressed by networks, there is a demand for more attention to capacity building in financial management, marketing, standards and quality, and the use of ICT tools. Much of the current extension practice is targeted at improving technical skills, not management skills.

The principle of federating upon graduation must be incorporated into the curricula of all FFSs. As FFS networks grow and take on more complex initiatives, networks will need investments to acquire and learn to use ICTs to bridge the information gap, enhance the diversification of business opportunities, and improve operational efficiency. Computer and Internet access and skills are high priorities. Revolving funds need to be developed into more sustainable and long-term investments by supporting networks in identifying viable income-generating activities.

INCAGRO: Developing a Market for Agricultural Innovation Services in Peru

John Preissing, Food and Agriculture Organization of the United Nations (FAO)

SYNOPSIS

The INCAGRO project seeks to establish a national agricultural science and technology system that is decentralized, pluralistic, demand-driven, and led by the private sector. The project achieves this objective by strengthening the market for agricultural innovation services, increasing strategic competencies in agricultural research for development, and promoting the institutionalization of policies, information, and the quality of innovation services. The most notable innovation of INCAGRO was the emergence of a demand-driven market for agricultural innovation services that was more extensive and inclusive than before. This achievement came about as a result of empowering clients to formulate, cofinance, regulate, implement, monitor, and evaluate extension services through the mechanisms and tools offered through two competitive funds. One fund increased the demand and supply of extension services through competitive bidding, and the other expanded the number and quality of extension providers. Another innovation of INCAGRO is that it provided effective national yet decentralized support through regional offices and a central headquarters. The potential long-term impact or sustainability of the model has not been established (more care is needed to document ex post impacts), but it is clear that with appropriate backstopping farmers can become authentic drivers of agricultural extension systems. The competitive grant funds owed their success to transparent policies and rigorous selection and monitoring. A small staff functioning as agricultural innovation brokers throughout Peru promoted efficiency and effectiveness within the agricultural innovation market. Organizational development is needed to work with underserved groups (and also larger groups of farmers to decrease administrative costs). The emphasis must shift to developing a more sustainable system based on private cost recovery, funding partners, and government support.

BASIC PROJECT DATA

The Innovation and Competitiveness Program for Peruvian Agriculture (INCAGRO, Innovación y Competitividad para el Agro Peruano) contributed to the development of a market for agricultural innovation services by paying close attention to how demand for such services is generated and how those services are supplied. INCAGRO has led to technical innovations that bolstered production and productivity and institutional innovations that fostered potentially sustainable models for delivering innovation services. INCAGRO's primary partner has been the Ministry of Agriculture and the national agriculture research and innovation institute (Instituto Nacional de Innovación Agraria) (financing details are shown in box 3.33).

Box 3.33 Sources of Support for the Innovation and Competitiveness Program for Peruvian Agriculture (INCAGRO)

The first phase of the INCAGRO research and extension program (November 1999–January 2005) was financed through a World Bank Adaptable Program Loan (US$9.6 million), the Government of Peru (US$1.44 million), and local counterparts (US$2.78 million), for a total of US$13.82 million. The second phase (October 2005–December 2010) was financed through a second Adaptable Program Loan (US$25 million) in addition to US$6 million from the Government of Peru and US$12 million from local counterparts, for a total of US$43 million. As of this writing, financing for a third phase remains under review by the Government of Peru and the World Bank.

Source: Author, based on INCAGRO project documents and World Bank 2005.

CONTEXT

Peru's public extension services grew considerably from the mid-1950s to the mid-1980s, but in the years that followed, a range of factors led to their decline. The number and range of services supplied by Peru's public extension system became financially unsustainable owing to government financial limitations, privatization trends, and the inhibiting presence of the Shining Path guerilla group (Ortiz 2006). The system was considered too top-down, too supply-driven in its focus on technology transfer, and too centralized. Although large-scale commercial producers could still obtain extension services, small and medium producers came to rely on sporadic support from NGOs. A common concern with the extension services supplied by NGOs was that those organizations were not well integrated with the AIS and its knowledge and information subsystem.

Because the provision of agricultural innovation services to Peru's small and medium-sized farmers was particularly weak, in 1999 the government signed a letter of intent with the World Bank to promote agricultural innovation through the Bank's Adaptable Loan Program in three phases: the establishment of the innovation system; scaling up the system; and a final consolidation phase (presently under discussion). A key feature of the resulting INCAGRO project is the use of competitive funding schemes to promote a market for agricultural innovation services.

DEVELOPING A MARKET FOR AGRICULTURAL INNOVATION SERVICES

INCAGRO's main objective is to establish a national agricultural science and technology system that is modern, decentralized, pluralistic, demand-driven, and led by the private sector. The project's three components are designed to achieve this objective by: (1) strengthening the market for agricultural innovation services; (2) increasing strategic competencies in agricultural research for development; and (3) promoting the institutionalization of policies, information, and the quality of innovation services.

Agricultural technology fund: Competitive grants to improve the demand and supply side of the market for agricultural innovation services

The Agricultural Technology Fund (Fondo de Tecnología Agraria, FTA) has financed projects developed by farmer organizations for support in agricultural extension. Project proposals are based on business plans and use standardized logframes. Independent, three-member panels of agribusiness leaders rate the proposals and determine which projects will be funded. The panels may also recommend changes in the content or size of the proposed projects. The INCAGRO team receives guidance from the evaluation panels for adjusting proposals with the farmer organizations. The exercise of developing a business plan, submitting proposals for competitive review, negotiating with INCAGRO "innovation brokers," and the follow-up monitoring and evaluation data demonstrates, particularly to farmers, that a positive return can be made on the investment in agricultural innovation services.

Proposals range from using innovation services to improve agricultural production and productivity to using them to improve agricultural products and agroindustry. To date, extension service projects covering 40 annual crops, 26 perennial crops, 10 kinds of farm animals, 11 kinds of fish, and 18 agroprocessing efforts have been funded. Crops include basic food crops as well as export crops, some raised organically.

An important aspect of the FTA model is that farmers own the project. They contract extension providers to complete a specified number of activities. Farmer groups are required to make a financial contribution in cash, plus any in-kind contributions. The cash contribution ranges between 15 and 30 percent of the total costs for extension projects. Farmers must form legal entities to sign contracts and receive government support. To meet these requirements, participants must be willing to collaborate, handle considerable legal paperwork, and have the capacity to manage and implement their projects.

The FTA fund makes it possible for farmer groups to gain organizational and project development skills by contracting an "ally" (*aliado*), a private individual or a public or private agency, to assist in formulating the project proposal, developing the corresponding business plan, identifying the right extension supplier, and managing project implementation on behalf of the farmer group (box 3.34 provides an example from the project). An ally is a new but critical innovation in the development of a functioning market for extension services. (For more on this concept, see the discussion of innovation brokers in TN 4.)

Competitive funds have expanded the market for extension service providers through various means. Producer organizations have hired their own extensionists, contracted individual private extension providers, signed agreements with NGOs, and partnered with cooperatives for the provision of extension services. The FTA guidelines for project proposals support a more holistic approach to agricultural innovation by including collaborating entities in the project proposal, such as private input and marketing firms in the

Box 3.34 An Ally Broadens Farmers' Skills to Articulate and Meet Their Demand for Innovation Services

Through the Agricultural Technology Fund, three cacao cooperatives in Huánuco Region partnered with an ally to acquire technical assistance and training in a project to increase the productivity of cacao, obtain certification for organic cacao production, and increase the marketing of their organic cacao. CAFÉ PERÚ (Central de Organizaciones Productoras de Café y Cacao del Perú) served as the ally and implementing agency for the project. The cooperatives received market analysis and specific training in cooperative management and product promotion. Starting from zero in the project's first year, more than 1,200 producers had obtained organic certification by the end of year three. Over the same period, cacao productivity rose from 340 to 600 kilograms per hectare, and the cooperative markets some 1,500 tons of organic cacao. Although cacao prices have risen overall, the productivity increases and the switch to organic production enabled producers' returns to rise from US$ 546 to US$ 1,543 per hectare. The total project cost was US$ 158,716 over the three years. INCAGRO staff, along with CAFÉ PERÚ, played an important role in brokering the arrangement between national and local partners.

Source: Author, INCAGRO (http://www.incagro.gob.pe).

value chain as well as public agencies. Together, these collaborators form a strategic alliance that is formalized in an Agreement of Participation. The agreement establishes the roles and responsibilities of each member of the alliance, their respective contributions to the project, and the final disposition of any items obtained as a result of the project. The idea is that a strong strategic alliance will raise the probability of success.

In addition to developing extension-based projects, farmer organizations can develop adaptive research projects to verify the technical and economic suitability of research findings in the local setting. The research is participatory, requiring the producer-clients to become involved in identifying problems or opportunities in their fields and contribute actively during all stages of the research. Producers can use the strategic alliance framework to include other actors in the value chain as part of the adaptive research

business plan and use the ally to assist them in implementing the adaptive research project. The research entity can be a public or private institution or an individual with specific competencies required for the project. In this case, the farmer organization must meet 5 percent of project costs.

Strategic Services Development Fund: Improving the supply side of the market for agricultural innovation services

The Strategic Services Development Fund (FDSE, Fondo para el Desarrollo de Servicios Estratégicos) uses competitive matching grants to promote basic and applied strategic research, focusing on genetic resources, biotechnologies, plant and animal protection, natural resource management, postharvest technologies, and conservation agriculture. It is a demand-driven system (box 3.35 provides examples). These strategic research areas were defined through national as well as decentralized workshops, with actors in the value chain for all of Peru's major agricultural products, and on the basis of in-depth studies.

Projects must involve strategic alliances of at least two major stakeholders in the research. A specific entity must make the proposal and lead the research; collaborating research organizations are encouraged and increase the competitiveness of a proposal. Funding is capped at US$125,000 per project. Grant recipients must match this funding by 50 percent or more through actual or in-kind resources. National, independent, three-member panels comprised of researchers evaluate proposals for funding.

A second purpose of the FDSE—to improve the supply of agricultural extension services—has proven useful for developing service providers. Competitive grants are awarded to train extension providers to establish extension services; use specific extension methods; learn particular crop and livestock practices, laboratory procedures, and postharvest storage practices; write and analyze business plans; and conduct market analysis. One example of how these funds are used is the training given by the Lambayeque Institute for Agricultural Development to extension providers in how to price and market their services. Another example is the value chains and improvements in value chain training on small livestock in Arequipa by the Institute for Development of the Informal Sector–Arequipa (IDESI, Instituto de Desarrollo del Sector Informal). Extension training providers have included cooperatives with their own staff, universities, national and regional research institutes, and national and regional NGOs. While this training has been useful

Box 3.35   Using Competitive Grants to Fund Multiple, Synergistic Innovation Services for a
New Oilseed Crop in Peru

Competitive grants awarded by the *Agricultural Technology Fund*[a] and Strategic Services Development Fund[b] for nine interrelated projects provided extension services for producers, adaptive research, and strategic research for the development of Sacha Inchi, a native Amazon oilseed plant with high levels of omega fatty acids. The projects involved actors all along the value chain, including the Ministry of Agriculture and Institute of Peruvian Amazon Research, rootstock providers, regional producers' associations, extension providers, and processing and marketing associations. Extension services were offered to community-based or regional producer organizations on improving and standardizing production, making production of organic Sacha Inchi more competitive, and improving producers' agribusiness skills. These projects reached 450 producers directly. Adaptive research projects included testing and validating two technology packages, one for organic production methods and the other for production and processing methods; these projects had 220 direct and 670 indirect beneficiaries. The four in-depth strategic research projects focused on integrated pest management methods, the identification and improvement of genetic lines of Sacha Inchi, the generation of elite lines, and asexual propagation methods.

Source: Author, INCAGRO (http://www.incagro.gob.pe).
a. Fondo de Tecnología Agraria; b. FDSE = Fondo para el Desarrollo de Servicios Estratégicos.

for improving the quality of the extension services market, it represented just 10 percent of the projects. According to the World Bank (2009), this low percentage reflects the small number of providers, the weakness of current and potential training institutions, and a lack of communication between the potential providers and suppliers.

Monitoring, evaluation, and policy development to support a high-quality market for agricultural innovation services

A key activity under the third component of INCAGRO has been to develop an effective project monitoring and reporting system that is agile and robust. For each project, baseline information is collected as part of the business plan proposal required for submitting the requests for funds. During the life of each project, data are collected during the "critical path" steps. At the end of each project, a final financial and technical report is prepared by the project executor. INCAGRO has developed strong, web-based tools (spreadsheets and templates) to aid in this analysis. Data for all of the projects are compiled for analyses by region, crop, type of intervention, gender, and other critical features.[1]

Periodic evaluations helped to improve how INCAGRO is implemented. For example, based on initial findings from the World Bank's Independent Evaluation Group, the competitive fund strategy was redesigned to ensure that disadvantaged groups would have greater access to the competitive funding cycles and have greater success in winning rounds. Stratifying the funds and providing direct assistance to targeted low-income groups were key features of INCAGRO's second phase.

RESULTS AND INNOVATIONS

The most notable innovation of INCAGRO was the emergence of a demand-driven market for agricultural innovation services that was more extensive and inclusive than before. This achievement came about as a result of empowering clients to formulate, cofinance, regulate, implement, monitor, and evaluate extension services through the mechanisms and tools offered by the two competitive funds (J. Ramirez-Gaston, personal communication, April 19, 2010). The demand and supply of extension services was increased through the FTA competitive bidding process, while the number and quality of extension providers increased through the FSDE.

Another innovation of INCAGRO is that it provides effective national yet decentralized support through seven regional offices and a central headquarters. Approximately 60 consultants led or supported the overall project, with an average of four per regional office (this number has fluctuated over the life of the project, based on competitive funding and activity levels). All INCAGRO staff members are consultants rather than permanent government employees. To improve their integration into the ministry, most INCA-

GRO offices are located at stations of the national agricultural research service, unless they are quite distant from a city center.

The impact of the individual projects funded by the grants has been significant, but challenges remain in ensuring a sustainable market for agricultural innovation services. There is no evidence yet that effective, sustained demand for technical assistance services exists or that the capacity to pay for these services (through increased incomes) will suffice to maintain them. In fact, one institutional challenge to the continued market for agricultural extension is the plethora of providers from NGOs and other government projects (some also funded by the World Bank) which provide technical assistance gratis or without competitive funding. Indicators of sustainability are discussed in the benefits section below.

BENEFITS, IMPACT, AND EXPERIENCE TO DATE

Two recent evaluations assessed the impact of INCAGRO (World Bank 2009; Ministry of Agriculture 2009). These and the INCAGRO database provide information on the project's near-term benefits and outcomes, but they are not conclusive on the potential long-term impact or sustainability of the model. Over eight years of INCAGRO's implementation, thousands of farmers demanded and received extension support (table 3.10). Over half of the funds used in the competitive grant projects came from financing provided by farmers and service providers, though much was in the form of in-kind contributions.

The two evaluation studies, using nonrandomized samples due to data limitations, reached positive findings on the project's impact. The Ministry of Agriculture study estimated that 56 percent of producers were likely to adopt the technology innovations, productivity increased by 86

percent, 77 percent of participants were willing to partially pay for extension services, the number of extension and research providers grew by 23 percent, and the diversity and quality of services increased by 25 percent. The same study calculated an internal rate of return between 23 and 34 percent, higher than typical returns for agricultural development projects. The benefit-cost ratio was estimated at two to one, and net present value at US$15 million (2009). Using a different sample, the World Bank study concluded that the economic rate of return for the FTA projects was 39 percent. The Bank's study also reports that Barrantes et al. (2004) calculated an average economic rate of return of 76 percent for a selected number of extension projects.

Based on these analyses, INCAGRO appears to have been a sound investment and successful project. Equity remains a concern, however: The greatest beneficiaries were medium- to large-scale producers rather than the most disadvantaged producers, including women. Strategies were adopted in the second phase to target more vulnerable groups with separate funding, more support, and training. The results of this effort are not clear, but it has led to a perception that costs per client have become higher. Finally, while it is possible to conclude with confidence that the competitive grant projects represented strong investments, INCAGRO itself must be judged against its broader goal of generating a sustainable model for an agricultural innovation market for extension services.

LESSONS LEARNED AND ISSUES FOR WIDER APPLICATION

Peru now has many new competitive funding schemes supporting agriculture, including at least two other schemes managed by the Ministry of Agriculture and others by science and technology and innovation units. In interviews, these funders credit INCAGRO for much of the success of competitive funding schemes, including their role in developing competitive funding strategies; in building the capacity of producer organizations to follow rigorous funding protocols and implement projects; preparing a cadre of professionals that now manage other funds; developing a pool of competent service providers; and creating competitive funding mechanisms and tools that were directly adopted by the new funding agencies. These are actually some of the strongest signs of INCAGRO's sustainability, if not as a program then as a concept. The long-term sustainability of individual extension service providers is not guaranteed through competitive funding schemes, but

Table 3.10 Measures of INCAGRO Project Output	
Projects supported through grants	**Outreach and funding**
Funding cycles: 36	72,000 farmers reached
FTA extension projects:	580,000 indirectly affected
34% to indigenous groups*	16% indigenous population*
12% to women's groups*	5.5% women farmers reached
1,211 proposals	directly*
330 approved	US$43.7 million expended
FDSE extension training:	53% from cofinancing
349 proposals	67% used strategic alliances
51 approved	

Sources: INCAGRO (http://www.incagro.gob.pe); Benites and Wiener 2008.

* Some of these beneficiaries may be indigenous women.

such schemes may represent a sustainable model for extension provision. Quite possibly, future markets for agricultural innovation services will include multiple competitive funding schemes that seek efficiency and responsiveness on the part of extension providers and effective demand from farmers. This model or market may grow, shrink, or change focus based on who funds it and on what producers demand of it.

In summary, major lessons and issues have emerged from INCAGRO. Farmers can become authentic drivers of agricultural extension systems, but initially they require professional backstopping (by "allies"). Transparent policies and rigorous selection and monitoring procedures are keys to the success of competitive grant funds. INCAGRO's operating procedures, information systems, and communication strategies were essential. A small number of staff

functioning as agricultural innovation brokers throughout Peru served to promote efficiency and effectiveness within the agricultural innovation market. Further, by basing staff throughout the country, INCAGRO supported national decentralization goals. Smaller projects dominated the market for extension services, incurring higher administrative costs. Organizational development is needed to work with underserved and also larger groups of farmers to decrease administrative costs (IAP 2). While it is important to focus on establishing funding cycles and tools, eventually the emphasis must shift to developing a more sustainable system based on private cost recovery, funding partners, and ongoing government support. Finally, more deliberate care is needed to document ex post impacts, including the careful and limited use of control groups to reach clearer conclusions on INCAGRO's impact.

Combining Extension Services with Agricultural Credit: The Experience of BASIX India

Vijay Mahajan, BASIX Group

K. Vasumathi, BASIX Group

SYNOPSIS

Since 1996, BASIX has piloted and scaled up livelihood promotion initiatives in which an integrated triad of services—financial services, agricultural, livestock and enterprise development services, and institutional development services—is delivered on a fee-for-service basis. BASIX has reached as many as 3.5 million poor households across 19 states in India. The emphasis has been less on increasing agricultural production than on reducing production costs through a variety of agricultural and business development services. A thorough analysis and understanding of particular subsectors (paddy, groundnut, dairy, and soy, among others) helped identify opportunities for increasing incomes and develop appropriate products and services. Facilitating linkages with high-end markets also resulted in a net increase in producers' incomes. Because products and services were tailored to customers' diverse needs, they were willing to pay for them. The mix of services enabled customers (primarily the rural poor) to increase their incomes from their crops, livestock, and other enterprises. While it is widely understood that financial services alone are insufficient for promoting livelihoods, BASIX provides an innovative strategy for offering such integrated services in a financially sustainable manner.

CONTEXT

More than 80 percent of India's roughly 90 million farm households operate on a small or marginal scale, farming less than two hectares. Most of them also usually have one or two buffaloes or cows, reared for milk and dung. Given that most small-scale and marginal farmers fall below the poverty line, any improvement in their income will help to reduce overall poverty in India. Credit can help farmers obtain yield-enhancing inputs (improved seed, fertilizer,

cattle feed) as well as irrigation pumps and crossbred cattle, but BASIX realized that financial services alone could not raise farmers' incomes. Farmers also needed awareness of better agricultural practices and preventive animal healthcare to reduce risk and costs. In other words, they needed a range of agricultural and livestock development services to gain the knowledge to improve crop and livestock production, mitigate risk, and develop stronger links to markets.

OBJECTIVE AND DESCRIPTION

BASIX (www.basixindia.com) began in 1996 "to promote a large number of sustainable livelihoods, including for the rural poor and women, through the provision of financial services and technical assistance in an integrated manner." BASIX, recognized as a pioneer in livelihood promotion, now operates in over 40,000 villages across 19 states in India with more than 3.5 million poor households.

For the first five years, BASIX delivered what it called "microcredit plus" services to clients. The "plus" components were Technical Assistance and Support Services (TASS), provided through various programs:

- *Dryland Agriculture Productivity Enhancement Program (DAPEP).* DAPEP introduced new crops or varieties; arranged for inputs such as seed, fertilizer, and pesticides (including biopesticides); provided extension support for new agronomic practices to cut costs and increase yields; and arranged for collective transport, processing, and purchase of produce by local agencies and companies.
- *Inter Borrower Exchange Program (IBEX).* IBEX focused on the exchange of expertise among borrowers on better management practices and local innovations.
- *External Resource Person Advisory Program (ERAP).* ERAP invited experts to serve as resource persons to

deliver advisory services on required topics and practices.

- **Self Help Group Quality Improvement Program (SHGQIP).** A microfinance agent model for promoting self-help groups was tested under this program.
- **Rural Infrastructure Revival (RIR).** Local community resources were activated to rehabilitate lift irrigation, the electricity supply, milk chilling plants, and other infrastructure.

These programs delivered services on a full grant basis or 50:50 or 75:25 cost sharing by customers and BASIX, depending on need and customers' willingness to pay. Under the triad strategy described in the next section, these services evolved into the services listed in table 3.11.

RATIONALE FOR BASIX'S TRIAD STRATEGY

In 2001, BASIX asked the Indian Market Research Bureau, an independent external agency, to assess the impact of BASIX among recipients of its services. Only 52 percent of customers who had received at least three rounds of microcredit had significantly improved their incomes, compared with a control group who received no credit. Income levels did not change among 25 percent of customers; 23 percent reported a decline.

BASIX carried out a detailed study of those who had experienced no increase or a decline in income and concluded that this outcome arose from unmanaged risk, low productivity, and limited access to markets, combined with poor terms for buying inputs and selling output. The analysis clearly identified several needs: to improve farmers' productivity, offer services to mitigate risk, improve producers' links to markets, and organize producers to gain a stronger bargaining position.

(See TN 2, which discusses the need for financial services, and IAP 2, which explains how to build strong federations of farmer groups to obtain better services.) In 2002, BASIX developed a "livelihood triad" strategy to provide comprehensive livelihood promotion services to poor rural households. Box 3.36 describes how services evolved for one particular subset of clients.

INNOVATIVE ELEMENT

As mentioned, the livelihood triad strategy includes the provision of financial inclusion services; agricultural, livestock and enterprise development services; and institutional development services detailed in table 3.11. Under Agricultural, Livestock, and Enterprise Development (AGLED) services, BASIX currently provides services to farmers growing several crops (cotton, groundnuts, soybeans, pulses, paddy rice, chilies, vegetables, mushrooms) (box 3.37) and lac (a form of organic resin) and producing milk and livestock (poultry, sheep, and goats).

Nonfarm business development services are also provided for selected activities such as tailoring, woodworking, bamboo work, retail stores, and niche handicrafts and handlooms. An example of the need for institutional development services beyond financial assistance is given in box 3.38.

ACHIEVEMENTS

Today BASIX works in more than 40,000 villages through a network of over 250 branches, each with five field executives under a team leader. Each field executive supervises five livelihood service advisers (LSAs). Each LSA covers about 10 villages, originating credit, selling insurance, collecting

Table 3.11   Services Included in the BASIX Livelihood Triad

Financial inclusion services	Agricultural, livestock, and enterprise development services	Institutional development services
Savings (directly in districts where BASIX has a banking license and through other banks elsewhere)	Improved productivity through higher yields from improved seed or practices	On an individual level, develop awareness, skills, and entrepreneurship
Credit: agricultural, allied, and nonfarm, short and long term	Improved productivity through cost reductions	Form producer groups, federations, cooperatives
Insurance for lives and livelihoods, including index-based weather insurance for crops	Risk mitigation (other than insurance), such as livestock vaccinations	Functional training in accounting and management information systems, using information technology
Money transfer, for migrant workers	Local value addition, such as processing cotton into lint (fiber) before selling	Build collaboration to deliver a wide range of services
Experimental products such as micropensions and warehouse receipts	Alternative market linkages: input supply and output sales	Sector and policy work: analysis and advocacy for changes and reforms

Source: Vijay Mahajan, BASIX.

Box 3.36 BASIX Services for Groundnut Farmers in Andhra Pradesh: From Financial Services to Livelihood Triad Services

In 2001–03, BASIX was working in Anantapur District, which had 700,000 groundnut farmers, with support from ICICI Bank. Based on recommendations from research institutes,[a] the program introduced drought-tolerant cultivars and agronomic practices to reduce the effects of drought, but three years of severe drought dashed efforts to raise groundnut yields. BASIX had greater success with improving groundnut marketing. BASIX formed a farmer cooperative and facilitated it to lease a local factory to shell groundnuts—an activity formerly done by intermediaries. This value-adding step enabled farmers to sell their produce at a higher price. Women's self-help groups bought groundnuts produced under irrigation in the *rabi* (winter) crop cycle and processed them for sale. Large hand-processed nuts were sold as seed for the upcoming *kharif* (summer monsoon) crop cycle, and smaller ones were sold for bird feed.

These interventions offered little for rainfall-dependent farmers affected by drought, however. Dairy production was promoted as an alternative livelihood strategy in 2003. BASIX identified villages to form milk collection routes, educated farmers in dairy farming, helped villagers grow fodder where some irrigation sources were available, and negotiated linkages with nearby bulk chilling centers established by the Andhra Pradesh Dairy Development Cooperative Federation. The market linkages facilitated with the federation led to the revival of chilling centers in Kalyandurg and Kannekal. Chilling center capacity increased from 2,000 liters to 10,000 liters. Outreach to women, for whom dairy became a primary livelihood activity, increased. Migration from the area declined. Many such efforts have led to the BASIX "livelihood triad" of services (financial inclusion services; agricultural, livestock, and enterprise development services; and institutional development services).

Source: S. Amarnath and K. Vasumathi, BASIX.

(a) The International Crops Research Institute for the Semi-Arid Tropics and the Central Research Institute for Dryland Agriculture.

Box 3.37 AGLED Services for Mushroom Cultivation

BASIX seeks to enhance clients' awareness of good practices related to their investments. One example is the fee-based Mushroom Development Services Package, which BASIX offers to customers who have sufficient space to raise mushrooms as an additional source of livelihood. The package includes:

- A review of the primary requirements for mushroom cultivation. A suitable location and the use of specific materials are key factors for good mushroom production.
- Mushroom bed installation (including straw-cutting techniques, soaking straw in water, preparing the bed using spawn and feeding material), aftercare for optimum production, identification of poisonous mushrooms, and precautions to be taken.

- An assessment of mushroom beds for progress of mycelium growth. Training is provided to conduct regular inspections of beds to ensure regular growth of mycelium, avoid losses, and enhance production.
- Training in measures to mitigate the effects (and risks) of high temperatures and low humidity.
- Training in producing two value-added products. It can sometimes be difficult to sell raw mushrooms. Value-added products such as mushroom pickles and soup always fetch higher prices and increase profits for producers.
- Input market linkages for spawn and polythene. BASIX helps customers identify sources of good spawn and facilitates the procurement of polythene and spawn.

Source: Tapaskumar Pati, BASIX.

Box 3.38   Contract Farming for Potato:
            The Need for Strong Farmer
            Organizations

BASIX became involved with financing the potato value chain in Jharkhand in 2005. PepsiCo was already buying potatoes from some farmers in Jharkhand but on a very limited scale. It approached BASIX to facilitate contract potato farming on a larger scale. BASIX had strengthened its capacity as a facilitator and subsequently helped negotiate contracts between PepsiCo and small-scale potato farmers to supply agreed quantities of potatoes at a specified quality and price to PepsiCo's chip-making factory in Kolkata. Although in the first years the contract resulted in increased yields and higher incomes for farmers, in later years farmers' yields were affected by problems with seed quality, heavy disease infestations, and unfavorable weather. Sometimes their potatoes were rejected at the factory because of poor quality, and other times they resorted to the open market, where they obtained better prices. Clearly the partners needed more than finance to continue their association productively. BASIX invested in strengthening the relationships with partner organizations and the farming community to ensure that contracts operated fairly.

Source: Mishra 2008.

payments, and selling AGLED services. Repayment schedules depend on the crop, cropping season, and household cash flows and range from 6 to 11 months and 1–3 installments. BASIX has more than 4,000 LSAs.

BASIX field executives identify and select villages or clusters of villages to receive services. A cluster is a group of villages within a radius of 6–8 kilometers, which offers a reasonable base for delivering services effectively and efficiently to customers. The branches start enrolling customers for services in villages where at least 30 borrowers engage in either crop or livestock activities.

BASIX has a cadre of over 1,000 livelihood services providers (LSPs). While LSAs function as salespeople, LSPs resemble extension agents. An LSP works with BASIX on a regular basis and is typically a high-school graduate trained as a para-extension worker or para-veterinarian. He or she covers 200–400 customers for one crop or activity. More than 10 percent of the LSAs and over 15 percent of the LSPs

are women. BASIX distributes product brochures in regional languages telling customers what services they can receive and explaining the service conditions. Customers pay 450 rupees (Rs) (US$10), including a service tax, for a year of AGLED services.

In 2010, AGLED services had over half a million customers. About half of them used agricultural and livestock services, and the remainder used services related to nonfarm activities. Among the agricultural services, BASIX provided a soil-testing service for more than 30,000 farmers, integrated pest management or integrated nutrient management services to nearly 160,000 crop customers, and field surveillance to more than 85,000 farmers. It connected most customers to input markets (seed, fertilizer, pesticide, and bio-inputs such as vermicompost and organic pesticide) and output markets. BASIX provided index-based weather insurance to more than 10,000 farmers for different crops in different agroclimatic zones in 2009, in collaboration with private insurance companies.

BASIX also conducted health checkups of nearly 450,000 animals, vaccinated nearly 165,000, and dewormed 200,000 animals. It trained more than 170,000 customers in feed, fodder, and other improved practices for dairying. More than 60,000 farmers were linked to milk marketing chains supported by cooperatives or private companies. More than 120,000 animals were insured in collaboration with private companies.

LESSONS LEARNED

It has taken BASIX about seven years to reach the scale described, and it has learned many lessons along the way. Lessons that may prove useful for similar initiatives are summarized here.

■ *It is vital to respond proactively to farmers' needs.* In its first two years, BASIX emphasized market research to identify which services farmers needed. The organization also conducted action-research through many pilot interventions. This research, which featured numerous field visits and group interactions with farmers, showed that small-scale farmers preferred cost-saving and risk-reducing interventions over yield-enhancing interventions requiring greater cash outlays. BASIX also learned that it was not possible to handle such interventions for a large number of crops, so it focused on a few crops grown by a large number of farmers, such as groundnuts in southern Andhra Pradesh, cotton in northern Andhra Pradesh, and soybeans in western Madhya Pradesh.

- *In designing services, focus on reducing costs.* Productivity can be increased by raising yields as well as by reducing the costs of producing the same amount of output. Local agricultural universities and research stations had developed many practices for increasing yields, so BASIX decided to focus on reducing costs. One example of this approach was to apply pesticide on cotton stems to reduce pest multiplication and reduce pesticide applications later in the season. Another example was the introduction of soil testing to enable more precise, economical fertilizer use. For dairy animals, simple practices like vaccination and periodic deworming were more cost-effective than procuring high-yielding crossbred animals.

- *Customized services enhance willingness to pay.* BASIX staff learned how to customize AGLED services to different agroclimatic zones, which enhanced farmers' willingness to pay for services. Customer satisfaction surveys conducted by independent audit teams found that the satisfaction level was nearly 80 percent; the main cause of dissatisfaction was inadequate visits from LSPs. To improve service, field executives introduced tighter monitoring of service delivery through passbooks, acknowledgement receipts, and service cards, but this practice was expensive. BASIX is piloting a strategy to monitor service delivery through mobile phones so farmers can report LSPs for poor service or missed visits.

- *Sustainability and extension of services to larger farmers.* The income from AGLED services in 2010 was nearly Rs 148 million (US$3 million). BASIX made a modest profit (nearly Rs 22 million or US$450,000) by providing these services to over half a million customers. With more LSPs reaching the breakeven number of customers, profitability is likely to improve. BASIX also plans to move some basic facilities like soil-testing labs and artificial insemination centers under its own control to improve its service to farmers.

Although BASIX agricultural credit operations are aimed at small-scale and marginal farmers, the organization plans to extend AGLED services to larger farmers to whom it does not extend credit in the same villages. In improving their yields, these farmers will generate additional production as well as employment opportunities for the landless poor, outcomes aligned with the BASIX mission. So far BASIX has worked mainly in poorer dryland districts. It is considering providing AGLED services in irrigated districts and for large-scale farmers where it has no credit operations through its new BASIX Krishi company. With these changes, BASIX is confident of reaching two to three million farmers with AGLED services by 2015.

NOTES

Module 3 Overview

1. It is useful to have a conceptual and analytical framework for implementing and analyzing extension reforms with a focus on pluralistic extension. Such a framework has been developed by a multidisciplinary group of researchers, and it distinguishes four sets of factors that need to be considered: the policy environment, the capacity of potential service providers, the type of farming systems and the market access of farm households, and the nature of the local communities, including their ability to cooperate. The analytical framework then "disentangles" the major characteristics of agricultural advisory services for which policy decisions must be made ("choice variables"): governance structures, capacity, management, organization, and advisory methods. Implementation of the resulting extension programs and reforms should focus on "best fit" rather than "best practice." For more information see Birner et al. (2009).

2. Many of the points in the table and this section are based on Birner et al. (2009).

3. These strategies and lessons are relevant for other advisory service programs; for more information, see Braun and Duveskog (2009) and Davis et al. (2010a).

4. See the discussion of the scaling-up tool developed by ZALF (the Leibniz Centre for Agricultural Landscape Research), described in Herberg and Schoening (2010).

Thematic Note 1

1. Farm Radio International (www.farmradio.org), an NGO that reaches millions of rural people, is one example.

Thematic Note 2

1. Others refer to these services as "value chain oriented services" (KIT, Faida Mali, and IIRR 2006; Webber and Labaste 2010), "Market-Oriented Agricultural Advisory Services" (Chipeta, Christoplos, and Katz 2008), and "marketing extension" (FAO, http://www.fao.org/ag/ags/agricultural-marketing-linkages/marketing-extension/en).

2. These actors have (re-)discovered the need to address small-scale farmers' demands; see, for example, the roundtable for a sustainable cocoa economy (www.roundtablecocoa.org) and the sustainable spices initiative (http://www.kit.nl/spiceconference).

3. "Republic of Mozambique: Country Programme Evaluation, IFAD http://www.ifad.org/evaluation/public_html/eksyst/doc/country/pf/mozambique/index.htm, accessed July 2011.

4. KIT and IIRR (2012). Women's skills and techniques can be assets for entrepreneurship, and services should be tailored to their circumstances as individual entrepreneurs and within groups. Capacity-building strategies and programs with a gender focus are helpful (Kahan 2007: 90-92).

5. A priority of AFAAS (www.afaas-africa.org).

6. Examples include sesame, peppers, and organic and Fair Trade produce (Jaleta, Gebremedhin, and Hoekstra 2009; Kristjanson et al. 2009).

7. AgriProfocus (http://www.agri-profocus.nl) fosters concerted efforts by public and private agencies to support smallholders' farming entrepreneurship.

Innovative Activity Profile 1

1. This IAP focuses primarily on agrodealers who continuously engage in supplying inputs. In reality, businesses that supply agro-inputs operate on a seasonal basis in many emerging markets. They often sell inputs as complementary products to their core general merchandise business. Subdealers, or "stockists," play a key role in improving farmers' access to inputs. This IAP focuses on agrodealers and only alludes to the development of subdealers/stockists.

2. Macroenvironmental factors (in particular the government's role in a country) substantially influence agrodealer participation in agricultural input markets.

3. The terms "value chain" and "agricultural input supply and marketing chain" are considered synonymous in this IAP. For consistency, this IAP uses the more contemporary "value chain" terminology.

4. Activities are varied and may include lectures, group discussions, role-playing, case studies, problem solving, hands-on analysis (such as the use of soil test kits), individual and group presentations, special studies to design promotional programs, observation/hands-on applicationn of agricultural inputs, product displays, and videos, among others.

Innovative Activity Profile 2

1. Federations in this paper are defined as farmer organizations that have emerged from bottom-up empowerment processes (such as the training of farmers' groups), leading to a higher level of organization at a given administrative or geographical level.

2. Growing from a nucleus outwards.

Innovative Activity Profile 3

1. Information on INCAGRO's subproject management and M&E tool (SIGES, Sistema de Gestión de Sub Proyectos) is available at http://www.incagro.gob.pe/WebIncagro/detalleArticulosBanner.do?c_codigoArticulo=000166.

REFERENCES AND FURTHER READING

Module 3 Overview

Alawy, A.S. 1998. Accessibility of Women's Groups to Agricultural Extension Services in Kenya: An Exploratory and Descriptive Study of Factors, Needs, and Problems. Unpublished PhD thesis, Ohio State University, Columbus.

Alston, J., C. Chan-Kang, M. Marra, P. Pardey, and T.J. Wyatt. 2000. "A Meta-analysis of Rates of Return to Agricultural R&D: *Ex Pede Herculem?*" IFPRI Research Report No. 113. Washington, DC: International Food Policy Research Institute (IFPRI).

Anderson, J.R. 2006. "Training and Visit (T&V) Extension Method." In the Sustainable Funding for Agricultural Innovation Systems Electronic discussion (http://www.dgroups.org/groups/worldbank/LAC-AgInnovations/index.cfm), which includes the Anderson video clip (http://www.gdln-lac.org/wms/Agriculture_Message_Series-08-15-2006-high%20bandwidth.WMV), accessed September 2011.

Anderson, J.R. 2007. "Agricultural Advisory Services." Background paper for *World Development Report 2008, Agriculture for Development*, World Bank, Washington, DC.

Anderson, J.R., and G. Feder. 2004. "Agricultural Extension: Good Intentions and Hard Realities." *The World Bank Research Observer* 19(1):41–60.

Benin, S., E. Nkonya, G. Okecho, J. Pender, S. Nahdy, S. Mugarura, and G. Kayobyo. 2007. "Assessing the Impact of the National Agricultural Advisory Services (NAADS) in the Uganda Rural Livelihoods." IFPRI Discussion Paper No. 724. Washington, DC: International Food Policy Research Institute (IFPRI).

Benor, D., and M. Baxter. 1984. *Training and Visit Extension*. Washington, DC: World Bank.

Birkhaeuser, D., R.E. Evenson, and G. Feder. 1991. "The Economic Impact of Agricultural Extension: A Review." *Economic Development and Cultural Change* 39(3):607–40.

Birner, R., K. Davis, J. Pender, E. Nkonya, P. Anandajayasekeram, J. Ekboir, A. Mbabu, D. Spielman, D. Horna, and S. Benin. 2009. "From Best Practice to Best Fit: A Framework for Analyzing Agricultural Advisory Services Worldwide." *Journal of Agricultural Extension and Education* 15(4):341–55.

Blackden, M., S. Canagarajah, S. Klasen, and D. Lawson. 2006. "Gender and Growth in Sub-Saharan Africa: Issues and Evidence." Research Paper No. 2006/37. Tokyo: United Nations University World Institute for Development Economics Research.

Braun, A.R., and D. Duveskog. 2009. The Farmer Field School approach: History, global assessment, and success stories. Unpublished report for the IFAD *Rural Poverty Report 2009*, International Fund for Agricultural Development (IFAD), Rome.

Chema, S., E. Gilbert, and J. Roseboom. 2003. "A Critical Review of Key Issues and Recent Experiences in Reforming Agricultural Research in Africa." ISNAR Research Report No. 24. The Hague: International Service for National Agricultural Research (ISNAR).

Christoplos, I. 2010. *Mobilizing the Potential of Rural and Agricultural Extension*. Rome: Food and Agriculture Organization of the United Nations (FAO) and the Global Forum for Rural Advisory Services (GFRAS).

Davis, K. 2008. "Extension in Sub-Saharan Africa: Overview and Assessment of Past and Current Models and Future Prospects." *Journal of International Agricultural and Extension Education* 15(3):15–28.

———. 2009. "The Important Role of Extension Systems." Brief 11 in "Agriculture and Climate Change: An Agenda for Negotiation in Copenhagen." 2020 Vision Focus 16. Washington, DC: International Food Policy Research Institute (IFPRI).

Davis, K.E., and B.K. Addom. 2010. "Sub-Saharan Africa." In *ICTs for Agricultural Extension: Global Experiments, Innovations, and Experiences*, edited by R. Saravanan. New Delhi: New India Publishing Agency.

Davis, K., E. Nkonya, D.A. Mekonnen, E. Kato, M. Odendo, R. Miiro, and J. Nkuba. 2010a. "Impact of Farmer Field Schools on Agricultural Productivity, Poverty, and Farmer Empowerment in East Africa." IFPRI Discussion Paper. Washington, DC: International Food Policy Research Institute (IFPRI).

Davis, K., B. Swanson, D. Amudavi, D.A. Mekonnen, A. Flohrs, J. Riese, C. Lamb, and E. Zerfu. 2010b. "In-depth Assessment of the Public Agricultural Extension System in Ethiopia and Recommendations for Improvement." IFPRI Discussion Paper No. 01041. Washington, DC: International Food Policy Research Institute (IFPRI).

Dercon, S., D.O. Gilligan, J. Hoddinot, and T. Woldehanna. 2008. "The Impact of Agricultural Extension and Roads on Poverty and Consumption Growth in Fifteen Ethiopian Villages." IFPRI Discussion Paper. Washington, DC: International Food Policy Research Institute (IFPRI).

Dixie, G. 2005. *Horticultural Marketing*. Marketing Extension Guide No. 5. Rome: Food and Agriculture Organization of the United Nations (FAO).

Engel, P., and M. Salomon. 1997. *Facilitating Innovation for Development*. Amsterdam: Royal Tropical Institute (KIT).

Frank, E. 1999. Gender, agricultural development, and food security in Amhara, Ethiopia: The contested identity of women farmers in Ethiopia. Unpublished paper, United States Agency for International Development (USAID), Washington, DC.

Gender and Governance Research Team. 2009. *Gender and Governance in Rural Services: Insights from India, Ghana, and Ethiopia*. Washington, DC: World Bank and International Food Policy Research Institute (IFPRI).

Government of Kenya. 2005. *National Agricultural Sector Extension Policy (NASEP)*. Nairobi: Ministry of Agriculture, Ministry of Livestock and Fisheries Development, and Ministry of Cooperative Development and Marketing.

Hafkin, N., and N. Taggart 2001. "Gender, Information Technology, and Developing Countries: An Analytical Study." Washington, DC: Academy for Educational Development and the Office of Women in Development, Bureau for Global Programs, Field Support and Research, United States Agency for International Development (USAID).

Hanson, J.C., and R.E. Just. 2001. "The Potential for Transition to Paid Extension: Some Guiding Economic Principles." *American Journal of Agricultural Economics* 83(3):777–84.

Haug, R., 1999. "Some Leading Issues in Agricultural Extension: A Literature Review." *Journal of Agricultural Education and Extension* 5(4):263–74.

Heemskerk, W., N. Lema, D. Guindo, C. Schouten, Z. Semgalawe, H. Verkuijl, B. de Steenhuijsen Piters, and P. Penninkhoff 2003. *A Guide to Demand-driven Agricultural Research: The Client-oriented Research Management Approach*. Amsterdam: Institut d'Economie Rural (IER), Department for Research and Development, and Royal Tropical Institute (KIT). http://www.kit.nl/smartsite.shtml?id=SINGLEPUBLICATION&ItemID=1500, accessed July 2011.

Heemskerk, W., and B. Wennink. 2004. "Building Social Capital for Agricultural Innovation: Experiences with Farmer Groups in Sub-Saharan Africa." KIT Bulletin No. 368. Amsterdam: Royal Tropical Institute (KIT).

Heemskerk, W., S. Nederlof, and B. Wennink. 2008. "Outsourcing Agricultural Advisory Services: Enhancing Rural Innovation in Sub-Saharan Africa." KIT Bulletin No. 380. Amsterdam: Royal Tropical Institute (KIT).

Herberg, L., and A. Schoening. 2010. "Sustainet: Scaling up Sustainable Agriculture." *Rural21: The International Journal for Rural Development* 44(2):21–23.

Hoffmann, V., M. Gerster-Bentaya, A. Christinck, and M. Lemma (eds.) 2009. *Basic Issues and Concepts.*

Volume 1 of *Handbook: Rural Extension*. Weikersheim: Margraf.

KIT (Royal Tropical Institute), Faida Mali, and IIRR (International Institute of Rural Reconstruction). 2006. "Chain Empowerment: Supporting African Farmers to Develop Markets." Amsterdam, Arusha, and Nairobi.

Kithuka, J., J. Mutemi, and A.H. Mohamed. 2007. "Keeping up with Technology: The Use of Mobile Telephony in Delivering Community-based Decentralised Animal Health Services in Mwingi and Kitui Districts, Kenya." FARM-Africa Working Paper No. 10. London: FARM-Africa.

Leeuwis, C., and A. van den Ban. 2004. *Communication for Rural Innovation (Rethinking Agricultural Extension)*. Oxford: Blackwell Science.

Nederlof, E.S., B. Wennink, and W. Heemskerk. 2008. "Access to Agricultural Services." Background paper for the International Fund for Agricultural Development (IFAD) *Rural Poverty Report* 2010. IFAD, http://www.ifad.org/rural/rpr2010/background/3.pdf, accessed September 2011.

Praneetvatakul, S., and H. Waibel. 2006. "Impact assessment of farmer field school using a multi-period panel data model." Presented at the 26th conference of the International Association of Agricultural Economists (IAAE), Brisbane, 12–18 August 2006.

Quisumbing, A.R. (ed.). 2003. *Household Decisions, Gender, and Development: A Synthesis of Recent Research*. Washington, DC: International Food Policy Research Institute (IFPRI).

Quizon, J., G. Feder, and R. Murgai. 2001. "Fiscal Sustainability of Agricultural Extension: The Case of the Farmer Field School Approach." *Journal of International Agricultural and Extension Education* (Spring):13–23.

Rahmato, D. 1993. "Land, Peasants, and the Drive for Collectivization in Ethiopia." In *Land in African Agrarian Systems*, edited by T.J. Bassett and D.E. Crummey. Madison: University of Wisconsin Press.

Rajalahti, R., J. Woelcke, and E. Pehu. 2005. "Monitoring and Evaluation for World Bank Agricultural Research and Extension Projects: A Good Practice Note." Agriculture and Rural Development Discussion Paper No. 20. Washington, DC: The World Bank.

Rivera, W.M., and G. Alex. 2004. "Extension Reform for Rural Development." Agriculture and Rural Development Discussion Paper No. 10. Washington, DC: World Bank.

Spielman, D.J., and R. Birner. 2008. "How Innovative Is Your Agriculture? Using Innovation Indicators and Benchmarks to Strengthen National Systems." Agriculture and Rural Development Discussion Paper No. 41. Washington, DC: World Bank.

Spielman, D.J., K. Davis, M. Negash, and G. Ayele. 2011. "Rural Innovation Systems and Networks: Findings from a Study of Ethiopian Smallholders." *Agriculture and Human Values* 28(2):195–212.

Sulaiman, S.V., and A. Hall 2002. "Beyond Technology Dissemination: Can Indian Agricultural Extension Reinvent Itself?" ICAR Policy Brief No. 16. New Delhi: International Council of Agricultural Research (ICAR).

Swanson, B., and R. Rajalahti. 2010. "Strengthening Agricultural Extension and Advisory Systems: Procedures for Assessing, Transforming, and Evaluating Extension Systems." Agricultural and Rural Development Discussion Paper No. 44. Washington, DC: World Bank.

Swanson, B.E., B.J. Farner, and R. Bahal 1990. "The Current Status of Agricultural Extension Worldwide." In *Report of the Global Consultation on Agricultural Extension,* edited by B.E. Swanson. Rome: Food and Agriculture Organization of the United Nations (FAO). Pp. 43–76.

Torero, M., S. Chowdhury, and A. Bedi. 2006. "Telecommunications Infrastructure and Economic Growth" Chapter 5 in *Information and Communications Technology for Development and Poverty Reduction,* edited by M. Torero and J. von Braun. Baltimore: Johns Hopkins.

Wennink, B., E.S. Nederlof, and W. Heemskerk (eds.). 2007. "Access of the Poor to Agricultural Services: The Role of Farmers' Organizations in Social Inclusion." KIT Bulletin No. 376. Amsterdam: Royal Tropical Institute (KIT).

World Bank. 2006a. "Enhancing Agricultural Innovation: How to Go Beyond the Strengthening of Research Systems." Washington, DC: World Bank.

———. 2006b (revised). "Investment in Agricultural Extension and Information Services." Module 3 in *Agriculture Investment Sourcebook.* Washington, DC: World Bank. www.worldbank.org/ard/ais, accessed july 2011.

———. 2011. *Information and Communication Technologies for Agriculture e-Sourcebook.* http://bit.ly/ICTinAG. Washington, DC.

World Bank, United Nations Development Programme (UNDP), and United Nations Development Fund for Women (UNIFEM). 2010. Presentation from Session 4: Governance and Fragility and Session 2. Challenges for Service Delivery for Women in Post Conflict States. Innovations and New Directions in the World Bank and the United Nations. New York.

Thematic Note 1

AFAAS–FARA (African Forum for Agricultural Advisory Services and Forum for Agricultural Research in Africa) and Neuchatel Initiative. 2009. Capacity development for market-oriented agricultural advisory services in sub-Saharan Africa: An innovative approach to rural development. Proposal for a Partnership Programme.

Blewett, T.J., A. Keim, J. Leser, and L. Jones. 2008. "Defining a Transformational Education Model for the Engaged University." *Journal of Extension* 46(3).

Chapman, R., and R. Tripp. 2003. "Changing Incentives for Agricultural Extension: A Review of Privatized Extension in Practice." ODI Agricultural Research and Extension Network Paper 132. London: Overseas Development Institute (ODI).

DNEA (Direcção Nacional de Extensão Agrária). 2005. "Programa Nacional de Extensão Agrária (PRONEA)." Maputo: Ministry of Agriculture. www.IIAM.gov.mz

———. 2007. *Extension Master Plan 2007–2016.* Maputo: Ministry of Agriculture. http://www.iiam.gov.mz/images/stories/pdf_files/%20plano_eg.pdf.

ECDPM (European Centre for Development Policy Management). 2008. "Capacity Change and Performance: Insights and Implications for Development Cooperation." Policy Management Brief 21. Maastricht: ECPDM.

FAO, KARI, and ILRI (Food and Agriculture Organization of the United Nations, Kenya Agricultural Research Institute, and International Livestock Research Institute). 2003. Farmer field schools: The Kenyan experience. Report of the FFS Stakeholders' Forum, March 27, 2003, Nairobi, Kenya.

Gêmo, H., C.K. Eicher, and S. Teclemariam. 2005. *Mozambique's Experience in Building a National Extension System.* East Lansing: Michigan State University.

Heemskerk, W., and B. Wennink. 2004. "Building Social Capital for Agricultural Innovation: Experiences with Farmer Groups in Sub-Saharan Africa." KIT Bulletin 368. Amsterdam: Royal Tropical Institute.

———. 2005. "Stakeholder-driven Funding Mechanisms for Agricultural Innovation: Case Studies from Sub-Saharan Africa." KIT Bulletin 373. Amsterdam: Royal Tropical Institute.

Heemskerk, W., S. Nederlof, and B. Wennink. 2008. "Outsourcing Agricultural Advisory Services: Enhancing Rural Innovation in Sub-Saharan Africa." KIT Bulletin 380. Amsterdam: Royal Tropical Institute.

Klerkx L., A. Hall, and C. Leeuwis. 2010. "Strengthening Agricultural Innovation Capacity: Are Innovation Brokers the Answer?" *International Journal of Agricultural Resources, Governance, and Ecology* 8(5–6): 409–438.

Nederlof, E.S., B. Wennink, and W. Heemskerk. 2008. Access to agricultural services. Background paper for the IFAD *Rural Poverty Report* 2010. http://www.ifad.org/rural/rpr2010/background/3.pdf

Spielman, D.J., K. Davis, M. Negash, and G. Ayele. 2011. Rural innovation systems and networks: findings from a study of Ethiopian smallholders. *Agriculture and Human Values* 28 (2): 195–212.

Swanson, B., and R. Rajalahti. 2010. "Strengthening Agricultural Extension and Advisory Systems: Procedures for Assessing, Transforming, and Evaluating Extension Systems." Agricultural and Rural Development Discussion Paper No. 44. Washington, DC: World Bank.

Wennink, B., and W. Heemskerk (eds.). 2006. "Farmers' Organizations and Agricultural Innovation. Case Studies from Benin, Rwanda, and Tanzania." KIT Bulletin 374. Amsterdam: Royal Tropical Institute.

Wennink, B., E.S. Nederlof, and W. Heemskerk (eds.). 2007. "Access of the Poor to Agricultural Services: The Role of Farmers' Organizations in Social Inclusion." KIT Bulletin 376. Amsterdam: Royal Tropical Institute.

Thematic Note 2

AFAAS. 2011. http://www.afaas-africa.org.

Benin, S., E. Nkonya, G. Okecho, J. Pender, S. Nahdy, S. Mugarura, and G. Kayobyo. 2007. "Assessing the Impact of the National Agricultural Advisory Services (NAADS) in the Uganda Rural Livelihoods." IFPRI Discussion Paper No. 724. Washington, DC: International Food Policy Research Institute (IFPRI).

Chipeta, S., I. Christoplos, and E. Katz. 2008. *Common Framework on Market-Oriented Agricultural Advisory Services*. Lindau: Agridea for Neuchâtel Group.

Christoplos, I. 2010. *Mobilizing the Potential of Rural and Agricultural Extension*. Rome: Food and Agriculture Organization of the United Nations (FAO) and the Global Forum for Rural Advisory Services (GFRAS).

DFID (UK Department for International Development) and SDC (Swiss Agency for Development and Cooperation). 2008. *Perspectives on The Making Markets Work for the Poor (M4P) Approach*. Bern: SDC.

Dixie, G. 2005. *Horticultural Marketing*. Marketing Extension Guide 5. Rome: Food and Agriculture Organization of the United Nations (FAO).

DNEA (Direcção Nacional de Extensão Agrária). 2007. *Extension Master Plan 2007–2016*. Maputo: Ministry of Agriculture. http://www.iiam.gov.mz/images/stories/pdf_files/%20plano_eg.pdf, accessed July 2011.

Friis-Hansen, E., and C. Aben. 2010. "Rise and Fall of Private Agricultural Service Providers in Sorti District Uganda." Presentation at the GFRAS Conference, Chile, http://www.rimisp.org/FCKeditor/UserFiles/File/documentos/docs/proyecto261/16.%20Rise%20and%20Fall%20of%20PSP2_Esbern%20Friis.pdf, accessed July 2011.

Hanlon, J., and T. Smart. 2008. *Do Bicycles Equal Development in Mozambique?* Rochester: James Currey.

Heemskerk, W., S. Nederlof, and B. Wennink. 2008. "Outsourcing Agricultural Advisory Services: Enhancing Rural Innovation in Sub-Saharan Africa." KIT Bulletin No. 380. Amsterdam: Royal Tropical Institute (KIT).

Hoffmann, V., A. Christinck, and M. Lemma (eds.). 2009. *Examples and Background Material*. Volume. 2 of *Handbook: Rural Extension*. Weikersheim: Margraf.

Hoffmann, V., M. Gerster-Bentaya, A. Christinck, and M. Lemma (eds.) 2009. *Basic Issues and Concepts*. Volume 1 of *Handbook: Rural Extension*. Weikersheim: Margraf.

Jaleta, M., B. Gebremedhin, and D. Hoekstra. 2009. "Smallholder Commercialization: Processes, Determinants, and Impact." Discussion Paper No. 18. Nairobi: International Livestock Research Institute (ILRI).

Kahan, D.G. 2007. "Farm Management Extension Services: A Review of Global Experience." Agricultural Management, Marketing, and Finance Occasional Paper No. 21. Rome: Food and Agriculture Organization (FAO).

KIT (Royal Tropical Institute). 2011 (forthcoming). "Gender in Value Chains." www.kit.nl.

KIT (Royal Tropical Institute), Faida Mali, and IIRR (International Institute of Rural Reconstruction). 2006. "Chain Empowerment: Supporting African Farmers to Develop Markets." Amsterdam, Arusha, and Nairobi.

KIT (Royal Tropical Institute) and IIRR (International Institute of Rural Reconstruction). 2008. "Trading Up: Building Cooperation between Farmers and Traders in Africa." Amsterdam and Nairobi.

———. 2010. "Value Chain Finance: Beyond Microfinance for Rural Entrepreneurs." Amsterdam and Nairobi.

———. 2012. *Gender in Value Chains*. Amsterdam, The Netherlands: Royal Tropical Institute; Arnhem, The Netherlands: AgriProFocus, and Nairobi, Kenya: International Institute of Rural Reconstruction (in press).

Kristjanson, P., R. Reid, N. Dickson, W. Clark, D. Romney, R. Puskur, S. MacMillan, and D. Grace. 2009. "Linking Research Knowledge with Action: Lessons from Sustainable Development Livestock Projects. PNAS, http://www.pnas.org/content/suppl/2009/03/16/0807414106.DC Supplemental/Appendix_PDF.pdf, accessed July 2011.

Lamers, J.P.A., P.R. Feil, N. Bayverdiyeva, Y. Guliyeva, and F. Djafarov. 2008. "From Kolchoz Systems to Fee-based Private Agricultural Extension: Achievements with a Client-oriented Training and Advisory Concept as Support for Private Farming in Azerbaijan." *Journal of Applied Biosciences* 8(1):262–71.

Malindi, G. 2011. "Agricultural Advisory Services under Value Chain Agriculture. A Case from Malawi." Presented at AFAAS conference, Accra, April 12, 2011, www.afaas-africa.org.

NAADS (National Agricultural Advisory Services). 2010. Agribusiness Development. Unpublished working paper, Kampala.

Odada, E.O., R.J. Scholes, K.J. Noone, C. Mbow, and W.O. Ochola (eds.). 2008. *A Strategy for Global Environmental Change Research in Africa: Science Plan and Implementation Strategy.* Stockholm: International Geosphere-Biosphere Programme (IGBP) Secretariat.

Pingali, P.L., Y. Khwaja, and M. Meijer. 2005. "Commercializing Small Farmers: Reducing Transaction Costs." FAO/ESA Working Paper No. 05-08. Rome: Food and Agriculture Organization of the United Nations (FAO).

Pingali, P.L., and M.W. Rosegrant. 1995. "Agricultural Commercialization and Diversification: Process and Policies." *Food Policy* 20(3):171–85.

Poitevin, B., and S. Hossain. 2006. "Marketing Extension: A Powerful Process in 6 Steps." Dhaka: Livelihoods, Empowerment and Agroforestry (LEAF) Project, Intercooperation, and Swiss Agency for Development and Cooperation (SDC).

Pyburn, R., F. van der Lee, and M. ter Heegde (2011, forthcoming). "Green Services: Getting Payment for Smallholder Contributions to NRM." Working Paper. Amsterdam: Royal Tropical Institute (KIT).

Roduner, R. 2007. "Donor Interventions in Value Chain Development." Working Paper. Community of Practice on Value Chains in Rural Development. Berne: Swiss Agency for Development and Cooperation (SDC).

Spielman, D.J., J. Ekboir, K. Davis, and C.M.O. Ochieng. 2008. "An Innovation Systems Perspective on Strengthening Agricultural Education and Training in Sub-Saharan Africa." *Agricultural Systems* 98:1–9.

Swanson, B., and R. Rajalahti. 2010. "Strengthening Agricultural Extension and Advisory Systems: Procedures for Assessing, Transforming, and Evaluating Extension Systems." Agricultural and Rural Development Discussion Paper No. 44. Washington, DC: World Bank.

UNCDF (United Nations Capital Development Fund). 2009. "The State of Local Economic Development in Mozambique." Prepared by P. Penninkhoff for United Nations Capital Development Fund. Development Policy and Practice. Amsterdam: Royal Tropical Institute (KIT).

UNDP (United Nations Development Programme). 2004a. *Business Development Services How to Guide.* Bratislava: UNDP Regional Centre.

———. 2004b. *Unleashing Entrepreneurship; Making Business Work for the Poor.* Commission on the Private Sector and Development, Report to the Secretary-General of the United Nations. New York.

USAID (United States Agency for International Development). 2009. "Building Partnerships on Higher Agricultural Education and Hatching Agribusiness Incubator in Mali." Washington, DC.

van Weperen, W. 2011. "Market-Oriented Advisory Services Approaches and Processes: Toward a Guide for Piloting MOAAS in AAS Systems." Presented at AFAAS conference, Accra, April 12, www.afaas-africa.org.

Webber, C.M., and P. Labaste. 2010. "Building Competitiveness in Africa's Agriculture: A Guide to Value Chain Concepts and Applications." Washington, DC: World Bank.

Wennink, B., E.S. Nederlof, and W. Heemskerk (eds.). 2007. "Access of the Poor to Agricultural Services: The Role of Farmers' Organizations in Social Inclusion." KIT Bulletin No. 376. Amsterdam: Royal Tropical Institute (KIT).

Wilk, E. de O., and J.E. Fensterseifer. 2003. "Towards a National Agribusiness System: A Conceptual Framework." *International Food and Agribusiness Management Review* 6(2):99–110.

World Bank. 2006. "Enhancing Agricultural Innovation: How to Go Beyond the Strengthening of Research Systems." Washington, DC.

———. 2010. "Designing and Implementing Agricultural Innovation Funds: Lessons from Competitive Research and Matching Grant Projects." Report No. 54857-GLB. Washington, DC.

———. 2011. Rural Alliances in Bolivia. A World Bank Case. http://web.worldbank.org/WBSITE/EXTERNAL/COUNTRIES/LACEXT/BOLIVIAEXTN/0,,contentMDK:22781102~pagePK:1497618~piPK:217854~theSitePK:322279,00.html, accessed July 2011.

Thematic Note 3

Alex, G., W. Zijp, and D. Byerlee. 2002. "Rural Extension and Advisory Services: New Directions." Rural Strategy Background Paper No. 9. Washington, DC: The World Bank.

BRAC (Bangladesh Rural Advancement Committee). 2010. Program information from www.brac.net, accessed July 2011.

Bhamoria, V. 2004. Tribal rehabilitation for livelihood enhancement: Experience of DHRUVA. Document produced for the IWMI-Tata Water Policy Programme, Anand.

DHRUVA (Dharampur Uththan Vahini). 2010. Program information from www.dhruva.org.in, accessed July 2011.

Echeverria, R. 2003. "Twenty Years of Reforming Extension in Latin America: Are We There Yet?" Presented at the Regional Workshop on Operationalising Reforms in Agricultural Extension in South Asia, New Delhi, May 6–8.

Farrington, J., I. Christoplos, A.D. Kidd, and M. Beckman. 2002. "Extension, Poverty, and Vulnerability: The Scope for Policy Reform: Final Report of a Study for the Neuchâtel Initiative." Working Paper No. 155. London: Overseas Development Institute (ODI).

Neuchâtel Group. 2002. "Common Framework on Financing Agricultural and Rural Extension." Lindau: Swiss Centre for Agricultural Extension and Rural Development.

Rivera, W.M., M.K. Quamar, and L.V. Crowder. 2001. "Agricultural and Rural Extension World Wide: Options for Institutional Reform in the Developing Countries." Rome: Extension, Education and Communication Service, Research, Extension and Training Division, Sustainable Development Department, Food and Agriculture Organization of the United Nations (FAO).

Sulaiman, R.V., and A.J. Hall. 2002. "An Innovation System Perspective on the Restructuring of Agricultural Extension: Evidence from India." *Outlook on Agriculture* 30(4):235–43.

———. 2004a. "Towards Extension-Plus: Opportunities and Challenges." Policy Brief No. 17. New Delhi: National Centre for Agricultural Economics and Policy Research (NCAP).

———. 2004b. "India: The Emergence of Extension-Plus: Future for Extension beyond Technology Transfer?" In W.M. Rivers and G. Alex (eds.), *Extension Reform for Rural Development*. Vol. 1. Washington, DC: World Bank. Pp. 19–29.

———. 2005. "Extension Policy at the National Level in Asia." *Plant Production Science* 8(3):308–19.

Swanson, B.E. 2006. "The Changing Role of Agricultural Extension in a Global Economy." *Journal of International Agricultural and Extension Education* 13(3):5–18.

Vaswani, L.K., V. Venkatakrishnan, R. Upadhyay, and J. Talati. 2003. "Agricultural Market Linkages: Evaluating and Evolving a Conceptual Framework in Indian Context." NABARD Occasional Paper 28. Mumbai: National Bank for Agriculture and Rural Development (NABARD).

VFPCK. 2009 "Progress Report, 2008-09." Vegetable and Fruit Promotion Council Kerala (VFPCK), Kochi.

XLRI (Xavier Labour Research Institute). 1999. "Report of the Impact Evaluation of Kerala Horticultural Development Programme." Bhubaneswar: Xavier Labour Research Institute.

Thematic Note 4

Alex, G., W. Zijp, and D. Byerlee. 2002. "Rural Extension and Advisory Services: New Directions." Rural Development Strategy Background Paper No. 9). Washington, DC: World Bank.

Anandajayasekeram, P., R. Puskur, and E. Zerfu. 2010. "Applying Innovation System Concept in Agricultural Research for Development: A Learning Module." Nairobi: International Livestock Research Institute (ILRI).

Berdegué, J., and G. Escobar. 2002. "Rural Diversity, Agricultural Innovation Policies, and Poverty Reduction." Agricultural Research and Extension Network Paper No. 122. London: Overseas Development Institute (ODI).

Burt, R. S. 2004. "Structural holes and good ideas." *American Journal of Sociology* 110(2):349–99.

Devaux, A., D. Horton, C. Velasco, G. Thiele, G. Lopez, T. Bernet, I. Reinoso, and M. Ordinola. 2009. "Collective Action for Market Chain Innovation in the Andes." *Food Policy* 34(1):31–38.

Devaux, A., J. Andrade-Piedra, D. Horton, M. Ordinola, G. Thiele, A. Thomann, and C. Velasco. 2010. "Brokering Innovation for Sustainable Development: The Papa Andina Case." ILAC Working Paper No. 12. Rome: Institutional Learning and Change (ILAC) Initiative.

Gêmo, H.R. 2006. *Recurcos humanos na extensão agrária pública em Mocambique (1987–2006). Estudos sobre investigação e extensão agrária.* Vol. 1. Maputo: Ministry of Agriculture.

Gildemacher, P.R., W. Kaguongo, O. Ortiz, A. Tesfaye, G. Woldegiorgis, W.W. Wagoire, M. Wakahiu, C. Leeuwis, and P.C. Struik. 2009. "Improving Potato Production in Kenya, Uganda, and Ethiopia: A System Diagnosis." *Potato Research* 52(2):173–205.

Granovetter, M. 1985. "Economic Action and Social Structure: The Problem of Embeddedness." *American Journal of Sociology* 91(3):481–510.

Gupta, A.K., R. Sinha, D. Koradia, R. Patel, M. Parmar, P. Rohit, H. Patel, K. Patel, V.S. Chand, T.J. James, A. Chandan, M. Patel, T.N. Prakash, P. Vivekanandan, and other members of Honey Bee Network. 2003. "Mobilizing Grassroots' Technological Innovations and Traditional Knowledge, Values, and Institutions: Articulating Social and Ethical Capital." *Futures* 35(9):975–87.

Hall, A., J. Clark, and G.C. Naik. 2007 "Technology Supply Chain or Innovation Capacity? Contrasting Experiences of Promoting Small-scale Irrigation Technology in South Asia." UNU-MERIT Working Paper No. 2007-014. Maastricht: United Nations University (UNU) and Maastricht Economic and Social Research and Training Centre on Innovation and Technology (MERIT).

Hartwich, F., V. Gottret, S. Babu, and J. Tola. 2007. "Building Public-private Partnerships for Agricultural Innovation in Latin America." IFPRI Discussion Paper No. 00699. Washington, DC: International Food Policy Research Institute (IFPRI).

Howells, J. 2006. "Intermediation and the Role of Intermediaries in Innovation." *Research Policy* 35(5):715–28.

Kibwika, P., A.E.J. Wals, and M.G. Nassuna-Musoke. 2009. "Competence Challenges of Demand-led Agricultural Research and Extension in Uganda." *Journal of Agricultural Education and Extension* 15(1):5–19.

Klerkx, L., N. Aarts, and C. Leeuwis. 2010. "Adaptive Management in Agricultural Innovation Systems: The Interactions between Innovation Networks and Their Environment." *Agricultural Systems* 103(6):390–400.

Klerkx, L., A. Hall, and C. Leeuwis. 2009. "Strengthening Agricultural Innovation Capacity: Are Innovation Brokers the Answer?" *International Journal of Agricultural Resources, Governance. and Ecology* 8(5/6):409–38.

Klerkx, L., and C. Leeuwis. 2008. "Matching Demand and Supply in the Agricultural Knowledge Infrastructure: Experiences with Innovation Intermediaries." *Food Policy* 33(3):260–76.

———. 2009. "The Emergence and Embedding of Innovation Brokers at Different Innovation System Levels: Insights from the Dutch Agricultural Sector." *Technological Forecasting and Social Change* 76(6):849–60.

Kristjanson, P., R.S. Reid, N. Dickson, W.C. Clark, D. Romney, R. Puskur, S. MacMillan, and D. Grace. 2009. "Linking International Agricultural Research Knowledge with Action for Sustainable Development." *Proceedings of the National Academy of Sciences* 9(13):5047–52.

Murthy, R. 2010. "India's Rural Inventors Drive Change." *Asia Times Online*, January 29, 2010, http://www.atimes.com/atimes/South_Asia/LA29Df03.html, accessed July 2011.

Obstfeld, D. 2005. "Social Networks, the Tertius Iungens Orientation, and Involvement in Innovation." *Administrative Science Quarterly* 50(1):100–30.

Pant, L.P., and H. Hambly-Odame. 2006. "Multi-stakeholder Deliberation on Dialectical Divides: An Operational Principle of the Systems of Innovation." *Knowledge Management for Development Journal* 2(3):60–74.

Rivera, W., and R. Sulaiman, V. 2009. "Extension: Object of Reform, Engine for Innovation." *Outlook on Agriculture* 38(3):267–73.

Smits, R., and S. Kuhlmann. 2004. "The Rise of Systemic Instruments in Innovation Policy." *International Journal of Foresight and Innovation Policy* 1(1/2):4–30.

Spielman, D.J., J. Ekboir, and K. Davis. 2009. "The Art and Science of Innovation Systems Inquiry: Applications to Sub-Saharan African Agriculture." *Technology in Society* 31(4):399–405.

World Bank. 2006. "Enhancing Agricultural Innovation: How to Go Beyond the Strengthening of Research Systems." Washington, DC.

Innovative Activity Profile I

IFDC (International Fertilizer Development Center). 2011. "Improved Livelihood for Sidr-Affected Rice Farmers (ILSAFARM) Project." Final Report, submitted to the USAID-Bangladesh. Muscle Shoals.

Thompson, T.P. 2003. A perspective on the IFDC strategy for agribusiness development. Unpublished paper, International Fertilizer Development Center (IFDC), Muscle Shoals.

———. 2005. "Agricultural Investment Note: Promoting Private Sector Fertilizer Markets." Washington, DC: World Bank.

USAID (United States Agency for International Development). 1996. "Privatizing Fertilizer Distribution: Bangladesh Case Study." USAID Evaluation Highlights No. 54. Washington, DC. http://pdf.usaid.gov/pdf_docs/PNABS524.pdf, accessed July 2011.

Innovative Activity Profile 2

These sources can be accessed through http://www.infobridge.org/ffsnet/; if not uploaded a single copy can be obtained from support@farmerfieldschool.net

Braun, A. 2006. Farmer Field School networks in Western Kenya: Evolution of activities and farmers' priorities. Unpublished report, FAO, Rome.

Braun, A., J.R. Okoth, H. Khaamala, and G.S. Khisa, 2007. "Building FFS Networks in East Africa." *LEISA Magazine* 23–1:18–19.

Braun, A.R., and D. Duveskog. 2009. The Farmer Field School approach: History, global assessment, and success stories. Unpublished report for the IFAD *Rural Poverty Report 2009*, International Fund for Agricultural Development (IFAD), Rome.

Davis, K., E. Nkonya, E. Kato, D. A. Mekonnen, M. Odendo, R. Miiro, and J. Nkuba. 2010. "Impact of Farmer Field Schools on Agricultural Productivity and Poverty in East Africa." Discussion Paper No. 00992. Washington, DC: International Food Policy Research Institute (IFPRI). http://www.ifpri.org/sites/default/files/publications/ifpridp00992.pdf, accessed July 2011.

FAO (Food and Agriculture Organization), 2006. Farmer Field School networks operational manual. Unpublished, FAO–Uganda, Kampala.

Gallagher, K., 2001. Self-financed field schools: Helping farmers go back to school in IPM/IPPM. Unpublished, http://www.share4dev.info/ffsnet/documents/3201.pdf, accessed July 2011.

———. 2002. "Self-financing Access to New Technologies: East African Farmer Innovations." SDI 7-23/1. SDI/UGF

World Summit 2002. Share4Dev, http://www.share4dev .info/kb/documents/3202.pdf, accessed July 2011.

Khisa, G., 2006. First Regional Meeting Report: Expansion of Farmer Field School Programme in Eastern and Southern Africa. Unpublished.

Khisa, G., and E. Heinemann. 2005. "Farmer Empowerment through Farmer Field Schools." In "Bright Spots Demonstrate Community Successes in African Agriculture," edited by F.W.T. Penning de Vries. Working Paper No. 102. Colombo: International Water Management Institute (IWMI). Pp. 71–83.

KIT (Royal Tropical Institute), Faida Mali, and IIRR (International Institute of Rural Reconstruction). 2006. "Chain Empowerment: Supporting African Farmers to Develop Markets." Amsterdam, Arusha, and Nairobi.

Okoth, J.R., G. Khisa, and J. Thomas, 2002. "Towards a Holistic Farmer Field School Approach for East Africa." *LEISA Magazine* 18–3.

———. 2003. "Towards Self-financed Farmer Field Schools." *LEISA Magazine* 19–1: 28-29.

Okoth, J., A.R. Braun, R. Delve, H. Khamaala, G. Khisa, and J. Thomas. 2006. "The Emergence of Farmer Field Schools Networks in Eastern Africa." Paper presented at the CAPRi Research Workshop on Collective Action and Market Access for Smallholders, 2–5 October, Cali.

Innovative Activity Profile 3

Barrantes, R., C. Trivelli, R. Morales S, and J.J. Miranda. 2004. "Análisis económico, social y financiero de la inversiones en innovación y evaluación ex ante de los retornos del proyecto INCAGRO." Lima: Innovación y Competitividad para el Agro Peruano (INCAGRO).

Benites, J.R., and H. Wiener. 2008. "INCAGRO: Converting ideas into values." Lima: Innovación y Competitividad para el Agro Peruano (INCAGRO).

Christoplos, I. 2010. *Mobilizing the Potential of Rural and Agricultural Extension.* Rome: Food and Agriculture Organization of the United Nations (FAO) and the Global Forum for Rural Advisory Services (GFRAS).

INCAGRO (Innovación y Competitividad para el Agro Peruano). http://www.incagro.gob.pe/WebIncagro/inicio.do, accessed May 2010.

Ministry of Agriculture. 2009. "Evaluación de impacto del INCAGRO: Proyecto de Investigación y Extensión Agrícola (PIEA)." Lima: Ministry of Agriculture.

Ortiz, O. 2006. "Evolution of Agricultural Extension and Information Dissemination in Peru: An Historical Perspective Focusing on Potato-related Pest Control." *Agriculture and Human Values* 23(4):477–89.

Swanson, B. 2008. *Global Review of Good Agricultural Extension and Advisory Service Practices.* Rome: Food and Agriculture Organization of the United Nations (FAO).

World Bank. 2005. "Implementation Completion Report on a Loan in the Amount of US$ 9.6 Million to the Republic of Peru for an Agricultural Research and Extension Project." Washington, DC.

———. 2006. "Enhancing Agricultural Innovation: How to Go Beyond the Strengthening of Research Systems." Washington, DC.

———. 2009. "Project Performance Assessment Report: Republic of Peru, Agriculture Research and Extension Project." Washington, DC.

Innovative Activity Profile 4

Vijay Mahajan, and K. Vasumathi. 2010. "Combining Extension Services with Agricultural Credit: The Experience of BASIX India." IFPRI Brief. Washington, DC: International Food Policy Research Institute (IFPRI). http://www.ifpri.org/sites/default/files/publications/focus 18_13.pdf, accessed March 2011.

Mishra, B.S. 2008. "Contract Farming for Potato: An Attempt to Include Poor Farmers in the Value Chain." *Enterprise Development and Microfinance* 19 (4): 331–43.

Agricultural Research within an Agricultural Innovation System

OVERVIEW

John Lynam, Consultant

EXECUTIVE SUMMARY

Investing in agricultural research within an AIS framework complements the traditional internal focus on capacity and research priorities with an external emphasis on better articulation of client demand and effective institutional partnerships. Agricultural research as a producer of new knowledge requires effective institutional arrangements to apply that knowledge. The types of organizations and nature of these partnerships in the generation of innovation will depend on the market orientation of the agricultural sector and private investment in agro-industry. In urban and transforming economies, these institutional partnerships will tend to focus on research linkages to agricultural input or processing industries, often within the frame of public-private partnerships, including technology transfer arrangements, and often facilitated by public financing arrangements. Such research linkages to the private sector and other actors will tend to be organized around clusters, and financing will often be in the form of competitive grants with cofinancing from the private sector.

In agrarian economies, on the other hand, external connectivity of research is primarily through bridging organizations, particularly extension services, farmer associations, trade associations, and NGOs, and farmer demand is articulated through nonmarket mechanisms with farmer representation. The latter tend to involve novel organizational arrangements, such as farmer councils and innovation platforms, new methodologies, organizational change within research institutes, and financing arrangements that support the increased transactions costs inherent in improved external connectivity. Farmer participation in the codesign of innovations is characteristic of these organizational arrangements, and it may be facilitated by innovation brokers. Financing is almost solely based on public sources and will tend to be organized around research foundations or agricultural research councils. There is an inherent tendency for research within an AIS to focus on market-driven applications, often within a value chain framework, and particular strategies are required to ensure that research continues to contribute to the reduction of rural poverty.

RATIONALE FOR INVESTMENT

As the globe enters a period of increasing constraints on land, water, and nutrient supplies, a tight balance between food supply and demand, and the certainty of climate change, new knowledge from agricultural research systems will be essential to maintain growth in agricultural productivity and in world food supplies. Locating agricultural research within an AIS is a means of heightening the performance of research systems through improved articulation with demand, more effective, better-differentiated institutional partnerships, and better market integration.

An innovation systems framework adds a set of new dimensions to the investment in agricultural science and technology. An AIS framework focuses attention on: (1) an expanded range of technologies (particularly postharvest and mechanical) provided by a differentiated set of suppliers; (2) demand responsiveness, particularly better connectivity and interaction of agricultural research with actors beyond farmers; and (3) adaptation to and facilitation of organizational innovations in credit, markets, insurance, farmer groups, and extension services.

The reframing of technological innovation coincides with the emphasis on market-led approaches for smallholder development, which have emerged in the wake of structural adjustment and market liberalization in the 1990s. Orienting research to markets, often through work in specific value chains, has become a principal vehicle for delivering new technologies and for combining them with the organizational and institutional innovations that so often accompany technical change in the agricultural sector.

In this sense, an AIS approach represents a relatively evolutionary form of institutional change in agricultural research, with a particular focus on enhancing the research system's external responsiveness. Yet improved responsiveness in the short term must be balanced with the investment strategies needed over the long term for a research system to be productive. In the process of balancing these short- and long-term imperatives, research organizations will arrive at a better alignment between internal research capacities and external partnerships and consortiums, increasingly with the private sector.

PAST EXPERIENCE IN ORGANIZATIONAL CHANGE IN AGRICULTURAL RESEARCH

Investment strategies for agricultural research over the past three decades have gone through a series of approaches, often requiring major organizational restructuring. Research restructuring has been much more pervasive in small countries than in large, which has given systems such as EMBRAPA in Brazil continuity in addition to significant levels of investment. The sections that follow describe these various approaches to investing in agricultural research.

Building national agricultural research institutes

Broadly defined, the 1980s and early 1990s were the period of the national agricultural research institute (NARI), when research units distributed throughout ministries of agriculture and other ministries were brought under a single, independent administrative structure. The assumptions ruling this restructuring were that economies of scale and scope could be achieved in agricultural research, budgetary resources allocated much more efficiently, and personnel policies freed from public civil service bureaucracy. Large countries such as Brazil and India developed a complex federal and state system of national research institutes, state agricultural universities, and state research institutes. Donor investment in agriculture was at an historical high, at least in percentage terms. The CGIAR network of international agricultural research centers expanded, and donors led by the World Bank and USAID funded programs that focused on training, infrastructure development, and program formulation within the newly formed NARIs.

From national research institutes to research systems

The dominance of the NARI gave way quite quickly in the late 1990s to reform based on the development of more pluralistic, decentralized systems, in which research funding and execution were separated, often through a competitive grants modality. The intent was to move away from reliance on a single research institution and toward the development of a broader-based national agricultural research system (NARS). At the same time, support for agriculture in aid budgets was declining, and domestic fiscal budgets came under pressure from structural adjustment. The rapidly growing Asian economies could support agricultural research from expanding tax revenues, and in Latin America market liberalization allowed greater participation of the private sector. In Africa, however, the reform of NARIs took place amid severely restricted budgets, often within a small-country context. The World Bank remained virtually the only donor investing in national agricultural research. Selective investment caused a few relatively strong NARSs to develop, leaving a majority of systems with limited capacities.

Decentralization and participatory research

This period also saw the rise of participatory research and the recognition that NARIs had to become more responsive to demand. Improved responsiveness was the principal justification for reforms that decentralized management within NARIs and created autonomous research councils. The councils, which often had farmer representation, now controlled the funding decisions (World Bank 2006), but their effectiveness varied greatly. Many had a narrow representation of stakeholders, consisting primarily of ministerial representatives or researchers, and their research prioritization process was not necessarily consistent and rigorous. The councils often had little influence on the policy process and how research was conducted. Many did not separate funding allocation and implementation effectively (see module 1). Decentralization was already a feature of large federal systems, but it involved significant tradeoffs in small systems. Scale economies in areas such as plant breeding were sacrificed, and operational budgets were directed to more adaptive research.

Shifting funding

Shifting funding to competitive grants had different impacts depending on the capacity of the research system. In larger systems with strong capacity, competitive mechanisms were used to improve research quality. Larger research systems used competitive grants to provide funding based on scientific peer review. They particularly aimed to: focus scientists' efforts on high-priority research or new fields of expertise; improve the relevance and quality of agricultural research, extension, and training; promote research partnerships and leverage research resources; and help to develop a more efficient and pluralistic research system (World Bank 2010). In smaller research systems, where financial constraints already limited core capacity, competitive grants were often used as a mechanism for farmers to articulate their demands for research more clearly, and farmers would participate in the grant review process. Such mechanisms reinforced the shift to adaptive research.

The use of competitive grants tended to undermine long-term strategic planning, however. A new crop variety can take ten years or more to develop; so does a locally adapted conservation agriculture system or a system to manage animal disease. The longer-term nature of agricultural research has produced significant debate on the extent to which research systems can respond to demand by allocating resources (often based on contestable funding mechanisms) to applied research as opposed to allocating resources (core funding) through longer-term strategic planning in relation to priority needs in the agricultural sector.[1] According to World Bank (2009, xii–xiii), a review of four World Bank projects in Latin America that employed competitive grant schemes for agricultural research,

A principal lesson concerns the importance of strengthening the capacity of research organizations, not just financing

research. Competitive funds can be an important vehicle for research financing and have a strategic role to play in piloting new ways of working, or focusing research on new topics; but they are most likely to make a sound and lasting contribution when they complement a relatively strong public sector framework for research (in this respect, prospects were brighter to begin with in Brazil and Colombia than they were in Nicaragua and Peru). Public funding is essential for agricultural innovation systems and private funding complements rather than substitutes for higher levels of public funding.

In Latin America, small-country systems became much more reliant on research outputs from CGIAR centers and spillins from private sector sources globally. Some countries even terminated public agricultural research altogether. Efforts to building core research capacities in sub-Saharan Africa, Central America, and the smaller South American countries got lost during this period of restructuring. In Africa, core research capacities remain largely underdeveloped after three decades of experimentation. While African governments committed themselves to increasing investment in agriculture to 10 percent of the overall national budget, by 2009 only a few countries had met the target.[2]

Limited operational funding

Operational funds are vital to research that results in interaction with rural communities outside the research station (on-farm adaptive research, multilocational testing networks, participatory plant breeding, disease surveillance programs, and soil fertility trials, for example). Because soil and biotic constraints are more severe in farmers' fields, carrying out research in farmers' conditions increases the relevance of the results.

The lack of operational funds is the first most binding constraint on the productivity of agricultural research, and the shortage of funds was particularly binding in Africa, where it limited the demand articulation that decentralization was supposed to provide. Ensuring access to operational funds is essential for research performance, but where budgets are highly constrained, such funding is the first to be cut to assure salaries, station running costs, and maintenance of core resources such as germplasm banks. Productive scientists are usually those who obtain external funds at the expense of the integrity of the overall research program.

The small-country problem

There remained a pervasive sense, especially among bilateral and multilateral investors, that agricultural research in the public sector was not meeting the performance standards that were expected. This perception tended to apply to research systems in smaller countries with little private sector capacity in agricultural research and with systemic market constraints.

Agricultural research in this context faces what is best termed a "small-country problem," in which limited market size, constraints on achieving economies of scope and scale, constrained fiscal budgets, and ineffective farmer demand for new technologies significantly limit the productivity of research. Research capacity is needed even to borrow technology, and in some conditions, technology can rarely even be borrowed. In sub-Saharan Africa, the potential for international spillins is very often limited because of the crops that are grown (they are not widely grown elsewhere) and the particular constraints on farmers' productivity (Pardey et al. 2007). Improving institutional performance under such circumstances is difficult, but the justification for investing in research in agrarian economies remains very strong, because agriculture remains the engine of growth for the overall economy.

Support to subregional research

One approach to the small-country problem, particularly in Africa, was to organize agricultural research at the subregional level to achieve scale economies and organize spillins efficiently. Since the late 1990s, many donors have shifted funding into subregional research organizations and the regional apex body (the Forum for Agricultural Research in Africa) and away from NARSs (see TN 3 on regional research). Initially subregional research organizations became a mechanism for coordinating regional research undertaken by CGIAR centers, but since the mid-2000s they have been a mechanism to "retail" research grants to national programs. The lost connection with international research centers left no framework to develop scale economies in these regional approaches, although two large projects funded by the World Bank (the East Africa Agricultural Productivity Project and West Africa Agricultural Productivity Project) were partially designed with that goal in mind. In Africa, the centralization represented by subregional organizations further diminishes farmers' ability to articulate demand (Sumberg 2005) in a context where the prospect of spillins is quite small, agroecologies extremely diverse, and the commodity structure of the food system very heterogeneous (Pardey et al. 2007).

Key lessons from reforms

Possibly the most important lessons drawn from attempts at restructuring agricultural research are:

- One size does not fit all. Context and path dependence matter in the design of an agricultural research system.
- The productivity of agricultural research differed significantly between large and small countries and between countries with well-functioning market economies and those without.
- Some period of consolidation was necessary for researchers to adapt to reorganization.
- Leadership was a crucial factor in NARIs that performed well.
- The lack of sustainable funding and weak capacity continued to limit the performance of agricultural research institutes in small countries.

The shift in investment in agricultural research from a focus on NARIs to a focus on NARSs and subregional research organizations took place as markets were liberalized, civil society expanded, and collective action increased in rural economies. To a significant degree, these developments were preconditions for investing in agricultural research within an AIS framework. One framework for needs assessment for agricultural research systems argues to (1) get the resources right, (2) get the priorities right, (3) get the linkages right, and (4) get the incentives right (Howard Elliott, personal communication). In an AIS, the investment framework shifts to focus on linkages and incentives and on identifying where further organizational change is oriented to external responsiveness. Whether this reorientation at this early period in the development of AISs can produce self-correcting change in both resource and capacity constraints is still largely untested, at least for research systems in small countries. The rest of this overview will focus on the evolving practice of undertaking agricultural research within an AIS framework.

RESEARCH WITHIN AN AIS

Knowledge and information are the engines of an AIS, and a market economy provides the incentives to search for improved products and processes that lead to overall gains in productivity within the agricultural economy. In many (if not most) agricultural economies in the developing world, however, resources, capacity, and market constraints retard the development and functionality of innovation systems for agriculture. The *World Development Report 2008* (World Bank 2007) differentiates between agrarian, transforming, and urbanized economies:

- *Agrarian economies* are almost always relatively small, mostly in sub-Saharan Africa, depend primarily on staple food crop production, and rely on agriculture for their economic growth. Their agricultural markets are not well integrated, transport and logistics are costly, and private investment in rural areas is still limited.
- *Transforming economies,* including China and India, are mostly in Asia, where economic growth is now led by the industrial sector and the economy is rapidly urbanizing. Growing urban demand, especially for higher-value products such as livestock and horticultural crops, is resulting in structural shifts in the agricultural economy. Even so, large areas of the rural economy still have high poverty rates and are not integrated into the growth process.
- *Urbanized economies* have most of their population in urban areas, are primarily located in Latin America, and the agricultural sectors are well integrated into global markets. Poverty is principally an urban problem.

Compared to the agrarian economies, transforming and urbanized economies invariably have better developed transport infrastructure and agricultural markets, a larger and more vibrant private sector, deeper R&D capacity, more effective agricultural institutions, and greater investment in ICT. These in turn are some of the preconditions for more functional innovation systems, particularly systems driven by expanding opportunities in more dynamic agricultural markets.

The sections that follow describe the contrasting roles of research and AISs in well-functioning market contexts and underdeveloped market contexts. A discussion of AIS approaches used to promote technological innovation is followed by an overview of key policy issues related to agricultural research in an AIS context, evolving areas of investment, and approaches for monitoring, evaluating, and scaling up agricultural research within an AIS.

Well-functioning markets and AISs

The functionality of an AIS rests on increasing connectivity within a widening organizational "matrix" in the agricultural sector. A growing private sector, increasing commoditization, and expanding market opportunities lead to an increasing array of organizations to promote their interests, most often in relation to government policy but also in relation to establishing norms of operation within their respective subsectors.

The proliferation of formal seed, chemical, and fertilizer associations, agroprocessing associations (for example, the Thai Tapioca Trade Association), animal feed milling associations, and commodity organizations (in module 1, the overview as well as TNs 3 and 4 provide examples) (from wheat and oilseed milling to horticultural exports) reflects the higher and more concentrated end of the value chain. In higher-value commodity chains, farmer organizations usually form around a particular commodity and often build off of cooperatives (dairy industries and the Colombian coffee federation are two examples). Such an organizational matrix balances competition with cooperation and organized collective action to further the interests of the subsector. Information flows are good: The subsector's needs are easily articulated, and appropriate institutional linkages and arrangements formulated, usually on a task basis.

At this stage of market development in the agricultural sector, an AIS is self-organizing. Public sector research has to be very responsive and flexible indeed, primarily to needs articulated through input-supply or agroprocessing firms, or it will quickly become an anachronism.

Innovation systems and the shift to a market context

As farmers integrate into the market economy, they rely more on inputs as a source of increased productivity and sell an increasing percentage of their production. Often they first diversify into higher-value crops and then specialize in particular production activities. Such intensification is facilitated by a widening array of innovations provided through markets for inputs and agricultural services. These innovations increasingly respond to changing urban demand, both for specific commodities and for specific quality characteristics in commodities.

Public agricultural research has continually wrestled with the issue of how to be more responsive to demand and, in a modernizing agricultural economy, how to balance farmers' needs with improved consumer acceptance. For example, in plant breeding and seed systems, seed companies are the translation point in the seed value chain between farmers' production constraints (how to produce a seed that farmers will buy) and consumers' quality requirements (if consumers and marketing agents want the product, then their preference feeds back to farmers in the form of a price premium for the commodity/variety). Plant breeding innovation by seed companies bridges the interests of producers and consumers, but only within a functioning, competitive seed market.

Private research capacity develops principally where agricultural markets function well, as in Latin America and Asia, especially in the large countries with their large markets. As agricultural economies modernize and the private sector becomes more active in funding its own agricultural research, innovation turns more to the application of frontier science, and public research tends to support private companies by developing new products (hybrid rice, for example) or supporting private sector research.

In agriculture, molecular biology and genomics represent this kind of frontier science, which is often supported through competitive grant schemes and in which universities often have a comparative advantage. For example, India's National Agricultural Innovation Project (NAIP) has a competitive grant scheme that funds innovation clusters around more basic research with potential applications of interest to the private sector. In Thailand, similar efforts are led by the Ministry of Science and Technology through its National Innovation Agency and BIOTEC program; they also focus on funding clusters of research and related applications. In addition, BIOTEC has set up two independent research programs, the Rice Gene Discovery Unit (applications of genomics and molecular biology in rice breeding) and the Cassava and Starch Technology Research Unit (molecular approaches to understanding starch synthesis and quality, with a particular focus on the Thai starch industry).

In these cases, public sector research is increasingly divorced from farmers as the primary clientele, relying instead on input markets as the mechanism for articulating farmer demand. Occasionally the interests of farmers and input companies do not coincide, however, as exemplified by the tensions surrounding pesticide use and the scaling up of integrated pest management programs in Asia.

Under these market-driven conditions, investments in public agricultural research tend to focus more on institutional innovations that reinforce the ties between research and the private sector. IPRs are emphasized, for example, often as much to ensure open access to publicly generated innovations as to protect innovations developed in the private sector. IPRs are often the basis for contractual arrangements in public-private partnerships. This connectivity can be reinforced by competitive grants that insist on public-private partnerships, brokers that can mediate between public research and subsector needs, science parks adjacent to research institutes that focus on areas of joint R&D, and venture capital funds that invest in developing products and markets based on research innovations.

Other areas of applied agricultural research are less well served by the private sector, however, and constitute more classical public goods, such as pulse and grain legume breeding, crop disease surveillance, development of forages for ruminants, and especially crop management and natural resource management research. Agricultural research institutes within dynamic agricultural sectors have to strike a balance between the more basic research that complements the private sector's interests and the more applied research that farmers need. This balance will become even more important with the increasing focus on using water and nutrients efficiently and reducing environmental externalities in production systems.

Even *transforming agricultural economies* have a role for public sector research in lagging rural areas with high poverty rates, usually associated with underdeveloped markets. External connectivity in these cases usually focuses on bridging organizations, particularly NGOs and extension services. In these cases, the AIS essentially reduces to the traditional partnerships between farmers, research, and bridging organizations, but those organizations provide a range of services beyond advisory services, including savings and credit schemes, farmer mobilization, and improved market access.

India's NAIP is an example of a funding program that stratifies its platforms or clusters based on relative market development and associated rural poverty. In more commercialized areas, NAIP's platforms involve public-private partnerships, and in lagging areas they involve traditional research and bridging organizations.

For smaller research institutes, this kind of stratification creates a dilemma. Should they focus on the more commercial areas and associated partnerships or focus on the lagging areas? The potential for innovation will be higher in the commercial areas, but the public interest may reside with the lagging areas. The tendency within an AIS will be toward the former, whereas the public role will in most instances lie in the latter.

Research within underdeveloped market contexts

The more agrarian economies, particularly those in sub-Saharan Africa, can be characterized as primarily dependent on smallholder, rainfed agriculture. Farmers face conditions of incomplete and unintegrated input and output markets, asymmetric information, and high transaction costs. Markets for insurance and credit are virtually nonexistent.

COMBINING TECHNICAL AND INSTITUTIONAL INNOVATIONS. For agricultural research to be effective under these conditions, technological innovations must usually be combined with organizational and institutional innovations, primarily to compensate for the lack of markets as an organizational impetus for innovations. Moreover, innovations tend to follow more orchestrated trajectories (Rajalahti, Janssen, and Pehu 2007)—in other words, the innovation process tends to be facilitated by external actors, and technical innovations are often integrated with organizational innovations.

For example, the deployment of improved sorghum varieties, microdosing of fertilizer, and pit technology for water harvesting in Sahelian countries becomes much more profitable for farmers to adopt if farmers form associations around warehouse receipt systems. In this way, technical innovations are combined with organizational innovations that compensate for incomplete input and credit markets and foster efficient bulking for output markets. Orchestrated trajectories are further facilitated by policy changes that legalize banks' acceptance of warehouse receipts, often initially with a loan guarantee program.

BRIDGING ORGANIZATIONS. Where markets are incomplete and unintegrated, bridging organizations—particularly extension services, farmer associations, trade associations, and NGOs—link the research and knowledge domain with the production and emerging market domain. These organizations deliver and adapt research products as well as develop supporting organizational innovations that provide greater access and efficiency in processing and marketing.

Bridging organizations are an imperfect mechanism for articulating consumer and farmer demand unless that demand is organized around a specific value chain. Value chains have become a dominant framework for orchestrated innovation platforms. Within an AIS, the emphasis on value chains will drive the organization of agricultural research back to a more centralized commodity approach. It will tend to drive the organization of farmer associations along similar lines. Where markets are already well developed, like markets for horticultural exports, organization along commodity lines is already evident. The Fresh Produce Exporters Association of Kenya (see module 1, box 1.5 in the overview) is but one example among many.

INCREASING PRIVATE SECTOR LINKAGES. In general, public agricultural research, especially within a NARI, is organized around core capacities involving some combination of plant breeding, disease and pest management, integrated crop management, soil and water management, livestock and fisheries, and potentially forest and rangeland management. Virtually all of these areas focus on improving land productivity and have farmers as their principal clients, either directly or through bridging organizations such as NGOs, farmer associations, and extension services.

The large change in organizing research within an AIS is the development of linkages between research and an emerging private sector. Research and bridging organizations are not well organized to effect such linkages (Larsen, Kim, and Theus 2009). In small agrarian economies, the private sector is not very prominent in rural areas. Large-scale processing tends to locate in major urban markets. Many nonfarm activities to generate income in rural areas are based in the household, such as brewing beer or processing root crops.[3] Haggblade (2009:A1-2) noted that "rural manufacturing remains limited across most of Africa . . . but, overall, local rural services, commercial and other business activity account for 80 percent of rural nonfarm earnings."

Innovations that have broad impact on the rural nonfarm economy thus tend to focus on organizational innovations that improve marketing efficiency or on small-scale processing where initial market conditions exist to specialize outside the household. The development of technology for small-scale processing relies primarily on mechanical innovations that usually come either from private industry, international borrowing, or occasionally (for specialized processing) from university engineering departments or industrial research institutes (Haggblade, Hazell, and Reardon 2007). The availability of specialized processing equipment depends, however, on the presence of local private capacity for manufacturing and distribution.

What then provides the basis for effective linkages between agricultural research and the private sector in agrarian economies? Strategies include vertical integration, brokering, and public-private partnerships.

In larger-scale processing, such as maize milling or feed production, a company's interest lies in procuring stable supplies of raw material of a certain quality at a competitive price. Where cost structures and margins permit, a company can vertically integrate across the value chain, as in export horticulture. Alternatively it can rely on imports, like the wheat milling industry of coastal West Africa. Under these conditions, there is no incentive for public-private partnerships.

If a company relies on domestic production supplied by market trading, however, it will be interested in combining higher farm productivity and lower costs with more efficient assembly, bulking, and marketing. An agroprocessing firm has little incentive to invest in increased farm-level productivity unless there is a quasi-monopoly on purchases of feedstock, such as for oil palm or sugarcane processing. Innovation platforms for staple food crops tend to focus on linking innovations in assembly and bulking by small-scale marketing agents with technologies for farmers to improve crop productivity. In value chains with significant returns to improved quality (for example, specialized coffee or dairy), coordination is usually needed from the farm to the final processing point. The mechanism for coordinating all of the actors in the value chain is usually provided by a specialized NGO with public funds (in other words, by a brokerage agency). The scope for public-private partnerships in agrarian economies is quite circumscribed, and enhanced farm productivity will generally continue to be the objective of partnerships or platforms.

Another principal sphere of interaction between research institutes and the private sector surrounds agricultural inputs. In this sphere, the potential interactions are more directly complementary. Public research provides varieties for seed companies, can target fertilizer blends and integrate them with organic sources, and can develop integrated pest management packages for safe horticultural production.

In crop breeding, the nature of public-private partnerships changes as competition and product development evolve in the seed market. Evolving private sector capacity in input markets, tied to responsive public research programs, can provide a direct channel for disseminating the products of public research. The public sector can focus its research on areas where input companies do not invest, which tend to be areas with high rates of poverty and ineffective farmer demand.

Technical innovation through AIS approaches

An evolving AIS has an increasing capacity to innovate, and agricultural research plays a key role in technical innovation. Kline and Rosenberg (1986) observed that "contrary to much common wisdom, the initiating step in most innovations is not research, but rather design" (quoted in Sumberg 2005). In this sense, *applied* agricultural research is probably a misnomer, as its central focus is on technology *design*, whether it is the design of an improved variety, an IPM system, a system to manage animal disease, or an agroforestry system. In such instances, the design process focuses to the extent possible on ensuring that farmers will be able to use the final product, principally to improve farm productivity. Ensuring that technical design is congruent with farmer utilization is a central tenet of farmer participatory research (including participatory plant breeding) and its antecedent, farming systems research, as the design process often entails iterations with farmers in testing and design modification—or rather codesign (see TN 4).

INNOVATION PLATFORMS. In practice, approaches to technical innovation within an AIS often take the form of project-based innovation platforms (TN 1; see also module 1, TN 2). Projects usually operate for a limited time; unsurprisingly they tend to focus on quick solutions to technical problems identified within the platform. This time frame orients research institutions to adapt existing knowledge and technology. Nevertheless, research institutes must balance this increasing demand for capacity to conduct shorter-term adaptive research with the longer-term research capacity required to expand the set of technical options. Farming systems research units were often successful locally but had difficulty achieving any significant scale in the adoption of adaptive research results. A value chain framework, which usually focuses on applied research, expands the specification of the problem and usually integrates technical innovation with institutional innovations in farmer organization and marketing to ensure that results are used throughout the value chain. Yet in many cases it is difficult to scale up results that are specific to particular value chains and contexts, as seen with warehouse receipt systems in underdeveloped grain markets such as those in Sahelian countries.

A research institute can also foster the development of innovation platforms by establishing a specialized unit or through funding arrangements such as those in Latin American research foundations or competitive grant programs. Through these units or funding arrangements, research institutes respond to some form of "articulated" demand, which may come only from a limited proportion of the farming population. Alternatively, the research institute may take a more proactive role and initiate an innovation platform itself, especially if it believes it has a technical innovation with a good chance of success and that an innovation platform is an effective means of ensuring its use. This approach can be described as "linear," but it is only another form of applying existing knowledge.

SCALE OF INNOVATION PLATFORMS. The scale of innovation platforms has not been much explored (Hall, Dijkman, and

Sulaiman V. 2010). In Africa, innovation platforms often operate at the district level, where the constraints lie in the heterogeneity of production systems and the location in relation to markets. In value chain platforms, scale is determined by the problem, market size, coordination requirements, and structure of the value chain. In Rwanda, for example, the conversion of the coffee subsector to high-value, specialty blends was done at a national, subsectoral level. In comparison, the development of organic horticultural value chains is often organized at the district level or lower. Scale is often a key design criterion in the development of a value chain platform. It determines whether the platform operates at the national, regional, or district level, whether it involves individual versus institutional partnerships, and whether policy change is a key element in the platform's activities.

Platforms that are not structured around value chains and that have poverty or food security objectives tend to approximate rural development programs and rely on public sector or civil society participants. Introducing an innovation perspective into what are essentially facilitation and service delivery platforms requires methodologies that are more participatory, focus more on the problem, and allow greater scope for experimentation. These requirements often limit the scale of operations and introduce higher implementation costs, however. Smaller-scale innovation platforms may be required, which will feed into and support broader rural development programs.

INNOVATION PLATFORMS WITH POVERTY OBJECTIVES. There is a tendency in an AIS to focus primarily on value chain approaches, as for example with the Kenya Agricultural Productivity Program (KAPP). After Kenya liberalized its markets in the 1990s and the private sector became increasingly active, Kenya generated successful market-led approaches to develop smallholder tea, horticulture, and dairy production as well as a fertilizer market. Despite these achievements, rural poverty remains very high in marginal agricultural areas and densely populated areas around Lake Victoria. Most smallholders are still net buyers of maize, the principal staple—only 18 percent of maize producers are net sellers and only 2 percent of producers account for 50 percent of sales (Jayne, Mather, and Mghenyi 2006). The Kenya Agricultural Research Institute, like many small-country agricultural research institutes, faces the dual challenge of innovating through market-led approaches while attempting to expand market participation or minimally improve household food security for the majority of smallholders.

As discussed, in India, NAIP (IAP 2) has managed the trade-off between market-driven innovation and innovation for lagging areas by creating two funding streams to support different types of innovation clusters. In lagging areas, funding supports the transaction costs involved in establishing innovation platforms and organizing partnerships. These platforms are bound by the duration of the grant, however, and either the problem focus has a short-term horizon, the results can be seen as a pilot for which larger-scale funding would be sought, or the funding is structured so that follow-up grants would be awarded to promising longer-term problem areas. However, a NARI such as the one in Kenya must internalize these trade-offs within the frame of its own research programming, the number of partnerships it can manage, the duration of the partnerships, and in the end the public good objectives that drive these decisions. In other words, it must prioritize whether to invest in more scientific capacity or collaborative capacity.

TAKING NOTE OF LONG-TERM NEEDS. Planning for longer-term needs in regard to technical innovation tends to take place in well-developed market economies and tends to focus on capacity needs in relation to basic and strategic research. At this stage in the development of the agricultural economy, there is a strong sense that new science will be a principal engine of continued agricultural growth and competitiveness in world markets. Scenario planning is often used to gauge alternative futures in relation to the evolving structure of the agricultural economy and to chart longer-term investment needs in agricultural R&D.

In developing countries, the private sector does not invest in scenario planning, although multinationals such as Monsanto do use such approaches, and a clearer division of labor emerges between public and private R&D. In developing basic research consortiums, NAIP used foresight scenario planning (see module 7, TN 3) to define longer-term needs. Tapping the potential of genomics and molecular biology as well as the potential of bioenergy often becomes a key focus in supporting continuing structural change in the overall economy.

KEY POLICY ISSUES

The key policy issues surrounding agricultural research in an innovation system range from the very broad (the need for an environment conducive to organizational interaction) to the very specific (the need for particular policy instruments that reduce the transaction costs of organizational collaboration). Policies that promote decentralization and democratization can also have a positive effect on rural innovation.

Incentives for organizational interaction

Within AISs, framework policies provide the broad incentives for increased organizational interaction in support of agricultural innovation: market policies, administrative policies, and financing mechanisms. Policies in general should enhance collaboration, collective action, and what might be termed "functional sufficiency" in innovation—the participation of a diversity of actors that bring a sufficient set of skills to the innovation process. As noted, market liberalization policies have been a critical driver of investments in the agricultural sector, market participation, and growth linkages in the rural economy.

With the trend toward market liberalization, government's role has shifted to improving marketing efficiency and defining standards for market participation, particularly the development of grades and standards in commodity markets and quality and safety standards in input markets. Seed certification, varietal testing and release, biosafety regulation, plant breeders' rights, and IPRs (see module 6) all set the rules under which product innovation takes place. They define the rules of competition (for example, between seed companies) and provide incentives for collaboration (for example, between integrated pest management firms and horticultural producers). The regulatory environment in turn provides incentives for the private sector to increase its investment in research, which often initially provokes competition with public research and then evolves into more collaborative modalities. The tension then becomes whether public research, especially if jointly funded by the public and private sector, is done under exclusive rights to particular companies or whether the research supports the whole agricultural sector. In general, public research turns its attention to more upstream research that has longer time horizons and addresses issues of less importance to the private sector, such as natural resource management, poverty reduction, gender, and equity.

Financing for transaction costs inherent in AISs

Financing mechanisms are possibly the most effective policy instruments for increasing organizational connectivity (box 4.1). Aside from providing incentives, financing mechanisms are also critical for supporting the increased transaction costs inherent in developing organizational linkages and platforms. Where benefits to collaboration are clear, particularly for the private sector, such transaction costs are a necessary component of investments in the innovation process. For public sector research, however, financing these costs requires explicit budgeting categories and clarity on how these expenditures will support institutional outcomes. To overcome the inherent risk and uncertainty and cover transaction costs for participation, special agricultural innovation funds (module 5, TN 2) have been developed (World Bank 2010).

Decentralization and equity issues

The trend toward democratization is often reflected in greater administrative and financial decentralization, in which the delivery of public services is managed at the district level and the public agencies involved become more accountable to local constituents, including women and men farmers. Decentralization has had a major impact on the organization of advisory services, animal and plant health services, and the development of infrastructure, especially for water and rural roads. Democratization is also accompanied by an expansion in civil society organizations, including the rebuilding of farmer organizations. In many respects, these processes are fundamental to creating capacity for rural innovation. In areas where local government is particularly responsive, they can facilitate innovations such as the Land Care movement in the Philippines.

EVOLVING AREAS OF INVESTMENT IN RESEARCH WITHIN AN AIS

How will research investments continue to evolve in an AIS context? Investments are expected to continue supporting the wider connections that lead to innovation. They will encompass new forms of collaboration and institutional structures as well as new technologies and systems that increase the flow of information among actors in an innovation system.

In some instances, greater collaboration and communication between actors in an innovation system will change where agricultural research occurs, who conducts and funds it, and its priorities. In Africa, however, public research agencies will be even more challenged to respond to the proliferation of small-scale innovation processes. Research investments will also be critical for counterbalancing the heavy market orientation of AISs and promoting greater inclusiveness and equity in innovation processes.

Box 4.1   Financing Agricultural Research and Innovation

Agricultural research in developing countries is characterized by significant underinvestment, especially in Africa. It remains very much a public activity that produces public goods. Government remains the largest contributor to public agricultural research, accounting for an average of 81 percent of funding. Internally generated funds, including contracts with private and public enterprises, account for only 7 percent. Because continuity of financing is so critical to the productivity of agricultural research, new approaches to funding have been piloted over the past two decades. As the precursors of funding strategies for AISs, they offer insight into the potential challenges involved.

The new approaches shared two fundamental characteristics. They separated funding decisions from the execution of research and they expanded the sources and sustainability of financing.

To ensure accountability and a research agenda that met users' needs, representatives of the private sector, farmers, and the public sector decided which research to fund, primarily through competitive grants. This separation required new organizational arrangements to manage the increased transaction costs, most often in the form of a research foundation, a national agricultural research council, or a government agency that managed competitive grants (see module 6, TN 2). Experience with research foundations yielded lessons on ensuring clients had a voice in funding decisions and on developing sufficient capacity to manage the funding within a strategic framework.

Research foundations and competitive grant funds have focused on funding projects that foster critical partnerships in the AIS, usually between public scientific institutes and the private sector. For example, Chile

organized clusters in which firms' market, processing, and management expertise were matched with expertise from public research institutes (IAP 3). Three government agencies, each with a slightly different mandate, managed the funding. In India a similar program was managed by a program office organized within a World Bank loan (IAP 2). The primary funding mechanism was competitive grants, awarded to specific types of partnerships, often with explicit contractual terms.

To diversify and sustain funding, the beneficiaries of the research are increasingly required to contribute. In Kenya, for example, levies have funded research on export crops (tea and coffee). In Uruguay, producer associations invested directly in national commodity research programs. More generally, tax incentives have encouraged companies to invest in R&D by hiring their own researchers or contracting with public agencies. In all of these arrangements, the research supported is defined much more specifically to reflect users' demands, ensure that appropriate products are developed, and promote accountability.

Such funding sources are still a small component of overall funding for agricultural research, however. Large parts of the agricultural sector still rely almost entirely on public funds to meet their needs, especially for plant breeding research in staple food crops and natural resource management research.

The *key lesson* is that a strong market and commercial orientation, if not bias, appears to exist in financing arrangements that move away from financing research and toward financing innovation within an AIS. A primary challenge is to ensure that research and innovation to generate public goods will be adequately supported.

Sources: Agricultural Science and Technology Indicators (http://www.asti.cgiar.org); Byrnes and Corning 1993.

Increasing connectivity between research and other innovation actors

In an AIS, research is chiefly oriented toward integration with the rest of the innovation actors, be they private, public, or civil society entities. In practice, most efforts have focused on interactions between the public and private sectors. Partnerships between public and private agencies involve a range of contractual arrangements, from informal to formal, under which research institutes provide products

or services to the private sector. For a public research agency, the degree of exclusivity in the use of its particular product or service, which is often under IP protection, often defines the contractual arrangement (see TN 2 in this module and module 5, TN 1).

Science parks create a useful nexus between commercial enterprises and research institutes by taking promising research products to market and providing backstopping in product modification. They function best where private investment capital and industrial engineering expertise coexist (see the overview in module 5).

The development of new markets for agricultural products often calls upon public expertise in agronomic and breeding research and relies on services ranging from brokering contracts and partnerships and assembling investment capital to developing capacity in the private sector. The costs of research for these niche markets often cannot be justified for public research institutes with their broad public good mandates. This issue is explored further in module 3, TN 4.

Value chain platforms, the most informal approach to linking publicly funded research and the private sector, seem to function best when coordination is needed to produce a specialized product like high-quality coffee. They require external funding and facilitation, and their effectiveness in generating innovation for a range of commodities and market conditions remains to be tested (see module 1, TN 2 and module 5, TN 4).

Enhancing access to information and communication

Enhanced information flows will improve the integration of an AIS. The facilitating role of ICTs is vital for researchers and other stakeholders in an AIS involved in gathering and manipulating data or interacting with global information resources (see box 4.2) (see World Bank 2011).[4] Information flows among actors in an AIS in developing countries have always been particularly costly, often asymmetric, and generally incomplete. New ICTs could significantly improve access to information and the availability of communication and collaboration tools, although access to information is only one component of the innovation process in agricultural systems. The design of technology in agricultural research oriented to increased farm productivity depends on a two-way flow of information between farmers and researchers. Farm-level innovation requires contextualized information (a particularly difficult problem, especially where farmers' level of education is limited), access to research products, and a significant learning-by-doing process. Researchers in turn need to understand the heterogeneity of farming systems and the constraints on farm productivity to inform their technology design and testing systems and ensure that they respond to farmers' needs. Some emerging areas of investment are discussed in the sections that follow.

MOBILE PHONES AS AN INFORMATION EXCHANGE DEVICE. Mobile phones are becoming pervasive in some rural areas.

Box 4.2 ICTs Make Agricultural Research More Inclusive

ICTs are making agricultural research more inclusive and at the same time more focused on development goals, because they change how, where, and to whom information flows. Information can flow in many directions; it can be highly dispersed and accessible; it can also be highly targeted and location specific. ICTs are significant in the research process but may be even more significant as a catalyst throughout the wider innovation system, in: collecting, storing, and analyzing data, with or without human interaction; geospatial applications; decision support and knowledge-based systems and robotics; embedded ICTs in farm equipment and processes (agrionics); connecting communities and enabling learning; collaboration with stakeholders across the research process; and the management of competitive innovation funds.

Sources: Agricultural Science and Technology Indicators (http://www.asti.cgiar.org); Byrnes and Corning 1993.

They are more than a mechanism for two-way flows of information. Kenyan farmers use them to obtain market prices, verify the certification of seed and fertilizer sellers, and obtain recommendations on which fertilizer and seed to choose. Equipped with GPS and cameras, mobile phones are becoming a very efficient means for researchers and farmers to collect farm-level information. For example, a network of sentinel farmers in the Great Lakes region of Africa monitors two cassava disease pandemics, the hybridized form of cassava mosaic virus and two species of cassava brown streak virus. A data template has been developed with the service provider, and farmers provide photographs of suspected new outbreaks. This effort could evolve into an interactive disease surveillance and control system. The potential of mobile phones for such interactive information flows between researchers and farmers will continue to evolve.

MARKET INTELLIGENCE UNITS. Within an AIS, research institutes will increasingly have to balance farmers' demands with those of the private sector, which will affect farmers indirectly through research that produces innovation elsewhere in the value chain. Within commercial agricultural economies, information flows are increasingly specific to the needs of each subsector. Market intelligence becomes an important public good in improving the efficiency of the market and directing investment in the subsector, including research investment. In countries such as Brazil, the Brazilian Institute of Geography and Statistics collects basic census and production data,[4] while agricultural economic research institutes collect information on prices, domestic trade volumes, agroprocessing output, and international markets and undertake analytical work in support of market development.

LEARNING ALLIANCES. Learning alliances provide a platform for networking often disparate R&D institutions around a common focal area involving program or project development and implementation. They focus especially on the learning associated with scaling up innovations, which usually involves the interaction of research, capacity building, project implementation, and evaluation. These platforms move beyond the scope of a traditional monitoring and evaluation (M&E) system and focus on broader learning objectives across institutions. Learning alliances are particularly intensive in the use of facilitation and information synthesis and require external funding to operate. See IAP 4.

Organizational change of research within an evolving AIS

Organizational change within a research system reflects increasing differentiation and specialization in the production of new knowledge, products, and services. This differentiation among R&D organizations is driven by improvements in their ability to respond to changing demands within the agricultural sector as organizational linkages within the sector improve (TNs 1 and 5; box 4.3). Closer interaction between publicly funded agricultural research and an expanding private sector and civil society has organizational implications for research because interaction shifts the locus of agricultural research, research priorities, and research financing.

This shift is most apparent in the rapidly growing economies of Asia, which are experiencing increasingly dynamic organizational change. Private investment in agricultural research is expanding rapidly, especially in plant breeding. Hybrid maize led the movement of seed companies into breeding, but they have since expanded into horticulture, hybrid rice, and plantation crops. In Asia and Latin America, public plant breeding has responded to these developments by either moving toward more prebreeding

and genomics research in support of the private sector or toward breeding for areas not covered by the private sector (box 4.4).

In sub-Saharan Africa the market and agricultural research context remains quite different, requiring NARIs to balance traditional research focusing on productivity with an emergent private sector that has almost no capacity to undertake research. Finding this balance, improving connectivity to the private sector to improve farmers' access to markets, and developing more effective linkages to bridging organizations that provide services to farmers are all on the agenda of how the NARI improves its connectivity within the AIS.

In many cases, approaches will vary depending on location and commodity, as research organizations have a role even in postconflict situations. In all cases, enhanced partnerships with a diversity of NGOs, farmer organizations, and civil society organizations are critical to effective performance.

With their persistent funding constraints, African NARIs face an increasing dilemma in organizing themselves to respond to an expanding range of innovation processes, which are often grouped around relatively micro market and development niches. As Hall, Dijkman, and Sulaiman V. (2010: 4) have argued, "Innovation diversity is central to research design, emphasizing that there is no optimal approach or way of organizing research into use for innovation and impact; rather it is context-specific and path-dependent." How African NARIs address the diversity question in their external linkages while organizing their limited internal resources around the biotic, natural resource, and impending climate change challenges facing African agriculture will be a persistent driver of organizational change within NARIs. Such change is best built on experimentation, piloting, and continuing enhancement of skills in developing external institutional arrangements, while evolving increased flexibility in designing research products and services.

Pro-poor innovation

The market orientation of AIS raises a number of challenges with respect to pro-poor innovation; in many ways these challenges echo the debate over the equity impacts of the Green Revolution. First, the rural poor are primarily found in contexts characterized by poorly developed markets, either because of inadequate transport infrastructure, distance to markets, or low population density. The poor often reside in marginal areas bypassed by private investment

Box 4.3   The Consultative Group on International Agricultural Research and the AIS

The institutional role of the 15 centers of the Consultative Group on International Agricultural Research (CGIAR) in agricultural innovation has been a source of debate and has changed significantly over time. The issue is most often framed in terms of where the centers operate within the research-to-development spectrum.

- For the first two to three decades, the centers' role was defined in terms of a division of labor with the NARIs (the centers would mostly develop technologies and NARIs would refine and disseminate them). Significant investment in training and capacity building within NARI research programs were made to develop a technology pipeline.
- Farming systems and participatory research in the 1980s and 1990s expanded the institutional matrix within which the centers operated (and the number of centers themselves expanded). The focus shifted from pushing the supply of technology to understanding farmers' demand for research and conducting more work on natural resource management. Core funding shifted to competition for project funding. The shift heightened the emphasis on achieving development outcomes—but within a less strategic, more project-driven modality.
- The 2000s saw Centers consolidate their downstream research and focus methodological develop-

ment under frameworks such as integrated natural resource management, agricultural research for development (AR4D), and knowledge to action programs—all forerunners or embodiments of AIS. Demand articulation was embedded in each of these approaches, and organizational and institutional innovation were seen as critical complements to technical innovation. The institutional matrix within which the centers worked expanded again.

In its current incarnation, the CGIAR intends for its 15 centers to function more as a system than as autonomous centers. They will operate through multicenter CGIAR Research Programs (CRPs). Financing will return to longer-term core funding allocated by funders through a Fund Council that is legally separate from the centers; centers will be governed by a Consortium Board. CRPs will be managed within a results-oriented framework and evaluated with respect to their contributions to four system-level outcomes. The new arrangements incorporate elements of an AIS approach, with a focus on measurable results (which are reflected in the contractual arrangements between each CRP and the Fund Council). This results orientation will have to differentiate between innovation-induced rural change and structured implementation of development projects.

Source: Author.

Box 4.4   Examples of Public-Private Engagement in Prebreeding and Genomics

Thailand has invested in higher-end genomics research and molecular breeding, not in the Ministry of Agriculture and Cooperatives but rather in the Ministry of Science and Technology. This high-end science is organized in clusters with private companies.

In Indonesia and the Philippines, the private sector accounts for one-fifth of agricultural R&D, partly because of the plantation structure of significant parts of the agricultural economy. In turn, the Indonesian Research Institute for Estate Crops generates a significant portion of its budget from contract research and commercial seed sales.

Source: Beintema and Stads 2008.

The same trends are apparent in Latin America, although not to the degree of Indonesia and Philippines. Colombia, Uruguay, and Argentina fund research on coffee, rice, sugar, and oil palm from levies on commodity sales. Chile has a number of specific competitive funds, all of which require collaboration with industry. These trends suggest that agricultural research is moving away from direct involvement with farmers and that private companies are increasingly the intermediaries between researchers and farmers.

until rural wages start to increase dramatically in other areas. The potential for developing sustainable links to markets under such circumstances is often limited.

Second, where market development is possible, the rural poor often lack the resources to access markets. Ensuring the participation of women is even more challenging. Women's roles in the value chain and the services that support those roles may need to be differentiated. Organizational innovations may also be needed to ensure that women participate in and benefit from the formation of farmer groups.

Finally, agricultural research linked to development agencies often has a significant role in improving food security for more subsistence-oriented households, including food-based nutritional approaches (especially for households affected by AIDS), limiting the "hungry season," and improving the resilience of the farming system. An AIS framework requires a very different institutional mix and methodologies to support innovation in such contexts, potentially extending into learning alliances. For AIS approaches, the incentives for researchers tend to come from more commercial producers, especially when researchers are working with value chains. A pro-poor innovation process requires more orchestrated investment approaches and in the end much more experience and evaluation of how the innovation process can be sure to include the poor.

MONITORING, EVALUATION, AND SCALING UP

Monitoring the performance of agricultural research within an agricultural innovation framework presents a number of conceptual and implementation issues. Optimally an M&E system will function at a project level, at a research institute and research system level, and at the level of the overall AIS, and indicators developed at lower levels will aggregate to higher levels. Virtually all of the experience with M&E in innovation systems is at the project level, however.

Spielman and Kelemework (2009) have developed an AIS performance index and tested it in Ethiopia and Vietnam. They divided the AIS into organizational domains, one of which was the knowledge and education domain, which includes agricultural research. Performance in this domain was measured in terms of knowledge production, reflected primarily in scientific publications.

When innovation outcomes are defined at the level of the overall AIS, the performance measure for agricultural research shifts from farmers' adoption of technology to the role of agricultural research as a producer of knowledge.

Impact at the farm level is attributed to the AIS rather than the agricultural research institutes. In the overall AIS, the monitoring of agricultural research organizations shifts toward process-oriented assessment (Daane 2010), the characterization of systemic linkages between agricultural research and other domains within an AIS, and their functionality. Ragasa et al. (2010) used this approach in developing an M&E system for agricultural research in Nigeria.

At the project level in an AIS, M&E in many ways is more comprehensive and tends to be organized around value chains, innovation platforms, public-private partnership arrangements, or competitive grants. In such contexts, agricultural research will be only one of many organizational partners in the M&E framework. The focus will be on understanding and adapting processes, often within an action-research modality (an iterative process of diagnosis, planning, action, evaluation, and reflection). The M&E system will serve several functions, particularly (1) monitoring project progress; (2) learning and change; (3) collecting data for testing hypotheses, often in relation to scaling up project results; and (4) project management (Njuki et al. 2010). Data will be collected on a combination of quantitative and qualitative indicators, the latter often collected through a participatory M&E process with project participants.

Defining the scope of innovation outcomes is often not straightforward. The outcomes reflect the application of new knowledge, products, or services within a context of organizational or institutional innovations. Investors will want to know the impact of these innovation outcomes and whether the organizational innovations are more cost effective than current institutional arrangements.

In a few cases, as in the sub-Saharan Africa Challenge Program of the CGIAR, experimental and quasi-experimental methods have been employed to test the comparable efficiency gains from using innovation system methods over traditional research and extension approaches (FARA 2009). Such tests are very expensive, however, and in the end not as rich as adaptive approaches employing M&E data in progressively improving the efficiency of AIS methods and processes.

Table 4.1 provides selected indicators for topics covered in this module.

ORGANIZATION OF THIS MODULE

Agricultural research within an AIS framework is still in quite early stages of development, and the experiences and applications described here have been drawn from a number of contexts and often within a piloting modality. This material has

Table 4.1 Schematic of a Monitoring and Evaluation Framework for Assessing the Performance of Agricultural Research within an AIS

Domain	Indicator
Innovation outcomes/research impact	– Technology adoption – Increased farm productivity and incomes – Increased efficiency within the value chain – Increased total factor productivity – Increased agricultural GDP
Demand articulation	– Forums for farmer voice – Joint priority setting – Participatory or codesign research – Number of public-private partnerships (PPPs)
Organizational interfaces/partnerships	– Number of innovation platforms, clusters, or consortiums – Number of PPPs – Value chain platforms – Scientist participation in networks
Organizational change	– Involvement of stakeholders in planning, priority setting, and evaluation – Presence of mechanisms or units for managing partnerships and for brokering innovation processes
Research productivity	– Number of competitive grants received – Number of peer reviewed articles published – Number of varieties released – Number of on-farm trials
Knowledge flows	– Number of hits on website – Articles or programs in mass media – Citation index – Extension bulletins produced and distributed

Source: Author.

Note: All indicators should be disaggregated by gender when possible.

been organized by typology in this overview and by theme in the notes that follow (except for TN 1).

The typology emphasizes that the market, organizational, and economic context needs to be understood in deciding on investment approaches for AIS. The themes covered in the notes enter into more detail on alternative areas of investment in important domains of AIS. They include a discussion of demand articulation and external organizational interfaces (TN 1); public-private partnerships (TN 2), regional research (TN 3), codesign in agricultural research (TN4), and management structures and organizational change (TN 5).

The TNs are followed by profiles of innovative approaches to agricultural research within an AIS framework. IAP 1 discusses the redesign of an international agricultural research center to align with an AIS approach. Two profiles focus on the implementation of agricultural innovation funds, especially in facilitating public-private partnerships, in India and Chile (IAPs 2 and 3). Another describes experiences with learning alliances (IAP 4)

Designing Agricultural Research Linkages within an AIS Framework

David J. Spielman, International Food Policy Research Institute (IFPRI)
Catherine Ragasa, International Food Policy Research Institute (IFPRI)
Riikka Rajalahti, World Bank

SYNOPSIS

Consensus is growing that new ways of conducting agricultural research are needed. To date, the operational implications of these changes and strategies for making them efficient, effective, and sustainable have been discussed very little. Lessons on strengthening the connectivity between agricultural research and other innovation system actors are viewed through the lens of three types of economies—agriculture-based, transforming, and urbanized—and two strategies: (1) investing in "demand articulation" mechanisms to better identify the needs of different user groups and (2) designing "organizational interfaces" that help transform research into real goods and services. There is a case for both market and nonmarket approaches to improving demand articulation and organizational interfaces. They include investment in formal mechanisms that provide stakeholder input to research organizations, more participatory mechanisms that bring researchers and farmers together to solve problems, innovation platforms that address larger, more complex challenges with diverse actors, commercialization programs that move research into the marketplace, and financing mechanisms that encourage collaborative research. Careful adaptation to the specific innovation contexts, strategies, and mechanisms is prerequisite for success.

BACKGROUND AND CONTEXT

Agricultural research needs to be examined within the broader analytical framework of an innovation system, which means recognizing that innovation in agricultural development may occur in collaboration with, separately from, or even in spite of agricultural research organizations. The challenge is to make public research organizations more responsive, dynamic, and competitive within this new landscape in agricultural development. To reach this goal, public research organizations will have to increase their relevance, their capacity to respond to a changing landscape, and their ability to produce goods and services that can be put to use in a socially or economically productive manner.

These statements are not a call for paying less attention to the quality of scientific inquiry and expertise in disciplinary fields. They are rather a call for greater interaction between researchers and other knowledge producers and users to maximize the quality of science and its impacts on society and the economy. Increased interaction means that public research organizations will continue to play a role in developing country agriculture but that their role must change. The key to this change will be flexible institutional arrangements that encourage dynamic, rapid responses to changing circumstances from public research organizations.

This TN examines specific investments in key design elements and approaches in three innovation contexts (box 4.5) similar to those discussed in the module overview. It focuses on key investments in articulating demand (identifying the needs of different user groups for the knowledge and information produced by research organizations) and designing organizational interfaces (modalities that help transform this knowledge and information into socially and economically relevant goods and services).

Research systems have undergone any number of reforms, ranging from rebuilding after a crisis to redesigning more complex and advanced systems. Little evidence points to which reforms actually work well in different types of research organizations and how these reforms might ultimately affect agricultural productivity and poverty. Without sufficient evidence, it is often difficult to provide conclusive insights into the returns on investing in large-scale reforms of research systems. The next best option is to examine

Box 4.5   The Three Innovation Contexts

(1) *Agriculture-based countries.* In these countries, farmers have limited access to agricultural markets, which in many cases do not function well. Most countries in sub-Saharan Africa are in this category. In agriculture-based countries, **research organizations** must develop an interface with their clients—primarily small-scale farmers, extension systems, and government decision makers—and with the rest of the national, regional, and global research system. The private sector engages mostly in licensing technologies to public breeding programs, multiplying improved seed developed by public breeding programs, distributing inputs such as chemical fertilizer, or providing other small-scale and localized agricultural products and services.

(2) *Transforming countries.* Transforming countries host innovation systems in which agricultural markets are expanding and developing. A subset of farmers gain from good connections to markets. Transforming countries can be characterized by an increased reliance on market-based approaches to

guide the contribution of agricultural research to the wider innovation system. *Many developing countries in South Asia, East Asia and the Pacific, the Middle East, and North Africa are in this category.*

(3) *Mature innovation countries.* These countries have innovation systems in which agricultural markets function relatively efficiently and farmers are effective market players. *Most countries in Latin America and the Caribbean and many in Europe and Central Asia are in this category.* In transforming and mature countries, **research organizations** should take greater notice of market demand and rely on market-based approaches to guide their contribution to the wider innovation system. Research organizations are required to interface with a wider set of clients—smallholders and commercial producers, diverse private sector actors (input suppliers, processors, wholesalers, retailers, industry associations, exporters), other service providers, and consumers—to create venues for them to express their needs and align national priorities to research agendas.

Source: Authors, based on World Bank (2007).

---

different reform processes to understand the impact pathways through which they are expected to work.

INVESTMENT NEEDED

This note describes *nonmarket* and *market-based approaches* to investment, starting with approaches that fit particularly well with agriculture-based contexts and moving to more commercial, market-oriented approaches. The note does not provide an exhaustive list of investment mechanisms but features the mechanisms that are most relevant for developing countries:

1. Strengthening information sharing and demand articulation in research systems through formal coordination organizations, enhanced communication, and ICTs.
2. Promoting greater participation of farmers and other clients in technology development processes.
3. Technology transfer and commercialization approaches.
4. Financing mechanisms for multistakeholder approaches.

Table 4.2 summarizes the approaches, their purposes, and the key knowledge assets used and exchanged as part of each approach. The approaches or mechanisms can be selected and combined to fit the particular need for innovation in a given context.

Strengthening information sharing and demand articulation in research systems through formal coordination organizations, enhanced communication, and ICTs

In many countries, formal organizations facilitate regular exchanges of information and identify research priorities. These organizations include committees, agencies, and other formal bodies that obtain farmers' input on research results (for example, their opinions of the performance of new cultivars), on longer-term priorities for research and/or competitive research funds, and on the wider policy issues associated with agricultural production and markets.

The public sector often leads and manages the process of setting up these formal organizations. Often they include representatives of farmers, extension services, the research system, and ideally other actors in the public sector, private sector, and civil society. Both centralized and decentralized approaches are applied. Organizations at the provincial/zonal level, such as the Research Extension–Farmer–Input–Linkage System in Nigeria or the Research and Extension Linkage Committees in Ghana (box 4.6), particularly fit agriculture-based contexts and tend to focus on consultation and receiving farmers' input on research results.

Organizations that operate at the national level use a more sophisticated set of tools for priority setting aside from stakeholder consultations, including tools for scenario and technology foresight, information databases, and M&E of research programs. Examples include the Senegal Agricultural Services and Producer Organizations Project

Table 4.2 Approaches to Strengthening the Articulation of Demand and Interfaces with the Agricultural Research System in Agriculture-Based, Transforming, and Mature Innovation Contexts

Approach	Purpose	Key assets	Examples (sources)
Formal coordination organizations	Information exchange, priority setting, coordination, fund allocation	Scientific information; extension and advisory services	*Agriculture-based:* Research Extension–Farmer–Input–Linkage System (REFILS) in Nigeria (Koyenikan 2008); Research and Extension Linkage Committees (REALCs) in Ghana (World Bank and IFPRI 2010); Senegal Agricultural Services and Producer Organizations Project *Transforming and urban:* Fund governance and national research/innovation councils or forums
Communication and ICT	Share information; demand articulation		See *Information and Communication Technologies for Agriculture Sourcebook* (World Bank 2011, forthcoming).
Participatory research	Engage farmers in research priority setting, selection, testing, and experimentation	Scientific information; extension services; capacity/methodology in participatory approach	Participatory plant breeding (Sperling et al. 2001; Morris and Bellon 2004) Central America Learning Alliance (Faminow, Carter, and Lundy 2009); CIALs in Colombia, Honduras, Ecuador, Bolivia, and Nicaragua (CIAT 2006; Quiros et al. 2004)
Codesign approaches	Engage diverse stakeholders in the entire R&D cycle	Scientific and local information; capacity in codesign approach	Liu (1997); Almekinders, Beukema, and Tromp (2009); Hocdé et al. (2009); Bernet et al. (2006, 2008)
Innovation platforms	Promote co-innovation; exchange information; identify opportunities and set priorities; promote policy change	Public and private technologies; capacity to reach commercial and underserved markets; private financing; farmer-private sector-policy maker linkages	*Agriculture:* Civil society partnerships: Papa Andina (Thiele et al. forthcoming; Devaux et al. 2009, 2010; Horton et al. 2010; Smith and Chataway 2007) *Transforming:* Agricultural innovation networks in Argentina (Ekboir and Parellada 2002; Trigo et al. 2009), Bolivia (Monge et al. 2008), Mexico (Ekboir et al. 2009), Andean South America (Devaux et al. 2009, 2010; Horton et al. 2010), and the Netherlands (Klerkx, Aarts, and Leeuwis 2010); Research consortiums: CLAYUCA on cassava (Patiño and Best 2002; see IAP 5 in module 1) *Urban:* Netherlands (Janssen and Braunschweig 2003; Klerkx and Leeuwis 2009a) *International and regional research networks:* CGIAR, FARA, ASARECA, APAARI
Consortiums			Australia; NAIP India (IAP 2)
Technology transfer	Acquire technology	Scientific information and tools; capacity for dealing with international agreements	*Agriculture:* Material transfer agreements between international and national research centers for wheat improvement (Dubin and Brennan 2010; Louwaars et al. 2005) and biotechnology (Byerlee and Fischer 2002) *Urban:* Agricultural biotechnology (Byerlee and Fischer 2002); drought-tolerant maize research (AATF 2011)

(Table continues on the following page)

Table 4.2 Approaches to Strengthening the Articulation of Demand and Interfaces with the Agricultural Research System in Agriculture-Based, Transforming, and Mature Innovation Contexts (continued)

Approach	Purpose	Key assets	Examples (sources)
Commercialization programs	Commercialize public research	Public technologies; capacity to commercialize new products	*Transforming:* ICSRISAT Hybrid Parents Research Consortia for sorghum and pearl millet (Gowda et al. 2004; Pray and Nagarajan 2009); Sustainable Commercialization of Seeds in Africa (SCOSA) (Jones 2006) *Urban:* Cooperative R&D agreements (Day-Rubenstein and Fuglie 2000); Plant genetic IP management (Louwaars et al. 2005)
Public-private research partnerships	Develop new products	Scientific information, tools, and materials; managerial capacity	East Coast fever vaccine development (Smith 2005; Spielman 2009); agricultural research (Spielman, Hartwich, and von Grebmer 2010); see also IAP 2 in module 6
Science parks and business incubators	Develop new products	Scientific information and tools; managerial capacity; private and public capital	CIAT and ICRISAT (Spielman, Hartwich, and von Grebmer 2010); see also TN 3 and IAP 1 in module 5
University-industry research collaborations	Promote co-innovation; commercialize public research	Public technologies; capacity to commercialize new products	Agricultural biotechnology (Ervin et al. 2003)
Alternative funding mechanisms	Farmer-funded research; finance research	Financing from financial markets and donors; specialized scientific services	Competitive grants and innovation funds (World Bank 2006, 2010; Gill and Carney 1999); Research prize schemes (Masters 2003); farmer levies (Klerkx and Leeuwis 2009b); market segmentation schemes (Kolady and Lesser 2008; Lybbert 2002)

Source: Authors.

Note: APAARI = Asia Pacific Association of Agricultural Research Institutions; ASARECA = Association for Strengthening Agricultural Research in Eastern and Central Africa; CIALS = Local agricultural research committees; CGIAR = Consultative Group on International Agricultural Research; CLAYUCA = Latin American and Caribbean Consortium to Support Cassava Research and Development; FARA = Forum for Agricultural Research in Africa.

Box 4.6 Research-Extension-Linkage Committees in Ghana: Experience and Lessons

In Ghana, Research-Extension-Linkage Committees (RELCs) include producers, researchers, and extension agents from the Ministry of Food and Agriculture (MoFA). The committees facilitate dialogue and elicit better guidance from producers about local research and extension efforts. Five RELCs were piloted, one in each of the country's major agroecological zones, under the World Bank–funded Agricultural Services Project. Eventually the committees were expanded to cover each of Ghana's 10 regions. Each regional RELC has 15 members, including two representatives of farmer organizations, one representative from a nongovernmental organization, one representative of agribusiness, and representatives from research and extension.[a] Under the Agricultural Services Project, the second call for proposals from the competitive research grant scheme was based on the RELCs' identification of farmers' problems. Thirteen research projects from seven regions were approved for funding.

Despite this effort at planning from the farm level up, the RELCs proved ineffective in strengthening links between research and others in the AIS. Funding for implementing RELC initiatives has been limited, partly because responsibility for allocating operating funds is divided between the national research institute (the Council for Scientific and Industrial Research) and MoFA. Perhaps owing to these financial constraints, the RELCs have not engaged greater numbers of farmers and end users and have had little influence on the research agenda. Sustainable financing for farmers' and end users' participation in the RELCs is likely to have made them more effective.

Source: World Bank 2007; Riikka Rajalahti, personal communication.
a. According to the project's 2002 procedure manual.

(box 4.7), research councils (module 1, TN 1), and competitive funds (module 5, TN 2).

Improved awareness of research programs, results, and applications—among research partners (national, international) and other stakeholders, including clients—are important for articulating demand in increasingly decentralized AISs and developing a platform for information sharing and collaboration. The key investment elements include development of a communications strategy and program; capacity building for staff on communications and ICTs; hardware and software for collecting and storing data, and a telecommunications and Internet platform. For details and examples, see World Bank (2011).

Promoting participation of farmers and other actors in technology development

Participatory research approaches, codesign, and innovation platforms offer pathways for farmers and other clients to develop agricultural technology with researchers. The next sections discuss these approaches and specific corresponding investments. The concluding discussion focuses on the potential for research consortiums to strengthen links between research and other actors in the AIS.

PARTICIPATORY RESEARCH APPROACHES. Participatory approaches identify farmers' demands and bring farmers' knowledge as well as researchers' knowledge to bear on

---

Box 4.7   Lessons from Senegal's Agricultural Services and Producer Organizations Project

The Agricultural Services and Producer Organizations Project (PASAOP) strengthens end-users' demand for services and public research institutions' ability to meet their demands. In its first phase (1999–2006), PASAOP established a network of producer organizations in 142 of 320 rural council areas, along with decentralized, demand-driven agricultural services. In its second phase (2006–11), the project further strengthens the institutional framework, extends the coverage of agricultural advisory services nationwide, supports the emergence of private service providers, strengthens research capacity and focus, and further empowers producer organizations, while increasing their social accountability and representation. Both project phases have built on the following approaches:

- *Restore the focus of ministries* active in agriculture on their core public functions: policy formulation, monitoring, and evaluation. Create specific directorates for policy analysis, forecasts, and statistics. Decentralize services with the creation of regional directorates.
- *Replace the traditional technology transfer model* with demand-driven support. Decentralized advisory services are managed jointly (including planning and evaluation) by a semipublic National Agency for Agricultural and Rural Advisory Services (ANCAR), producer organizations, and private agribusiness.
- *Establish transparent, competitive financing* for research on agriculture and agroprocessing through

the National Fund for Agricultural Research (FNRAA).
- *Engage producer organizations* in decision making as genuine advocates of proposals. Producers also chair the management committee of FNRAA to ensure that research programs are relevant to their needs.
- *Link producers* through a network of rural consultative forums (CLCOPs) in 152 rural council areas so producers contribute fully to defining, implementing, and evaluating research and extension programs. Producer organizations have also established and manage their own Demand Driven Rural Services Fund, which allocates resources to microprojects prepared by producer organizations.

Benefits

To date, PASAOP has helped improve the quality and selling price of groundnuts, level and quality of community seed stocks, beneficiaries' incomes (12 percent higher), and nonfarm household income. Producer satisfaction with services is 80 percent against a target of 100 percent. Food security increased among 62 percent of producers against a target of 60 percent. In producer organizations, 45 percent of members adopted at least one technology in their production systems against a target of 50 percent. The agricultural research system generated 22 technologies. Cofinancing of FNRAA by other

(Box continues on the following page)

Box 4.7 Lessons from Senegal's Agricultural Services and Producer Organizations Project (continued)

stakeholders (government, other donors, commodity organizations, and the private sector) is expected to reach 30 percent.

Lessons

■ **Invest in producer organizations.** Local institutions can responsibly and efficiently implement their activities and limit the need for a project to establish an implementation unit. Producer federations with adequate project support improved their efficiency and internal governance (quality of records, meetings, actions taken, satisfaction of members). Demand-driven funds that supported physical investments and equipment were more effective than those focused only on soft investments.

■ **Two channels for research proposals (from researchers and from users) provided flexibility.** They made it possible to respond to producers' immediate concerns as well as opportunities identified by scientists.

■ **Complement core funding with competitive funding.** Together, these two mechanisms guarantee that institutional development continues. Funding for operating costs goes directly to research teams working on projects relevant to users, to whom they are accountable. Core funding for developing human resources is essential to elicit relevant proposals of good quality.

■ **Specific pro-poor strategies must be designed into the project.** The project's second phase seems to have had a greater impact in richer households than poorer households.

Sources: Diaw, Samba, and Arcand 2009 on impact assessment of Phase 2 of PASAOP; World Bank Project Appraisal Documents for Phases 1 and 2 of PSAOP; World Bank Implementation and Completion Report for PSAOP.
Note: PSAOP = Programme d'Appui aux Services Agricoles et aux Organisations Paysannes; ANCAR = Agence Nationale de Conseil Agricole et Rural; FNRAA = Fonds National pour la Recherche Agricole et Agro-Alimentaire; CLCOP = Cadre Local de Consultation des Organisations de Producteurs.

agricultural problems. Farmers (and others) participate in monitoring and evaluating the results. Some participatory research is done in farmers' fields. This approach is particularly suited to *agriculture-based countries* in which resources are at a premium and farmers are often isolated from others in the AIS. The approach allows research organizations to complement their programs in cultivar improvement and crop management with work on more integrated and natural resource management issues, such as common resource management of pastures, shared water resources, fisheries, and communal forests, and incorporate gender and community-based development perspectives through farmer organizations, forest user groups, and local savings and credit associations.

CODESIGN APPROACHES. Codesign approaches (discussed in detail in TN 4) seek better articulation between the supply of research (from researchers) and demand for research (from users). Researchers engage systematically with a heterogeneous set of actors, which may include farmers, input suppliers, traders, processors, researchers, NGOs, and government officials in the iterative, adaptive, and flexible process of developing innovations. The core principles of codesign include joint planning, implementation, and decision making related to all activities that foster innovation; close coordination among stakeholders at all strategic and operational levels; and combining scientific, other technical, and local knowledge and other resources.

Codesign is often used *when problems are complex and/or the scale involved is challenging.* Examples include the shared management of a dwindling natural resource held in common (a forest or water source, for example); the period of adjustment to new policies or market operations; the development of shared understanding of problems and their solutions, when there is potential to do so; and problems for which previously designed solutions or scientific and technical knowledge are not available. Given the issues of scale involved in such a large group of actors and their numerous concerns, codesign relies on at least some of the concerned stakeholders to have the experience and skills to facilitate, coordinate, and negotiate multistakeholder efforts (module 1). The Papa Andina program implemented in Bolivia, Ecuador, and Peru successfully combined and applied codesign and innovation platform approaches (TN 4, box 4).

INNOVATION PLATFORMS. Innovation platforms (or networks and forums) assemble stakeholders to share information, identify opportunities, discuss problems, and agree on joint activities related to a shared interest, often with a specific commodity/cluster focus. They usually provide a means for many participants to exchange opinions but tend to imply less commitment to addressing the needs identified, compared to codesign approaches, consortiums, or competitive grant schemes. Innovation platforms focus on all kinds of innovation, not necessarily research alone, and they may be led by actors other than researchers. Even so, they present an important venue and opportunity for many research organizations to engage with other AIS actors, improve their understanding of how they can best fit into the AIS, and develop partnerships. In *transforming countries,* innovation platforms are likely to be more mature than in *agriculture-based countries,* where public support and funding are prerequisites for success. The key assets or contributions by each actor in the interface may be explicit (for example, they may consist of scientific or market information, tools, and materials, both proprietary and nonproprietary) or more implicit (such as the capacity to manage complex projects, move technologies through regulatory processes, or market and distribute new products).

Examples of innovation platforms include the Central America Learning Alliance, a multistakeholder network that promotes rural enterprise development (IAP 4), and the innovation network that promoted zero-tillage cultivation practices in Argentina (module 1, IAP 1). Papa Andina (TN 4) and the client-oriented research management approach (box 4.22 in TN 5) apply both nonmarket and market-based strategies.

INVESTMENT NEEDS IN PARTICIPATORY AND CODESIGN APPROACHES AND INNOVATION PLATFORMS. Specific investments improve the likelihood that these approaches and platforms will function more successfully.

- **Invest in researchers' capacity to work in innovation systems.** Researchers must have the capacity to diagnose innovation systems and the ability to participate in and sometimes facilitate group processes involving people with diverse stakes in a commodity or value chain. Aside from their technical and scientific expertise, they will need the skills involved for organizing actors, coordinating activities, and consulting, negotiating, monitoring, and evaluating.

- **Invest in other partners' skills.** Farmers, universities, NGOs, the private sector, and others will need skills in designing partnerships, building trust, and effective communication. Farmer organizations often need help in learning how to articulate their demands, establish links to local government, and engage in social learning and experimentation to innovate rather than simply demonstrate or accept technological "fixes."

- **Invest in bringing people together.** Operational funds are needed to run committees and cover the costs of face-to-face, facilitated group meetings (coordination, facilitation) and the collective action that are inherent to collaboration at all stages of the codesign process.

- **Invest in innovation brokers.** A good facilitator or a project team is required to take an initiative forward. Innovation brokers can limit the failures that occur when different interests and conflicting agendas frustrate initiatives designed to foster partnership. They can also reduce competition between the public and private sectors, creating a more coordinated approach to problem solving. Innovation brokers do not often emerge of their own accord. Their facilitation role needs to be funded, supported, and linked to activities in research, extension, and the broader innovation system.

- **Invest in incentives for participation.** These incentives often take the form of funding that makes partnerships work: operational costs and costs of joint R&D.

- **Invest in value chain analysis and development.** Investments in value chain development are a key entry point for research organizations in *transforming countries* to contribute solutions that enhance the benefits (and lower the costs) to actors along the value chain. Tools such as value chain analyses—including participatory approaches to such analyses—can identify constraints and market opportunities at different stages of the value chain as well as entry points for support.

RESEARCH AND INNOVATION CONSORTIUMS. Consortiums are more formal mechanisms than networks or innovation platforms. They bring together diverse partners around a specific and common problem requiring research investment, jointly define R&D strategies, and finance and implement the subsequent research-innovation project. They often—but do not necessarily—focus on applied R&D. Consortiums often require multidisciplinary teams consisting of private, public, civil society, and producer actors. Most consortiums have a lead organization, and each partner has a specific role and commits resources. Contributions from a range of actors, including private enterprises, cover various aspects of R&D (demand identification, R&D investment, technology transfer and adoption). Consortiums are often funded through competitive grants (which match funds to resources mobilized by partners) for a limited period.

Australia (box 4.8) and the Netherlands (box 1.14 in module 1, TN 1) are examples of *mature urban innovation* contexts where a consortium approach helped R&D meet specific challenges. Consortium approaches have shown promise in *transforming countries*; see the discussions of approaches in India (IAP 2) and Chile (IAP 3).

Technology transfer and commercialization approaches to integrating private actors

Technology transfer is the foundation of many research programs in *agriculture-based* countries and prevalent in *transforming* and *urbanized* countries. Transforming and more mature innovation contexts increasingly rely on formal transfers of technology from public research organizations, universities, and the private sector. Such technology may require IP protection and/or other legal agreements that transfer property rights to commercial or international partners. Many of the technology transfer and commercialization approaches in these countries build on approaches introduced earlier, but they require a higher level of capacity with respect to advanced science and technology, complex regulatory systems, IP protection, sophisticated

---

Box 4.8   Design of the Australian National Agricultural Innovation System

Australia's AIS is one of the most dynamic and successful in the world. Direct engagement of producers through their financing and oversight of commodity-focused (mainly applied) research was the primary mechanism for gaining insight into the needs and demands of key user-groups. Sharpened priority setting, increasingly involving ex ante economic analysis of competing proposals along with ex post impact assessments, has been the hallmark of the approach. Agricultural research intensity has been maintained at nearly 0.04 of agricultural GDP, among the highest levels in the world, and total factor productivity for agriculture has been close to 2 percent per year since the major reforms in the agricultural research system began in the mid-1980s.

A key feature of the reforms is the creation of Cooperative Research Centres (CRCs), which are joint agreements between research providers to undertake R&D in particular areas. CRCs must comprise at least one Australian end-user (either from the private, public, or community sector) and one Australian institution of higher education (or research institute affiliated with a university). These institutions work for a limited period (generally seven years) to resolve technological problems in a multidisciplinary fashion. The involvement of universities and their disciplinary expertise is especially important for linking industry

demand to academic centers of excellence in joint problem-solving.

Traditionally Australia has invested relatively heavily in agricultural research through a blend of public and private (producer levy) funds, which were largely used by federal and state government agencies with some producer oversight through farmer membership on various advisory committees and an institutional watchdog (the Productivity Commission for institutional learning and ensuring accountability). Producer funding was matched equally by federal government support of up to 1 percent of respective commodity GDP.

A *key lesson* is that a charismatic change leader with a relevant vision is critical. In this case, it was a minister of primary industries, who was insightful and effective (originally a farmer, then a research agricultural economist and a politician). The strong (albeit less than perfect) accountability mechanisms built into the new processes, such as the CRCs, surely helped greatly. A major lesson for other countries is that, given the inherent complexity of the AIS, it is critical for public policy analysts to keep pursuing their understanding of the realities and opportunities in agricultural research as it evolves and to keep a sharp eye on the effectiveness of institutional arrangements.

Source: Jock Anderson, personal communication.

---

markets and market infrastructure, and international trade considerations.

The capacity to manage formal technology transfer mechanisms is critical to engage effectively in public-private partnerships and, increasingly, to transfer technologies that can be disseminated through market channels. *Technology transfer offices* are special units affiliated with a research organization or university with a mandate to identify and protect as well as facilitate the use and commercialization of research results. These offices can expand the recognition of the research organization's work (thereby strengthening public perceptions of its value), move technologies to end-users (seed companies, farmers) on an exclusive or nonexclusive basis, and generate revenues to fund continuous research.

Technology transfer offices can provide special expertise on IP protection and/or legal agreements and contribute to formal transfers of technology from public organizations or universities or from the private sector to commercial or international partners (see box 6.20 in TN 3 of module 6 and TN 5 in module 5. Several examples of this interface have been used successfully to disseminate hybrid parent lines of pearl millet and sorghum in India, with substantial improvement in the availability of improved seed and yields for small-scale farmers in semiarid and arid tropics (Gowda et al. 2004; Pray and Nagarajan 2009). Aside from

technology transfer offices, other pathways to technology transfer may be applied (summarized in box 4.9).

Some technology transfer offices also host *incubators* to help technology-oriented firms (often established by researchers) commercialize new technology. Incubators provide hands-on management assistance, access to financing, business and technical support services, shared office space, and access to equipment. For details, see module 5, TN 3.

SCIENCE PARK APPROACHES. Science parks (also called technology or research parks) are a mechanism for fostering public-private partnerships in *more mature innovation contexts*.[1] Science parks are organizations managed by specialized professionals, whose main aim is to increase local wealth by promoting a culture of innovation and improving the competitiveness of local businesses and knowledge-based institutions. A science park stimulates and manages the flow of knowledge and technology among universities, R&D institutions, companies, and markets; facilitates the creation and growth of innovation-based companies through incubation and spin-off processes; and provides other value-added services together with high-quality space and facilities.

Science parks function best where there is investment capital from the private sector, industrial engineering expertise, and a sufficient knowledge and technology base.

Box 4.9 Technology Transfer Pathways

Technology transfer agreements. The classic example of technology transfer agreements is the formal exchange of breeding materials for crop improvement, typically from international research centers or universities in industrialized countries to national research organizations in developing countries. Scientists and research managers in developing countries require additional skills to understand the increasingly complex material transfer and intellectual property agreements that govern technology transfer; they must also expand their linkages to international and regional science networks. Great success has been achieved with technology transfer programs (for example, for wheat and rice improvement in Asia, NERICA rice in Africa, and orange-fleshed sweet potato in several postconflict countries in Africa).

Commercialization programs. These programs create windows for private companies or entrepreneurs to access public research outputs and move them into commercial use. Often this approach is used to move improved breeding material from public research organizations to private seed companies. For example, the Hybrid Parents Research Consortiums of the International Crops Research Institute for the Semi-Arid Tropics have provided more than 35 Indian companies with improved sorghum, pearl millet, and pigeonpea lines for commercial use. The program for Sustainable Commercialization of Seeds in Africa, the Eastern and Southern Africa Seed Alliance, and the West Africa Seed Alliance are also designed to improve the private sector's access to breeding materials and strengthen its seed marketing capacity.

Source: Authors.

As discussed in the module overview, they are a useful nexus between the private sector and research institutes (particularly universities), taking promising research products to market and providing backstopping for product modification. Their diverse services include facilitating the creation of public-private partnerships for research, providing infrastructure, and providing other services, including business development. The scope of this note does not allow the numerous science parks to be discussed in detail (including China's agricultural demonstration and technology parks; CIAT's Agronatura, and France's Agropolis); see module 5, IAP 1 on the incubator affiliated with the Agri-Science Park of the International Crops Research Institute for the Semi-Arid Tropics (ICRISAT).

Financing mechanisms for multistakeholder approaches

Transforming and mature countries often demonstrate a higher level of complexity and capacity when it comes to funding research organizations and activities, especially where functioning markets exist alongside an organized agricultural sector. Aside from public core funding for research, a great number of funding mechanisms or other arrangements (such as levies on sales, public-private partnerships, cofinancing with farmer organizations and trade associations,[2] R&D tax deductions, joint ventures, or research partnerships, see IAP 3 on Chile) incentivize and reduce the transaction and risk management costs associated with collaborative research (for a summary on financing agricultural innovation, see module 5, TN 6). This TN briefly describes the two main mechanisms—competitive research grants and matching grants—which are described in detail in module 5, TN 2.

Competitive research grants are a common mechanism for funding basic, strategic, and applied research through competition based on scientific peer review. The aim is to focus scientists' efforts on high-priority research or new fields of expertise, improve the relevance and quality of agricultural research, promote research partnerships, and leverage research resources (from the public or private sector). See IAP 2 for an example of a competitive research grant scheme to promoting multistakeholder consortiums in India (World Bank 2010). Funds for competitive grant schemes usually come from the public sector and are managed by a public or semiautonomous organization. Competitive grants have been used to fund consortiums working on specific research themes.

Matching grants are used for financing near-market technology generation, technology transfer and adoption, or business-related innovation, often by including multiple stakeholders (see module 6, IAP 1 for a matching grant scheme to develop agribusiness in Zambia). Matching grants require a financial commitment from the beneficiaries (farmers, entrepreneurs) and therefore may be more effective than competitive research grants at enhancing the dissemination and use of knowledge and technology. They are also better suited for funding overall innovation and activities requiring private sector engagement.

Both competitive research grants and matching grants involve short- to medium-term funding arrangements. They should complement, never substitute for, stable funding for long-term research, private sector development, human resource development, and infrastructure maintenance and development.

POTENTIAL BENEFITS

The immediate benefits of these investments are straightforward. Research organizations gain greater relevance and responsiveness, ultimately leading to greater impact on agricultural development, food security, and poverty reduction. In many agriculture-based countries, these impacts are measured in terms of increased yields (output per unit of land) and production (total output).

Where markets operate with some degree of efficiency, potential benefits may include higher returns to crop cultivation (Kaaria et al. 2009; Thiele et al. forthcoming; Devaux et al. 2009, 2010; Cavatassi et al. 2009). Potential benefits also extend to improvements in gender aspects of agricultural development, such as changes in the household assets owned by men and women.

Beyond the immediate benefits to productivity, output, and welfare, these approaches carve out a niche for research organizations within a rapidly changing agricultural landscape. They provide research organizations with new clients and markets as well as access to new resources and assets. In urbanized systems particularly, an improved interface between research and other AIS actors may accelerate the rate of innovation by bringing the best science to bear on real problems and ensuring that sufficient resources are allocated to solving problems. The research system will become more responsive to the demands of society because users such as farmers and consumers have many different pathways to express their needs.

POLICY ISSUES

Most issues related to the policies and governance structures that enable research institutions to participate more fully and successfully in the AIS are detailed in module 6. A few key issues should be mentioned here, however.

- **Sustainability requires managerial and structural reforms.** Research organizations often organize their personnel and assets by discipline, but this form of organization makes it costly to bring personnel and assets together to resolve problems in agricultural value chains. Management and structural reforms are vital to overcome this barrier; see the discussion in TN 5.
- **Institutional change and reform require stable, long-term support.** Efforts to encourage research organizations to interface with other user-groups and respond to their demands require considerable time, effort, and resources. Policy makers must commit the time, space, and funding to implement reforms and build the related capacity.
- **The participation of civil society, including women, may require specific policy initiatives.** Farmer associations and community-based organizations cannot operate in their members' interests in an environment hostile to grassroots and women's participation. Policies to foster equitable participation and social mobilization can (for example) provide operational funds to build marginalized groups' capacity to participate, cover the costs of their participation, and require that financing mechanisms have specific criteria to promote inclusiveness.
- **Are public funds used where they are most needed?** A value chain approach with a focus on multiple stakeholders can lead public research organizations to serve those who need their services least. Research organizations typically struggle with such trade-offs. For example, should they develop technologies for high-potential agricultural areas where the gains are likely to be high, or should they concentrate on technologies suited to both high- and low-potential areas? Decisions on how to address these tradeoffs require strong leadership from policy makers to ensure that public funds are used as intended.
- **Foster a conducive investment environment.** The key policy issue for a mature innovation system is to create a climate that supports private sector participation and development. Policies are needed for public research to contribute to private participation (through sound regulatory frameworks, for example) and also to ensure that women and the poor are included in the activities and benefits of innovation.

LESSONS LEARNED

The following lessons related to designing agricultural research linkages within an AIS are grouped into general lessons, lessons on the approaches that are best in particular innovation contexts (agriculture-based, transforming, and mature urban countries), and lessons related to particular mechanisms linking research to other AIS actors.

General lessons:

- **While large structural reforms are a good investment, smaller, more evolutionary, and incremental approaches to systemic change sometimes work best.** Invest in stepwise efforts to engage diverse user-groups, define problems collectively, build joint action plans, develop internal capacity, and learn through iterative processes. Such interventions sometimes involve only short-lived projects, marginalized administrative units, short-term bridge financing, or small team initiatives, but they foster responsiveness, dynamism, and competitiveness. Often they are more grounded in a specific innovation challenge.
- **Experiences from industrialized countries can prove instructive.** For example, Australia's approach to formalizing joint public and industry funding for its rural research program, and its regular and broadly consultative review of progress, could be effective in other contexts.
- **Invest in a mix of integrated approaches.** The appropriate mix depends on the specific circumstances of a country's agricultural research system, but it could involve a combination of formal research/innovation governance arrangements, participatory or codesign research approaches, and more commercially oriented approaches and financing mechanisms.
- **Approach capacity strengthening more comprehensively and iteratively than in the past.** Bench scientists require management training to interact effectively with other AIS stakeholders and ultimately improve the quality and impact of their research. Develop courses and learning materials based on experimentation and rigorous assessments of approaches that work or do not work in different contexts. To create a critical mass of researchers with skills suited to the AIS, integrate participatory processes and innovation network techniques into agricultural education systems.
- **Organizations also need new capacities and incentives to reform.** The ability of researchers and research organizations to leverage constructive interactions at some lower experimental level depends on the signals—authorization,

encouragement, or financing—from higher levels. In designing and implementing strategies to facilitate interactions and linkages, incentives and motivating factors among staff and leaders of research organizations (and other organizations with which they interact) must be assessed with care. Organizations need to enhance support for risk-taking managers and collaborative teams experimenting with learning approaches—but coupled with periodic external evaluations. Change of the kind described here requires strong, long-term leadership and political commitment in addition to incentives.

Lessons specific to particular innovation contexts:

- *In agriculture-based countries, improve researchers' responsiveness to farmers' needs and increase access to global science and technology through a diversified, cross-cutting approach to participatory research and technology transfer.* Strengthen individuals' capacity to use participatory approaches by building skills in facilitation, negotiation, conflict prevention and resolution, building relationships and trust, and developing the rules of the game. Broaden research organizations' access to technology by expanding their links to international science networks and their understanding of complex material transfer and IP agreements.
- *In transforming countries, use combined market/non-market approaches (making use of the skills just described) to engage the private sector more actively and encourage opportunity-driven entrepreneurship.* Public sector orchestration and financing are key to addressing transaction costs (of collective action and negotiation, for example), reducing risk, and providing incentives. Innovation networks and platforms are vital tools to identify opportunities, set priorities, and influence the research agenda.
- *In urban countries, policy makers and practitioners can invest in sophisticated competitive funding mechanisms.* The channels through which user groups articulate their needs should be advanced enough to ensure that science, technology, and innovation respond to market opportunities and that public research organizations work alongside the private sector and other stakeholders.

Lessons related to specific mechanisms:

- *Pay careful attention to the design of multistakeholder approaches and platforms, because they do not work in all contexts.* These platforms need good facilitation to bring stakeholders (with their potentially divisive power relationships, capacity differences, and levels of interest) together. To sustain these programs, enhance negotiation and conflict management, improve the representation of poor and marginalized farmers, fully fund communication and knowledge management, and clearly define roles and functions of advisory committees, secretariats, and members. Engagement of high-level policy makers is often crucial.
- *It takes time to form and sustain networks or platforms.* These interfaces require clear priorities, roles, and milestones. Substantive capacity strengthening of all partners in partnership design, trust-building, and effective communication is required for these approaches to work, along with incentives for participation.
- *Consortium approaches have the advantage of a problem-oriented focus.* This focus permits the definition of partners' objectives, goals, and responsibilities, which in turn permits better management and evaluation of the collaborative effort. The disadvantage is that the reason for collaboration ends the moment that the problem ceases to need attention.
- *Innovation brokers play an important role in facilitating change in an innovation system.* More formal approaches to innovation brokering include the use of research coordination councils, committees, and other bodies (see module 1, TN 2).
- *Analyze the pros and cons of new funding mechanisms carefully before introducing them.* Matching grants may better suit innovation contexts where private sector engagement is crucial and where dissemination requires significant attention. Competitive research grants can develop high-quality research portfolios, but they tend to have high operational costs and have been ineffective in engaging the private sector and disseminating knowledge and technology. Small research systems may not allow sufficient scope for real competition.

Building and Strengthening Public-Private Partnerships in Agricultural Research

Frank Hartwich, United Nations Industrial Development Organization (UNIDO)

SYNOPSIS

Often the challenges of developing country agriculture can be met only by pooling the limited public and private resources available and unleashing innovative potential out of critical interaction. Investments that support the building and conduct of research partnerships can lead to more substantial and viable research that fosters social and development needs while not getting overshadowed by private interests. Before providing financial support to public-private partnerships, public and development agencies must determine whether the partners' interests and objectives are sufficiently mutual and whether the partnership will generate synergies from joint use of knowledge and resources. If these conditions are met, brokering a partnership arrangement is a powerful means to foster agricultural research.

RATIONALE FOR INVESTING IN PUBLIC-PRIVATE RESEARCH PARTNERSHIPS

The innovation system approach extends the conventional argument for public investments in agricultural research[1] by saying that a single organization, be it private or public, may not be able to assemble the necessary resources, capacities, and knowledge to generate and diffuse innovations. (See also module 5, TN 1.) The increasingly blurred roles of the public and private sectors in agricultural research can be seen in the rise of public-private partnerships (PPPs). The two sectors are redefining their traditional division of labor as their overlapping interests and the benefits of combining resources lead both of them to engage in both types of research (Hall et al. 2001; Hartwich and Negro 2010; Muraguri 2010). Some of the implications of this mixed panorama in agricultural research are summarized in box 4.10.

In PPPs, at least one public and one private organization share resources, knowledge, and risks to achieve a match of interests and jointly deliver products and services.[2] In agricultural research, PPPs can be seen as arrangements that bring together partners with different skills and knowledge to contribute jointly to the generation, adaptation, and/or diffusion of an innovation. Usually the partnership agreement is in the form of a contract that establishes each partner's commitments and the distribution of benefits.

PPPs in agricultural research can be set up not only to generate knowledge via research but to foster the diffusion and application of knowledge among private actors (agribusinesses, farmers) as well as public actors (universities, research institutes, and extension agencies). In this respect, PPPs can be distinguished by their contributions to the research-development continuum (table 4.3). Some partnerships also engage in a mix of types of research and development; research-based activities precede the stage of product development.

PPPs can be distinguished further according to the partners engaged. Common partners are research institutes, universities, extension agencies in the public sector, and producer associations, businesses, and individual producers in the private sector. Many partnerships involve a public research organization that has the main responsibility for conducting the research, but in others a private entity conducts the research and public agencies diffuse and/or fund the research. Other types of organizations engaged in partnerships include farmer and community groups, private associations, investment and sector development promotion bodies, and funding agencies. Often more than two parties are engaged; depending on the leading partner one can distinguish PPPs led by private enterprises, research institutes and universities, government (ministries), and so on. For an example of PPPs among public research organizations, the private sector, and farmer organizations, see module 5, IAP 2.

Box 4.10 Public-Private Partnerships and the Changing Roles of Public and Private Agents in Agricultural Research

- *PPPs are more than an organizational solution* between market and hierarchy; they are arrangements that maximize benefits by using the creativity and synergy of collective action to respond simultaneously to public and private needs.

- *Public-private partnerships (PPPs) generate social benefits* in the way that they leverage resources from the private sector and combine them with public resources. In this way, additional resources and capacities become available to address pressing research issues.

- *PPPs can be instrumental in increasing the overall volume of agricultural research,* although developing countries may not replace public funds substituted by private funds (and match levels of private investment in more developed countries).

- *To develop and improve their image* in line with public goals and respond to corporate responsibility criteria, private companies will partner with the public sector. Corporate social responsibility alone is not a good motivation for PPPs that want to promote sustainable businesses, however.

- *It is simplistic to reduce the private sector's interest in agricultural research* to the development of private goods that can be protected by IPRs. In fact, private partners in only a small fraction of 124 PPPs in agricultural and agro-industrial research in Latin America were interested in protecting property rights; the few examples occurred only in seed companies. Apparently private companies benefit more from the synergistic use of resources than from protecting the IP of research results.

Source: Hartwich and Tola 2007.

Table 4.3 Types of Research Subject to Public-Private Partnerships

Type of research	Example
Creative research partnerships that generate and explore new scientific finding	Biotechnological exploration of compounds in pharmaceutical plants
Applied research partnerships that use an existing research methodology to generate new solutions	Development of a plant variety that resists a new pathogen
Development-oriented research partnerships that focus on the development of products	Development of a new potato chip
Diffusion-oriented partnerships that promote the dissemination of developed knowledge and technology	A seed multiplication program supporting dissemination of a particular plant variety

Source: Author.

INVESTMENT OPTIONS FOR SUPPORTING PUBLIC-PRIVATE RESEARCH PARTNERSHIPS

PPPs become interesting subjects for investment in two ways. First, they constitute product development mechanisms that foster collaboration for innovation that otherwise would not occur. For example, a private company and a public research institute may not collaborate in the development and diffusion of a new pest-management technology simply because they are not informed about their common interest and complementary capacities. In fact, potential partners from the public and private sector often fail to collaborate because each lacks knowledge about the other sector and the potential benefits of a partnership. A funding agency could improve information flows by financing meetings and a platform where these players can start interacting and identify common interests to develop a common work agenda.

Second, PPPs constitute funding mechanisms in their own right that enable a funding agency to engage with actors in agricultural innovation, particularly the private sector. (See also IAP 2 and module 5, TN 2 and IAP 2.) For example, a development agency or donor can use a PPP to support the efforts of an international agribusiness specializing in dairy products to work with small-scale dairy producers who initially lack the capacity to participate in the dairy value chain. The funding agency would ensure that activities under the partnership would also generate social benefits, particularly for the small-scale producers.

Quite a number of PPPs have been set up over the past ten or more years in developing country agriculture (Hartwich and Tola 2007; Spielman and Hartwich 2009). In many cases the public and private organizations involved found that partnering in research is to their benefit. Examples include partnerships where research reduces the costs of processing primary products, partnerships that improved product quality to access higher-value markets, or partnerships to exchange planting material and outsource seed multiplication to the private sector (boxes 4.11, 4.12, and 4.13.)

The failure of PPPs to fulfill their potential provides the context for a range of public investment opportunities to build and improve PPPs for agricultural research. But in which PPPs should development agents invest? In general, a precondition for investment in PPPs in agricultural research is that they should count on financial support from government and development agencies; they should address problems of public concern that require collective action; and they should pool capacities for innovation. Specific investment opportunities include:

- *Investments in bringing partners together*—specifically, in identifying partners, their common interests, available resources and competencies, and potential synergies. One option is to invest in the organization of platforms that allow public and private organizations interested in developing and adopting agricultural innovations to meet, exchange information, and develop joint projects. Often a neutral broker is needed to facilitate the platform; see module 1 and module 3, TN 4 for examples.

- *Investments in fostering better planning.* Investments can support the design of partnerships that facilitate efficient work and fair distribution of benefits within the partnership framework.

- *Investments to partly fund the operation of the partnership, complementing the public contribution.* An example of such investments is a fund that provides grants to research projects conducted in partnership between various organizations. (See also IAP 2 on NAIP in this module and TNs 1 and 2 on PPPs and innovation funds in module 5.) The fund's contribution to the partnership usually depends, among other considerations, on the partnership's ability to respond to certain public interests.

- *Investments to set clear legal rules and framework conditions for partnering.* Many public institutions still lack clear rules to determine when and how they can work with the private sector; in the end, individual initiatives determine whether partnerships are set up. Many research institutes leave it to their legal departments to screen and approve collaborative research projects.

- *Investments in coaching partners to ensure that public and private benefits reach the partners and society.* Often it is not enough to help organizations to set up a promising partnership. During the partnership, conflicts may arise, partners fail to comply with promises, and certain framework conditions can change. Support during the partnership (for example, through a partner-

---

Box 4.11   Public-Private Partnership for Participatory Research in Potato Production in Ecuador

An international potato chip producer operating in Ecuador could not procure enough potatoes of suitable quality to use all of its local processing capacity. The company tried to provide incentives to farmers through contract farming and higher prices, with unsatisfactory results. The company finally determined that on its own it could not persuade small-scale farmers to deliver more and higher-quality potatoes. It partnered with the National Agricultural Research Institute (INIAP, Instituto Nacional Autónomo de Investigaciones Agropecuarias) to focus on identifying, developing, and distributing potato varieties with the required processing qualities. The partnership had an extension component in which INIAP helped farmers adopt the new varieties, grow them with appropriate management practices, and increase productivity and output.

After some months of negotiation, the partnership was set up and a formal agreement signed. INIAP provided research and extension capacities while the private company provided funding. A national research fund contributed additional public funds. The partnership ended after some years because the company had achieved its objectives.

Factors that enabled the partnership to succeed included a good match of interests (INIAP sought to support small-scale farmers, from whom the company wanted to buy quality potatoes). The partners' competencies were also well matched: INIAP was expert in developing and disseminating improved potato varieties, and the company had expertise in evaluating the cost-reduction and product-improvement potential of the potato varieties.

Source: Author.

Box 4.12   Public-Private Partnership for Processing Cashew Nuts in Northern Brazil

In the late 1990s, Brazil's cashew nut sector was in ruins. Competition and price variation on the world market, along with poor product quality and mismanagement, caused the processing industry to go bankrupt, taking a large number of small-scale producers with it. Producers' crop remained without buyers, and they had few alternatives to cashew production.

Brazil's national agricultural research institute, EMBRAPA, partnered with a cashew-exporting company and a number of organizations of small-scale growers to develop and diffuse microprocessing units for cashews that would be owned and managed by the grower organizations. Researchers adapted large-scale processing technology to develop much smaller processing units they called "minifactories" (*minifabricas*). The export company provided knowledge of market demand, product quality, and processing requirements, as well as some quite limited funding for training farmers. EMPRAPA covered the main costs of the partnership, which involved laboratory research and on-site testing of growers' processing units.

EMBRAPA constantly exchanged information with growers and the exporter about the appropriateness and readiness of the technology. Growers slowly became familiar with the technology and started to adapt it to their own needs. Eventually farmers gained substantial capacity in cashew processing and the commodity was once again, through the exporting partners and soon through competing exporters, sold internationally. Farmer incomes benefited significantly as farmers regained a market for their primary product and profited from the value added by their primary processing units. The exporting company also increased its profits, which benefitted employees and shareholders. The partnership succeeded because EMBRAPA's advance project identification and planning were good, the market for cashews remained strong, and EMBRAPA and the exporting company had complementary knowledge and skills (processing technology in EMPRAPA and market information in the exporting company).

Source: Author.
Note: EMBRAPA = Empresa Brasileira de Pesquisa Agropecuária.

Box 4.13   Public-Private Partnership for Research on New Wheat Varieties in Argentina

A private European company specializing in wheat, sorghum, and maize seed wanted to benefit from Argentina's rapidly expanding market for cereal seed but knew little of Argentina's wheat breeding program over the years. For example, it did not know which specific resistance and tolerance traits had been introduced (and failed) and which varieties, lines, and breeding strategies had been used to raise wheat yields. The national agricultural research institute (INTA, Instituto Nacional de Tecnología Agropecuaria), partly funded through a levy on the agricultural sector, had downsized its wheat breeding program and was looking for a partner with whom it could exchange information and genetic material. It was also interested multiplying

certain varieties that it had developed but not tested, released, or multiplied. The two organizations entered a partnership that envisioned the exchange of certain genetic materials to develop and multiply seed of new varieties. The IP for the resulting varieties would remain with the public sector, but the seed company would benefit from seed sales and pay royalties to the public institute. The partnership was phased out after a number of years and the company (among others) now caters to a well-established market for wheat seed. Factors that contributed to the partnership's success included their complementary skills (INTA's excellence in wheat breeding and the company's advantages in seed multiplication and marketing).

Source: Author.

ship support unit) may be useful. The unit could also help partners set up an efficient system for monitoring and evaluating the partnership. For example, CIAT's partnership development and facilitation unit has played this role.

POTENTIAL BENEFITS OF INVESTING IN PUBLIC-PRIVATE RESEARCH PARTNERSHIPS

For a number of important reasons, public and private organizations participate in research partnerships and governments encourage them to do so. First, supporting PPPs in agricultural research provides an opportunity to improve the performance of agricultural research operations. Research partnerships can also broaden the scope of research activities, increase efficiency and synergy, access complementary resources, and promote organizational learning among the partners (Hagedoorn, Link, and Vonortas 2000). The potential benefits from funding measures that initiate PPPs and help them operate successfully can include:

- *Better use of existing research capacity* by allowing public and private partners to form and draw from a greater critical mass of scientific capacity.
- *The research process becomes more creative* when allowing public and private partners to join complementary competencies.
- *Agricultural research productivity and results improve* when public and private partners develop synergies through the combined use of resources.
- *Cost-efficiency in agricultural research improves* when public and private partners share costs and benefit from more efficient private sector management practices.
- *Research results can arrive more quickly* owing to the private partner's drive for more immediate results.
- *Investments in agricultural research for social benefits increase* by identifying compatible private and public interests and combining private and public sources of funding.

Boxes 4.11 and 4.12 illustrate how the efficient collaboration of public and private agents in the Ecuadorian potato industry and the Brazilian cashew nut industry yielded some of these benefits.

It is important to note that these are *potential* benefits. They will be achieved only if a number of conditions are met. In certain situations partnerships may even "block competition" and "create monopolies" (Hagedoorn, Link,

and Vonortas 2000:568). In others, particularly in small developing countries, partnerships may be instrumental in bringing together the scarce but necessary research resources.

Aside from potential benefits, potential risks exist. Risks related to uncertainty and failure are inherent in any research endeavor, but every partnership also carries a risk that the partners may not contribute to the partnership as initially negotiated. Partnership agreements may not be met for any number of reasons, including changes in the market and business environment for which the partnership's products are geared. The legal and government framework may prove too inflexible for the partnership to progress, or the framework can change to prevent the public partner from fulfilling its commitment. Finally, the relationship between the partners can run into difficulties because of misunderstandings, a clash of cultures, and distrust. For this very reason, partnerships often benefit from investments in building trust, brokering, and partnership development—both initially and indeed throughout the partnership.

(POLICY) ISSUES OF OUTSOURCING, SUSTAINABLE FUNDING, AND PUBLIC GOODS

Governments often use PPPs to *outsource* service provision to private companies (improving service provision through "bringing in private sector efficiency"). In agricultural research in developing countries, where research capacity is insufficient both in the public and private sectors, outsourcing is hardly relevant. Instead PPPs aim to maintain public research facilities and strengthen them with contributions from and collaboration with the private sector to attain *a critical mass* in research.

Partnerships must not persist over time. They are agreements that help partners reach an end but they are not an end in themselves. *Sustainability over the duration of the PPP* depends on the partnership's capacity to cover the related costs of human resources, infrastructure, and equipment as well as operations, which in itself is positively related to the negotiations and the setting of binding contractual relationships at the start. More than other contractual relationships, however, partnerships are prone to financial risks. If one partner loses interest, changes the strategic focus, or becomes insolvent, the partnership will lose part of the anticipated contributions despite any prior commitments.

Core funding is a prerequisite for PPPs established in the public interest. Competitive grants that require public and private collaboration (or vice versa) are one instrument to develop PPPs. Evidence from competitive grant schemes for agricultural research in Latin America indicates that without a solid core budget, public organizations cannot enter such partnerships from a sufficiently strong position to negotiate an agreement that reflects their interests (Echeverría 1998a).

LESSONS LEARNED

Despite the numbers of successful PPPs, the concept is still under development. In many situations PPPs fail to meet their potential to improve agriculture and rural livelihoods in developing countries, for two reasons. First, the number of PPPs is still very low and cannot match the opportunities to conduct research and develop innovations by means of PPPs. Second, many efforts to build partnerships among public research organizations and the private sector fail to bring sufficient benefits to both partners; usually one partner profits at the expense of the other (Hartwich and Tola 2007). Nor do most PPPs use their complementary resources effectively to produce substantial social benefits. Planning can be weak. Prior to initiating their partnership, partners do not engage in the necessary negotiations to clarify their common interests, each partner's commitments, and the redistribution of benefits. Usually these partnerships collapse as soon as resources become scarce and/or accrued benefits are not shared equally.

Experience on the best strategies for supporting collaborative agricultural research and PPPs is mixed (Hall 2006). Specialists in plant breeding and biotechnology research stress the value of strengthening public institutions' awareness of and capacity to manage IPRs (Lewis 2000; Byerlee and Fischer 2002). Other analysts emphasize the importance of negotiating not only IP issues but other aspects of benefit sharing and resource commitment (Rausser, Simon, and Ameden 2000; Vieira and Hartwich 2002; Hall et al. 2003). The following sections discuss some of the most important lessons from various studies and experiences.

Public-private partnerships as funding opportunities

PPPs are a mechanism through which funding agencies can build collaborative research as well as an arrangement into which they can inject public funds. In a partnership to develop wheat varieties such as the one in Argentina described in box 4.13, public research funds (for example, a competitive grant) would support the partnership financially while ensuring that social benefits are met.

Partnering may not always be the best option

Public research organizations and development funding agencies that seek to support partnerships need to separate the wheat from the chaff. Promoting and financing partnerships simply for their own sake is not useful. Partnerships may not produce good results, and in many situations another solution is more appropriate. For example, a public research agency may consider contracting the services of a private laboratory. A private company may prefer to set up its own research unit rather than collaborate with public researchers.

Motivation matters

Governments and funding agencies that want to support partnerships need to assess prospective partners' motivations for entering a partnership. According to Spielman and Hartwich (2009), one can distinguish between:

- **Representational partnerships,** in which one partner joins the partnership for prestige without contributing knowledge and/or resources.
- **Outsourcing partnerships,** in which one partner seeks to outsource research and diffusion activities to another.

Box 4.14   Indicators for Evaluating Public-Private Partnerships in Agricultural Research

- Costs of interaction among the partners, including time for negotiation as well as cost of communication and monitoring compliance.
- Contribution of each partner to the partnership in terms of human resources and expenditure.
- Strengths, weaknesses, opportunities, and threats of the partnership as perceived by partners, beneficiaries, and other stakeholders.
- Mapping of communication and information flows among individuals in the partnership.
- Anecdotal information on the history and functioning of the partnership as perceived by partners, beneficiaries, and other stakeholders.
- Joint authorship of publications by partners.
- Attribution of patents and user rights.
- Intensity of use of research results by each of the partners.

Source: Author.

- *Competency-led partnerships,* in which one partner is trying to tap into the competencies of another.
- *Finance partnerships,* in which one partner aims to secure complementary funding.
- *Co-innovation partnerships,* in which one or both partners try to reach a critical mass in research and synergy in developing innovations.

This last type of partnership, with its focus on adding value through collaboration, is the partnership in which public development agencies will mostly want to invest.

Importance of brokers

Evidence from Latin America indicates that PPP-promoting agents, acting as brokers in the public interest, play a crucial role in building partnerships, particularly to motivate potential partners, build trust among them, and provide credibility to partnership initiatives (Hartwich et al. 2007). Gradually, as common interests are identified and partnerships are formalized, the roles and contributions of partners need to be negotiated to ensure that partnership arrangements are in alignment with partners' interests, their capacities, and the prevailing technological and market opportunities.

Lack of planning

Through inappropriate planning and management, PPPs often fail to use their complementary resources effectively and benefit both the public and the private sector. It helps to develop a clear contract specifying the resources each partner will commit over the project period as well as the use of the research results, including IPRs and payment of royalties (boxes 4.12, 4.13). It can also be useful to strengthen partners' capacities in negotiating partnerships as well as in planning and implementing activities under partnership agreements.

As noted, for many PPPs, *protection of IP is not always a priority.* IPRs often are of minor or no concern to the private partners, who are interested instead in public sector support to obtain primary materials for newly developed and/or quality products and eventually market them to consumers. Research on new varieties may be the exception here. In the partnership described in box 4.13, any varieties developed remained public goods, and the seed company had to pay royalties on sales of those varieties. The main aim of the seed company was to market very good varieties (which it could not develop on its own) to complement its existing product portfolio. (See module 6, TN 3 on IPR and module 5, TN 5 on technology transfer offices.)

Learning and evaluation

M&E is a particular challenge in partnerships. Many criteria can be used to evaluate the outcomes and impacts of agricultural research, but evaluating collaborative processes is different and onerous for several reasons. First, partnerships can involve multiple and heterogeneous partners and stakeholders with different goals. Second, it is difficult to monitor the joint use of knowledge and other resources by partners. Third, partnerships change over time and in response to internal and external conditions. Finally, there is the inherent difficulty in attributing benefits to the collaboration itself and not to partners' individual activities (de Bruijn and van der Voort n.d.). A unified framework for the evaluation of research partnerships does not exist, and little progress on this issue has been reported in the literature (Hagedoorn, Link, and Vonortas 2000; Serafin, Bustamante, and Schramm 2008). Box 4.14 lists some indicators that, despite these difficulties, may help in monitoring and evaluating the success of collaborative research processes.

RECOMMENDATIONS TO BUILD SUPPORT FOR PUBLIC-PRIVATE RESEARCH PARTNERSHIPS

Public funding agencies that want to invest in PPPs should consider five minimum conditions that must be met beforehand (Hartwich et al. 2008):

- The existence of a common interest which, during initial negotiations, needs to be agreed upon by the partners. For example, setting up a partnership for developing a new pesticide could unleash substantial synergies among the partners, but the pesticide itself could be so harmful to people and the environment that public involvement could not be justified.
- Each partner must show a clear commitment that goes beyond shared interest. Commitments must come in the form of time and resources allocated to the partnership.
- The individual benefits of each of the partners must outweigh their individual costs; otherwise partners have no incentive to engage in the partnership.
- Benefits must be distributed proportionally. One partner cannot reap a large share of the benefits while the other receives only a marginal payoff.
- Overall benefits must outweigh overall costs. Without this synergy, there is no justification to engage in any partnership. The costs of collaborating would outweigh the benefits, and for one of the partners it would be better to pursue the goal alone. This last condition puts

PPPs in the context of innovation systems, suggesting that mechanisms of collective action, joint learning, and complementary use of resources are all operational in the development and diffusion of innovation.

In conclusion, investing in PPPs makes sense where there is agreement on objectives, strong commitments, some added value through partnering, and a fair distribution of benefits. PPPs that are badly designed, fragile, and/or result in a public subsidy to private organizations do not represent value for public investment. Partnerships should also have a limited time period. Funding agencies should particularly exclude from funding any partnerships that evolve spontaneously in the search of funding opportunities or prestige.

A public funding agency may find few of these partnerships in place, however, and could consider helping to build adequate partnerships. If after a period of support and brokerage a PPP shows signs of complying with the criteria just mentioned, it can be considered for further funding.

Finally, any PPP that is up and running can benefit from continuous support to its management that helps sustain it until the end. Given the complex relations prevalent in partnerships, tension is to be expected, and a funding agency should consider the support for partnership management as a promising opportunity.

Regional Research in an Agricultural Innovation System Framework: Bringing Order to Complexity

Howard Elliott, Consultant

SYNOPSIS

Spontaneous self-organization will not be sufficient for a strong regional AIS to emerge. The emergence of a regional AIS will require compatible organizations and institutions that can bond, a timely and adequate concentration of resources, and a sustained infusion of resources. Proposed initiatives should be assessed against six basic criteria for their contributions to national and regional innovation systems. First, participants in a regional arrangement must developed and share a vision of mutual benefit and make a formal commitment prior to concrete investments. Second, the regional arrangement should lead to greater efficiency in research for development, especially through economies of scope and scale. Participants must understand that the gain from research is not where the expenditure is made but where the benefits are used. Third, the research design and approval process must ensure research quality and relevance. Investment in strong M&E is necessary to gain political support. Fourth, regional arrangements must improve national capacity, especially in the smaller countries. Fifth, a regional AIS should foster collective action on shared challenges that cannot be addressed by a single nation, especially transboundary problems of a regional public good nature. At some point, an innovation system may emerge as participants become aware of their connectedness and begin to coordinate through information flows, markets, and new organizations. A sequence of many small investments leading towards collaboration, collective action, and stepwise integration is needed. Finally, sustainable regional organizations will require a higher-order political and financial mechanism committed to a regional strategy. Self-sustaining regional centers can emerge when a host-country institution accepts a regional role and has a business model that sustains it through a national core commitment, research grants, service fees, and projects.

BACKGROUND AND CONTEXT

Regional agricultural research has a long history rooted in colonial initiatives and commercial crops (in West Africa or South Asia, for example). In most instances these initiatives came to an end with independence. Regional and subregional organizations emerged again in the 1980s, when several regions sought to address challenges they would or could not address alone. They also wanted to strengthen their influence over the networks of international centers operating in their regions. The new arrangements were expected to lead to greater regional ownership of research and economies of scale and scope. Not all regions were organized in the same way. All evolved in path-dependent ways.

- *Asia and the Pacific* had APAARI,[1] which included both giants like India and China and the microsystems of the Pacific without any subregional associations.
- *The Near East and North Africa* was covered by a similar organization, AARINENA,[2] without differentiation.
- *In sub-Saharan Africa*, the rationale for subregional programs focused on economies of scale and scope, the development of regional public goods, and the need to create regional capacity to compensate for weak NARSs. In addition, the subregional organizations emerged politically to bring coordination and ownership to networks run by the CGIAR research centers. The pioneer subregional organization for Africa was SACCAR,[3] created in 1984 to serve the frontline countries surrounding then-apartheid South Africa. It was followed by CORAF in 1987 (as a Francophone group), ASARECA in 1994,[4] and CORAF/WECARD in 1997 (to include Anglophone countries).[5]
- *In Latin America,* PROCISUR[6] served as the forerunner and model for other subregional organizations: PROCIANDINO in the Andes, PROCITROPICOS in the

Amazon,[7] and in Central America and the Caribbean. Successful regional commodity programs also developed.

Every subregional organization evolves within its unique combination of constraints and driving forces. Latin America offers several rich experiences from which sub-Saharan Africa can benefit. Latin American countries have used their middle-income status to invest in agriculture as a dynamic sector. Strong rural organization and fiscal systems that leave resource-based revenues in the regions are often important elements in this strategy (box 4.15).

Many regional and subregional initiatives have strengthened particular components of an innovation system at the national level, but they have been faced variously with national free-riding, avoidance of responsibility or inability to carry out commitments to collective action, and sensitivities with respect to "equitable sharing of the benefits."

Despite these issues, the need for collective regional activity is greater than ever, because more problems and opportunities are transnational.

On the demand side, investment is needed in the capacity to exploit market-driven agricultural opportunities and to negotiate rules and standards for better competition. On the supply side, investment in knowledge for addressing emerging challenges, such as transboundary diseases, climate change, water scarcity, and increased price volatility in global markets. All of these challenges have significant implications for the knowledge system and do not recognize country borders. Moreover, opportunities provided by advances in biotechnology require concentrated investments in infrastructure, advanced computing, and scarce human capacity that call for central hubs and platforms that offer economies of scale (as in the example in box 4.16).

Box 4.15 Subregional Organizations in Latin America: Strong National Capacity, Commitment to Research, and Alignment with an Emerging Regional Economic Community as Drivers of Success

Created 1980, PROCISUR has been an autonomous regional organization since 1981, located with IICA in the headquarters of Mercosur (the Southern Common Market). In its evolution, PROCISUR came to focus on themes affecting all countries of the region, including biotechnology policy and intraregional trade. PROCISUR's success in generating political and financial support for itself and its members reflects the influence of two strong research systems, Argentina and Brazil, as well as a functional regional economic community, Mercosur. PROCISUR has been adept at championing projects that bring together national universities and national research institutes with regional and international partners, thus closing a gap at the national level that may exist. (LOTASSA, a research project on forage legumes, is one example.) As the regional community, MERCOSUR has several redistributive measures to support the smaller members of the community. The integration of markets and realignment of production have created incentives for investment and relocating

economic activity in the region. Despite this success, in recent years, the smaller members of Mercosur have increasingly resisted deeper economic and regulatory integration. Finding a common policy has become more difficult.

South America has also seen the emergence of commodity consortia funded by the region to address special needs for research or integration with the rest of the innovation system. The Latin American Fund for Irrigated Rice (FLAR) collaborates with CIAT but takes responsibility for irrigated rice in temperate countries. CONDESAN was created with the help of CIP to involve more than 30 actors in the potato innovation system. After 25 years of public research on cassava, CLAYUCA's role has evolved into organizing public and private organizations to promote uses for cassava in addition to food. The concept of a regional innovation system seems to be taking shape. Convergence of policies, market integration, and some equalization measures are important factors.

Source: Author.
Note: CLAYUCA = Latin American and Caribbean Consortium to Support Cassava Research and Development; CONDESAN = Consorcio de Desarrollo Andino/Consortium for Andean Development; FLAR = Fondo Latinoamericano para Arroz de Riego; IICA = Instituto Interamericano de Cooperación para la Agricultura; LOTASSSA = Lotus Adaptation and Sustainability in South America; Mercosur = Mercado Común del Sur, Mercado Comun do Sul; PROCISUR = Programa Cooperativo para el Desarrollo Tecnológico Agroalimentario y Agroindustrial del Cono Sur.

Box 4.16   Creating Sustainable Scientific Hubs: An Example of the Biosciences in Eastern and Central Africa

The entry into biosciences research usually begins by developing capacity in a central location to serve a number of clients, such as university scientists, commodity breeding programs, and often nonagricultural research as well. Major indivisibilities in laboratory and computing facilities for genomic research, for example, mean that there are significant economies of scale in creating a hub. Over time, and with the falling cost of more applied operations, decentralized capacity is developed. One new platform for biosciences, the BecA-Hub (Biosciences Eastern and Central Africa), benefited from major support from the Canadian International Development Agency to upgrade facilities at the International Livestock Research Institute (ILRI) in conjunction with the New Partnership for Africa's Development (NEPAD).

Located at ILRI's Nairobi campus, BecA is one of three regional initiatives championed by NEPAD. The hub's sustainability depends on whether it builds on its role as an international facility, serves its clients in East and Central Africa, and generates strong international collaborative research program on animal disease. The location of a regional hub at an international research center brings many logistical and administrative facilities in addition to economies of scale and scope.

From the earliest days of biotechnology in developed countries, the location of a biotechnology facility has been controversial. Should it be based in a separate center, faculty of science, or faculty of agriculture? The business models for sustainable research in biotechnology differ in universities and national agricultural research institutes, and given scale considerations, stable funding arrangements will be a determining factor. BecA estimates that it will have to generate at least 50 percent of its funds from hosted research projects to sustain its capacity. Growth will come through an expansion of target countries, broadening the scope of activities, a good synergy of African and international partners, and incubation/innovation projects.

BecA currently supports staff from 6 international centers, has helped 60 MSc and PhD students advance their research, and has hosted 41 projects (17 in crops and 24 in livestock). A NEPAD "BecANet" provides resources for national agricultural research systems and university scientists to use these resources to support their own priorities. In West Africa, the Alliance for a Green Revolution in Africa supports the West Africa Center for Crop Improvement, a regional postgraduate training program at the University of Ghana, Legon. The center, which benefits from a faculty institute for biosciences and backstopping from Cornell University, has attracted scientists from francophone countries as well.

Source: BecA website, http://hub.africabiosciences.org/.

This TN discusses two central questions. First, how can investments in regional programs and institutions better strengthen national AISs? Second, how can they support the emergence of a regional AIS? This TN argues that if more complex systems are to emerge, they will require (1) compatible organizations and institutions that can bond, (2) a timely and adequate concentration of resources, and (3) a sustained infusion of resources over a sufficient period (Elliott 2010). The agenda, science, and methods are changing rapidly at all levels, and it is difficult to design organizations to deal with a moving target. Spontaneous self-organization will not be sufficient for a strong regional AIS to emerge.

INVESTMENT NEEDED

While there is no model for a regional research system that is well integrated within an AIS, a few main investment elements are needed to ensure the necessary concentration, connectivity, and continuity for a self-sustaining regional innovation system to emerge:

- *Support consensus on a regional vision and agreement.* The most important element is that participants in a regional arrangement share a vision of mutual benefit and make a formal commitment. Support for this dialogue and agreement on vision, research priorities, and cost and benefit sharing are needed prior to concrete investments. A mutual accountability framework must be developed and provide incentives for all partners to deliver on their commitments.[8]

- *Select a committed host organization.* A regional research system must be led by a committed and capable host institution, preferably the most effective research organization in the region. The host institution needs a business model that sustains the effort through a national core commitment, research grants, service fees, and projects. In other words, the host organization should have a sustainable research program that ensures its continued ability to serve a regional demand.

- **Support national partners.** A successful regional research system requires sufficient national capacity for research, dissemination, and coordination (national and regional) to avoid competition and duplication and to reach synergies. Support is needed to address the human resource gaps, perhaps by investing in MSc programs at the national and regional levels to give government, NGOs, and aid organizations the staff they require to participate in the initiative.

- **Design a governance structure.** The regional initiative requires a board with representatives from participating countries, including financial decision makers, political allies that ensure continuity, and other relevant stakeholders. The board is expected to recruit strong leadership, ensure a strategy is in place, oversee its implementation, and support the regional mission. Governance structures have to be capable of changing strategies and adapting governance itself to new conditions.

- **Base a strong secretariat in the host organization.** A secretariat performs the day-to-day administrative functions of a regional network, platform, or hub as mandated. The secretariat must have staff and skills to support efforts to meet clients' needs, generate and communicate new information to members, and raise funds.

- **Maintain regional funding authority.** A regional research arrangement requires a regional funding authority to maintain its scale of operations and commitment to a regional role. National commitment and support from the host government are required. Reliance on donor funding is not a sustainable solution.[9]

- **Develop procedures for selecting, monitoring, and evaluating regional research projects.** Clear and transparent procedures, taking notice of stakeholders' concerns, are needed to ensure both quality and relevance of research. Such procedures would permit the balanced articulation of demand, base the selection and approval of projects on strong and independent scientific evaluations, specify the financial and reporting requirements, and outline requirements for objective ex ante evaluations, program monitoring, and ex post evaluations. See module 5, TN 2 on innovation funds for details on sound procedures.

Box 4.17 presents an example of the challenges presented in developing a regional research system.

POTENTIAL BENEFITS

Regional research efforts may create different types of benefits. Often such efforts are designed to gain efficiency in the production of research for development in a given priority area, primarily through economies of scale and scope. Participants can share costs for infrastructure, software, laboratory equipment, and highly specialized scientists, as in BecA (box 4.16). Regional efforts also put institutional arrangements into place that facilitate the use of research results. Their financial sustainability requires special attention, however.

POLICY ISSUES

The policy issues that are particularly relevant to whether and how regional research might contribute to the emergence of a regional innovation system are discussed in the following points. Perhaps the central issue is the tension between national and regional priorities and the considerable challenges of collective action at a regional level.

- **Achieving balanced growth of innovation system components through regional collaboration.** Coordination among NARIs, universities, and technology dissemination is often weak or even resisted at the national level. The source of this problem may be unbalanced growth among the components, their sequential development, different scales of operation, and different bases for organization (agroecological, political administrative, or centralized for economies of scope and scale). Regional facilities may allow a national knowledge system to operate at a higher scale by avoiding the need for large fixed investments in one or more sectors through regional borrowing.

- **Partnerships in producing regional public goods.** National research systems, subregional organizations, and international agricultural research centers all contribute in complementary ways to the production of public goods. Pingali (2010) warns that an unintended consequence of the CGIAR research centers' shift toward product adaptation and dissemination relative to innovation and product development is a potential break in the R&D pipeline that supplies public good research and technologies to raise agricultural productivity in developing countries. Improved partnership is the elusive solution to issues of competition and crowding out, but it requires behavioral and institutional change (Horton, Prain, and Thiele 2009).

The Regional Universities Forum in Africa (RUFORUM), initiated in 2004, grew out of the Rockefeller Foundation project (FORUM) to help 5 countries and 10 universities specialize around particular themes, create university centers of excellence, and train graduate students from participating countries. Its challenge was "to build a national and regional research for development (R4D) innovation system based on partnerships and dialogue between all actors, impact-oriented and well-coordinated, where universities are an integral part." RUFORUM is recognized as an international nongovernmental organization and is supported by the European Community and the Bill and Melinda Gates Foundation.

Experience to date. RUFORUM planned to roll out seven regional thematic programs, each assigned to a lead university, and train more than 800 MSc and 150 PhD students by 2013. Students from participating countries would receive local tuition and living allowances. The attractiveness of the concept (and the absence of another body with a similar mandate) caused membership to grow from the original 12 to 25 universities in Eastern and Southern Africa alone. In 2010 a strategic reflection meeting identified the challenging trade-off between a sharp focus and an expansion in scope and scale, including a proposal for RUFORUM to expand to West Africa. A ministerial Conference on Higher Education in Agriculture called for universities to (1) create a pan-African university, (2) contribute fully to NEPAD-CAADP processes,[a] (3) link to the private sector as a potential source of funding and internships, (4) engage in policy formulation and implementation, and (5) coordinate with ministries in the national innovation system. Another resolution that emphasized the AIS was to include investment in agricultural higher education within the NEPAD-CAADP 10 percent budget allocation (provided it did not dilute the commitment to agriculture).

Lessons. RUFORUM faces at least three large challenges arising from its initial success. First, how can it meet the needs of its 25 university member-owners? Second, should it consolidate activities at a meaningful subregional scale, with tangible results for participating countries, or should it follow donor pressure to expand and assume additional functions? Third, how can it manage the sustainable addition of functions as it moves from network to advocacy platform to possibly more intensive service functions?

RUFORUM's Business Plan 2011–2016 describes how balanced growth would allow it to take on an expanded function as a continental forum. Documentation of lessons from the "RUFORUM model" in eastern and southern Africa would create legitimacy and credibility beyond its current region. Continental action would require both changes in governance and full cost recovery by new initiatives to avoid competition with delivering RUFORUM's flagship MSc programs, which serve national and regional AIS needs directly. As seen with the effort to create a hub for the biosciences in East and Central Africa (box 4.16), sustainable financing for a platform or hub depends on many things: a subsidy from the host country, continuous donor commitment, or integration of national and regional activities through products and services. Regional efforts often founder when the national host cannot or will not continue its regional role, so selecting the host for a center of excellence is a critical decision. Many major universities were created with support from Ford and Rockefeller Foundations and bilateral aid programs. A handful emerged to play sustained regional roles, whereas others' regional presence diminished in the wake of purely national decisions and crises or a host institution's inability to sustain regional commitments beyond donor funding.[b] The Conference on Higher Education in Agriculture pushed for a continental forum rather than a consolidation of action at the subregional level or a definition of subregional communities that are natural partners.

Source: RUFORUM 2005, 2010, 2011.

a. Under NEPAD (the New Partnership for Africa's Development), African heads of state committed to allocating a 10 percent share of their national budgets to agriculture. CAADP (the Comprehensive Africa Agriculture Development Programme) is a NEPAD program with wide support from international, regional, and national actors that is designed to increase Africa's agricultural productivity.

b. The breakup of the East African Community broke up the University of East Africa. The creation of dozens of new state universities in Nigeria siphoned experienced staff from the University of Ibadan, who gained promotion to top management, professorships, and department heads just when Ibadan's renaissance was attracting students from neighboring Francophone and Anglophone countries.

- *Sustainable funding.* A corollary of the previous point is that R&D mandates have to come from funding bodies that can commit to sustained funding of regional activities. Regional centers of excellence will be sustainable only if it is national policy to play a regional role and maintain that role through resources from the government or institute. It helps if the governance mechanism for agricultural research includes people with access to decision makers if not the decision makers themselves. Regional initiatives must be seen to generate tangible benefits with clear value added to national efforts.
- *Shared regional goals and a policy-making framework.* Regional policies presume there must be some regional policy maker with the power to fund regional challenges and organize regional collective action. The current overlapping of regional economic communities and lack of coherence with R&D structures needs to be resolved. Regional agricultural R&D priorities must be embedded in the strategic agendas of regional economic communities.
- *Subsidiarity and equity considerations.* Regional collective action to deal with inequality, gender, and social unrest across national borders is extremely difficult to organize. Border populations are sometimes marginalized within their own country or even in opposition to the central government,[10] so transboundary problems sometimes may not attract central policy makers' concern. Gender mainstreaming is a widely accepted concept at the regional level but can be implemented only at the national level. Finally, regional collective action takes place in an environment of overlapping mandates and resource competition; the result is that sometimes a national priority is underfunded because it is perceived as something that others will fund.
- *Policies for international markets.* Whether regional research contributes to the emergence of a regional innovation system is linked to the development of the wider environment: regional economic communities, removal of trade barriers, and mechanisms to compensate the losers for their participation in the wider system. Seed policy harmonization, common standards, and certification requirements have fostered seed industries within regions. Regional seed associations emerge as seed markets grow; some national companies become international; and many small companies flourish by finding a niche in areas where multinationals have little interest. Policies and institutions that widen regional markets can facilitate these developments. Investment in

creating these institutions rather than designating coordinators is justified.

LESSONS LEARNED AND GUIDANCE FOR FUTURE INVESTMENTS

The long experience with regional research efforts offers a considerable number of lessons and other guidance. The issues with the greatest relevance for regional research and innovation systems are discussed in the sections that follow.

High-level policy support

High-level policy frameworks such as the NEPAD-CAADP commitments are important if they succeed in raising real financial contributions by national governments. Such high-level organizations have not previously been successful in implementing scientific programs, but if they raise national commitments or create frameworks for regional economic communities to finance regional collective action, they play their role.

Appropriate level of coordination

Ensuring that regional research programs integrate with the national AIS is a matter of how well the national system is integrated with its own subregional organization or benefits from alternative mechanisms. Contracts, networks, strategic alliances, partnerships, and mergers are all ways of achieving coordination. Researchers can coordinate around tasks, problems, target communities, commodities, and scientific challenges. Their linkages with the farmers, the private sector, and advisory services will differ among activities and countries. For this reason, innovation usually emerges through self-organization. Subregional organizations can design mechanisms to coordinate the activities of international centers through regional programs but wisely do not fight activities that escape their control.[11] Attempts to impose "coordination" run the risk of imposing costs that exceed efficiency and relevance gains.

Particular outcomes will be context-specific and often path-dependent

The examples presented here describe a diversity of approaches designed to deal with issues of centralization and decentralization, governance and ownership of a program, political and financial sustainability, responsiveness to demand, and quality of science. It is possible to analyze proposed investments in light of experience with these approaches.

Analyze the AIS prior to investment in individual components

First, recognize that an AIS of some type already exists in a country. It may not be consciously defined, and it may have missing pieces, but it can be described in terms of a system's five elements: its objectives, its components, its resources, its means of coordination, and its environment (context). Second, analyze both external and internal drivers of change. The external drivers include the global and regional economic environment, opportunities for innovation that trade may open for small countries, the global advance of science and technology, and external finance. Other critical factors may be partly or fully endogenous, such as the national capacity to profit from new technology, regional arrangements for accessing and using technology, and ease of access to different markets. Third, since institutions evolve, formulate scenarios around their most likely life cycles. Understanding path dependency allows one to build on it or to overcome it. Finally, to overcome undesirable path dependency, institutional innovation requires leadership, financial commitment, and sustained effort by national decision makers, investors, and donors.

Assess and design interventions against a checklist of six basic criteria

Proposed initiatives can be assessed against six basic criteria for their contributions to national and regional innovation systems. The six criteria are summarized in table 4.4 along with an interpretation of their meaning. Readers should look for examples of their application in the boxes that accompany this TN or apply them to cases with which they are familiar. The six criteria are not exclusive but serve as a checklist of common elements in the success or failure of regional action. They are elaborated in the sections that follow with potential actions to facilitate innovation.

FIT WITH REGIONAL VISION, CONTEXT, AND PRIORITIES. A national innovation system either works within its regional environment and constraints or attempts to overcome them. The most important element is that participants in a regional arrangement share a vision of mutual benefit. Threats of loss may be more compelling promoters of collective action than opportunities for expanded production

or potential gains from cost reduction. Regional efforts designed to share the cost of "lumpy" infrastructure, costly software or laboratory equipment, and highly specialized scientists are logical arguments but may not be seen as compelling reasons for collective action by policy makers. Agreement on regional priorities that create an "effective demand" for research output by pooling the demand of small users assumes that markets and mechanisms exist for research outputs to be used and final agricultural products to be sold. This outcome requires all components of the system to move in a convergent way, however. Economic integration among countries, at least in the targeted sectors, helps achieve wider markets and greater demand for research. Important investments for innovation could focus on (1) strengthening the ability of the regional economic community to create wider markets, (2) creating regional seed associations, (3) harmonizing policy, and (4) building regional buffer stocks.

EFFICIENCY IN THE PRODUCTION OF RESEARCH FOR DEVELOPMENT. The regional arrangement should lead to greater efficiency in the production of research for development in a given priority area. Some gains may be made by eliminating duplication of low-technology research, but much larger gains in research productivity will come through economies of scale and economies of scope,[12] and often the hard decision of where to locate a regional center of excellence will have to be made. A regional arrangement may also bring better access to global knowledge and state-of-the-art technology, along with an understanding of how to use it and better share its benefits across recommendation domains cut by national borders.

Given the heterogeneity of small systems, allocating problems at the regional level through breeding networks will allow better use of limited capacity. In some cases, less complexity may be called for. Commodity programs on water-efficient maize and orange-fleshed sweet potato have regional specificity that is organized along a narrow value chain.

Mechanisms for benefit sharing must be put in place. Participants in the arrangement must understand that the gain from research is not where the expenditure is made but where the benefits are used.

QUALITY AND RELEVANCE OF SCIENCE. The processes by which research is designed and approved need to ensure both the quality and relevance of that research. A proper balance is needed between the articulation of demand (by technology users) and the response from scientists as

Table 4.4 Six Criteria for Assessing Prospective Regional Research Initiatives for Their Contributions to Regional and National Innovation

Criteria and indicators	Interpretation or specific examples
(1) Fit with regional vision, context, and priorities	
Shared vision at national and regional levels	– Do countries have the same perception of potential gains and losses from an initiative? – Is it a shared challenge with the same sense of priority or urgency?
Potential for technical spillovers and spillins	– How much of the recommendation domain is found in each member country? – How much of the targeted area is found in each country?
Economic impact and incentives for regional action	– Are spillovers beyond the research area highly likely to occur without frictions? – Do mechanisms exist to share benefits of regional investments?
Evidence that policies, standards, and regulations encourage spillovers and spillin benefits	– Is there clear analysis of spatial distribution of benefits along the value chain? – Do regulations favor cross-border investment in markets, transportation, and freedom of trade?
Components of innovation system are evolving in compatible way and in line with new challenges	– Are institutional strategies addressing new issues such as climate change, adaptation needs, water scarcity? – Do educational institutions produce future researchers or competent technical staff?
(2) Efficiency in the production of research for development	
Research located in the most effective site in the region	– Is the network hub located where research capacity and infrastructure, ICT, and administrative depth are present? – Does it cluster with other AIS components to serve as a center of excellence? – Is it located in the appropriate agroecological zone if relevant?
Economies of scale can be gained in research from lumpy capital, specialized software, scarce human resources	– Can a centralized capacity draw on a wide testing network for rapid feedback on adapted or adaptable cultivars/results? – Does the hub have a sustainable program of research that ensures its continued ability to serve a regional demand?
Economies of scope can be realized from locating the program in multi-interest center or cluster	– Is there a mixture of disciplines and a balance of strategic and applied research to serve multiple commodities, themes, or functional groups?
Research taps into global knowledge cost-effectively	– Do research contracts and partnership arrangements ensure cost-effective access to global science and technology? – What incentives linked to location, special skills, or environment attract foreign partners? – Can strong ICT compensate for size?
(3) Quality and relevance of science	
Project approval based on strong and independent scientific evaluation	– Are program and project review subject to formal evaluation according to agreed criteria and processes? – Do governance and management respect scientific evaluation?
Mechanism for project approval of projects includes key stakeholder concerns	– How are the private sector's and farmers' concerns included in analysis, consultation, and decision-making processes? – Do processes help build participation and demand by key stakeholders?
The balance between long-term strategic and short-term problem-solving research is appropriate	– Is scientific quality and feasibility ensured by peer review? – Is relevance ensured by governance mechanisms or financial support by client?
Access to advanced breeding, biotechnology, or analytic techniques helps problem-solving research	– How do regional platforms remain open to partners across public, private, and NGO sectors? – What internships, training awards, and fee structures make advanced facilities accessible to downstream demand?
Mechanisms for objective ex ante evaluation, program monitoring, and ex post evaluation	– Does a functional M&E system provide objective evidence of performance against development objectives at the national level? – What peer review from regional experts exists for regional program? – What mechanism exists for subregional partners to ensure performance?
(4) Contribution to national capacity and functioning of national AIS	
Regional initiative provides support for long-term staff development at national level	– Does regional program build national capacity in sustainable way? – How does regional program reinforce gender mainstreaming? – Do standards for regional awards raise the standard for national research? – How does program reduce net regional brain drain?

(Table continues on the following page)

Table 4.4 Six Criteria for Assessing Prospective Regional Research Initiatives for Their Contributions to Regional and National Innovation (continued)

Criteria and indicators	Interpretation or specific examples
Gap-filling by regional initiative is temporary, linked to training, and does not compete for funds	– Does regional staff contribute to training and refrain from crowding out national staff? – Where regional program provides training or equipment, does the national authority assume responsibility for maintenance and replacement?
Regional program enhances ability of national research-for-development institutions to generate political and financial sustainability	– Does regional research meet a national as well as regional need? – Do regional projects enhance the reputation and credibility of national scientists? – Do regional projects connect with the national agricultural innovation system or do they work only on regional and global public goods?
(5) Facilitation of collective action	
Increased ability and incentives for collaboration	– Does initiative identify and address problems that partners could not address on their own? – Does collaboration reduce research costs through complementary skills, resources, or access to partners? – Does addressing transboundary issues enhance regional food security?
Planning and priority setting processes identify "regional public goods"	– Do processes help identify regional problems that can be addressed by collective research for development? – Will initiative bring in issues that were previously believed to be beyond research, such as postdisaster or postconflict recovery?
Compensatory or redistributive mechanisms to ensure sense of "fairness" of arrangements	– Do countries that are not part of the hub have funds to access transborder services from neighbors? – Do their scientists participate equally in regional missions (travel, per diem, consultancy fees) written into initiatives?
(6) Political and financial sustainability	
Design of program and equity in benefit sharing builds financial commitment by governments of the region	– Is host country of regional program willing to maintain commitment to regional collaboration in the absence of donor funding? – Is it able to sustain a regional role on basis of core funding and generation of research contracts? – What sanctions are there for governments who evade their commitments?
Governance structure of initiative brings in financial decision makers or political allies that ensure continuity	– Are relevant ministries represented in governance of the initiative if necessary to ensure political and financial support? – How are key stakeholders maintained as allies of the initiative? – How are nongovernmental regional stakeholders included in the governance?
Regional hierarchy supportive of initiative?	– To what degree can a regional innovation system emerge without a hierarchical structure at the regional level? – What arrangements by regional economic communities are needed to ensure a favorable environment for uptake of research results?

Source: Author.

Note: The criteria in this table were identified inductively by the author from experience with strengthening agricultural research systems and used deductively in applying an AIS perspective to regional research arrangements that are currently being initiated.

suppliers of knowledge and technology. Not all countries in a regional organization will have the same priorities, and how priorities are funded is important enough to be a separate criterion. Investment in strong M&E is necessary to gain political support. The success stories of CGIAR research centers usually focus on two things: either genetic breakthroughs leading to increased productivity or successful control of plant diseases and pests (especially cassava pests). Better methodologies are needed to evaluate returns to natural resource management to ensure proper attention to this aspect of research at the national level.

CONTRIBUTION TO NATIONAL CAPACITY. National capacity is often the scarcest factor. A regional program with large and small members may help to build capacity in the smaller units, if only to enable them to play their role in a regional effort. It is important to identify whether regional efforts compete among themselves or compete with the national programs for scarce scientific capacity and whether there is sustainable national support for this area of research. The politics of higher education make it intensely national. Investment in MSc programs at the national and regional level is needed to give government, NGOs, and aid organizations the staff they require (as discussed in box 4.17 on RUFORUM).

FACILITATION OF COLLECTIVE ACTION. Collective action—in the very specific sense that parties assemble to address shared challenges that they would not or could not address alone—should increase. It may often address transboundary problems of a regional public good nature, such as regional economic policies, transborder watersheds, diseases and pests, and postconflict/postdisaster rehabilitation. At some stage, an innovation system may begin to emerge from the interaction of components and reach a point where components become aware of their connectedness and begin to coordinate through information flows, markets, and new organizations. The system may become "emergent" and create new complexity. Various governments and donor programs may consciously attempt to strengthen linkages or create human and institutional capacities to overcome binding constraints that exist at a given time, but the system may become conscious and self-correcting. A sequence of many small investments leading towards collaboration, collective action, and stepwise integration is needed. In a changing environment it is unlikely that a fully designed system can be imposed from above.

POLITICAL AND FINANCIAL SUSTAINABILITY. Sustainability of regional arrangements depends on the incentives for participation by all parties. A regional role allows a host country to move to a higher level of complexity in its science. The collapse of regional organizations is precipitated most often by national decisions by a member country that free rides, shirks responsibility, becomes unable to carry out commitments to the collective action, or determines it is not sharing equitably in the benefits.

These problems need to be addressed by governance and financing mechanisms, methods of ensuring access to and sharing of benefits, and some formal policing of commitments. Sustainable regional organizations will require a higher-order political and financial mechanism committed to a regional strategy. Self-sustaining regional centers can emerge when a host-country institution accepts a regional role and has a business model that sustains it through a national core commitment, research grants, service fees, and projects. A mutual accountability framework that provides incentives for all partners to deliver on their commitments is necessary.

Box 4.18 describes the continuing challenges of subregional organizations in sub-Saharan Africa to secure funding.

Box 4.18   Subregional Organizations in Sub-Saharan Africa: Challenges with Secure Funding and Collective Action

In Africa, the three subregional organizations—ASARECA, CORAF/WECARD, and SADC-FANR—are quite different.[a] They shared a common goal of increasing African ownership and control of research funded by donors and the centers of the Consultative Group on International Agricultural Research (CGIAR). Their separate evolution reflects the composition of their initial memberships, the role of donors under the Special Program for African Agricultural Research (SPAAR), and path dependency imposed by historical structures and objectives. Different SPAAR donors had different motivations for creating subregional organizations, and the original memberships reflected their interest and funding.

ASARECA. ASARECA was an association of 10 directors of national agricultural research institutes; its geographic spread was defined by countries included in the regional offices of the United States Agency for International Development or the World Bank. Freedom from regional political direction facilitated research collaboration, even among countries whose unstable borders provoked political conflict, but the absence of political coverage limited financial sustainability and commitment to action. In 2007 ASARECA created a mixed board (representatives of donors, national research institutes, and stakeholders), disbanded its networks, and consolidated their work into seven programs with donor funding through a trust fund held in the World Bank. The national agricultural research institutes have not divided up management of the new programs, which remain centralized in Entebbe for administrative rather than scientific reasons. Unlike Latin America, in sub-Saharan Africa no single regional economic community is present to create the environment for wider market integration and policy harmonization.

(Box continues on the following page)

CORAF/WECARD. In West and Central Africa, the SPAAR strategy of base centers and regional programs was a familiar model, especially in the Francophone countries. The need to bridge Anglophone/Francophone structures involved integration on a North/South (interior/coastal country) basis within ECOWAS.

Southern Africa. The Southern African Center for Cooperation in Agricultural Research and Training (SACCAR), an autonomous center composed of frontline states, brought the CGIAR's regional programs under a mixed board of national research institute directors, university representatives, and a donor committee. Donors coordinated their efforts under Cooperation for Development in Africa, subsumed under SPAAR when the World Bank became the dominant donor to research. With the end of apartheid and opening of South Africa, the frontline states remained wary of SPAAR's message of strong "base centers" and "regional programs," which implied centralization around a scientifically dominant South Africa. When SACCAR disbanded in 2001, its functions were brought under a directorate of the SADC Secretariat. It has taken almost ten years for a new subregional research organization, similar to ASARECA and CORAF, to replace SACCAR. The Center for Coordination of Agricultural Research and Development in Southern Africa (CCARDESA) was established in 2010.

Lessons from Sub-Saharan African subregional organizations. No subregional organization has achieved significant financial support from its member governments. Political fragmentation, overlapping regional economic communities, and membership in multiple regional units has prevented the subregional organizations from taking on significant collective challenges such as climate change, postconflict rehabilitation, and development of regional markets. It would be a true example of "collective action" if the governments in each region had assumed responsibility for funding. Through their scientific networks, however, the subregional organizations have successfully addressed some transboundary emergencies and challenges. For example, ASARECA helped Burundi overcome a banana wilt attack with replacement cultivars. The cultivars were developed and maintained in Uganda and multiplied with help from Kenya and Uganda under reallocated funding from regional donors. Like PROCISUR, their next step (and an essential element in regional innovation) is to strengthen technologies, institutional linkages, and policies of the regional economic communities that integrate markets.

Source: Author.

a. ASARECA = Association for Strengthening Agricultural Research in Eastern and Central Africa. CORAF/WECARD = Conseil Ouest et Centre Africain pour la Recherche et le Développement Agricoles (West and Central African Council for Agricultural Research and Development). SADC-FANR = Southern Africa Development Community–Food, Agriculture, and Natural Resources.

Codesigning Innovations: How Can Research Engage with Multiple Stakeholders?

Bernard Triomphe, Agricultural Research for Development, France (CIRAD)

SYNOPSIS

Codesign aims at achieving better articulation between research supply and user-driven demand for problem-solving. It implies that researchers engage systematically with multiple stakeholders in the iterative, adaptive, flexible, and nonlinear process of developing innovations. Core codesign principles include: (1) joint planning, implementation, and decision making related to activities aiming to foster innovation and (2) close coordination among stakeholders at the strategic and operational levels, combining scientific, technical, and local knowledge and other resources. Codesign may be implemented at any scale, depending on the nature of the problem, the innovation being developed, and the types of stakeholder involved. Investing in codesign approaches implies covering the costs of coordination, facilitation, and collective action inherent to working collaboratively at all stages of the codesign process. Investment is also needed to build capacity required by different stakeholders, including researchers.

BACKGROUND AND CONTEXT

Research is not necessary for innovation to take place, but often research is an important part of an innovation process. Innovations usually result from a process of networking and interactive learning among a heterogeneous set of actors, which may include farmers, input suppliers, traders, processors, researchers, NGOs, and government officials. Many research organizations and researchers in developing countries have some experience in research collaboration with other public sector professionals and with farmers (often within the framework of competitive research grants), yet they generally lack the more wide-ranging exposure, related skills, and attitudes to engage effectively in collaborative research with more diverse public and private stakeholders. Individual researchers and research systems need to change and expand the scope of their research, methodologies, and core skills.

Since the 1990s, "new" research approaches have emerged and been consolidated to ensure that researchers interact and collaborate effectively with users in identifying and producing the knowledge and innovations to respond to a rapidly changing local, national, or international environment, be it biophysical or socioeconomic. These codesign approaches include well-tested and documented "branded" approaches (table 4.5; boxes 4.20 and 4.21). Each "brand" has its specificities but also borrows more or less explicitly from related approaches, so the boundaries between approaches are blurred.

"Codesign" is a generic term referring to any approach by which researchers contribute explicitly to developing innovations together with other stakeholders (or users) at all or most stages of the innovation process by making use of four underlying principles (adapted from Liu 1997):

1. The processes of producing knowledge and solving users' problems have equal strategic importance and run concurrently during codesign.
2. All stakeholders involved have the right and a fair opportunity to take part in all stages of the codesign process. Research does not have an inherently stronger input than other stakeholders have. Researchers do not necessarily lead or even initiate the process.
3. Goals, objectives, ethical values, and the way that they are effectively translated into approaches, governance, operating structures, and activities are subject to explicit initial and periodic negotiations and formal agreements.
4. Periodic reflection on the progress achieved and the consistency of the approach in relation to the stated goals is integral and essential to an effective codesign process.

Table 4.5 Key Approaches or Brands Fitting under the Codesign Umbrella

Approach	Key references	Key features and focus	Examples
Participatory technology or innovation development (PTD/PID)	Veldhuizen, Waters-Bayer, and de Zeeuw (1997); Sanginga et al. (2008)	Systematized steps and methods to develop production or natural resource management innovations, with a strong focus on local people, knowledge, and resources	PROLINNOVA program (www.prolinnova.net)
Participatory action-learning and action-research (PAR)	Liu (1997); Almekinders, Beukema, and Tromp (2009); Hocdé et al. (2009); Faure et al. (2010)	Negotiating common goals and setups, combining production of knowledge and problem-solving	ASOSID (box 4.20)[b]
Participatory market chain approach	Bernet et al. (2006, 2008)	Add value by creating interactions and coordination among stakeholders along an existing or new value chain	Papa Andina Box 4.21
Companion modeling (ComMod)[a]	Bousquet, Trébuil, and Cerf (2005); Béguin and Cerf (2009)	Combining and representing different types of knowledge and exploring scenarios for collective action in natural resource management	New irrigation arrangements in northern Thailand

Source: Author.

a. See http://cormas.cirad.fr/ComMod/en/index.htm.

b. Asociación para la Agricultura Sostenible en base a Siembra Directa (Association for Sustainable Agriculture Based on Direct Seeding).

In codesign, the concurrent and explicit application of these four principles contrasts strongly with more conventional R&D and other so-called participatory or collaborative approaches, even though the latter may appear to share some characteristics of codesign.

While this module treats general issues related to the role of research in AIS, this TN focuses on how research may engage in AIS *in practice,* drawing from recent experience with codesign approaches to improve agricultural productivity, develop market chains, and manage natural resources in developing countries.

INVESTMENT NEEDED

Ideally, codesign approaches should be applied only when the context, the problems to be solved, and the stakeholders are well suited to such approaches (table 4.6). Usually the circumstances are right when one or several of the following conditions are met:

- Concerned stakeholders face changes in their socioeconomic or biophysical environment, or problems that typically cannot be solved by one of them alone, because of the complexity and/or scale involved. Examples include managing a dwindling common natural resource or adjusting to new policies or to changes in how markets function.
- A shared understanding of the problems and their solutions can be reached.
- Current scientific and technical knowledge and pre-designed solutions to address the issues at hand are inadequate or inaccessible.

- At least some of the concerned stakeholders have prior experience with the skills necessary for the negotiation, facilitation, and coordination of multistakeholder efforts.
- Codesign requires a positive, open attitude, motivation, and sufficient degrees of freedom among individuals and institutions toward multistakeholder collaboration. This condition implies, among other things, that stakeholders can recognize the legitimacy of all other stakeholders in being part of the process, an incentive structure compatible with codesign exists, and participants have the ability to operate outside the established rules and paradigms within each institution.

Conducting an effective codesign innovation process typically involves organizing three main interlinked and overlapping phases: (1) exploratory phase, (2) implementation phase, and (3) a dissemination and exit phase. Investing heavily in capacity-building of all stakeholders involved, with regard to specific thematic issues as well as the principles and approach of codesign, will be necessary (Triomphe and Hocdé 2010). Box 4.19 summarizes the associated costs and investments.

Exploratory phase

Three goals may be pursued: (1) diagnosing the situation faced by stakeholders; (2) identifying and characterizing concerned stakeholders, their demands, and needs; and (3) negotiating the overall goals of the codesign process and the related institutional and operational mechanisms and arrangements, all of which influence the effective implementation of codesign activities during the next phase.

Table 4.6   Examples of Problems, Corresponding Potential Innovations, and Key Potential Components of a Codesign Approach Adapted to Address Those Problems

Types of problems or issues	Types of innovations	Examples of components of an adapted codesign approach
– Reducing costs or increasing profitability of cropping or farming systems, making farming more environmentally friendly	– New cropping or farming systems – New arrangements and institutions	– Joint experimentation – Creation of multistakeholder alliances and platforms – Multiscale networking
– Unequal access to irrigation among different types of farmers in a given watershed – Conflicts for common resource use (such as pastureland)	– New irrigation or grazing techniques – New rules or institutions at community watershed level	– Joint experimentation – Role-playing games – Participatory and simulation modeling
– Reducing deforestation – Managing erosion and natural resources (such as biodiversity)	– New farming systems – New land uses and new policies and/or regulations for land use	– Territorial multistakeholder committees – Participatory land use and policy planning – Creation of multistakeholder alliances, platforms, and similar mechanisms
– Responding to farmers' and consumers' needs related to crop and food quality	– New germplasm	– Participatory plant breeding – Geographical indications
– Limited access of poor farmers to high-value markets	– Infrastructure development – New food processing techniques – Farmer organization for marketing	– Approaches for pro-poor market chain innovation (such as development of geographical indications) – Supply chain coordination – Public-private partnerships

Source: Author.

Box 4.19   Costs and Investments Associated with Codesign

Codesign approaches require that proper funding (or cofunding) be made available for a number of specific expenses, especially:

1. Holding all necessary initial negotiations among concerned stakeholders: travel and meetings.
2. Enlisting researchers from several disciplines (biophysical and social sciences), as required.
3. Ensuring sufficient staff involvement from key stakeholders, and funding the time of those who are not in a position to support themselves.
4. Expenses related to the proper functioning of multi-stakeholder coordination instruments and mechanisms, such as multistakeholder platforms, steering committees, and facilitation costs, without forgetting the funding needed for corresponding communications strategies.
5. Resources to hire a full-time or at least part-time facilitator or innovation broker (these resources might be especially critical).
6. A multifaceted capacity-building program directed at strengthening the capacities and skills of each and every stakeholder on a variety of topics over the duration of the codesign process.

7. The establishment and implementation of a formal, rigorous, and participatory monitoring and evaluation system, which is necessary to provide feedback to guide the direction and content of the codesign process.
8. Sufficient funding should be made available to document the codesign process and its major outcomes in diverse media, from classical scientific and technical publications to videos, Internet-based products, and policy briefs. There should be as many formats and products as types of stakeholders involved or concerned by the problem.

Other costs typically associated with a codesign approach may not differ much from the costs of other approaches: the cost of running a multisite, on-farm experimentation scheme, costs of a large-scale dissemination strategy, costs of specific research activities, and so forth. It may be possible to share some of the costs among the partners, and as encouraging results are generated, it may be possible to leverage further investment.

Source: Triomphe and Hocdé 2010.

The perspective and perceptions of each stakeholder about problems and opportunities need to be brought shared, understood, and recognized by others as legitimate so that suitable solutions may be identified collectively. Diverse *diagnostic methods* can be used to achieve an accurate collective representation of the situation; they usually involve developing some sort of a conceptual model (or simplified representation) of the problem, as proposed in the ARDI method (actors, resources, dynamics, interaction) described in Etienne (2005).

In *characterizing stakeholders,* the key is to understand the actual motivation and goals of each stakeholder, its history and trajectory, its strengths and weaknesses, its actual political clout, and its past and current interactions with other stakeholders. Specific methods and tools have been developed to elicit this understanding, such as stakeholder mapping or the analysis of sociotechnical networks.

Intense *negotiations* in bilateral and multilateral arenas are needed to identify the mutually acceptable overall goals and objectives of the codesign process. Negotiations also revolve around identifying and agreeing on the roles and functions of each stakeholder and on the resources that each must commit or find. Negotiations need to come up with effective mechanisms for managing the codesign process during implementation at the strategic/governance and operational levels, such as steering and implementation committees. *At the strategic level,* goals and objectives need to be reassessed dynamically and adjustments made periodically to refocus the collective energies and to solve any tensions or conflicts, which frequently arise during multiple stakeholder endeavors. *At the operational level,* a key concern is effective implementation and dealing successfully with technical, logistical, and financial issues. Who is selected to represent the various stakeholders in these committees will greatly influence their eventual effectiveness.

Implementation phase

All activities are conducted in effective multistakeholder fashion, and the implementation is expected to reflect the overall goals, governance, and operational mechanisms established as an output of the exploratory phase. Joint experimentation and participatory M&E are two important activities occurring in this phase.

In *joint experimentation,* the nature of the experimentation depends on the types of innovations sought:

- When new cropping or farming systems are being designed, *agronomic trials* codesigned by farmers,

researchers, and extension agents are often conducted. These same trials support field visits and hands-on training during which stakeholders may be invited to share their experiences and assess the results.

- When the goal is pro-poor market chain innovation, the implementation phase will usually involve *developing and testing new products,* or *finding safe paths* for small-scale farmers to enter high-value markets. This work involves not only technical research and innovation (to develop both pre- and postharvest technology) but also economic and organizational innovation (for example, to test the viability of new products or organize stakeholders more effectively along the value chain).

- Successful technical and commercial innovation often requires *changes in organizations and institutional arrangements.* Institutional innovation may require experimentation with new coordination or collective-action mechanisms, such as a new farmer organization in charge of collecting products for subsequent joint marketing.

In a codesign approach, *participatory M&E* focuses on outputs, the process itself, and on providing the elements needed to assess the continued relevance of the goals and the methods used to solve the problems identified during the exploratory phase. Participatory M&E thus provides strategic inputs for guiding and dynamically adjusting the overall codesign process.

Dissemination and exit phase

An important task is to conclude the codesign process in a way that will lead to sustained and scaled-up application of the innovations developed or enable the process to be used for other issues, in other settings. A priority is to document and take stock collectively of what was achieved, both the expected and unexpected outcomes. The outcomes can take many forms; examples include innovations of different types, new knowledge, individual and collective learning, strengthened capacities, and new institutional norms and behaviors. In this phase, some activities also aim at sharing some of the results as well as scaling them up or out. Successful codesign processes often pave the way for launching activities or programs pursuing one or more of the following goals:

- Consolidating and expanding the use of pilot innovations to achieve more significant and sustainable impacts (scaling up).
- Tackling new problems in the same area, with the same stakeholder group.

- Expanding the codesign/innovation process to new areas and new stakeholder groups, or institutionalizing the corresponding approaches within existing or new institutions.

At times, a codesign approach may also need to be terminated before a satisfactory outcome has been achieved, because conditions for continuing are no longer favorable. Some stakeholders might not wish to collaborate further; they may view the costs and time as too high or too uncertain. What is important under such circumstances is to minimize the potential long-term damage that an unmanaged failure (usually in the form of a conflict or crisis) might cause to future collaboration.

For examples of codesign processes in Mexico and the Andean Region, see boxes 4.20 and 4.21.

Box 4.20 Applying Codesign for Conservation Agriculture in Central Mexico

How it started: identifying a common problem and a potential solution. In early 2000, a small group of international researchers met with representatives from the private sector and the government of Guanajuato State in Central Mexico. They discussed the opportunities and challenges of a joint effort to develop and diffuse conservation agriculture in the Bajío (lowland) region, where a crisis in environmental sustainability was underway. Thousands of mechanized smallholders produced high-yielding cereals (wheat, barley, maize, and sorghum) in the Bajío using large amounts of fertilizer and irrigation water. Production costs had soared, making the profitability of grain production uncertain. Competition for scarce irrigation water was increasingly fierce. Conservation agriculture, internationally heralded for its potential to reduce costs and save water, was seen as a relevant solution. Conservation agriculture had been successfully tested for years in the region, but residual technical problems and poor coordination among stakeholders prevented its significant adoption.

Launching the codesign process. Researchers did not add to the pile of experiments showing the potential or drawbacks of conservation agriculture. Nor did they develop a participatory scheme to fine-tune and test even more locally adapted conservation agriculture with a handful of farmers. Instead, research broke away from well-established paradigms. A flurry of negotiations followed with key local or national stakeholders in the Bajío, including farmer-managed local and regional water associations, the state extension agency, high-level policy makers from the state government, the private sector (input providers, no-till equipment manufacturers), national research, and funding agencies. A major objective was to establish a means of coordinating and implementing a multifaceted conservation agriculture program in the Bajío. It soon became apparent that such a program would be complex, involving numerous diagnostic activities, trials and demonstrations, the introduction of new equipment, training of technicians and farmers, and monitoring and evaluation, among other activities. In 2002, all stakeholders agreed that the rather informal coordination started in 2000 had to make way for a more formal one in the shape of a nonprofit association, ASOSID AC.[a] The key stakeholders were the founding members of ASOSID, which became the recognized agency for implementing the program.

The role of research and the results obtained. Throughout the process, research played key roles. Unusually, it was a major innovation champion in the initial stages, assuming the role of innovation broker by enrolling key stakeholders and tirelessly negotiating (bilaterally, multilaterally) the common objectives and approach. At the same time, it produced technical, economic, and social knowledge about the situation, developed new cropping systems in close collaboration with farmer innovators, and helped shape the enabling environment to establish ASOSID. It strengthened stakeholders' capacity—an investment that continues. Aside from the creation of ASOSID (a major institutional innovation), key results included the increasingly wide adoption of conservation agriculture and related water-saving techniques. The capacity and reach of farmers' local and regional water-user associations expanded. The associations decided to venture beyond their original mandate to manage irrigation water and gradually got involved in advisory services and alternative crop marketing. Although it has gone through several stages, ASOSID is at work ten years later, long after international research ceased to be a major force in its agenda and activities.

Source: Author, based on Triomphe, Hocdé, and Chia 2006 and www.asosid.com.
a. Asociación para la Agricultura Sostenible en base a Siembra Directa (Association for Sustainable Agriculture Based on Direct Seeding).

Box 4.21   Features of Papa Andina's Partnership Programs

Papa Andina is a regional partnership funded by the Swiss Agency for Development and Cooperation and other donors. It involves the International Potato Center (CIP) and national agricultural research organizations in Bolivia, Ecuador, and Peru. Since its inception in 1998, Papa Andina has shifted its focus from implementing a regional research agenda to developing a regional learning agenda and strengthening national capacities for innovation.

Organizational interface. Papa Andina's participatory market chain approach (PMCA) features facilitated, face-to-face meetings that involve diverse market chain actors, researchers, and other agricultural service providers in exploring options for market chain innovation. The facilitating R&D organization then conducts or arranges for R&D on specific innovations. A project team based at CIP (with external funding) has continued to facilitate interactions between researchers, market actors, and decision makers at various policy levels (in theory this task should eventually pass to the market actors). This group sees itself as an innovation broker and plays a lead role in a Learning Alliance that has been established to promote exchanges among different groups working on market chain innovation and development in Peru. Full-time facilitators and innovation brokers reduce some of the transaction costs and coordination issues related to partnerships and networks. Financial sustainability beyond donor funding remains controversial and problematic for Papa Andina, both in CIP and in national research organizations.

Outcomes. PMCA and stakeholder platforms have achieved higher prices for native products, increased farmers' revenues, developed more stable markets for producers of native potatoes (partly through successful branding and marketing), and increased farmer's self-esteem. In Bolivia, new potato products sold to supermarkets enable farmers to receive 30–40 percent higher prices than they received in traditional markets. The innovation network in Ecuador (Plataforma) enabled farmers to raise yields by 33 percent, improve input-output ratios by 20 percent, and increase gross margins per hectare fourfold.[a]

New products and markets. Other key outcomes include the creation of a new brand of high-quality fresh potatoes for the wholesale market, a new native potato chip product and brand, and the first brand of high-quality native potatoes to be marketed in Peruvian supermarkets. Technological innovations improved pest and disease management and the selection of harvested produce. A national platform, CAPAC-Peru,[b] was established to promote the marketing of quality potato products and innovation, in which local actors are gradually taking more responsibility as their capacity and trust increases. CAPAC helped organize small-scale farmers to supply potatoes meeting the more demanding market requirements. When a multinational entered the market, Papa Andina began to work on corporate social responsibility to balance corporate interests with the interests of community suppliers and the environment. Other indirect results include the popularization of native potatoes in Peru's urban cuisine and the establishment of Peru's annual National Potato Day, which caused the United Nations to declare 2008 the International Year of the Potato.

Key lessons

- Approaches such as PMCA require substantial time and resources for capacity development if they are to strengthen linkages between researchers, economic actors, and policy makers.
- Traditional evaluation approaches based on objectives and logical frameworks do not work for innovation processes and innovation brokers' performance. The processes and tasks involved are too complex and results often take some time to be apparent.
- A pro-poor focus is vital to market chain approaches and innovation networks, which run the risk of benefiting those who are better able to take advantage of new market opportunities and innovations. In Peru, native potato varieties have evolved from "poor peoples' food" to a source of national pride, and the main beneficiaries have been the smallholders from the high Andes who preserved and grew them over thousands of years.

Source: Devaux et al. 2009, 2010; Horton et al. 2010; author.
a. Impact statistics from Cavatassi et al. 2009. (b) CAPC = Cadenas Productivas Agrícolas de Calidad en el Perú (Quality Agricultural Productivity Chains in Peru).

POTENTIAL BENEFITS

Benefits of codesign are diverse and depend greatly on the specific goals of each codesign experience. Generally speaking, benefits include a mix of:

- More suitable and diverse innovations that are more appropriate, easier to adopt, and developed more rapidly than innovations generated through conventional approaches.
- Involved stakeholders, whose individual and collective capacities for action, research, and problem-solving are strengthened. If attention has been duly paid to the weakest stakeholders, their technical, social, and at times political endeavors may be empowered.
- Institutions develop better routines and capacities to implement their respective missions and goals, owing to their involvement in codesign.
- New institutional arrangements allowing better coordination and synergies among stakeholders.
- A virtuous, sustainable circle through which, at the end of the codesign process, the various stakeholders are more willing and able to keep innovating as needs or opportunities arise. In short, a greater capacity for stakeholders to take their destiny into their own hands.

For research, experience and skills in applying codesign approaches can result in several additional benefits, such as a greater ability to work in an interdisciplinary fashion and to think systemically. The approach helps to renew and open the research agenda and to reduce the typical divide between research and societal needs.

POLICY ISSUES

Strengthening the capacities of stakeholders involved in codesign is an essential part of the process and improves their ability to interact with each other and with their institutional and socioeconomic environment. Such interactions allow the visions and concerns of a specific stakeholder group to become visible and legitimate to other stakeholders, and hence may eventually influence the scope and nature of the innovations being developed, the distribution of benefits among stakeholders, and other outcomes.

Practitioners must be prepared to deal with the strong ethical and political dimensions of codesign processes. Large power asymmetries can prevail among stakeholders. Codesign processes frequently deal with or uncover conflictive situations. The process can have different consequences for different stakeholders' livelihoods or their respective places and influence in the institutional and political landscape. In some cases, codesign empowers the weakest stakeholders, but empowerment is not automatic; it may be necessary to overcome tensions arising from the resistance and inertia of traditionally powerful stakeholders.

Codesign may affect the strategic positioning of research and its role in the innovation landscape. Research may decide it is legitimate to go beyond its traditional role as a producer of neutral knowledge and invest explicitly in such areas as innovation brokering or documenting stakeholders' positions and rationales in conflicts over resource management.

Public research organizations may also need to become better acquainted with innovation development initiatives and modes of collaboration led by private stakeholders, who often exhibit more responsiveness than public stakeholders to emerging opportunities and who possess the skills and tools to deal with consumers and markets. On the other hand, research may play a vital role in innovation programs by ensuring that public goods are identified, produced, and protected and that political agendas and concerns relating to sustainability, poverty reduction, and equity are duly reflected in the collaboration.

LESSONS LEARNED

Experiences with codesign show that efforts to innovate are most successful when they tackle innovation in its broad sense and diverse dimensions, including technological, organizational, and institutional dimensions. In ASOSID and Papa Andina, innovations ranged from production techniques (new pest management techniques) to new institutions. A narrow focus on predefined solutions is not likely to yield effective innovation.

By necessity, codesign is a highly iterative, dynamic approach. The unfolding of an actual innovation process is by nature highly iterative and dynamic. It typically involves overlapping and interlinked phases and activities, including participatory identification of demands and problems, stakeholder mapping and enrollment, the development of rules and modalities for collective action, joint experimentation on different innovative solutions, capacity building, participatory monitoring and evaluation, joint learning among stakeholders, and the sharing and dissemination of results and outcomes among stakeholders.

Codesign is really a set of guiding principles. It is not a blueprint or a ready-built, standard approach, method, or toolbox that can be implemented "as is." Codesign is only one of many possible and complementary approaches that researchers may need and decide to use, after having thought carefully about its justifications, advantages, intrinsic complexities, and the limits of what it can or cannot contribute.

Innovations may be codesigned on a small to large scale, depending on the issue at hand and the stakeholders involved. While most experience with the approach has been gained at a local or limited scales (in problems affecting a few communities at a time, for example), proponents of codesign now tend to work at multiple scales and/or involve institutions, supply chains, and networks operating at a regional, national, or even international scale.

Codesign requires strong bottom-up, participatory processes shaped by interactions among key individuals ("champions") acting autonomously rather than along existing hierarchical and institutional channels and routines.

Different stakeholders need to be involved. As noted, each stakeholder has a unique legitimacy, role, knowledge base, and contribution to the codesign process, depending on its own stake in the outcome, demands, desires, needs, and previous experience, capacities, and skills in relation to the specific issues and objectives being addressed. For their part, end-users of innovations need to be given a fair opportunity to play a central role throughout the innovation process. Mapping and analyzing stakeholders is an effective way to increase the chances that a codesign approach will be realistic as well as successful.

Demonstrable, early progress with tangible (visible) innovations is important for keeping stakeholders motivated and actively engaged throughout the codesign process. It also increases their sense of ownership.

There are no theoretical limitations to the types and number of stakeholders that can or should be involved. In practical terms, however, the ability to effectively coordinate multiple stakeholders and maintain "reasonable" transaction costs can reach a limit. An important lesson for process management is to keep stakeholders' involvement as flexible and dynamic as possible, with stakeholders entering or exiting the process, or becoming more or less active, depending on the phase of the codesign process.

Stakeholder coordination needs to be formalized, however, via specific instruments (such as steering committees and multistakeholder platforms) to allow joint strategic decision making and effective joint implementation of activities.

Approaches for codesigning innovations still represent a novel field of investment, requiring investments in developing new roles and new methods. To increase the chances that researchers and their organizations will be successful in such endeavors, they need to critically assess the roles they usually play in an innovation process and take appropriate steps to develop new roles and skills that may be needed, such as skills in facilitation, negotiation, building and nurturing partnerships, understanding power relationships and how to deal with them, and reflection. They also need new conceptual frameworks and tools for assessing problems; understanding the diversity of stakeholders' objectives, perceptions, and criteria; exploring new scenarios and innovations effectively and ex ante; and assessing impact in its tangible and intangible dimensions.

Allow flexibility with project proposals. Donors need to adapt their guidelines for acceptable proposals to co-design approaches. Donors usually require clear proposals and funding plans that outline several years of activities, with detailed explanations of what, where, when, and how activities will occur and at what cost. In contrast, when funding "true" codesign approaches, donors should allow the maximum level of flexibility by those submitting proposals. This flexibility includes giving proper consideration to conducting a true exploratory phase (whose outcome, by definition, cannot be known beforehand), allowing a significant margin of freedom for plans and budgets to be developed and adjusted as and when needed, and allocating sufficient funding for the typically significant transaction costs and other specific costs related to operating a codesign approach. Since conditions for an effective codesign process are not always suitable, one outcome of the exploratory phase may be a decision not to engage in codesign after all. Far from being a negative or undesirable outcome, a decision of this kind should be viewed as an excellent way of preventing limited resources from being wasted on a potentially ill-fated process.

Organizational Change for Learning and Innovation

Douglas Horton, Consultant

SYNOPSIS

Earlier sections of this module identified strategies for agricultural research organizations to work more effectively with partners in an AIS. This note discusses the changes that research organizations may need in their management practices, structures, and incentives if they are to use these strategies effectively. It emphasizes the need for agricultural research organizations to become "learning organizations" that are responsive to changes in their environment and innovative in their policies, management practices, and structures. Becoming a learning organization frequently requires: shifting from closed innovation strategies to more open ones; shifting from simple, hierarchical organizational designs to more complex ones that feature multidisciplinary teamwork and multi-organizational collaboration; shifting from traditional planning and implementation systems to adaptive management; expanding evaluation functions to encompass both accountability and learning; and incorporating societal concerns and priorities into performance incentives.

BACKGROUND AND CONTEXT

This thematic note outlines changes in organizational structures, management systems, and incentives that agricultural research organizations can make to improve their performance in AISs. It develops a number of themes introduced previously and offers practical suggestions for improving research-action linkages and partnering in the day-to-day work of agricultural research organizations. The changes discussed may apply to organizations operating at the local or zonal research level, national agricultural research institutes, regional or subregional bodies, and international agricultural research centers. Whatever the case, the aim is to make agricultural research organizations more open and responsive to changes in the external operating environment and more effective in linking research to the practical needs of clients and society more broadly.

INVESTMENTS NEEDED

This section highlights the importance of change management and capacity development. It identifies the main areas in which agricultural research organizations may need to change and where investments may be useful. Priorities for investment to support organizational change are summarized in table 4.7.

Capacity development and change management strategies

The shift from a linear model of technology generation and transfer to an innovation system model, which involves a range of stakeholders from the start, requires new competencies related to communication, participatory planning, facilitation of teamwork, and learning-oriented evaluation. Conventional capacity development has concentrated on developing the knowledge and skills of individuals, but research organizations that perform effectively in innovation systems also require changes in policies, management systems, and incentives.

There are two basic approaches to organizational change: top-down (deliberate) change and bottom-up (emergent) change. In recent years, agricultural research organizations with weak leadership have implemented numerous but disconnected strategies, often driven by external donors. This type of bottom-up change fragments decision-making and jeopardizes the coherence of an organization's programs. Transforming an agricultural research organization into a learning organization requires that bottom-up initiatives be complemented with strong leadership from the top, to ensure that organizational learning takes place and that useful organizational innovations are mainstreamed.[1] Leadership within

Table 4.7 Priorities for Investment to Support Organizational Change

Organizational element	Priority for investment and change
Capacity development and change management	Develop new competencies related to communication, facilitation, and mediation needed to work with diverse stakeholders in identifying and developing new opportunities for technical and institutional innovation
Strategy formulation	Shift from production of research outputs to fostering innovation processes that contribute to broad socioeconomic goals
Accountability and governance	Include representatives of diverse stakeholders, including smallholders, market agents, and consumers, in governance bodies
Partnership policies	Formulate policies for working with partners, including the objectives and types of partnership and principles for decision making, communication, and sharing of costs and benefits
Planning and priority setting	Develop practical procedures for systematic planning and priority setting, which combine stakeholder inputs with analysis of costs and benefits
Monitoring and evaluation (M&E)	Develop learning-oriented M&E systems that clarify "impact pathways," monitor progress in relation to these markets, and use results to improve the design and implementation of ongoing and future work
Incentives for change	Reward teamwork and partnerships that produce practical results. Develop competitive grant schemes for innovation projects
Administration and finance	Increase flexibility in arrangements to allow adaptive management and responsiveness to emerging needs and opportunities
Organizational arrangements	Develop mechanisms or units to manage inter-organizational partnerships with multiple lines of accountability
Beyond the agricultural research organization	Develop specialized innovation brokerage units outside of the national agricultural research organization

Source: Author.

the organization is needed to formulate appropriate goals and strategies, improve policies and management systems, and adjust organizational structures where appropriate.

Organizational change is a highly political process, because there are winners and losers. For that reason, change initiatives need to have a powerful guiding coalition and local support. Effective capacity development and change management require experts, local and international, who possess not only technical expertise but skills in coaching, process facilitation, and management development.

For agricultural research organizations to shift their focus from *doing research* to *using research to foster innovation*, they are likely to need changes in the following areas: strategy formulation; accountability to end-users and beneficiaries; partnership policies; planning and evaluation systems; incentives; administration and finance; and organizational arrangements. Box 4.22 describes how a Tanzanian capacity-development program addresses some of these needs.

Strategy formulation

Reformulation of the basic goals, values, mission statements, and strategy documents of agricultural research organizations can be a crucial area for investment, because this information influences the motivation and guides the behavior of all the organization's members and programs. The mission and strategy of agricultural research organizations traditionally focused on producing research for major

crops and livestock. In an AIS, the mission and strategy should focus on fostering innovation processes that address broader social goals, including poverty reduction; improved food security; improved health and nutrition; and sustainable management of natural resources.

Reformulating the strategy of an organization is not the job of a management consultant or senior manager. It needs to be done in a participatory fashion, involving representatives of a cross-section of management and staff, to build a strong coalition for change within the organization.

Accountability and governance

Accountability refers to the processes and practices that an organization uses to keep its stakeholders informed, take into account and balance their interests, and ensure adequate responses to their concerns (Blagescu and Young 2005) (box 4.23). In agricultural research institutes, accountability procedures usually focus on meeting the needs of funding bodies (the public treasury and donors) and pay little attention to the needs and interests of partners and intended users of research products and services (development programs, farmers, market agents, consumers). In particular, small-scale farmers have little voice and influence in decision making.

Investors can encourage agricultural research organizations to include representatives of different stakeholders, including smallholders, in their governance bodies (see module 1, TN 1) and to incorporate principles of good

Tanzania's Ministry of Agriculture and Food Security wanted to internalize a strong orientation to clients in all of its services. The ministry's Client Service Charter, adopted in 2002, required the ministry to establish service contracts with clients. The national agricultural research system adopted a Client-Oriented Research and Development Management Approach (CORDEMA) in 2003. Under the multidonor Agricultural Sector Development Programme (2006–13), public and private providers of agricultural research use the CORDEMA approach to provide more relevant and effective services. Funding for services comes from performance-based contracts and Zonal Agricultural Research and Development Funds (ZARDEFs) (competitive grants).

Organizational change involves training public researchers and their partners, who also develop a collaborative, market-focused agenda for R&D. Funding is available to plan and competitively fund collaborative R&D. Change management focuses on developing capacities in (1) human resource management, (2) financial management, (3) partnership and linkage management, (4) planning, monitoring, and evaluation, and (5) output orientation, dissemination, and information management.

The capacity development program includes all 24 research managers (national and zonal), 30 zonal CORDEMA facilitators, and more than 280 researchers and partner agencies competing for grants. It began with the development of a National Facilitation Team and subnational training teams. The curriculum was developed collaboratively by the Ministry of Agriculture, a university, a farmer networking organization, and the Royal Tropical Institute of the Netherlands.

Results and benefits. Although it is too early to verify whether agricultural innovation has increased and influenced the impact of research on development, the CORDEMA program has improved awareness that research should be managed as a performance-oriented, demand-driven service. Among other results, the program increased interaction between research, the private sector, and farmers. These "innovation triangles" benefited from research funds available through the grant program and activities related to

District Agricultural Development Plans. Researchers participated in Farmer Field Schools and farmer groups, including farmer research groups. The capacity to develop effective research proposals for national and international research funds improved.

Lessons. Lessons from this capacity-strengthening program include:

- *A comprehensive framework is needed for change management.* It should include regular reflection based on close monitoring of the framework, institutional support, flexibility, and links between all management areas.

- *Build awareness of the need for a client and service orientation at all levels*, among staff of the research organization at the national level, ministerial decision makers, and national policy makers. Broad, effective awareness cannot be achieved by zonal and district champions alone.

- *Provide resources for training and sustained learning with follow-up funding.* To maintain the momentum for change created during training, fund at least part of the subsequent organizational change activities as well as win collaborative research proposals. Otherwise the momentum is lost.

- *Durable change in research organizations concerns not only researchers but all staff.* It includes service units and financial administrators as well as support staff and field assistants.

- *Trained and competent facilitators are needed nationally and locally.* To avoid conflicts of interest, facilitation should not be combined with resource management or implementation.

- *Monitoring organizational and institutional change is essential* to maintain momentum and inform decision-makers.

- *Research organizations need autonomy for full institutionalization of CORDEMA.* In Tanzania, implementation was slowed by a hierarchy of policy makers, many of whom were not directly involved with or committed to the change process.

Sources: Personal communication from Willem Heemskerk (KIT), Ninatubu Lema (Department of Research and Training), and Zainab Semgalawe (World Bank); www.agriculture.co.tz; www.kit.nl; DRT 2008; Hawkins et al. 2009; Heemskerk et al. 2003; Lema, Schouten, and Schrader 2003; Schrader et al. 2003.

Box 4.23   Principles of Accountability

The One World Trust has formulated four principles that agricultural research organizations can apply to improve their accountability, not only to funding bodies but to key stakeholders and partners in innovation processes:

- **Participation.** The organization should involve stakeholders in its decision-making processes and activities.
- **Transparency.** The organization should make information about its aims and activities available to its stakeholders.
- **Evaluation.** The organization should reflect on and learn from past experiences and provide evidence for reporting on progress and impact.
- **Feedback management.** The organization should invite and respond to feedback, comments, and critiques of its activities.

Source: One World Trust, www.oneworldtrust.org.

accountability into the rules and procedures of their boards of trustees and other governing bodies.

Partnership policies

Although agricultural research organizations already work with partners, whose numbers will only grow as AIS approaches become mainstreamed, few agricultural research organizations have formal partnership policies, leading to frequent confusion, inefficiency, and conflict. A priority for organizational reform and for investment in this area is to support the formulation of appropriate policies for working with partners in research and innovation processes. Partnership policies need to define:

- The objectives the organization seeks by engaging in partnerships.
- The main types of partnerships employed by the organization.
- Principles for working in partnership, including decision making, communication, and sharing of costs and benefits (including intellectual property).

Since there is little experience in this area to date, the payoff to investment projects that support development of

innovative partnership policies can go beyond local benefits and include valuable contributions to general knowledge. As indicated in TN 2, procedures for selecting partners and managing relationships are quite different for upstream research partnerships than they are for downstream partnerships to promote innovation. These differences need to be built into partnership policies and management systems.

Planning and evaluation systems

Planning and evaluation in agricultural research organizations are often ad hoc and externally driven. An important area for investment is the development of simple but effective institutional systems and procedures for planning, priority setting, monitoring, evaluation, and impact assessment. The goal should not be to introduce the most sophisticated methods available but to develop local institutional capacity and commitment to continue with systematic planning and evaluation after the initial investment project has been completed.

PLANNING AND PRIORITY SETTING. Planning and priority setting assume even greater importance in the changing context for public agricultural research. As competitive grants come to replace core funding and block grants for agricultural research, as the goals of agricultural research organizations proliferate and become more complex, and as agricultural research organizations work with more partners, they need more systematic planning and priority-setting procedures that combine stakeholder inputs with analysis of the research costs and benefits. Module 7 in this sourcebook provides guidance on planning and assessment approaches and bodies (such as research-innovation councils, research networks, and subsector networks, platforms, or associations) that can perform planning and priority setting. The sourcebook on planning agricultural research (Gijsbers et al. 2001) and the collection on prioritizing agricultural research (Raitzer and Norton 2009) provide useful approaches and methods for planning and priority setting. The value of well-facilitated priority-setting exercises goes beyond their empirical results. By engaging partners and external stakeholders in a process of shared reflection over the validity of assumptions underpinning impact pathways, priority-setting exercises also promote collective learning and strengthen relationships among stakeholders. (Raitzer and Norton 2009, 2).

MONITORING AND EVALUATION. As noted, agricultural research organizations typically employ monitoring and evaluation to satisfy accountability requirements and report to external funding bodies. M&E programs are seldom designed to draw lessons from experience to improve the design and implementation of programs. Similarly, ex post impact assessment is done mainly to document results of past investments and justify future funding, rather than to learn from experience with a view to improving future programming (Kelly et al. 2008; see also module 7).

Improving M&E systems is an important area for investment. Innovative evaluation approaches are needed to:

- Articulate "theories of change" and "impact pathways" for projects and programs, with clearly defined progress markers.
- Monitor and evaluate progress and results in relation to these markers.
- Use the results to improve the design and implementation of ongoing and future research and research-related activities.

Module 7 provides guidance on M&E, and useful approaches are also presented by Raitzer and Norton (2009) and Walker et al. (2008). Issues and ideas for developing institutional evaluation systems for agricultural research organizations are discussed in the inception report for establishing a CGIAR independent evaluation arrangement (Markie and Compton 2011).

Incentives for change

Another priority for investment projects is to ensure that researchers have adequate incentives to communicate and work effectively with others—other researchers in other disciplines, development professionals, agricultural service providers, and farmers and other market chain actors—to promote agricultural innovation. There are two main ways to provide such incentives. The first is through human resource management policies and practices that reward teamwork producing practical results. The second is through competitive grant schemes for innovation projects. Without such incentives, individuals are more likely to continue producing research publications, regardless of their relevance, than to work with partners to ensure that research results are relevant and useful. Module 5 describes various approaches for implementing competitive grants schemes; box 4.24 below provides an example from Peru.

Box 4.24   Promoting Agricultural Innovation through a Competitive Funding Scheme in Peru

During the 1990s, Peru took important steps to liberalize the economy, but smallholders did not share in the benefits. Public expenditures on research represented only 0.2 percent of agricultural GDP. In 1999, the Government of Peru and the World Bank initiated a program to increase the competitiveness of the agricultural sector through the adoption of environmentally sound technologies generated and disseminated by a decentralized technology innovation system led by the private sector. The new program relied on a competitive funding scheme to promote innovation. A Fund for Technical Assistance supported extension services for producer groups and field-level adaptive research. A Strategic Services Development Fund supported research as well training for extension providers. Further support aimed at developing institutional capacity in the public sector to formulate and implement agricultural innovation policy in conjunction with the private sector. INCAGRO, the program coordination unit, resided in the Ministry of Agriculture and was administratively independent of the national agricultural research institute (INIA, Instituto Nacional de Innovación Agraria). INCAGRO's innovative features included:

- *An approach and tools for rigorous and transparent funding* (clear rules, an independent vetting committee, and an efficient and transparent monitoring system).
- *Empowerment of producer groups* to plan and carry out agricultural innovation projects and demand extension services.
- *Use of business plans* to estimate expected economic benefits.
- *Cofinancing of innovation projects* through producer groups and/or alliances.
- *Establishment of regional offices* with staff to facilitate and coach the producer groups and alliances.
- *Innovation project budgets* that include funds for hiring technical experts as innovation brokers.

(Box continues on the following page)

■ *A database management system* that covers all aspects of the project cycle.

Results. The program reached most regions of the country and appears to have increased the volume and quality of available extension services. The research fund also appears to have contributed to setting national priorities for research funding, developing a research agenda, strengthening researchers' capacities for innovation, strengthening producer organizations, and achieving research impacts. An ex post cost-benefit study found that a sample of 171 agricultural extension subprojects financed through INCAGRO had an average internal rate of return of 54 percent.

Not unexpectedly, establishing the competitive funding mechanism outside of INIA created resistance, and a 2008 law placed INCAGRO under INIA. Even so, INCAGRO had helped create institutional capacity in the regions, including producer organizations and public-private alliances which could compete for new cross-sectoral innovation funds established by the government. The largest fund was based on the INCAGRO model and designed by former INCAGRO staff.

A recent law requires INIA to promote the establishment of a national, pluralistic, demand-driven agricultural innovation system. Though many practical aspects of these institutional innovations remain to be worked out and the future of INIA is uncertain, the competitive funding model to promote agricultural innovations has proven effective.

Lessons. The main lessons from this experience include:

■ Institutional success did not depend on designing an institutional model for agricultural innovation for the country but on sound implementation of the competitive funds themselves (in other words, the power of the model was that it created institutional capacities on the demand side).
■ Despite the significant risks involved in promoting innovative approaches from outside the predominant institutional structures, specific circumstances can justify such an approach.
■ Transparent policies and selection and monitoring procedures are keys to successful competitive funds. INCAGRO's operating procedures, information systems, and communication strategies were important in implementing the program effectively.
■ Placing INCAGRO staff throughout the country supported national decentralization goals. Other competitive funding programs in Peru noted the strength of this decentralized approach.

Sources: Klaus Urban, FAO (personal communication); Días Avíla, Salles-Filho, and Alonso 2010; INCAGRO 2010; Fresco 2010; Vargas Winstanly 2010; López Heredia 2010; http://www.incagro.gob.pe/WebIncagro/inicio.do.
Note: More nuanced findings on the Peruvian case are presented in World Bank Independent Evaluation Group (2009). http://lnweb90.worldbank.org/oed/oeddoclib.nsf/InterLandingPagesByUNID/DB83D0B3CC8500D085257 8330014721A.

Administration and finance

As innovation processes are inherently dynamic and unpredictable, working to promote pro-poor innovation requires considerable flexibility in administrative and financial arrangements. Yet flexibility can be difficult to achieve in traditional public administrative systems, which demand considerable forward planning of activities and expenditures. Working to promote innovation processes also requires extensive work "off campus" that is difficult to monitor.

Agricultural research organizations that support or engage in pro-poor innovation may need to adjust their administrative and financial procedures to gain more flexibility and responsiveness to unanticipated needs and opportunities. Decentralizing decision making to units with regional, thematic, or value chain mandates can help bring administration and finance "closer to the field." Another option is to establish semiautonomous bodies (such as foundations) that may use administrative and accounting procedures that are simpler and more flexible than those which are common in the national agricultural research organization within the public sector. PROINPA in Bolivia is a successful example of a foundation dedicated to agricultural research and development (Gandarillas et al. 2007).[2]

Organizational arrangements

By themselves, changes in organizational structure are unlikely to produce the behavioral changes for agricultural research organizations to promote innovation more effectively, although some new organizational arrangements can be useful. They include specialized units responsible for spanning institutional boundaries and for innovation brokering.

BOUNDARY MANAGEMENT. Working across institutional boundaries is essential for the effective operation of innovation systems. The development of units responsible for managing interorganizational relations and partnerships is a new and promising area for investment in agricultural research for development. Agricultural research organizations have tended to partner exclusively with other research entities, but they require a more diverse set of partners to promote innovation. The idea is not to partner for the sake of partnering, but to partner when and where it is essential to achieve impact. One way to achieve this outcome is for specialized bodies to manage communication and relations between agricultural research organizations and other key stakeholders in innovation processes.

It is essential that boundary management units have lines of responsibility and accountability to groups on both sides of the organizational boundary and not report only to the agricultural research organization (Cash et al. 2003:8086). Because the benefits of boundary management units may accrue to several organizations, there are limited incentives for any one organization to take the initiative and bear the full costs. For this reason, boundary management constitutes an important area for investment by national governments or external donors that wish to promote pro-poor innovation.

INNOVATION BROKERAGE. The establishment of sustainable innovation brokerage units (see module 3, TN4) is another potentially high-payoff area for public investment. Innovation brokerage is less concerned with linking researchers in a research organization to external stakeholders than with fostering innovation processes and bringing in needed research products or services from wherever they reside. Brokering innovation processes is related to boundary management but focuses on articulating demands for research products and services, forming stakeholder networks for innovation, and managing innovation processes (Klerkx, Hall, and Leeuwis 2009, 413). In performing these functions, innovation brokers enable other organizations to work together and innovate. These functions may be performed by independent bodies or by units within or attached to agricultural research organizations. Innovation brokerage units have played useful roles in stimulating and facilitating innovation processes, but the participants (particularly small-scale farmers and market agents) have been reluctant to cover the full cost of their operations—hence the key role for public investment.

Looking beyond the agricultural research organization

As the previous paragraphs indicate, the changes required for agricultural research organizations to contribute more productively to innovation systems are not all within agricultural research organizations themselves. To establish legitimacy and trust, it might be necessary to assign innovation brokerage functions to organizations that are independent from the main participants in agricultural innovation processes. Peru's competitive grant scheme (see also module 5, TN2) to promote pro-poor innovation was established in the Ministry of Agriculture and operated independently from the national research institute (box 4.24).

POTENTIAL BENEFITS

Effective capacity development and change management in the areas highlighted previously are essential for implementing the approaches recommended in other thematic notes of this module (linkages, public-private partnerships, regional programs, and codesign of new technologies). The benefits that can be expected from these measures are described in the other thematic notes and innovative activity profiles. Box 4.22 describes the likely direct benefits from capacity strengthening (CORDEMA, Tanzania), box 4.24 describes benefits from changes in incentive and funding schemes (INCAGRO, Peru), and box 4.25 describes benefits arising from sweeping organizational change (NAIP, India).[3]

Given the complex nature of organizational change and the emergent nature of the results, it is notoriously difficult to evaluate and measure the benefits of organizational change processes. The types of benefits that may be expected to result from well-managed organizational change processes carried out under favorable conditions can be summarized as follows:

- Greater awareness on the part of researchers of the importance of working in coalitions with other stakeholders in innovation processes.
- Improved relations between agricultural researchers, policy makers, and economic actors (producers, market agents, consumers).
- Changes in the research portfolio to emphasize research with higher short-term potential impact.

The National Agricultural Innovation Project (NAIP) is a major reform initiative implemented over six years (2006–12) by the Indian Council of Agricultural Research (ICAR), with funding from the Government of India and the World Bank. NAIP seeks to accelerate the collaborative development and application of agricultural innovations involving public research organizations, farmers, the private sector, and other stakeholders. ICAR is responsible for catalyzing institutional change through initiatives in the areas of policy, strategy, governance, financial management, and accountability mechanisms and through a massive human resource development initiative. NAIP has established 51 market-oriented collaborative research alliances of public, private, and nongovernmental organizations as well as farmer groups and international organizations. Applied research focuses on technological innovation in disadvantaged rural areas. Basic/strategic research focuses on such areas as biotechnology, nanotechnology, and postharvest technology. Nearly 60 percent of the research funding provided under NAIP is channeled through competitive grants to research partnerships involving public, private, and nongovernmental organizations.

ICAR faced several challenges in implementing NAIP because of the scale of the project, its broad vision of joint technology development by public and private organizations, and the formation of coalitions with a wide range of partners. It addressed these challenges through extensive awareness campaigns prior to initiating the grant program; a helpdesk to support potential project partners; a sophisticated monitoring and evaluation system; and partnership guidelines that provide for management of intellectual property and sharing of capital expenditures.

Results. The project's main results and benefits relate to organizational and institutional change. NAIP gave public sector scientists direct experience of the challenges involved in partnering with a wide range of nonconventional partners critical for innovation. Lessons from this experience have not been systematically documented to date, but interactions with scientists involved in NAIP suggest the following organizational and institutional outcomes:

- Greater appreciation of the range of skills needed for innovation and the complementary roles of diverse partners.
- The importance of broad consultations with a range of actors before conceiving project ideas and developing concept notes.
- Better understanding of how to develop large-scale projects with multiple partners, which can achieve significant impact.
- More frequent project reviews with partners at regular intervals.
- Improved facilitation of partnerships and brokering of innovation processes.
- Increased confidence of scientists to work with private and nongovernmental organizations.

Lessons. Externally funded programs such as NAIP provide useful opportunities for researchers to learn how to work with the wide range of actors needed for innovation, but such partnerships are challenging for organizations with a long history of working in isolation. Aspects of research project management (review, financing, procurement, and so forth) in public organizations must change to provide the support and flexibility for partnerships to flourish. Systematic assessment of and reflection on the experience gained through NAIP will yield valuable insights to further reform the national research system and enable good practices fostered by NAIP to take root and multiply.

Sources: Rasheed Sulaiman V, Centre for Research on Innovation and Science Policy (CRISP), Hyderabad (personal communication); Mruthyunjaya 2010; NAIP 2010.

- Greater influence on policy processes.
- Better mobilization of resources to support research and innovation processes.
- Improved uptake and use of research results.
- Expanded socioeconomic and environmental benefits.

POLICY ISSUES

Organizational change directed at enabling public research institutions to participate more fully in the innovation system must give particular attention to three policy issues. They include gender and equity issues, the national commitment to change, and the possibility that broader reforms may be needed to support innovation.

Dealing with gender and equity issues in organizational change

Investment projects are convenient mechanisms for supporting agricultural research organizations in addressing important and complex issues, such as gender, equity, and empowering poor people. In fact, having access to external resources and legitimacy can often be crucial for agricultural research organizations to begin working on these issues. Gender and empowerment issues are especially important for pro-poor agricultural innovation, because women feature so prominently in the target population. Additionally, empowering farmers and strengthening their organizations may be essential for the success of multistakeholder processes. The *Gender in Agriculture Sourcebook* (World Bank, FAO, and IFAD 2009) presents principles and approaches for introducing gender work into agricultural research organizations.

Key role of national leadership and commitment to change

In any organizational change effort, it is important to keep in mind that while external agents can motivate and support change, local commitment and leadership are essential for the initiative to succeed. A key role for external change agents is to assess and cultivate local commitment and leadership among policy makers for the organizational changes that will enable the agricultural research organization to promote agricultural innovation more effectively. The local commitment of resources to the change process is a key indicator of commitment. Where local leadership and financial commitments for change are not forthcoming, *the appropriate decision might be to not proceed with the investment project.*

Need for broader public sector reform

As public agricultural research organizations are part of larger public administrative systems, successful efforts to introduce changes needed to perform more effectively in the AIS may require changes in the broader system of public administration. For this reason, modernizing agricultural research organizations is best viewed as part of a broader initiative of public sector reform. A recent evaluation of World Bank experiences with public sector reform summarizes useful lessons on what works and why (World Bank Independent Evaluation Group 2008).

LESSONS AND RECOMMENDATIONS

A number of lessons and recommendations provide guidance for initiating organizational change that makes it possible for public research organizations to participate more fully in the AIS, especially as proponents of pro-poor innovation. They are discussed in the sections that follow.

Investment projects may be useful vehicles for initiating and mainstreaming changes

Organizational change efforts are fraught with difficulties and often lose momentum or veer off course. Participants within the organizations concerned may lack or lose legitimacy and can benefit from external support and guidance. For these reasons, it is useful to organize organizational change efforts as "institutional projects" with defined leadership, goals, capacity development strategies, budgets, timelines, evaluation procedures, and lines of accountability. In this sense, investment projects play a useful role in organizational change. One thing to keep in mind, however, is that while external agents can play useful roles in supporting and legitimizing change processes, leadership for organizational change must come from within the organization. Experience shows that it is easier to introduce changes through externally funded projects than to mainstream changes in agricultural research organizations. On the other hand, a large-scale project such as NAIP (box 4.25), which has allowed ICAR to develop, fund, and operate a large number of consortia, can help change efforts to reach a critical mass and become embedded in organizational routines. From the very start of a project that aims to promote organizational change, it is important to develop strategies for mainstreaming innovations.

Adaptive management Is needed for organizational change projects

Change processes are highly dynamic and inherently unpredictable. For this reason, it is inappropriate to attempt to plan organizational change projects in great detail and then implement them rigidly as planned. Adaptive management is needed for change projects, and project managers in external funding bodies may need to be creative to maintain the required accountability without imposing undue limitations on the local teams who need to spearhead organizational change.

Developing internal capacity for organizational innovation is crucial for agricultural research organizations

To keep up with rapid alterations in the social, economic, environmental, and technological landscape, agricultural research organizations need the capacity to adjust their policies, management practices, and structures as conditions change. The continuing success of an agricultural research organization will depend on its capacity for organizational innovation. Agricultural research organizations often fail to develop this capacity because of the lack of continuity in policies and leadership in agricultural research.

Bringing about cultural change takes time and multipronged approaches

Organizational changes (institutional change) may take considerable time to take root; withdrawing support too quickly can jeopardize the sustainability of results. Change agents and donors need to keep this in mind. In many cases, the changes that are needed to allow agricultural research organizations to contribute more effectively to pro-poor innovation amount to a significant shift in culture, and organizational cultures are notoriously resistant to change.

Past organizational change has focused on formal structural issues—position titles, reporting relationships, and the titles and boundaries of various organizational units—yet day-to-day activities are more often influenced by formal and informal rules and norms enshrined in an organization's "standard operating procedures." Deep changes in organizations also produce losers as well as winners, and the potential losers often fight long and hard to retain their status and privileges. The high rate of turnover of managers and researchers in many agricultural research organizations presents another important challenge. Innovation system approaches are seldom taught in agricultural universities and must be introduced to new staff members when they join the organization. If staff turnover is high, these "new approaches" may need to be introduced again and again.

Strengthening capacity needs to be skillfully meshed with managing organizational change

Public organizations are often perceived as resisting change. Many seek capacity—the ability to get things done—but not change—a different way of doing old and new things. In working with agricultural research organizations, it is important to understand which aspects of the status quo are amenable to change and which ones are not, so that an appropriate capacity development initiative can be designed and appropriate alliances can be forged with progressive elements in agricultural research organizations, NGOs, the policy community, and the private sector. The case of EMBRAPA in Brazil illustrates how an agricultural research organization can successfully embrace both capacity development and organizational change.

Redesigning a Livestock Research Institute to Support Livestock Development within an AIS Approach

Ranjitha Puskur, International Livestock Research Institute (ILRI)
Peter Ballantyne, International Livestock Research Institute (ILRI)
Patti Kristjanson, World Agroforestry Centre (ICRAF)

SYNOPSIS

The International Livestock Research Institute (ILRI) has undertaken significant institutional change as it attempts to shift its research from a linear, science-driven approach to an innovation systems approach through the adoption of a research-for-development strategy, fostering of new partnerships and knowledge brokering roles, reorganizing and refocusing its research focus, and by adjusting its skills and human resource needs. To illustrate these changes and their rationale, this profile draws on lessons from a wide range of projects and research into linking knowledge with action.

CONTEXT

The imperative to invest in agricultural research that provides more benefits to more poor people is driven by the increasing numbers of poor throughout the world, the global food crisis, and evidence of mounting climate change, among many other forces. Livestock are a key asset for poor households, especially women, and they often contribute to better health and livelihoods. Yet the capacity of livestock-related research to produce measurable reductions in poverty has been questioned, indicating that perhaps other, less linear approaches merit evaluation.

Livestock systems and mixed crop-livestock systems are inherently complex and diverse. National agricultural research and extension systems in developing countries are relatively weaker at working on livestock than on crops. Public services for animal breeding, health, and market information are grossly underinvested and often underdeveloped, and private participation remains quite limited. These conditions make it especially challenging to develop livestock innovation systems.

PROJECT OBJECTIVE AND DESCRIPTION

The objective—redesigning ILRI to support livestock development within an AIS approach—was attempted by developing an approach that uses innovation systems and value chain perspectives to design and implement an expanding portfolio of research-for-development (R4D) projects with an emphasis on developing innovation capacity among system actors with external funding (box 4.26). Adopting this approach has required considerable innovation in how livestock research is done and in the issues it addresses. That innovation was supported by developing a new research strategy and approach, impact orientation, changes made to human resource and discipline mix, partnership management, and strategic communication (described in more detail under "innovative elements").

INNOVATIVE ELEMENTS

Over the past decade, ILRI's organizational structure and research approaches have evolved to pursue the new agenda. Some of the key and innovative elements are:

- *A research-for-development strategy and approach.* ILRI's research strategy features a holistic systems perspective extending from production and markets to institutions and policy. Its R4D projects balance technical and process issues through orchestrated innovation networks, in which coalitions of actors along particular value chains (such as those for dairy or small ruminant production) identify knowledge required by target groups and test options to address them. The networks emphasize joint action and learning.
- *A knowledge brokering role.* ILRI engages as a knowledge partner that integrates or bundles complementary knowledge and technologies to promote pro-poor livestock development. It uses knowledge to influence

A starting point for building capacity in livestock innovation systems is to make it easier for the actors to innovate—through organizing, partnering, and linking in a number of ways. This form of capacity building is a major element of recent projects by the International Livestock Research Institute (ILRI) and its partners.

Networking actors. The Fodder Innovation Project (funded by the UK Department for International Development) works with partners within and outside government in India and Nigeria to form innovation networks that enable actors in local livestock systems to organize for innovation. Working with a diverse set of actors in the system made it possible to address broad system constraints rather than narrow technical issues. Partner organizations have started to institutionalize this approach (see box 4.27 for details).

Designing interventions around a service hub. The East Africa Dairy Development Project, initiated in late 2007, has built upon ILRI's experiences in Kenya and elsewhere to design interventions around a "service hub" that develops a network of actors to introduce, test, and offer a range of services, technical options, and information.

Forming public-private partnerships to reach clients. Several projects engendered new partnerships between public and private agencies:

- The Livestock, Livelihoods, and Markets project (LiLi), started in 2007 with the International Crop Research Institute for the Semi-Arid Tropics, promotes better delivery of livestock services and market participation of smallholder goat and cattle keepers in Southern Africa. Through an innovation platform facilitated by the Namibia National Farmers Union in the LiLi Project, AGRA (a commercial agricultural cooperative) sponsors a veterinary outlet as part of its Social Responsibility Outreach Program.

- Partners from public, private, and nonprofit organizations developed and tested index-based livestock insurance in Kenya's Marsabit District in a project initiated in 2009. The insurance proved commercially viable and will be scaled up for use in Ethiopia. The first pilot involved Equity Bank of Kenya, UAP Insurance, and Swiss-Re as commercial partners. More than 2,000 contracts have been issued, covering livestock worth over US$1 million and attracting premiums exceeding US$77,000.

- A vaccine against East Coast fever has existed for more than three decades. Highly effective and in great demand, the vaccine has been produced in ILRI's laboratories, but more widespread distribution would require an effective cold chain. To scale up production and make the vaccine more widely available, the Global Alliance in Livestock Veterinary Medicines is partnering with ILRI and private companies to establish viable commercial production and delivery systems (module 6, IAP 2).

New partnerships broaden the participation of the poor. Several new projects respond to zoonotic diseases such as avian influenza by building the capacity of veterinarians and public health officials in early detection, diagnosis, and response. The projects mitigate disease risk by improving coordination at the national level. An "ecohealth" project initiated in 2009 in Southeast Asia develops community-led options to prevent and control emerging zoonoses. A group of regional health and disease surveillance networks and institutions catalyzed this effort. Under the Safe Food, Fair Food Project, initiated in 2008, ILRI and its partners promote risk-based approaches to improve food safety and the participation of the poor in informal markets for livestock products in West Africa.

Source: Authors; www.fodderinnovation.org; http://www.ilri.org/ibli/; www.GALVmed.org; http://www.ilri.org/EcoZd; http://www.ilri.org/SafeFoodFairFood; http://www.slideshare.net/ILRI/using-hubs-to-increase-smallholder-farmers-access-to-services-experiences-from-the-east-africa-dairy-development-project; http://mahider.ilri.org/bitstream/10568/1787/1/InnovationPlatformMozambique.pdf.

actions globally and in target regions, facilitates and convenes livestock R&D actors around pro-poor livestock issues, and identifies gaps in knowledge and technologies to fill.

- *Reorganizing and refocusing research.* There is no perfect way to organize human resources, but ILRI recognized that impact depended on replacing discipline-based and geographically specific research projects with a new culture of working across disciplines and thematic areas. ILRI has built up multi- and transdisciplinary research capacities and added capacity for poverty and gender research and impact assessment. Its Science Advisory Panel helps to assure research quality, renew intellectual capital, and provide an informal, outside evaluation of ILRI's work. Links between strategy, planning, and research implementation have been strengthened.

- *Aligning human resources and skills.* ILRI began to change its culture by building a new mix of capacities (scientific, managerial, and business and partnership management) and testing staff evaluations that reflected the multifaceted roles involved in its new way of working. Financial and other systems were reformed in parallel. ILRI emphasizes larger projects to improve the efficiency of human resources, finance, and administrative support.

- *Forging strategic R&D partnerships.* As indicated in box 4.26, ILRI engages proactively with key enabling partners (policy makers, regulators) and implementing partners (farmers, market agents' organizations, private companies, NGOs, and government) and provides incentives to research managers to do so. Its R&D partnerships benefit from complementary competencies and capacities and its work is aligned with broader government, NGO, and private initiatives. A partnership strategy and guidelines (ILRI 2008) support these efforts.

- *Strategic communications and knowledge management* play a key role in engaging and supporting partnerships, influencing the global and regional livestock agenda, and making ILRI's research outputs accessible. Its communications strategy differentiates the information needs of its stakeholders and networks of influence.

BENEFITS, LESSONS, AND ISSUES FOR WIDER APPLICATION

Lessons and issues for wider application from the new approaches, as well as from the "Linking Knowledge with Action" study (Kristjanson et al. 2009), are summarized in the sections that follow. Box 4.27 presents detailed lessons from the Fodder Innovation Project, which was instrumental in the design of subsequent projects and reflects an AIS approach.

Designing livestock innovation systems and processes

Localized innovation systems (often established around sharply defined value chains) are most effective and have the greatest potential to create impact. Extracting meaningful, practical principles and lessons from these context-specific, path-dependent innovation processes makes it complicated to scale up successful approaches or replicate them. Practitioners must give rigorous attention to learning what works where and how, and then make judicious use of the elements. Building on the social capital of actors, their history of collaboration, successes, and current initiatives can overcome some of these challenges.

One question is whether innovation system approaches need to be expressed through projects. Is it possible to build coordinating mechanisms into innovation systems without bureaucratizing and immobilizing them in a top-down program? Where should ownership be located? In the national agricultural research system? Are the most appropriate contexts for an innovation system approach in commodity research, in regional programs, or in programs based on agroecological areas? These questions are part of a long list of queries that practitioners will need to consider in deciding on the best means of achieving their goals.

Orchestrating coalitions and building innovation capacity

Spaces and environments have to be created to facilitate interaction and communication among actors in the innovation system. They need a place to articulate demands and promising solutions and create the "pull" that elicits innovation. The scale (national, regional, district, or village) of such platforms depends on the problem being addressed, coordination requirements, and structure of the value chain. Experience shows that effective projects require a year simply to lay the groundwork. This important issue needs to be highlighted and negotiated with donors. For example, the Fodder Innovation Project created loose networks of actors around the issue of fodder scarcity at the district level, but as the networks emerged, it became clear that the focus needed to be at the broader level of the livestock value chain.

Are partnerships among individuals or organizations? Often projects identify like-minded individuals in organizations, with personal contacts and relations playing a big role. This means of operating is inherently unstable in the longer term. It is critical to test and learn how networking can become a routine in the organizations involved and not be limited to select individuals.

Boundary spanning and brokerage functions are critical. Brokers by definition have to be good communicators who

Box 4.27 Lessons and Operational Issues from the Fodder Innovation Project

The Fodder Innovation Project provided practical guidance for implementing other projects with an innovation systems orientation:

- *Fodder was too narrow a theme for building a network.* It is more appropriate to build networks and innovation capacities around crop-livestock value chains that mobilize wider coalitions of partners and more interest. Appropriate technology introduced through partnerships that ILRI had made prior to the project proved to be a useful catalyst to involve new stakeholders and raise and address broader system constraints.

- *Building true partnerships, facilitating stakeholder platforms, and building innovation capacities take time.* These processes and projects need longer time frames to mature and gain currency in policy debates and organizational change.

- *Innovation processes need one or more organizations or individuals to assume the critical roles of broker, connector, and catalyst.* An organization's ability to do this depends on its particular situation. The history of the partners and stakeholders, their social capital, and the legitimacy and credibility they bring are all critical factors.

- *Monitoring and evaluating the processes and resulting changes are essential but far from trivial tasks.* Traditional logframes and monitoring and evaluation systems are inadequate for measuring many of the indeterminate outcomes of innovation systems (see module 7, IAP 6 and Lilja et al. 2010).

- *Financial management and planning must be flexible* and adept at accommodating emerging opportunities and challenges.

- *Engage policy actors from the beginning* to identify windows for influence and for ownership of research results. Policy stakeholders have observed that the evidence of impact is very valuable but the evidence base is too small.

In its examination of the Fodder Innovation Project (among others), the "Linking Knowledge with Action" study concluded that projects are more likely to link knowledge with action when they (1) recognize that scientific research is just one "piece of the puzzle," (2) apply systems-oriented strategies, and (3) engage the partners who are best positioned to transform knowledge jointly created by all project members into actions (strategies, policies, interventions, technologies) leading to better and more sustainable livelihoods. The knowledge flows both ways between practitioners/implementers/policy makers and researchers—making the emphasis on linking *with* action rather than linking *to* action an important one.

Sources: Authors; Kristjanson et al. 2009; de Haan et al. 2006.

are skilled at supporting collaboration and interactive processes that involve different types of stakeholders. A critical function of brokers is to manage and deal with large asymmetries of power among actors. Brokering roles can be played by local government, extension services, CSOs, national research systems, or even the private sector, depending on the constellation of skills, capacities, social capital, legitimacy, and credibility they possess. Engaging nontraditional partners like the private sector is still a challenge for NGOs and government as well as public research institutes. ILRI has found that assuming these brokerage roles is not always the best solution, given that these intensive, long-term, and local processes demand continuous engagement that is rarely supported by short project periods.

It seems much more logical that ILRI should focus instead on building the capacities of key partners to play these roles. This approach poses its own challenges: The skills required cannot be mastered easily in formal training alone. They require substantial coaching and mentoring on the job. Two vital questions for project designers to answer are who is best placed to play this role, and who is responsible for setting arrangements in motion.

Researchers will need to hone their skills to assume the roles that the innovation system requires, and researchers from complementary disciplines rarely found in traditional research organizations will need to be engaged: anthropologists, political economists, communications specialists, and project managers, among others. New institutional arrangements will make it possible to work more effectively with partners to capitalize on the competencies and human resources they bring.

Monitoring, evaluation, and impact assessment: new approaches urgently needed

In research-for-development projects, finding the right balance between the production of international public goods and the achievement of local development impact is a recurring challenge for international organizations like ILRI. When such projects attempt to build innovation capacity, they often must choose between a capacity-building approach aiming for sustainability, with much wider and deeper potential impacts, or an approach that seeks the rapid "adoption" of research products to raise productivity and incomes (often only for a short time). Projects are needed that blend the implementation of a good development strategy with rigorous scientific research.

Research projects are much more likely to link knowledge with action when they are designed as much for learning as for knowing. These projects are openly experimental, embracing failures so as to learn from them throughout the project's life. This kind of learning does not occur unless risk-taking managers are funded and rewarded; these managers also must be evaluated regularly by external experts (Kristjanson et al. 2009).

It is important to develop M&E frameworks to track both processes and outcomes and serve the twin objectives of learning and accountability. In the typical three- to four-year research-for-development project, it is difficult to demonstrate change, because it often depends on complex processes and interactions among diverse organizations and individuals coming together for the first time. In the Fodder Innovation Project, technical, institutional, and organizational changes appeared to reinforce each other and generate improvements in livestock systems that could improve livelihoods. In reality, it was difficult to draw clear causal links in these complex adaptive systems and attribute specific changes to specific interventions. Impact assessment frameworks and methodologies are still imperfect tools for demonstrating impacts and proof of concept. More appropriate tools must be developed and tested.

Pro-poor partners help make innovation pro-poor

The Fodder Innovation Project showed that working through and with partner organizations that had an explicit pro-poor mandate and agenda helped target interventions better and ensure pro-poor outcomes through negotiation in the networks. Although service delivery generally improved in Fodder Project areas owing to the networks' actions, better service delivery did not guarantee that the poor would benefit. The possibility that they would benefit increased only when champions in the network negotiated the conditions to ensure that outcome. It is vital to gain greater clarity on which alternative mechanisms will ensure pro-poor outcomes if partner organizations are not specifically committed to those outcomes.

Policy engagement

Innovation is more likely to occur if it is fostered by specific policies and institutional arrangements. The evidence and learning from research projects can inform policies so that they result in better outcomes and impacts. Although many organizations tend to assume that they understand policy makers' needs for information and knowledge, in practice research organizations often seem to lack an adequate understanding of policy processes and the best mechanisms to incorporate evidence and knowledge into policy decisions. Engaging policy actors from the outset is one strategy for enabling policy makers to influence and own research results. Engagement in policy processes demands special expertise and targeted, strategic communication.

An Innovative Approach to Agricultural Technology Development and Transfer in India

Mohinder S. Mudahar, Consultant

SYNOPSIS OF PROJECT DATA

Project: National Agricultural Innovation Project
Cost: US$250 million (IDA amount: US$200 million)
Date: September 18, 2006 to December 31, 2012 (now extended to June 30, 2014)
Contact: Paul S. Sidhu, World Bank

DEVELOPMENT CONTEXT

Almost 85 percent of India's poor live in rural areas and depend on agriculture for their livelihood. Despite the large potential of India's agricultural sector, overall growth in agricultural production and productivity is low. Agricultural practices for managing natural resources (land, water, and biodiversity) remain unsustainable, and the transition to a market-oriented, globally competitive agricultural sector has been slow. The Government of India views agricultural R&D as a critical means of improving agricultural productivity and increasing agricultural growth. Strengthening institutional capacity in the NARS, improving coordination among institutions within and outside the agricultural research system, and promoting partnerships between national agricultural research institutions, the growing private sector, and NGOs are essential to speed the transition to a more competitive agricultural sector. The National Agricultural Innovation Project (NAIP), fourth in a series of projects funded by the World Bank to improve agricultural research and technology in India, addresses these challenges by changing the way in which scientists, farmers, and agricultural entrepreneurs interact in the national AIS.

PROJECT OBJECTIVES, DESCRIPTION, AND EVOLUTION

NAIP incorporates lessons from the three earlier projects (representing almost 25 years of experience), including the need to develop public-private partnerships, integrate technology development and transfer mechanisms, and finance research through competitive research grants. The project's development objective is "to contribute to sustainable transformation of agriculture and accelerate the collaborative development and application of agricultural innovations between public research organizations, farmers, the private sector, and other stakeholders."[1]

The project seeks to achieve this objective by strengthening the role of the Indian Council for Agricultural Research (ICAR) in catalyzing and managing change in the NARS and by financing agricultural research through innovative consortiums of public organizations, universities, private enterprises, NGOs, and other stakeholders. Activities pursued under the project include:

- Strengthening ICAR's technology foresight and policy analysis to envision and plan for future needs.
- Strengthening ICAR's communications and information capacity through better dialogue and interaction with the public at large, farming community, private sector, and within the ICAR system itself.
- Reinforcing the research system's capacity through national and international training.
- Improving technology transfer and commercialization through business planning and development units and IPR management.
- Adopting organizational and management reforms such as M&E, procurement, and financial management throughout the ICAR system.

The project also promotes the development of three kinds of multistakeholder, multidisciplinary consortiums of public and private organizations, universities, NGOs, and others focusing on three high-priority research themes. *Market-oriented, collaborative research alliances* focus on sustainably improving the productivity, profitability, and sustainability of selected agricultural value chains. *Livelihood research alliances* focus on strategies to sustain secure rural livelihoods in about 110 disadvantaged districts, mostly in rainfed, hilly and mountainous, dryland, tribal, and coastal areas. *Basic and strategic research alliances* focus on well-defined areas of frontier science with potential applications for problems in Indian agriculture.

Promising consortiums and research alliances are selected through a competitive process, and NAIP funds their proposed research. Members of each consortium are jointly responsible for the governance, design, and implementation of their research programs; maintaining satisfactory fiduciary and safeguard arrangements; applying the resulting innovations; and disseminating new knowledge through conferences, innovation marketplaces, networks, and communications strategies.

INNOVATIVE PROJECT ELEMENTS

India has one of the world's largest public agricultural research systems.[2] In this context, NAIP produced three critical innovations: scenario planning, new kinds of partnerships, and the Helpdesk.

- *Scenario planning.* ICAR and World Bank teams conducted a scenario planning exercise that identified and analyzed critical policy and institutional challenges facing the agricultural sector and identified corresponding reforms that would strengthen the research system's ability to meet those challenges. Scenario planning enabled ICAR management to assess the consequences of alternative reform scenarios, including their likely benefits and impact, and identify specific reforms to be supported through NAIP. This process has not only increased government ownership of the reforms but also its commitment to implement them.
- *Expanding capacity and resources via partnerships.* The underlying principle of NAIP is the formation and management of consortiums that bring research institutions together with those who use research results. The collaborative arrangements developed by these stakeholders optimize the use of research resources in an enhanced process of innovation, value addition, commercialization, and technology transfer that solves specific agricultural development problems. Research proposals prepared by the consortiums clearly define the roles of the consortium leader and the other partners, including the budget for each institute. The broad array of participants and clarity about their specific roles and

contributions lead to significant synergies and value addition in the design and implementation of their research projects. ICAR and NAIP management have invested considerable time in building partnerships and providing support through meetings, workshops, and the Helpdesk.

- *Helpdesk.* The Helpdesk was established to support the new and more challenging partnerships that the consortiums represent. The project outsourced Helpdesk functions to one national institute and informed the prospective consortiums that it was there to help them in a number of ways: by providing guidance for preparing concept notes and full research proposals, assisting in matching consortium partners, and helping to overcome initial problems in managing the consortiums. The Helpdesk does not charge for its services. The experts managing the Helpdesk understand all of the process and details involved in forming consortiums; developing, selecting, and approving concept notes and full proposals; and the priority research themes. Users found the Helpdesk effective in facilitating the proposal selection process and forming consortiums. The Helpdesk used a number of tools in its work: the Helpdesk portal, e-learning and multimedia modules, databases of potential partner institutions and organizations, case studies of agricultural projects using direct e-mail responses to potential consortium members.

BENEFITS, IMPACT, AND EXPERIENCE

NAIP was approved in April 2006 and the approval of consortiums was completed only in December 2009. The sections that follow describe some of the early results. Readers interested in tracking the project's progress are directed to the NAIP website (http://www.naip.icar.org.in).

Overwhelming national interest in the consortium approach

The number of consortiums was three times the number anticipated, far exceeding expectations. The overwhelming response enabled consortium leaders to assemble the consortium partners with relative ease. From a total of 188 consortiums, 142 were selected through a two-stage competitive review, and the remaining 46 were sponsored. The average consortium budget is about US$1.4 million, of which approximately 62 percent of the committed amount went to the 188 consortium leaders and 38 percent to the

646 consortiums' partners. Table 4.8 summarizes the types of participating institutions, their relative commitments, and their budgets.

EXPANDING THE INSTITUTIONAL BASE FOR AGRICULTURAL RESEARCH

NAIP introduced greater pluralism into agricultural research, given that 38 percent of consortium institutes (leaders and partners combined) come from outside the ICAR–state agricultural university system. Through NAIP, ICAR has promoted public-private partnerships on a large scale for the first time. The project has provided the opportunity for ICAR and the state universities to collaborate with the Indian Institutes of Technology (IITs), the Indian Institutes of Management (IIMs), general universities, and institutes in the Council for Scientific and Industrial Research. The main motivations for these institutions to participate in NAIP are the opportunity to work with agricultural scientists and access their skills and facilities, the opportunity to work in ICT and biotechnology applications in agriculture, and the availability of research grant funds.

Positive experience from partnerships

As of this writing, field visits, workshops, and supervision missions indicate that the consortiums have been working smoothly. There is consensus in ICAR and among consortium partners that the consortium approach has promoted pluralism, synergy, teamwork, value addition, learning, and better research; they believe that as a result the development impacts will be much larger. On average, each consortium has four partners. Box 4.28 summarizes the main issues raised by partners.

Focus on high-priority agricultural research themes

Through the competitive selection process, a number of potentially high-quality and high-impact subprojects were selected to address the research themes described earlier. The focus on value chains is intended to solve practical problems in commercial agriculture and agribusiness (box 4.29 describes achievements in research for two value chains). At the other end of the spectrum, work in 110 disadvantaged districts with NGOs has given researchers opportunities to address poverty and growth problems.

Improved quality and relevance of research

The synergy created through partnerships has improved the quality and usefulness of research results. Agricultural scientists now have access to unique skills (in ICT and biotechnology, for example) and research facilities provided by the scientists who had generally never worked with scientists from the national agricultural research system. Through their work with the private sector and NGOs, more public sector researchers have been exposed to the perspectives of these partners and their sense of urgency for solving clients' practical development problems. Box 4.30 summarizes the project's preliminary results and likely impact.

Institutional development

By sponsoring formal training and, even more important, developing new kinds of partnerships, the project has strengthened the institutions that serve agriculture, agribusiness, and livelihood security and is preparing them to deal with the development challenges of 21st century agriculture. A vital element of institutional development is the continuous interaction between public, private, and NGO sectors and the willingness of ICAR institutes to work

Type of institution	Institutions as consortium leaders (%)	Institutions as consortium partners (%)	Share of participating institutions in the NAIP budget (%)
ICAR institutes	46.5	37.2	50.8
State and central agricultural universities	30.9	22.9	26.3
International institutes	2.7	1.4	1.9
Central institutes	8.0	7.1	6.6
State institutes	–	2.0	0.6
Private agencies	3.2	10.4	3.9
Other universities	3.2	2.3	2.4
NGOs	2.7	13.3	6.3
Other institutions	2.7	3.4	1.2

Table 4.8 Consortium Leaders and Partner Institutions in the National Agricultural Innovation Project, India

Source: NAIP Project Implementation Unit.

The 188 consortiums consist of 188 consortium leaders (one leader for each consortium) and 646 partners, coming from about 370 public, private, and nongovernmental organizations. (Together, the consortium leaders and their partners comprise 834 project-implementing units.) Many consortium partners are working together for the first time in addition to participating in a World Bank-funded project for the first time. In this situation, implementation challenges are expected. They are being addressed as part of the learning process for the consortium approach.

The coordination of the consortiums generally appears to be working well, but some partners have experienced problems arising from poor coordination. The performance of 188 consortiums was rated in 2010 using a scorecard system, and 15 (8 percent) were rated Not Satisfactory. Milestones were developed to upgrade their performance. Instances of problems with staff commitment, staff skills, and the flow of funds to partners have occurred. Most consortium partners had problems with World Bank fiduciary requirements (procurement and financial management), environmental and social safeguard requirements, and monitoring and evaluation. They find the requirements too rigid, especially the reporting requirements. Often funds have been delayed. Most of these issues have been or are being addressed. Finally, capacity has been strengthened through training, workshops, the Helpdesk, and manuals. Although some problems continue, the project implementation team is committed to addressing them and learning how to make the consortium approach more sustainable for all involved.

Source: Based on surveys conducted by NAIP Project Implementation Unit.

A value chain model for producing bioethanol from sweet sorghum in rainfed areas through collective action and partnerships. This consortium is led by the International Crops Institute for the Semi-Arid Tropics (ICRISAT) and six partners (five research institutes, one private company). Although it is only partway through its research program, the consortium has increased awareness of sweet sorghum's multiple uses in the project area (food, feed, fodder, and fuel) and organized farmer groups to produce the sweet sorghum crop. Farmers increased sweet sorghum yields by growing improved varieties and using better management practices, and new farm equipment provided under the project reduced the drudgery of farm operations. Farmers now use sorghum grain for food and feed, stalks for syrup production, and bagasse for fuel and fodder. A crushing unit (owned by the private partner) using ICRISAT's bioethanol conversion technology was established to produce syrup and the organizational structure of the unit is designed to manage the supply chain. One ton of stalks produces 269 liters of juice and 50 kilograms of syrup. Syrup production costs were reduced by increasing juice recovery, syrup recovery, and labor efficiency. The syrup is sold to the bioethanol, food, and pharmaceutical industries. Although farmers who cultivate sweet sorghum and raise livestock have benefitted from this arrangement, the profitability of bioethanol production ultimately depends on crude oil prices.

A value chain of banana pseudostem for fiber and other value-added products. This consortium is led by a university and four other partners participate (one research institute and three private business partners). Banana pseudostem is generally regarded as a waste product and source of pollution after bananas are harvested. In seeking to develop a value chain for banana pseudostem, the university and research institute have focused on the backward linkages (to banana farmers)

(Box continues on the following page)

and the private sector has focused on the forward linkages (developing useful, marketable products from banana pseudostem). Consortium members have already been able to produce fiber, yarn, paper, candy, woven fabric, artificial leather, cellulose powder, vermicompost, and liquid fertilizer from banana pseudostem; the machinery needed to produce these value-added products has already been designed, fabricated, and commissioned under this project (although not yet commercially available in India). Having demonstrated that these products are technically feasible, the consortium is currently analyzing the financial and economic prospects for various products (and conducting experiments on producing yarn from banana and synthetic fiber). Farm income has improved because farmers have been able to sell pseudostem through the project, and the conversion of banana pseudostem into various products has generated employment.

These examples demonstrate achievements that would not have been possible without the complementary resources provided by each consortium partner. These examples also demonstrate substantial economic and social impact through value addition for all the stakeholders and the country.

Source: Based on information provided by NAIP Project Implementation Unit and the consortiums.

Catalyzing and managing change in the national agricultural research system. Knowledge can be a powerful change agent, and NAIP has given considerable attention strengthening research capacity through the provision of knowledge. Students, teachers, and scientists now have access to a much deeper knowledge base. Over 2,000 scientific and professional journals can be accessed in 124 libraries; 155 of 368 e-courses have already been developed; 6,000 PhD theses have been digitally uploaded; electronic information on agriculture is available through an "agro-web"; and 10 business planning and development units have been established. Significant formal training should have a long-term effect on the human resources available to the national research system (and wider AIS): already 1,611 experts have been trained (1,441 nationally, 170 internationally).

Research to strengthen value chains. The project supports research for value chains representing a wide spectrum of potentially high-value agricultural products: banana pseudostem, briquettes from industrial residues, industrial agroforestry, oceanic tuna, potato and potato products, natural dyes, bioethanol, coconut, seed of spices and flowers, millet foods, sorghum foods, and maize and maize products—among others. Preliminary results for most of these value chains appear promising with respect to the new technologies used and the potential economic gains. For example, it may be possible to develop a new industry involving a large number of small-scale entrepreneurs in producing and distributing briquettes.

Research on more sustainable and secure rural livelihoods. Subprojects encompass a wide range of topics, including: scaling up crop production technologies; increasing water storage capacity through improved natural resource management (extremely important in drought-prone parts of India); backyard poultry production; generating employment through various natural resource management interventions; expanding irrigated area; improving grain storage capacity through storage bins; vermi-composting units; rice-fish-poultry farming; drought mitigation measures; and water harvesting. A few consortiums report early results in improving yields of maize (by 30 percent), rice (37 percent), soybeans (22 percent), wheat (32 percent), sorghum (24 percent), and cotton (126 percent). Improved resource-use efficiency and increased productivity are likely to raise incomes

(Box continues on the following page)

Box 4.30   Preliminary Results of the National Agricultural Innovation Project, India (continued)

among marginal and small-scale farmers. To mainstream the best practices they develop, some consortiums are establishing sustainability funds.

Basic and strategic research (frontier science). Results from these subprojects are expected to advance scientific knowledge and yield economic benefits for farm families, the agricultural sector, and the economy as a whole. Subprojects focus on natural resource management innovations as well as innovations that improve agricultural productivity, quality, and value added in staple crops, horticultural products, livestock, and fish. Promising results include the identification of 30 genes specific to cotton fiber development; 10 herbal extracts to control ticks in cattle; a chip-based biosensor and a micro-well chip platform to detect ultra trace concentrations of pesticides and adulterants in milk; and a prototype rubber dam for small-scale watersheds.

Source: Information provided by NAIP Project Implementation Unit and consortiums.

outside their system. ICAR has started to mainstream the consortium approach and competitive selection process throughout its institutes. If these actions continue on a large scale, the process, interventions, and impacts initiated through NAIP are likely to be sustained and substantial. The consortium approach has encouraged partner institutes to consider new strategies for solving real agricultural problems. The technical solutions emerging from the consortiums appear to have benefited from increased interaction and creativity, given their quality, economic potential, effectiveness, and level of appropriateness to clients' needs.

LESSONS LEARNED AND ISSUES FOR WIDER APPLICATION

The formation of consortiums, partnerships, and the competitive selection process has been a major but time-consuming achievement. Innovation is needed to shorten this process without sacrificing quality and pluralism. All heads of research organizations must be encouraged to reach out to scientists whose work complements the project's research themes and overall benefits—a strategy that should be pursued for all agricultural research, irrespective of the funding source. Based on experience to date, the consortium approach is likely to emerge as a best practice for agricultural research, if indeed it ensures high returns to investments in agricultural research and promotes collaboration throughout the wider AIS.

Project teams must develop a strong sense of the number of proposals that are likely to be submitted and plan accordingly. Subproject selection and review in NAIP were delayed by 18 months by the high number of proposals submitted for consideration. The proposals were not only numerous but complex to develop, review, revise, and approve. The involvement of new partners from diverse backgrounds and inclusion of research issues that the public research system is formally pursuing for the first time (value chains and sustainable livelihood security) also extended project selection and approval. If future projects fund a larger-than-expected number of subprojects, they must make appropriate arrangements to handle the increased workload for management, procurement, financial management, monitoring, evaluation, and safeguard management. Otherwise implementation will be delayed.

From the start, all consortiums must be aware that they are required to follow agreed procurement procedures and receive appropriate procurement training. The procurement of goods (especially scientific equipment and supplies), services, and works must keep pace with the implementation of the project and research subprojects. Procurement under NAIP was very slow, especially in the beginning. Most consortium partners had never worked with Bank-funded projects. It took some time to convince and train them to use World Bank procurement procedures.

The Helpdesk seems to be a best practice to adopt in collaborative programs for agricultural research. In NAIP, the Helpdesk portal has been extremely useful in forming consortiums and preparing proposals, and it should be available to address implementation problems as they arise, especially considering the large array of consortiums and organizations involved in the project. It could also be useful for disseminating success stories.

At the beginning, establish an effective M&E system for internally tracking the project's progress and performance as well as its likely impact. A good M&E system provides regular feedback to project management about potential problems as well as progress, results, and impact. Supplement the internal M&E system with an independent external M&E system for the benefit of the project management team.

Establish an effective outreach and communications system from the start. This system should promote emerging best practices, share success stories, disseminate knowledge and experience about consortiums that perform well, and share the emerging outcomes of the project with policy makers, the scientific community, and general public. The communications system will use a range of media and formats (not simply print, or scientific journals, but electronic and visual media in appropriate languages) if it is serious about reaching stakeholders and the public.

The Agricultural Technology Consortium Model in Chile

Rodrigo Vega Alarcón, Consultant, formerly with the Foundation for Agricultural Innovation (FIA)

SYNOPSIS

This IAP presents and discusses the main objectives, results, and lessons from Chile's experience with agricultural technology consortiums. These formal alliances promote joint work between industry and science within a market framework. Chile is regarded as a pioneer in using this model to focus public policy on innovation. Consortiums are a good option when industry is strongly committed to the process and the partners possess the technological capabilities to develop the kinds of products they seek. When these conditions are not met, a strictly corporate model such as the one used in Chile may not work. Special programs may be needed, for example, if consortiums are intended to include small-scale producers.

BACKGROUND AND CONTEXT

Chile's economy has grown at a remarkable pace since the mid-1980s. The country has pursued consistent economic reforms, including market liberalization and important free trade agreements, and it has developed an export model relying on its abundant natural resources and commodities. These reforms have benefited from political and social stability, underpinned by sound democratization.

An important warning signal emerged in the second half of the 1990s, however, when economic growth began to flag. The decade that followed saw near-zero growth in total factor productivity.[1] In response, the government adopted a new innovation policy to make Chile's economy more competitive. The National Council on Innovation for Competitiveness (Consejo Nacional de Innovación para la Competitividad) was formed to advise the president on strategies and policies to foster innovation and increase competitiveness in the medium and long term. The resulting national innovation strategy has three pillars: science, human capital development, and business innovation (CNIC 2007a,

2007b). In conjunction with this strategy, a new institutional framework was developed to give priority to economic sectors exhibiting the highest growth potential, create appropriate instruments to realize that potential, and allocate more public and private resources to research, development, and innovation.[2]

New policies and instruments have been designed and implemented to further innovation. They include the strategic use of mining royalties; tax deductions for research, development, and innovation; the organization of clusters (high-priority areas for innovation); and the establishment of technology consortiums. This note discusses the main objectives, results, and lessons from Chile's experience with agricultural technology consortiums, which promote joint research and innovation by industry and science within a market framework.

Chile is considered a pioneer in using consortiums to align public research with national innovation policies, and the government issued its first tender for a Technology Business Consortium in 2004. The Chilean consortiums were initially modeled on Australia's Cooperative Research Centres, established in 1990 (see box 4.4 in the overview of this module). The Cooperative Research Centres linked academia, public research institutes, business, and producers through a forum for dialogue, which proved to be the key to developing trust and a meaningful relationship between the parties. This process allowed researchers to understand the problems and needs of private industry and producers and seek specific solutions. It also allowed industry and producers to understand the importance of research and innovation to their business strategies.

OBJECTIVES AND DESCRIPTION OF CONSORTIUMS

As noted, Chile's technology consortiums are expected to strengthen links between research communities and local and global business communities, thereby improving competitiveness and opening new business opportunities (Álvarez et al. 2010). Their specific objectives are to:

- Increase national competitiveness by encouraging companies to invest in research, development, and innovation.
- Implement programs for research and development that target long-term profits.
- Encourage links between science and industry.
- Develop and strengthen scientific skills and techniques.

A consortium is defined as a technology company in which one or more companies agree to carry out joint projects with universities, institutes, and/or technology centers to develop new technologies that can improve and add value to production processes and products (Álvarez et al. 2010). The theoretical justification for this type of instrument is to discover solutions for market failures limiting innovation by an enterprise and encourage partnership strategies to incorporate knowledge externalities, coordinate the use of complementarities, and share the risk of investment in technology innovation.

Public agencies[3] fund and manage the consortium program. The maximum contribution of the public sector to a single consortium is US$6 million, and each consortium can operate for up to five years. The maximum yearly public contribution to a consortium's total budget is 25 percent for research activities, with an additional contribution of 10 percent for human capital development and 15 percent for research infrastructure. This funding is matched by cofinancing of 50 percent from nonpublic consortium members. Box 4.31 describes the conditions that consortiums must meet to receive public financing.

Like a corporation, a consortium is run by a board composed of representatives from academia and industry in a number according to their capital ratio. Their responsibility is to define the consortium's strategic aims, determine which research projects to pursue, and allocate funding to each. A general manager reporting to the board is responsible for consortium administration and management. The manager's main responsibility is to coordinate all activities of the consortium. Several consortiums have also established technical committees (appointed by the board) composed of researchers and business professionals who are not involved directly with the research; their role is to monitor

---

Box 4.31   Characteristics and Conditions for Business-Technology Consortiums to Receive Public Financing

The consortiums and the projects under them must:

- Produce results that contribute to economic growth in Chile.
- Create permanent capacity for research, development, and innovation in Chile.
- Use mechanisms to generate, transfer, and adopt knowledge that will achieve the anticipated impacts.
- Identify suitable participants for collaborative work, based on their capacity (including managerial skills), the proposed mode of collaboration between universities, research institutes, and private companies, and the level of commitment to achieve the desired results.
- Provide sufficient cofinancing. The commitment of the partners is expressed in the proportion of cofinancing they are prepared to commit and the returns they envision as a result.

The following conditions are expected to be met:

- The objective justifies the need for different enterprises and research institutions to undertake a sustained R&D effort through the consortium.
- The consortium's work will significantly strengthen existing industries and stimulate the emergence of new ones.
- The projected work requires public funding to be performed.
- The resources requested are consistent with the projects and their anticipated results.

Consortiums are generally formed by enterprises that seek to use cutting-edge research to satisfy the needs of their particular productive sector. An integral part of the business model for these enterprises (and usually a prerequisite for their participation in a consortium) is the development of *patentable results* that can be *licensed* and generate *spinoffs* (in the form of new businesses).

Source: Author, based on information from CONICYT.

---

research and ensure that it is being conducted as planned, evaluate new research proposed by the consortium, make related recommendations to the board.

INNOVATIVE ELEMENT

Technology consortiums have several novel aspects. The institutional setup of the consortiums was a major innovation because it gave a market orientation (from the companies) to research on technical solutions (to the science-based institutions). This intermediate approach does not imply that research and business should change in any fundamental way; it is simply a means to strengthen the relationship between private enterprise and science. Companies and research centers forge links by pursuing applied research for profit. Members act together "upstream" when they develop technology but act in competition "downstream" in the marketplace. Companies determine the priorities and corresponding research projects, which are executed by research institutions (either as consortium members or externally contracted agencies). When a group of companies or an industry identifies common priorities, problems, or opportunities and sets priorities, resources are allocated more efficiently to address them. Greater trust and understanding among the actors will increase their propensity to share information and achieve synergies for industry and advances for research.

BENEFITS AND IMPACT

The previous discussion has given an idea of the kinds of benefits, tangible and intangible, that can emanate from technology consortiums. As the Chilean experience with this model is so recent, however, the empirical evidence of impact is limited. Most of the consortiums studied have focused on improving the competitiveness of productive sectors rather than on improving capacity to pursue innovative activities. The consortiums' main contributions are improved access to technological and other kinds of knowledge (such as marketing, international market regulations and requirements, and staff with specific kinds of expertise) and joint technology development by researchers with companies. On-going assessments have identified the following trends:

- The large public investment in these consortiums leveraged significant private resources for research, development, and innovation. These private investments have

risen considerably since the consortiums were established.

- Consortiums have demonstrated the capacity to identify and implement projects relevant to business or industry and to achieve economies of scale in applied research.
- Industries' and companies' demands for specific applied research spurred the formation of appropriate institutions and stronger research teams to develop technology.
- Participants gained access to knowledge that otherwise would have been very difficult for them to acquire.
- In some initiatives, the participation of experienced and emerging businesses allowed the companies to learn from one another. In the wine industry, for example, most developing vineyards work with more established companies.
- The trust engendered in the course of the research is likely to have important long-term implications for the actors involved, increasing the potential for further collaboration.

Given the consortiums' short duration and the applied research they generally conduct, applications for IP protection are still very low. For the same reason, the companies in the consortiums have not yet achieved major technological breakthroughs.

LESSONS LEARNED AND GUIDANCE FOR THE FUTURE

Studies in Chile and in Latin America more broadly suggest that various factors influence consortium partners' willingness to exchange knowledge, collaborate in R&D, and thus produce innovations (Álvarez et al. 2010):

- The greater the number of partners, the less profitable the results.
- The more that consortium partners compete directly in the final market for goods or services, the more they produce results of limited impact. Not all firms that compete in final markets produce results of limited value when they partner in consortiums, however. Firms that negotiate clear agreements for managing IP before they join a consortium produce better results, because their fears of losing trade secrets or failing to recover R&D costs are alleviated.
- The greater the geographical distance between partners, the less profitable the results.
- The greater the partners' experience, the better the results.

- The greater the number of employees in R&D or other technical areas in the consortium, the better the results.
- The more frequent the interaction between partners, the better the results.
- The more satisfied the partners are with the contractual provisions to protect IP and resolve conflicts, the better the results.
- The greater the trust and goodwill between partners, the better the results.

Empirical evaluations of consortiums have yielded other lessons. One lesson is that the long startup times of consortiums (about one year) in relation to their duration acts as a disincentive to participation. Startup delays usually involve uncertainty over who will participate, what kind of formal relationship they will adopt, what resources each will provide (social capital, physical capital, counterpart funding), and the IP arrangements.

Part of the application process should be a formal presentation by the parties of the consortium's objectives, the issues it will address, and its projected business model, governance model, IP arrangements, legal status, national and international partners, and other important features. The consortium can operate for a startup period of up to one year with financial support to assess whether to continue or terminate the relationship. This "pre-consortium" period can be useful to define both the business and governance models that the consortium will adopt. It is essential that the business and governance models be consistent with one another. The final structure chosen for any single consortium will not necessarily work for another because in each case the business model is likely to vary.

The preconsortium stage is also the time to define IP arrangements. It is better to define these arrangements before any revenue is generated to avoid conflicts later when revenue begins to flow. The alternatives are for the IP to be owned by the joint venture or consortium or for each partner to own a share of the IP. In that case, the consortium acts as a technology broker, charging a fee for administering any royalties or commissions.

Another lesson is that governance and management make a difference. An active, committed board and technical committees as well as fluid dialogue between participants and high-level management are essential. One concern expressed by companies and research centers alike is that the difference in their objectives is the greatest obstacle to their success in consortiums (box 4.32 provides an example from the wine industry). To address this problem, the consortium must have a very good full-time manager and constant support from government agencies *with a voice— but no vote.*

A corporate model may not be suitable for all types of consortiums. Consortiums are a good option when industry is strongly committed to the process and the partners possess the technological capabilities to develop the kinds of products they seek. When these conditions are not met, a strictly corporate model may not work.

For example, if consortiums are intended to include small-scale producers, a number of other considerations become important, such as the potential social and economic importance of smallholders' participation in the subsector; government's commitment to support their participation in a consortium; guidance for smallholders to form and function in organizations; and advisory services that enable smallholders to manage new technology and practices successfully. Potato production in Chile is dominated by smallholders with few assets and thus limited capacity to participate in consortiums. A special program could establish better links between these farmers and research institutions (box 4.33), fostering the trust and experience that could make them more effective partners in a consortium, or a less formal association could be devised.

Two strategic elements should be taken into account when implementing a consortium (Lavados 2009):

- *Business model.* From the outset, the business model needs to be clear. Is the consortium geared to develop a single product or multiple products? (To date, most Chilean consortiums have attempted to develop a portfolio of products.) Are local or global markets targeted? How will the partners acquire their medium-term revenues (royalties, licensing, product sales, or some other mechanism)?
- *Portfolio of programs and projects.* Research projects usually deliver results in the medium and long term. To attract private sector partners, consortiums will need to develop a portfolio of projects that is balanced between short- and long-term marketable products.

Another lesson is that Chilean scientific and technological capacity is not always sufficient to address companies' or industries' increasingly complex needs for R&D. A proactive plan must be implemented to generate local knowledge and/or import knowledge and expertise from abroad (immigration of specialists, alliances with international R&D centers).

The quality and prestige of Chilean wine have risen to the extent that Chile now ranks among the top five nations in global wine sales. Two consortiums, Tecnovid and Vinnova, have joined forces to maintain and enhance this position and achieve the economies of scale required for successful R&D. The R&D requirements are set by the companies and executed by university research teams. The consortium ultimately seeks to become a viable enterprise that manages continuous innovation in the wine industry, from processing to marketing.

Through their merger, the consortiums will improve the productivity of the wine industry, develop programs for disseminating and adopting knowledge that will enable the wine industry to use research results quickly and efficiently, and develop the human capital needed to innovate at high scientific and technological levels. The research programs focus on improving the quality, productivity, and differentiation of wines and on developing sustainable production practices. For example, researchers are studying which wines are most competitive among consumers, nationally and internationally; strategies to ensure that planting material is free of disease and pests; the origins of certain physio-logical disorders in Merlot grapes; the fermentative capacity of various strains of yeast, and winemaking processes that yield higher levels of antioxidants in the final product.

The anticipated benefits and results of the collaboration include: positive and synergistic interaction between academia and the wine industry (all teams consist of business professionals and university researchers); a competent professional team attuned to corporate and business requirements channels the industry's needs for research and innovation; and research that not only solves scientific problems but produces results that can be incorporated into an enterprise's management and marketing.

Lessons include the importance of sound leadership, of working with mature enterprises, and of focusing on innovations that are important to the national economy and the industry. To meet global challenges, the wine industry increasingly needs collaboration upstream in R&D, needs to acquire additional technology and knowledge, and needs to increase its competitiveness in the wine market. If the best expertise to reach those goals does not exist in Chile, industry must seek it abroad.

Sources: Vinnova and author.

Box 4.33   A Business-Technology Consortium for Potato

Potatoes, an important part of the Chilean diet, are grown by 15,000 producers on 50,000 hectares, mostly on small farms. The potato subsector contributes only a fraction of agricultural GDP. A business-technology consortium has been formed to improve the competitiveness of Chile's potato industry through stronger participation in the international market for potato seed, potato for consumption, processed products, and potato varieties. The Consortium has taken the form of a new company, Consorcio Papa Chile SA, through which a large part of the potato industry (mainly small-scale producers) entered into a formal association with technology institutes. The company has 17 shareholders, of which 15 are producer organizations (representing 1,500 small- and medium-scale farmers who operate 4,500 hectares in various locations) and 2 are research entities: the Instituto de Investigaciones Agropecuarias (Institute of Agricultural Research) and Los Lagos University.

The Consortium has five lines of action: developing technology, strengthening human capital, adding value, management, and transferring and diffusing technology. It pursues research to develop production technologies adapted to particular agroecological conditions and market requirements and is developing new potato varieties with better postharvest characteristics and traits that match consumers' preferences. It has implemented a market intelligence system as well.

(Box continues on the following page)

Box 4.33   A Business-Technology Consortium for Potato (continued)

The main benefit of the Consortium, aside from assembling commercial and public partners to produce innovations requested by the industry, is that it is likely to increase private investments in science and technology for the subsector.

One lesson from the potato Consortium that may prove useful elsewhere is that it is quite difficult to launch a consortium in an industry with multiple, diverse, geographically dispersed, and heterogeneous actors, for which the products generated are essentially public goods. A second important lesson is that a corporate structure does not seem to be the best match for a subsector with these characteristics. Finally, given the characteristics of the subsector and heterogeneity of the partners, a special program is needed to foster trust between partners (producers, companies, and technology institutions) if they are to work toward a common goal.

Source: Potato Consortium and author.

Finally, the role of public agencies in designing, implementing, and evaluating the work of consortiums must be reconsidered. The specific capacities required in personnel charged with tracking and monitoring consortiums as a whole must be reviewed. The technical, administrative, and financial procedures used in consortiums should also be analyzed with a view to learning which practices enable consortiums to produce the best research outcomes.

Linking Research and Development Actors through Learning Alliances

Mark Lundy, International Centre for Tropical Agriculture (CIAT)
María Verónica Gottret, Tropical Agriculture Research and Education Center (CATIE)
Rupert Best, Catholic Relief Services (CRS)

SYNOPSIS

Learning alliances view research and development outputs as inputs to processes of rural innovation that are place- and time-specific. Methods and tools will change as users adapt them to their needs and realities. Understanding why adaptations occur, the extent that these lead to positive or negative changes in livelihoods, and documenting and sharing lessons learned are key objectives. The learning alliance approach differs substantially from the common practice of attempting to train development practitioners in new methods through short, one-off training courses. Learning alliances rely on an iterative learning process jointly undertaken among multiple stakeholders with a common interest or goal through a series of learning cycles, typically over 12–24 months. The Central American learning alliance improved connectivity between organizations working on similar topics, provided better access to information and knowledge on rural enterprise development, and access to improved methods and tools. Attitudes have shifted from competition to collaboration as partners.

CONTEXT

Millions of dollars are spent each year on R&D initiatives to improve rural livelihoods in the developing world. Despite this expenditure, rural poverty remains an intractable problem in many places. Among the multiple causes of this situation is the limited collective learning that occurs between researchers, development workers, cooperation agencies, policy makers, and private enterprise. As a result, useful research does not benefit the poor, lessons learned do not influence research, cooperation and policy agendas are less relevant than they could be—and development falters. Starting in 2003, a group of actors in Central America came together to explore how to improve the links between

research and development actors through learning alliances (defined in box 4.34).

OBJECTIVE AND DESCRIPTION

The chosen thematic focus of the Central America learning alliance, consisting of international and local NGOs, an international agricultural research center, a national university, and the International Development Research Centre (IDRC), was rural enterprise development.[1] Learning alliances were proposed as a process-driven vehicle through which the effectiveness of investments in rural livelihoods could be augmented.

During the first phase of the learning alliance (starting in 2003), process facilitation and knowledge management were managed by CIAT. In 2008, these functions were devolved to the Tropical Agriculture Research and Education Center (CATIE, Centro Agronómico Tropical de Investigación y Enseñanza) in Central America based on a regional agreement and cost-sharing strategy. In its second phase, the Central American Learning Alliance continues to promote sustainable rural economic development, considering the following objectives:

- Facilitate joint learning processes between development agencies, technical and financial cooperation, universities and research centers, and state and private actors, based on the development, validation, and dissemination of strategies, concepts, methodologies, and tools for developing effective rural economic development.
- Enhance knowledge management to influence the design of public and private policies that provide incentives for the development of sustainable rural commodity chains, based on the implementation of research activities relevant to sustainable rural economic development.

Box 4.34 Advantages and Impacts of Learning Alliances

Learning alliances rely on an iterative learning process jointly undertaken among multiple stakeholders with a common interest or goal. Typically, stakeholders might include research organizations, development and cooperation agencies, universities, policy makers, and private businesses. Learning alliances facilitate the development of cumulative, shared knowledge between these stakeholders about what works, what doesn't, and why in temporal and spatial contexts. Shared and accessible knowledge in this sense contributes to improved development outcomes as lessons are quickly identified and learned. Improved links among research and development actors improve both research focus and development practice. As funds diminish, increased efficiency becomes paramount in achieving positive livelihood change. Finally, jointly developed proposals are also more attractive for funding agencies as they have a higher potential for scaling out and up and therefore to achieve broader impact.

A well-functioning learning alliance achieves the following outputs:

- Cumulative and shared knowledge about approaches, methods, and policies that work in different places, cultural contexts, and times (as well as those that do not), and the reasons for success or failure.
- Learning opportunities across organizational and geographical boundaries through the establishment and support of communities of practice around specific topics.
- Synergy among multiple actors by providing a vehicle for collaboration, helping to highlight and develop diverse solutions to problems that may appear intractable to the individual actors.
- Contribute to healthy innovation systems by building bridges between islands of experience, helping to assess how these results were achieved, and what others can learn from these experiences.
- Capacity development for implementing, scaling out, and improving innovative approaches and methods.

Table B4.34 Types of Learning Alliances

Type	Need	Focus
1	Building capacity and skills base	Training and learning to use concrete, practical approaches and proven methods
2	Developing new methods, tools, and approaches	Action-research that generates methodological guides based on good practice, which is then validated through capacity-development learning cycles
3	Generating information that can lead to policy influence	Conventional socioeconomic research to understand principles and lessons across experiences

Sources: Authors and Best, Ferris, and Mundy 2009.

Common principals applied by the market-oriented alliance

Collaborative processes require agreement on certain basic principles to govern collective work. Table 4.9 lists and describes these principles.

Agenda setting

The selection of themes and topics in the learning alliance is based on dialogue among partners to identify knowledge or skill gaps that limit the success of their interventions. Once a topic has been selected, the interested partners define the central learning questions, which may range from basic development issues to research hypotheses. Partners select a small number of these options for development as full learning cycles. Attempts are made to target areas where partners have both an interest and ongoing projects, to align the learning process with concrete results that are useful for improving existing projects or contribute to the development of new proposals.

Learning cycles: A shared process of documentation and reflection

The learning alliance approach differs substantially from the common practice of attempting to train development

Table 4.9 Key Principles for an Effective Multipartner Learning Alliance

Principal	Description
Clear objectives	Multiple stakeholders have different objectives and interests. A learning alliance is based on the identification and negotiation of common interests, needs, and capacities of participating organizations and individuals. What does each organization bring to the alliance? What complementarities or gaps exist? What does each organization hope to achieve through the collaboration? How can the alliance add value to partner activities?
Shared responsibilities, costs, and benefits	Organizations and individuals participate in learning alliances when: (1) they perceive benefits from this association, (2) transaction costs are lower than expected benefits, (3) benefits from collective action are perceived to be greater than those obtained individually, and (4) results do not conflict with other key interests. Learning alliances seek to benefit all parties. Therefore, transaction costs and responsibilities, as well as benefits and credit for achievements, are shared among partners in a transparent fashion.
Outputs as inputs, documentation, and experience sharing	Rural communities are diverse and no universally applicable recipe for sustainable development exists. Learning alliances view research and development outputs as inputs to processes of rural innovation that are place- and time-specific. Methods and tools will change as users adapt them to their needs and realities. Understanding why adaptations occur, the extent that these lead to positive or negative changes in livelihoods, and documenting and sharing lessons learned are key objectives.
Differentiated but linked learning mechanisms	Learning alliances have a diverse range of participants. Identifying each group's questions and willingness to participate in the learning process is critical to success. Flexible but connected learning methods are needed.
Long-term, trust-based relationships	Rural development processes stretch over many years or decades. To influence positive change and understand why that change has occurred requires long-term, stable relationships capable of evolving to meet new challenges. Trust is the glue that cements these relationships, but develops gradually as partners interact with each other and perceive concrete benefits from collaboration.

Source: Lundy and Gottret 2007.

practitioners in new methods through short, one-off training courses. It involves establishing a series of "learning spaces," typically over 12–24 months (Best, Ferris, and Mundy 2009).

The development of feedback loops and space for reflection as a way to improve practice is the final method used by the learning alliance. It is implemented through face-to-face meetings as well as web-based tools.

- *Capacity-strengthening workshops* are used to train local partners for implementing new approaches and methods developed by learning alliance partners based on best practice and action-research. Follow-up is provided by learning alliance partners. Results are documented and feedback given to improve these new approaches and methods. This process is critical for action-research and strategic research results to be scaled out and up.
- *Backstopping* is carried out by an alliance member with more experience with a specific tool or approach who pays periodic visits to other partners who are adapting the tool to their needs.
- *Write shops* are used to help distil lessons learned into documents. They are especially useful with partners who have difficulties finding time to write up results from their work. The learning alliance adapted methods and tools developed by IFAD (Berdegué et al. 2002) and Douthwaite et al. (2007) for this purpose.

- *National learning fairs* are based on the Most Significant Change method developed by Davies and Dart (2005).
- *Web-based tools* are used principally for documentation and dissemination among partner agencies.

INNOVATIVE ELEMENT: LEARNING ALLIANCE PROCESS AND TOOLS

The learning alliance approach is made up of four interrelated strategies:

1. *Capacity development activities* seek to strengthen or improve partners' capacities in the selection, use, adaptation, and improvement of specific approaches, methods, and tools. This process is directly linked to specific learning cycles. Capacity building is not limited to training workshops but focused on practical, field-level use, follow-up, adaptation, and improvement, with continuing support as partners implement the prototype. As a result, partners strengthen their ability to use specific tools and approaches, adapt them to their needs, and discern when specific methods might or might not be useful.

2. *Targeted action research* responds to specific knowledge gaps identified with partner agencies. In this strategy, key research questions are identified and fieldwork designed and implemented in collaboration between research and development agencies. Outcomes and findings are shared with other partner agencies, selected decision

makers, and the public in general in workshops and in electronic formats.

3. *Connectivity and knowledge management* strive to increase the relationships that form the basis of the learning alliance. The "densification" of networks and personal connections is critical to the success of the alliance. To achieve this, the alliance makes use of face-to-face meetings, training, and exchange visits, as well as virtual tools such as a website and a list server.

4. *Evidence-based decision-making* in partner organizations, public entities, cooperation agencies, and private firms. This strategy has been markedly less successful than the previous three in engaging nonpartners. Despite this difficulty, the learning alliance partners feel that this is a critical capacity that should be developed to leverage higher-level change based on field results. Work in this direction is being piloted with the public sector in Honduras and Nicaragua (Swisscontact, Catholic Relief Services, and CATIE) and Colombia (CIAT); with cooperation agencies in Guatemala and the Dominican Republic (CIAT, Oxfam Great Britain, Sustainable Food Lab); and with the private sector in Guatemala (Oxfam Great Britain and the Sustainable Food Lab).

BENEFITS AND IMPACT

By 2007, the Central American Learning Alliance had contributed to significant changes in partner knowledge, attitudes, and practices. Evidence showed improved connectivity between organizations working on similar topics, better access to information and knowledge on rural enterprise development, and access to improved methods and tools. Attitudes shifted from competition to collaboration as organizations witnessed that working together enhances their capacity to serve rural communities' needs and receive cooperation funds rather than undermining it. Rural enterprise development practices and knowledge management have improved, as shown by increased effectiveness in existing projects and more strategic new projects. These shifts in turn contribute to a more efficient innovation system in favor of rural enterprise development, as evidenced by the shared use and generation of information, joint capacity building programs, and large-scale, collaborative projects.

The first phase of Central American learning alliance worked with a total of 25 direct partner agencies and through their networks influenced 116 additional organizations. In total the learning alliance contributed to change in organizations working with 33,000 rural families (approximately 175,000 people) in Honduras, Nicaragua,

Guatemala, and El Salvador. With a four-year budget of US$499,000, the alliance leveraged an additional US$990,000 in in-kind and additional funding, exceeding budgeted counterpart funding by a factor of ten. The alliance website (www.alianzasdeaprendizaje.org) is a key site for practitioners focused on rural enterprise development in Latin America.

Strengthened networks and knowledge management contribute to improved processes of collaboration between partners. Partner agencies report the use of methods and tools from the alliance in 46 occasions in ongoing projects. Community-level assessment in 2007 led to the identification of 30 cases of most significant change that highlight the positive impact of these tools on income generation, natural resource management, and the role of women.

This initiative started a new phase in July 2009 when five of the organizations that participated in the Learning Alliance during its first phase—Catholic Relief Services, the Netherlands Development Organization (SNV), the Swiss Foundation for Technical Cooperation (Swisscontact), Oxfam Great Britain, and CATIE signed a Cooperation Agreement for five years to support a Coordination Unit that is being facilitated by CATIE.

This new phase started with the first jointly developed project cofunded by IDRC, "Leveraging Information and Knowledge for Inclusive and Sustainable Agricultural Value Chains Development" (K4ValueChains) with the participation of CATIE, CIAT, Catholic Relief Services, SNV, and Swisscontact. This action-research project aims to leverage the development of sustainable and inclusive value chains through the strategic access and use of information and communication technologies to improve the participation of the poor in overall chain governance and decision making, contributing to value chain competitiveness, sustainability, poverty reduction, and food security.

With the collaboration of the World Bank's Agricultural Risk Management Team (ARMT) and the Regional Unit for Technical Assistance (RUTA), the learning alliance has also started a pilot learning cycle on price risk management with seven coffee cooperatives in Nicaragua. With a five-year (2009–13) core budget of US$25,000, the regional learning alliance has already leveraged an additional funding of US$940,310 for strategic learning, and US$206,400 for capacity development in Nicaragua, where the Deutsche Gesellschaft für Internationale Zusammenarbeit (GIZ) and Lutheran World Relief (LWR) are also participating.

Participation in learning alliances has transformed the work of development partners and has broadened the work of research centers. The changes that the learning alliance

The learning alliance approach has been successfully adopted within Catholic Relief Services' (CRS) Agriculture and Environment Program. It highlighted the interdependence of development actors and how, through partnership and collaboration with appropriate research and development actors, real gains can be made in achieving common goals. The reasons that led to CRS's adoption of the learning alliance approach included:

- A desire to demonstrate the benefits of market-led approaches in relief and development.
- Frustration with the effectiveness of traditional training programs.
- A disappointing record of adoption of innovations in methodologies, technologies, and partnerships based on traditional training methods.
- A lack of feedback in more typical learning processes.
- Insufficient impact assessment and follow-up.

Prior to adopting the alliance approach, almost all the farmer training undertaken by CRS and its partners dealt with a single skill set (for example, how to increase production of a particular commodity, or basic business and marketing skills). CRS and the International Center for Tropical Agriculture (CIAT) came to understand that self-sustaining growth and development of farmers requires multiple sets of skills. The learning alliance partners also realized that these skill sets required the integration of several sectors (e.g., microfinance, agriculture, water, and enterprise development).

A research-development partnership was established between CRS, CIAT, and the International Institute of Tropical Agriculture–FOODNET in which they pooled agroenterprise skills and expertise with the methodologies and tools that they had been developing to help smallholder farmers link to markets. Together they mapped out an iterative process of learning, putting into practice what had been learned, and then analyzing and reflecting on the results. From small beginnings in East Africa and Central America in 2002–04, CRS is now involved in agroenterprise development learning alliances in 5 regions, with participation of around 30 countries.

In countries where the learning alliance has been most active, CRS agriculture programming has undergone a radical change. Where formerly CRS's attention was narrowly focused on a low-input/low-output, subsistence farming-oriented approach to food security, now CRS programs integrate the goal of enabling small-scale producers to enter competitive markets by identifying market opportunities, strengthening rural enterprise, and converting poorly coordinated supply chains into value chains. Many learning alliance participants sum up this change by saying that they are helping farmers transition from "struggling to sell what they have produced, to producing what sells."

Sources: Author and Best, Ferris, and Mundy 2009.

experience brought to the work of CIAT and Catholic Relief Services are detailed in box 4.35.

LESSONS LEARNED AND ISSUES FOR WIDER APPLICATION

Several important lessons can be extracted from the learning alliance for innovation systems work. A key lesson is the need to increase connectivity and information flows between individuals in a transparent fashion, facilitated by an honest broker. This process is critical to build trust among participants. The experience highlights the importance of individuals as opposed to organizations as well as the need to avoid organizational standard bearers who feel threatened by open

systems where "the competition" also participates. Second, clear and shared objectives, applicable results, personal commitment, and flexibility are key elements. An effective innovation system adds value to individual participants in diverse ways by leveraging a collective motivation to work smarter, learn, and share with others. Finally, the facilitation of an innovation system is an art in itself. The learning alliance taught the partners to value diversity of opinion and tension as a crucible of creative ideas.

In the hopes of contributing to more effective innovation systems in the future, there are also several critical issues and/or errors that can be taken from the learning alliances. These include: (1) the difficulty of selling a process in a project- and outcome-driven context; (2) a lack of causality in many of the results, which makes evaluation and reporting difficult in more formal channels; (3) an initial excessive emphasis on web-based tools when what seems to work best is face-to-face exchanges to build trust and innovation; (4) the difficulty in securing on-going funding for a sometimes "fuzzy," demand-driven process; (5) the need to proactively involve more members of the overall food system (such as public policy makers and the private sector) from the outset; and (6) the need for eventual buy-in from key decision makers in the organizational "home" of the innovation system.

Specific interventions highlighted by partner agencies to support innovation systems projects or programs focus on connectivity and information. Key interventions in connectivity include face-to-face exchanges and trust building.

Once a minimum level of trust exists, the use of ICT-based communication tools is useful. Sequencing is important here. Trust is the basic building block without which technical fixes are of limited use. A second key type of intervention focuses on increased information access and flow. Participants in the learning alliance value new ideas about how to resolve constraints, short case studies illustrating the application of these ideas in diverse contexts, access to people with experience using these tools, and feedback mechanisms to share their experiences with others. Investments in simple process documentation (for example, contracting local reporters), knowledge-sharing fairs, and web-based platforms as well as support for write shops to make sense of outcomes are useful here.

NOTES

Module 4 Overview

1. New Zealand reformed its agricultural research system in the early 1990s, creating separate research institutes under a research funding foundation based on short-term contestable grants. By 2005 government policy was reformed: "The aim appears to be a move away from short-term contestable funding and a move toward long-term commitment of resources to individual providers to plan their own priorities" (Johnson 2006:8).

2. By 2009, only Chad, Ethiopia, Madagascar, Malawi, Mali, Namibia, and Niger had met the target, but for many if not most of these countries (Malawi, for example), the new funds were directed more to short-term interventions, especially fertilizer subsidy programs, rather than to longer-term investments in agricultural research.

3. In Sub-Saharan Africa nonfarm rural income is about 34 percent of total income, compared to 51 percent in Asia and 47 percent in Latin America (Reardon, Berdegué, and Barrett 2007).

4. IBGE, Instituto Brasileiro de Geografia e Estatística.

Thematic Note 1

1. For other definitions and functions, see the International Association of Science Parks, http://www.iasp.ws/publico/intro.jsp.

2. Examples include cofinancing through trade organizations and farmer associations to pay for research and steer priorities toward the needs of farmer-members. Examples include research interfaces with such organizations as La Fundación Nacional del Arroz (FUNDARROZ, National Rice Foundation) in Venezuela. For such arrangements to work, the associations must develop the capacity to articulate their needs. The provision of research and extension services will need to be pluralistic (that is, involving many potential providers with the skills to provide those services).

Thematic Note 2

1. The conventional rationale for investing in agricultural research suggests that governments and development agents should finance agricultural research when the private sector "underinvests." This condition is usually given when technology is a public good—in other words, when others cannot be excluded from its use and the private sector cannot recover the costs of developing it (Anderson 1998; Pray and Umali-Deininger 1998; Day-Rubenstein and Fuglie 1999).

2. Horton, Prain, and Thiele (2010) discuss PPPs in the broader context of partnerships.

Thematic Note 3

1. Asian and Pacific Association of Agricultural Research Institutes.

2. Association of Agricultural Research Institutes in the Near East and North Africa.

3. The Southern African Center for Cooperation in Agricultural Research and Training.

4. Association for Strengthening Agricultural Research in Eastern and Central Africa.

5. CORAF/WECARD = Conseil Ouest et Centre Africain pour la Recherche et le Développement Agricoles (West and Central African Council for Agricultural Research and Development); ASARECA = Association for Strengthening Agricultural Research in Eastern and Central Africa.

6. Programa Cooperativo para el Desarrollo Tecnológico Agroalimentario y Agroindustrial del Cono Sur (Cooperative Program for Technological Development in Agrifood and Agroindustry in the Southern Cone).

7. Cooperative Agricultural Research and Technology Transfer Program for the Andean Subregion (Programa Cooperativo de Innovación Tecnológica Agropecuaria para la Región Andina) and the Cooperative Program on Research and Technology Transfer for the South American Tropics (Programa Cooperativo de Investigación y Transferencia de Tecnología para los Trópicos Suramericanos).

8. The ASARECA priorities assume potential benefits across frontiers are counted equally. They may not be realized if trade is banned during periods of drought and market access is restricted by informal barriers.

9. Regional economic communities need to assume responsibility for funding regional public goods, whereas donor funding may be channeled through funding authority for specific activities/projects.

10. Efforts by CGIAR research centers in Eastern and Southern Africa to develop a postdisaster and postconflict program classified 17 of 25 countries as belonging to these categories, but countries were very resistant to becoming involved in their neighbors' internal problems or found border areas too difficult to target.

11. Many well-focused, tightly organized initiatives by CGIAR research centers and the Bill and Melinda Gates Foundation do not come through subregional organizations.

12. Economies of scale occur when the cost per unit of research output fall with the number of units of output produced, usually through better use of major fixed investment or specialized skills. Economies of scope occur when the cost of a desired output falls with an increase in the number of different research outputs being produced. These gains occur when there is cross-commodity or cross-disciplinary learning as an external economy.

Thematic Note 5

1. The World Bank Institute's *Capacity Development Results Framework* (Otoo, Agapitova, and Behrens 2009) presents a useful step-by-step guide and associated methods and tools for planning, implementing, and evaluating interventions to develop capacity.

2. It should be noted, however, that PROINPA's autonomy from the national government has sometimes led to strained relations with public-sector bodies responsible for agricultural research and development.

3. For a more extensive discussion of NAIP with a somewhat different emphasis, see IAP 2.

Innovative Activity Profile 2

1. For a detailed description of NAIP, see World Bank (2006) and http://www.naip.icar.org.in.

2. India's national agricultural research system consists of ICAR as well as a large network of state and central agricultural universities. ICAR is an autonomous organization under the Ministry of Agriculture. At present, there are over 90 ICAR institutes and 50 state (45) and central (5) agricultural universities.

Innovative Activity Profile 3

1. Recent studies have shown that TFP [total factor productivity] growth depends heavily on at least two factors that are very important to this Council: the quality of human resources and spending on research and development (Bitran Colodro and González Urrutia 2010).

2. The government plans to raise the contribution to research, development, and innovation from 0.7 percent of GDP to 2.5 percent of GDP by 2025. At the same time, it will reallocate a large share of the public investment in research to the private sector. Two-thirds of public expenditures on research currently go to the public sector and one-third goes to the private sector; in the future, the allocations will be just the reverse.

3. InnovaChile-CORFO is the most important multisectoral public innovation agency (CORFO is the Corporación de Fomento de la Producción) along with CONICYT (Comisión Nacional de Investigación Científica y Tecnológica) and FIA (the Foundation for Agricultural Innovation / Fundación para la Innovación Agraria).

Innovative Activity Profile 4

1. "Rural enterprise development" is a catch-all phrase that includes methods and tools to facilitate market linkages for small producer associations. In the first phase of the learning alliance, topics covered included the identification of market opportunities with producer groups, participatory supply chain analysis and upgrading strategies, and inclusive innovation in products and processes. For a current list of themes, see http://www.alianzasdeaprendizaje.org.

REFERENCES AND FURTHER READING

Module 4 Overview

Beintema, N., and G.J. Stads. 2006. "Agricultural R&D in Sub-Saharan Africa: An Era of Stagnation." Washington, DC: International Food Policy Research Institute (IFPRI).

———. 2008. "Agricultural R&D Capacity and Investments in the Asia–Pacific Region." Research Brief No. 11. Washington, DC: International Food Policy Research Institute (IFPRI).

Byrnes, K., and S. Corning. 1993. *Programming for Sustainability: Lessons Learned in Organizing and Financing Private Sector Agricultural Research in Latin America and the Caribbean.* Washington, DC: Bureau for Latin America and the Caribbean, United States Agency for International Development (USAID).

Daane, J. 2010. "Enhancing Performance of Agricultural Innovation Systems." *Rural Development News* 1:76–82.

FARA (Forum for Agricultural Research in Africa). 2009. "Sub-Saharan Africa Challenge Programme: Research Plan and Programme for Impact Assessment." Accra.

Haggblade, S. 2009. "Bringing the Poor into a Growth Agenda: What Role for Africa's Rural Nonfarm Economy?" East Lansing: Michigan State University.

Haggblade, S., P. Hazell, and T. Reardon (eds.). 2007. *Transforming the Rural Nonfarm Economy: Opportunities and Threats in the Developing World.* Baltimore: Johns Hopkins.

Hall, A., J. Dijkman, and R. Sulaiman V. 2010. "Research Into Use: An Experiment In Innovation." LINKLook, www.innovationstudies.org, accessed March 2011.

Jayne, T., D. Mather, and E. Mghenyi. 2006. "Smallholder Farming under Increasingly Difficult Circumstances: Policy and Public Investment Priorities for Africa." MSU International Development Working Paper. East Lansing: Michigan State University.

Johnson, R. 2006. "Whither the Crown Research Institutes? Funding Issues." Paper presented at the New Zealand Agricultural and Resource Economics Society Conference, Nelson, August 25–7.

Kline, S.J., and N. Rosenberg. 1986. "An Overview of Innovation." In *The Positive Sum Strategy: Harnessing Technology for Economic Growth,* edited by R. Landau and N. Rosenberg. Washington, DC: National Academy Press.

Larsen, K., R. Kim, and F. Theus. 2009. *Agribusiness and Innovation Systems in Africa*. Washington, DC: World Bank.

Njuki, J., P. Pali, K. Nyikahadzoi, P. Olaride, and A. Adekunle. 2010. Monitoring and Evaluation Strategy for the Sub-Saharan Africa Challenge Program. Accra.

Pardey, P., J. James, J. Alston, S. Wood, B. Koo, E. Binenbaum, T. Hurley, and P. Glewwe. 2007. *Science, Technology and Skills*. Minneapolis: International Science and Technology Practice and Policy (InSTePP), University of Minnesota.

Ragasa, C., S. Babu, A.S. Abdullahi, and B.Y. Abubakar. 2010. "Strengthening Innovation Capacity of Nigerian Agricultural Research Organizations." IFPRI Discussion Paper No. 01050. Washington, DC: International Food Policy Research Institute (IFPRI).

Rajalahti, R., W. Janssen, and E. Pehu, E. 2007. *Agricultural Innovation Systems: From Diagnostics toward Operational Practices*. Washington, DC: World Bank.

Reardon, T., J. Berdegué, and C.B. Barrett. 2007. "Household Income Diversification into Rural Nonfarm Activities." In *Transforming the Rural Non-farm Economy*, edited by S. Haggblade, P.B.R. Hazell, and T. Reardon. Baltimore: Johns Hopkins.

Spielman, D., and D. Kelemework. 2009. *Measuring Agricultural Innovation System Properties and Performance: Illustrations from Ethiopia and Vietnam*. IFPRI Discussion Paper 851. Washington, DC: IFPRI.

Sumberg, J., 2005. "Systems of Innovation Theory and the Changing Architecture of Agricultural Research in Africa." *Food Policy* 30:21–41.

World Bank. 2006. *Agriculture Investment Sourcebook*. Washington, DC.

———. 2007. *World Development Report 2008: Agriculture for Development*. Washington, DC.

———. 2009. "Agricultural Research and Competitive Grant Schemes: An IEG Performance Assessment of Four Projects in Latin America." Washington, DC.

———. 2010. "Designing and Implementing Agricultural Innovation Funds: Lessons from Competitive Research and Matching Grant Projects." Report No. 54857-GLB. Washington, DC.

———. 2011. *Information and Communication Technologies for Agriculture e-Sourcebook*. http://bit.ly/ICTinAG. Washington, DC.

Thematic Note 1

AATF (African Agricultural Technology Foundation). 2011. "Water Efficient Maize for Africa." http://www.aatf-africa.org/wema/, accessed February 2011.

Almekinders, C.J.M., L. Beukema, and C. Tromp. 2009. *Research in Action: Theories and Practices for Innovation and Social Change*. Wageningen: Wageningen Academic Publishers.

Bernet, T., A. Devaux, G. Thiele, G. López, C. Velasco, K. Manrique, and M. Ordinola. 2008. "The Participatory Market Chain Approach: Stimulating pro-poor market-chain innovation." ILAC Working Paper No. 21. Rome: Institutional Learning and Change Initiative (ILAC).

Bernet, T., G. Thiele, and T. Zschocke, 2006. *Participatory Market Chain Approach (PMCA) - User Guide*. Lima, Peru: CIP-Papa Andina. http://papandina.cip.cgiar.org/fileadmin/PMCA/User-Guide.pdf.

Byerlee, D., and K. Fischer. 2002. "Accessing Modern Science: Policy and Institutional Options for Agricultural Biotechnology in Developing Countries." *World Development* 30(6): 931–48.

Cavatassi, R., M. Gonzalez, P. Winters, J. Andrade-Piedra, P. Espinosa, and G. Thiele. 2009. "Linking Smallholders to the New Agricultural Economy: An Evaluation of the Plataformas Program in Ecuador." ESA Working Paper No. 09-06. Rome: Food and Agriculture Organization (FAO).

CIAT (International Center for Tropical Agriculture). 2006. The CIALS at a Glance. http://webapp.ciat.cgiar.org/ipra/ing/glance.htm, accessed February 2011.

Day-Rubenstein, K., and K.O. Fuglie. 2000. The CRADA model for public-private research and technology transfer in agriculture. In *Public-Private Collaboration in Agricultural Research: New Institutional Arrangements and Economic Implications*, edited by K.O. Fuglie and D.E. Schimmelpfenning. Ames: Iowa State University.

Devaux, A., J. Andrade-Piedra, D. Horton, M. Ordinola, G. Thiele, A. Thomann, and C. Velasco. 2010. "Brokering Innovation for Sustainable Development: The Papa Andina Case." ILAC Working Paper No. 12. Rome: Institutional Learning and Change Initiative (ILAC).

Devaux, A., D. Horton, C. Velasco, G. Thiele, G. Lopez, T. Bernet, I. Reinoso, and M. Ordinola. 2009. "Collective Action for Market Chain Innovation in the Andes." *Food Policy* 34:31–8.

Diaw, A., M. Samba, and J.L. Arcand. 2009. Analyse de l'impact du Programme des Services Agricoles at Organizations de Producteurs-Phase 2 (PSAOP II) sur la sécurité alimentaire des ménages ruraux. Unpublished study prepared by team from University of Gaston Shepherd, Senegal, and CERDI-Université d'Auvergne, France.

Dubin, H.J., and J.P. Brennan. 2010. "Combating Stem and Leaf Rust of Wheat: Historical Perspective, Impacts, and Lessons Learned." In *Proven Successes in Agricultural Development: A Technical Compendium to Millions Fed*, edited by D.J. Spielman and R. Pandya-Lorch. Washington, DC: International Food Policy Research Institute (IFPRI).

Ekboir, J.M., G. Dutrénit, V. Martínez, A.T. Vargas, and A.O. Vera-Cruz. 2009. "Successful Organizational Learning in the Management of Agricultural Research and Innovation: The Mexican Produce Foundations." Research Report No. 162. Washington, DC: International Food Policy Research Institute (IFPRI).

Ekboir, J., and G. Parellada. 2002. "Public-Private Interactions and Technology Policy in Innovation Processes for Zero Tillage in Argentina." In *Agricultural Research Policy in an Era of Privatization,* edited by D. Byerlee and R. Echeverría. Oxon, UK: CABI.

Ervin, D., T. Lomax, S. Buccola , K. Kim, E. Minor, H. Yang, L. Glenna, E. Jaeger, D. Biscotti, W. Armbruster, K. Clancy, W. Lacy, R. Welsh, and Y. Xia. 2003. *University-industry Relationship: Framing the Issues for Academic Research in Agricultural Biotechnology.* Philadelphia: Pew Initiative on Food and Biotechnology.

Faminow, M.D., S.E. Carter, and M. Lundy. 2009. "Social Entrepreneurship and Learning: The Case of the Central America Learning Alliance." *Journal of Developmental Entrepreneurship* 14(4):433–50.

Gill, G., and D. Carney. 1999. "Competitive Agricultural Technology Funds in Developing Countries." *Natural Resources Perspective* 41:1–10.

Gowda, C.L.L., B.V.S. Reddy, K.N. Rai, and K.B. Saxena. 2004. "ICRISAT Collaboration with the Seed Industry in Asia." Paper presented at the Asian Seed Congress 2004, September 13–17, Seoul. In *APSA Technical Report* No. 38. Bangkok: Asia and Pacific Seed Association (APSA).

Hocdé H., B. Triomphe, G. Faure, and M. Dulcire. 2009. "From Participation to Partnership—A Different Way for Researchers to Accompany Innovation Processes: Challenges and Difficulties." In *Innovation Africa: Enriching Farmers' Livelihoods,* edited by P.C. Sanginga, A. Waters-Bayer, S. Kaaria, J. Njuki, and C. Wettasinha. London: Earthscan. Pp. 135–50.

Horton, D., B. Akello, L. Aliguma, T. Bernet, A. Devaux, B. Lemaga, D. Magala, S. Mayanja, I. Sekitto, G. Thiele, and C. Velasco. 2010. "Developing Capacity for Agricultural Market Chain Innovation: Experience with the 'PMCA' in Uganda." *Journal of International Development* 2:367.

Janssen, W., and T. Braunschweig. 2003. "Trends in the Organization and Financing of Agricultural Research in Developed Countries: Implications for Developing Countries." ISNAR Research Report 22. The Hague: International Service for National Agricultural Research (SNAR).

Jones, R. 2006. "The Concept of Seed Enterprise Enhancement and Development Services (Seeds)." Electronic Newsletter of the African Seed Trade Association, http://www.seed quest.com/seed/associations/international/afsta/news letters/pdf/06dec_e.pdf, accessed March 2011.

Kaaria, S., J. Njuki, A. Abenakyo, R. Delve, and P. Sanginga. 2009. "Enabling Rural Innovation: Empowering Farmers to Take Advantage of Market Opportunities and Improve Livelihoods." In *Innovation Africa: Enriching Farmers' Livelihoods,* edited by P.C. Sanginga, A. Waters-Bayer, S. Kaaria, J. Njuki, and C. Wettasinha. London: Earthscan. Pp. 167–85.

Klerkx, L., and C. Leeuwis. 2009a. "The Emergence and Embedding of Innovation Brokers at Different Innovation System Levels: Insights from the Dutch Agricultural Sector." *Technological Forecasting and Social Change* 76: 849–860.

———. 2009b. "Operationalizing Demand-Driven Agricultural Research: Institutional Influences in a Public and Private System of Research Planning in the Netherlands." *Journal of Agricultural Education and Extension* 15(2):161–75.

Klerkx, L., N. Aarts, C. Leeuwis. 2010. "Adaptive Management in Agricultural Innovation Systems: The Interactions between Innovation Networks and their Environment." *Agricultural Systems* 103(6): 390–400.

Kolady, D., and W. Lesser. 2008. "Can Owners Afford Humanitarian Donations in Agbiotech: The case of Genetically Engineered Eggplant in India." *Electronic Journal of Biotechnology* 11(2):1–8.

Koyenikan, M.J. 2008. "Issues for Agricultural Extension Policy in Nigeria." *Journal of Agricultural Extension* 12(2):52–62.

Liu, M. 1997. *Fondements et pratiques de la Recherche-Action.* Paris: L'Harmattan.

Louwaars, N.P., R. Tripp, D. Eaton, V. Henson-Apollonio, R. Hu, M. Mendoza, F. Muhhuku, S. Pal, and J. Wekundah. 2005. *Impacts of Strengthened Intellectual Property Rights Regimes on the Plant Breeding Industry in Developing Countries: A Synthesis of Five Case Studies.* Wageningen: Centre for Genetic Resources.

Lybbert, T.J. 2002. "Technology Transfer for Humanitarian Use: Economic Issues and Market Segmentation Approaches." *IP Strategy Today* 5:17–25.

Masters, W.A. 2003. "Research Prizes: A Mechanism to Reward Agricultural Innovation in Low-income Regions." *AgBioForum* 6:71–4.

Monge, M., F. Hartwich, and D. Halgin. 2008. "How Change Agents and Social Capital Influence the Adoption of Innovations among Small Farmers: Evidence from Social Networks in Rural Bolivia." IFPRI Discussion Paper No. 761. Washington, DC: International Food Policy Research Institute (IFPRI).

Morris, M.L., and M.R. Bellon. 2004. "Participatory Plant Breeding Research: Opportunities and Challenges for the International Crop Improvement System." *Euphytica* 136(1):21–35.

Patiño, B.O., and R. Best. 2002. "Strategic Alliances of Cassava Farmers with Private and Public Sectors: A New

Approach for Development of the Cassava Crop in Latin America." Paper presented at the 9th Japan International Research Center for Agricultural Sciences (JIRCAS) International symposium, Value Addition to Agricultural Products, October 16–17, Ibaraki.

Pray, C.E., and L. Nagarajan. 2009. "Pearl Millet and Sorghum Improvement in India." IFPRI Discussion Paper No. 919. Washington, DC: International Food Policy Research Institute (IFPRI).

Quiros, C.A., B. Douthwaite, J.I. Roa, and J. Ashby. 2004. "Colombia, Latin America, and the Spread of Local Agricultural Research Committees (CIALS): Extension through Farmer Research." In *Demand-Driven Approaches to Agricultural Extension.* Vol. 3 of *Extension Reform for Agricultural Development,* edited by W. Rivera and G. Alex. Washington, DC: World Bank. Pp. 10–19.

Smith, J. 2005. "Context-bound Knowledge Production, Capacity Building, and New Product Networks. *Journal of International Development* 17(5): 647–59.

Smith, J., and J. Chataway. 2008. "Learning from the Periphery: The CGIAR and Civil Society Partnerships." Innogen Working Paper No. 72. Edinburgh: Innogen.

Sperling, L., J.A. Ashby, M.E. Smith, E. Weltzien, and S. McGuire. 2001. "A Framework for Analyzing Participatory Plant Breeding Approaches and Results." *Euphytica* 122(3):439–50.

Spielman, D.J. 2009. "Public-Private Partnerships and Propoor Livestock Research: The Search for an East Coast Fever Vaccine." In *Enhancing the Effectiveness of Sustainability Partnerships: Summary of a Workshop,* edited by D. Vollmer. Washington, DC: National Academies Press.

Spielman, D.J., F. Hartwich, and K. von Grebmer. 2010. "Public-Private Partnerships and Developing-country Agriculture: Evidence from the International Agricultural Research System." *Public Administration and Development* 30(4):261–76.

Spielman, D., and D. Kelemework. 2009. "Measuring Agricultural Innovation System Properties and Performance: Illustrations from Ethiopia and Vietnam." IFPRI Discussion Paper No. 00851. Washington, DC: International Food Policy Research Institute (IFPRI).

Thiele, G., A. Devaux, I. Reinoso, H. Pico, F. Montesdeoca, M. Pumisacho, J. Andrade, C. Velasco, P. Flores, R. Esprella, A. Thomann, K. Manrique, and D. Horton. Forthcoming. "Multi-stakeholder Platforms for Linking Small Farmers to Value Chains: Evidence from the Andes." *International Journal of Agricultural Sustainability.*

Trigo, E., E. Cap, V. Malach, and F. Villarreal. 2009. "The Case of Zero-tillage Technology in Argentina." IFPRI Discussion Paper No. 915. Washington, DC: International Food Policy Research Institute (IFPRI).

World Bank. 2006. "Institutional Innovation in Agricultural Research and Extension Systems in Latin America and the Caribbean." Washington, DC.

———. 2007. *World Development Report 2008: Agriculture for Development.* Washington, DC.

———. 2010. "Designing and Implementing Agricultural Innovation Funds: Lessons from Competitive Research and Matching Grant Projects." Washington, DC.

———. 2011 (forthcoming). *Information and Communication Technologies for Agriculture e-Sourcebook.* Washington, DC.

World Bank and IFPRI (International Food Policy Research Institute). 2010. "Gender and Governance in Rural Services: Insights from India, Ghana, and Ethiopia." Washington, DC.

Thematic Note 2

Anderson, J.R. 1998. "Selected Policy Issues in International Agricultural Research: On Striving for International Public Goods in an Era of Donor Fatigue." *World Development* 26(6):1149–62.

Byerlee, D., and K. Fischer. 2002. "Accessing Modern Science: Policy and Institutional Options for Agricultural Biotechnology in Developing Countries." *World Development* 30(6): 931–48.

Day-Rubenstein, K. and K.O. Fuglie. 1999. "Resource Allocation in Joint Public-Private Agricultural Research." *Journal of Agribusiness* 17(2):123–34.

de Bruijn, J.A., and H.G. van der Voort. n.d. "Public-private Partnership in Scientific Research: A Framework for Evaluation." Consultative Committee of Sector Councils for Research and Development (COS), the Netherlands.

Echeverría, R.G. 1998a. "Agricultural Research Policy Issues in Latin America: An Overview." *World Development* 26(6):1103–11.

———. 1998b. "Will Competitive Funding Improve the Performance of Agricultural Research?" ISNAR Discussion Paper No. 98-16. The Hague: International Service for National Agricultural Research (ISNAR).

Hagedoorn, J., A.N. Link, and N.S. Vonortas. 2000. "Research Partnerships." *Research Policy* 29:567–86.

Hall, A. 2006. "Public-Private Sector Partnerships in an Agricultural System of Innovation: Concepts and Challenges." UNU-MERIT Working Paper No. 2006-002. Maastricht: United Nations University (UNU) and Maastricht Economic Research Institute on Innovation and Technology (MERIT).

Hall, A., G. Bockett, S. Taylor, M.V.K. Sivamohan, and N. Clark. 2001. "Why Research Partnerships Really Matter: Innovation Theory, Institutional Arrangements, and

Implications for Developing New Technology for the Poor." *World Development* 29(5):783–97.

Hall. A., R. Sulaiman V., N. Clark, and B. Yoganand. 2003. "From Measuring Impact to Learning Institutional Lessons: An Innovation Systems Perspective on Improving the Management of International Agricultural Research." *Agricultural Systems* 78(2):213–41.

Hartwich, F., M.V. Gottret, S.C. Babu, and J. Tola. 2007. "Building Public–private Partnerships for Agricultural Innovation in Latin America: Lessons from Capacity Strengthening." IFPRI Discussion Paper No. 699. Washington, DC: International Food Policy Research Institute (IFPRI).

Hartwich, F., and C. Negro. 2010. "The Role of Collaborative Partnerships in Industry Innovation: Lessons Learned from New Zealand's Dairy Industry." *Agribusiness* 26(3):1–25.

Hartwich, F., O. Quirós, and J. Garza. 2009. "Partnerships for Agroindustrial Research and Development in Costa Rica and El Salvador." In *Fuelling Economic Growth: The Role of Public-Private Sector Research in Development*, edited by M. Graham and J. Woo. London Practical Action Publishing. http://www.idrc.ca/en/ev-135830-201-1-DO_TOPIC.html.

Hartwich, F., and J. Tola. 2007. "Public–private Partnerships for Agricultural Innovation: Concepts and Experiences from 124 Cases in Latin America." *International Journal on Agricultural Resources Governance and Ecology* 6 (2). 240–55.

Hartwich, F., J. Tola, A. Engler, J.J. Espinoza, G. Ghezan, C. Gonzalez, J.A. Silva, and J. Vásquez. 2008. "Guidelines for Building Public-private Partnerships for Agricultural and Agroindustrial Innovation." *Food Policy in Practice.* Washington, DC: International Food Policy Research Institute (IFPRI).

Horton, D., G. Prain, and G. Thiele. 2010. "Perspectives on Partnership: Highlights of a Literature Review." ILAC Brief 25. Rome: Institutional Learning and Change Initiative (ILAC).

Lewis, J. 2000. "Leveraging Partnerships between the Public and Private Sector: Experience of USAID's Agricultural Biotechnology Program." In *Agricultural Biotechnology and the Poor: Proceedings of an International Conference*, edited by G.J. Persley and M.M. Lantin. Washington, DC: Consultative Group on International Agricultural Research (CGIAR).

Muraguri, L. 2010. "Unplugged! An Analysis of Agricultural Biotechnology PPPs in Kenya." *Journal of International Development* 33(3):289–307.

OECD (Organisation for Economic Co-operation and Development). 1998. "Special Issue on Public/Private Partnerships in Science and Technology." *Science Technology Industry Review* 23. Paris.

Pray, C.E., and D. Umali-Deininger. 1998. "The Private Sector in Agricultural Research Systems: Will It Fill the Gap?" *World Development* 26(6):1127–48.

Rausser, G., L. Simon, and H. Ameden. 2000. "Public–Private Alliances in Biotechnology: Can They Narrow the Knowledge Gaps between Rich and Poor?" *Food Policy* 25(4):499–513.

Serafin, R., C. Bustamante, and C. Schramm. 2008. "What Is Current Practice in Evaluating Cross-sector Partnerships for Sustainable Development?" TPI Working Paper No. 1/2008. London: The Partnering Initiative–International Business Leaders Forum.

Spielman, D., and F. Hartwich. 2009. "Public-private Partnerships and Developing-country Agriculture." In: I. Scoones and J. Thompson, Farmers First Revisited. Bourton on Dunsmore, Rugby, U.K. Practical Action

Vieira, L-F., and F. Hartwich. 2002. "Approaching Public-Private Partnerships for Agroindustrial Research: A Methodological Framework." Coronado, Costa Rica: International Service for National Agricultural Research (ISNAR).

Thematic Note 3

Beintema, N., and H. Elliott. 2009. "Setting Meaningful Investment Targets in Agricultural Research and Development: Challenges, Opportunities, and Fiscal Realities." Paper prepared for the Expert Meeting on How to Feed the World in 2050, Food and Agricultural Organizations, June 24–26, Rome. FAO, ftp://ftp.fao.org/docrep/fao/012/ak978e/ak978e00.pdf, accessed March 2011.

CGIAR (Consultative Group on International Agricultural Research). 2010. "Embracing Change." CGIAR, http://www.cgiar.org/changemanagement/embracing change2/embracing_change_june8_2010.html, accessed March 2011. *Report on outcome of Global Conference on Agricultural Research for Development. Update on development of Consortium Mega-programs and strategic results framework. Progress report on creation of the Donor Fund.*

Coleman, J.S., with D. Court. 1993. *University Development in the Third World: The Rockefeller Foundation Experience.* New York: Pergamon.

Cox, T. Paul. 2010. "Cassava Development in the Networking Age." *New Agriculturalist*, http://www.new-ag.info/en/focus/focusItem.php?a=1586, accessed August 2011.

de Janvry, A. 2010. "Agriculture for Development: New Paradigm and Options for Success." In *The New Landscape of Global Agriculture.* Proceedings of the 27th Conference of the International Association of Agricultural Economists. New York: Wiley.

Elliott, H. 2010. Regional Agricultural Innovation Systems: Emergence or Intelligent Design? Draft manuscript and PowerPoint presentation for the RUFORUM Strategic Reflection Meeting, Bellagio, June. *Discussion of the conditions under which a regional agricultural innovation system may emerge and the historical lack of coordination of efforts in time, emphasis, and duration.*

Horton, D., G. Prain, and G. Thiele. 2009. "Perspectives on Partnership: A Literature Review." CIP Working Paper No. 2009-3. Lima: International Potato Center (CIP). *A good review of partnership activity, both conceptually and in the way the CGIAR research centers have tried to implement it. Draws on global literature, center policies, and CGIAR Science Council discussion.*

IFPRI and ASARECA (International Food Policy Research Institute and Association for Strengthening Agricultural Research in Eastern and Central Africa). 2006. "Strategic Priorities for Agricultural Development in Eastern and Central Africa." IFPRI Research Report No. 150. Washington, DC. *Identification of priority crops by development domains and spillovers among countries for regional investment in eastern and central Africa.*

Johnson, S. 2010. *Where Good Ideas Come From: The Natural History of Innovation.* New York: Riverhead Books (Penguin).

North, D.C. 1990. *Institutions, Institutional Change, and Economic Performance.* Cambridge: Cambridge University Press. *Classic work by Nobel Prize-winning new institutional economist.*

Pardey, P., J. James, J. Alston, S. Wood, B. Koo, E. Binenbaum, T. Hurley, and P. Glewwe. 2007. *Science, Technology and Skills.* Minneapolis: International Science and Technology Practice and Policy (InSTePP), University of Minnesota. *Discussion of innovation, incentives for collective action, and "technological distance" to argue that geographical proximity may not necessarily translate into spillover potential and that regional cooperative agreements may not be the most efficient way to capitalize on spillovers.*

Pingali, P. 2010. "Global Agricultural R&D and the Changing Aid Architecture." In *The New Landscape of Global Agriculture.* Proceedings of the 27th Conference of the International Association of Agricultural Economists. New York: Wiley.

RUFORUM (Regional Universities Forum for Capacity Building in Africa). 2005. "Ten Year Strategic Plan 2006-2015: Fostering Innovation and Adaptive Capacity of Universities to Develop and Sustain High Quality in Training, Impact-oriented Research, and Collaboration."

http://www.ruforum.org/category/documents-taxonomy/ruforum-policy-documents/, accessed August 2011.

———. 2010. Realigning the Regional Universities Forum for Capacity Building in Agriculture (RUFORUM) to a Changing Mandate and Emerging Challenges. Report of the Strategic Reflection Meeting of RUFORUM, Bellagio, April 2010. Unpublished.

———. 2011. "Business Plan 2011-2016: Consolidating Lessons and Success for Balanced Growth." http://www.ruforum.org/category/documents-taxonomy/ruforum-policy-documents/, accessed August 2011.

World Bank. 2007. Project Appraisal Document for West Africa Agricultural Productivity Program (WAAPP) Support Project. Internal document, Washington, DC. *Detail of adaptable program loan for West Africa.*

———. 2009. Project Appraisal Document for East Africa Agricultural Productivity Program. http://web.worldbank.org/external/projects/main?menuPK=51521804&pagePK=51351007&piPK=64675967&theSitePK=40941&menuPK=64154159&searchMenuPK=51521783&theSitePK=40941&entityID=000350881_20091124093008&searchMenuPK=51521783&theSitePK=40941, accessed March 2011. *Detail of adaptable program loan to Ethiopia, Kenya, and Tanzania for creation of Regional Centers of Excellence and outreach.*

Thematic Note 4

Almekinders, C.J.M., L. Beukema, and C. Tromp. 2009. *Research in Action: Theories and Practices for Innovation and Social Change.* Wageningen: Wageningen Academic Publishers.

Béguin, P. 2003. "Design as a Mutual Learning Process between Users and Designers." *Interacting with Computers* 15(5):709–30.

Béguin, P., and M. Cerf (eds.). 2009. *Dynamique des savoirs, dynamique des changements.* Toulouse: Octares.

Bernet, T., G. Thiele, and T. Zschocke, 2006: Participatory Market Chain Approach (PMCA). User's Guide. International Potato Center (CIP), Lima, Peru.

Bernet T., A. Devaux, G. Thiele, G. Lopez, C. Velasco, K. Manrique, and M. Ordinola, 2008: "The Participatory Market Chain Approach: Stimulating Pro-poor Market-driven Innovation." ILAC Brief No. 21. Rome: Institutional Learning and Change Initiative (ILAC).

Bousquet F., G. Trébuil, and B. Hardy (eds.). 2005. Companion Modeling and Multi agent Systems for Integrated Natural Resource Management in Asia. Los Baños, Philippines: International Rice Research Institute (IRRI) and Centre de Coopération Internationale en Recherche Agronomique pour le Développement (CIRAD).

Cavatassi, R., M. Gonzalez, P. Winters, J. Andrade-Piedra, P. Espinosa, and G. Thiele. 2009. "Linking Smallholders to the New Agricultural Economy: An Evaluation of the Plataformas Program in Ecuador." ESA Working Paper No. 09-06. Rome: Food and Agriculture Organization (FAO).

Devaux, A., J. Andrade-Piedra, D. Horton, M. Ordinola, G. Thiele, A. Thomann, and C. Velasco. 2010. "Brokering Innovation for Sustainable Development: The Papa Andina Case." ILAC Working Paper No. 12. Rome: Institutional Learning and Change Initiative (ILAC).

Devaux, A., D. Horton, C. Velasco, G. Thiele, G. López, T. Bernet, I. Reinoso, and M. Ordinola. 2009. "Collective Action for Market Chain Innovation in the Andes." *Food Policy* 34:31–38.

Etienne, M. 2005. "Co-construction d'un modèle d'accompagnement selon la méthode ARDI : guide méthodologique." Avignon: Institut National de la Recherche Agronomique (INRA). ComMod, www.commod.org, accessed March 2011.

Faure, G., P. Gasselin, P., B. Triomphe, L. Temple, and H. Hocdé. 2010. "Innover avec les acteurs du monde rural : la recherche-action en partenariat." [Innovate with rural stakeholders: Action-research in partnership.] Versailles: Quae.

Hocdé H., B. Triomphe, G. Faure, and M. Dulcire. 2009. "From Participation to Partnership—A Different Way for Researchers to Accompany Innovation Processes: Challenges and Difficulties." In *Innovation Africa: Enriching Farmers' Livelihoods,* edited by P.C. Sanginga, A. Waters-Bayer, S. Kaaria, J. Njuki, and C. Wettasinha. London: Earthscan. Pp. 135–50.

Horton, D., B. Akello, L. Aliguma, T. Bernet, A. Devaux, B. Lemaga, D. Magala, S. Mayanja, I. Sekitto, G. Thiele, and C. Velasco. 2010. "Developing Capacity for Agricultural Market Chain Innovation: Experience with the 'PMCA' in Uganda." *Journal of International Development* 2:367.

Klerkx, L., N. Aarts, and C. Leeuwis. 2010. "Adaptive Management in Agricultural Innovation Systems: The Interactions between Innovation Networks and Their Environment." *Agricultural Systems* 103(6):390–400).

Liu, M. 1997. *Fondements et pratiques de la Recherche-Action.* Paris: L'Harmattan.

Sanginga, P.C., A. Waters-Bayer, S. Kaaria, J. Njuki, and C. Wettasinha (eds.). 2008. *Innovation Africa: Enriching Farmers' Livelihoods* London: Earthscan.

Triomphe, B., and H. Hocdé. 2010. "Financer une recherche-action en partenariat : stratégies et pratiques." In *Innover avec les acteurs du monde rural : la recherche-action en partenariat,* edited by G. Faure, P. Gasselin,

B. Triomphe, L. Temple, and H. Hocdé. Versailles: Quae. Pp. 197–203.

Triomphe, B., H. Hocdé, and E. Chia. 2006. "Quand les agronomes pensent innovation et les institutions transfert: des malentendus sur la forme ou des visions différentes sur le développement? Le cas du Bajio guanajuatense (Mexique)." In *Agronomes et innovations,* edited by J. Caneill. 3ème édition des entretiens du Pradel. Actes du colloque des 8–10 septembre 2004. Paris: L'Harmattan. Pp. 247–66.

Veldhuizen, L. van, A. Waters-Bayer, and H. de Zeeuw. 1997. *Developing Technology with Farmers: A Trainer's Guide for Participatory Learning.* London: Zed Books.

Thematic Note 5

Blagescu, M., and J. Young. 2005. Partnerships and accountability: Current thinking and approaches among agencies supporting civil society organizations. Working Paper 255. London: Overseas Development Institute.

Cash, D., W. Clark, F. Alcock, N. Dickson, N. Eckley, D. Guston, J. Jager, and R. Mitchell. 2003. "Knowledge Systems for Sustainable Development." *Proceedings of the National Academy of Sciences* 100(14):8086–91.

CGIAR (Consultative Group on International Agricultural Research). 2011. "A Strategy and Results Framework for the CGIAR." Washington, DC.

Días Avíla, A.F., S. Salles-Filho, and J.E. Alonso. 2010. *Impacto de la I&D Agraria en el Perú: La experiencia de INCAGRO.* Lima: INCAGRO.

DRT (Department of Research and Training). 2008. "Enhancing Service Delivery for Agricultural Sector Development: CORDEMA Training Modules." Morogoro: DRT, Sokoine University of Agriculture, MVIWATA, and Royal Tropical Institute.

Fresco, H.W. 2010. "Promoviendo el Mercado de Servicios de Extensión Agraria en el Perú: La experiencia de INCAGRO." Lima: INCAGRO.

Gandarillas, A., J. Blajos, G. Aguirre, A. Devaux, and G. Thiele. 2007. "Changing Paradigms for Organising R&D: Agricultural Research and the Creation of the PROINPA Foundation in Bolivia." *International Journal of Agricultural Resources Governance and Ecology* 6(2):256–76.

Gijsbers, G., W. Janssen, H. Odame, and G. Meijerink. 2001. *Planning Agricultural Research: A Sourcebook.* Wallingford, UK: CABI Publishing in association with the International Service for National Agricultural Research (ISNAR).

Hawkins, R., W. Heemskerk, R. Booth, J. Daane, A. Maatman, and A.A. Adekunle 2009. "Integrated Agricultural Research for Development (IAR4D): A Concept for the Forum for Agricultural Research in Africa (FARA)

Sub-Saharan Africa Challenge Programme (SSA-CP)." Accra: FARA. ICRA, http://www.icra-edu.org/objects/anglolearn/IAR4D_concept_paper.pdf, accessed April 2011.

Heemskerk W., N. Lema, D. Guindo, C. Schouten, Z. Semgalawe, H. Verkuijl, B. de Steenhuijsen Piters, and P. Penninkhoff. 2003. "A Guide to Demand-driven Agricultural Research: The Client-oriented Research Management Approach." Amsterdam: KIT Publishers. Royal Tropical Institute (KIT), http://www.kit.nl/smartsite.shtml?id=SINGLEPUBLICATION&ItemID=1500, accessed April 2011.

Kelly, T., Ryan, J., and Gregersen, H. 2008. Enhancing ex post impact assessment of agricultural research: the CGIAR experience. *Research Policy* 17 (3):201-212.

López Heredia, D.A. 2010. "Mejorando la oferta de extensionistas agrarios en el Perú: La experiencia de INCAGRO." Lima: INCAGRO.

ILRI (International Livestock Research Institute). 2008. "ILRI's Partnership Strategy and Management System." Nairobi.

INCAGRO. 2010. "Proyectos de I&D+ Innovación Agraria: Productos y Resultados en Cifras." Lima.

Klerkx, L., A. Hall, and C. Leeuwis. 2009. "Strengthening Agricultural Innovation Capacity: Are Innovation Brokers the Answer?" *International Journal of Agricultural Resources, Governance, and Ecology* 8(5/6):409–38.

Lema N.M., C. Schouten, and T. Schrader (eds.). 2003. "Managing Research for Agricultural Development." Proceedings of the national COR workshop, Moshi, 27–28 May 2003. Dar es Salaam: Department for Research and Development (DRD) and Royal Tropical Institute (KIT).

Markie, J., and J. Compton. 2011. Establishment of a CGIAR Independent Evaluation Arrangement (IEA): Inception Report for Consideration of the CGIAR Fund Council, 5–6 April 2011. Unpublished report.

Mruthyunjaya, Dr. 2010. "Context, Design and Approach of NAIP: Experience of Implementation." Paper presented at the Workshop on Enhancing Capacity for Innovation: Learning from Practice, 5–6 May, Hyderabad.

NAIP (National Agricultural Innovation Project). 2010. "Half Yearly Progress Report (April–September 2010)." Delhi: Project Implementation Unit, National Agricultural Innovation Project, Krishi Anusandhan Bhawan II, Indian Council of Agricultural Research.

Otoo S., N. Agapitova, and J. Behrens. 2009. "Capacity Development Results Framework: A Strategic and Results-oriented Approach to Learning for Capacity Development." Washington, DC: World Bank Institute. http://wbi.worldbank.org/wbi/about/capacity-and-results, accessed April 2011.

Raitzer, D., and G. Norton (eds.). 2009. *Prioritizing Agricultural Research for Development*. Wallingford, UK: CABI.

Schrader T., E. van Poelje, G. Sempeho, A. Cissé, and J. Kampen, 2003. "Business Unusual: Making Public Sector Agricultural Research Organizations More Responsive to Clients' Needs." Amsterdam: Royal Tropical Institute (KIT) and Directorate General for International Cooperation (DGIS).

Vargas Winstanly, S. 2010. "Investigando para innovar, innovando para investigar: La experiencia de INCAGRO." Lima: INCAGRO.

Walker, T., M. Maredia, T. Kelley, R. La Rovere, D. Templeton, G. Thiele, and B. Douthwaite. 2008. "Strategic Guidance for Ex Post Impact Assessment of Agricultural Research." Rome: Standing Panel for Impact Assessment, Consultative Group on International Agricultural Research Science Council.

World Bank, FAO (Food and Agriculture Organization), and IFAD (International Fund for Agricultural Development). 2009. *Gender in Agriculture Sourcebook*. Washington, DC.

World Bank Independent Evaluation Group. 2008. "Public Sector Reform: What Works and Why?" Washington, DC.

———. 2009. *"Agricultural Research and Competitive Grant Schemes: An IEG Performance Assessment of Four Projects in Latin America"* Report No. 49149. Washington, DC.

Innovative Activity Profile 1

de Haan N, Romney D, Bezkorowajnyj P, Olufajo O. 2006. Feeding livestock through partnerships. *Knowledge Management for Development Journal* 2(3):123–35.

ILRI (International Livestock Research Institute). 2002. "Livestock—A Pathway out of Poverty: ILRI Strategy to 2010." Nairobi.

———. 2008. "ILRI's Partnership Strategy and Management System." http://hdl.handle.net/10568/566, accessed April 2011.

Kristjanson, P., R.S. Reid, N. Dickson, W.C. Clark, D. Romney, and R. Puskur. 2009. "Linking International Agricultural Research Knowledge with Action for Sustainable Development." *Proceedings of the National Academies of Science* 106(13):5047–52.

Lilja, N., P. Kristjanson, and J. Watts. 2010. "Rethinking Impact: Understanding the Complexity of Poverty and Change: Overview." *Development in Practice* 20(8): 917–32.

Innovative Activity Profile 2

Avila, A.F.D., and R.E. Evenson. 2005. Total factor productivity growth in agriculture: The role of technological capital. Unpublished draft, Economic Growth Center, Yale University, New Haven.

Evenson, R.E., C.E. Pray, and M.W. Rosegrant. 1999. *Agricultural Research and Productivity Growth in India*. Research Report No. 109. Washington, DC: International Food Policy Research Institute (IFPRI).

Pal, S., and D. Jha. 2007. Public-private partnerships in Agricultural R&D: Challenges and Prospects. In *Institutional Alternatives and Governance of Agriculture*, edited by V. Ballabh. New Delhi: Academic Foundation.

World Bank. 1998. "India: National Agricultural Technology Project, Project Appraisal Document." Report No. 17082-IN, Rural Development Sector Unit, South Asia Regional Office, Washington, DC.

———. 2006. "India: National Agricultural Innovation Project, Project Appraisal Document." Report No. 34908-IN, Agriculture and Rural Development Sector Unit, South Asia Region, Washington, DC.

———. 2007. *World Development Report 2008: Agriculture for Development*, Washington, DC.

Innovative Activity Profile 3

Álvarez, R., J.M. Benavente, C. Contreras, and J.L. Contreras. 2010. "Consorcios Tecnológicos en América Latina: Una primera exploración de los casos de Argentina, Chile, Colombia y Uruguay." Technical Note No. IDB TN-127. Washington, DC: BID (Banco Interamericano de Desarrollo). http://idbdocs.iadb.org/wsdocs/getdocument.aspx?docnum=35242324, accessed March 2011.

Benavente H., J.M. 2009. Untitled presentation for the Primer Encuentro de Consorcios Agropecuarios, April, Puerto Varas, Chile. Slide share, http://www.slideshare.net/cnicchile/consorcios-tecnolgicos-agropecuarios-1400911, accessed March 2011.

Bitran Colodro, E. and C.M. González Urrutia. 2010. "Productividad Total de Factores, Crecimiento e Innovación: Documento de Referencia. Santiago de Chile: Consejo Nacional de Innovación para la Competitividad (CNIC). http://biblioteca.cnic.cl/content/view/1021165/Productividad-Total-de-Factores-Crecimiento-e-Innovacion.html#content-top, accessed August 2011.

CNIC (Consejo Nacional de Innovación para la Competitividad). 2007a. *Hacía una estrategia nacional de innovación para la competitividad*. Vol. 1. Santiago de Chile: CNIC. http://bligoo.com/media/users/3/181868/files/18144/Presentacion(5).pdf, accessed March 2011.

———. 2007b. *Hacía una estrategia nacional de innovación para la competitividad*. Vol. 2. Santiago de Chile: CNIC. http://www.cnic.cl/content/view/472445/Presentacion.html, accessed March 2011.

———. 2010. *Evaluation Report of National Innovation Strategy for Competitiveness*, Chile. International Evaluation Panel, March 2010. http://biblioteca.cnic.cl/media/users/3/181868/files/18813/Reporte_PANEL.pdf, accessed August 2011.

Intelis Center, University of Chile. 2010a. "Análisis Institucional de Instrumentos Consorcios Tecnológicos Empresariales (CTE) y Evaluación de Medio Término de Consorcios Conicyt e Innova Chile." Final report. Santiago de Chile.

———. 2010b. FIA Analysis Consortium. Final report. Santiago de Chile.

Lavados, G., J. 2009. "Instrumento Consorcios en Chile: Transitando hacia la Segunda Etapa." Presentation for the Primer Encuentro de Consorcios Agropecuarios, April, Puerto Varas, Chile. FIA, http://www.fia.cl/difus/notici/InstrumConsorcioChile170409VF.pdf, accessed March 2011.

Innovative Activity Profile 4

Berdegué et al. 2007. Sistematización de experiencias locales de desarrollo rural: Guía metodológica. FIDAmerica and PREVAL. Mayo, 50p. http://preval.org/documentos/guia_metodologica.pdf.

Best, R., S. Ferris, and P. Mundy (eds.). 2009. "Working Together, Learning Together: Learning Alliances in Agroenterprise Development." Baltimore: Catholic Relief Services. USAID, http://pdf.usaid.gov/pdf_docs/PNADQ286.pdf, accessed August 2011.

Davies, R., and Dart, J. 2005. "The 'Most Significant Change' (MSC) Technique: A guide to its use." CARE International, United Kingdom; Oxfam Community Aid Abroad, Australia; Learning to Learn, Government of South Australia; Oxfam New Zealand; Christian Aid, United Kingdom; Exchange, United Kingdom; Ibis, Denmark; Mellemfolkeligt Samvirke (MS), Denmark; Lutheran World Relief, United States of America. April, 104p. http://www.mande.co.uk/docs/MSCGuide.pdf

Douthwaite, B., Alvarez, B. S., Cook, S., Davies, R., George, P., Howell, J., Mackay, R. and Rubiano, J. "Participatory impact pathways analysis: a practical application of program theory in research-for-development". *Canadian Journal of Program Evaluation*, 22(2), Fall 2007.

CRS and RII-CIAT (Catholic Relief Services and Rural Innovation Institute–International Center for Tropical Agriculture). 2007. "Preparing Farmer Groups to Engage Successfully with Markets: A Field Guide for 5 Key Skill Sets." Baltimore: Catholic Relief Services.

Ferris, S., P. Mundy, and R. Best (eds.). 2009. "Getting to Market: From Agriculture to Agroenterprise." Baltimore: Catholic Relief Services.

Lundy, M. 2008. Diversified livelihoods through effective agro-enterprise interventions: creating a cumulative learning framework. Unpublished final project report presented to the International Development Research Centre (IDRC), Ottawa.

Lundy, M., and M.V. Gottret. 2007. "Learning Alliances: Building Multi-Stakeholder Innovation Systems in Agro-enterprise Development." In *Learning Alliances: Scaling up Innovations in Water, Sanitation, and Hygiene*, edited by S. Smits, P. Moriarty, and C. Sijbesma. Technical Paper No. 47. Delft: IRC International Water and Sanitation Centre. Pp. 37–57.

Lundy, M., M.V. Gottret, and J. Ashby. 2004. "Building Multi-stakeholder Innovation Systems through Learning Alliances." ILAC Brief No. 8. Rome.

USEFUL WEBLINKS

Thematic Note 3

AGRA (Alliance for a Green Revolution in Africa) http://www.agra-alliance.org/

ASARECA   http://www.asareca.org/

Bill and Melinda Gates Foundation http://www.gatesfoundation.org/Pages/home.aspx

BecA-Hub (Biosciences Eastern and Central Africa) http://hub.africabiosciences.org/

CLAYUCA   http://www.clayuca.org/

CONDESAN   http://www.condesan.org

CORAF/WECARD   http://www.coraf.org/

PROCISUR   www.procisur.org

RUFORUM   http://www.ruforum.org/

SADC-FANR   http://www.sadc.int/fanr/

WACCI www.wacci.edu

Innovative Activity Profile I

CGIAR Research Program on Livestock and Fish http://livestockfish.wordpress.com

East Africa Dairy Development Project www.eadairy.org

Ecosystem approach to managing emerging zoonotic diseases http://www.ilri.org/EcoZd

Fodder Innovation Project www.fodderinnovation.org

Index Based Livestock Insurance (IBLI) Project http://www.ilri.org/ibli/

Global Alliance in Livestock Veterinary Medicines www.GALVmed.org

Safe Food, Fair Food http://www.ilri.org/SafeFoodFairFood

Innovative Activity Profile 3

Vinnova-Tecnovid www.vinnova.cl

FIA www.fia.cl

CORFO-Comité Innova Chile www.corfo.cl

Incentives and Resources for Innovation Partnerships and Business Development

OVERVIEW

Josef Ernstberger, Consultant

EXECUTIVE SUMMARY

Governments in developing countries increasingly *intervene actively in supporting private sector development* through diverse means. Public investments in business development can *direct* private investments towards areas of significant public interest and areas where the private sector alone would generally underinvest. They can *facilitate* or *stimulate* private investment through a conducive policy, legal, and institutional environment. Public investments for business development can also *complement* private investments (for example, by funding services or basic research). Such public-private partnerships need to become a *strategic element* of the agricultural development agenda. Formal technology transfer mechanisms (IPRs, licensing) offered through specialized *technology transfer offices* are critical to engage effectively in PPPs and disseminate technology through market channels.

The appropriate *funding mechanisms* to support innovation by collaborating public institutions, private entrepreneurs, and other actors depend on the public good to be produced and the role of the public sector. Useful alternatives include specialized innovation funds and matching grants to provide incentives for collaboration and risk taking. The use of *venture capital funding* has been limited in developing countries, but small and medium agricultural enterprises require risk capital to capture opportunities presented by agricultural innovation.

Support for *business incubation* helps to scale up small and often newly formed enterprises that bring innovative technologies and services to market. Developing countries require broader, less intensive, and more diverse incubator services to develop entrepreneurial, innovative cultures and business environments. *Agricultural clusters* foster innovation through proximity; they encompass interdependent firms in a value chain, service providers, and associated institutions. Cluster-based approaches have increased agricultural productivity, innovation, and business formation.

The *key policy issues* for agricultural business development and PPPs involve their potential for *altering development priorities*, the potential *welfare effects* of agricultural innovation and growth driven by private interests, *welfare concerns* related to gender and social equity, and prospects for building a "*shared responsibility system*" capable of balancing the sometimes divergent interests of the public sector, private sector, and civil society. Finally, in an environment characterized by increasing private involvement in agricultural innovation, very clear criteria will be needed to determine when public intervention is justified and at what level. Every publicly supported partnership or business development program must have a clear *time frame* and *exit strategy*. If the temporary nature of public involvement is not clear at the outset, private investors' decisions and business plans will be biased. The sustainability of social and environmental services, on the other hand, is often assured only through long-term public support.

RATIONALE FOR PUBLIC INVESTMENT

The power of the private sector to innovate and foster economic growth is a critical driver of long-term, sustainable development in agriculture. Development programs apply two major instruments to engage with the private sector for this purpose: *agricultural business development,* which aims to stimulate general economic growth, and *public-private partnerships* (PPPs), which aim to address development issues in conjunction with the private sector.

Agricultural development aims to achieve three objectives: national *food security, income* for rural people, and the *sustainability* of natural resources. These interdependent, sometimes conflicting objectives must be finely balanced; for example, food security must not be attained regardless of the cost to the environment. The private sector's primary goals are to generate *income* and *economic growth,* and its involvement in agricultural development carries the risk that development will be inequitable. The challenge is to find the common interests that will enable the private sector to use its many advantages to encourage balanced agricultural development and innovation.

Public sector support for and cooperation with the private sector are generally considered justified, valuable, or even necessary to: (1) compensate for market failures that prevent or hinder necessary private investments, (2) stimulate growth and help businesses become established; (3) generate and/or direct innovation in areas considered important for society; or (4) reduce some of the risk inherent in commercializing new technologies.

Support for agricultural business development

For agricultural business development, the most important objectives are to generate *qualitative* and *innovation-led* economic growth and income opportunities. "Qualitative growth" is associated with a range of additional public goods that especially reduce extreme poverty, provide food security, narrow structural inequalities, protect the environment, or sustain the growth process itself (Thomas et al. 2000). "Innovation-led" growth is based on innovative technologies, processes, products, markets, or organizational arrangements rather than on large additional uses of natural resources.

Public investments for business development are important because they can accelerate and improve the quality of growth in several ways. They can *direct* private investments towards areas of significant public interest and areas where the private sector alone would generally underinvest. They can *facilitate* or *stimulate* private investment through a conducive policy, legal, and institutional environment. Public investments for business development can also *complement* private investments (for example, by funding services or basic research).

The agricultural sector is characterized by specific market failures that are less prominent in other sectors, including the problem of scale, the time lags, and the multitude of partners. To overcome these and other initial obstacles to private investment and enable new products or technologies to be introduced, the public sector can provide incentives such as tax incentives, grants, and guarantees. Many governments support investments in new agricultural products or production systems until they can be commercialized or support the commitment of actors throughout an industry (in production, processing, and marketing) until a secure and mature business foundation develops. Most of the recently subsidized biofuel production programs were justified on these grounds.

A typical problem in developing agribusinesses is that the number and diversity of market players in a given value chain is often high. A chain's production base often consists of large numbers of small-scale, unorganized, geographically scattered producers. In circumstances such as these, which the market alone cannot improve, it makes little sense to introduce new production processes and products.

Public funds are needed to facilitate the vertical and horizontal linkages that will make the value chain efficient—for example, to organize farmers into cooperatives and associations and create platforms for institutional cooperation. For example, China's government supports *farmer-company* or *farmer-company-researcher arrangements,* in which farmers organize to partner with investors in processing/marketing industries, contract research institutions to develop certain products, or move into processing and marketing their products themselves (see IAP 2).

Support for public-private partnerships

Public support for PPPs in agriculture moves beyond business development and facilitation and makes direct use of individual *private sector actors to generate public goods.* Many governments and development agencies recognize that it can be more effective and sustainable for the public sector to work with the private sector to generate public goods in ways that enable each sector to build on its comparative advantages. Such direct cooperation can benefit both partners and is particularly useful in the following areas:

- **Social services.** Governments and development agencies use cooperation with private partners to invest directly in the delivery of social services. Public support ranges from providing favorable conditions for private investments in targeted poor or remote areas to contractual agreements between public institutions and private investors to deliver specific social goods and services. Such partnerships can offer a number of benefits, not only to small, local companies but to large multinational firms, which have come under increasing scrutiny to ensure that they operate in socially responsible ways. For example, under a PPP supported by GIZ and the multinational food company Tchibo GmbH, factories in Bangladesh, China, and Thailand are trained to implement production and labor standards with worker participation and sustainably improve employees' working conditions (GIZ 2009). In this way, PPPs pave the way for the development and adoption of international social standards.

- **Environmental services.** The protection and sustainable use of natural resources in agricultural production are important public goods. To preclude private entities from externalizing environmental costs or the costs of using natural resources, governments traditionally apply two sets of instruments: (1) regulations and controls and (2) incentives and disincentives in the form of financial instruments, such as subsidies for water-saving technologies or planting trees on eroded hillsides. The second set of instruments has recently become much more diverse and has created a new line of business in agriculture. Payments for environmental services involve not only payments for avoiding environmental costs (for example, payments for not cropping in watershed areas of reservoirs) but payments for actively generating environmental benefits. Farmers in the European Alps are subsidized to continue livestock production to maintain the characteristic alpine landscape, for example. Many countries use carbon sequestration funds to support farmers' efforts to plant trees, protect grasslands, use biogas, or pursue similar activities that reduce greenhouse gas emissions.

- **Innovation and technology adoption.** Agriculture in particular suffers from significant underinvestment in R&D (see module 4, TN 2) because of the high degree of externalities (such as benefits not being captured by the investor). PPPs lower the risk for individual private partners; at the same time, they foster the adaptation and dissemination of new research results or existing knowledge and technologies. This kind of collaboration helps to overcome problems of underinvestment and accelerates technological progress.

Notably in agricultural PPPs farmers play an important role as partners and providers of public services. This involves the mobilization and organization of farmers in formal or informal associations, cooperatives, or groups as a first step; then these organizations can partner with public services organizations or participate in wider partnerships with public and private organizations. Most commonly this partnering involves training and extension services but can also include adaptive research and technology testing or social and environmental services.

Table 5.1 summarizes some of the instruments used in business development, including PPPs.

PAST EXPERIENCE

Agricultural development agencies traditionally have been ambivalent about business development. Attitudes range from seeing business as an obstacle to agricultural development—at worst, the "evil middleman" that must be controlled—to regarding business as a necessary link between farmers and markets and finally to regarding business development as a driving force for agricultural innovation, growth, and development.

Business development is a relative latecomer to the agricultural development agenda, but starting in the mid-1990s the business sector came to be widely recognized as an important driver of agricultural development. Agricultural development programs and projects started to address the development of the private sector, particularly small and medium enterprises (SMEs) and value chains, and to seek ways of improving the business environment and facilitating business operations. Table 5.2 presents a more schematic view of the relationship between business development and partnership objectives and instruments.

Business development and partnership instruments

Recognition of the business sector's role in agricultural innovation and growth was accompanied by business development services (BDS) similar to those employed in nonagricultural sectors for many years. While most of these instruments are applied for general business development

Table 5.1 Business Development Instruments Used in Nonagricultural Sectors (and Later Adapted to Agriculture)

Type of instrument	Target firms	Key features
Tax incentives (for R&D)	– All firms (generally more attractive for larger firms)	– Motivate companies to invest in R&D and innovation – R&D tax credit to reduce a firm's tax liabilities, based on the amount spent to develop new products or improve existing products – Large enterprises paying more tax will benefit more than small firms
Business advisory services	– Small and medium enterprises (SMEs)	– Broad business support, including training and advisory services – Acts as primary service provider – Basically supply driven
Business development services	– SMEs	– Broad business support, including training and advisory services provided to individual businesses (more on a demand-driven basis than advisory services) – Often coordinates other service providers – Focus on building capacity within the business development service industry
Business incubation	– Startups and SMEs with high growth potential (dynamic enterprises)	– Integrated mix of intensive strategic and operational support provided to entrepreneurs and businesses selected for their growth potential – Focus on helping firms manage risk and build competitiveness through early, high-risk growth stages – Support typically ends when clients "graduate" by reaching particular milestones – May be linked with educational or research institutions
Science and technology parks[a]	– Emerging and established technology businesses, but may target specific industries	– Focus on helping relatively mature businesses accelerate growth – May use incubation as way to source future clients – May be linked to national, cluster-driven development strategies
Industry clusters[b]	– Related and supporting businesses and other organizations linked by a shared value chain (vertical) or shared final market (horizontal) – Concentrated in technology industries	– May be linked with educational or research institutions – May use incubation to source future clients – May be linked to national competitiveness strategies
Public-private partnerships (PPPs)	– Mature, strong, experienced companies (sharing of responsibilities) – Private partners can be diverse, from small to multinational	– Addresses delivery of public goods more directly – Addresses diverse set of public goods (social, environmental goods and technologies) – Can have sustainability problems (e.g., if social or environmental goods are targeted)

Source: infoDev Monitoring, Evaluation, and Impact Assessment Study, n.d.
a. For more information about technology parks, see infoDev, http://www.infodev.org/itparks. b. A practical application of Michael Porter's industry cluster theory, explored in *The Competitive Advantage of Nations* (1990).

---

or private sector partnerships, they can also be tailored to particularly stimulate and direct private sector development for R&D and innovation.

TAX INCENTIVES FOR R&D AND INNOVATION. R&D tax incentives have been used to encourage more spending on R&D. Such tax incentives are usually provided in the form of tax deductions based on the amounts spent in financing agricultural *research and innovation*. They can be a suitable instrument to overcome market failure resulting in underinvestment in R&D as they motivate companies to invest in innovation. Although such incentives may not always be limited to large corporations but include small and medium enterprises, clearly large enterprises that pay more tax

than small firms will benefit proportionately (World Bank 2006). Even though the use of tax incentives is widespread (over two-thirds of OECD members have tax incentives, as well as many developing countries), evidence of their cost effectiveness is not clear. Box 5.1 describes the experience in Chile, which has yet to undergo a closer impact evaluation. In general, tax incentives tend to benefit larger companies with large-enough revenue streams. Small and start-up companies may have difficulties benefiting from tax deductions because their revenue base is limited.

Other forms of tax incentives used in some countries include personal income tax deductions for individuals investing in startup businesses (effective only if the investor is paying any substantial income tax) and tax relief on

Table 5.2 Objectives, Instruments, and Financial Support Mechanisms for Business Development and Partnerships

		Business development and partnership instruments					
		BDS	**Incubators**	**Science parks**	**Clusters**	**TTOs**	**PPPs**
Objective — Overall		Qualitative, innovation-led *economic growth and income opportunities*					Specific *public goods* (e.g., social, environmental, innovations with high public good content)
Objective — Intermediate		Business startups		Business growth			Generation of innovations of public interest
		Business growth		Commercialization of new technologies			Transfer and application of technologies
					Efficiency gains (e.g., through synergies, reduced transaction costs, among other means)		
Financial support mechanisms		Often initial public funding, but service fee collection important for sustainability

Matching grants | Most incubators subsidized, fee collection and cost recovery difficult | Complementary public funding (e.g., infrastructure, public sector institutions)

Tax breaks and other financial incentives (e.g., subsidies, public sector credit guarantees) are common

Risk capital

Commercial financing | | Initial public funding replaced by service fees | Matching grants

Competitive grants |

Source: Author.

Note: BDS = business development services; TTO = technology transfer office; PPP = public-private partnership.

Box 5.1 Research and Development Tax Incentive Law in Chile

Chile was one of the first Latin American countries to introduce competitive funding programs for agricultural research. These programs have helped to increase the volume and quality of Chilean agricultural and nonagricultural research significantly. Another step was to introduce tax incentives.

The main objectives of the incentives are to:

- Increase private investment in R&D.
- Strengthen the link between research centers and companies.

The R&D tax incentive works as follows:

- Companies hire registered R&D centers through a previously approved R&D contract.

Source: Adapted from Noe 2007.

- Thirty-five percent of the payments private companies make to the research center against an R&D contract are considered a credit against corporate taxes.
- The remaining 65 percent of the payment is automatically considered expense for tax purposes.

The main requirements to qualify are:

- Research centers need to be registered. The criteria to be included in the registry include years of operation, research capabilities, and good accounting practices to ensure appropriate enforcement.
- Contracts need to be approved by CORFO, the Chilean development and innovation agency.
- The company and research center cannot be related.

donations to research foundations or endowments. Infant firms can benefit from incentives such as tax grace periods while they grow, which may be an important strategy in countries that tax firms even before they start production. Reducing taxes on importing equipment and supplies needed in R&D can help innovative projects. Innovation can be further encouraged by establishing special economic zones offering tax and regulatory relief, especially to stimulate cooperation with foreign partners.

BUSINESS DEVELOPMENT SERVICES. Business development services (BDS) comprise a wide range of nonfinancial services provided by public and private suppliers (BDS providers) to entrepreneurs who use them to operate more efficiently and expand their businesses. BDS thus may include training, consultancy, and advisory services, marketing assistance, information, technology development and transfer, and business linkage promotion. The high cost and low impact of many BDS approaches has caused government and international donors to shift from providing highly subsidized BDS toward a demand-led, market-based approach based on undistorted private service markets. The most frequently mentioned weaknesses of the former approach include (Altenburg and Stamm 2004):

■ *Lack of financial sustainability.* Because most services are highly subsidized, service providers come to depend on continuous public support. Most countries are cutting back on public expenditures, including business support measures.
■ *Insufficient outreach.* Even in times marked by high government revenues, deficit spending, or substantial inflows of foreign aid, business services usually only reach a relatively small percentage of the target group.
■ *Lack of business orientation.* Public service providers can share some of the less admirable tendencies of publicly administered programs. Budget allocations are not linked to program performance, employees do not act in a business-like fashion, and beneficiaries are not treated as clients. Incentives—both for support institutions and the people working within them—often are not designed for them to actively seek links with the business sector and strive for client satisfaction.
■ *Poor quality.* Service providers and clients work under different incentive systems, operational routines, and even mindsets, with the result that service supply often is not tailored to clients' needs. In addition, products delivered at low cost or for free may induce a debilitating dependency and cynicism over quality and value.

■ *Crowding out private competitors.* Delivering services at highly subsidized rates distorts markets and hampers the emergence of commercially viable service providers.

In contrast, the new BDS market paradigm highlights the need to deliver services at cost-covering rates and for providers to operate in a demand-driven, business-like manner. Service providers should either be private companies or public entities organized like firms with respect to their incentive systems, personnel, culture, and attitudes. Services should be regarded as commercial products, and the companies that receive services should be regarded as customers rather than beneficiaries. Providers should always charge fees high enough to secure the provider's financial sustainability (box 5.2).

BUSINESS INCUBATORS. Incubation first emerged in developed countries in the 1980s, operating alongside many other generic business development services and evolving to provide narrow and deep services for a small, select group of companies. Developing countries picked up the concept, and today more incubators are based in developing than developed countries. Observers and the "global business incubation community" estimate that of about 5,000 business incubators worldwide, at least 1,000 are based in Asia (approximately half in China), 1,000 in North America, 900 in Europe, and close to 400 in Latin America (with a sizeable and robust industry in Brazil).

As the name implies, incubators nurture young firms, helping them to survive and grow during the startup period when they are most vulnerable. Incubators provide hands-on management assistance, access to financing, and business and technical support services; they frequently also provide shared office space and access to equipment. Although they work with a broad spectrum of business development models, the vast majority of business incubators fall into two general categories: *technology* (focusing on commercializing new technology and transferring technology) or *mixed use* (serving a wide range of clients). Hybrid models, combining outreach, virtual, and broader services with more traditional incubation for new and existing businesses, have emerged in many developing countries, particularly those with smaller economies, limited generic business support services, weak cultures of entrepreneurship, difficult business environments, and limited resources to support innovation. In these circumstances, the demand for intensive, narrowly focused, and deep incubation services is minuscule. Instead, broader, less intensive, and more diverse services are needed to extend impact and develop

Box 5.2   Main Aspects of Developing and Implementing Demand-Driven, Sustainable
          Business Development Services

***Market assessment and reasoned justification of any
public intervention.*** Well-intended government interventions may be harmful for the long-term development of markets for business services and business
development services, because they may distort
prices, create bad habits, and crowd out private competitors. For this reason, intervention, rather than
nonintervention, in markets for business development services has to be justified on the basis of a clear
analysis of the situation.

Separating funding from service delivery. Major
problems arise with subsidized service provision
when the service provider and the organization
managing and administering the funds are identical.
Without relatively complex external supervision
arrangements, it is nearly impossible to commit this
"system" to an efficient and cost-sensitive execution of
its tasks. Inefficiencies often result from an explicit or
implicit obligation to spend funds in a given period,
disregarding careful targeting and the best possible
cost-benefit ratio. Separation of funding and delivery
functions will reduce the risk of crowding out private
suppliers, especially when private companies can apply
for public funding to provide services. This kind of
competition increases the transparency of service markets and provides additional information on whether
there still is a case for public intervention.

Improved accountability. Many service providers
offer a more or less ample set of services, often without
having established an accountability system to measure
the cost and the income generated by each service
offered. It is highly important to improve accountability and enable service providers to monitor market
success and cost-related aspects of each and every service offered. Service providers with a public function or
mission may then decide to cross-subsidize different
services to maintain important services that cannot be
provided on a cost-covering basis.

Monitoring and evaluating performance. Public
service provision must be continuously, transparently,

and independently monitored and evaluated. The two
subsystems of service provision (funding and delivery)
need to be evaluated according to different performance criteria. Within the organization that manages
the funds, the cost-benefit ratio of the previously
established objective(s) should guide the evaluation.
At the level of service delivery, the evaluation should
focus on the proven impact and efficiency of links
between the provider and small and medium enterprises (SMEs).

***Establishing a direct link between performance
and resource allocation.*** Monitoring and evaluation
are not objectives in their own right but should be
used for continuously improving the system. The
most effective way to ensure the system's responsiveness is to link the allocation of funds directly to the
performance of those who supply services. The success of effective service providers is largely based on
incentive systems that provide financial rewards for
good performers.

Compulsory cofinancing. To ensure that SMEs feel
some ownership of the services they receive, every
transaction should be partly financed by the customer.
The proportion of cofinancing will depend on the
character of the service and the final objective of service provision. Additional factors to consider include
the business environment and changes in the institutional setting. Services with predictable and appropri-
able outcomes should be largely financed by the
customer, while in some strategic areas it will be necessary to step up the share of funds transferred. When
SMEs operate under conditions of economic growth or
at least stability, a higher proportion of private financing should be expected. On the other hand, when
macroeconomic conditions are volatile and competition is becoming life-threatening for many companies,
governments or other funding organizations may opt
for a higher share of subsidies, stressing short-term
impact and outreach and temporarily sidelining
aspects of financial sustainability.

Source: Altenburg and Stamm 2004.

entrepreneurial and innovative cultures and business environments. For more information on how an incubation
approach may contribute to business development and
innovation in agriculture, see TN 3.

TECHNOLOGY PARKS AND CLUSTERS. Technology parks (also
known as science parks or research parks) are usually linked
with educational or research institutions and provide infrastructure and support services for businesses, particularly

real estate and office space. Technology parks can foster business-driven development and innovation because they focus on creating links and collaboration among diverse businesses, many of them large, established businesses for technology transfer and economic development. Business incubation shares some of the features of technology parks and industry clusters, in that each involves a "place," processes, and companies with potential for growth. The main difference is that business incubation focuses on startup companies that eventually graduate from the incubator service, whereas companies in clusters or parks usually intend to maintain their association. Technology parks often have their own business incubators dedicated to "growing" tenants for the park, and some clusters have a business incubation component as well.[1]

TECHNOLOGY TRANSFER OFFICES. Technology transfer translates agricultural research innovations into applications. The vast majority of agricultural innovations in developing countries arise from publicly sponsored research centers that typically are inexperienced and unprepared to engage in formal mechanisms of technology transfer. Technology transfer offices (TTOs), which are usually affiliated with research organizations or universities, have the mandate to identify and protect research results[2] with a view to facilitating their use and commercialization. Some TTOs also host incubation services for businesses to commercialize technology. TN 5 offers a detailed discussion of TTOs.

INDUSTRY CLUSTERS. Since 2000, industry clusters have become an increasingly popular model for organizing strategies and policies to promote regional development. Clusters are agglomerations of strongly interdependent firms (including specialized suppliers) linked to each other in a value-adding production chain, service providers, and associated institutions in a particular field. Some clusters encompass strategic alliances with universities, research institutes, knowledge-intensive business services, bridging institutions (brokers, consultants), and customers. Cluster-based approaches for business development and innovation have increased agricultural productivity, innovation, and business formation (Gibbs and Bernat 1998; Andersson et al. 2004; World Bank 2009b). TN 4 summarizes key lessons from cluster-based business development and innovation.

PARTNERSHIPS. At first, private partners in agricultural development initiatives focused on developing and strengthening businesses, but growing appreciation of the private sector's role in agricultural development has led to innovative partnerships in which the public and private sectors are true business partners, producing public goods. The partners bring their complementary skills to a program or project, with varying levels of involvement and responsibility (TN 1). Partnerships range from global strategic alliances to specific forms of cooperation in individual development projects. In true PPPs, the financial and managerial strength and experience of the private partners is important (box 5.3), but many development programs have difficulty working with financially strong and experienced companies and are concerned that the partnership will weaken their focus on poverty. The objective of PPPs is not to support weak businesses through public contracts, however, but to engage with the private sector as a reliable partner that can deliver a public good efficiently. A comparison with the procurement of public works is useful, because it follows requirements for bidders based on size, past experience, financial strength, and reliability.

Funding mechanisms

A range of innovative funding mechanisms can be used to support innovators and their links to public institutions, private entrepreneurs, and other actors.[3] Rather than funding innovation through block grants, many countries use specialized innovation funds to provide incentives for collaboration and risk taking. For example, competitive research grants target research-related activities to mobilize public and private research capacity. Matching grants are widely and increasingly used to stimulate engagement between the private sector and farmers in activities related to technology generation, technology dissemination, and innovation processes.

Matching grants show greater promise than competitive research grants in fostering business-driven innovation development, as they tend to be better at promoting pluralism in applied technology development, transfer, and adoption by enhancing ownership among actors. They are also well suited to overall development of agribusiness because they can be adapted to support productive partnerships, provide technical assistance and other services, promote productive activities by farmer groups, support value-added activities, and build small-scale infrastructure (World Bank 2010). TN 2 reviews experiences with grant schemes and the various opportunities they present; IAP 4 provides examples.

Venture capital funding is explicitly designed for investment in a high-risk business or security of some type. It has

Box 5.3 Critical Choices for Public-Private Partnerships

Before entering into any partnership project or program, the rationale for the investment must be fully understood, along with the problem(s) to be fixed or outcomes to be achieved. These considerations have critical implications for the choice of:

- **Institutions and partners.** If delivery of a public good is the main objective of the partnership, it must involve organizations that are capable of generating the good (or that can undergo institutional development to acquire this capability). This issue is explored in the discussion of institutions and partners later in this module.
- **Financing instrument.** The appropriate financing instrument depends on the type of public good to be produced and the role of the public sector. When governments act as brokers in these sorts of partnerships, common instruments include government guarantees, risk or venture capital for developing and introducing new technologies, or competitive and matching grants to "buy" public goods from the private sector. For environmental and social services, temporary or permanent subsidies are more common instruments. For creating an enabling business environment, legislative instruments are complemented by direct investments in services, infrastructure, and facilities.
- **Exit strategy.** Every support program must have a clear time frame and exit strategy. The public sector intervenes to support private investment until industries mature or technologies are proven and adopted. If the temporary nature of public involvement is not clear at the outset, private investors' decisions and business plans will be biased. The sustainability of social and environmental services, on the other hand, is often assured only though long-term public support.

Source: Author.

been used widely outside agriculture to support business-driven development, but its application in agriculture has been limited. Small and medium agricultural enterprises require risk capital to fully capture the opportunities presented by agricultural innovation, however. TN 6 summarizes experiences with and applicability of risk capital investment models for agriculture in developing and middle-income countries.

KEY POLICY ISSUES

The key policy issues for agricultural business development and PPPs involve their potential for altering development priorities, the potential welfare effects of agricultural innovation and growth driven by private interests, welfare concerns related to gender and social equity, and prospects for building a "shared responsibility system" capable of balancing the sometimes divergent interests of the public sector, private sector, and civil society. Finally, in an environment characterized by increasing private involvement in agricultural innovation, very clear criteria will be needed to determine when public intervention is justified and at what level.

Food supply versus rural incomes

Building on business development and business-driven innovation, agricultural development is likely to see priorities shifting toward growth and income generation and possibly away from regional and global food security. Business development is driven by the interest of producers, processors, or traders of agricultural products in generating profits (for example, by increasing factor productivity, adding value, or developing new products and markets) and capturing a meaningful proportion of these gains as additional income rather than passing them on to consumers.

Increasing factor productivity in primary agricultural production is the main approach to increase food production, but increased factor productivity does not always translate into higher profits for producers over the long run. Historical trends show that after early adopters of a technology achieve their initial high gains, most of the ensuing productivity gains are passed on quickly from producers to consumers in the form of absolutely or relatively lower prices for foods. This scenario offers little to interest key business actors, with the possible exception of the input supply industry (box 5.4).

Box 5.4   Different Trajectories of Agricultural Growth and Producer-Consumer Welfare Distribution

From an economic welfare perspective, additional agricultural economic growth and ultimately additional welfare in a society are generated by higher levels of productivity, which in turn depend strongly on innovation. Technologies that increase the volume of production (such as a higher-yielding crop variety) would shift the supply function to the "right" and increase welfare, but this shift would largely come at the expense of producers because of a price-decreasing effect. Technologies that reduce the cost of production (such as insect-resistant Bt cotton, which reduces the use of costly pesticides) would shift the supply function "downward" again, with a significant welfare gain that would in this case be shared more equally between consumers and producers.

An alternative option for growth is possible by actively addressing the demand function or creating new demands for agricultural food and nonfood products. For example, agricultural raw materials like maize, sugarcane, or cassava could be used in new ways for energy production; consumers' changing preferences could be served through market differentiation, brand-naming of products, and so on; an agricultural raw material could serve as the basis for chemical or pharmaceutical products; or agricultural products could be designed as functional foods. While the vast majority of agricultural production is still destined to meet the world's basic requirements for food, market differentiation and new products and uses will increasingly play a role in raising the overall value of agricultural production, especially in middle- and higher-income countries, with production coming from developed and developing countries.

Source: Author.

Distributional effects of business-driven agricultural growth

The distributional aspects of agricultural innovations are related to the food security issue just mentioned. As explained in box 5.4, the nature and type of innovations and technologies influence both absolute growth and welfare gains as well as the distribution of welfare gains between producers and consumers. Agricultural innovations and growth driven by business interests not only open new growth opportunities but are likely to move agricultural development in a different direction. Although innovation generated by the private sector can be important in securing continued gains in agricultural productivity, particularly from the agricultural input side (in the form of new seed or chemicals, for which royalties can be sufficiently protected), the private sector will probably give much more attention to adding value in agricultural production and production systems. This emphasis will favor agricultural industry (companies supplying inputs and processing or marketing products) and possibly farmers as the primary producers, but it could put poor consumers of staple foods at a disadvantage.

Gender and social equity

Compared to the public sector, private interests are not equally sensitive to gender or other social equity considerations, with the risk that a growing role of the private sector could further erode social equity in agricultural development. For example, women make up most of the rural workforce (70 percent), but in comparison to men, women still have far fewer resources (land, capital, and access to knowledge) to carry out their tasks (World Economic Forum 2011). The growing influence of civil society and shared responsibility for social and gender issues by the public and private sector are positive, continuing developments, but on their own they will not ensure that gender and equity considerations are reflected in agricultural development. Governments need to incentivize the private sector and promote and safeguard gender and social equity in all business development activities, PPP contracts, services, and access to resources and knowledge. Specifically targeted PPPs can make an important contribution to gender and social equity goals (for an example, see IAP 2 on the China Technology Transfer Project).

A shared responsibility system with new roles for stakeholders

A well-functioning society provides a range of checks and balances to keep the agricultural development agenda in line with the desired priorities. Many companies have started to integrate social and environmental responsibilities in their long-term business concepts because they understand that this strategy is important for long-term commercial success. A balance between private commercial interests and the interests of society is not attained automatically, however. It is the product of smooth interaction among governments, civil society, and the private sector, under a shared responsibility system (for example, see Henckes et al. 2004).

While PPPs and business development activities can be interesting for most countries, these activities are unlikely to deliver the desired results in the absence of a minimum level of governance and sufficient control mechanisms in civil society (box 5.5). A capacity analysis is a prudent step to take before investing in PPPs and business development in a given setting. The analysis should assess risks and risk mitigation measures and determine whether supportive capacity building is needed. To work well, a shared responsibility system requires a high level of transparency, sensitization, and sufficiently effective mechanisms for society to influence policies as well as private sector decisions—conditions that are anything but perfect in most countries.

Market distortion

Market failure is the main justification for public sector interventions, but efforts to address market failures often distort markets unduly. Two key questions need to be answered: First, is public investment justified? Second, what is the right amount or level of support? The generation of a public good is a necessary condition for intervention, but it is not sufficient justification. Public funds need to generate *additional* positive economic net returns that would not have been generated without public investment. In other words, the public sector should not pursue investments that the private sector is likely to undertake on its own. For practical purposes, this determination can be difficult to make, however. Nor is it easy to determine the right level of public sector support (box 5.6). TN 1 describes methods for determining additionality.

NEW DIRECTIONS, PRIORITIES, AND REQUIREMENTS FOR INVESTMENT

Given that business development and PPPs are becoming important drivers for innovation and growth in agriculture, policy makers and development agencies need to consider a range of issues related to supporting partnerships and business. Among these considerations, *developing a vision* of the priorities and strategic choices related to agricultural development is perhaps the most fundamental. Agriculture provides more than food. It produces essential commodities that increasingly include energy, environmental services, and a wide range of social goods. Given that three-quarters

---

Box 5.5   Consumers Want to Be Engaged

Civil society itself has developed a wide range of instruments to express its interests and preferences and influence governments, development agencies, and private companies. "Mature and sensitive" societies significantly influence the direction of technology development and business opportunities. For significant numbers of consumers in these societies, the value of a food product lies not only in its taste, nutritional value, and chemical and physical properties but in the resources used to produce it, the impact on the environment, the contribution to global warming, and the social conditions and safety of farm or factory workers.

According to the 2010 Cone Shared Responsibility Study, 84 percent of Americans believe that their ideas can help companies create products and services that are a win for consumers, business, and society, yet only 53 percent feel that companies effectively encourage them to speak up on corporate social and environmental practices and products. A majority of consumers want to be engaged on four key pillars of responsible business, including how a company conducts its business (85 percent), its products and packaging (83 percent), its support of social and environmental issues (81 percent) and its marketing and advertising (74 percent). Consumers are prepared to dedicate time and money to help influence corporate social/environmental practices through surveys and research (70 percent), buying or boycotting a company's products (44 percent), or through email, phone, or employee communications (32 percent), among other activities.

Sources: Author; Cone 2010.

Box 5.6   Additionality Criteria to Use in Deciding Whether to Provide Public Funding

Public investment decisions need to be based on an *economic* and a *financial* analysis. The economic analysis determines whether an investment is likely to generate additional benefits at minimum rate of return for public capital investments. The financial analysis shows whether public funds are required at all. It helps to ascertain, for example, if an investment is unattractive to private investors, and it determines the level or amount of public funds needed (the share of public grants or level of subsidy, for example).

Source: Author.

Additionality is a key consideration in programs to support private sector development. Certainly companies will enjoy public support and will give any assurance to donors that they would not have invested without such support. This assertion needs to be proven. Especially in matching grant programs, the risk is high that public funds will simply replace private funds. This substitution is not only inefficient but disturbs and biases competitiveness among producers or businesses.

of the world's poorest people live in rural areas, agricultural development goals are often torn between delivering low-cost food using increasingly scarce and more expensive resources and providing income for poor farmers. Effective partnering and business development must be guided by a vision for agricultural development that sorts through these issues, sets priorities, and makes strategic and often difficult choices.

The roles of the public and private sector must also be defined clearly. Business-driven agricultural development, combined with PPPs, will require a *definition or redefinition of the roles and responsibilities of the public and private sectors.* Many governments already find it challenging to define and implement clear policies of what the public sector should do in relation to the private sector. Ministries of agriculture and their subordinate institutions often still engage directly in enterprises related to agricultural production, such as input supply or food processing facilities. They will need to reassess these enterprises with a view to facilitating participation by the private sector and partnering with private entities.

A fundamental role of government is to create conditions enabling the private sector to generate economic growth through innovation and the development of new businesses. The lifeline of an *enabling business environment* is a strong legal and institutional framework capable of protecting investors and intellectual property. Module 6 gives examples of strategies to develop an enabling environment for agricultural innovation.

The public sector also *intervenes actively in supporting private sector development* through business development and advisory services, business incubators, technology parks, and cluster development. *Formal technology transfer* mechanisms, such as intellectual property protection and legal agreements that transfer property rights to commercial or international partners, are becoming critical to engage effectively in PPPs and disseminate technology through market channels. TTOs are an interesting option, especially if they are affiliated with research organizations or universities or host incubation services—in other words, if they focus on nurturing businesses that aim at commercializing technology.

Working in isolation, public institutions and private companies are less and less likely to address global, regional, or even local agricultural development challenges. *New partnerships need to be encouraged,* from global strategic partnerships and alliances to innovative PPPs and individual stakeholder partnerships. Such partnerships need to shift from being ad hoc initiatives to becoming a strategic element of the agricultural development agenda.

With the growing diversity of partners and institutional arrangements, the demand for more *innovative funding mechanisms* is growing as well. The significant experience with some of these mechanisms, such as competitive research grants and matching grants, can be built upon; at the same time, new applications for other funding mechanisms, such as risk capital funding, are beginning to emerge.

MONITORING AND EVALUATING INVESTMENTS

Tracking and attributing the results of a business innovation or partnership program is highly challenging. The particular challenges relate to the timeframe, complexity of the processes, great array of stakeholders, and external influences. Key aspects of M&E include clarifying the program's objectives, identifying appropriate indicators, establishing appropriate M&E arrangements, following common monitoring practices, and evaluating impacts.

For complex business innovation or partnership programs, a multistage grouping of indicators may be useful. For example, the PAID framework includes process indicators (P), used to track the first stage of a program; action indicators (A), used to track activities and inputs provided by the program; investment indicators (I), used to track investments and co-investments by the private entities receiving support; and delivered results (D), used to measure final outcomes.[4] Table 5.3 provides examples of indicators for designing monitoring systems.

Table 5.3 Possible Monitoring and Evaluation Indicators

Type	Indicator	Program type			
		Grant	PPPs	Incubator	Cluster
Process indicators	Analytical activities (competitiveness, availability of infrastructure, capacity of institutions)	X	X	X	X
	Memoranda of understanding	X	X	X	X
	Agreed schedules and levels of participation (including aspects of gender or other social equity)	X	X	X	X
	Other milestones in process for designing and establishing a business support scheme (for example, private sector mapping)				
Action indicators	Technical assistance provided	X	X	X	X
	Completion of strategies			X	X
	Completion of action plans with responsibility split among representatives			X	X
	Delivery: Number of subprojects terminated within a year after the planned date	X		X	X
	Success rate: Number of subprojects that have achieved the planned milestones	X	X		
	Punctuality: Ratio of realized and planned time for subproject execution	X	X		
	Length of subproject cycle (number of months)	X	X		
	Number of incubatees supported			X	
Investment indicators	Outsourcing for efficiency: Share of contracted research within subproject activities (percentage of total)	X			
	Additionality of resources attracted by mechanism (from clients, government, private sector, and partners)	X	X		
	Business formation: Number of new business registered			X	X
	Value of newly registered capital			X	X
	Accessed financing during the incubation process			X	
	Business retention: Percentage of graduates staying in the community in which they were incubated			X	
	Business success: Percentage of incubator graduates staying in business			X	
	National and foreign investment attracted				X
Delivered results	Factor productivity (crop yields, labor productivity)	X	X		X
	Trends in natural resource degradation (soil erosion rates)	X	X		X
	Social rate of return to research (percentage)	X			
	Absolute and relative poverty rates (percentage)	X	X		
	Scientific quality and spillover benefit (publications, citations, peer evaluations)	X			
	Increase in the value of sales, farmer value-added, the quality of produce of farmers engaged in partnerships	X	X		X
	Increase in the income/profitability or competitiveness of target actors (agribusiness, farmers, and others)	X	X	X	X
	Increase in innovation (technical, organizational, and other) among the target actors	X	X	X	X
	Public return on investment (e.g., tax revenue versus public spending)			X	X
	Sector or subsector growth rates				X
	Employment				X

Sources: World Bank 2010; World Bank 2009b; author.

Foundations for Public-Private Partnerships

Josef Ernstberger, Consultant

SYNOPSIS

Governments and bilateral and multilateral development agencies are engaging the private sector to deliver a range of services traditionally delivered by the public sector. Such PPPs involve new arrangements under which governments obtain services through contracts with direct payments or various forms of subsidization. PPPs are also increasingly valued as a means of unleashing the private sector's capacity to generate innovation in the rural sector. This note discusses the opportunities and constraints of PPPs, including their institutional settings, the capacities and skills on both sides, partnership arrangements (especially the need for contracts that clearly define outcomes and ensure accountability), and the need for independent supervision and monitoring of PPPs.

BACKGROUND AND CONTEXT

Public-private partnerships (PPPs) bring the complementary skills of the public and private sectors to a program or project in which each partner has a different level of involvement and responsibility, with the objective of providing public goods or services. These partnerships leverage the strength and reflect the interests of the individual partners. These partnerships are not intended primarily for business development, for which business development programs are a better alternative, but for bringing about a public good outcome in partnership with the private sector. The public sector is interested in cooperating with the private sector to use technologies, capital, and know-how and (ideally) to benefit from the comparatively greater flexibility, innovative capacity, and efficiency of private companies. For the private partners, the incentive is generally that the collaboration opens an interesting new line of business.

PPPs in agriculture mainly target opportunities for environmental, social, or equitable growth. They range in scope from cooperating on an individual project, to generating ideas and innovations in a specific field or for a specific purpose, to major strategic alliances that address major concerns in the development of the agricultural sector.

PPPs implemented through individual development projects are characterized by a relatively clear and specific expected outcome or service for which the private partner provides know-how and technical solutions. In infrastructure projects this kind of arrangement is generally known as a "design-build," "design-bid-build," or "design-build-operate" contractual arrangement. For example, an IT company may be invited to develop and implement a tailored search engine for a local Internet-based agricultural extension system.

When PPPs are used to generate ideas and innovations, the public sector defines more or less specific fields and objectives in which it seeks progress (environmental health, social welfare, agricultural growth, and so forth) and issues a call for proposals to public and private entities, who submit their ideas for collaborative work to generate the desired innovations. The proposals are submitted for review, and if they are selected for funding, the public and private partners implement the programs.

Strategic alliances usually involve long-term cooperation (ten or more years), multinational companies, or groups of companies. Examples include the development and introduction of minimum social and environmental standards for agricultural or forestry products, fair trade arrangements, and similar ambitious programs.

In agriculture, PPPs are more promising in some areas than others. They can be quite effective for introducing environmental and/or social production and processing standards, which then become national and sometimes even international standards (box 5.7). Examples of collaborative projects that subsequently resulted in national legislation include projects on standards for organic and fair trade food production, control of child labor, and the protection of forest biodiversity.

Box 5.7 A Public-Private Partnership to Implement Labor Standards in Asia

Many international companies produce goods in developing countries where international labor standards are not followed and in any case are challenging to monitor. A public-private partnership run by GIZ and Tchibo GmbH engages workers in factories in Bangladesh, China, and Thailand in implementing the International Labour Organization's core labor standards and improving working conditions. Since the project started, 13 local training providers have acquired the specialized knowledge and tools to advise suppliers on setting up company structures for dialogue and implementing social standards. Forty firms are taking part in the pilot, and the number is rising. All have designed and implemented action plans, including plans for electing staff representatives, avoiding forced labor, and reducing overtime.

Source: GIZ 2009.
Note: GIZ = Deutsche Gesellschaft für Internationale Zusammenarbeit.

PPPs have also helped governments or development organizations direct innovation toward specific areas of public interest. Governments or development organizations either acquire innovations directly from the private sector or, through competitive research grant schemes or idea competitions, challenge the private sector to pursue innovative ideas and technologies of public interest. Efforts could be as specific as identifying a solution to a particular plant disease or as general as improving energy use in agricultural production systems. (See table 5.1 in the module overview; see also module 4, IAPs 2 and 3, for examples from India and Chile.)

A number of countries have started to use private organizations or NGOs to provide specialized services (such as agricultural advisory services) that public agencies once provided. Development organizations have turned to commercial or noncommercial NGOs to deliver extension or training to farmers or train cooperatives and agricultural processors. For more detail, see module 3.

With technologies and production processes becoming more complex and technically demanding, governments increasingly use private sector capacity for regulatory controls and enforcement. Partnerships have been formed with leading enterprises or specialized private companies and

laboratories to regulate biosafety hazards, control animal diseases, detect genetically modified ingredients in food, and enforce restrictions on seed imports or exports, among other activities.

IMPLEMENTATION ARRANGEMENTS AND INVESTMENT NEEDED

Countries require a *governance framework* that brings together the capacities for designing and executing PPPs. In most countries, line ministries or lower-level government institutions execute PPPs, which may require public officials to master new skills. They must know how the private sector operates and design collaborative programs accordingly. They must assess the risks and incentives that make collaboration attractive and do not waste taxpayers' money. They need skills to negotiate and manage contracts, and to avoid disturbing markets, they need to be able to assess the long-term implications of their partnership programs for industry structure and competitiveness. A new way to provide these capacities is to establish PPP units within cross-sectoral ministries such as finance or planning (Dutz et al. 2006). These units can operate in several ways. For example, they can provide information and guidance to other government departments for designing and preparing PPPs, provide advisory support and funding to line departments or agencies, or directly approve PPPs.

The *execution of PPPs entails numerous steps*, beginning with the selection of private sector partners. For most PPPs, specially convened committees—whose membership comes from government and the private sector but should not be dominated by the government representatives—select the private partners. Depending on the nature of the partnership envisaged, the committee may also include farmers or representatives of farmer organizations, agricultural education, extension, or NGOs. Competitive application and selection procedures are common. The objectives of the partnership program, the conditions for application, and the funding arrangements are publicly advertised. One or more rounds of proposals follow. In evaluating proposals, the selection committee often calls upon additional expertise in the form of a technical expert group or individual experts. These experts might prepare a short list of candidates, but the final decision remains with the committee. After selecting the private and public partners, it is a good practice to formalize the partnership through a contract or signed memorandum of understanding.

Generally PPPs are *managed by project implementation units* established for this purpose and supported by a secretariat with appropriate technical and administrative capabilities. Every PPP also requires a framework for M&E to ensure that the program is on track and make adjustments as needed. It is vital for the partners to agree beforehand on the M&E framework and the practical arrangements for implementing it. The framework must be flexible enough to handle an unpredictable timeframe and outcomes, given that many of the key variables will have a significant amount of uncertainty. An independent but mutually accepted monitoring agency or organization should implement the framework.

Particularly in the agricultural sector, PPPs require *supplementary funding to build capacity in most or all private partners.* Capacity building can extend from the application process to proposal development and program implementation.

POTENTIAL BENEFITS

As mentioned, the major benefits of PPPs derive from using the complementary strengths of the public and private partners to:

- *Attain efficiency gains.* Many public goods can be delivered by private partners more cost-effectively, especially if contracts are output oriented and give the private partners the flexibility to identify the most cost-effective technical solutions.
- *Mobilize resources and investments.* PPPs have significant potential to mobilize additional resources and funding. Sometimes only limited or even no public funds are needed to trigger significant private investment (box 5.8; IAP 3).
- *Develop innovative solutions.* PPPs can encourage the private sector to come forward with creative ideas.
- *Reduce risk.* Transferring part of the project risk to private partners can be one of the key benefits of PPPs and result in better control over public spending, the service delivery time frame, and quality of service. Output- or delivery-based payments can be effective for reducing risks.

POLICY ISSUES

PPPs often require the public and private sector to *redefine their respective roles and responsibilities.* This issue requires particular attention from policy makers. Many governments already find it difficult to separate the functions of the two sectors; they may subsidize private enterprises without clear justification or operate what appear to be commercial agri-

cultural ventures such as plantations or seed companies. For these reasons, PPPs can aggravate rather than resolve uncertainty over what each sector should properly do.

Many PPPs also suffer from a *lack of suitable public institutions and legislative arrangements.* Donor programs often bridge this gap by setting up their own program- or project-specific implementation units and operational manuals. While project-specific arrangements may be necessary, as long as an overall institutional and legislative framework is lacking, it will be challenging to implement PPPs as a regular feature of government procedures and programs.

As discussed, poorly designed PPPs can easily *distort markets,* and many PPPs probably do so. Major market distortion occurs when public funds are invested in activities that the private sector would pursue in any case (crowding out private investment) or when private companies participating in PPPs achieve an unfair and unjustified market or production advantage over their competitors.

To attract public investment, private companies tend to emphasize their social or environmental responsibility, but their ultimate (and legitimate) interest is to generate revenue. PPPs will remain *sustainable* and private partners will meet their obligations in the long run only if this interest is appropriately factored into partnership arrangements acceptable to all participants. The simplest PPPs use public funds to pay private enterprises to deliver public goods. More complex arrangements pay for the delivery of public goods by incorporating the price of the public goods (such as social or environmental services) into output prices.

Although PPPs can reduce the gap between technology development and adoption and make public research institutions more effective, responsive, and demand driven, they do have risks. Private interests can supersede public interests in public research agendas. Governments can *lose the public research capacity* that is critical for developing technologies that may not be commercially attractive but are in the public interest.

LESSONS LEARNED

The experience with PPPs can be distilled into guiding principles to help practitioners develop and invest in partnerships while avoiding problems such as market distortion and poor sustainability. Over the years, important lessons have been learned about which private enterprises and farmer organizations make successful partners for the public sector and which strategies work best for selecting them. Issues related to contracts, the partnership's time frame, the

Box 5.8   A Public-Private Partnership to Conserve Genetic Resources in China

The Chinese government maintains important genetic stocks for animal breeding at a number of stations throughout the country. In Anhui Province, the government contracted a medium-scale company engaged in pig breeding and processing to protect an endangered pig variety called Wei Pig, which was close to extinction. The company undertook to commercialize the variety as a niche product while continuing with its mainstream pig production business. The market strategy for Wei pig involved a special "near nature" production process, in which poor, small-scale farmers in mountainous areas raised pigs under contract for the company, which operates a breeding center that also functions as a park for visitors. The higher prices of meat from these pigs compared to conventionally produced pigs reflect the higher costs of the more environmentally and socially oriented process used to produce them. The government provided a share of the initial financing for the company and its contract farmers. This partnership made it possible to close a government facility and sustainably protect genetic resources without public funding, once the cofinancing ended. The partnership succeeded owing to a combination of private technical know-how and experience, an innovative business and marketing concept, and significant private financial resources, complemented by public startup funds.

Source: Author.

capacity gaps that may need to be filled, and arrangements for M&E are also important to consider before the partnership begins. Details of these guidelines, lessons, and issues follow.

Guiding principles for partnership

Guiding principles for PPPs involve the clarity of their political objectives, potential mutual benefits, additionality, competition and transparency, and sharing of risks and responsibilities:

- *Consistency with political objectives.* The public partner must clearly define the larger political objective that motivates the partnership. For example, government may want to stimulate agricultural growth as an instrument to address rural poverty and reduce increasing income disparities in a society (equitable growth). It may want to protect specific natural resources and future livelihood systems (such as forest products for traditional medicine), support specific vulnerable groups (by improving labor conditions, promoting pro-poor growth, or improving gender equity). It may want to bolster national food security.
- *Mutual benefits.* As discussed, all partners must benefit sufficiently from the partnership to honor their commitments for as long as required. The public sector benefits if the PPP proves to be effective and efficient at generating the desired public goods. Benefits to the private sector can be indirect (for example, its participation will improve its reputation by demonstrating social or environmental responsibility), but in most cases the desired benefits are legitimate financial profits. It is important to understand the benefits required by all partners to prevent the partnership from being abused. For example, the real incentive for private partners may be to gain closer ties to government to pursue a hidden agenda such as influencing political decisions or obtaining an unfair advantage over competitors.
- *Additionality.* Public funds should support PPPs only when the private sector would not undertake a similar activity to achieve the same outcome, either on its own or as required by law.
- *Competition and transparency.* Like other forms of public investment, PPPs must give all competent private partners an equal opportunity to compete for the business opportunities supported. Setting priorities for funding, selecting partners and programs, allocating resources to partnerships, implementing and monitoring partnership programs, and all other decisions must be transparent.
- *Sharing risks and responsibilities.* In PPPs the commitment of partners is generally demonstrated by an appropriate sharing of risk and responsibilities. A major difference between PPPs and traditional public contracts such as infrastructure contracts is that the participating private companies contribute financial and/or human resources (for example, through cofinancing, matching funds, and other arrangements).

Selecting the right partners with the right capacity

The financial and managerial strength and experience of private partners are important for the success of partnerships. As mentioned in the module overview, PPPs have to be clearly distinguished from programs to support SMEs: SME development focuses on the enterprises themselves by helping them to become fully established or providing venture capital, but PPPs focus on achieving public good outcomes. Partnerships with financially strong, experienced companies are more likely to succeed in delivering those public good outcomes efficiently and reliably. As discussed in the module overview, properly constructed PPPs do not support weak business with lucrative public contracts; nor do they weaken the public sector's or donor agency's focus on poverty. These problems can be avoided if the public sector performs due diligence and thoroughly assesses prospective partners beforehand on the basis of minimum criteria related to their financial strength, management capacity, and demonstrated ability to deliver the kinds of public goods required.

Farmers can be effective partners in PPPs and improve their impact in rural communities, yet very few farmers have the capacity to perform in PPPs without some form of assistance. Most PPPs involving farmers have a strong component for organizing farmers and providing the skills they need to perform their role in the partnership. Partnerships with farmer organizations work best when the organizations have a clear and narrow interest or focus. For example, water user associations have become strong public partners in many countries and have successfully assumed traditional public service functions. Farmer-managed grazing associations or other natural resource management organizations have successfully regulated access to and use of common resources.

Farmer organizations are increasingly important commercial partners, given that the most efficient agricultural value chains emanate from a strong, organized producer community. Forming these organizations will help to balance the power among the partners and is practically the only feasible way for government and companies to interact with producers, because organization substantially reduces the cost of interacting with large numbers of smallholders.

Preparing the way for successful partnerships

The administrative steps in selecting partners and implementing programs were discussed earlier. The next points describe practical steps to lay the groundwork for successful partnerships.

- *For PPPs, the public good outcome is the primary objective, and it must be clearly defined* (an innovation, an environmental or social service, the performance of a traditional public service by a private partner, and so on). Confused objectives are a common defect of many programs, which fail to clarify whether their primary objective is to support or develop businesses or to deliver specific public goods.

- *Clearly define the criteria that a private partner must meet to participate,* especially the type of partner (private company, farmer group, and so on), legal status (perhaps farmer organizations must incorporate to participate), size, previous experience, and technical, financial, and managerial capacity.

- *Clearly specify the contribution expected from each partner.* Contributions can take many forms: financial resources, human resources, risk-sharing arrangements, sharing of innovations, or access to confidential or internal information (financial data, income, cash flow, technical processes). It is rarely in the interest of companies to share information related to innovative technologies, business concepts, or financial status, but if this information is necessary to achieve the partnership's objectives, this requirement must be very clear before potential partners apply for funds. This point reinforces the earlier point that absolute clarity at the outset can prevent conflicts from derailing a partnership.

- *Provide transparent information on modes of public financing and decision making.* Information about the size of and conditions for public financing is generally straightforward. The problem lies more with the lack of clarity over government procedures and the time they require. Private companies are especially frustrated by complex and lengthy public procurement procedures and bureaucratic clearances, but if they are aware of procedural requirements beforehand they may cope better.

Contracts

Like any business relationship, each PPP should be based on a signed contract between the partners, usually representatives of the government partner, company, and farmer organization, as relevant). At a minimum, the contracts must:

- Define each expected outcome, the corresponding indicators (measurable and conducive to monitoring), and time-bound targets. This level of specificity may seem challenging for PPPs intended to develop innovations, because of the organic nature and unpredictability of the process. The intended outcome is already clear, however, for specific partnerships based on a proposal to provide an innovative business idea, technology, or public good or service, and the partners need to be assessed against their progress in achieving that outcome.

- Specify the resources committed by the partners, along with a detailed financing and implementation plan.
- Include a monitoring framework with agreements on the monitoring arrangements and contract supervision by an accepted third-party monitoring institution or individual (see the concluding part of this section).
- Define the exit strategies and follow-up arrangements (as discussed next).

Exit strategies and sustainability

PPPs generating social or environmental goods may end once a good has been delivered, or they may need to continue for as long as the public good is required (ideally without public support). For example, if the social or environmental value of a particular good becomes a critical part of the marketing concept (as with many organic and fair trade products), the *level of commercial interest* may be sufficient for the government to withdraw public support. If the partnership pilots environmental technologies or social standards that are eventually embodied in *mandatory regulations*, they will be sustained if compliance with regulations is assured.

PPPs generating innovations require an exit strategy based on *transferring and adopting the innovation.* Many PPPs are designed to include elements of technology transfer. For example, World Bank-supported grant programs in Albania and Armenia required PPPs to earmark a certain percentage of funds for publicizing and/or transferring their innovations to farmers or small businesses. Because it is not in the interest of private companies to share innovative ideas or new technologies with potential competitors without compensation, future ownership of any innovation developed by the partners and any obligations to share information must be spelled out in the PPP design and clarified in the contract. Many arrangements have been developed to protect financial and other incentives to share technology and information.

PPPs that shift public service provision to private partners can be sustained by *institutionalizing partnership arrangements.* For example, private partners can be organized and authorized to collect user fees, which must be high enough to assure financial sustainability and/or maintain the business interest of the private partner. Many water-user organizations have used this kind of arrangement once initial public support for their activities ends. Private food quality testing laboratories receive a fee for performing public-service functions.

Capacity building

A precondition for involving farmers in PPPs is to form organizations and equip the members with the skills to be effective partners, as discussed earlier. Aside from needing to acquire management skills and an understanding of managerial procedures, farmer organizations may also require training related to agricultural production and processing, quality standards, participating in adaptive research, testing technology, and providing social and environmental services. The best practice for PPPs is separate supporting activities in capacity building very clearly from the actual funding of PPPs. One option, for example, would be to include capacity building as a separate component of the project.

Monitoring and evaluation

If a PPP encompasses more than a single project, the M&E system should provide information corresponding to several levels of activity. First, at the level of *individual projects* (often called "subprojects" when they are part of a large program that funds numerous small projects), the M&E system should provide information on adherence to contractual arrangements and outcomes from each subproject. Second, at the *program level,* the M&E system should provide information on the program's processes, outcomes, and attractiveness. Third, at the *policy level,* the M&E system should generate feedback on the program's broad effects on economic growth (the causal link between the program and the resulting innovation and factor productivity) and wider social benefits.

PROJECT OR SUBPROJECT LEVEL. As mentioned, project-level M&E combines *contract supervision* as well as *outcome monitoring and evaluation,* and it should be done by an independent, third-party M&E institution that is acceptable to all partners. The basis for M&E of individual projects (subprojects) is established when partners enter the proposal development and approval process. No proposal should be approved for funding unless it possesses a set of clearly defined objectives linked to measurable and monitorable outcome indicators and time-bound targets. The indicators should distinguish between the public good objectives (such as the provision of an environmental or social good or an innovation for scaling up) and the financial or commercial objectives (such as target figures for production, sales, or income and benefit sharing). These detailed indicators and targets are included in the contract between the individual partners. The contract also specifies detailed input and output parameters, such as the financial and other resources that the partners have each committed to provide.

PROGRAM OR PROJECT LEVEL. Program/project-level M&E should aggregate the achievements of individual subprojects and devote most of its attention to examining the effectiveness and efficiency of the PPP program. Many programs report outstanding achievements of individual subprojects but offer very little information, for example, on whether the program proved attractive for investors, partners, and beneficiaries; on any design flaws that emerged; or on whether investments were taken up slowly or quickly. Few programs use "with-program" and "without-program"

scenarios to assess their effectiveness, because designing control groups or control scenarios is difficult. Program M&E should also provide information on the overall use of funds, factors driving the use of funds, and mechanisms for delivering funds. All of this information will improve how similar projects are designed and implemented.

POLICY LEVEL. Some of the most difficult issues in monitoring and evaluating PPP programs relate to whether and to what extent a PPP program has influenced overall economic growth, whether it experienced or caused interference with other parts of the economy, what its wider social or environmental implications (positive and negative) may have been, the sustainability of its impacts, and finally the policy measures that should be taken as a consequence of these findings. The list of failed government interventions in the private sector is long, and the effects have sometimes been significant. On the other hand, sound analyses of the effects of PPPs will be invaluable for formulating policy (box 5.9).

Box 5.9 The Importance of Policy-Level Monitoring, Evaluation, and Analysis

The economic effects of public-private partnerships (PPPs) can be very large. For example, many governments responded to skyrocketing energy prices by launching programs to generate and support innovations in renewable energy. Some of these programs related to biofuels have had far-reaching consequences for agriculture and food production that remain controversial and are not yet completely understood. Another example of the economic effects of PPPs comes from China, where the government promoted PPPs (among many other strategies) to develop a modern dairy industry. The result was an enormous overstimulation of milk production. The subsequent collapse of the dairy market and bankruptcy of many producers was accelerated by scandals over contaminated milk but was probably inevitable, given the inappropriate level of market interference.

Most PPPs will not have such vast economic effects, because they are small and locally confined, but the positive or negative macroeconomic and policy implications even of small programs should be analyzed and understood. Policy-level M&E and analysis are

also important for determining whether PPPs are the right instruments for achieving certain objectives (for example, taxes or regulations could be used instead of PPPs to induce compliance with social or environmental goals).

Policy-level evaluations of PPPs can also determine whether more extensive policy measures might be warranted. For example, the China Agricultural Technology Transfer Project supports a number of PPPs with private companies to develop and test new technologies for manure treatment. Cattle manure treatment plants of different sizes, using different fermentation processes, are being tested in Heilongjiang Province; special technology to treat duck manure was developed in Anhui Province. Aside from introducing and testing technical solutions, partners in these projects are analyzing their commercial and financial parameters to learn, for example, how the competitiveness of the livestock industry would be influenced if manure treatment were to become mandatory. In other words, aside from stimulating technical innovation, these PPPs are preparing the ground for decisions about future policy measures.

Source: Author.

Innovation Funds

Josef Ernstberger, Consultant
Riikka Rajalahti, World Bank

SYNOPSIS

This note summarizes experiences with the two main mechanisms used to fund agricultural innovation, competitive research grants (CRGs), and matching grants (MGs). It offers lessons and guidelines for designing and implementing innovation funds, drawing on information and analysis published in World Bank (2010). That report primarily analyzed experience with World Bank investments, but the lessons are relevant in other contexts. The decision to use a grant scheme to fund innovation and the choice of scheme require a rigorous decision on the objective, which must be embedded in a strategic vision of the innovation system. CRGs are often used in the larger context of agricultural research system reform programs. MGs tend to work best when combined with complementary investments (infrastructure, financial services). Significant costs can be associated with setting up and administering a grant program, and the cost-effectiveness of procedures to keep overhead low must be balanced against the need to ensure accountability and transparency of operations. Funding priorities must be set with stakeholders, who should participate in the governance of the grant or program and provide continuous feedback on implementation. Transaction costs associated with participation in a grant scheme may be significant and reduce the pool of private applicants. Grant recipients are often not adept at fulfilling M&E requirements and will benefit from specific training and hands-on support.

BACKGROUND AND CONTEXT

Institutional core funding from public resources has been the most common means of funding research and innovation, as many core research activities require long-term support. Core funding also facilitates the development of integrated research programs to address major problems, commodity needs, or geographical areas. Many core research activities, such as plant breeding, have only one research provider, so competitive funding is irrelevant (World Bank 1999a). Institutional core funding does not easily lend itself to innovation processes that engage diverse stakeholders, however, even though much innovation can result from tapping the vast pool of creative ideas they possess. In many instances, those ideas never translate into innovation because the incentives and human and financial resources that enable collaboration and innovation are lacking.

To compensate for this failure, many countries are using innovation funds in the agricultural sector to strengthen innovation-promoting links among public institutions, private entrepreneurs, and other actors, such as groups of rural producers. These funds—often in the form of grants—create platforms for innovative activity by providing incentives to improve research collaboration and quality. They complement traditional core funding allocated annually to specific public research institutions to pursue their core research agenda.

Competitive research grants (CRGs) fund research based on national competition and scientific peer review. Transparent procedures are used to select the proposals that will receive funding, based on rigorous criteria. Grants can accomplish objectives that may be difficult to achieve through core funding, such as innovation in specific areas of research (perhaps through adaptive on-farm research projects that require organizations and farmers to work together) or innovation in a target region. If they are well designed, grants can bring greater contestability to the innovation process; the funding may not necessarily flow to the traditional recipients. Grants can promote research partnerships, leverage research resources, and help develop a more efficient, demand-driven, and pluralistic research system by involving clients in setting priorities and financing, executing, and evaluating research (World Bank 2009, 2010). For an example of using competitive grants to promote multistakeholder collaboration in India, see module 4, IAP 2.

Matching grants (MGs) can be used to finance research but increasingly promote near-market technology generation, technology transfer and adoption, private economic activity, and overall innovation, often by including multiple stakeholders. By focusing greater attention on demand and use from the very beginning, basically by attracting users of technologies and knowledge in partnerships (and requiring a matching commitment), MGs may be more effective than CRGs at enhancing the use of technology and knowledge by farmers and other entrepreneurs. Funds from the granting organization (usually a public agency) are matched with funds from the beneficiary. Table 5.4 compares the three financing modes.

The use of MGs in particular warrants caution. They should be used for public good investments, such as generating technical or institutional innovations with positive economic, social, or environmental externalities or reversing market failures. They can thus be used for addressing knowledge gaps and reducing the risk for investment when the market fails to produce a necessary or desirable good, but at the same time other public sector instruments may be more effective and less costly.

The following list indicates when grants may or may not be a good choice (van der Meer and Noordam 2004; Donovan 2006):

- **The lack of public goods such as infrastructure, legislation, or information.** In this case, the appropriate solution is to invest in these public goods. It will not help to give grants to reduce the high costs of production caused by their absence.
- **The lack of economies of scale.** No single enterprise is large enough to make the lumpy investments needed to overcome this problem. In this case, grants do not automatically help, although it may help for governments to support collective action for making lumpy investments.
- **High risk,** arising (for example) from the long gestation periods for certain investments, political instability, lack of transparency in government policy, or natural disasters. Private insurance schemes can handle some risks, and governments should first deal with any deficiencies in their own policies and performance before considering grants.
- **High costs of protecting property rights.** In general, governments should establish and protect property rights and provide subsidies (grants) only where the costs of enforcing those rights are too high.
- **Lack of commercialization of the economy.** In such an economy, the development of financial services is especially slow. Grants should not be used in these cases for subsidizing credit, but they may be justified for training, developing management information systems, or helping to expand rural outreach of credit providers and install new technologies.
- **Lack of technology, information, or trained staff.** Grants may be useful to solve these problems.

Table 5.4 Comparison of Competitive Research Grants (CRGs), Matching Grants (MGs), and Core (Block) Funding

Issue	CRGs	MGs	Core (block) funding
Primary objectives and activities	Basic, strategic, and adaptive research (and extension)	– Demand-driven, near-market technology development, dissemination, and adoption and overall innovation processes – Private sector activity, including agribusiness and productive partnership creation	Long-term strategic research and institution building and strengthening
Key stakeholder	Primarily research system actors, increasingly also private sector	Farmers, private sector, NGOs, research institutes, extension services, other service providers based on objective	Public research organizations
Capacity requirements for success	– Requires a critical mass of staff and a steady operational budget to allow true competition and result in improved research quality – Capacity to compete and administer (including technical review) crucial for success	– More flexible with capacity requirements of participants or the administrative burden – Business understanding, ability to partner, and monitoring and evaluation requirements great	Little additional capacity required
Cost sharing	Limited, due to ownership issues	High for demand-driven activities	Limited to in-kind resources
Overhead and transaction costs	Significant, due to management and monitoring and evaluation	Varies; can be significant depending upon the degree of decentralization of the program and the overall purpose and actors involved	Low or nonexistent
Incentives for partnership	Fair, depending on criteria	High	Limited
Sustainability	Limited unless complement long-term funding	Limited unless complement long-term funding	High

Source: Authors, adapted from World Bank 2010.

INVESTMENT OR ACTION NEEDED

A number of investments and actions are needed for innovation funds to perform successfully. They are briefly explained in the sections that follow. See World Bank (2010) for detailed explanations and examples.

Establish a fund for financing projects

A grant scheme requires a fund that provides financing for the selected subprojects. The fund is usually situated within a ministry of agriculture and managed and disbursed to the winning participants by the fund secretariat (which is independent of the ministry). The actual size of the fund depends on the anticipated portfolio of subprojects (that is, the size of subprojects, the size of the grant subsidy, and the anticipated number of subprojects). Successful grant schemes supporting subprojects may range from several hundred dollars for small farmer groups and initiatives to more than US$1 million.

Establish rules regulating the terms of the competition and implementation

Transparency of management, institutional arrangements, eligibility of applicants, projects and expenditures, selection criteria and processes (calling for proposals, approval, contracting, disbursement, financial management, audits, procurement, safeguard management requirements, and M&E procedures) are the most critical elements for a successful grant scheme. Significant preparatory work is required to identify and define appropriate operational procedures.

Conduct a communications campaign

Successful grant schemes require a rigorous awareness raising and communications campaign, managed by the grant administrator (secretariat), targeting the potential applicants. A communications campaign (either a nationwide information campaign or a more targeted marketing approach) guarantees that potential grant applicants learn about the innovation fund and related capacity-building opportunities. Communications campaigns may use diverse means, such as mass media, a specific Internet site, and/or more targeted communications (such as stakeholder meetings and face-to-face contact).

Build capacity of participants and service providers

Most grant schemes need to build capacity in grant-implementing units (related to the role of implementing units, communication, administrative procedures, and requirements of the grants), potential applicants (understanding the procedures, proposal development, partnering, financial management, and M&E), and potential service providers (proposal development skills, technical skills, and M&E capacity).

Establish a governance and management system

Grant schemes require effective governance and strong management. A good practice is to maintain separate units for policy setting, technical evaluation, management, and impact evaluation. The main governing responsibility resides with a *governing board* (sometimes also called an "advisory board" or "coordinating committee"). Table 5.5 describes the typical governance and management structure for a grant scheme. Grant schemes require a secretariat to handle day-to-day administrative functions such as communication, processing, coordination, and M&E. The capacity and placement of the secretariat is of significant importance for the success of a grant program. The options range from placing the secretariat within a public institution managing the fund or outsourcing it to private or civil society.

POTENTIAL BENEFITS

The major benefits of grant schemes arise from their capacity to tap into a vast pool of *creative ideas*. For designers and implementers of grant schemes, the emerging innovations (technological and organizational) are one of the most exciting and often surprising features. By mobilizing new partners, grant programs facilitate significant opportunities for *synergies* in technology and institutional innovation development. Some schemes stipulate particular partnership arrangements, such as company-research or company-farmer partnerships (as in China and Vietnam). In Turkey, a matching grant program helped to promote scientific and technical collaboration and technology development (box 5.10).

CRGs in particular, with their clear, outcome-oriented design, can significantly increase the chances that *research will succeed and that research resources will be used effectively*. When grants focus on specific themes or activities, the approach of inviting proposals provides a good opportunity to receive the best ideas and select only proposals with a high likelihood of achieving the desired outcome.

MGs in particular have a strong business orientation (expectation of revenue) from the outset and result in more *business-driven innovations*. Because users determine what

Table 5.5 A Generalized Governance and Management Structure for Grant Schemes

Unit	Composition	Responsibilities
Governing board	Often consists of representatives of key stakeholders associated with the grant scheme, such as government, farmers, agribusiness, and finance. A nonvoting representative of the secretariat usually participates.	Responsible for overall program policy. Oversees operations, establishes program priorities, awards grants, and represents program with funding agencies. Ensures close connection between the selection criteria used to evaluate proposals and the system-level objectives to which the competitive research grant scheme should contribute.
Secretariat	Composition depends on the type and size of grant scheme. The secretariat should have administrative capacity, including capacity to manage contracts and procurement; technical expertise (for example, in agribusiness); and M&E experience. The capacity and stability of the secretariat are often crucial for the success of the grant scheme.	Responsible for managing programs and carrying out daily operations. Provides support for governing and technical bodies and facilitates communications about program operations.
Technical advisory committee	This committee may be a subcommittee of the governing board or may be combined with the technical review panel described below.	Provides technical input for planning programs and setting priorities, advises on peer reviewer selection, and monitors technical quality of research subprojects.
Technical review panel	Often composed of 3 members selected from a pool of approved experts, including technical and financial experts. The size and complexity of the proposals will determine the number of experts required. Proposals for small subprojects may require one reviewer, whereas larger or technically new or more complex proposals may require 2–3 reviewers.	Responsible for evaluating, scoring, and ranking proposals. Makes funding recommendations.
Appeals body	Often managed by the grant secretariat. Appeal decisions are made by steering committee or governing council associated with the project or host institution.	Responsible for handling any petitions that may arise from the decisions by the reviewers or the approval committee.

Source: Adapted from World Bank 2010.

Box 5.10 Getting the Most from Matching Grant Schemes: The Turkey Technology Development Project

Perhaps the most notable legacy of the Turkey Technology Development Project (a US$100-million project approved in 1991) was the construction of the Technology Development Foundation of Turkey (TTGV). The Foundation provided funds for Challenge Programs, which stimulate applied research and technology development by industry through a mix of matching grants, income notes, and conditional loans. TTGV stimulates private investment in the development of industrial technology by providing seed capital (matching funds) for market-driven research and development (R&D) subprojects in a host of critical industrial sectors, including agro-industry.

In 1992, TTGV began to cofinance R&D subprojects in the private sector with the assistance of the World Bank. Proposals were solicited twice yearly. Of 273 proposals submitted by April 1998, 103 (37.7 percent) had been approved for funding. The funded projects elicited US$99 million in funding—US$44 million from TTGV and the remainder from private matching funds. The majority of the approved subprojects (84) were for technology development; 67 of these subprojects had concluded by the time the project's Implementation Completion Report was submitted in 1998. A large majority of subprojects funded through those grants succeeded technically as well as commercially.

Source: World Bank 1999b, 2006.
Note: TTGV = Türkiye Teknoloji Geliştirme Vakfi.

---

kind of innovation they require, the resulting innovations often have fewer problems with adoption or transfer. The innovations must pass the test of commercial usefulness, which increases the attractiveness and efficiency of the funds used in such grants.

POLICY ISSUES

The policy issues related to innovation funds resemble those for many of the other investments discussed in this sourcebook (sustainability, equity), but special concerns are related to the potential fragmentation of the research agenda and distortion of markets.

CRGs can be an effective component in a portfolio of funding mechanisms, but they must complement rather than substitute for long-term public funding for strategic research through block grants. When institutional block grants fall below 40–50 percent of the funding portfolio, the viability of long-term research may be compromised (EMBRAPA, IDB, and World Bank 2000). Competitive grant schemes—either for research or other activities—may introduce instability into the funding structure of institutions that compete for grants. Sustainability is also threatened by the fact that most grant schemes funded by donor organizations provide financing for only a limited number of years, and donor-initiated schemes are rarely mainstreamed into government programs. Inequitable access poses a challenge for grant schemes, particularly competitive ones. Grant schemes may be inaccessible for administrative reasons, lack of capacity to participate, or

limitations inherent in the funding, which could favor particular themes or areas and unwittingly discriminate against certain groups of applicants (World Bank 2010). In Colombia, MGs were used to incentivize collaboration between smallholder groups and the private sector (box 5.11). For other examples of MG schemes, see IAP 2 on China in this module, IAP 6 in module 1, and IAP 1 in module 6.

The risk of "projectization" and the accompanying failure to build capacity are acute in grant schemes, especially in CRGs, which do not require the counterpart funding implicit in MGs and PPPs. Competitive grants may be used as stopgaps to gain resources that cannot be obtained through the national research system or financial services. This inappropriate use of competitive grants yields a set of ad hoc research projects that contribute to no overriding research strategy (World Bank 2010).

Finally, many grants by definition interfere in markets as they try to resolve market failures. As observed, poorly designed grants can easily distort markets by directing

---

Box 5.11   Colombia Productive Partnerships Project: Incentivizing Market Inclusion through Matching Grants

The Colombia Productive Partnerships Project creates favorable conditions for large buyers and small sellers to establish mutually beneficial and sustainable relationships. It offers matching grants to complement producers' own resources and/or funding from other sources (local governments, municipalities, commercial partners). Producer organizations use the grants to obtain technical assistance and build their capacity (for example, to meet quality standards, bargain, or enhance their entrepreneurial and negotiating skills). Through the grants, producer organizations gain the ability and incentives to invest in collective goods such as storage facilities and packing facilities. The grants also enable individual small-scale producers to invest in productivity-enhancing infrastructure and gain startup capital to meet buyers' requirements. The types of partners have varied: over half have been food processors, one-third wholesalers, and the remainder supermarkets and retailers (for domestic and international markets).

By the end of the project's first phase, of 136 partnerships financed initially, 118 were sustainably operating in a wide range of markets. The average income of small-scale producers had increased by 77 percent and their employment by 70 percent. Success varied, but the relationship between the buyer and producer

was terminated only in 13 percent of partnerships. A particular set of incentives, infrastructure, and market conditions is needed to create and sustain well-functioning, productive partnerships. The key lessons for success were:

- *A stronger producer organization yielded a more successful partnership.* Social cohesion and business skills were difficult to achieve and are emphasized more strongly in the second phase of the project.
- *A rigorous, transparent, and competitive selection process* ensured the credibility and integrity of the grant scheme.
- *Technical service providers as facilitators* were fundamental to building trust with the commercial buyers.
- *Management and support of partnerships should be outsourced* to local service providers at the end of the project.

This project was the first World Bank project of its type in Latin America and the Caribbean. Since its inception, similar projects have been initiated in Brazil, Bolivia, Guatemala, Honduras, Jamaica, Panama, and Peru.

Source: Collion, forthcoming.

money to activities that would have attracted private funding anyway and by conferring undue advantages on some companies at the expense of others (see TN 1).

LESSONS LEARNED

The lessons learned from implementing innovation funds are briefly and partially discussed here. For a more comprehensive discussion, see World Bank (2010).

The decision to use a grant scheme to fund innovation and the choice of the scheme require a rigorous decision on the objective, which must be embedded in a strategic vision of the innovation system. *Weak objectives lead to incoherent priorities and funding rules and an inefficient use of resources.* Caution is advised with CRGs, given the tendency to distort long-term research agendas and markets (box 5.12). *No short-term grant scheme can substitute for long-term funding* for research, private sector development, human resource development, or infrastructure maintenance and development.

CRGs are often used in the larger context of *agricultural research system reform* programs. MGs on the other hand tend to work best when combined with *complementary investments* such as infrastructure, financial services, collective action, and market development. They also tend to benefit from *complementary policies* that are similarly introduced to provide enabling conditions that make overall investment more effective, and more attractive to prospective investors.

Significant costs can be associated with setting up and administering a grant program. Many programs set overhead costs at 10 percent of the budget, but one analysis found overhead costs of competitive grant schemes to be 25 percent or more in some cases (World Bank 1999a). The cost-effectiveness of procedures to keep overhead low must be balanced against the need to ensure accountability and transparency of operations.

Funding priorities must be set with stakeholders, who should participate in the governance of the grant or program

---

Box 5.12   Lessons from Competitive Grant Programs in Latin America

Do not just finance research; strengthen the capacity of research organizations. Competitive funds can be an important vehicle for financing research, piloting new ways of working, or focusing research on new topics, but they are most likely to make a sound and lasting contribution when they complement a relatively strong public sector framework for research. Public funding is essential for agricultural innovation systems. Private funding complements rather than substitutes for higher levels of public funding.

To compete for funding, research institutions require a minimum core budget and critical mass of staff. All institutions need some core funding to maintain and improve their physical and human resources. The sustainability of the public research apparatus and competitive funding alike will depend on continuous public funding. Grants usually fund operating costs over two to three years. A grant model is unlikely to flourish in a climate of fiscal austerity because there is only limited scope for private funds to substitute for public money.

The competitive fund model is more likely to strengthen the strongest agencies providing research and extension than it is to reduce disparities between

the strong and the weak. Competition between alternative service providers breaks down when the range of providers is limited and many potential providers lack the skills to prepare viable proposals.

The competitive model itself has not spurred large growth in the role of the private (for-profit) sector as a provider of agricultural research and extension, but it has contributed to the broader process of private sector development. Commercial firms have played a smaller role as providers relative to public and private nonprofit agencies. To the extent that commercial firms have played a role, they have largely done so *outside* the framework of competitive grant schemes. On the other hand, under the competitive schemes, through the medium of subproject copayments, producers have provided private funds as a complement to public sector grants; they have received training in the preparation of business plans; and they have become more market-oriented owing to partnerships with producer associations that have been facilitated by competitive funding agreements. In this sense, the competitive fund model has contributed to the broader goal of private sector development without entailing a major role for commercial firms as service providers.

Source: World Bank 2009.

and provide continuous feedback on implementation. Qualifying stakeholders to fulfill these vital roles entails an investment in capacity building, as discussed earlier.

Board composition and *transparency of activities* require careful attention. Boards can easily be taken hostage by one interest group that skews decisions in favor of its constituency at the expense of others (World Bank 2010). The capacity and institutional location of the secretariat is also crucial (box 5.13).

If the *absorptive capacity is low*, grant schemes will have fewer participants, lower-quality proposals, and a tendency to favor well-established research or private business entities. In CRGs this situation will limit the chances for innovative ideas, and in MGs established businesses may abuse grants to strengthen their business positions and dominate a market.

Calls for proposals can target an entire nation, specific regions, and/or stakeholders. A *two-stage selection process*, starting with concept notes and followed by full proposal development, is recommended to winnow out ineligible applications at an early stage. Given that most MGs target farmer groups or enterprises, often all proposals that meet the minimum eligibility criteria will be funded. *A good practice is to carry out a field appraisal before accepting a concept note for further development.* The field appraisal is helpful for verifying the information and identifying needs for technical assistance (for example, for developing the full proposal). In large and/or new and technically complex projects

that use MGs, a *second field appraisal* may facilitate progress with the grant scheme. In some cases eligible proposals are selected by ranking.

Consider limiting the grant contribution in the overall business portfolio of an enterprise and in the budget of a public research institution. Businesses receiving MGs should not use them as the main source of financing for a startup business. Grants can easily be abused in this way, replacing commercial financing (risk or venture capital funding) or core public funding.

Transaction costs associated with participation in a grant scheme may be significant and reduce the pool of private applicants. To retain applicants' interest in a grant scheme, the secretariat needs appropriate administrative and communication skills. Procedures for application and implementation need to be as streamlined and clear as possible. Simplified procurement methods (shopping or commercial practices) are often more suitable for business grantees, since it is in their interest to use funds efficiently. Tracking and documenting the outcomes of innovation funding are too often neglected in grant schemes, yet a sound M&E system enables grant schemes to identify and address problems as they arise. Specialized M&E personnel can be employed in the grant secretariat; the responsibilities can be outsourced to independent experts; or a combination of both options may be used. Grant recipients often are not adept at fulfilling M&E requirements and will benefit from *specific training and hands-on support.*

Box 5.13   Recommended Options for Grant Program Secretariats

- An existing public institution, such as a ministry or local government agency, can generate additional political buy-in, institutional sustainability, and leverage. The same public institution may also be the key stakeholder within the grant and innovation system and provide a higher chance of sustainability or institutional mainstreaming. Disadvantages arise from the tendency of such institutions to be bureaucratic, interfere politically, and lack commercial acumen and understanding.
- An existing private entity that brings an aptitude for business and less bureaucracy but could be affected by conflicts of interest or unwillingness to meet donor requirements (procurement and fiduci-

ary practices or social/environmental safeguards, for example).
- Creating an autonomous public/private unit.
- Using the services of a NGO when independence and flexibility are most important and local capacity is very low. NGOs bring knowledge about donor requirements but may entail higher costs, problems with long-term institutional sustainability, and a lack of business acumen.
- If outsourcing to a competent body is not feasible, it is usually best to provide sufficient capacity building to the staff of the hosting organization, and/or supplement it with technically and administratively appropriate staff.

Source: Authors.

Accelerating the Development of Agribusiness Enterprises by Using Business Incubators

Seth Ayers, World Bank Institute

SYNOPSIS

This note explores business incubation as a mechanism for effectively and sustainably accelerating the growth of startup enterprises that bring innovative technologies and services to market. Typically startup enterprises face a number of challenges in growing, including the lack of technical assistance, appropriate financing, networks of partners and customers, and infrastructure. Business incubators have effectively supported the growth of enterprises across many sectors, including agriculture. Although the number of agribusiness incubators is limited, interest is growing in expanding them and learning from experience in other sectors. This note summarizes experiences with two agribusiness incubators and offers key lessons learned from others around the world, including business plan development (for example, a ten-year financing and sustainability plan) based on a detailed market assessment and structuring an effective management team.

BACKGROUND AND CONTEXT

Especially in emerging economies, SMEs play a major role in economic development. Formal SMEs contribute up to 45 percent of employment and up to 33 percent of GDP in developing economies; these numbers are significantly higher when the estimated contributions of SMEs in the informal sector are included (IFC 2010). Enterprises are particularly important for generating innovations in response to market demand.

For these reasons, creating links with business development is one of the most important challenges facing agricultural innovation in developing countries. The agricultural sector has consistently struggled to bring new technologies, services, and business models to market at a scale that can have substantial impact. Efforts have been limited by many factors, including the difficulty of coordinating the diverse partners required to develop and deploy new technologies successfully; market failures that discourage private investment; lack of access to financing; limited engagement of end-users in product development; and poor alignment of complementary roles for the public and private sectors (Boettiger and Alvarez 2010). It is critical to find effective ways to help SMEs to sustainably scale up their operations.

Business incubators occupy the space between mechanisms such as business development services (BDS), technology parks, and other platforms for business development. The core of business incubation is a focus on emerging enterprises. Services can range from mentoring to seed financing to influencing policies (box 5.14).

Since 2000, as awareness of their usefulness has grown, the number of business incubators has expanded substantially in developing countries.[1] Countries such as Brazil and India have established large domestic networks of incubators, facilitated by a national association. Table 5.6 provides

---

Box 5.14   Services That Incubators Can Provide

- Capacity-building, training, and mentoring services.
- Technology testing, demonstration, and certification facilities.
- Technology transfer and IP policy advisory services.
- International networking and collaboration.
- Policy advocacy and market intelligence.
- Links to investors and other financing sources.
- Infrastructure (IT and office facilities, for example).

Source: infoDev (www.infodev.org).

Table 5.6 Business Incubator Typology and Overview

Context and features	Strengths (broad)	Challenges (broad)
Mixed portfolio business incubation		
– Targets high-growth firms in a range of sectors – May select sectors that align with the overall regional or national competitiveness strategy – May exist in environments with little entrepreneurial activity	– Can align with regional and national strategies, germinate new areas of competitive capacity, and provide a locus for innovation in this regard – In environments with little entrepreneurial activity, may include extensive "pre-incubation" and education to source entrepreneurs and businesses	– Where new competitive sectors are under development, time to achieve impact and scale may be long – Where there is little entrepreneurship, attracting clients with high growth potential can be a challenge; it may not be possible to operate at the scale necessary to support the incubator's business model
Technology business incubation		
– Targets high-growth technology firms – Requires foundation of strong technology and human capital infrastructure – Where this infrastructure and human capital are weak, may require extensive pre-incubation activities – May exist in economies in transition	– Can be an economic resource by attracting and developing research, skills, and businesses – Can develop technology as a new source of competitive capacity	– May be challenging to scale businesses beyond seed stage because of lack of financing and difficulties entering international markets
Business incubation with university relationships		
– Frequently the university or academic institution has a role as founder and is a source of resources such as research, expertise, space, and/or funds – Typically targets technology firm, but may work with other sectors	– Opportunity to bridge the gap between research and commercialization or technology transfer – Access to intellectual property and the potential to develop competitive businesses from it – Often provides financial stability for incubators	– Can create "cultural" tensions if academics seen as good researchers but poor managers or if the university is seen as too bureaucratic or risk-averse
Agribusiness incubation		
– Targets firms in the agricultural sector – Aim is to commercialize innovative practices or transform sector firms from slow growth to growth	– Can often have significant economic and social impact by improving the livelihoods of communities – Can have an agritechnology focus and focus on commercialization	– Requires both business and community development skills – May be challenging to enter markets beyond local communities

Source: infoDev, "Incubation Models," http://www.idisc.net/en/Page.MEIA.Incubator.Models.html.

an overview of the types of business incubators that can be deployed in different markets.

INVESTMENT NEEDED

The incubator must be designed based on market demand, which is reflected in a detailed ten-year business model that outlines how the incubator will be sustainable. In establishing a business incubator, the main issues and areas of investment include selecting the appropriate model, establishing a successful management and governance apparatus, and accumulating the appropriate physical assets.[2]

The broad choice in *selecting which type of incubator* to develop is between a mixed-portfolio incubator and a sector-specific incubator, such as an incubator for technology or the agricultural sector (subsector) (table 5.6). For the agricultural sector, both models can be relevant and can support businesses that provide services, products, or new technologies contributing to agricultural innovation.

■ *In a mixed-portfolio model*, the incubator will incubate a business in any sector, pursuing any business activity from technology to agriculture, as long as it believes that the business is scalable. (Scalability is key to ensuring that the incubatee will be able to pay for rent, services, and possibly royalties.) A mixed-portfolio incubator is often a more appropriate choice in developing countries, because the pool of scalable SMEs in a specific sector is limited, and a sector-specific model is not viable without a substantial and ongoing subsidy. For examples, see Fundación Chile, which is a unique, one-stop business incubator, and Technoserve of Mozambique, which leverages BDS to transform entire sectors (infoDev 2011).

■ *A sector-specific incubator* is appropriate if there is a sufficient pool of clients (incubatees) and demand for agribusiness development. It will be important to leverage other actors in the sector, including well-established enterprises that can be potential clients for the enterprises being incubated. For examples, see box 5.20 on

Rutgers University later in this note and infoDev (2011) on Fundacion Jalisco in Mexico.

- *A technology-oriented incubator* is particularly appropriate when technology transfer and commercialization are desired. In that case, the incubator supports actors (often affiliated with research centers or universities) in commercializing a technology. For an example, see box 5.19 on ICRISAT later in this note and infoDev (2011) on CENTEV/UFV (the Federal University of Vicosa in Brazil).

The choice of *business model* determines how the incubator will sustain its operations—in other words, what types of services the incubator will offer to attract the targeted clients and generate revenue.[3] Broadly speaking, three revenue models have been developed for different business incubation environments (box 5.15). Most business incubation environments will combine elements of each model. For agriculture, revenue from tenants and other clients, complemented by public support, may be the most appropriate model, as used in the Agri-Business Incubator.

As with any business, a key factor for the success of business incubator is *strong leadership and management*, which can develop a sustainable business plan, adjust the plan as needed, and ensure that business meets the desired objectives (infoDev 2009). A committed and skilled manager can guide entrepreneurs through the development process and act as their on-site management counselor. This person should have business experience, the ability to analyze the issues facing client entrepreneurs as they develop their businesses, the ability to develop networks that will serve clients, and the ability to work with stakeholders to retain their support for the program. Box 5.16 describes typical staff and tenant levels.

In a business incubator, the *board of directors provides strategic guidance* to management and helps build complementary relationships in the community where the incubator operates. The board is composed of representatives from the management of the incubator (often the incubator manager) and from external partners, including those that provided the financial and material resources to establish the incubator (Medeiros et al. 1992). Board members should be selected from every area in which the incubator has needs.

Finally, the *physical assets* associated with an incubator range from the availability of IT equipment to real estate with sufficient space for the incubator office, offices for incubatees, and other services/equipment. The level of physical assets is determined by the business model, client pool, and incubator type. The two most common choices

---

Box 5.15 Key Features of Revenue Models for Business Incubators

Revenue from tenants and other clients. Rent (40–60+ percent) is the most common source of revenue in this model, but fees for the business support provided (business incubation fees) and for the use of facilities and other services can be just as important. Hot-desking fees (renting a desk and computer connected to the Internet by the hour) can be important for broader incubation models. This model is financially self-sufficient, given that the incubator relies on "free" buildings, has minimum economies of scale, and often has anchor tenants.

Revenue from sharing in clients' success. This model is based on small equity positions or royalty agreements on gross sales and brokerage fees on raising finance. For example, ParqueSoft in Colombia[a] requires clients to pay 20 percent of their sales as commission.

This model can help ensure the incubator's sustainability while aligning both the incubator and client business to growth of the business and its revenue. The model requires stakeholders to have a long-term vision, because it can take ten years to develop revenue streams that will sustain operations into the future. The model also requires managerial sophistication, a well-developed business environment (to form and protect an investment), and functioning capital markets (if it relies on brokerage fees from finance raised).

Ongoing government or donor funding. A long-term commitment from government, a donor, and/or other organization finances the incubator. This model is potentially risky, because it has no additional revenue streams. If funding is discontinued, the incubator is likely to close.

Source: infoDev's Online Incubator Toolkit (www.infodev.org/idisc).
(a) Described in box 5.21.

Box 5.16   Typical Numbers of Incubator Staff
            and Tenants

The typical incubator will have 20 or more client companies (tenants), some of whom may be virtual clients who are not physically located in the incubator but receive technical assistance and other services. A building of about 2,000 square meters should easily handle 20 tenants. At a minimum, staffing should include a manager with business experience who has been trained in incubator operation, possibly an administrative assistant, secretary/receptionist, and at least one business counselor who provides technical services directly to tenants. A minimal maintenance staff is probably also required, but numbers will vary by location.

Source: infoDev's Incubator Toolkit (www.idisc.net).

are to establish a stand-alone business incubator in an existing building or use a complementary organization already in operation to house and operate the incubator. The ICRISAT incubator, for example, is housed within the ICRISAT facility, which offers both physical space and access to new technologies and equipment.

POLICY ISSUES

The policy issues[4] that arise most often with business incubators include their sustainability, the public sector's role in creating an environment in which business incubators can operate successfully, and adapting incubator models to address social concerns.

Sustainability is a key factor in designing the business model for an incubator. The design must consider the potential client pool (what is the existing market for prospective tenants?) and mix of services that will create a sufficient income base and cover operating costs.

The more business-friendly the market environment is, the more likely a business incubator is to succeed. The World Bank's annual *Doing Business* report ranks countries based on the extent to which their market environments facilitate common business operations and transactions. The report and corresponding website (www.doingbusiness.org) provide recommendations for creating a supportive market environment for business, including business incubators. Specific policy instruments that governments can use to support incubators include tax incentives and early-stage

soft funding such as grants (for an example from India, see box 5.17). Since most incubators operate as nonprofit organizations, the public sector can play an important role in providing physical space for the incubator as well as financing to cover operating costs.

Business incubators can target specific sectors such as agriculture or certain segments of society, such as women entrepreneurs. Incubators that have a specific focus may require additional public support, given that they aim to serve a small subsegment of the market and not the market as a whole, which means that they are more challenging to sustain. In Tianjin, China, an incubator for women's businesses has had considerable impact on the growth and sustainability of enterprises that women own and manage (box 5.18).

BENEFITS

The benefits of business incubation range from direct financial benefits in terms of tax revenues to significant improvements in SME sustainability, through which new technologies, services, and business models can be delivered and scaled up. In addition, business incubators raise awareness of entrepreneurship. They create a cluster of entrepreneurial activity around a particular sector, such as agriculture, as the incubator becomes a primary point of contact for actors working in the sector.

The National Business Incubator Association (NBIA), based in the United States, estimates that over 7,000 incubators operate around the world, with more than 1,100 in the United States alone (of which 94 percent operate as nonprofit organizations). In the United States, the impact of business incubation has been well documented (University of Michigan et al. 1997; Knopp 2007). For example, every US$1 of public investment in an incubator has yielded US$30 in local tax revenue; 84 percent of incubator graduates stay in the community where they were incubated; and 87 percent of incubator graduates remain in business. The economic impact and investment return from business incubators, as demonstrated by the experience in the United States, indicates the opportunities for agriculture, particularly in developing countries, to use business incubation for developing and mainstreaming new agricultural technologies.

LESSONS LEARNED AND RECOMMENDATIONS FOR PRACTITIONERS

The experience gained from using business incubators in a number of settings, within and outside agriculture, offers

Box 5.17 The Government of India's Incentives to Support Business Incubators

- Incubators and small and medium enterprises (SMEs) that are clients of incubators are exempt from service tax and corporate tax.
- Foreign equity ownership of incubated SMEs can be as high as 100 percent.
- Seed funds for SMEs (US$230,000 per fund) are managed by selected incubators.

Source: Department of Science & Technology, Government of India (http://dst.gov.in).

Box 5.18 The Tianjin Women's Business Incubator

The Tianjin Women's Business Incubator (TWBI) is China's first women's incubator, started with a grant of US$300,000 contributed by infoDev, a building contributed from the Tianjin Municipal Government, and a cash investment from the Tianjin Women's Federation along with three other local government authorities. As of 2009, TWBI worked with 48 on-site tenants and 10 off-site tenants. At that point, the incubator had graduated 16 companies, creating new jobs for more than 3,000 people; assisted 2,000 women entrepreneurs to obtain microfinance; and provided business training to more than 20,000 entrepreneurs. Operating near full capacity, TWBI has almost reached financial self-sufficiency through charges for office rent, business services, and external training courses.

Source: Author.

many useful guidelines on incubator design, management, and finance. They are summarized in the sections that follow; practitioners interested in developing their own incubators will also want to make use of the resources listed in the references for more comprehensive advice.

Choose the right incubator model for the context

As discussed, in a few markets, particularly in middle-income and larger economies, sector-specific incubators are viable, but they will not be viable in many developing countries without a substantial subsidy. To sustain their operations, most incubators in developing countries must operate as mixed-portfolio incubators by providing services to scalable SMEs in all sectors.

An example of a successful sector-specific incubator is the agribusiness incubator located within ICRISAT in Hyderabad, India (box 5.19). The incubator benefits not only from a large market but also from the substantial pool of R&D available to be commercialized (see IAP 1 for details). The incubator played a key role in building a business around a new seed technology and providing shared equipment for processing sweet sorghum into ethanol.

Develop a comprehensive business plan that captures lessons and evolves to suit changing needs

As noted in the case study of the Rutgers Food Innovation Center (box 5.20), preparing a comprehensive business plan, including a seven-year financing strategy, was essential to success. The Center's iterative approach to design and implementation allowed it to test the business plan before investing substantially in building the center. Incubator development must allow for iteration; ideally, the investment should permit modifications over time.

Incubator management and board must be strong

Strong management will attract clients and help the incubator become viable. For example, a key aspect of the business model used by iPark, an incubator in Jordon, was to recruit successful (serial) entrepreneurs to bring a mix of entrepreneurial experience to the incubator. iPark maintains a strong management relationship with tenants, pursues a flexible approach to solving problems, and recognizes that the incubator will not succeed if its clients do not succeed.

As an incubator evolves, it may need to change the composition of its board.[5] Management must determine which skill set will be most useful in enabling board members to support the incubator's operations. For instance, in many developing countries access to finance for SMEs and incubatees is a significant issue. Therefore, many incubators select Board members who represent local banks as way to sensitize these banks to the challenges that incubatees face in accessing capital.

Ensure access to finance for clients

Figure 5.1 shows that enterprises require different amounts and types of financing (for example, more patient capital

Box 5.19   The Agri-Business Incubator@ICRISAT

The Agri-Business Incubator@ICRISAT (ABI), launched in 2003, estimates that its various programs have benefited more than 40,000 farmers. ABI is an initiative of the International Crops Research Institute for the Semi-Arid Tropics (ICRISAT) in partnership with the Department of Science and Technology (DST), Government of India. It promotes technologies developed exclusively by ICRISAT, jointly developed with collaborators, or agricultural technologies developed by R&D centers of excellence, universities, and other institutions. ABI develops agricultural enterprises by providing various services and facilities:

■ *Entrepreneurship development.* ABI offers entrepreneurs support from concept to commercializa-

tion, with a focus on seed ventures, biofuels, and farm systems solutions.
■ *Technology commercialization.* ABI is a platform for commercializing technologies developed by public institutions through the creation of agribusinesses. ABI also facilitates the commercialization of technologies and services that have been developed by entrepreneurs and benefit the agricultural sector. Products and technologies incubated by ABI include sweet sorghum for ethanol production, insect-resistant transgenic cotton, and pesticide-free crops produced through organic farming.
■ *Services and facilities.* ABI offers technology consulting, business development, and training services, as well as office space, laboratories, and agricultural land to test new technologies and services.

Source: Agri-Business Incubator@ICRISAT.

Box 5.20   The Rutgers Food Innovation Center

The Rutgers Food Innovation Center, based in New Jersey, provides business and technology expertise to small and midsize food and agribusiness companies in the Mid-Atlantic and Northeast regions of the United States and, through its outreach capacity, to food and agribusinesses throughout the world. The Center created over 1,000 new jobs by incubating new businesses and over US$200 million in revenue growth for clients, in addition to millions of dollars in local tax revenue. It has assisted more than 1,200 companies and entrepreneurs since it began operations in 2000, including:

■ *Farmers and agricultural cooperatives* desiring to create new businesses based on value-added agricultural products and/or developing new markets for their existing commodities.
■ *Startup food companies* coping with challenges such as financing, technology, regulations, market development, and infrastructure requirements.
■ *Existing small and midsize food companies* seeking to access new technologies, upgrade quality assurance

capabilities, enter new markets, train their workforce, and expand and improve their operations.
■ *Retail and food service establishments* seeking to improve their operations and purchase locally grown New Jersey products.

An extensive feasibility study in 1999 helped identify the prospective client base. The center developed a detailed business plan based on significant primary and secondary market research, a national benchmarking study on best practices in food business incubation, a comprehensive strategic plan, and a seven-year financial pro forma. Based on this plan, the Food Innovation Center began operations in 2000. *A very important consideration was that the center did not begin its program with a dedicated facility but held back until its program was fully developed.* The center operated out of a rented office for eight years before moving into a full-scale facility. During this period it tested its model, fully developed its programs and services, gradually hired staff, and established a network of resources to meet the needs of a

(Box continues on the following page)

Box 5.20   The Rutgers Food Innovation Center   (continued)

broad clientele. It ensured that sufficient demand existed for its services before investing in a building (funded entirely through grants).

The Center's facility consists of a *Client Services Area*, where clients receive marketing, development,

and analytical support from concept to commercialization. A *Shared-Use Processing Area* uses an array of food-processing technologies to produce a broad range of value-added agricultural and food products.

Source: Rutgers' Food Incubator, http://www.foodinnovation.rutgers.edu/incubatorlinks.html.

Figure 5.1   Financing Gap for Small Enterprises

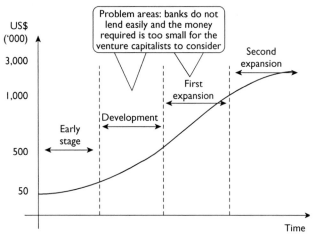

Source: infoDev 2008.

early on)[6] at different stages of development. If these amounts and types of financing are not available in the market, enterprises will struggle to scale, and the incubator's impact will suffer. It is important that an assessment of the financing available in the market is made before establishing an incubator to determine if financing gaps exist and can be addressed effectively.

Lessons on real estate investment

The incubator's client base will determine the array of services and physical assets required, including real estate. The client base must consist of a relatively large number and mix of small enterprises, because not all small enterprises are growth-oriented. The type and number of enterprises will determine the space, equipment, and other resources required, as well as where the incubator needs to be located. If the incubator provides only business planning and strategy

assistance, only office space is required. If the incubator also provides office space and shared equipment (office equipment and agricultural processing equipment, for example), it needs an appropriate space to house them. If the incubator plans to offer access to the latest agricultural information and/or to enable virtual mentoring and collaboration between enterprises, it will need the appropriate IT equipment. For an example, see box 5.21.

Lessons from financing arrangements

Lessons related to financing incubators differ depending on their source of funding. A model based on revenue from incubatees or from equity may become self-sufficient with time, and the only continuing support it may need will be the use of buildings, free of charge or at cost. In more challenging markets, as noted, a proportion of ongoing government funding is likely to be required, especially if the social and economic return on investment is greater than for other BDS interventions. The financial strain on the public sector may make it very difficult for it to afford long-term funding, however. If a model based on long-term support from an external donor is considered, it will be important to conduct a rigorous cost-benefit analysis to compare incubation to BDS and other business development support mechanisms before committing.

Business incubators that rely upon *rent and client fees* as their main revenue sources cannot, as a general rule, be financially self-sustainable in commercially leased accommodation or where they pay the capital costs of a building, unless they have other, substantial forms of support. Generally this support consists of ongoing financial subsidies by a third party, which can be an unreliable and unpredictable source of funding. It is very hard for business incubators to achieve adequate margins in commercially rented

Box 5.21   Real Estate and Management Arrangements Associated with Parquesoft Centers, Colombia

Colombia's network of ParqueSoft Centers comprises 14 incubators situated throughout the country and managed from a headquarters in Cali. The founder is a highly charismatic individual who used his expertise in information and communications technology to establish ParqueSoft and promote social development through business creation. More than 270 businesses participate in ParqueSoft. These interlocking business entities provide internal as well as external consulting, marketing, and training support. Member companies pay a 20 percent commission on each sale to ParqueSoft, which also receives grants, in-kind donations, rents, and service fees. This revenue allows ParqueSoft to pay for advertising, maintain its building in Cali, and cover other overhead expenses. Local universities provide office space in each location except Cali. ParqueSoft is largely self-sustaining, though it must seek grants and contributions in addition to revenues. Its growth to 14 centers proves it is scalable, and it has required very limited public sector support other than real estate.

The ParqueSoft brand is known internationally and is associated with high-quality services. An association with ParqueSoft allows the startup companies to obtain larger contracts that are effectively subcontracted to the member companies. The use of cubicles in ParqueSoft buildings permits more intense networking and interaction by the companies. Although this setup builds the ParqueSoft brand and makes each company seem competitive with larger, more established companies, it should be pointed out that it also diminishes the efforts of individual companies to establish their own identities and brand equity.

Source: infoDev 2009.

accommodation. They run the risk of either failing financially or having to cut costs until they are really nothing more than real estate operations. More commonly, business incubators attempt to rent buildings at a purely symbolic rate (for example, US$1 per year) or to obtain funds to purchase or construct their own facilities, both of which are more reliable strategies for securing on-going support.

Taking a small proportion of *equity* or a *royalty on gross sales* for a period can be a very good way for an incubator to receive payment for its value-adding services once the company being assisted has succeeded (not up front, when the company is short of cash). This success-sharing strategy aligns the business incubator's mission with that of its clients. Increasing numbers of technology incubators take a small equity position or negotiate royalty agreements as a condition for membership in an incubator. Realistically, taking equity applies only to high-growth and generally intellectual property-driven companies, in situations where clear exit mechanisms (such as initial public offerings or trade sales) exist. This option makes little sense for most service companies or for locations where exit mechanisms are unclear. In that case, royalties may be a better approach.

Business models that rely on sharing clients' success have proven somewhat problematic. Returns from the equity position, royalties, and brokerage on finance cannot be relied upon for financial sustainability in the short term. It can take ten years to realize returns, and a portfolio of at least 20 companies is required to spread the risk. A high level of management expertise is also required.

Aside from a sound financing model, incubators need to "walk the talk" and demonstrate *high levels of financial management capability* to incubated enterprises. Financial management consists of planning, overseeing, and controlling the incubator funds, whether they are brought in through services or provided by partners and investors. Mechanisms need to be developed that allow the incubator manager to know accurately the amounts available, the needs, and the investment capacity. Additionally, financial management should focus on bringing in new sources of funds for the incubator.

Agricultural Clusters

Florian Theus, World Bank Institute
Douglas Zeng, World Bank

SYNOPSIS

A cluster-based approach helps to identify policy and institutional impediments to competitiveness and innovation. Through dialogues at the cluster level, new partnerships can be forged between cluster leaders and various public organizations (such as those working on industrial development, infrastructure development, research, innovation, and training) to formulate and expedite policy reforms. Given that developing countries have very limited financial resources, selectivity and efficacy are important guiding principles of competitiveness strategies. Focusing on a cluster approach mirrors these principles, helps in scaling up to the industry level, and fosters regional innovation systems. A government may choose to pursue cluster initiatives along with simultaneous policy reforms, because the two approaches may create positive externalities and help government develop a compelling case for policy reform. Based on a broad range of successful cases, the most important factors in the success of a cluster program or initiative appear to be demanding markets, positive joint action, and institutions capable of moderating and focusing power imbalances—complemented by the push factors of education and prior technological knowledge. Perhaps most important from a policy perspective is the ability to adopt flexible and coevolving policies designed to foster cluster emergence, which reflect the development of institutions, technologies, and firms in a dynamic and self-organizing process. It is critical to find a balance between carrying capacities, the institutional setup, and the individual incentive design.

CLUSTERS AND THE RATIONALE FOR INVESTING IN THEM

The cluster approach focuses on networks of production and value chains rather than on nationwide initiatives to foster innovation. Clusters are agglomerations or networks of production populated by strongly interdependent firms (including specialized suppliers) within a value-adding production chain as well as service providers and associated institutions in a particular field. In some cases, clusters also encompass strategic alliances with universities, research institutes, knowledge-intensive business services, bridging institutions (brokers, consultants), and customers. These entities are linked by externalities and complementarities and are usually located near each other. Agricultural clusters often form geographic and sectoral agglomerations of enterprises (Schmitz 1992). The most dynamic clusters spring up spontaneously, without direct intervention by external actors.

Why and when to use a cluster approach

Cluster-based policy aims at removing the imperfections of innovation systems by enabling them to function more efficiently and avoid coordination failures. A cluster-based approach is a realistic way to identify the policy and institutional impediments to competitiveness and innovation. When a critical mass of firms moves simultaneously to function as an initial cluster, they become an effective vehicle for catalyzing reform. Through dialogues at the cluster level, new partnerships can be forged between cluster leaders and various public organizations (such as those working on industrial development, infrastructure development, research, innovation, and training) to formulate, effectively sequence, and expedite policy reforms.[1] A government may choose to pursue cluster initiatives along with simultaneous policy reforms, because the two approaches may create positive externalities and help government develop a compelling case for policy reform.

The cluster perspective provides a number of advantages over the traditional sectoral approach in both analyzing and promoting competitiveness, innovation, and innovation networks. Given that developing countries have very limited financial resources, selectivity and efficacy are important guiding principles of competitiveness strategies. Focusing on a cluster approach mirrors these principles, helps in scaling up to the industry level, and fosters regional innovation systems. A value chain may be too narrow a domain when the surrounding innovation ecosystem is underdeveloped. On the other hand, the national innovation system may be too broad a domain, and its top-down approach (unlike the bottom-up approach possible with a cluster strategy, discussed in "Lessons Learned") would increase the risk of making mistakes on a large scale.

Clusters versus value chains

While a value chain approach[2] can be used in the absence of a cluster approach, *value chains must be supported for a cluster approach to work.* Cluster development and value chain enhancement must go hand in hand if a cluster aims to promote innovation-based competitiveness in developing countries, where (1) value chains are often very unstructured throughout their segments (transportation, distribution, enabling environment), thus requiring intervention by numerous stakeholders who cannot resolve these problems alone; (2) trust among stakeholders is weak, and a special effort is needed to build social capital; and (3) obstacles need to be addressed by multiple stakeholders and value chain segments (USAID 2008).

Spontaneous versus orchestrated clusters

The most dynamic clusters spring up spontaneously, without direct intervention by external actors (McCormick and Mittulah 2005). On the other hand, agricultural clusters in developing countries may be fostered by local and/or national government as well as donor support.

In supporting the transformation of clusters into innovation systems, evidence points to the importance of complementary policies, programs, and financial mechanisms that can foster new linkages and create opportunities for sustained growth. Where these links and opportunities are not established, stagnation and decline in the face of crisis and challenges often followed (Zeng 2010). Regulatory frameworks and extension agencies also play important roles in agricultural cluster development.

It is obvious that cluster development is a long-term, multifaceted approach that is unsuited to short-term investment projects. External support is often provided during later phases of cluster development, when clusters have demonstrated their potential.

ENABLING INVESTMENTS TO SUPPORT AGRICULTURAL CLUSTERS

The cluster approach involves many actors whose roles are always evolving, which makes the role of the state complex and location specific. Roles of the public and private sectors are becoming increasingly blurred. The private sector in many developing countries is providing quasi-public goods, such as training smallholders in the use of technology, instituting quality control, or providing finance (Larsen, Kim, and Theus 2009). Cooperation and collaboration schemes have been implemented in virtually all spheres that originally might have been the domain of the public sector (see the discussion of PPPs later in this note).

Three broad areas of investment are commonly needed to support the development of agricultural clusters. They include investment in infrastructure and the policy environment, investment in a regulatory framework, and investment in formal and informal institutions (and their coordination).

Importance of improved infrastructure and policy environment

Cluster programs and investments are effective only where minimum conditions of macroeconomic and physical stability, hard and soft infrastructure for doing business, and basic institutions for supply-side functions are met. Government plays an important role in this regard, often supported by donors. Economic reforms such as deregulation of domestic markets, removal of explicit and implicit trade barriers, ending distortions in exchange rates and taxation, as well as the development of a sound property rights regime are some of the measures that need to be taken. For instance, the Kenyan cut flower cluster succeeded owing to the enactment of legislation setting up promotional schemes (such as manufacturing under bond, export compensation, and export promotion zones for horticultural exports), protecting intellectual property rights, and enforcing quality standards (Zeng 2010). The public good character of infrastructure—especially the transport, hygiene, and cooling facilities critically important for many food products—makes government and donor involvement imperative.

Regulations, quality assurance, and standards

Local governments often try to improve services and regulations in ways that help to generate business, enable clusters to operate normally, and maintain dynamic growth. In addition, governments enact specific regulations, especially related to types of investments, product quality, and standards to ensure that products made in clusters have a market future.

Persistent capacity constraints can cause standards and quality management regimes to fall short of implementation, however. Targeted assistance to governments has proven successful in removing this constraint. In the Ugandan fish cluster, the government provided leadership and coordination to develop local standards and, through its fisheries inspection service, regular monitoring. The development and enforcement of standards helped sustain the pressure on the clusters to keep up with improved process-related standards (the same could be done for product upgrading) (Kiggundu 2005).

The crucial role of institutions

Institutions are crucial in agricultural clusters (figure 5.2). They may be public (state agencies, regional entities on competitiveness and innovation, and educational institutions, among others) or private (banks, business organizations, and companies) and formal or informal (networks, learning through transactions with local and external agents, and so on).

GENERATING AND ACCESSING FINANCIAL RESOURCES. Throughout developing countries, the lack of financial services forestalls process and product upgrading and cluster development. The high risks, uncertainty of collateral, informality of many SMEs, and high transaction costs limit financial services in rural areas. Where financial services are available, banks often provide expensive credit under stringent repayment schedules. Financing innovators and startups becomes extremely difficult under these conditions.

Developing capacity in financial service providers as well as their cluster clients (firms) can be an important step in reducing information asymmetries and risk. Banks benefit from capacity building to improve credit checks. Firms, on the other hand, benefit from expertise in developing business plans. The list of mobilizing co-investments may include commercial bank financing, equity financing, development bank financing, an industry tax or levy, venture capital investment, government investment, and voluntary industry investments[3] (see the discussion in TN 6 on risk capital). The approach should factor in the consideration that venture capital, as cases such as Silicon Valley show, generally lags behind cluster formation. Venture capital firms are attracted to new clusters once they show substantial economic activity with the expectation of future profits (Braunerhjelm and Feldman 2007).

LEARNING AND TRAINING. Various industry-specific modes of learning and training—formal and informal—are

Figure 5.2   Institutions with Crucial Roles in Agricultural Clusters

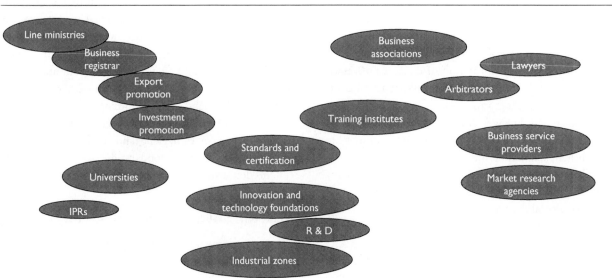

Public institutions   Private institutions

Line ministries · Business registrar · Export promotion · Investment promotion · Universities · IPRs · Standards and certification · Innovation and technology foundations · R & D · Industrial zones · Business associations · Lawyers · Arbitrators · Training institutes · Business service providers · Market research agencies

Source: World Bank 2009.

important means of acquiring and disseminating knowledge and technology in clusters, because they influence the clusters' potential for innovation and competitiveness. Modes of learning and training generally include: apprenticeship, on-site training at suppliers' factories, on-the-job training, expert contracting, support mechanisms provided by public institutions, learning through transactions with local and external agents, and learning-by-doing in the areas of production and maintenance (Zeng 2006).

On-the-job-training, a major mode of learning for high-tech-intense clusters, is often absent or insufficient in agro-clusters in developing countries. Support programs for training institutes, expert contracting, and training of masters are means to strengthen training and learning capacity. For example, in the South African wine cluster, capacity was improved through the newly created Elsenburg Training Institute, with vineyards and a cellar. Most established producers exchanged production knowledge through this institute, which performed a function similar to viniculture forums and root stock associations (Wood and Kaplan 2006). (See box 5.22.)

Governments can also strengthen major educational institutions to meet cluster needs. A critical element of curriculum design is to foster links with the private sector so that the skills of graduates meet the agribusiness requirements of the cluster (see examples in module 2). The same point applies to research institutes and universities involved in R&D: They need to focus on applied research and areas of prime importance for the respective cluster. Finally, location is key to cluster development. As major sources of technology and knowledge spillovers, as well as technology commercialization, agricultural universities and possibly their incubators are major sites of innovation and thus significant for cluster development.

INSTITUTIONS OF TECHNOLOGY TRANSFER, KNOWLEDGE GENERATION, AND SHARING. Progress in technology, innovation, and product diversity are linked to institutions of technology transfer, knowledge generation, and sharing, mostly supplied or supported by the public sector or by donors (TN 5). A prime example of best practice and long-term development of such institutions is Fundación Chile, the key institution enabling technology transfer to Chile's renewable natural resource clusters. Fundación Chile started out as a specialized service provider and began to incubate companies to nurture a demonstration effect for new technologies (see module 1, IAP 3).

INSTITUTION(S) FOR COORDINATION AND STRATEGY. Nurturing a coordinating body may be important. Forming competitive clusters requires collaboration among multiple levels of government, companies, teaching and research institutions, service providers, standard-setting bodies, and private organizations. A strong coordinating body, usually a public agency, is needed for that role. Through unambiguous policy and with the involvement of all relevant actors, a national competent authority on technological upgrading may provide leadership in organizing background research across carefully selected sectors, developing standards and performance targets for technological change, and developing reward systems and support and enforcement mechanisms. An example is the South African Wine and Brandy Company, which was formed to represent the local wine industry cluster and to implement the vision of innovation-driven and market-oriented production and exporting of wine. After restructuring, it incorporated divisions focusing on basic and applied research, international market development, and social and political transformation (Wood and Kaplan 2006).

POTENTIAL BENEFITS

Evidence from rural industry clusters in the United States shows that clusters contribute positively to regional economic growth and higher wages for rural workers (Henry and Drabenstott 1996; Gibbs and Bernat 1997). Because of their proximity, cluster constituents enjoy the economic benefits of several location-specific externalities and synergies. A well-developed concentration of related agribusiness spurs three important activities:

- *Increased productivity* (through specialized inputs, access to information, synergies, and access to public goods).
- *More rapid innovation* (through cooperative research and competitive striving). Nothing sparks productive innovation better than competition in proximity—whether technological innovation, as seen in the IT clusters of Bangalore or Korea, or creative innovation, as in the fashion design clusters in New York and Paris.
- *New business formation* (filling in niches and expanding the boundaries of the cluster map). In clusters, new businesses form as a consequence of competition, demand for services, and the attraction of investors; this dynamic in turn spurs innovation.

The cluster and its location-specific externalities and synergies accrue a range of benefits:

- *Better and more efficient access to infrastructure, specialized human resources, and inputs, including capital.* Firms readily obtain access to vital inputs such as suppliers, information, technology, financing institutions, and institutions of higher education.
- *Reduction of costs.* Transaction costs are considerably lowered because of proximity in the cluster. Proximity offers vital advantages for the agricultural sector in developing countries, in particular for SMEs. Often firms can source products and services from inside the cluster and forgo the (greater) cost of having to develop or produce the product or service. Costs are also being reduced through economies of scale and scope, as in the case of joint marketing and bulk purchasing. Costs related to hiring talented employees are reduced as well, provided talent is made available in the cluster.
- *Access to information and services.* Being in a cluster provides members with preferred access to extensive market, technical, and competitive information that accumulates in the cluster. For example, through a close relationship with sophisticated buyers within a cluster, suppliers are more attuned to their specific needs. Business organizations and also business fairs function as information and service hubs; informal day-to-day contact with similar companies plays an equally important role.
- *Attraction of foreign investment.* If clusters are leading centers for their industries, they will attract all the key players from home and abroad. In fact, foreign-owned companies can enhance the leadership of the cluster and contribute to its upgrading, as experienced in the Ugandan fish cluster where this function was played by European firms.
- *Better recognition and marketing.* For small and developing businesses, locating in a cluster near competitors and related industries may help them to grow, gain recognition, and attain status more rapidly within the market. The South African wine cluster is indicative. Only when a company was established to market wine for a pool of producers did they gain the clout to export and market on a large scale. One producer alone could not have done it.

One cluster often seeds or enhances others as it disperses activities in the value chain to reduce risk, access cheaper inputs, or better serve particular regional markets. Export-oriented clusters usually generate above-average wages, productivity, and innovation.[4] The South African wine cluster demonstrates those benefits and shows how the development of a cluster can spur innovation and economic growth in an industry (box 5.22).

POLICY ISSUES

With regard to agricultural cluster projects, the most prominent policy issues concern power imbalances, social and environmental issues, as well as a need to define public versus private sector roles.

Some actors in clusters and innovation systems, such as SMEs and their associations, can encounter enormous difficulties in trying to grow (Parrilli 2006). The strongest agents use the market to maintain control over resources and decision-making and invest in innovative and costly activities (such as R&D or promotion campaigns) that strengthen their position compared to weaker competitors. Leaders are needed for a cluster, but programs and projects need to guarantee equal opportunity of access and competition with a view to sustaining the cluster. The commercial interests of a cluster can infringe upon local communities and their way of life, leading to social tensions and hampering the cluster's prospects. For example, social tensions with local residents erupted after cut flower farms privatized the public beaches used for public recreation around Lake Naivasha (Bolo 2006). Horticulturists and fishers around the lake have come into conflict over diminishing fish numbers (Bolo 2006) (box 5.23). Potential externalities have to be factored into policies and programs and should reflect the interests of all potential stakeholders.

Clusters whose productive processes rely heavily on natural resources encounter challenges arising from their side-effects on the production base, growth, or changing environmental conditions (such as climate change). At the onset it is difficult to predict the size, concentration, and output a cluster might attain, but the environmental implications can be severe (box 5.23). A common analytical tool is an environmental impact study. Scenario-based forecasting is an additional tool to hedge against the potential problems outlined in an impact study (see module 7, TN 3). A specific challenge to agriculture relates to common-pool resources, in which exclusion of beneficiaries through physical and institutional means is especially costly and exploitation by one user reduces the availability of the resource for others (Ostrom et al. 1999).

Box 5.22　The Wine Cluster in South Africa: Outcomes and Success Factors

The South African wine industry has grown significantly since the early 1990s in an extremely competitive global market. Exports rose from 20 to 177 million liters between 1992 and 2002. Over the same period, table wine production increased by 33 percent. This phenomenal change came about through a combination of institutional, structural, and market factors.

Two policy initiatives underlay the wine cluster's success. First, the abolition of the quota system precipitated a shift toward varieties for which global demand was increasing. Incentives promoted extensive new planting and replanting. In 2002, 37 percent of vineyards were less than eight years old, and grape quality had increased. Second, broad macroeconomic policies aided the growth of the cluster, including the liberalization of agricultural trade, deregulation, land reforms, reduction of direct subsidies, and the introduction of a minimum wage for farmers.

South Africa's wine producers can be divided into four segments: established producers, new producers, cooperative producers, and wholesalers (some of which produce wine in addition to their primary role of marketing, sales, and distribution). Technical support came from the Agricultural Research Council (ARC), which is partially funded by the state. The Wine Industry Network for Expertise and Technology (Winetech) plays

a key coordinating role as the hub of an extensive network of industry actors and scientists and technicians from the universities and ARC. Winetech serves as an advisory council to the South African Wine and Brandy Company (SAWB), a nonprofit company that represents wine producers, workers, and wholesalers. It has an explicit commitment to enhance the competitiveness of every aspect of the industry through innovation. SAWB designated the private organization Wines of South Africa (WOSA) to act on behalf of some 320 South African wine exporters to promote South African wines internationally. The competitive realities of the global market reinforced the roles of SAWB and WOSA, because marketing the country's brands had become too big of a job for individual firms.

Greater cooperation and collaboration among producers and other institutions increased innovation and helped to overcome market imperfections and inefficiencies. Innovation altered marketing, wine-making processes, the choice of varieties, and production practices. Producers exchange knowledge through viticultural and vinicultural forums, a root-stock association, two varietal associations, Winetech, and the Elsenburg Training Institute. Extensive use of local and international consultants and information about international market trends was a key component of success.

Source: Wood and Kaplan 2006.

Box 5.23　Environmental Challenges for Cluster Development: Examples from Kenya and Tanzania

Kenya cut flower industry (a scale challenge). Kenya's cut flower industry needs water, but the extent to which the cluster around Lake Naivasha would evolve was not foreseen. Horticultural farms in the lake region have encroached on riparian ecosystems, leading to pollution and excessive withdrawals of water from Lake Naivasha. The lake continues to recede, with a concomitant loss of aquatic life, threatening the livelihoods of local fishers and the food supply. The challenge is to guarantee the necessary inputs into production on a sustained basis while limiting environmental externalities.

Lake Victoria fisheries (a common-pool resource challenge). A combination of poverty in the fishing com-

munity, heightened competition, and an ineffective regulatory regime threatened to transform Lake Victoria into an open-access resource, with dire consequences for fishers' livelihoods, fish populations, and the quality and quantity of lake water. Through efforts by the government, international organizations, and civil society, all stakeholders have been involved in managing the fisheries resource. Fishers have been involved in planning and managing fisheries, which is expected to increase their access to Lake Victoria fisheries and help reconcile the potentially conflicting goals of sustainable fishery management and the livelihoods of communities that catch fish using improper gear and methods.

Sources: Bolo 2006; McCormick and Mitullah 2005.

LESSONS LEARNED AND RECOMMENDATIONS FOR SUPPORTING CLUSTER DEVELOPMENT

In developing a cluster initiative and identifying supporting investments, practitioners need to know which emerging clusters have potential and which analytical tools they will need to develop a program or project. Successful clusters develop on the basis of specific combinations of capabilities, incentives, and opportunities. It is hard to predict which of the locations identified in the first phase of this process will become the dominant location in the final phase. Policy makers and development practitioners should promote clustering in sectors that already show comparative advantage (Rodriguez-Clare 2005) rather than distorting prices by promoting the development of sectors with high clustering potential. It is important to include several key clusters in the long list of candidates for inclusion in the initiative. This limits the danger of trying to create clusters where none exist (World Bank 2009).

Identifying emerging clusters with potential

Before selecting specific clusters for intervention, industry specialists should carry out *broad cluster mapping* at the national level. The regional economies in a country are specialized, with each region exhibiting competitiveness in a different mix of industry clusters. It is very important not to ignore interdependencies between regions and between clusters. Cluster mapping will help assemble a detailed picture of the location and performance of industries with a special focus on the dynamics, linkages, or externalities across industries that give rise to agroclusters.[5]

The analysis of emerging clusters begins by identifying either a driving, export-oriented industry or a fast-growing "emerging" industry. Export-oriented industries can be identified using the *Employment Concentration Factor,* an *Input-Output model, Cluster Dependency Factors, Specialization analysis,* and the *San Diego cluster analysis.*[6] All are described in detail in SANDAG (2005) and Goetz et al. (2004).

Analytical tools for developing a cluster initiative and informing investments

Once potential clusters are selected for inclusion in an initiative, several analytical tools can guide cluster initiatives and investments. *Product and market segmentation* will identify the products and markets in which each cluster competes. These products and market segments can give direction to *Porter's five-forces* and *competitiveness position*

analyses, which offer more rigorous approaches to the question of where to compete (for details on all of these tools, see World Bank 2009).

It often is vital to the success of a cluster for a donor program to help *assess the potential of target markets,* in particular those in developed countries. Focus first on existing export contacts and identify specific buyers. Include detailed monthly pricing information from United States and European markets to develop a strategy to enter market segments offering the greatest opportunities. Focus initially on customer demand and work back to what value chains and clusters produce and how they must change to meet customers' needs. Agro-industries may have different market segments, with different demands and competitors; ensure that demand exists not only for the industry in general but also for the specific product and its value chain. Demanding markets can also be local markets, which can serve as an essential first step in the cluster's development.[7]

Benchmarking of the sector is used to examine gaps between the performance standards of domestic firms and the standards required by multinationals, new markets, and new buyers. Cost benchmarking improves the awareness of internal costs and enables comparisons with competitors. It also helps in determining which niches may need less improvement than others and therefore inform product and market differentiation.[8]

Value chain analysis can contribute to strategic decisions about which products to target in cluster development. The analytical framework is based on three major functions in the value chain: source, make, and deliver. In addition, the chain's performance is measured, benchmarks established, and performance gaps analyzed, taking into account government and market failures.[9] Finally, *institutional mapping* can identify which public and private institutions exist in the cluster and determine which institutions may be reformed, abolished, or strengthened.[10]

An incremental approach to supporting and designing cluster development: some lessons

The *initial phase* of cluster development involves sowing the seeds of institutional reform, creating new forms of property rights, and setting out strategic programs. It also involves establishing trust among key stakeholders and convincing them to collaborate (which will also be a continuous activity). In *subsequent phases,* the focus shifts to providing incentives to assist startups, attracting entrepreneurs from the diaspora abroad back to the country, and developing regional clusters, in some cases around special economic zones (Braunerhjelm and Feldman 2007; Zeng 2010). Many agricultural clusters might be remote from urban centers, and linking them to support structures such as banks or universities will be difficult.

Financing is another major issue in subsequent phases of cluster development, once startups and investors consider the basic framework conditions to be in place. As access to capital and finance increases for firms, they can increase their activity and productivity. Often, new investors such as venture capital firms cause new businesses to form, which in turn may alter the cluster.

A common learning trajectory among firms in developing countries starts off with process upgrading, followed by product upgrading. For this reason, an incremental approach to investments and programs seems promising. It may also be worth focusing first on local markets, if their potential is high enough, as a first step in upgrading. Early accomplishments are essential in fostering trust and need to be incorporated in early action plans. To create such trust, the project and the cluster should focus on a series of smaller value chain initiatives that create confidence in cluster activities while addressing constraints to innovation and competitiveness in various segments of the value chain. These initiatives may include technical assistance on low-cost production and postharvest handling, observation trips to see best practices in other countries, technical assistance for packaging, and trade missions to major international markets to meet potential clients.

SEQUENCING, PRIORITIZATION, AND THE INCLUSION OF LEAD FIRMS. Given limited resources and the ambition to achieve the biggest impact for the competitiveness of the agricultural cluster, the major and most urgent areas of public investment must be identified. The analytical information mentioned above should be combined with market trend and segmentation analysis to address major shortcomings in ways that support the strategic development of the cluster.

A parallel dialogue with major industry agents needs to be initiated, their feedback factored in, and a common vision built. The outcome should be an informed decision on where to invest and under what time frame, in line with the vision for the cluster. It is pivotal to identify stakeholder leaders when the project is being designed or first implemented and to identify lead firms and make them central in efforts to improve the value chain and develop the cluster. Lead firms in successful clusters often spearhead collaborative efforts, interact with government, and attract follower

firms. In projects where lead firms were missing, cluster development and value chain enhancement encountered serious obstacles.

In agriculture, most lead firms are major buyers with a stake in ensuring quality and price competitiveness and thus an important stake in the production process. Obtaining strong commitment from a lead firm is not always possible at the outset, and the implementation team will need to assess how much can be done to improve the value chain without cooperation from a lead firm. In successful cases that exhibited these characteristics, the cluster identified segments of the value chain where it could work without active participation from a lead firm, such as promoting micro or small and medium enterprises or dealing with environmental issues (USAID 2008). Ultimately, however, the support of lead firms is required for cluster promotion and value chain enhancement to be effective.

GUIDING PRINCIPLES ON TARGETED PUBLIC-PRIVATE PARTNERSHIPS. In agricultural innovation and clustering, PPPs (discussed in detail in TN 1 and in module 4, TN 2) can be seen as arrangements that assemble partners with different skills to generate, adapt, and/or diffuse innovation, build infrastructure, export and market products, and pool financial resources. Usually PPPs are formalized through contractual agreements specifying the partners' commitments and sharing of benefits. Potential partners include regional industry promotion agencies, research institutes, universities, extension agencies, market promotion agencies in the public sector, producer associations, and businesses and individual producers in the private sector. PPPs must satisfy the condition that overall benefits outweigh the overall costs. Without this synergy, there is no justification to partner.

Agents that promote PPPs, such as donor agencies, play a crucial role in motivating potential partners, building trust among partners, and providing credibility for PPPs in general (box 5.24). Matching grants are a successful mechanism for making these partnerships work, especially if the objective is to foster links between markets and the public sector (see TN 2, the discussion in the module overview, and World Bank 2010).

DIRECT AND INDIRECT WAYS OF FACILITATING AND INDUCING COOPERATION. When collaboration and joint action provide obvious benefits to all major stakeholders, as in the South African wine cluster, they may arise spontaneously. In most cases, cooperation among competing firms and producers does not occur naturally; it is likely to require deliberate and sustained action (Wood and Kaplan 2006).

Box 5.24   Public-Private Partnership Supports Cluster Development in Uganda's Fish-Processing Industry

After fish exports from Uganda were banned by the European Union owing to concerns about bacterial contamination, the government, donor agencies, fish-processor association, and private firms worked together closely and swiftly to help the industry improve its processing practices. Standard operating procedures were devised for inspectors, and a voluntary code of conduct on good manufacturing principles for fish-processing firms was established through the Uganda Integrated Program of the United Nations Industrial Development Organization (UIP-UNIDO) and the Uganda Fish Processors and Exporters Association. UIP

provided technical assistance to the government for timely and effective communication with the European Commission. It identified and paid private consulting firms (based in Europe) to strengthen the audit systems of the government's Department for Fisheries Resources and train fisheries inspectors as well as quality assurance managers across all firms that processed and exported fish. The Lake Victoria Environmental Management Program, supported by the World Bank, provided duty allowances, transportation, and other logistical support critical to implement the revitalized inspection and law enforcement system.

Source: Kiggundu 2005.

Both the provision of incentives and direct facilitation (often funded and implemented by the public sector) seem to be of value in cluster development.

Incentives for joint action can take the form of tax deductions for dues paid to business associations, access to infrastructure, or financial resources for firms agreeing to work together. In Cambodia, a PPP for food processing supported by GIZ enabled five companies producing mineral water, ice, and soy sauce to upgrade their manufacturing facilities, machinery, and technology. To receive financial and technical support, the companies agreed to invite interested SMEs to visit to exchange knowledge, skills, and experiences. As a result of this induced knowledge sharing, there was more demand than available spots to function as lead companies, and collaboration in the cluster and industry intensified.[11]

Creating value along the supply chain and enhancing collective efficiency in the cluster requires trust among all stakeholders. *Direct facilitation* by the government and other organizations, including donors and NGOs, is often needed to stimulate and sustain joint action among key cluster agents (box 5.25). The "honest broker" role is essential, especially for uniting small producers and linking buyers and sellers. On the other hand, project staff must ensure that stakeholders eventually assume this role themselves in the interest of sustainability. Experience indicates that trust can be established more effectively by creating a flow of successful small activities (training, initial transactions between small and large producers, and so on) that lead to more significant transactions, such as joint exporting (USAID 2008).

DEVELOPING CAPACITY IN RELEVANT INSTITUTIONS THROUGH TECHNICAL ASSISTANCE. As the discussion has indicated, institutions, public and private, are crucial in agricultural clusters (figure 5.2). Through targeted programs, training (including study tours), and sharing international best practices, donors play a vital role in strengthening institutions and promoting cluster development. For example, donor support can improve the *capacity of state agencies in monitoring and enforcing quality standards*, as seen in the example from Uganda in box 5.24. Donors' involvement may be necessary to help incipient producers meet complex international standards. Buyers cannot play this role, because the standards embody specialized knowledge that most buyers do not possess (McCormick and Mittulah 2005). Adding capacity to an enforcing institution, and making sure the respective officers (and possibly some critical buyers) gain the necessary knowledge of sanitary and phytosanitary measures, appear to be vital components of cluster development and value chain enhancement in developing countries.

Donors can also strengthen the intermediary institutions (such as professional and business organizations) that act as important mechanisms for tapping into foreign knowledge, coordinating activities in a cluster, and lobbying government. Successful efforts to build capacity have focused on creating a service mentality, training staff, establishing proper financing schemes (such as a fee structure), and providing key business organizations with the necessary infrastructure (including a website). In extension systems, donors add capacity by training officers to use new

Box 5.25   A Joint Action Project in Paraguay Improves Competitiveness of a Sesame Value Chain
            through the Cluster Approach

Successful joint action for cluster development and alleviating rural poverty occurred under the United States Agency for International Development's (USAID's) Vende ("Sell!") project in Paraguay. In 2003, sesame processors in north-central Paraguay were receiving orders from Japan and Korea that they could not fill, because regional suppliers could not provide the right quantity and quality of raw, industrial-grade sesame. The processors required a type of sesame that could be harvested only by hand. Some processors considered moving their facilities to other countries. Although sesame farmers and processors distrusted one another and had never collaborated closely, Vende brought them together. The project helped them to set clear expectations and develop a realistic understanding of the financial returns to cooperation. It educated participants about the value chain and players involved. Its emphasis on increasing sales constantly reminded people that they were working toward a mutually rewarding goal. After two years, sesame production had doubled in Paraguay. Exports for the four processors involved in the program rose by approximately US$8 million. Thousands of farmers in Paraguay's impoverished north-central zone and elsewhere increased their incomes by growing a more viable cash crop. Vende technicians are building local capacity to ensure that these efforts are sustained.

Source: USAID 2008.

methods to explain and demonstrate appropriate technology for cluster firms. In credit institutions, donors can add capacity by improving credit checks; the financial capacity of firms will improve if they learn how to write business plans. Donors may also assist in developing innovative financial services and credit schemes for banks.

Technical assistance has strengthened institutions and promoted cluster development by establishing national standards based on international food standards such as GlobalGAP. A cluster's competitiveness and capacity to innovate particularly benefits from programs that link firms within value and supply chains in ways that benefit all actors involved, including small-scale, geographically scattered producers. Subcontracting schemes are a particularly successful way to link small-scale producers with processors and buyers. They ensure reliable, better-quality production for buyers and access to credit, training, and economies of scale for producers.

MONITORING AND EVALUATION. Cluster M&E begins by mapping three main areas that require monitoring: the *resources* of the cluster (natural resources, firm absorptive capacity, linkages, and human capital, for example), the *activities* to support the cluster (these can be private or public initiatives or donor programs), and the *finance* and *funds* available. The evaluation should also be designed to capture *lessons on process*. Results of cluster initiatives may not mature for a long time. To deal with the inherent challenges, it could be useful to devise an M&E assessment similar to the *management effectiveness tracking tool* used by the Global Environment Facility.[12] The tool categorizes indicators by context, planning, inputs, processing, outputs, and outcomes, all of which are relevant to agricultural cluster projects. Instead of absolute values, a score is used, so results can be plotted to facilitate comparison (for example, across agricultural clusters). The tool can be applied as a self-monitoring and external monitoring tool, and it gives immediate feedback and suggestions for improvement.

Apart from tracking process and results through a sound M&E system, a *good governance framework* is needed to minimize the risks associated with government failure related to misinformation or capture by the industry/cluster.

Technology Transfer Offices: Facilitating Intellectual Property Protection for Agricultural Innovation

Alan B. Bennett, University of California, Davis
Riikka Rajalahti, World Bank
Andrea Pape-Christiansen, Consultant

SYNOPSIS

Technology transfer is a critical process in transforming agricultural research innovations into applications for end users. The vast majority of agricultural innovations in developing countries arise from publicly sponsored research centers or universities, which typically are unprepared to engage in formal mechanisms of technology transfer. These mechanisms may require intellectual property protection and/or legal agreements for transferring intangible and tangible property rights to other public, commercial, or international partners. A sustained investment in capacity building is essential for technology transfer programs to have an impact. It is critical to assess whether an institution has a broad base of research assets and a culture that will support a technology transfer program. If so, investment in a technology transfer office requires paying attention to: (1) identifying an appropriate business model, (2) establishing an institutional policy framework consistent with national laws to clarify responsibilities of the institution and its employees, (3) building the capacity to address intellectual property protection and to negotiate legal agreements to transfer intangible and tangible property, (4) gaining experience to develop business strategies to effectively disseminate technology and work with private partners, and (5) communicating both internally and externally the aspirations (building appropriate culture) and successes of the institution in technology transfer.

BACKGROUND AND CONTEXT

Agricultural innovations reach farmers largely through direct transfers of knowledge, agricultural practices, devices, or seed from research centers to farmers. Extension services in many countries aim at bridging the gap between public researchers and farmers. These traditional approaches are insufficient in strong market economies and for advanced technologies that may require a more formal hand-off to permit commercial development or may require regulatory approval from governmental authorities. Increasingly, more formal mechanisms of technology transfer are required, involving intellectual property (IP) protection and legal agreements to transfer both intangible and tangible property for further development and distribution. These new technology transfer requirements prevail within and between developed and developing countries.

Although it is clear that public research organizations play important roles in all countries, more developed countries have a strong private innovative sector that is virtually absent from developing countries. For this reason, public research organizations in developing countries remain a primary source of local innovation and are likely to play a role further along the innovation pipeline. They require internal capacity to deliver innovations to commercial partners either for further development or for dissemination through market channels. In most cases, public research institutions (centers, universities) have focused their programmatic development on scientific research capacity and not on the development of expertise to engage in the legal transactions needed to translate their research into applications. Moreover, many public research institutions have historically managed their technology transfer programs by passively waiting for potential licensees to knock on their door and seek out new technology.

International technology transfer can be particularly complex and is highly bi-directional. Advanced genetic technologies are increasingly transferred from multinational private companies to public research centers in developing countries, at the same time that germplasm or other genetic resources move in the other direction. These transfers require carefully balanced agreements to ensure that the partnership achieves its objectives and that both parties receive appropriate recognition for their proprietary contributions. Developing this capacity is not easy and typically requires both educational as well as institutional investment.

This thematic note explores the steps in developing that capacity in public research institutions by establishing a technology transfer office (TTO) to assist in the legal transfer of technology. Investing in and building institutional capacity for technology transfer within a public organization may cover a broad range of activities. These activities may include the protection and licensing of IP but are more likely to focus on the support of public/private partnerships, the development of business strategies, or the transfer of tangible property directly to commercial channels such as seed companies or agricultural or veterinary product suppliers. These activities share a range of skills and experience related to an understanding of intangible and tangible property rights and transfers, of legal contracts and agreements, and of business activities and strategies that are relevant for the local region.

The technology transfer function may be addressed in different manners and must fit local needs and resources. The four main business models for TTOs are summarized in table 5.7. For developing countries, a TTO operated jointly among many institutions (a consortium/network TTO) may be the best solution to attain economies of scale

as well as a critical mass of research assets and expertise in technology transfer.

INVESTMENT NEEDED

An effective technology transfer program or department, based on a proper assessment and policy alignment (see "Lessons Learned"), requires major investments. The most important are staff, infrastructure, capacity, and governance.

Staff and site

A TTO will require a general director or manager (usually a scientist with extensive business experience rather than a lawyer) who is at a minimum responsible for business development, communication, and negotiation. The office will also require administrative personnel. The physical assets associated with a TTO can range from office equipment (IT, furniture, and so on), an IP library with access to online legal databases, to real estate encompassing sufficient space for the TTO.

Minimum training and core skills

Table 5.8 describes the minimum training requirements for different groups in the institution and IP management

Table 5.7	The Four Main Technology Transfer Office (TTO) Business Models			
	Independent TTO department within an institution	**Network-based TTO**	**Subsidiary company**	**Outsourcing**
When?	When prospects for technology transfer and commercialization are high within one institution and sufficient resources allow establishment	When individual institutions lack resources and critical mass (research base) but institutional culture is conducive for entrepreneurial activity	When research and resource base are sufficient but institutional culture is not conducive for entrepreneurial activity	Suitable particularly when institution(s) generate technology suitable for high-value, income generating opportunities
Key benefits associated with the option	– Alignment with institutional objectives – Revenue to institution	Sharing of costs and expertise	– May encourage a positive perception of technology transfer and demonstrate seriousness – More operational flexibility and the ability to structure staff remuneration packages	– Minimizes investments and risks for the institution – More operational flexibility and the ability to structure staff remuneration packages
Disadvantages	Investment requirements per institution may be high	Requirement for shared procedures and agreements on revenue sharing often challenging	Lack of alignment with institutional objectives and unresponsive to policy constraints of the institution such as publication or conflict of interest	Overhead costs (fees) reduce revenue to institution Less geared toward technology transfer for the broader public good

Source: Authors, adapted from Campbell 2007.

Table 5.8 Generalized Intellectual Property (IP) Training Needs of Different Groups of Staff in a Technology Transfer Office (TTO)

Group	Minimum training requirement
Researchers	– Maintaining good laboratory records – A basic understanding of the types of IP agreements, especially in the context of exchanging research material and information – The importance of confidentiality, especially with respect to publishing and delivering academic presentations – When to disclose IP guidelines and procedures
Research managers and institution directors	– The importance of IP management and management functions – IP protection processes and procedures; the investments required to manage IP effectively (including decisions required at different stages of IP and research development) – Implementing IP policies, processes, and procedures – An appreciation of the role of technology in addressing socioeconomic needs
IP managers	– Overview of IP management from the generation of IP property to its exploitation and application – Awareness building – Understanding of science (understanding of certain fields of science an added benefit)
Operations	– Finance: Understanding IP policy guidelines, namely, systems and processes to handle IP payments and receipts (for example, royalties); the administration of benefits to researchers and the institution – Human resources: IP policy guidelines and interface with other institutional policies such as conditions of service, recruitment, conflicts of interest and commitment, and contracting with clients – Legal services: IP policy guidelines, IP contracts and agreements, understanding what constitutes IP and the different forms of IP protection, and IP negotiation – Grant and contract research: IP contracts and agreements, especially clauses regarding IP ownership, and IP policy guidelines

Source: Pefile and Krattiger 2007.

office. Technology transfer programs will need the capacity to develop business strategies around new technologies. These strategies can serve as a tool to market innovative technologies to existing companies or become the basis for starting new companies to implement the strategy. The secondary role of a TTO is to establish and maintain connections to national and international investment communities as well as to other providers of services for business development. These connections facilitate the development of startup companies to commercialize technology. The TTO will also need the ability to communicate—internally and externally—its aspirations and successes in technology transfer. Consistent communication within the TTO is critical to develop and sustain a culture of entrepreneurship and engagement in the technology transfer process.

The TTO must have the core skills to manage IP protection and the capacity to negotiate and execute legal agreements to transfer intangible and tangible property. The main functions of TTO staff include: (1) evaluating invention disclosures and deciding whether to file patents or other forms of IP protection, (2) managing or monitoring patent prosecution, (3) developing, with business development staff, a commercialization strategy, (4) negotiating and executing technology transfer agreements ranging from options to licenses, and (5) once an agreement is concluded,

monitoring technology developments and compliance with the terms and conditions of the option or license.

Reporting and governance structure

A TTO requires reporting and governance functions. The TTO management will be accountable to a governing body that may consist of faculty or research center members, administrators, and external business leaders. An advisory group from inside and outside the institution is expected to bring new experience to the organization and act as internal and external champions.

Often a departmental TTO reports to a senior university staff member, whereas a TTO company will be responsible to a board, which may be chaired by a university senior staff member. Because the TTO represents a linking function between a university or research institute and business, governance and advisory arrangements to support both its internal linkage to the university or research institute and its external linkage to the business community are important. The TTO can also become the "face" of the research organization to the business community and as such needs to have strong governance and advisory relationships to ensure that it maintains a high degree of integrity and credibility with the outside. The TTO will be expected to produce at least annual reports of activity which provide financial information to its governance board(s) and also provide accounts of nonfinancial public benefits that may have been realized, such as development of a new medicine or SME.

POTENTIAL BENEFITS

Investment in technology transfer and IP management capacity for public research institutions and universities will almost certainly have an impact on the way the institution looks at its research outcomes and the extent to which it can partner with companies to either codevelop or commercialize its research results. In developed countries where universities have adopted robust technology transfer programs, the resulting impact of those universities on the adoption of new technologies, rise in entrepreneurship, synthesis of new medicines and other products, develop-

ment of jobs, and increased prosperity through regional and national economic development is legendary.

In contrast to developed countries, where most research capacity resides in private companies, the vast majority of the research capacity in developing countries resides in public research institutions and universities. For this reason, one might expect an even greater relative impact of technology transfer from the public sector in developing countries.

Within universities (see box 5.26 on China), robust technology transfer programs also have many important benefits that are quite separate from royalty income, such as (Campbell 2007):

■ **Productive interaction** with the industrial community. Ideas shuttling back and forth between the academy/ research institute and private sector often increase the quality of research.

Box 5.26   Intellectual Property Management at Tsinghua University, China

Intellectual property (IP) is a relatively new legal and social concept in China. Formal legislation was introduced in the 1980s and subsequently strengthened. Universities now usually own the IP emanating from government-funded research, but technology transfer and commercialization remain low. Most universities lack IP policies and independent offices for IP management.

Tsinghua University is an exception. Its IP Office develops IP policies and manages university IP, which includes patents, trade secrets, know-how, trademarks, copyrights, and any related rights. The university's IP policy clearly states what constitutes employee work. The policy requires an investigator to disclose all results of a finished project to the administrative department, which then decides whether to apply for a patent. If results appear to have commercial value but are not suitable for a patent, they remain a trade secret. An industry-sponsored research agreement must have a clause on ownership of resulting IP, allocation of patent costs, and sharing of revenue made from the IP, among other arrangements, and the IP Office examines the contract before it becomes effective. When a faculty member or other employee goes to another domestic or foreign university or institute to conduct research, any resulting IP should be assigned, or at least jointly

assigned, to Tsinghua University, unless another agreement takes precedence. Under the university policy, at least 25 percent of revenue generated by a piece of IP is shared with the inventor(s) as cash or equity.

Tsinghua University spared no effort to educate its faculty members and students about IP and the university's IP policy. It implemented procedures for examining collaborative research agreements and sponsored research agreements between the university and other institutions or companies, for which it designed a standard contract. A special fund covers patent costs, including application fees, examination fees, agency fees, and maintenance fees for the first three years after a patent is issued.

Together, these measures caused Tsinghua University to own more patents than any other Chinese university. From 1985 to 2000, Tsinghua University filed 1,587 patent applications. Since 2001, the average annual growth rate of the university's patent filings has been 26 percent. In 2004, the university filed 43 foreign applications (including Patent Cooperation Treaty filings). The numbers of patents issued to the university rose from 121 in 1999 to 537 in 2004. Other universities with a similar level of IP management include Peking University (University of Beijing) and the Chinese University of Technology.

Source: Heher 2007.

- *Increased industrial support* of research at universities or public research facilities.
- *More willingness from central and local governments* to support research for economic development.
- *Students' or public researchers' exposure* to the world of industry and the commercial opportunities of research (including training in entrepreneurship), which influence their career aspirations and the national economy.
- *Financial support* from grateful alumni and other entrepreneurs who have grown wealthy from companies started from university/institute research.

POLICY ISSUES

A critical and well-understood caveat is that the positive economic and public benefits of technology transfer programs take many years or even decades to be fully realized. Sustained investment in capacity building is essential to achieve the potentially broad impact that technology transfer programs are very likely to have, and sustained support—fiscal and otherwise—is needed from senior administration to set the program's mission, policies, and priorities.

Technology transfer is a multifaceted process with important policy, economic, and managerial ramifications. The public sector's role is particularly to address issues related to the enabling environment, such as the removal of technical, legal, and administrative barriers to technology transfer, sound economic policy, regulatory frameworks, and transparency, which have implications for transferring private and public technology and the success of a TTO. A public subsidy to establish a TTO may also be an appropriate incentive in developing countries where experience is limited (Fernandez 2007).

LESSONS FOR IMPLEMENTATION

The general lessons for implementing TTOs have evolved in many contexts. They involve the need for long-term institution building, the prerequisite of a robust research base, the choice to develop a TTO for a single institution or adopt a network approach, the value of an institutional framework for technology transfer, and the skills, cultural environment, and incentives to support a TTO.

Long-term institutional building

Building capacity for technology transfer requires serious programmatic planning and a long-term institutional commitment to the activity. The early investment may yield financial returns, but only after several years. The immediate returns are measured in terms of increased PPPs, an enhanced capacity to engage international partners, and an enhanced capacity to deliver research results to commercial channels. Nelson (2007) describes five economic "lessons" for institutions to consider before deciding to establish a technology transfer program:

- *Technology transfer will not make your university rich.* A successful program will make a small profit but will not support the university. It will, however, provide many other benefits to the institution and the community.
- *Building a robust technology transfer program takes sustained financial investment.* Investments are required to develop a patent portfolio, attract expert talent, and train office professionals.
- *It will likely take eight to ten years before your program stops losing money*—and it may never make your institution any substantial amount. It takes time to build an IP portfolio, establish contacts, and develop skills in technology transfer. Once these conditions are met, the TTO *may* begin to make money.
- *It may take two decades or more* before a university technology transfer program (including entrepreneurial spinouts) substantially affects the local and regional economy. Expecting substantial returns in a few years leads to underinvestment and disappointment.
- *The ultimate impact may be very large*, however—both economically and culturally—for the university, its graduates, and the community.

A robust research base

Technology transfer programs require a robust research base, with the capacity to develop new technologies with significant commercial applications. As mentioned, a good practice prior to investing in technology transfer programs is to assess the research assets. This assessment should include the research assets and research capacity to supply a steady stream of innovations to be "transferred." This assessment should consider the scientific staff of the institution, its laboratories and facilities, its existing and potential base of research funding, and its existing or potential international collaborations. In addition, this assessment should look at the culture of the institutional leadership, its scientists, and students (if applicable). If measured only by royalty income, an institution with a smaller research base will have a more difficult time breaking even. Less research means fewer inventions. An example of successful

TTO at a large university with a large research base is Unicamp Brazil (box 5.27).

Individual or network-based technology transfer office

If an institution has sufficient research assets to develop a steady stream of innovations and a culture that would support an active technology transfer program, it can move forward and establish a TTO. If the research base or institutional culture is weak, it may be useful to broaden the assessment to encompass a set of regional research institutions. Building a technology transfer program to serve a cluster of institutions is the more difficult option, however, and requires several institutions to agree formally on strategies. In many cases it may be the only way to assemble a sufficiently large research base to justify the investment. An assessment in Chile concluded that an effectively staffed technology transfer office should receive somewhere between 20 and 30 innovation disclosures per year (Fernandez 2007). In Chile, this level of activity could be achieved only if TTOs served a cluster of universities.

An institutional policy framework

Another good practice, following the research base assessment, is to assess and to develop an institutional policy framework for technology transfer. The framework must

---

Box 5.27   From University to Industry: Technology Transfer at Unicamp in Brazil

Brazil has dramatically increased technology transfer and innovation through Inova, the technology transfer office established by the State University of Campinas (Unicamp) in 2003 and the first technology transfer office established in a Brazilian university. A multidisciplinary university with more than 31,000 students and 20 research units, Unicamp pursues a variety of technologies in many fields. By 2007, Inova had become the most frequent patentor and licensor in Brazil. In only two-and-a-half years it signed 128 technology transfer agreements, licensed 45 technologies to private companies and the government, and applied for 153 new patents, 22 trademarks, and 24 software registrations. Its technology transfer agreements will last for more than ten years, and they have already generated royalties for the university ranging from 1.5 percent to 10 percent of the net income from the licensed technology. Unicamp grants inventors 33 percent of royalty and licensing income. The greatest contributor to the patent and licensing portfolio is the Chemistry Institute (48 percent); the corresponding figure for agribusiness and food is 16 percent. Inova's patent database is available online.

Under Brazilian law, Unicamp owns 100 percent of its professors' and researchers' results. The law permits public institutions to give up ownership to the inventor, but Inova has not taken this route; its inventors lack commercial expertise and find it more attractive for Inova to commercialize the technology and give the inventor part of the licensing fee. Unicamp also

commonly practices sponsored research. In such cases, ownership rights are normally split 50/50.

Factors in Inova's success include:

- **Inova is driven by market demand.** Instead of selecting Unicamp's technologies and offering them to the market, Inova examines market demand and seeks solutions inside the university.
- **The technology transfer team comes from private institutions and has business skills.** They are not researchers.
- **The government provides many incentives to companies,** such as tax benefits to companies that pay royalties; tax benefits to companies that invest in research and development, within or outside the company; compensation for taxes on royalties paid abroad during the execution of technology transfer contracts; tax exemptions for fees paid to maintain patents, trademarks, and cultivar registrations abroad; and sponsorship/subsidy of 60 percent of the salary of a scientist hired by a company.

These coordinated efforts will increase patenting and technology transfer in Brazil, strengthen the relationship between public institutions (where Brazilian research is mainly concentrated) and private companies, and contribute strongly to innovation. Other public universities and research centers have been studying Inova's model to emulate it.

Source: Di Giorgio 2007.

be consistent with national employment and IP laws and any exemptions that may exist for public researchers or for universities and their faculty. This framework can clarify the responsibilities of the institution and its employees prior to investing in the TTO.

The first step is to assess the national legal framework, including the IP policy, and ensure that the institution's IP policy and objectives for its technology transfer program are consistent with the national legal framework as well as its own mission and policies. The institution's IP policy should provide the basis for structuring the technology transfer program as well as the basis for transferring tangible property developed within the institution.[1] Support from senior administration is critical (Nelson 2007). Clear mandates will help technology transfer professionals choose among competing priorities and the ever-present trade-offs between private sector and public sector values. A well-understood review and appeal process needs to be put in place early.

In addition to establishing the broad aspirations and objectives, an institutional IP policy must address several key issues (box 5.28). (See module 6, TN 3). These issues are ownership, researchers' obligations, the institution's obligations, and administrative responsibilities.

Importance of core skills to manage IP protection and capacity to manage legal agreements

The required expertise is complex. Technical knowledge is needed to clearly understand a range of new innovations and how they might meet standards of patentability; legal skills must be sufficient to craft complex legal agreements. Many TTOs have found that technical expertise is indispensible and difficult to learn, whereas legal knowledge can be acquired "on the job" and supported by the judicious use of external legal counsel. Many staff members are not needed to support this activity, but their skills should reflect the major scientific disciplines of the institution.

Importance of business skills and communications strategy and skills

The long-term success of the TTO will depend on its ability to expand the impact of the institution's research by effectively transferring technologies (either intellectual or tangible property) to commercial partners and by supporting new research partnerships. Its long-term success will also depend on its ability to effectively communicate the results of its activities to the institutional stakeholders, including its leadership and sponsors. The profile for staff in this area includes business training, such as a master's degree in business administration, significant work experience in business development in private companies, and preferably experience in starting a company. Overall, staff in this area must have the interpersonal skills to interact easily with a range of business professionals and have experience in selling new ideas in a wide range of contexts. The number of staff needed in this area is small. For most operations, one person is enough, but that person must be skilled and experienced.

Culture and incentives for technology transfer and commercialization

A passionate interest in technology transfer within the institution, the TTO management, but particularly the most senior management of the university or institute is a prerequisite for success. An essential ongoing activity is to identify and foster relationships with stakeholders, including academics, representatives of the business and user community, and regional and governmental offices. The most important group at the outset is the internal community that must be supported and encouraged to engage in technology transfer and entrepreneurial behaviors (Campbell 2007). Faculty reward systems such as professional advancement or revenue for engaging in technology transfer, along with removing cultural barriers and staffing the TTO, are key factors for success in technology transfer (Campbell 2007). For details, see Siegel et al. (2003).

Critical steps in establishing a TTO

The following minimal activities should be undertaken *once a decision is taken to establish* a new TTO (Young 2007):

■ *Assist researchers in identifying results that have commercial value and document the discoveries through a disclosure process.* The disclosure-of-invention form should be simple (a complex form deters disclosure). More detailed information can be obtained through subsequent interviews with the inventor.

■ *Evaluate the commercial potential of disclosed innovations.* A TTO exists to find commercial applications for technology and partners to realize the commercial potential, not to judge the value of the science. Such evaluations may be the most difficult of all tasks for a TTO.

Box 5.28   Key Issues to Be Addressed by an Institution's Intellectual Property Policy

Ownership. An institution's intellectual property (IP) policy should clearly address who owns IP developed in the institution (for example, the inventor/researcher owns the IP; the research institution owns it; a company providing research funds owns it; the government agencies providing research funds own it; or no-one owns it, and all IP is committed to the public domain). In most cases, institutions cannot manage IP effectively unless they own all IP developed within their walls regardless of funding source, but this condition may not always be possible. Whatever is decided regarding IP ownership, it must be very clear to prevent any ambiguity over who has the legal ability to transfer technologies.

Researchers' obligations. The policy needs to clearly describe the obligations of research staff. When the institution owns all IP, its researchers typically are required to

disclose possible inventions before publication, to assign ownership to the employer/institution, to assist in evaluation and patenting, and to report potential conflicts of interest.

The institution's obligations. The policy also needs to clearly describe the institution's obligations in managing IP. When the institution owns the IP, it typically is required to manage IP effectively, to pay patenting costs, and to share revenue with inventors. Of particular interest to researchers is the actual share of revenue that will go to the inventors, which can range widely from place to place but is typically 25–50 percent of net revenue after expenses.

Administering the policy. The policy should identify who in the institution is responsible for administering the policy and procedures for compliance.

Source: Authors.

There are many approaches to invention evaluation. The evaluation process lays the foundation for future decisions about IP protection and marketing.

- *Determine whether to protect IP in the innovation.* If needed, secure funding for filing patent, trademark, or copyright applications, and manage the protection process. The challenge of securing funding for IP protection internationally—especially when seeking protection in highly industrialized countries, where the primary markets for the expected products lie—is often overwhelming and perhaps even impossible in many developing economies because of the tremendous expense. Yet there may be very small or nonexistent commercial markets for the innovation in the country of origin, which can present a serious dilemma. The only solution in many cases is to secure protection in the country of origin first, thereby "buying time" under the requirements of the Patent Cooperation Treaty to find a corporate partner to pay the patent costs internationally as a business expense in the license agreement.

- *Conduct market research* to identify potential industry partners, and then market the innovations. Research has shown that in the United States, the primary source for identification of licensees is the inventor. In industrialized countries, inventors typically are familiar with the marketplace in their area of scientific expertise; through

professional networking, they may even know their counterparts in industry (potential licensees) on a personal basis.

- *Once one or more industry partners are identified for an innovation, negotiate legal contracts* (license agreements) with these industry partners to transfer IP rights in the innovation in exchange for royalties or other considerations. The goal is to negotiate a fair arrangement that facilitates and assists the commercial partner in successfully developing and marketing the product, rather than simply seeking to negotiate the absolute highest fees and royalties in the agreement. Developing industry partnerships can lead to many unexpected benefits, such as sponsored research, student employment opportunities, consulting opportunities, and even philanthropic donations to the institution.

- *Maintain and manage administrative functions in support of the primary functions of IP protection and technology transfer.* These functions can include accounting, royalty distributions, licensee performance management, and patent application management.

- *If the TTO decides not to pursue IP protection and commercialization of an innovation, implement a process* to ensure that others have an opportunity to pursue protection and commercialization if they choose. The "others" will most often be inventors.

Risk Capital for Agriculture in Developing and Middle-Income Countries

Alistair Brett, Consultant

SYNOPSIS

Risk capital is money explicitly available for investment into a high-risk business or a security of some type—typically those which are not publicly traded on any national stock exchange. In this note, "risk capital" refers to investment in a company or project at an early or high-risk stage. Because interest in agricultural risk capital investment is relatively new (although investment at other stages is already in place), many of the investment vehicles cited in this note are still relatively unproven. This note draws on lessons from using risk capital in other sectors, especially in innovative technology. It describes current and possible future investment models, their benefits, and potential applications in agricultural investment. Global lessons from developing investment vehicles and investments in both agriculture and other asset classes are presented with recommendations for policy makers and practitioners. Because capital for investment does not exist in isolation, an enabling environment must be in place or under development. Traditional venture capital is not appropriate for countries lacking essential features for venture funding, such as a strong flow of investable opportunities (which can be stimulated by the public and private sectors), access to domestic or foreign stock markets, a large business sector for trade sales of companies, and an entrepreneurial culture where risk and failure are acceptable. It is always critical to ask the question: What problems are to be solved or needs to be met? Only then can it be known whether the provision and use of investment capital could help accomplish the expected outcomes.

BACKGROUND AND CONTEXT

Investment in agriculture is growing because of improved profitability projections and the interest of development agencies and governments to increase investment in the sector to achieve food security and economic growth. Investment is essential for the agricultural sector to grow; it is estimated that net investments of US$83 billion per year must be made in the agricultural sector in developing countries if there is to be enough food to feed the world population of 9.1 billion in 2050. Private investors need to be the major sources of this capital. Public investment cannot meet the needs, but it can be effective in stimulating and leveraging private investment in the sector. "...[A]gricultural investment growth has not only been spurred by increased agricultural prices and food security concerns, but also importantly because of innovation and experience in *risk mitigation* of investment. One manifestation of this phenomenon is the proliferation of funds set up to target the agricultural sector—*agricultural investment funds*" (FAO 2010, xv).[1]

Investing in agricultural innovation is an important part of overall agricultural investments. Increasingly agricultural innovation is seen as a sector that offers profitable investment opportunities for private investment funds as well as alleviating poverty and increasing food security (World Bank 2007). Investments are being made in SMEs that are developing innovative agricultural technologies to improve the quality of crops, reduce risk and losses, and improve efficiency to increase competitiveness. It is expected that the level of innovative technology used in agriculture in developing countries will significantly increase through new applications of biological and information technologies. The availability of "risk capital," money explicitly intended for investment into a high-risk business or a security of some type (typically those which are not publicly traded on any national stock exchange), is warranted. Gaps in the provision of finance for agriculture and agricultural innovation for SMEs and early-stage firms are evident, however, as illustrated by data on Africa, which show that the "meso-finance stage financing is the most difficult to obtain" (table 5.9) (the finance gap for small enterprises is also noted in TN 3). The choice of a risk capital investment model depends on the growth stage of the project or company that will receive the funds.

Innovation funds, including competitive grants or matching grants as described in TN 2, may be used to make a small enterprise in the early stages of development "investment-ready" for the types of investment shown in figure 5.3.[2] Innovation funds may, for example, support moving an idea through the feasibility and proof of concept phased to a stage that is much more attractive to angel and seed capital. Incubators (TN 3) may provide similar support.

Typical equity investment levels, which do not have clearly defined boundaries, are:

1. *Angel investment.* An angel investor provides backing to very early-stage businesses or business concepts. For example, a business may have little more than a business

concept and perhaps a plan for growing the business.[3] Angel investors may group together to form angel investment pools or come from the category referred to as FFFs (friends, family, and fools—see figure 5.3).

2. *Seed fund investment.* A pool of money used to back companies that are too small to attract venture firms but require too much money for angel investors. For example, a business may have a prototype product or service but few sales.

3. *Venture capital fund investment.* Venture capital funds pool and manage money from institutional investors, such as pension funds and insurance companies, as well as from other venture funds and wealthy individuals. They take equity stakes in SMEs with strong growth potential.

This note focuses on the venture capital model of risk capital and pre-venture capital. Because much of the interest in risk capital investment for agricultural innovation is relatively new, although investment at other stages is already in place, *many of the investment vehicles discussed here have a limited track record for investment in agricultural innovation and are consequently unproven.* For this reason, the discussion that follows draws on lessons learned in the use of risk capital in other more traditional investment sectors, especially in innovative technology.

RISK CAPITAL INVESTMENT MODELS

The elements needed to provide risk capital for innovation are: (1) an adequate number of opportunities for investment (referred to as "deal-flow"); (2) a structured vehicle that provides a source of funds; (3) defined criteria for investments made by the investment vehicle; (4) a methodology for evaluating and selecting projects to be supported according these criteria; and (5) a fund management entity and governance to monitor and manage funded projects.

Venture capital in agriculture: The venture capital fund model

A suitable investment model is required.[4] Although traditional venture capital may not always be appropriate for many developing countries, it is a helpful model to which alternative financing models such as angel, seed, or meso-level investment can be compared. Venture capital[5] is a form of private equity provided for early-stage and more mature companies with substantial market potential. Returns on venture capital investment are from a trade sale (sale to, or merger with, another company) or an initial

Table 5.9 The Gap in Access to Enterprise Finance in Africa

Funding level	Funding category	Availability
US$10 million	Project financing	Yes
US$1–10 million	Venture capital/private equity	Some
US$50,000–1 million	Startup/seed capital/growth capital/meso-finance	No
US$100–50,000	Microfinance	Yes

Source: Adapted from Ashley, Warner, and Romano 2005.

Figure 5.3   Typical Financing Stages for Company Growth

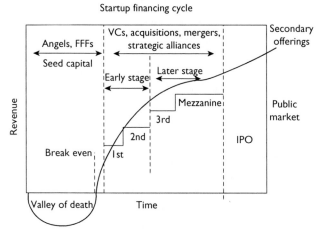

Startup financing cycle

Source: Wikipedia (http://en.wikipedia.org/wiki/Startup_company), accessed September 2011.
Note: FFFs = family, friends, fools; VC = venture capital; IPO = initial public offering.

public offering (IPO) in which the company becomes authorized to sell its stock to the general public on a stock exchange. Venture capital funds will not only provide money but will mentor their investee firms. Venture capital funds are very selective in making investments and may review many hundreds of business plans before investing in one opportunity. By their nature, venture capital investments are high risk and investments may fail. On average, about one in 10 venture capital investments will provide a substantial return on investment. Others may fail or provide insufficient returns to justify the investment. Venture capital funds usually invest at several stages of a business's development: *startup* (funding for businesses at the early stage of product or service development), *first-round* (funding for businesses that have initial sales), *second-round* (working capital for early-stage companies that are selling product, but not yet turning a profit), *third-round* (expansion funds for a profitable company; also called mezzanine financing), and *fourth-round* (financing for exit preparation such as a trade sale or an IPO).

The venture capital fund management structure

Venture capital firms are typically structured as limited partnerships ("limited" because they limit the liability of investors in the fund, who are referred to as the "limited partners"). Limited partnerships have "general partners," which serve as the managers of the venture capital fund and investment advisors for the venture capital funds raised. The limited partners have no decision-making authority for the investments being made. These limited partnerships are legal entities, which hold the funds from the limited partners and have a limited lifetime, typically around ten years. This means that the fund must cash in (exit) their investments in, say, five to seven years, and this need will be a factor in selecting opportunities in which to invest.

Venture capital fund management team

Venture capital fund management teams receive a combination of management fees and a share of the profits. Compensation in the form of a percentage of the fund's capital means that there is a lower limit on the amount of capital necessary to support qualified management teams. Thus this venture fund model cannot function if only small amounts of capital are available (for example, seed funds have to consider other compensation systems).

The *fund manager* is responsible for the overall financial and administrative management of investments, including:

assessments on potential investments (known as "due diligence"); representation of the fund to the potential investment's board of directors or equivalent; defining exit strategies; monitoring investments and taking corrective action when needed; and communication with all investment parties. An *advisory board* composed of independent members and investor representatives should: provide guidance on the implementation of the fund's investment strategy; ensure adherence of the fund to its investment charter; and resolve conflicts of interest. An *investment committee* composed of fund representatives and/or other investors should: approve all investment decisions of the fund, on the basis of reports of evaluations and due diligence performed by the fund manager; conduct postinvestment monitoring; and review progress of the fund's portfolio and fund manager performance.

Funding criteria

Key criteria may include the following:

- A strong and committed core management team with a demonstrated performance track record, commitment, enthusiasm, and energy, although in some cases a reason to invest may be to grow the potential of new businesses.
- Sales of products or services locally or in other markets.
- Potential for scaling up the business.
- Potential for sustainable high growth for the business.
- Expectation of sustainable long-term competitive advantage.
- A viable business model (overall business concept) followed by a viable business plan (a detailed plan for growing the business) delivering an attractive return on investment.
- A clear strategy for a cashing in their investment within a reasonable time period (known as the investment exit).

A selection of agricultural investment funds and the type of instrument and investment preferences is shown in table 5.10. A few of these funds, mostly the new ones, support agricultural innovation, but overall they focus on agribusiness, value chain development, and food processing. Box 5.29 describes one representative fund in more detail.

POTENTIAL BENEFITS

The impact of providing risk capital for investment in innovative agricultural ventures will depend on the reasons for

Table 5.10 Representative Agricultural Investment Funds

Fund (founding date) Purpose and capital base	Countries/target investments	Financial instruments
Actis Africa Agribusiness Fund (2006) www.act.is US$92.7 million private equity. This is a specialized fund from Actis, a leading private equity investor with sixty years of experience in emerging markets.	*Côte d'Ivoire, Kenya, South Sudan, Tanzania, Zambia* Agribusiness across the supply chain.	*Equity and quasi-equity investments* Deal size: US$5–15 million.
African Agricultural Capital (2005) www.aac.co.ke/web/ Venture Capital Fund (US$8 million), which is fully invested in 16 ventures. In response to the absence of an investment facility that focuses on the development of private initiatives in agriculture in East Africa, the Rockefeller Foundation, the Gatsby Charitable Foundation, and Volksvermogen NV set up the fund to invest in agriculture-related SMEs in East Africa.	*Kenya, Tanzania, Uganda* – Small and medium-sized agricultural enterprises. – Provision of risk capital to seed companies operating in agricultural value chains.	*Equity, quasi-equity, and debt investments* – Most likely there are no investments below US$100,000. – Objective is to earn a minimum gross return of 12%/yr on funds invested.
African Agribusiness Investment Fund (2008) www.agrivie.com/index.html US$100 million private equity.	*Botswana, Ghana, Kenya, Nigeria, South Africa, Tanzania, Uganda* Agribusiness sector across sub-Saharan Africa along the value chain.	*Equity and quasi-equity investment.*
Agribusiness Partners L.P. (United States of America/Russian Federation)/Agribusiness Partners International Fund partnership (1995) www.burlingtoncg.com/api.shtml US$100 million. The fund had an initial guaranty from the Overseas Private Investment Corporation.	*Georgia, Kazakhstan, Moldova, Russian Federation, Ukraine* Agribusiness and food-processing companies.	*Private equity, venture capital*
Omnivore Capital (2010) The Godrej Group US$50 million (target). Investment in scientists and entrepreneurs who are innovating to improve agricultural productivity. Investment in small and medium-sized companies focused on agricultural innovation.	*Canada, India, USA*	*Seed and venture capital*

Source: FAO (2010, 164 ff).

Box 5.29 African Agriculture Fund

The African Agriculture Fund was established in 2009 by AfDB, AGRA, BOAD, IFAD, and AFD. The fund has an initial target size of US$150 million and expects to raise additional commitments up to an aggregate capital amount of US$500 million. Investment objectives of the fund, chosen because they are assessed as future high-growth sectors, include:

■ Food production industries (or provide financial services to small agribusiness operators).
■ Invest in the value chain to reduce transaction costs of producers/processors and in storage/marketing.
■ Grow their markets within the region or develop export opportunities.
■ Main investment sectors: cereal production, roots and tubers, livestock and dairy products, fruit products, seed production and fertilizers, fats and oils,

investment financing, equity, and quasi-equity products, Technical Assistance Facility (TAF).

The fund has two windows of financing for enterprises. A small–medium company financing window offers investment between US$0.15 million and US$4 million to help bridge the typical early-stage financing gap. A large company financing window is designed for investment up to US$15 million for more mature firms.

Fund terms and exit strategy are: a five-year commitment period; seven- to ten-year investment maturity; the exit strategy shall be, as the case may be, to provide for the option for local agricultural producers to acquire interests in the targets; and average Internal Rate of Return per target shall be around the mid-teens.

(Box continues on the following page)

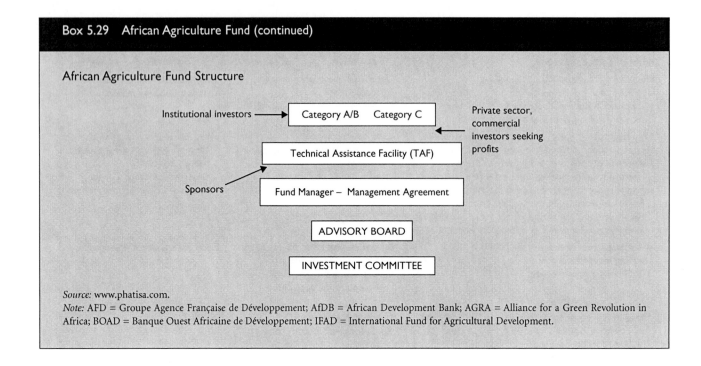

Box 5.29   African Agriculture Fund (continued)

African Agriculture Fund Structure

Institutional investors → Category A/B   Category C

Private sector, commercial investors seeking profits

Technical Assistance Facility (TAF)

Sponsors →

Fund Manager – Management Agreement

ADVISORY BOARD

INVESTMENT COMMITTEE

Source: www.phatisa.com.
Note: AFD = Groupe Agence Française de Développement; AfDB = African Development Bank; AGRA = Alliance for a Green Revolution in Africa; BOAD = Banque Ouest Africaine de Développement; IFAD = International Fund for Agricultural Development.

investing: Is the intention to earn a financial return on the investment, support innovation, create economic development, or promote social good? For typical venture capital investments, the Internal Rate of Return (IRR) is used as a measure of the cash return on an investment. The IRR can be thought of as the effective rate of interest earned by the investment.[6] A project may be a good investment if its IRR is greater than the rate of return that could be earned by alternative investments of equal risk. A venture capital fund in a developed country has traditionally expected an IRR of at least 20–25 percent (before the 2008/09 crisis).

The African Agriculture Fund cited in box 5.29 is an example of a mix of public and private funds. The presence of public support can be attractive to private investors. Although agriculture may have a lower IRR than other types of investment, the fund is designed to reduce risk where possible by adhering to strict investment criteria. Funds that provide training and technology transfer to farmers further reduce investment risk. Another benefit is the provision by some funds (such as African Agricultural Capital) of equity investment with possible debt financing in cases where taking an equity position in a business is not inappropriate (for example, when there may be too much dilution of value for the shares held by existing investors).

POLICY ISSUES

Sustained innovation and commercialization need consistent policies for the long term. Obtaining early-stage financing for SMEs with science- and technology-based products or services remains difficult even in developed markets. Investment capital is moving to later stages to avoid risk in developed markets and even in some developing markets such as India. SMEs do not have the tangible assets required by banks and other lenders as collateral; their assets may be intangible, in the form of patents or know-how. These firms may also have an unreliable revenue stream and inexperienced management. Before investment capital can be deployed effectively and efficiently, it may be necessary to make improvements in a country's IP regime and improve the support ecosystem. But it is also necessary to be realistic about what improvements can be made in the short term. Waiting until there is a fully functioning support system in place will mean delaying action indefinitely.

The frequent public policy response is (Hodgson 2009):

■ *The creation of some venture funding entity* that provides, often with public sector participation, risk capital for new, knowledge-based businesses.

- *Mentoring and training entrepreneurs* in how to prepare their ideas to be investment-ready and thus attract funding by the right presentation of evidence on the opportunity and potential rewards to those looking to make investments.

This response may be necessary, but it is not sufficient as a capacity-building policy. For example, this note emphasizes the importance of creating the ecosystem for innovation, including the role of government policy in improving existing systems, such as reducing barriers to success, as well as making more radical changes. For agribusiness support this may include physical infrastructure such as transport systems as well as education.

Contributions of public and private sectors to both finance and knowledge generation should be coordinated to achieve maximum benefits. Public sector support in reducing barriers to business, for example, by easing the process of business registration and creating a more favorable tax regime and tax inspection systems, can help combine public and private sector investments. A related issue is to assure that foreign investors will be able to transfer their capital gains out of the country.

IMPLEMENTING RISK CAPITAL SYSTEMS: LESSONS LEARNED AND RECOMMENDATIONS FOR PRACTITIONERS

Although the provision of risk capital equity investment funds for agricultural innovation is a new trend, lessons have been learned from both agricultural investment and from investing in nonagricultural sectors.

Investment fund management

The fund management (general partners and support staff) must have the necessary skills to carry out the management functions described earlier, including in-country experience of working under the constraints of a frequently underdeveloped ecosystem to support the selection and growth of the fund's investments. Experience has shown that nonresident management teams and lack of experience in the country or in similar environments may lead to failure.

Advisory boards, which are responsible for guiding the fund's investment strategy and making sure that the fund adheres to its investment objectives, all too frequently in developing countries do not in fact provide effective advice, owing to a lack of experience or understanding of investment processes. Advisory boards composed of government officials or representatives from large enterprises do not have the skills to carry out the duties of an advisory board.

Even more important, the investment committee, which must approve all investment decisions of the fund as recommended by the management, must be fully informed and experienced in such decision making, which is not always the case in developing countries. There is a need to provide training for advisory boards and investment committees working with agricultural innovation projects.

Challenges for a developing country or region

The challenges specific to developing countries or regions include:

- *Setting the right balance* between (1) investing in technology commercialization and (2) investing in technology absorption and adaption, together with investment in building the capacity to support technology translation and development.
- *Understanding that capital for investment does not exist in isolation.* It is critically important to have an ecosystem (enabling environment) that includes but is not limited to such features as: provision of advisory services, a supportive IP regime, access to markets, an effective governance structure, availability of investment exits as a way to realize a return on investment, and the existence of the rule of law to provide investor confidence.
- *Recognizing that traditional venture capital is not appropriate for many countries* lacking the features necessary for venture funding, such as a strong flow of investable opportunities, stock markets, a large business sector for public listing or trade sales of companies, and an entrepreneurial culture where risk and failure are acceptable.[7]
- *Building the ability to provide small amounts of funding quickly* for very early-stage technology development or technology translation funding, and knowing how to invest these funds in a "smart" fashion, to attract private sector participation and resources needed to scale up investments.
- *Maintaining close contacts* with the private sector and other investment funds in regions as well as globally.
- *Finding partners* to provide access to public and private financing outside the country when only limited financing is available within the country.
- *Having a locally based investment fund manager* (that is, one who resides within the country).

A cautionary lesson from developing countries is that investment decisions can be distorted by political influence. Government officials typically lack the experience to evaluate and manage investment opportunities. In India, for example, the selection of government-funded projects that were candidates for further investment was contracted to a private sector group.

Practical issues for risk capital use in early-stage development

Policies should focus on the practical rather than the ideological. A practical problem for many developing countries in negotiating financing agreements to either acquire a technology or license IP to others is that someone has to take the first step and agree to provide initial funding, which could be matched later by others. Sometimes this funding can be in the form of a grant for early-stage development, as noted. Should these grants take too long to be approved (as is often the case), the deal may be lost.

Note that usually angel and seed funding rely on the availability of later-stage capital to get a return on their investment by having their shares bought out. Of special interest to developing countries is that some investment models specifically attempt to address the "investment gap" (also referred to as the "Valley of Death") which occurs when private and public funding are either unavailable in the first place or run out, and where the company's net cash flow does not close the funding gap. Many businesses—frequently those based on research discoveries—continue to reside in the Valley of Death because they lack the financial support and skilled management teams to progress into the "proof of relevancy" phase.

Recommendations for practitioners

A few recommendations should be considered:

■ *Ask the question: What problems are to be solved or needs to be met?* The answer will determine if the provision and use of investment capital will help accomplish the expected outcomes. Businesses and governments often believe that what they need most is investment capital when in fact it may not be the *immediate* critical need; a more pressing need might be finding business partners or gaining access to markets.

■ *Set the right balance* between (1) investing in technology commercialization and (2) investing in technology absorption and adaption, together with investment in building the capacity to support technology translation and development.

■ *Determine the financing objective.* Is it to earn a financial return on the investment, to foster economic development or social welfare, or achieve another purpose? The purpose of an investment or investment strategy should be decided and made clear to all involved at the start of the process.

■ *No investment can be made without a sufficient "deal flow"*—a continuous source of investment possibilities. The public and private sectors can stimulate deal flow.

■ *When developing an investment fund or other investment vehicle, decide how much money will be dedicated to the fund or other vehicle and under what conditions or constraints.* This decision is critically connected to the question of what is the purpose and investment strategy of the fund. Some public investment funds have sizeable amounts of money but have not succeeded because of a poorly conceived or implemented investment strategy. Other funds, with limited capital, have not been able to support businesses to become self-sustaining.

■ *Structure public funding to attract private funds,* either initially or later (for example, as matching funds or a guarantee for the private investment).

■ *A majority of nongovernment representatives* should be appointed to management boards of investment programs using government funds. Government representatives should provide guidance but may want to create maximum good by funding too many projects.

Developing Entrepreneurs through an Agribusiness Incubator at ICRISAT

Kiran K. Sharma, International Crops Research Institute for the Semi-Arid Tropics (ICRISAT)

S.M. Karuppanchetty, ICRISAT

S. Aravazhi, ICRISAT

SYNOPSIS

The Agri-Business Incubation (ABI) Program at ICRISAT, launched in 2003, is an initiative of the International Crops Research Institute for the Semi-Arid Tropics (ICRISAT) in partnership with India's Department of Science and Technology (DST). ABI promotes agricultural technologies developed by ICRISAT, other R&D centers of excellence, universities, and other institutions, separately and jointly. Its approach features a dual service and outreach strategy. The service strategy focuses business development on five strategic areas, building on the expertise of ICRISAT and its partners: seed, biofuels, ventures to develop particular innovations (products or services), farming (high-value crops), and agricultural biotechnology. The outreach strategy involves collaborative business incubation to bring a wider range of expertise and resources to bear on business development to foster agricultural development in other regions.

CONTEXT: INCUBATORS IN INDIA AND AGRIBUSINESS

Business incubators are gaining a foothold in India. A recent survey found that their numbers had grown from 10 in 2000 to 30 business incubators and science and technology parks involved in the commercialization of software and other engineering technologies in 2009 (NSTEDB and ISBA 2009). Of the 495 ventures that graduated from the business incubators in India, 387 remained in business. More than 10,000 jobs were created through these ventures. These incubators have stakeholders in government agencies, financial institutions, and venture capital operations. Only three were involved in agribusiness in 2008, although various government departments, which recently created entrepreneurship promotion programs, have expressed an interest in establishing agribusiness incubators.

Agribusiness incubators can take the form of comprehensive occupational schools, offering rural producers and workers sufficient knowledge, experience, infrastructure, and means to become agribusiness entrepreneurs. This endogenous movement can have far-reaching effects, promoting the overall modernization of primary production, industrialization, and marketing and development of rural areas.

More specifically, however, an agribusiness incubator creates a mechanism to assist in the identification, adaptation, and commercialization of products from public and private agricultural research institutions and universities. From a development perspective, the goal of agribusiness incubation programs is to develop and commercialize new products, technologies, and services to improve productivity in farmers' fields and increase the practical impact of research conducted in India's academic and research institutions. Incubators provide a means of leveraging the significant resources invested in R&D and infrastructure, generating employment and income in India's rural areas, and ultimately creating wealth to support the livelihoods of the poor.

AGRI-BUSINESS INCUBATION PROGRAM OBJECTIVE AND DESCRIPTION

The Agri-Business Incubation (ABI) program, launched in 2003, is an initiative of the International Crops Research Institute for the Semi-Arid Tropics (ICRISAT) in partnership with the Department of Science and Technology (DST), Government of India. ABI promotes agricultural technologies developed by ICRISAT, other R&D centers of excellence, universities, and other institutions, separately and jointly. The incubator was set up as part of ICRISAT's Agri Science Park (later the Agribusiness and Innovation Platform). ABI is governed by a board of advisors headed by the Director General of ICRISAT and by a standing advisory committee that counsels the board on strategy and client intake and exit.

ABI represents a new resource to promote enterprise development in agriculture and facilitate business among entrepreneurs and technology developers. The pillars for high-performance incubation are R&D, business planning, business development, and access to capital (figure 5.4). The framework encompasses all the services and support systems offered to an agribusiness venture, such as technology transfer, business facilitation, and technical guidance, especially those in ABI's focal areas of seed, biofuel, and farm systems. ABI also facilitates the commercialization of services that benefit farmers.

INNOVATIVE ELEMENT

ABI is the only incubator with an inclusive, market-oriented development plan that seeks to improve farmers' livelihoods through business incubation. Based on the experience gained in the years since ABI's inception, the approach has evolved to benefit the farmers through a vertical strategy (*service strategy*) and a horizontal strategy (an *outreach strategy* based on partnerships in collaborative business incubation).

The *service strategy* focuses development on strategic areas related to the mandates of ICRISAT and its partners:

- **Seed ventures.** Rural entrepreneurs receive support in developing a seed business to meet the demand for high-quality seed of open-pollinated crops. Through partnering with public and private entities, entrepreneurs are assisted in seed production, processing, and marketing (box 5.30).
- **Biofuel ventures.** ABI promotes industries involved in producing ethanol from sweet sorghum and other agricultural materials.
- **Innovative ventures.** Innovative agribusiness ventures are based on proprietary products or novel services with good market potential.
- **Farm ventures.** Contract farming, organic farming, and precision farming are among the commercial farming ventures promoted through ABI.
- **Agribiotech ventures.** ABI enables seed companies to engage in the emerging area of agricultural biotechnology by developing genetic transformation protocols for commercial crops, molecular markers for traits of interest for seed producers, and tissue culture methods for producing medicinal, horticultural, and tree crops.

The *outreach strategy* of ABI is to collaborate with organizations globally in business incubation (cobusiness

Figure 5.4 Framework for Business Incubation in ABI

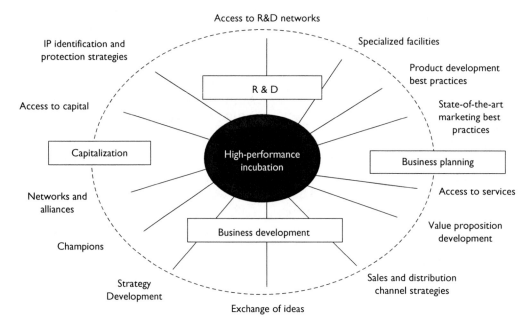

Source: ABI Strategic Business Plan 2008–13.

Box 5.30 Aakruthi Agricultural Associates: An Incubator Graduate

Aakruthi Agricultural Associates of India (AAI) was launched in 2004 as a commercial alternative to government agricultural extension services in Anantapur Province of Andhra Pradesh. AAI joined forces with Agri-Business Incubation (ABI) in 2005 and in 2006 became the second venture to graduate from the incubator.

Through the incubator, AAI developed a successful business model based on partnership with seed providers like ICRISAT, national and state agricultural research centers, and farmer franchisees. AAI designed replicable seed business ventures and proprietary methods for recruiting farmer entrepreneurs for local, low-cost, high-productivity seed multiplication. These ventures create seed delivery systems that offer an alternative to government-supported channels and allow more rapid introduction of superior varieties. AAI has built a network of 70 farmer entrepreneurs in Andhra Pradesh. These entrepreneurs pay franchise fees and receive working capital advances. They engage more than 300 farmers in seed multiplication and currently have 4,000 acres under production.

The target crops include groundnuts, chickpeas, pigeonpeas, and rice. Oil and legume seeds, in particular, offer limited commercial opportunity for multiplication and distribution due to their inherent low multiplication ratios. Through its franchise concept, AAI can sell seed of these crops in markets where demand significantly exceeds supply. Demand for groundnut seed in Anantapur, for example, exceeds

supply by 80 percent. The gap for chickpeas is 30 percent; for pigeonpeas, 70 percent.

When AAI graduated from the incubator in 2006, it became a full business partner with ICRISAT for distributing ICRISAT seed in Andhra Pradesh. AAI has compressed the time between the release of new seed and market acceptance in Andhra Pradesh from eight years to less than three years. The advantage of more rapid market penetration is significant for both ICRISAT and farmers. In 2009 the company generated revenues of 27 million rupees. Its net profit margin was 2 percent and is expected to reach 20 percent.

ABI assisted AAI with several critical elements of its development, including the creation of a business plan; provision of technical knowledge and seed science backstopping; introductions to multiple stakeholders and potential sources of financing; and introductions and links to the national research system and other public providers of technology. Arguably the most significant assistance that ABI provided to AAI consisted of increasing its credibility with government officials.

Despite these achievements, AAI's growth is constrained by a lack of external financing. In this area of development, ABI has not been able to assist its clients as successfully as it would wish, although it helped to secure financing for the company's seed processing plant, and 12,000 square feet of warehouse capacity was provided by the Department of Marketing. ABI has also assisted AAI in renting numerous local seed storage and distribution centers.

Source: AAI Annual Report 2009 (unpublished document).

incubation). The benefits of cobusiness incubation are that it provides enhanced support and services to a greater number of entrepreneurs; enables complementary business and technology development in a greater number of regions; fosters cross-border ventures and business development; provides access to a greater range of physical, technical, and other facilities for clients; improves access to a greater range of markets; offers common branding that can make clients' businesses more marketable; and maintains an inclusive, market-oriented development strategy.

Cobusiness incubation services with other institutional partners include: planning, development, and implementation of a business incubator; facilitating coordination

and operations; capacity building in business incubation operations; business consultancy support services; access by Technology Development Board entrepreneurs to seed capital; development and implementation of incubation services in the focal areas (seed, agricultural biotechnology, biofuel, other innovation, farms, and potentially other areas); and making the system successful and self-sustaining through M&E. To date, key partners for cobusiness incubation have come from the Network of Indian Agri-Business Incubators (NIABI) and from Mozambique.[1] ABI is the coordinating body for NIABI, which is implemented by ICAR under the World Bank-funded NAIP project.

IMPACT: TECHNOLOGIES COMMERCIALIZED

ABI has supported more than 158 ventures in agribusiness since 2003. Among ABI's clients, 62 percent are seed entrepreneurs, 13 percent are incubatees located on site, 30 percent are cobusiness incubatees, and 4 percent are biofuel entrepreneurs. To date, agribusiness products and technologies incubated through ABI have included sweet sorghum for ethanol production; insect-resistant transgenic cotton (box 5.31); a biofermentor for biopesticide production; a drought-tolerant groundnut variety; better-yielding chickpea varieties; biopesticide formulations; and organic farming methods. Businesses supported by ABI are estimated to have benefited 40,000 farmers.

In Andhra Pradesh and Maharashtra, 4,000 acres have been brought under sweet sorghum cultivation for ethanol production. (See other details in module 4, box 4.29 in IAP 2.) The drought-tolerant groundnut variety is used by 1,500 farmers on more than 5,000 acres; a new chickpea variety is planted in 100,000 acres in Anantapur District by 20,000 farmers.

LESSONS LEARNED

ABI has chosen a fairly risky strategy of combining new entrepreneurs with new technology—a risk that is partly offset by close linkages with world-class scientists. ABI's access to capital and commercial expertise is also somewhat less developed than that of most other incubators. Some observations on sustainability and challenges follow.

Sustainability

ABI operates as a business. An initial startup grant of US$444,444 as capital for infrastructure and US$111,111 as a recurring grant was provided by the National Science and Technology Entrepreneurship Development Board (NSTEDB). On average, ABI is generating US$250,000 per year; its annual operating expenses are around US$2 million. Since its inception in 2003, ABI's average annual growth rate has been 30 percent. ABI works on two financial models:

- *Under the capital gains model,* ABI takes an equity share in companies it incubates. This model is useful for startup companies with highly proprietary technologies and strong entrepreneurship capabilities. It requires less management support, but the new technology must be very strong.
- *The revenue-generation model* is a franchisee model in which revenue for ABI is generated through service fees, royalties, rental fees, and one-time fees. It is useful for small-scale entrepreneurs who need significant management support but do not require strong technology inputs; their business is based on incremental technologies or pure services.

Challenges in business incubation

Table 5.11 summarizes challenges that ABI has encountered since its inception and strategies and solutions used to deal with them.

Box 5.31 Agri-Biotech Incubation with Bioseed Research India

In 2003, Agri-Business Incubation (ABI) client Bioseed Research India licensed Bt gene technology from Monsanto to develop and commercialize insect-resistant cotton varieties. ABI provided its client with training in Bt breeding techniques, biosafety consultancy, and lab and greenhouse facilities. These services helped the mid-level seed company to enter the high-end seed business and gain a strong market share through early entry into the market for Bt cotton seed. The incubation service benefitted from commercialization of Bioseed's Bt technology. About 525,000 packets of Bt cotton seed were sold during the past two years. The company's varieties are grown by 200,000 farmers on 500,000 acres.

Source: ABI.

Table 5.11 Challenges Faced by Agri-Business Incubator Since Its Inception and Solutions

Segment	Challenge	Strategy/solution
Goal and objectives	– Balancing technology commercialization and agricultural development. – Profit or nonprofit entity.	– Focus on incubating enterprises oriented to commercializing technology for agricultural development. – Nonprofit status ideal under society or section 25 clause of Indian Companies Act, 1956.
Target segments	Identifying and retaining innovators is difficult.	Target niche areas to retain clients.
Agricultural technology	– Availability of technology limited to what ICRISAT can provide. – Appropriate technology not available at right time.	Provide incentives to scientist at ICRISAT for technology transfer to increase the pool of technology.
Innovations	– Limitations to innovations in R&D institutes because right enabling environment is lacking. – Individual innovators handicapped by ICRISAT's intellectual property (IP) policy.	– Organize regular innovation camps and motivational programs to scout innovations in the institute. – Encourage individual innovators through official innovation camps and recognition in a public forum by employers.
Markets	– Low pricing of end products limits attractiveness of market for envisaged enterprises. – Ag-biotech market segment is not buoyant and is constrained by ICRISAT IP policy.	Work for alternative end products to increase market opportunities.
Enterprise initiatives	– Risk inherent in technology-based businesses deters entrepreneurs and incubation. – Rural enterprises have less risk-taking ability and need high amount of management support.	– Provide entrepreneurs internships on innovation and incubation with part-time options. – Provide rural enterprises a high level of management support and a service package that includes risk coverage.
Organization and policy	– ICRISAT's intellectual property right (IPR) policy and standard material transfer agreement (SMTA) deter technology commercialization and incubation. – Stringent entry and exit procedures limit intake of incubatees.	– Liberalize IPR and SMTA policies. – Offer single-window clearance mechanism for incubation.
Operations	– Security systems prevent private clients from working late at ICRISAT and limit client retention. – Access to facilities and services for the incubatees is not always available.	– The security systems need to be flexible for private clients. – The host institute can allocate and list the facilities and services available for enabling effective service to clients.
Human resources	Limited availability of incubation professionals reduces opportunities to incubate successful ventures.	Certified short-term incubation training program offered by acclaimed universities.
Financial	– Capital investment provided by the donor must not be redundant. Allow flexible fund transfers across cost centers to fund utilization as required by industry. – Rent must be billed to incubatees on full-cost rather than partial-cost recovery basis.	– Donors' capital investment can be more flexible (confirmed clients can then request the facilities they need). – Rental revenues need to sustain the incubator.
Sustainability	– Innovative agricultural entrepreneurs mostly cannot pay for incubation services. – Institutional realignments and changes will erode the sustainability of ABI.	– Revenues from innovators must be packaged in the investment, either upfront or based on margin of profits. – Maintain a reserve of 12 months of operating funds.
Cobusiness incubation partnerships	Leadership vacuum and poor follow-up/initiative of national agricultural research systems (NARSs), primarily in Africa (Mozambique).	– Strategize with non-NARS organizations along with NARSs as consortium partners.

Source: ABI.

The China Technology Transfer Project

Josef Ernstberger, Consultant

SYNOPSIS OF PROJECT DATA

Country:	China: Shaanxi, Anhui, Hunan, Heilongjiang Provinces
Project:	Agricultural Technology Transfer Project
Financing:	About US$207 million, of which US$100 million is a World Bank loan
Implementing agency:	State Office for Comprehensive Agricultural Development
Dates:	Approved April 28, 2005; original closing date (December 31, 2010) extended to December 31, 2011
Includes support for:	About 120 subprojects.

CONTEXT

Since the end of the 1990s, China's agricultural sector has entered a phase of urgent and challenging structural transformation dictated by five major developments:

- *Lagging agricultural output and incomes.* Farm income growth had fallen alarmingly behind overall income growth. Policy priorities have shifted from concerns over food self-sufficiency and low consumer prices toward serious concerns about income growth for the farming population and the widening disparity in rural and urban incomes.

- *Natural resource pressure.* As agricultural output expanded rapidly in the 1990s, production encroached into more and more fragile ecological environments, and farmers adopted unsustainable production practices.

- *Changing demand and consumer preferences.* Rapid growth in the nonagricultural economy and growing urbanization caused changes in food preferences and consumption patterns. Demand for meat, fruit, vegetables, and other high-value commodities, including "green" and organic food, rose rapidly.

- *Demand for advanced food processing and marketing.* Increasing urbanization and differentiation of the food consumption structure demanded an advanced processing, marketing, and catering industry. Enterprises in this sector have mushroomed throughout the country. Most of these enterprises were naturally competing with small-scale farmers on profit margins, or entirely new agricultural markets emerged in which smallholders often had a relatively weak position.

- *New market challenges and opportunities.* China's accession to the World Trade Organization required rapid liberalization of trade practices and further opening up of the border. Internal production and food quality standards had to be improved and adjusted to international standards.

A critical bottleneck for the transformation required in the agricultural sector was the slow transfer and adoption of modern science, technology, and knowledge-intensive agriculture. The prevailing farming environment in China was characterized by a highly fragmented production structure dependent on a multitude of very small farms. This farm structure made it difficult: (1) to expand the use of new technologies and/or supply high-value markets that need a critical mass, farm size, or contain other critical elements to reach economies of scale, (2) to reach farmers by the traditional extension system, because the national extension system model is poorly suited to reach large numbers of farmers and meet their increasingly individualized demand for knowledge and information, and (3) for farmers to know about and respond effectively to market signals. The government-based research and extension system was not sufficiently responsive to the new challenges and opportunities presented by agricultural technologies, markets, and farmers' demands. It was supply-oriented, engaged in extending the government's programs and production targets, and had no effective means of dealing with the constraints small-scale farmers encountered in adopting new technologies.

The Technology Transfer Project responded to strategic concerns in China's agriculture by providing a learning platform for developing innovative models for public sector facilitation and support, including fostering better PPPs in agriculture. The primary addressee of the project was the public sector's agricultural support and development system, in particular the State Office for Comprehensive Agricultural Development (SOCAD), which is China's main funding institution, disbursing about US$2 billion each year for agricultural development in the country. A significant amount of this budget supports agribusinesses. Any improvements in targeting these funds and improving funding modalities would therefore have a tremendous impact.

PROJECT OBJECTIVES AND DESCRIPTION

The project's overall development objective was to develop and test innovative models for agricultural technology transfer and application aimed at generating additional farm income with potential for scaling up. In this way the project sought to give poor farmers a chance to participate in high-value agricultural markets, domestically and internationally. This objective was to be achieved by supporting viable models for restructuring and modernizing agricultural production, processing, and marketing through various forms of vertical and horizontal integration, the introduction of innovative technologies, and new institutional arrangements and PPPs.

Since mid-2005, the project has been underway in four provinces (Anhui, Hunan, Heilongjiang, and Shaanxi), which includes the Yangling High Technology Agricultural Demonstration Zone. The initial closing date of December 2011 was extended by one year to provide more time for analysis and to learn from experience. The main components of the project are:

1. *Technology transfer and information markets and services,* which includes the building of technology transfer markets and exhibition and demonstration facilities.
2. *Promotion of commercially attractive key technologies and new institutional arrangements,* including (1) researcher-investor-farmer technologies, which involved the partial financing of typically tripartite joint ventures with the objective of designing successful

investment models in which researcher-investor-farmer partnerships are tested; (2) targeted technology transfer, which financed technology transfers targeting farmer groups who did not have adequate access to information, capital, or decision-making power to adopt technologies on their own; and (3) public support programs, which involved financing activities that enabled the private sector to realize its role in commercializing innovative technologies (such as food testing and certification or use of IPRs) and technologies that did not appeal to the private sector on commercial grounds but had a clear public good nature (such as water-saving technology or waste treatment).

PROJECT INNOVATIONS

The *first innovation* of this project was that, unlike traditional public sector support projects, it combined public funding for research, extension, training, and institution building with private investment. The government funded the development and dissemination of public goods, but these activities were implemented by the private sector to foster the integration of public and private investments (World Bank 2010).

The *second innovation* of this project was to focus on technologies that increase smallholders' incomes. Often public investments in agricultural research raise smallholders' productivity but not necessarily their incomes (World Bank 2010).

The *third innovation* was to develop and fund institutions such as farmer associations as part of the "technology package," provided they would improve the dissemination of new technologies.[1] In addition, under all subproject proposals involving commercial enterprises it was made mandatory that at least 50 percent of the subproject funding would be used to directly support farmers in providing the raw material for the enterprises (production base or farm outreach).

The *fourth innovation* was to design this project specifically as a learning platform. Recognizing that a single project could have only a limited impact in a country as big as China, the project was designed as a stepping-stone to improve the effectiveness of public investment and partnership arrangements by partnering with private players. For this purpose a framework for M&E was designed to extract and disseminate lessons from this project, particularly in two areas: (1) the understanding and rationale for public funding (why and in what areas is it justified to use public funds for agricultural development) and (2) the improvement of procedures for the use of public funds (such as contracts with private partners, working with farmer cooperatives and associations, and similar arrangements) (World Bank 2005–10).

BENEFITS, IMPACT, AND EXPERIENCE

Many subprojects supported under the project generated remarkable benefits in terms of value added, additional income, or entire new lines of business with new products and markets. These achievements were based on: (1) innovative institutional and partnership arrangements (for example, public sector research with companies or farmer organizations or research-farmer or research-company arrangements); (2) innovative funding arrangements; and (3) innovative technologies fostered by these new partnerships. However, the project was designed as a learning project. As such its achievements in terms of value-added or additional income cannot be seen as an end but provide only the tool for learning and drawing more generic lessons. The project developed an analysis and lessons learning framework (World Bank 2005–10), but the evaluation process continues, with a consolidation of results yet to come. Even so, some key outcomes have been identified.

National and provincial agricultural support programs have already adopted several design elements tested under the project. A number of policy documents have been prepared, which influenced SOCAD's funding policies, including:

- The combination of company and farm outreach support under a joint investment.
- Importance of farmer associations as new and valuable partners in PPPs.
- PPP arrangements to be guided by a clear understanding of public good outcomes to be specified in contractual arrangements with objectives, indicators, and milestone-based targets.
- Monitoring of PPP contracts and performance as a crucial success factor.

Models have been developed with innovative management approaches and technologies for women farmers. The outmigration of male labor from many rural areas in China increases not only the demand for technologies suitable for women but the need for women to develop their managerial skills as farm operators and entrepreneurs. The project developed a model partnership with the All-China Women's Federation that successfully addressed these needs.

Technology barriers caused by fragmented production and small farm size were overcome by partnerships between companies and farmers or farmer associations. For example, organic food production or integrated pest management technology were difficult for individual famers to adopt because their small holdings prevented them from exploiting commercial opportunities. However, in groups and in combination with a strong processing or marketing partner, smallholders could be linked with commercial opportunities.

New products and markets, which require the commitment of multiple stakeholders over time to develop and reach, were opened by involving the public sector as a broker. A typical example was the development of camphor production in one area, where an investor used the support of the local government and the project to organize farmers to plant sufficient trees and build a critical mass of production required for a viable processing facility.

Many subprojects show that organizing farmers into associations or under company-farmer arrangements enables them to enter into higher-value production through branding, product certification (green or organic certification, for example), or accessing new markets (especially export markets). In some instances, farmers were organized because companies were the driving force, but in others farmer associations alone achieved these objectives.

Contractual arrangements between the government and private entities introduced under the project showed that private institutions could successfully deliver public goods and services. Many subprojects involve farmer associations or companies providing animal health services, training farmers, or providing other extension services. Most project proposals target poor farmers or disadvantaged groups. Several proposals have involved innovative technologies and environmental services, such as the treatment of manure, waste, or crop residues.

Although the private sector is assuming many functions traditionally performed by government institutions, the project has demonstrated that the government remains central to areas such as food safety, protection of property rights, regulations, and policy direction. Project investments in those areas (for example, in food quality standards testing and certification in Yangling, or testing for the presence of genetically modified ingredients and certification in Anhui) show the importance of the government in relation to the private sector in agricultural development.

LESSONS LEARNED AND ISSUES FOR WIDER APPLICATION

The summary of impact and experience gives some indication of the lessons that the project has already provided.

The sections that follow explain some of these lessons in greater detail. They concern the importance of aligning project objectives to the context, supporting multiple stakeholders at the same time to attain such goals as a more equitable sharing of benefits, and the importance of assessing and matching potential partners with care.

Importance of objective and context

One of the most critical success factors for this project was that it was based on a critical analysis of the sectoral context and a vision for the sector's development. For example, the decision to aim for income and value generation versus quantity and productivity increases had far-reaching implications for the design and approval of individual subprojects. Similarly, the decision to emphasize farmer-based investments for farmer groups or in combination with enterprise investments was based on the recognition that farm incomes had high priority given the wide urban–rural income disparity.

Supporting multiple stakeholders under one investment

Traditionally a kind of a "trickle-down" effect was expected when either companies or research institutions were supported through project funding, with the expectation they would have a positive influence on farm incomes. In many cases, this effect did not occur. The project has taken a probably unique approach in combining funding for different groups of stakeholders in a single investment, often under a joint business plan. Different arrangements such as company-farmer, research-farmer, or research-company-farmer were supported. The main advantages of this approach were as follows:

- *It reduced the risk* that companies would use public funds to crowd out farm production and important sources of farm household income for high-value products (for example, that a meat processor would run its own pig farm). In line with the overall rural development objective, the project forced companies to enter agreements with farmers or farmer organizations to source their raw material from farm households and allocated at least the same amount of funding to support farmers (with technical and management training, basic infrastructure, and other resources).

- *It controlled the sharing of benefits.* Supporting companies and farmers under a joint investment makes it possible to monitor and control the distribution of benefits.
- *It facilitated the commercialization of research results.* Linking public research (individuals as well as institutions) with companies or farmer groups and associations in a joint business undertaking helps to ensure that research results are directly relevant to business. Different arrangements have been tested successfully, such as contracts between researchers and companies or farmer associations, or participating researchers and research institutions as shareholders.

Another lesson from the project is that the Chinese political and administrative environment made it difficult to select proposals based on competitive procedures. A matching grant system was chosen as the preferable approach to select private partners and suitable proposals. The selection and approval of proposals was often combined with a "negotiation process," in which many proposals were modified to sharpen the public good outcomes and develop them into PPPs (in other words, elements of a solicitation process were incorporated).

Choice of companies and private partners

Initially several companies applying for support perceived the project as an opportunity to receive public funds to resolve a difficult business situation—startup companies or companies that could not obtain additional commercial financing. Many of these companies could not afford to wait until their proposal had been vetted and funds became available, so they dropped out of the process. The *key lesson* from this experience is that a project needs to provide sufficient design clarity on whether it aims to act as a business support and development project or a project aimed to support PPPs, in which case:

- *Private partners need to be thoroughly appraised* for their financial capacity, management skills, technical know-how, and experience.
- *The PPP should not be the primary determinant* of a partnering company's financial success.
- *Companies with a track record of successful businesses* in a field relevant to the PPP objective provide the highest chance for a successful partnership.

Agricultural Cluster Development in Nicaragua

Nuria Ackermann, United Nations Industrial Development Organization (UNIDO)

SYNOPSIS

This Austrian-funded UNIDO project (total project budget of approximately €1.5 million) was implemented between 2005 and 2008 in Nicaragua to foster the sustainable development of 14 agricultural and agro-industrial clusters. The project's innovative element was its strategy for delivering technical assistance, in which local institutions were subcontracted to act as cluster brokers. This approach considerably reduced the risk that cluster activities would falter after project closure because the capacity to sustain such activities was lacking. Support institutions can more easily cater to the needs of a well-organized cluster than of isolated actors, because the cluster enhances the effectiveness and client orientation of their services. In Nicaragua, cluster development turned out to be a strong catalyst for mobilizing additional resources for private sector development, especially for infrastructure and the strengthening of local productive capacities and organizations.

CONTEXT

Small and medium-sized agricultural and agro-industrial production units account for a significant share of employment and income in Nicaragua, but their small size and limited resources trap them in a vicious circle of low productivity, cutthroat competition, and low or even decreasing incomes. UNIDO's current strategy to overcome these problems is to promote clusters, which are geographically defined business systems where producers specialize in the same or related production activities. UNIDO's entry point for fostering systemic competitiveness in a specific cluster was to facilitate the generation of trust and social capital.

PROJECT OBJECTIVES AND DESCRIPTION

UNIDO's approach placed particular emphasis on the role of cluster brokers, who could mediate between stakeholders and trigger synergies between various territorial initiatives, such as infrastructure development, capacity building and skill development, adjustments of the regulatory framework, access to capital, technology transfer, and promotional activities. Technical assistance sought to facilitate partnerships as well as to link existing or planned initiatives so that they would not be implemented in a parallel or uncoordinated manner.

In the project's first phase (2003–05), 11 pilot clusters received assistance. Best practices identified during this period helped adapt the cluster development methodology to the Nicaraguan context. In the final phase described here (2005–08, with €1.5 million from Austria), the project focused on strengthening participatory processes in target clusters, defining joint strategies, and implementing actions that contributed to competitive growth and the exploitation of market opportunities. More strategically, it reinforced local stakeholders' capacity to lead cluster development and provide continuity once the project ended. It also helped to increase the application of the knowledge of productive linkages gained through the project by developing a critical mass of academics and policy makers to apply this learning.

Indirect assistance to clusters through brokering institutions

The innovative element in this phase of the project was that UNIDO personnel stepped back from direct intervention as cluster brokers and selected a local civil society entity to assume this role in each cluster. UNIDO provided comprehensive classroom and on-the-job training based on its cluster development methodology to the employees of these brokering institutions. In this way, the brokering institutions could obtain the skills required to continue their work as facilitators after the project ended.

Clusters were selected for development based on field surveys that gathered information about the concentration of enterprises in the cluster, existence of a local identity and shared culture among cluster stakeholders, the degree of organization within the productive sector, the motivation of entrepreneurs and local institutions, the existence of complementary projects in the locality, the potential for third-party funding for cluster activities, and market demand and trends for products produced by the cluster, among other criteria. The 14 clusters eventually selected belonged to a number of subsectors, including the cocoa, dairy, livestock, coffee, and banana subsectors.

Brokering institutions were selected through competitive bidding. Selection criteria included being active in the cluster but perceived as neutral by stakeholders, having strong ties with or a good understanding of the productive sector, and having qualified, experienced human resources. After training in the cluster development methodology, the brokering institutions implemented it in their respective clusters. UNIDO provided backstopping throughout this on-the-job training, which was critical to capacity building and to adapting the approach to the local context. Brokering institutions were enabled to take stock of parallel activities to support the SMEs in the target clusters and align actors and activities in ways that would enhance each cluster's performance.

UNIDO mainly financed activities to facilitate collaboration and synergies in territorial initiatives. In a few cases it cofunded competitiveness-enhancing activities to trigger imitation effects. The search for public and private funding was a key responsibility of the brokering institutions. An advantage was that UNIDO could draw on or leverage a number of complementary initiatives in the target clusters (box 5.32).

Capacity building for representatives of political and academic institutions

UNIDO organized specific seminars for university teachers and political actors at the national level to increase their awareness of the importance of fostering systemic territorial competitiveness. The aim was to ensure that a critical mass of opinion and decision makers would take ownership of the cluster approach and continue promoting activities after the project's closure. Between June 2006 and November 2007, UNIDO organized seven seminars on theoretical aspects of cluster promotion for 25 professors of the economics departments of 8 prioritized universities and 20 representatives of the public sector. The training topics were established in a participatory manner with the academic counterparts, since the main objective was to build the capacity of university professors to lecture on cluster development and motivate them to establish closer links between the universities and the productive sector.

BENEFITS, IMPACT, AND EXPERIENCE

The benefits, impact, and sustainability of indirect assistance with brokering and capacity building for academicians and policy makers are described in the sections that follow. The discussion is supported by case studies.

Box 5.32   Complementary Cluster Development Activities Reinforced the UNIDO Project

- *A conducive policy framework,* including a National Development Plan focused on sector-specific and territorial enhancement of competitiveness and the Presidential Competitiveness Commission, an executive body for public-private dialogue.
- *Infrastructure development,* including transport, water, and electricity infrastructure developed by the National Energy Commission, Inter-American Development Bank, and the Government of Japan.
- *Institutional strengthening,* including the establishment or development of local cooperatives as well as the strengthening of national producer organizations.

- *Other technical or financial assistance projects,* focusing on upgrading quality, facilitating market access, and strengthening design and technological capacities among producers and local authorities (variously funded by Germany, Finland, the United Kingdom, and European Union).
- *Activity by nongovernmental and civil society organizations* engaged in strengthening local cooperatives, production processes, and product quality (through training and diagnostics) as well as market access (promotional activities), including Oxfam, HORIZONT3000, and the Interchurch Organization for Development Cooperation.

Source: Author.

Indirect assistance to clusters through brokering institutions

The brokering institutions performed their new roles well in the 14 clusters. Although the activities varied by cluster, the UNIDO methodology was generally followed. In most clusters, the brokering institution conducted a participatory diagnostic study to identify the main bottlenecks of the cluster. By involving all of the main local stakeholders, this exercise helped to create an initial base of trust. The brokering institution usually went on to promote the establishment of a cluster commission or territorial governance board that integrated public and private actors. The brokering institution also provided training and backstopping to members of the cluster commission and helped them design and implement a joint action plan to improve cluster performance.

Most brokering institutions started promoting joint activities among cluster stakeholders even before the commission had been set up to achieve "quick wins," to sustain the actors' sense of motivation, and to increase social capital over time. Examples include the establishment of joint farmer sales networks, vertical networks between primary producers and processors, the organization of local trade fairs, the adaptation of financial services to smallholders' needs, and the introduction of environmentally and economically sustainable irrigation systems (boxes 5.33 and 5.34 provide specific examples).

After the project ended, public-private collaboration continued or increased in 8 of the 14 clusters. In the other clusters, communication between stakeholders remained fluid, and levels of trust were preserved, but the cluster commissions/governance boards and brokering institutions stopped performing their leading roles and the cluster lost momentum.

Beyond the specific assistance provided under the project, cluster initiatives unlocked a number of investment and economic opportunities. By engaging in business partnerships and networking, firms could pool assets, increase their advocacy, and mobilize local, national, and international stakeholders

---

Box 5.33   The Livestock Cluster in Chontales

In the Nicaraguan department of Chontales, around 5,350 producers raise livestock for meat on a small scale. The brokering institution for this cluster was the Center for Entrepreneurial Management (CEGE), which belongs to the National Union of Agriculture and Livestock Farmers (UNAG). UNAG had around 3,000 members in Chontales alone and was very active in supporting farmers both technically and financially.

In 2006, after a participatory diagnostic study, CEGE facilitated the creation of a cluster commission, which was composed of representatives of the various cattle farmers' associations, local universities, the biggest local slaughterhouse, various municipalities, the police, and other public institutions. Under CEGE's guidance, these representatives prepared a joint action plan and implemented concrete activities to overcome the bottlenecks. Project participants achieved the following results:

■ *Small-scale cattle farmers were struggling with the nontransparent grading and pricing system* applied by local slaughterhouses. CEGE helped to mediate a stable supply agreement between five producers'

associations and the biggest local slaughterhouse. The agreement specified the quality standards and a premium for cattle meeting those standards. Producers raised their incomes; the slaughterhouse gained a consistently good source of supply.

■ *Cluster actors collaborated against widespread livestock theft.* Three municipalities engaged livestock guards (typically volunteers trained by the local police and financed by the collaborating producers).

■ *The largest slaughterhouse (87 percent), UNIDO (7 percent), and the local university (6 percent) sponsored research by three university students on slaughtering and boning capacities.* The results motivated the slaughterhouse to fully finance the expansion of its slaughtering capacity from 350 to 550 cattle per day.

CEGE continues to operate as the local brokering institution. The cluster commission remains active and has expanded its scope. Because most cattle farmers in Chontales are also milk producers, CEGE and the cluster commission are fostering public-private linkages in both the local livestock and dairy sectors.

Source: Author.
Note: CEGE = Centro de Gestión Empresarial, Universidad Centroamericana de Nicaragua; UNAG = Unión Nacional de Agricultores y Ganaderos.

Box 5.34 The Banana Cluster in Rivas

A significant share of Nicaraguan banana production is concentrated in the department of Rivas, where around 4,000 small-scale banana producers operate. The Banana Producers Association in Rivas (APLARI) was the brokering institution for this cluster. APLARI provided assistance to its 500 members to improve production and marketing. In this case the cluster commission included members of banana producer organizations, local representatives of the Ministries of Agriculture and Transport, the police, the army, local universities, financial institutions, and international cooperation agencies operating in the area. APLARI helped the commission members prepare a joint action plan, in which the value of the planned activities amounted to US$850,000. Of this, 5 percent was financed by UNIDO and the remainder came from APLARI, the government, the European Union, the United Kingdom, Inter-American Development Bank, and others. Some of the main results included:

■ *APLARI helped institutions to adapt their support services to producers' technical requirements.* One result of this realignment was that producers started to use the local university's soil testing service.

■ *The Ministry of Agriculture had tried to introduce "best agricultural practices" among banana producers in Rivas but had failed because of a lack of direct communication with the productive sector.* Thanks to APLARI's intervention and European Union support, 18 leading producers agreed to invest in these practices on their farms. APLARI and the Ministry of Agriculture provided training and technical assistance to the farmers. These producers received higher prices for their bananas and served as a model for other farmers.

■ *APLARI supported four small producer cooperatives to establish a sales network to increase their supply volume and bargaining power.* The cooperatives started selling jointly to a Honduran banana chip company and a high-end supermarket chain in the capital. The incomes of producers involved in these sales agreements had increased by up to 50 percent by the end of the project.

■ *Banana producers on the lake island of Ometepe found it difficult to sell to clients on the mainland.* The only private boat company charged extremely high fares for transportation and did not respect the time schedule. Intervention by public authorities (in particular the National Port Operator) fixed transportation fees and restored adherence to schedules. Port facilities were upgraded and expanded. The improved boat connection with the mainland benefited not only the producers in Ometepe but facilitated the arrival of tourists to the island.

When the project ended, APLARI and the cluster commission continued promoting public-private activities in Rivas to strengthen the performance of the banana cluster.

Source: Author.
Note: APLARI = Asociación de Plataneros de Rivas.

whose support would be out of reach to individual stakeholders. The agglomeration of local private enterprises around a clear set of development objectives facilitated public-private dialogue, as it provided the private sector with a clear agenda and legitimate leaders. Support institutions could more easily cater to a well-organized cluster than to isolated actors, so the effectiveness and client orientation of their services improved.

Capacity-building for representatives of political and academic institutions

At the cluster level, municipalities and other public actors were actively involved in cluster commissions and began to implement activities with the productive sector to foster systemic competitiveness. At the national level, however, newly elected authorities preferred terminology associated with a value chain approach, which may alter perceptions of the cluster approach.

It had been envisaged to target university headquarters for seminars on the cluster approach, as they could establish the curriculums of their regional branches and foster wider dissemination of the approach. In practice, the regional university branches were much more receptive to the cluster approach, mainly because as cluster members they felt more committed to local economic development.

Several regional branches of universities trained by the project established links with the productive sector in their territory by becoming active members of the cluster commissions. Seven of eight universities participating in the seminars included cluster topics in their curriculums and had introduced 683 undergraduate and 121 graduate students to the cluster approach by the end of the project.

LESSONS LEARNED AND ISSUES FOR WIDER APPLICATION

A *key lesson* was that when the brokering institution was a local producer association, facilitation continued after the project ended. These types of organizations were well embedded in the local context and benefited from a high level of legitimacy with those they represented. Producer associations sustained their role as facilitators because they perceived it as beneficial to members and organizational objectives. In contrast, brokering NGOs often stopped working once the subcontract with UNIDO had ended and funds to continue their work were not forthcoming. Unlike the producer associations, the NGOs had a broader scope of objectives, and their missions were less linked to a specific territory and its producers.

Other lessons include:

- *UNIDO's long-term engagement in Nicaragua,* combined with its wide-ranging experience in business network and cluster development in Nicaragua and the region, were instrumental in developing a tailor-made training methodology and ensuring an innovative and sustainable project implementation strategy.

- *Given the limitations of time and funding, cluster development must target a limited number of clusters* that become instrumental for showcasing the benefits of the approach and transferring competencies to local institutions that can replicate the approach. The selection of target clusters—based on clearly specified and agreed upon criteria—is critical, because they can have a demonstration effect extending beyond their immediate concerns. Careful selection of the brokering institutions and the specific employees that would act as brokers on the ground also contributed significantly to success.

- *Social capital can develop in a cluster only if its stakeholders clearly perceive the advantages of joint action.* It is important to focus from the very beginning on identifying and implementing "quick win" joint actions to increase motivation and expand cooperation.

- *The combination of classroom training, learning by doing, and mentoring proved crucial* for learning the approach and adapting it to specific local needs. Training in fund raising was not sufficient to guarantee sustainability once the project ended, however.

- *The involvement of local university branches* helped to disseminate the cluster development approach and helped universities form an integral part of the cluster.

A final lesson is that the UNIDO approach (based on the generation of trust, establishment of business linkages, and formation of PPPs) can unleash a cluster's growth potential but cannot create potential where none exists. Social capital can help to leverage limited resources, but local development will still be hampered if a territory lacks sufficient funds to mobilize.

Providing Farmers with Direct Access to Innovation Funds

Bernard Triomphe, Agricultural Research for Development, France (CIRAD)

Mariana Wongtschowski, Royal Tropical Institute (KIT)

Anton Krone, SaveAct, Prolinnova South Africa

Ann Waters-Bayer, ETC Foundation

David Lugg, FAO

Laurens van Veldhuizen, ETC Foundation

SYNOPSIS

Specific grant programs—referred to here as "Farmer Innovation Funds"—can be designed to provide direct, fairly simple competitive access to small grants or loans for individual farmers or farmer groups, businesses, or other stakeholders who wish to adapt, develop, or adopt innovations and business initiatives on topics and issues of their own choosing. Access to such funding allows a wide range of innovations to be tackled, and under proper conditions may expand enthusiasm and innovation capacity among smallholders, other rural stakeholders, and those who support them. Different funding schemes have been tested and adapted in several countries throughout Africa, Asia, and Eastern Europe, with specific objectives, types of farmers, setups, grant sizes, and screening and support mechanisms. The funding schemes are highly dynamic, evolving in response to changing circumstances and experience gained. Farmer Innovation Funds work better if and when decentralized settings are used and when support institutions have the necessary skills and experience to implement them. Funding mechanisms can be made more sustainable by linking them with savings and credit schemes and structures (should they exist) and/or by embedding them within existing agricultural R&D institutions and mechanisms for fostering innovation. Farmer Innovation Funds are most powerful when they are not implemented in isolation but as part of systemic, long-term efforts to promote and strengthen sustainable farming, participatory innovation development, and dynamic innovation systems and processes, in which the roles and skills of various stakeholders (particularly smallholders) are recognized and supported.

CONTEXT: WHY PROVIDE INNOVATION FUNDS DIRECTLY TO FARMERS?

Although efforts have been made to provide public funding to foster innovation among a diverse group of stakeholders through competitive bidding (see, for example, World Bank 2010), such funds still tend to be allocated primarily to research and extension institutions or other formal actors in the agricultural sector (such as large NGOs), partly as a result of the high administrative and technical requirements for accessing the funds. Consequently, such institutions and actors retain an overwhelming influence and control over the main decisions related to who should benefit from such funding, how the innovation process is organized, what types of activities are implemented, and by whom. Conversely, farmers and other stakeholders involved rarely have direct access to (and hence have little to say about) funding to implement their own ideas about which innovations o explore. In most cases, farmers receive limited financial support to compensate them for the cost of their participation in specific activities being funded (such as working on experiments, linking with other actors, and so on) or to motivate them to try out new technologies developed by others.

PROJECT OBJECTIVES AND DESCRIPTION

This profile examines how funding schemes to support farmer innovation ("Farmer Innovation Funds," FIF) can be designed and what lessons can be drawn, based on two sets of experiences:

■ *Local Innovation Support Funds (LISFs)* were initiated under the Prolinnova[1] network to test if and how research and innovation funding could be channeled to, governed by, and accessed by small-scale farmers through small grants (typically a few hundred dollars or less) for developing innovations of their own choosing. LISFs specifically target poor and vulnerable households and focus more (but not exclusively) on local ideas and technologies (existing or new), depending on what farmers actually want to achieve. LISFs have been operating on a pilot basis in several low-income countries in Africa and Asia over the past five years.

- **_Competitive Grant Programs (CGPs)_** focus on commercially oriented, small- to medium-size farmer groups and small rural businesses. Although CGPs work with poor farmers, they prioritize commercially oriented ones. The CGP focuses on business and market-oriented activities and emphasizes adapting and adopting existing technologies (but not exclusively). CGP grants are generally much larger than LISF grants (typically US$10,000 or more). Grants include funding for investments to set up the innovative activity, for external technical assistance, and for technology transfer and demonstration to other farmers and stakeholders. A CGP initially operated in Albania, and similar schemes are being implemented in Armenia, Azerbaijan, and Kazakhstan.

Both funds have some generic features. They are both designed to provide a diverse spectrum of beneficiaries or clients (including farmers, land-users, and rural businesses, either as individuals or as members of groups) with easy access to relatively modest grants or loans allowing them to develop, invest in, and strengthen initiatives and innovations that they consider worth pursuing and which also have the potential to produce public goods such as economic growth and rural employment, social equity, and ecosystem services.

The funds can be used for various purposes and types of innovations, including technical ones (natural resource management, for example, or improved production, processing, or transformation of produce), organizational ones (such as better access to input, service, and produce markets), and institutional ones (such as creating new institutions and rules or transforming existing ones). To fulfill this purpose, funds may be used for implementing diverse types of activities: experimenting on a smallholder's own farm, engaging in joint experimentation and other activities by farmers and other stakeholders (researchers, extension agents, and so on), transferring existing technology, or sharing and disseminating successful experiences. In doing so, the aim is also to strengthen the individual and collective

capacity of the fund recipients to innovate and to increase their overall contribution to and participation in the innovation process.

Funds are administered by small multistakeholder committees or secretariats (usually with 5–10 members but no more than 2 or 3 in the case of the CGPs) in charge of organizing the calls for proposals and creating sufficient awareness about the fund, clarifying the funding modalities (grant size and cofunding share, interest rate if a loan is involved, and so on), screening applications in a formalized and transparent way, and overseeing the effective disbursement of funds (adapted to the financial services and circuits available to the applicants).

In most cases, field days or innovation fairs and/or commercial radio or TV programs (in the case of CGPs) are organized to share the results obtained by farmers through their fund-supported activities. The intention is to increase awareness about the funds and motivate more farmers to apply for the next cycle of funding.

Other key activities typically include capacity building for those who handle the fund at the local level as well as representatives of organizations supporting farmers' innovation. A typical fund program also seeks to establish an enabling environment for implementing grants, allowing careful M&E, and ensuring effective learning and sharing with members of the FIF committees and with relevant agricultural R&D institutions and policy makers. These efforts are aimed at creating awareness and support for the fund program's longer-term sustainability.

INNOVATIVE ELEMENT

Farmer innovation funds present a handful of innovative elements:

- They are designed to be easily accessible to small-scale farmers and other stakeholders through simple application forms and procedures, simple fund disbursement modalities, support provided to farmers to fill in application forms and meet eligibility criteria, and the possibility of applying as individuals or groups.
- They are meant to solve problems and to test innovations defined and chosen freely by the applicants themselves.
- Some FIFs (such as LISFs) strive to give farmers a prominent role in fund governance, including setting up criteria for selecting applicants, screening proposals, and M&E.
- FIFs have a relatively light administrative structure, so that over time the corresponding costs are reduced and

bureaucracy is minimized, allowing timely response to applicants and disbursement of grants.

BENEFITS AND IMPACTS

The funds have reached several thousand farmers to date. Table 5.12 summarizes information on numbers and amounts of grants made in several countries in 2005–10.

Impact assessments of the LISF program are being carried out. Initial evidence in Ghana and Ethiopia indicate that

many "new" innovations are emerging as a result of LISFs. Not only the farmer innovators but also other farmers who benefitted from sharing results are reporting higher crop and livestock productivity and increased savings and incomes. In both Ethiopia and Ghana, different stakeholders state that more use is being made of participatory approaches to extension work in the zones where the LISFs operate. For examples of innovations explored through FIFs, see box 5.35.

CGPs have proven effective in supporting farmer groups and emerging rural businesses to introduce, test, and

---

Table 5.12 Key Characteristics of Farmer Innovation Fund Grants Made in Several Countries, 2005–10

Country	Period covered	Applications received	Percent approved	Loan or grant?	Award size (US$)	Who are the applicants?
LISF scheme						
Cambodia	2005–09	193	69	Loan	10–100	Individuals filtered by group
Ethiopia	2005–09	109	43	Grant	<100–300+	Both individuals and groups of 4–5 persons
Ghana	2008–09	80	43	Grant	30–300	Mostly individuals
Kenya	2008–09	103	22	Grant	50–250	Mixed/unisex groups and individuals
Nepal	2004–09	63	38	Grant	50–750	Mostly individuals
South Africa	2005–09	65	23	Grant	700–2300+	Mixed/unisex groups and individuals
Tanzania	2008–09	25	64	Grant	500–1000	Group applications only
Uganda	2005–08	98	68	Mostly loan	25–120	Initially groups, later also individuals
CGP scheme						
Albania	2002–08	656	22	Grant	3,000–15,000	Groups and associations
Armenia	2006–09	276	20	Grant	Up to 20,000	Groups, associations, small businesses
Azerbaijan	2007–10	279	22	Grant	10,000–30,000	Groups, associations, small businesses

Source: Authors.

Box 5.35 Innovation Themes Explored in the Local Innovation Support Funds and Competitive Grant Programs

Crop and animal husbandry. Examples include devising inexpensive animal rations by replacing externally bought feed with locally available feed, treating animal disease with local plants, selecting germplasm adapted to local conditions, controlling bacterial wilt in enset (false banana), devising effective water-harvesting methods, improving apple and peach production technologies, and using plastic mulches in vineyards.

Processing and storage. Examples include vegetable preservation, improved sheep cheese production and brand marketing, and improved onion storage.

Improved quality and marketing. Examples include collection and standardization of olive oil, improved

lean-meat pig production and marketing, improved packaging of aromatic and medicinal herbs, and improved packaging and marketing of honey.

Development of niche markets. Examples include production of honeybee feed, production of saplings for forest and ornamental trees, and production of aromatic and medicinal plants.

Sustainable natural resource management. Examples include increasing biodiversity and combating deforestation through regeneration of an endangered native tree species of economic value.

Social innovation. Examples include organization of groups for developing innovations and improving savings and credit schemes.

Source: Compiled from several LISF and CGP reports.

demonstrate innovative technologies to a broader audience of potential rural entrepreneurs and beneficiaries. About 85 percent of the direct grantees (for example, 700 farmers in Albania) experienced an increase in yearly income and were likely to continue their activities after completion. Over 20,000 farmers were directly exposed to new technologies through the technology transfer activities, with an estimated 3–5 emulators per grant at completion and an additional number likely to adopt and possibly adapt the technologies in subsequent years (boxes 5.36 and 5.37 provide examples of a CGP and an LISF case).

LESSONS LEARNED AND ISSUES FOR WIDER APPLICATION

A number of lessons from the Prolinnova experience with FIFs may be useful in designing similar interventions. They are summarized in the sections that follow.

Ensure that funds are used for their intended purpose

Farmer funds are meant to support innovation or promote the adoption of new, relevant technologies (rather than

---

Box 5.36   An Example of a Competitive Grant Program Grant: Improved Onion Storage and Marketing in Albania

A local farmer association asked for a competitive grant to build a small onion storage facility to lengthen the marketing period and obtain higher prices during the off-season. During the grant period, the association increased from 7 to 32 members, each with about 0.15 hectares of onions and total production of around 350 tons per year, and signed a contract with a trader in Tirana to purchase the onions. Around 37 tons could be stored at a time, with further investments planned to increase storage capacity and to purchase a vehicle for distributing the produce. Other activities included purchase of a sprayer and irrigation pump for use by members and drying onion seed for planting. During the grant period, dissemination included two workshops, five training days, three publications, and a local TV broadcast focusing on various aspects of onion production and marketing. This grant eventually resulted in linking production with markets and contributed substantially to the development of a viable farmer association in a remote corner of Albania.

Source: World Bank 2011.

Box 5.37   An Example of a Local Innovation Support Fund Grant: Propagating *Podocarpus* in Ethiopia

In the highlands near Ambo in Ethiopia, communities rear livestock, produce crops, and plant trees for food and income. One tree genus of socioeconomic importance is *Podocarpus*, a conifer that produces good timber. These trees are becoming extinct because of high demand and the long dormancy of the seed, which takes up to a year to germinate. A farmer, Jifara Workineh, applied for and obtained an LISF grant to test various germination methods with the aim of shortening the dormancy period and regenerating the tree population in his community. The LISF grant provided him with the required material inputs. Jifara eventually developed a successful method of reducing the dormancy period by placing seed mixed with soil in a polybag, burying it in a hole, and providing sufficient regular water. The method resulted in a high germination rate (85 percent) and reduced the dormancy period from over a year to three weeks. Based on these results, Jifara received an award from the government, which raised his self-esteem; his income increased from selling seedlings; and the community's stocks of *Podocarpus* have increased. In addition, researchers' and especially extension workers' attitudes towards farmers changed, as they now recognized the contributions of local farmers to local solutions using mostly local resources.

Source: Prolinnova–Ethiopia, personal communication.

---

purchasing inputs alone). They generally achieve the intended objective if good overall management of the program is ensured, local community-based organizations and farmer organizations are strongly involved, proposals are selected according to clear criteria and procedures, and grants are formalized through signed contracts.

With respect to LISF schemes, greater clarity about fund use emerged gradually, after some initial confusion about whether LISF grants could cover input costs. With respect to CGPs, three elements are taken into consideration in the grant: investment, technical assistance, and technology transfer. Operating costs and purchase of inputs are normally part of the beneficiary's contribution, unless they are clearly related to the demonstration function of the grant.

Target the funds carefully

The best results are observed when funds are set up to target preexisting community-based organizations, farmer groups, and institutions that have prior experience with participatory approaches, and when good support institutions or service providers are selected.

Support institutions often face significant challenges in terms of their ability to provide sufficient initial mentoring to farmers and other grantees and then to step back and allow fuller appropriation of the scheme by local actors. Support institutions also need to develop the capacity to attract and involve major "conventional" agricultural R&D actors, a recurrent challenge for NGOs engaged in FIF schemes. In particular, it is critical to bring local and national extension and research partners on board to facilitate the scaling-up of the process and the results of such schemes.

Value added of farmer innovation compared to joint innovation by farmers and agricultural R&D

Localized innovation with minimal support from outside generally yields results that can be readily understood and available to neighbors of farmers benefitting directly from FIF grants. It is also a good way of empowering farmers and strengthening their capacity to engage with the formal agricultural R&D environment.

For its part, joint (multistakeholder) innovation deriving from structured and systematic interactions between farmers, researchers, and other actors in agricultural R&D is geared more toward generating results that can be scaled up with greater certainty. It is more costly and riskier than local (farmer) innovation, as it invariably takes time to assemble the right mix of partners and skills and to ensure it is

adequately resourced. It may result in delayed implementation of activities and "hijacking" of the process by researchers or extension agents, which may lower the motivation of the farmers. But provided things are done properly, that the diverse stakeholders perceive the value of working together, and that trust develops among the parties, several advantages emerge as the potential synergies between these actors come into play. Farmers, scientists, extension agents, and private business owners learn and improvise together—which is the full expression of an effective innovation system. Joint innovation also usually implies improved research design, more rigor in implementation, and better documentation of results, all of which increase the prospects of wider application and dissemination of innovation.

Costs associated with implementing an effective FIF program

One key objective in establishing a fund program is to keep administrative, support, and supervision costs as low as possible, compared to the amount invested in experimentation by grantees. Given the fairly experimental nature of the LISF and CGP programs profiled here (implying that new mechanisms and setups had to be designed), and the fact that some expenses are fixed (such as those for M&E or for establishing and operating committees), the relative amounts devoted to such costs versus the amount of the grants themselves may be quite high, especially in the initial stages.

Experience with LISFs over the past five years indicates that, in the start-up phase, about two-thirds of the program costs are associated with capacity building for farmers and support institutions, operational costs (making calls, screening proposals, reviewing progress, and so on), creating awareness about the fund, technical external backstopping by service providers and research, sharing and disseminating process and results, and M&E and impact assessment. About one-third of program costs are for the grants themselves. Over time, the costs associated with running an FIF gradually decrease relative to the cost share of the grants, to about two-thirds grants and one-third running costs. Overall, the absolute costs for the grant component in the FIFs vary greatly according to size of awards, scale of the program, and level of cofunding by grantees.

Effectiveness and dynamics of fund setups

The appropriate setup (local versus institutional, decentralized versus centralized) for governing and managing the fund depends on the specific context, experiences, opportunities, and the scale at which implementing a FIF program makes the most sense. FIF programs have generally been moving towards more farmer-led governance mechanisms and structures. This shift requires support organizations to move away from managing funds directly and to become more involved in building the capacity of farmer groups to manage funds autonomously. Support institutions also play an increasingly important role in ensuring the quality of proposals and integrity of fund use. The case from Cambodia (box 5.38) illustrates the evolving nature of the setup.

Ideally, both local and central approaches could be implemented at the same time. This strategy would speed the learning process and the way the entire agricultural R&D system operates and responds to farmers' needs and desires.

Sustainability and scaling up of FIF programs: Advances and challenges

At the *community level*, considerable progress has been achieved in giving farmers access to innovation resources and in building their capacity to collaboratively manage funding schemes at their level through specific training and mentoring by support organizations. In financial terms, sustainability and a sense of responsibility can be enhanced by putting payback arrangements into place within community organizations, as done in the case of the LISF Cambodia and Uganda. Not all stakeholders are keen on a loan-type mechanism for funding research and innovation, however, as this approach is easily confused with a classic microcredit or loan scheme.

At a higher level, efforts to *institutionalize the FIF concept within the country's agricultural R&D systems* are incipient. In Tanzania, a local government has agreed to help replenish the LISF in one district. In other countries, some government agencies participate actively in implementing LISF pilots. In Cambodia, the government and donors are interested in supporting the recently established central institution running the LISF over the long term (box 5.38). In Albania, the Ministry of Agriculture, Food and Consumer Protection built on the initial CGP approach by creating an investment grant model in the

---

Box 5.38 Dynamics of the Local Innovation Support Fund Setup in Cambodia, 2006–10

An initial structure for the Local Innovation Support Fund (LISF) in Cambodia was designed based on a feasibility study carried out in 2005–06. From the start, it was decided to operate the LISF as a revolving fund. Farmers could apply to the LISF for a loan, on which interest was charged. This setup was seen as the best way to make farmers feel more responsible for carrying out LISF activities and to replenish and expand the initial fund, linking it to existing community-based savings and credit schemes. From 2005 to 2008, LISF pilots were established in three provinces, each with a different organization playing the leading role, overseen by an LISF National Steering Committee coordinated by the Cambodian Center for Study and Development in Agriculture (CEDAC), which also coordinates Prolinnova–Cambodia. Although LISF operations were highly decentralized in operational terms, the three provinces followed a common procedure. Fund requests by individual farmers were first sent to a farmer association, which compiled and forwarded

them to the lead LISF partner in the province. After a preliminary review of the proposals, this partner forwarded them to the LISF National Steering Committee for a final decision.

In 2008, the LISF scheme was expanded to 11 provinces involving a total of 20 NGO members of Prolinnova–Cambodia, but it proved too difficult to ensure the necessary capacity building and the quality of the proposals and ensuing experimentation. It was also challenging to handle the varying degree of ownership by farmers and local support institutions as well as to monitor the results. Moreover, it was difficult to attract funding from the national government and from international donors.

In response, starting in 2011, a new structure was designed to implement the LISF through a farmer-governed, centralized national fund under an existing farmer organization at the national level, Farmer and Nature Net (FNN). CEDAC and other Prolinnova–Cambodia partners play solely an advisory role to the FNN.

Sources: Vitou 2008; FAIR workshop reports 2009, 2010.
Note: CEDAC = Centre d'Etude et de Développement Agricole Cambodgien.

---

official government agricultural support program. In Central Asia, CGP-like schemes have been set up to emulate the success obtained in Albania.

When integrating innovation funds into a wider framework, is it highly desirable to implement parallel activities contributing to farmer empowerment and capacity strengthening. When such integration has been achieved, the scope and opportunities for an effective and sustainable funding scheme are greatly expanded, at least in contexts where poverty and serious social and economic problems prevail.

With respect to choosing the appropriate grant size, small grants are extremely relevant for very poor, risk-averse farmers working with little outside help and little need for investment in equipment or infrastructure, and on innovations that have a strong location-specific character or an inherently small niche. They may also be a good way to start and experiment with the FIF concept and process, before institutionalization takes place.

Larger grants are relevant for better-off farmers, for group applications, when costs of external research and advisory services are factored in, and for supporting increased commercialization in rural areas. If and when actors from formal agricultural R&D are ready to integrate FIFs into their activities, they will usually be more willing to go with bigger grants than with small grants.

CONCLUSIONS

FIFs are a valuable, vital component of a wider approach to strengthening innovation capacities and systems. By making innovation funds more readily available to farmers and other relevant stakeholders, FIFs, as illustrated by the LISF and CGP experiences, are performing an essential role in strengthening innovation and promoting a greater role and voice for farmers and other rural stakeholders in governance of agricultural R&D.

FIFs are most powerful when they are not implemented in isolation but form part of systemic, long-term efforts to promote and strengthen sustainable farming, participatory innovation development, and dynamic innovation systems and processes, in which the roles and skills of various stakeholders (particularly smallholders) are recognized and supported. An FIF program should be complemented with investments in a number of related areas that create an enabling environment for agriculture: improving research and extension (to make them more responsive to demand), nurturing the emergence and consolidation of a vibrant private sector capable of providing services and inputs and processing the produce, establishing effective coordination mechanisms among these stakeholders, designing and funding policies that will favor rather than restrain innovation, improving education and training, and favoring market linkages, among others.

NOTES

Module 5 Overview

1. See the infoDev Monitoring, Evaluation, and Impact Assessment study at http://www.idisc.net/en/Page.MEIA.Study.Overview.html.

2. For example, through intellectual property rights (IPRs) to promote greater impact of the research and innovations emanating from an institution (for example, by licensing technology to partners).

3. Financing instruments for agricultural innovation are discussed in module 6.

4. See World Bank (2009b), which suggests this approach for cluster evaluation; with some modification, it is suitable for various business development programs.

Thematic Note 3

1. The Agribusiness Community of Practice recently launched by infoDev (www.infodev.org) raises awareness of the utility and need for agribusiness incubation. For more details on agribusiness incubators, see infoDev (2011).

2. Examples of incubator models included in this section are drawn from infoDev (2009), which assesses the best international practices for ICT incubators and includes case studies from Argentina, Chile, Colombia, Jordan, and South Africa.

3. This section is drawn from infoDev's Incubator Toolkit (www.idisc.net).

4. This section is drawn from infoDev's Toolkit on Business Incubation. The forms noted in the section can be accessed via the infoDev site: www.idisc.net.

5. Sally Hayhow, referenced in infoDev's Incubator Toolkit in the section on "The Role of the Incubator Board" (http://www.idisc.net/en/Article.163.html).

6. Patient capital is long-term capital invested without expectation of a quickly realized profit.

Thematic Note 4

1. For details on a cluster policy approach, see World Bank (2009).

2. For details on a value chain approach, see FIAS (2007).

3. For a concise but detailed discussion, see World Bank (2009).

4. For analytical work on this issue see Guiliani, Pietrobelli, and Rabelotti (2004, 2005) and World Bank (2009).

5. For a detailed description of advantages and disadvantages of the cluster mapping method, see World Bank (2009).

6. Another method would be to look at trade statistics.

7. See in particular USAID (2008).

8. World Bank (2009) provides a data-capture template that helps to implement this diagnostic tool.

9. For further information on value chain analysis, see FIAS (2007) and World Bank (2009).

10. For further information on institutional mapping, see World Bank (2009).

11. This project continues; the preliminary results described here were transmitted in an interview with the author.

12. See GEF (n.d.) and Negeli-Ganz (2008).

Thematic Note 5

1. Many published IP policies can provide guidance for developing these policies; Kowalsky (2007) addresses the role and structure of institutional policies.

Thematic Note 6

1. This report, based on research into 31 agricultural investment funds, is a more general overview of investment than this thematic note and is a useful and up-to-date source of trends. It also contains seven case studies and models of agricultural investment funds. The report focused primarily on sub-Saharan Africa and transitional economies and identified over 80 investment funds working in agriculture and rural development.

2. Probably the best known example, which has been replicated by many countries, is the US Small Business Investment Research program. See http://www.ncbi.nlm.nih.gov/books/NBK9607/.

3. For examples of angel and seed investing in agriculture, see de Kleene (n.d.) and the presentation "Angel Investing in Agriculture" (docstoc. www.docstoc.com/docs/10759/Angel-Investing-and-Agriculture).

4. An important model is Yissum Technology Transfer at Hebrew University of Jerusalem, which is responsible for commercializing an array of successful products that generate over US$2 billion in worldwide sales every year. See http://www.yissum.co.il/.

5. There are many descriptions of venture capital; see, for example, Wyse (2007).

6. The technical definition is that the IRR of an investment is the discount rate that makes the Net Present Value (NPV) of the investment's cash flow stream equal to zero. There are many ways to measure economic development and social good which are beyond the scope of this note.

7. There is a persistent myth that a large percentage of new businesses in the developed world receive bank loans and venture capital funding. In the United States, the

Kauffman Foundation's New Firm Longitudinal Study estimates that 5 percent of new firms obtain business bank loans. Venture capitalists invested in 440 startup/seed-stage companies; angels invested in 20,000 early stage firms ("Funding for the Rest of Us Entrepreneurs," http://www.nyew.org/2010/05/funding-for-the-rest-of-us-entrepreneurs/, accessed August 2011). That leaves approximately 95 percent of all new businesses not funded by these entities. Interest in venture capital continues because venture-backed firms have generated enormous economic impact.

Innovative Activity Profile 1

1. Institute of International Agriculture, Mozambique (IIAM); STEP, Thapar University, Patiala, India; NIABI members: Tamil Nadu Agricultural University (TNAU), Coimbatore, India; Anand Agricultural University (AAU), Anand, India; Birsa Agricultural University (BAU), Ranchi, India; Central Institute of Fisheries Technology (CIFT), Cochin, India; Central Institute for Research on Cotton Technology (CIRCOT), Mumbai, India; CCS Haryana Agricultural University (CCS HAU), Hisar, India; Indian Agricultural Research Institute (IARI), New Delhi, India; Indian Veterinary Research Institute (IVRI), Izatnagar, India; Jawaharlal Nehru Krishi Vishwa Vidyalaya (JNKVV), Jabalpur, India; National Institute of Research on Jute and Allied Fibre Technology (NIRJAFT), Kolkata, India.

Innovative Activity Profile 2

1. World Bank (2005), in particular Annex 4, describes the important role of farmer organizations in this project.

Innovative Activity Profile 4

1. Promoting Local Innovation in ecologically oriented agriculture and natural resource management (www.prolinnova.net).

REFERENCES AND FURTHER READING

Module 5 Overview

Andersson, T., S.S. Serger, J. Sorvik, and E.W. Hansson. 2004. *The Cluster Policies White Book*. Malmö: International Organisation for Knowledge Economy and Enterprise Development (IKED). http://www.iberpyme online.org/Documentos/TheClusterPoliciesWhitebook .pdf, accessed April 2011.

Aghion, P., and P. Howitt, P. 2005. "Growth with Quality-Improving Innovations: An Integrated Framework." In *Handbook of Economic Growth*, Vol. 1B, edited by P. Aghion and S.N. Durlauf. San Diego: Elsevier.

Altenburg, T., and A. Stamm. 2004. "Towards a More Effective Provision of Business Services." Discussion Paper No. 4. Bonn: German Development Institute.

Cone. 2010. Shared Responsibility Study Fact Sheet. http://www.coneinc.com/stuff/contentmgr/files/0/4b6d5 2e9ecfa4eb96b6ea2a801e48cc6/files/cone_2010_shared_ responsibility_survey_fact_sheet.pdf, accessed April 2011.

David Noe, Chilean Ministry of Finance, December 2007, www.oecd.org/dataoecd/59/62/40023795.pdf, accessed March 2011.

GIZ (Deutsche Gesellschaft für Internationale Zusammenarbeit). 2009. "Public Private Partnerships (PPP): Methods and Instruments." http://www.gtz.de/de/doku mente/gtz2009-en-ppp-advisoryservice.pdf.

Henckes, C., et al. 2004. Shared Responsibility: Promoting Pro-Poor Growth in Disadvantaged Regions. Draft Approach Paper, Eschborn, Bonn, New York, and Washington, DC.

infoDev. n.d. Monitoring, Evaluation, and Impact Assessment Study. http://www.idisc.net/en/Page.MEIA.Study .Overview.html, accessed April 2011.

———. 2008. "International Good Practice for Establishment of Sustainable IT Parks: Review of Experiences in Select Countries." Washington, DC: World Bank. http://www .infodev.org/en/Publication.557.html, accessed April 2011.

Noe, D. 2007. "R&D Tax Incentive in Chile." Presentation, December 2007. OECD, www.oecd.org/dataoecd/59/62/ 40023795.pdf, accessed March 2011.

Thomas, V., M. Dailami, A. Dhareshwar, R.E. López, D. Kaufmann, A. Kishor, and Y. Wang. 2000. *The Quality of Growth*. New York: Oxford University Press.

World Bank. 2006. *Agriculture Investment Sourcebook*. Washington, DC.

———. 2009a. "Agricultural Research and Competitive Grant Schemes: An IEG Performance Assessment of Four Projects in Latin America." Washington, DC.

———. 2009b. "Clusters for Competitiveness: A Practical Guide and Policy Implications for Developing Cluster Initiatives." Washington, DC.

———. 2010. "Designing and Implementing Agricultural Innovation Funds: Lessons from Competitive Research and Matching Grant Projects." Report No. 54857-GLB. Washington, DC.

World Economic Forum. 2011. "Realizing a New Vision for Agriculture: A Roadmap for Stakeholders." Geneva.

http://www3.weforum.org/docs/IP/AM11/CO/WEF_Ag
ricultureNewVision_Roadmap_2011.pdf, accessed April
2011.

Thematic Note 1

Biggs, T. 1999. "Microeconometric Evaluation of the Mauritius Technology Diffusion Scheme (TDS)." RPED Paper No. 108. Washington, DC: World Bank.

Dutz, M., C. Harris, I. Dhingra, and C. Shugart. 2006. "Public-Private Partnership Units." Public Policy for the Private Sector Note No. 311. Washington, DC: World Bank.

Ferroni, M. 2010. "Can Private Sector R&D Reach Small Farms?" In World Food Security: Can Private Sector R&D Feed the Poor? Proceedings of the 15th Annual International Conference, Parliament House, Canberra. Deakin, ACT: Crawford Fund. Pp. 3–12. http://www.crawfordfund.org/assets/files/conference/proceedings/Crawford_Fund_2009_Conference_Proceedings.pdf, accessed March 2011.

GIZ (Deutsche Gesellschaft für Internationale Zusammenarbeit). 2009. "Public Private Partnerships (PPP): Methods and Instruments." http://www.gtz.de/de/dokumente/gtz2009-en-ppp-advisoryservice.pdf, accessed April 2011.

Hartwich, F., C. Gonzalez, and L.-F. Vieira. 2005. "Public-Private-Partnerships for Innovation-led Growth in Agrichains: A Useful Tool for Development in Latin America?" International Service for National Agricultural Research Systems Discussion Paper No. 1. Washington, DC: International Food Policy Research Institute (IFPRI).

World Bank. 2005. China Agricultural Technology Transfer Project. Project Appraisal Document, World Bank, Washington, DC.

World Economic Forum. 2011. "Realizing a New Vision for Agriculture: A Roadmap for Stakeholders." Geneva. http://www3.weforum.org/docs/IP/AM11/CO/WEF_AgricultureNewVision_Roadmap_2011.pdf, accessed April 2011.

Thematic Note 2

Collion, M-H. (forthcoming). "Lessons from Latin America Productive Partnerships Support Project." ARD Note. Washington, DC: World Bank.

Donovan, G. 2006. "When Markets Do Not Work, Should Grants be Used?" Agriculture and Rural Development Policy Note. Washington, DC: World Bank.

EMBRAPA, IDB (Inter-American Development Bank), and World Bank. 2000. Competitive Grants in the New Millennium: A Global Workshop for Designers and Practitioners. Proceedings of the international workshop held in Brasilia, 16-18 May 2000. Available at: http://www-wds.worldbank.org/external/default/WDSContentServer/WDSP/IB/2005/07/21/000012009_20050721110629/Rendered/INDEX/319020rev0Competitve0grant0wkshp.txt (accessed January 2011).

van der Meer, K., and M. Noordam. 2004. "The Use of Grants to Address Market Failures: A Review of World Bank Rural Development Projects." Agriculture and Rural Development Paper No. 27. Washington, DC: World Bank.

World Bank. 1999a. "Agriculture Technology Notes: Competitive Research Grant Programs, Financing Agriculture Research Within World Bank Loans." Rural Development Department. No. 24. Washington, DC.

———. 1999b. "Turkey Technology Development Project." Project Appraisal Document. Washington, DC: World Bank.

———. 2006. "Turkey Technology Development Project." Implementation Completion Report. Washington, DC: World Bank.

———. 2009. Agricultural Research and Competitive Grant Schemes: An IEG Performance Assessment of Four Projects in Latin America." Washington, DC.

———. 2010. "Designing and Implementing Agricultural Innovation Funds: Lessons from Competitive Research and Matching Grant Projects." Report No. 54857-GLB. Washington, DC.

Thematic Note 3

Boettiger, S., and S. Alvarez 2010. "Getting Better Technologies to the Poor: A Landscape of Commercialization Challenges and Opportunities." Berkeley: Public Intellectual Property Resources for Agriculture (PIPRA). http://www.pipra.org/documents/PIPRA_CommercializationReport_photos.pdf, accessed August 2011.

IFC (International Finance Corporation). 2010. "Scaling-Up SME Access to Financial Services in the Developing World." Washington, DC.

infoDev 2008. "Financing Technology Entrepreneurs & SMEs in Developing Countries." (www.infodev.org/financing), accessed April 2011.

———. 2009. "A Model for Sustainable and Replicable ICT Incubators in Sub-Saharan Africa." Washington, DC: World Bank. infoDev, http://www.infodev.org/en/Publication.734.html, accessed April 2011.

———. 2011. "Agribusiness Incubation: Good Practice Assessment and Training Module." Washington, DC.

Knopp, L. 2007. *2006 State of the Business Incubation Industry.* Athens, Ohio: National Business Incubation Association (NBIA).

Medeiros, J.A., L.A. Medeiros, T. Martins, and S. Perilo. 1992. "Pólos, parques e incubadoras: A busca da modernizaçao e competitividade." Brasilia: Conselho Nacional de Desenvolvimento Científico e Tecnológico (CNPq), *Instituto Brasileiro de Informação* em Ciência e Tecnologia (IBICT), and Serviço Nacional de Aprendizagem Industrial (SENAI).

University of Michigan, NBIA (National Business Incubation Association), Ohio University, and Southern Technology Council. 1997. *Business Incubation Works.* Athens, Ohio: NBIA.

Thematic Note 4

Albaladejo, M. 2001. "The Determinants of Competitiveness in SME Clusters: Evidence and Policies for Latin America." In *Small-Scale Enterprises in Developing and Transitional Economies*, edited by H. Katrack and R. Strange. London: Macmillan.

Bolo, M.O. 2006. "The Lake Naivasha Cut Flower Cluster in Kenya." In *Africa: Knowledge, Technology, and Cluster-Based Growth*, edited by D.Z. Zeng. World Bank Institute Development Study. Washington, DC: World Bank. Pp. 37–52.

Braunerhjelm, P., and M.P. Feldman (eds.). 2007. *Cluster Genesis: Technology-based Industrial Development.* Oxford: Oxford University Press.

Ceglie, G., and M. Dini. 2000. *SME Clusters and Network Development in Developing Countries: The Experience of UNIDO.* PSD Technical Working Paper. Vienna: United Nations Industrial Development Organization (UNIDO).

Dini, M. 2003. Consideraciones sobre politicas para el desarrollo de cluster. Agora 2000 mimeo for Inter-American Development Bank.

FIAS (Foreign Investment Advisory Service). 2007. "Moving Toward Competition: A Value Chain Approach." Washington, DC: World Bank.

GEF (Global Environment Facility). n.d. "Management Effectiveness Tracking Tool." World Database on Protected Areas, http://www.wdpa.org/ME/PDF/METT.pdf, accessed April 2011.

Goetz, S.J., M. Shields and Q. Wang. 2004. "Agricultural and Food Industry Clusters in the Northeast US: Technical Report." Regional Rural Development Paper No.26. University Park, PA: The Northeast Regional Center for Rural Development, Pennsylvania State University.

Gibbs, R.M., and G.A. Bernat 1997. "Rural Industry Clusters Raise Local Earnings." *Rural Development Perspectives* 12(3):18–25.

Giuliani, E., C. Pietrobelli, and R. Rabellotti. 2004. "Upgrading in Clusters in Latin America and the Caribbean: The Role of Policies." Washington, DC: Inter-American Development Bank. http://siteresources.worldbank.org/INTEXPCOMNET/Resources/Pietrobelli_and_Rabellotti_2004.pdf, accessed August 2011.

Giuliani, E., C. Pietrobelli, and R. Rabellotti. 2005. "Upgrading in Global Value Chains: Lessons for Latin American Clusters." *World Development* 33(4):549–73.

Hausman, R., and D. Rodrik. 2003. "Economic Development as Self-Discovery." *Journal of Development Economics* 72(2):603–33.

Henry, M., and M. Drabenstott. 1996. "A New Micro View of the US Rural Economy." Economic Review, Federal Reserve Bank of Kansas City, 2nd Quarter 1996. Pp.53–70.

Kiggundu, R. 2005. "Learning to Change: Why the Fish Processing Clusters in Uganda Learned to Upgrade." In *Industrial Clusters and Innovation Systems in Africa: Institutions, Markets, and Policy*, edited by B. Oyelaran-Oyeyinka and D. McCormick. New York: United Nations University Press. Pp. 159–86.

Kiggundu, R. 2006. "The Lake Victoria Fishing Cluster in Uganda." In *Africa: Knowledge, Technology, and Cluster-Based Growth*, edited by D.Z. Zeng. World Bank Institute Development Study. Washington, DC: World Bank. Pp. 87–96.

Larsen, K., R. Kim, and F. Theus. 2009. *Agribusiness and Innovation Systems in Africa.* Washington, DC: World Bank

McCormick, D., and W. Mittulah. 2005. "Global Markets and Local Responses: The Changing Institutions in the Lake Victoria Fish Cluster." In *Industrial Clusters and Innovation Systems in Africa: Institutions, Markets, and Policy*, edited by B. Oyelaran-Oyeyinka, and D. McCormick. New York: United Nations University Press. Pp. 211–40.

Munnich, L.W., G. Schrock, and G. Cook. 2002. "Rural Knowledge Clusters: The Challenge of Rural Economic Prosperity." Reviews of Economic Development Literature and Practice No. 12. Washington, DC: US Economic Development Association, United States Department of Commerce.

Mytelka, L. 2005. "From Clusters to Innovation Systems in Traditional Industries." In *Industrial Clusters and Innovation Systems in Africa: Institutions, Markets, and Policy*, edited by B. Oyelaran-Oyeyinka and D. McCormick. New York: United Nations University Press. Pp.39–62.

Negeli-Ganz, B. 2008. "Monitoring and Evaluation of Clusters." Paper presented at the 3rd International Clustering Conference, December 4, Istanbul.

Ostrom, E., J. Burger, C.B. Field, R.B. Norgaard, and D. Policansky. 1999. "Revisiting the Commons: Local Lessons, Global Challenges." *Science* 284(5412):278–82.

Oyelaran-Oyeyinka, B., and D. McCormick (eds.) 2005. *Industrial Clusters and Innovation Systems in Africa: Institutions, Markets, and Policy.* New York: United Nations University Press.

Parilli, M.D. 2006. "Cluster Trajectories in Developing Countries: A Stage and Eclectic Approach Applied to Survival Clusters in Central America." In *Clusters and Globalisation: The Development of Urban and Regional Economies,* edited by C. Pitelis, R. Sugden, and J.R. Wilson. New York: Edward Elgar.

Porter, M. 1998. "Clusters and Competition: New Agendas for Companies, Governments, and Institutions." In *On Competition,* edited by M. Porter. Boston: Harvard Business Review. Pp. 197–288.

Rodriguez-Clare, A. 2005. "Coordination Failures, Clusters, and Microeconomic Interventions." Washington, DC: Inter-American Development Bank (IDB).

SANDAG (San Diego Association of Governments). 2005. "Understanding Cluster Analysis." San Diego Association of Governments, http://www.sandag.org/rta/transfer/cluster_analysis.pdf, accessed August 2011.

Schmitz, H. 1992. "On the Clustering of Small Firms." *IDS Bulletin* 23(3):64–69.

USAID (United States Agency for International Development). 2003. "An Assessment of Cluster-Based Approaches." An Evaluation Undertaken by The Mitchel Group. Washington, DC.

———. 2008. "Value Chains and the Cluster Approach: Transforming Relationships to Increase the Competitiveness and Focus on End Markets." MicroReport No. 148. Washington, DC.

Wood, E., and D. Kaplan. 2006. "The Wine Cluster in South Africa." In *Africa: Knowledge, Technology, and Cluster-Based Growth,* edited by D.Z. Zeng. World Bank Institute Development Study. Washington, DC: World Bank. Pp. 109–18.

World Bank. 2009. "Clusters for Competitiveness." http://siteresources.worldbank.org/INTEXPCOMNET/Resources/cluster_initiative_pub_web_ver.pdf, accessed April 2011.

———. 2010. "Designing and Implementing Agricultural Innovation Funds: Lessons from Competitive Research and Matching Grant Projects." Report No. 54857-GLB. Washington, DC.

Zeng, D.Z. (ed.) 2006. *Africa: Knowledge, Technology, and Cluster-Based Growth.* World Bank Institute Development Study. Washington, DC: World Bank.

———. 2010. "Building Engines for Growth and Competitiveness in China: Experiences with Special Economic Zones and Industrial Clusters." Directions in Development No. 56447. Washington, DC: World Bank.

Thematic Note 5

Campbell, A.F. 2007. "How to Set Up a Technology Transfer Office: Experiences from Europe." In *Intellectual Property Management in Health and Agricultural Innovation: A Handbook of Best Practices,* vol. 1, edited by A. Krattiger, R.T. Mahoney, L. Nelsen, J.A. Thomson, A.B. Bennett, K. Satyanarayana, G.D. Graff, C. Fernandez, and S.P. Kowalski. Oxford, UK, and Davis, California: Centre for the Management of Intellectual Property in Health Research and Development (MIHR) and Public Intellectual Property Resource for Agriculture (PIPRA). Pp. 559–66. www.ipHandbook.org, accessed April 2011.

Di Giorgio, R.C. 2007. "From University to Industry: Technology Transfer at Unicamp in Brazil." In *Intellectual Property Management in Health and Agricultural Innovation: A Handbook of Best Practices,* vol. 1, edited by A. Krattiger, R.T. Mahoney, L. Nelsen, J.A. Thomson, A.B. Bennett, K. Satyanarayana, G.D. Graff, C. Fernandez, and S.P. Kowalski. Oxford, UK, and Davis, California: Centre for the Management of Intellectual Property in Health Research and Development (MIHR) and Public Intellectual Property Resource for Agriculture (PIPRA). Pp. 1747–54. www.ipHandbook.org, accessed April 2011.

Fernandez, C. 2007. "How to Set Up a Technology Transfer System in a Developing Country." In *Intellectual Property Management in Health and Agricultural Innovation: A Handbook of Best Practices,* vol. 1, edited by A. Krattiger, R.T. Mahoney, L. Nelsen, J.A. Thomson, A.B. Bennett, K. Satyanarayana, G.D. Graff, C. Fernandez, and S.P. Kowalski. Oxford, UK, and Davis, California: Centre for the Management of Intellectual Property in Health Research and Development (MIHR) and Public Intellectual Property Resource for Agriculture (PIPRA). Pp. 567–74. www.ipHandbook.org, accessed April 2011.

Guo, H. 2007. "IP Management at Chinese Universities." In *Intellectual Property Management in Health and Agricultural Innovation: A Handbook of Best Practices,* vol. 1, edited by A. Krattiger, R.T. Mahoney, L. Nelsen, J.A. Thomson, A.B. Bennett, K. Satyanarayana, G.D. Graff, C. Fernandez, and S.P. Kowalski. Oxford, UK, and Davis, California: Centre for the Management of Intellectual Property in Health Research and Development (MIHR) and Public Intellectual Property Resource for Agriculture (PIPRA). Pp. 1673–82. www.ipHandbook.org, accessed April 2011.

Heher, A.D. 2007. "Benchmarking of Technology Transfer Offices and What It Means for Developing Countries." In *Intellectual Property Management in Health and Agricultural Innovation: A Handbook of Best Practices,* vol. 1, edited by A. Krattiger, R.T. Mahoney, L. Nelsen, J.A. Thomson, A.B. Bennett, K. Satyanarayana, G.D. Graff, C. Fernandez, and S.P. Kowalski. Oxford, UK, and Davis,

California: Centre for the Management of Intellectual Property in Health Research and Development (MIHR) and Public Intellectual Property Resource for Agriculture (PIPRA). Pp. 207–28. www.ipHandbook.org, accessed April 2011.

Kowalsky, S. 2007. "Making the Most of Intellectual Property: Developing an Institutional IP Policy." In *Intellectual Property Management in Health and Agricultural Innovation: A Handbook of Best Practices,* vol. 1, edited by A. Krattiger, R.T. Mahoney, L. Nelsen, J.A. Thomson, A.B. Bennett, K. Satyanarayana, G.D. Graff, C. Fernandez, and S.P. Kowalski. Oxford, UK, and Davis, California: Centre for the Management of Intellectual Property in Health Research and Development (MIHR) and Public Intellectual Property Resource for Agriculture (PIPRA). Pp. 485–94. www.ipHandbook.org, accessed April 2011.

Nelson, L. 2007. "Ten Things Heads of Universities Should Know about Setting Up a Technology Transfer Office." In *Intellectual Property Management in Health and Agricultural Innovation: A Handbook of Best Practices,* vol. 1, edited by A. Krattiger, R.T. Mahoney, L. Nelsen, J.A. Thomson, A.B. Bennett, K. Satyanarayana, G.D. Graff, C. Fernandez, and S.P. Kowalski. Oxford, UK, and Davis, California: Centre for the Management of Intellectual Property in Health Research and Development (MIHR) and Public Intellectual Property Resource for Agriculture (PIPRA). Pp. 537–43. www.ipHandbook.org, accessed April 2011.

Pefile, S., and A. Krattiger. 2007. "Training Staff in IP Management." In *Intellectual Property Management in Health and Agricultural Innovation: A Handbook of Best Practices,* vol. 1, edited by A. Krattiger, R.T. Mahoney, L. Nelsen, J.A. Thomson, A.B. Bennett, K. Satyanarayana, G.D. Graff, C. Fernandez, and S.P. Kowalski. Oxford, UK, and Davis, California: Centre for the Management of Intellectual Property in Health Research and Development (MIHR) and Public Intellectual Property Resource for Agriculture (PIPRA). Pp. 597–615.

Young, T.A. 2007. "Establishing a Technology Transfer Office." In *Intellectual Property Management in Health and Agricultural Innovation: A Handbook of Best Practices,* vol. 1, edited by A. Krattiger, R.T. Mahoney, L. Nelsen, J.A. Thomson, A.B. Bennett, K. Satyanarayana, G.D. Graff, C. Fernandez, and S.P. Kowalski. Oxford, UK, and Davis, California: Centre for the Management of Intellectual Property in Health Research and Development (MIHR) and Public Intellectual Property Resource for Agriculture (PIPRA). Pp. 545–58.

Thematic Note 6

Al-Rifai, T., and A. Khan. 2000. "The Role of Venture Capital in Contemporary Islamic Finance." http://alhaqq society.org/downloads/VCContemporaryIslamic Finance.pdf, accessed April 2011.

Ashley, C., M. Warner, and J. Romano. 2005. "Directions for Private Sector Development Instruments in Africa: 8 Strategies for the Policy Maker." London: Overseas Development Institute (ODI). www.odi.org.uk/resources/download/1438.pdf, accessed April 2011.

Boettiger, S., and S. Alvarez. n.d. "Getting Better Technologies to the Poor: A Landscape of Commercialization Challenges and Opportunities." PIPRA, http://www.pipra.org/documents/PIPRA_Commercialization Report_photos.pdf, accessed April 2011.

Capital for Enterprise Ltd. n.d. "Enterprise Capital Funds." www.bis.gov.uk/files/file45483.pdf, accessed April 2011.

de Kleene, S. n.d. "Agricultural Growth Corridors." Presentation. Beira Corridor and Southern Agricultural Growth Corridor of Tanzania. http://brusselsbriefings.files.word-press.com/2010/08/de-cleene-pp-en1.pptx, accessed April 2011.

Department of Biotechnology, Ministry of Science and Technology, Government of India. n.d. "Biotechnology Industry Partnership Programme (BIPP)." http://dbtindia.nic.in/AboutBIPP.pdf, accessed April 2011.

FAO (Food and Agriculture Organization). 2010. *Agricultural Investment Funds for Developing Countries.* Rome. www.fao.org/fileadmin/user_upload/ags/.../investment_funds.pdf, accessed April 2011.

Hodgson, B. 2008. "Accessing Finance: Intellectual Capital Readiness of Entrepreneurs and MSMEs." Abstract of a paper prepared for the WIPO Italy International Convention on IP and Competitiveness of MSMEs, Rome, December 10 and 11. WIPO, www.wipo_smes_rom_09 _c_theme03_3-related1.pdf, accessed April 2011.

Lerner, J. 2009. *Boulevard of Broken Dreams: Why Public Efforts to Boost Entrepreneurship and Venture Capital Have Failed – and What to Do about It.* Princeton: Princeton University Press.

National Committee for International Cooperation and Sustainable Development (NCDO). 2008. "Venture Capital and Private Equity Funds for Development Index 2008." 2008. Amsterdam: NCDO. www.bidnetwork.org/download.php?id=158389, accessed April 2011.

Tabor, S.R., W. Janssen, and H. Bruneau (eds.). 1998. "Financing Agricultural Research: A Sourcebook." The Hague: International Service for National Agricultural Research (ISNAR). ftp://ftp.cgiar.org/isnar/publicat/pdf/fsb/fsb-f.pdf, accessed April 2011.

World Bank. 2007. *World Development Report 2008: Agriculture for Development.* Washington, DC.

Wyse, R. 2007. "What the Public Should Know about Venture Capital." In *Intellectual Property Management in*

Health and Agricultural Innovation: A Handbook of Best Practices, vol. 1, edited by A. Krattiger, R.T. Mahoney, L. Nelsen, J.A. Thomson, A.B. Bennett, K. Satyanarayana, G.D. Graff, C. Fernandez, and S.P. Kowalski. Oxford, UK, and Davis, California: Centre for the Management of Intellectual Property in Health Research and Development (MIHR) and Public Intellectual Property Resource for Agriculture (PIPRA). Pp. 1281–88. www.ipHandbook.org, accessed April 2011.

Innovative Activity Profile 1

NSTEDB (National Science and Technology Entrepreneurship Development Board and ISBA (Indian STEPs and Business incubators Association). "First Status Report on Technology Business Incubation in India, 2009." http://www.nstedb.com/fsr-tbi09/index.html, accessed April 2011.

Innovative Activity Profile 2

State Office for Comprehensive Agricultural Development. 2009-2010. Various monitoring and evaluation reports (internal documents).

World Bank. 2005–10. Aide-memoires prepared by World Bank supervision missions (internal documents).

———. 2005. "China Agriculture Technology Transfer Project." Project Appraisal Document. Washington, DC: World Bank.

———. 2010. Designing and Implementing Agricultural Innovation Funds: Lessons from Competitive Research and Matching Grant Projects." Report No. 54857-GLB. Washington, DC: World Bank.

Innovative Activity Profile 3

UNIDO (United Nations Industrial Development Organization). 2008a. "Evaluación Independiente NICARAGUA; Fortalecimiento y Difusión del Desarrollo de Conglomerados en Nicaragua (UE/NIC/05/001); Fortalecimiento y Difusión del Desarrollo de Con-glomerados en Nicaragua – Creación de capacidades y "Cluster-to-Cluster" (UE/NIC/05/003)." http://www.unido.org/index.php?id=o49971, accessed April 2011.

———. 2008b. "Proyecto Fortalecimiento y difusión del desarrollo de conglomerados en Nicaragua (UE/NIC/05/001) 2005–2008: Informe Final Consolidado, Junio 2008." http://www.unido.org/index.php?id=o4307, accessed April 2011.

Innovative Activity Profile 4

van Veldhuizen, L., A. Krone, M. Wongtschowski, and A. Waters-Bayer. 2008. "FAIR: Farmer Access to Innovation Resources." *Rural Development News* 2:47–52.

Wongtschowski, M., B. Triomphe, A. Krone, A. Waters-Bayer, and L. van Veldhuizen. 2010. "Towards a Farmer-governed Approach to Agricultural Research for Development: Lessons from International Experiences with Local Innovation Support Funds." In *Symposium Proceedings, Innovation and Sustainable Development in Agriculture, June 28–30, 2010, Montpellier,* edited by E. Coudel, H. Devautour, C. Soulard, and B. Hubert. http://hal.archives-ouvertes.fr/ISDA2010, accessed August 2011.

World Bank. 2010. "Designing and Implementing Agricultural Innovation Funds: Lessons from Competitive Research and Matching Grant Projects." Washington, DC.

———. 2011. "Albania Agriculture Services Project: Implementation Completion Report." Washington, DC.

ADDITIONAL RESOURCES

Thematic Note 3

infoDev's Business Incubation Toolkit, and links to developing country regional and national incubator associations: www.infodev.org/idisc

The United States National Association of Business Incubators: www.nbia.or

Creating an Enabling Environment for Agricultural Innovation

OVERVIEW

Johannes Roseboom, Consultant

EXECUTIVE SUMMARY

The "enabling environment" for agricultural innovation encompasses factors that influence agricultural innovation positively but are controlled by policy domains other than agricultural innovation policy. An agricultural innovation policy seeks coordination with these other domains to ensure that together they enable agricultural innovation. Cross-cutting policy issues affecting agricultural innovation include policies to reduce poverty and sustain the environment, to foster collaboration between the public and private sectors, and to build social capital more generally.

Three clusters of enabling factors for agricultural innovation appear to require attention and investment in most developing countries: (1) *innovation policy* and corresponding governance structures to strengthen the broader framework for agricultural innovation policies; (2) *regulatory frameworks* that stimulate innovation directly (such as IPRs) or indirectly (standards that stimulate trade) or steer innovation towards certain preferred outcomes (safer food); and (3) *accompanying agricultural investments* in rural credit, infrastructure, and markets.

Innovation policy is a new area, and in most countries the governance structure for innovation is only starting to emerge. A particular challenge is where to assign responsibility for innovation policy within the government structure. Some countries delegate this task to the ministry in charge of science and technology, while others establish a higher-level entity that brings relevant ministries together to coordinate national innovation policy. In most countries, the overall objective of the national innovation policy is to facilitate the transition toward a *knowledge economy*, resulting in increased competitiveness and sustainable economic growth. A national innovation policy defines the roles and functions of actors and stakeholders within the national innovation system (NIS), provides an overall framework for innovation policies specific to particular sectors, and sets priorities across sectors and technologies. It creates positive conditions for innovation by investing in public goods essential for an innovative knowledge economy.

Regulatory frameworks important for agricultural innovation include those for IP; biosafety; and standards and technical regulations related to agricultural health and food safety and quality aspects. Countries will need assistance to develop legislation, assess the options from which they can choose, develop their regulatory agencies, and invest in standards-related infrastructure.

Better coordination of agricultural innovation investments with accompanying rural investments should lead to greater synergy and impact. Investments in rural financing systems will adopt a more holistic approach to financial services, including credit, savings, money transfers, leasing, and insurance. Investments in roads and market institutions and infrastructure help to improve agricultural productivity, reduce marketing costs, increase profit margins, and open up new opportunities for innovation.

These policies, investments, and regulatory reforms will trigger significant changes, such as improving the access of agricultural products to foreign markets, increasing private investment in agricultural R&D, and fostering the use of more sustainable agricultural practices. Policy measures will be needed to ensure that people are not left behind and make the transition to more promising economic activities.

RATIONALE FOR INVESTMENT

A key characteristic of the innovation systems approach is its holistic perspective on innovation as a multifaceted, iterative process that is very much shaped by the context within which it takes place. For that reason, national innovation policies are usually formulated as overarching policies trying to coordinate a wide spectrum of policy domains— science and technology policy, education policy, economic policy, industrial policy, infrastructure policy, taxation policy, and justice policy, among others—in such a way that together they create an environment that enables and stimulates innovation in the most positive way. Such overarching coordination is only possible with strong, high-level political support, often in the person of the prime minister or president chairing the council in charge of national innovation policy.

Sector-specific innovation policies (such as the policy for agricultural innovation) more or less replicate the national innovation policy's overarching and coordinating nature, but they will often have considerably less political clout to influence policies outside their domains. For example, a sector-specific innovation policy will have little influence over the adoption of a tax regime for R&D. Such a matter is more often dealt with at the national level.

One problem with the holism of the innovation system approach is that it tends to incorporate its enabling environment. Because innovation systems (or for that matter any soft system) do not exist "out there" as objective entities or realities but rather exist only "in the minds of those who define them" (Daane 2010), there is no natural delineation between what is core to an innovation system and what should be considered its enabling environment. An artificial but potentially practical solution to this problem is to define the "enabling environment" as those factors that influence agricultural innovation positively but that are controlled by policy domains other than the domain of agricultural innovation policy per se. An agricultural innovation policy will

have to interact and seek coordination with these other policy domains to ensure that together they enable agricultural innovation in the most positive way. This interaction may possibly lead to collaboration in the form of joint projects or programs, although not necessarily.

The agricultural innovation policy landscape

Figure 6.1 sketches the most relevant policy domains shaping agricultural innovation. At the most aggregate level, *political stability* is by far the most critical, overarching factor for any innovation system. Without such stability, investments in innovation activities (particularly those with long time horizons, such as plant breeding) are too risky to be attractive. Moreover, war and civil unrest often affect the knowledge infrastructure (research stations are destroyed, libraries plundered, and so on) and, by uprooting people, lead to a loss of knowledge and experience of agricultural practices and trade relations.

At the same time, it is important to realize that innovation in itself can be very destabilizing, because it comes with what Schumpeter labeled "creative destruction." New products and new production methods take over from old ones and in that process destroy old jobs, vested interests, and sometimes whole industries. As a result, innovations may encounter much opposition and catalyze social unrest. Creating new opportunities for those who lose their jobs

Figure 6.1   Policy Spheres Shaping the Environment for Agricultural Innovation

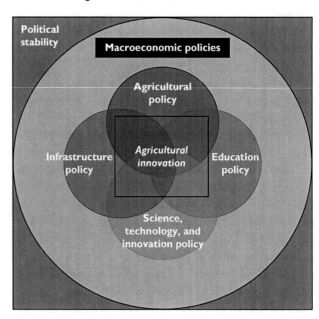

Source: Author.

(for example, by offering training to acquire new skills) is one of those measures that should accompany innovation to help reduce friction.

At the next level, *macroeconomic policies* dealing with taxation, exchange rate, market, and trade policies and similar matters can have an important impact on the relative competitiveness of agriculture in regional and global markets. In developing countries, many of these policies work against agriculture. They result in a net taxation of agriculture that hampers agricultural growth and innovation. Economic reform programs for the past twenty-five years have addressed this macroeconomic imbalance with some success (World Bank 2007b), but it remains a cause for concern and policy attention, especially considering that developed countries heavily subsidize agriculture to the detriment of developing countries. Consensus is growing (amid concern over rising food prices) that agriculture has been relatively neglected in developing countries by both donors and governments and that agricultural budgets have to be raised. The economic reform agenda focused initially only on improving the productivity and competitiveness of the agricultural sector. Over the years, however, poverty reduction and environmental sustainability have become equally important objectives. This expanded policy agenda also affects public investment decisions related to agricultural innovation. The big challenge for policy makers is to decide how to handle trade-offs between the different objectives.

At the meso level, four policy domains have the most direct influence on agricultural innovation: (1) agriculture; (2) science, technology, and innovation (STI); (3) education; and (4) infrastructure. These domains overlap considerably, and it is not always clear which domain should lead when it comes to investments. For example, agricultural research can be paid out of the agricultural budget or the science and technology budget. The scope of these different policy domains is also usually broader than agricultural innovation per se. An important task for an agricultural innovation policy is to influence and coordinate these policy domains (including investment decisions in those domains) so that they create the best environment for agricultural innovation to prosper. Each policy domain is described in detail in the sections that follow.

AGRICULTURAL POLICY. An agricultural policy usually comprises a wide range of topics, including agricultural health, research and extension, input supply, rural credit, land reform and improvement, rural infrastructure, market regulation and development, trade promotion, and sector organizations (farmer organizations, cooperatives, commodity boards, and the like). Some of these topics fall fully within the agricultural innovation policy orbit—agricultural research and extension, for example—whereas others partially overlap. Coordination between these various topics is important, because many of them complement each other. For example, the lack of rural credit often restrains the exploitation of market opportunities and new technologies.

A crucial element in agricultural transformation is farmers' integration into markets. With farmers' increasing integration into markets, market institutions (commodity chain organizations are one example) and regulations (such as product and health standards) become more important and require attention and investment. While agriculture-based economies depend on self-sufficiency for food security, urban economies depend on markets (including international markets) for food security.

EDUCATION POLICY. Agricultural education and training are core components of an AIS (see module 2), but they are also part of a broader national education policy that plays an enabling role. There is a strong positive correlation between primary education enrollment rates and agricultural productivity. The effectiveness of agricultural extension and training programs depends strongly on the basic skills that farmers acquire through primary education. At the same time, basic educational skills are important for those who seek employment outside agriculture.

In many developing countries, *vocational education* at the secondary level is virtually nonexistent (UNESCO education statistics),[1] and job specialization starts only after secondary school. Elsewhere a long tradition of vocational education at the secondary level equips the next generation of farmers with skills and knowledge. Despite the considerable debate about the disadvantages of forcing young people to make career choices early in life, the problem with waiting too long is that most students never reach the tertiary level. Tertiary education usually targets the more specialized jobs in agriculture, which may not necessarily be the best preparation for an all-round farmer (module 2). Promoting the introduction or expansion of vocational training in agriculture at the secondary level should advance agricultural innovation, but it will require many countries to rethink their national education policies.

SCIENCE, TECHNOLOGY, AND INNOVATION POLICY. In recent years, many developing countries—especially middle-income countries—have started to recognize the crucial role of innovation in economic growth and are aiming to make the transition toward a knowledge economy. As a consequence, STI policy is rapidly gaining importance in these countries.

This new emphasis on STI significantly affects the public agricultural research and extension services that pre-date STI initiatives by several decades. After many decades of relative isolation within ministries of agriculture, these agencies must now interact with new STI agencies that have a far wider scope that requires agricultural research and extension to compete with nonagricultural topics for resources.

INFRASTRUCTURE POLICY. Innovation opportunities often depend strongly on infrastructure such as roads, railways, utilities, and irrigation systems. High transportation costs are notorious for cutting heavily into the prices farmers and agribusinesses receive for their products and raising the costs of the agricultural inputs they purchase. When farmers and agribusinesses find it unprofitable to produce for the market, agricultural production often remains below its potential. It is affordable access to markets that makes it worthwhile and feasible to adopt new technologies, specialize, and raise production. The economic impact of lower transportation costs and improved market access can be quite dramatic (see box 6.2 later in this overview). As illustrated in IAP 1 for Zambia, investments in improved feeder roads can be an essential component of efforts targeted at enhancing agricultural innovation, value addition, and competitiveness.

Key enabling factors

The agricultural innovation policy landscape depicted in figure 6.1 comprises a wide range of enabling factors that are critical to agricultural innovation. It is impossible to cover them all, but the more important ones can be clustered as follows:

- Investments in innovation policy and corresponding governance structures that strengthen the broader framework for an agricultural innovation policy.
- Investments in regulatory frameworks affecting agricultural innovation, such as IPRs, sanitary and phytosanitary (SPS) measures, and biosafety regulations, aim to stimulate private actors to invest in innovation, improve the quality and safety of their products, and/or facilitate trade.
- Accompanying rural investments such as investments in rural credit, rural infrastructure (irrigation, roads, and utilities), and agricultural markets.

PAST EXPERIENCE

Experiences with investments in enabling factors have been quite mixed. Some types of investment have been around for decades (if not centuries—witness rural infrastructure), whereas others have emerged only very recently (biosafety regulations). Hence these interventions have quite different track records, as discussed next.

Innovation policy and governance structures

Investment in a national innovation policy and corresponding governance structures strengthens coordination across policy domains on innovation issues, addresses issues relevant across sectors (such as IPRs or tax deductions for innovation), provides a framework for more sector-specific innovation policies (including an agricultural innovation policy), and, not unimportantly, prioritizes public innovation investments across sectors. In many instances, national innovation policy has generated a substantial influx of new ideas and instruments into the agricultural innovation domain, including such concepts as business incubators and risk capital (see module 5). Embedding agricultural innovation policy in the national innovation policy may provoke inevitable complications and frictions, but at the end of the day it should result in a stronger AIS.

Innovation policies were first implemented in developed countries in the 1990s and have been emerging in developing countries only in the decade since then. In most countries, innovation policies and their accompanying governance structures are still very much in flux; in fact, the large majority of developing countries, particularly the smaller ones, still lack an innovation policy. In this sense, the historical record of innovation policy is still very short, both in developing and developed countries.

Regulatory frameworks

Except for environmental standards, the other three regulatory frameworks (IPRs, SPS standards, and product standards) have been around for decades, if not centuries. The international standardization and mutual recognition of these frameworks have been on the political agenda for quite some time. The Paris Convention for the Protection of Industrial Property, launched in 1883, was one of the first international treaties on IPRs. It has been revised many times and gained numerous signatories since then. Globalization and intensified trade have put increased pressure on countries to adopt these frameworks; the international community supports them strongly, and several international initiatives provide technical assistance and build capacity to implement them.[2]

One type of IPR that is quite specific to agriculture is plant variety rights (PVRs). In developed countries in particular, PVRs have been instrumental in developing a private seed industry and enabling public plant breeding to be funded through royalties. Only 68 countries are currently members of the International Union for the Protection of New Varieties of Plants (UPOV)[3], indicating an absence of PVR legislation consistent with UPOV standards. Membership is especially weak in Africa and Asia.

Considerable debate surrounds the introduction of PVR legislation in developing countries, many of which are under pressure to introduce legislation to meet the deadline (originally 2005, now 2016) set by the TRIPS[4] agreement. The debate focuses particularly on two issues. The first issue is farmers' rights in relation to breeders' rights: To what extent can farmers re-use, exchange, or sell PVR-protected seed? The second issue is the role of farm communities as custodians of genetic diversity: Should seed companies compensate communities for their services? With respect to the second issue, UPOV takes the position that farmers' customary role as curators of genetic resources is best regulated separately from PVR legislation. On the issue of farmers' rights, UPOV has moved over time toward a more restrictive standard favoring plant breeders. The criticism of developing countries is that UPOV is pushing for the adoption of developed country standards that are not necessarily adequate for developing countries (box 6.1).

Box 6.1   Plant Variety Rights Legislation in Africa

In 1998, the Heads of State of the African Union (AU) adopted the "African Model Law for the Protection of the Rights of Local Communities, Farmers, and Breeders, and for the Regulation of Access to Biological Resources." This watershed document addresses two issues—plant variety protection and access to biological resources—based on the premise that both issues are closely linked. The AU model law strongly favors farmers' rights over breeders' rights: PVRs can be withheld or nullified for reasons such as food security, health, biological diversity, and any other requirement of the farming community for propagation material of a particular variety. The model law also emphasizes the protection of Africa's biological resources and traditional knowledge.

The model law was criticized heavily by UPOV and WIPO.[a] Discussions between the AU, UPOV, and WIPO in 2001 did not reconcile their differences. Yet the AU member states did not hold a unified position on the issues. Some members (Egypt, Kenya, South Africa, and Tunisia) belonged to UPOV many years before the AU developed its model law. In 2002 the African intellectual property organization OAPI,[b] comprising some 16 Francophone African countries, approved a plant variety protection (PVP) chapter largely in line with UPOV standards as part of the 1999 Bangui Agreement. This decision was taken despite major opposition by international nongovernmental organizations.

In more recent years, the discussion in Africa regarding PVP has moved from the AU to the subregional economic communities, such as ECOWAS, SADC, and EAC.[c] Their strategy is to harmonize the (emerging) seed regulatory frameworks within their communities to facilitate trade and to join forces where possible to reduce regulatory costs. For example, ECOWAS and SADC each recently adopted the idea of setting up a common variety release system in their respective communities. Both communities have initiatives to work toward an integrated, regional PVP system. SADC, for example, developed a draft protocol for national PVP legislation. In other developing regions, regional economic communities are keen promoters of standardizing PVP systems. Most African countries seem to be moving toward adopting a PVP system that is compatible with the international UPOV standard—but only after much heated debate.

Source: Author.
a. UPOV is the International Union for the Protection of New Varieties of Plants and WIPO is the World Intellectual Property Organization. b. OAPI (Organisation Africaine de la Propriété Intellectuelle) was created in the early 1960s to replace the French institute in charge of IPRs prior to independence. It manages a single IPR system across 16 countries. c. Economic Community of West African States, Southern African Development Community, and East African Community.

Accompanying rural investments

Most countries have a long history of addressing enabling factors such as rural credit, rural infrastructure, and agricultural markets. In many instances, government interventions in these factors in the form of direct investment or facilitating private or mixed investment have been far from adequate and often rather fragmented, lacking attention to coherence among interventions. To cite a recent World Bank evaluation report on agricultural investments in sub-Saharan Africa):

> ...the lending support provided by the Bank has not reflected the interconnected nature of agriculture activities. Rather, the lending has been "sprinkled" across an array of activities in rural space, including research, extension, marketing reform, drought relief, seed development, and transport, but with little recognition of the relationships among them and the need for all of these areas to be developed at the same time, or at least in an optimal sequence, to effectively contribute to agricultural development. While the Bank's broader rural focus from the mid-1980s was justified, an unintended result was that it led to less focused attention on the need for various activities that are critical for agricultural development in rural space to come together at the same time or to take place in some optimal sequence. (World Bank 2007a, xxv)

RURAL CREDIT. The lack of working capital and access to affordable credit often prevents farmers and agribusinesses from buying modern inputs and equipment and fully benefiting from proven technological opportunities. Despite many attempts to address this issue, lack of affordable rural credit remains a major bottleneck in many countries. The formal banking sector is still largely absent from rural areas, because it perceives the risks and transaction costs to be too high to make business attractive. Popular rural credit schemes run by governments from the 1950s to the 1980s did little to attract commercial banks and proved unsustainable because of poor management and high default rates. The microfinance movement that emerged in the late 1990s tried to bridge the rural finance gap through self-help groups, which absorb the high costs inherent in small transactions and use social control to reduce risks (IAP 4). This approach has its limitations, and the model has not succeeded everywhere.

What is needed is a more active involvement of commercial banks in agriculture. Previous approaches tended to isolate financing for agriculture from the development of the wider financial system and overemphasized credit as opposed to savings and other financial services. Within a financial systems approach, however, financing for agriculture is viewed as part of the wider rural finance market.

Underpinning this approach is the fact that institutions adhering to commercial principles are more likely to achieve outreach and sustainability. The public sector's role is to concentrate on ensuring that the environment is conducive to the emergence and growth of such institutions (World Bank 2006).

For example, the Innovative Finance Initiative of the Alliance for a Green Revolution in Africa tries to mobilize commercial banks to provide more credit to the agricultural sector through a loan guarantee scheme. By absorbing some of the risks that commercial banks run when lending to agriculture, the initiative has managed to leverage some US$4 billion from commercial banks in the form of affordable loans for farmers and agribusinesses (www.agra-alliance.org).

Another practice that has fallen out of favor is to use subsidized credit to introduce new technologies. Such schemes have often undermined farmers' repayment discipline because farmers considered the subsidies to be gifts rather than loans. Jump-starting the introduction of a new technology is best done through a direct subsidy (starter packets at reduced costs, for example).

RURAL INFRASTRUCTURE. Early research on economic growth illustrated the importance of infrastructure, provided that: (1) a good balance was maintained with other investments and (2) infrastructure and related services were run efficiently. More recent econometric research suggests that infrastructure investment and improvement may have received too little attention in the lowest-income countries. There are also signs that rapidly growing middle-income countries have underinvested in infrastructure, leading in some cases to geographic patterns of development that hamper economic growth (Willoughby 2002). Other recent studies of infrastructure investments conclude that:

- *Institutional reforms are needed* to strengthen the capacity of local and regional governments to formulate and implement an infrastructure policy and to strengthen the capacity of infrastructure organizations to provide customer-responsive services.
- *Institutional reforms in the more advanced countries led to greater involvement of the private sector* in investing and managing infrastructure, which requires improved capacity at the government level to run transparent tender procedures and maintain open competition.
- *Decisions to invest in infrastructure should focus* on regions that lag in economic development.

Box 6.2   Economic Impact of Rural Roads in Bangladesh

A detailed econometric study of the impact of investments in rural roads in Bangladesh found substantial savings in household transport expenses, averaging about 36 percent in villages participating in the Rural Development Project (RDP) and 38 percent in villages participating in the Rural Roads and Market Improvement and Maintenance Project (RRMIMP). Road improvement also significantly affected men's agricultural wages (which rose by 27 percent in RDP villages), fertilizer prices (which fell by about 5 percent in RDP and RRMIMP areas), and aggregate crop indices (prices increased by about 4 percent in both project samples,

whereas production increased by about 38 percent in RDP and 30 percent in RRMIMP villages). The road effects are substantial for adult labor supply in RDP villages and schooling of both boys and girls. The overall effect of road improvement on per capita consumption was estimated at 11 percent in both project areas.

This study clearly shows that investment in rural roads unleashes the agricultural production potential of rural areas. The supply response to what looks like modest input and output price changes is quite dramatic in the study areas. Much of this additional production found its way to the market.

Source: Khandker, Bakht, and Koolwal 2006.

Despite broad agreement about the importance of rural roads for linking farmers with markets, surprisingly little statistical evidence exists on the size and nature of the benefits of rural roads or their distributional impacts. Isolating the impact of investments in rural roads on agricultural productivity from other enabling factors is not only challenging (Walle and Cratty 2004), but many other benefits must be considered—higher wages, better access to schooling and health services, and so forth (see box 6.2).

Another rural infrastructure investment, irrigation, is considered an innovation in its own right as well as an important enabler of agricultural innovation more generally. A key reason cited for the limited impact of Green Revolution technology (improved varieties in combination with modern inputs) in sub-Saharan Africa is the very limited area under irrigation in comparison to other regions, particularly Asia. The underlying problem is that investment costs per irrigation unit are many times higher in sub-Saharan Africa than in Asia. Irrigation investment projects also tend to fail more often in sub-Saharan Africa. The World Bank's Operations Evaluation Department identified specific weaknesses in irrigation investment projects, including irrigation system design, operation and maintenance, cost recovery, and user groups. For example, cost-recovery schemes did not improve operation and maintenance because revenues went into the general treasury. Despite these weaknesses, World Bank irrigation projects report good returns on average, but these projects require above-average preparation and oversight because of their complexity.

MARKET INSTITUTIONS AND INFRASTRUCTURE. Investment in market institutions and infrastructure was greatly affected by the market liberalization ideology that dominated the economic policy debate during the 1980s and 1990s. During the 1960s and 1970s, many governments played an active, direct role in agricultural markets, and donors provided significant direct investment in state-owned companies, government-controlled cooperatives, and public marketing agencies. When these government-dominated systems fell into disgrace because of their poor performance, donor support for them evaporated. Difficult, lengthy, and sometimes disruptive processes of privatization and market liberalization marked the ensuing transition to private market-based systems. It took some time to realize that well-functioning markets would not inevitably emerge (and foster agricultural innovation); some form of government assistance is often needed. Attention has recently focused on strengthening a new architecture for agricultural market institutions and incentives, promoting private commercial activity, and reorienting state activity to providing enabling regulatory and physical infrastructure; as a result, donor investments in market institutions have begun to increase again (World Bank 2006).

KEY POLICY ISSUES

Aside from the more thematic policies that shape agricultural innovation, discussed previously, several cross-cutting policy issues affect agricultural innovation. They include policies to reduce poverty and sustain the environment, to foster collaboration between the public and private sectors, and to build social capital more generally.

Poverty reduction and environmental sustainability

The impact of investments in agricultural innovation has been measured mainly in terms of improvements in agricultural productivity (see the numerous rate-of-return studies). Over the years, however, environmental sustainability and poverty reduction have assumed equal importance as outcomes of agricultural innovation. This changed perspective affects not only the orientation of investments in agricultural innovation but investments in enabling factors. An environmental sustainability assessment is standard procedure for major investment projects in most countries. The poverty alleviation impact of new technologies is often difficult to assess ex ante, however. In this sense, innovation in itself is a rather crude poverty alleviation instrument, in contrast to enabling factors such as investments in rural infrastructure or rural credit, which can be targeted far more specifically to the poor.

Public-private collaboration

With widespread adoption of the market-economy model, many governments are minimizing direct intervention in the economy and, where possible, leaving things to the private sector. When government intervention is unavoidable, governments are delegating or contracting implementation to the private sector as much as possible. For example, in closing the rural finance gap, the preferred approach now is to involve commercial banks (often by subsidizing them to take on less profitable rural loans) or microfinance schemes rather than to establish government-owned rural banks. The construction of rural infrastructure is contracted out to the private sector, which is increasingly contracted to handle infrastructure operations and maintenance as well.

A primary objective of many national innovation policies is to create the right incentives for private investment in innovation. Governments can use five important instruments to stimulate private investment in innovation: (1) IPR legislation; (2) tax deductions and subsidies for R&D; (3) antitrust legislation (because a competitive environment stimulates innovation); (4) subsidized risk capital (either directly or through tax deduction facilities) and business incubators; and (5) restraining bureaucratic procedures for introducing new products and technologies.

Social capital

When it comes to strengthening the various enabling factors that stimulate agricultural innovation, social capital (the institutions, relationships, and norms that shape the quality and quantity of a society's social interactions) often stands out as critical to success. Examples of social capital's important role include the management of irrigation schemes, self-help groups in microfinance initiatives, communal road maintenance, the establishment of value chains, and similar efforts.

NEW DIRECTIONS, PRIORITIES, AND REQUIREMENTS FOR INVESTMENT

The previous section described the policies that shape the enabling environment for agricultural innovation, which can range from the very generic to the very specific. This section explores concrete investments that will stimulate agricultural innovation by creating a more positive enabling environment.

Investments in innovation policy and governance structures

A national innovation policy should:

- *Provide guidance to actors and stakeholders* within the NIS regarding their roles and functions in the system, and give a sense of direction (in other words, describe what they want to achieve together).
- *Provide an overall framework* within which innovation policies specific to particular sectors—agriculture, health, energy, and so on—and particular technologies—ICT, biotechnology, nanotechnology, and so on—should fit. The national policy sets priorities across sectors and technologies (and the public resources allocated to them), whereas the more specific policies set priorities within a particular sector or technology field.
- *Create the best possible conditions for innovation* by investing in a range of public goods essential for an innovative knowledge economy. A functional analysis of the NIS is a good starting point to identify which functions of the system are particularly weak and require additional attention and investment. Aside from the more traditional investments in the generation and exchange of scientific knowledge, investments are needed to support the application of scientific and industrial knowledge throughout the economy. Although this responsibility primarily belongs to the private sector,

government can stimulate innovative behavior in the business sector by providing incentives for private innovation (subsidies, tax breaks, and recognition, for example) and by stimulating the startup of new, knowledge-intensive companies using business incubators, venture capital, and similar measures (module 5).

Since innovation policy is such a new area, in most countries the governance structure for innovation has only very recently started to emerge. "Governance" concerns the systems and practices that governments use within their NISs to set priorities and agendas, design and implement policies, and obtain knowledge about their impacts (OECD 2005).

A particular challenge is where to assign responsibility for innovation policy within the government structure. A considerable number of countries have delegated this task to the ministry in charge of science and technology, while others have opted to establish a higher-level entity that brings relevant ministries together to coordinate national innovation policy. TNs 1 and 2 discuss investments in innovation policy and innovation governance structures in greater detail.

Investments in policy and regulatory frameworks that affect agricultural innovation

The most important regulatory reforms underway at present that affect agricultural innovation include:

- *IPRs.* Like PVRs and patents, trademarks, certification marks, and geographic indications are IPR instruments that, applied correctly, support private investments in product quality and distinctiveness that go beyond minimum standards. The TRIPS agreement places considerable pressure on countries to comply. These issues are discussed in TN 3 and IAP 3.
- *Biosafety.* Frameworks (including instruments and activities) that analyze and manage risks in the sectors on food safety, animal life and health, and plant life and health, including associated environmental risks (which came together under the so-called biosecurity framework). For example, the establishment of proper legislation and enforcement capacity regarding genetically modified organisms (GMOs) is a prerequisite for regulating their adoption (or prohibition). Signatories to the Cartagena Protocol on Biosafety assume the obligation to put the necessary legislation and enforcement capacity in place.[5] A large number of countries in Africa and Asia have yet to adopt biosafety legislation and enforcement regimes. Investments are needed to put regulations in

place, establish oversight structures, and train personnel. TN 4 discusses the ins and outs of biosafety regulation.

- *Technical regulations and standards.* TN 5 explores the introduction and upgrading of technical regulations and standards related to food safety, animal life and health, plant life and health, and quality-related attributes. The past several decades have seen a tremendous expansion of the number of technical regulations and standards emerging in these areas. This momentum reflects the intensification of regional and global trade and heightened concerns over accompanying threats to food safety and animal and plant health, as well as consumer concerns on the environmental impacts of agriculture production. It also reflects a wider set of innovations in science and technology that permit very sensitive detection and analytical methods, as well as improved knowledge of the quality and associated health hazards of agrifood products. But these emerging technical regulation and standards are also defining the focus of agricultural innovation. For example, plant breeding can be steered toward developing products that attain a preferred quality attribute (size, color, taste, and so on), while the prohibition of certain pesticides (due to stricter regulations) will induce research on alternatives for the control of pests and diseases.

For some time, countries may need assistance to develop the necessary legislation and assess the options from which they can choose. They will also need support to build and strengthen the related regulatory agencies and invest in standards-related infrastructure.

The regulatory reforms currently being implemented are expected to trigger all kinds of changes, such as improving the access of agricultural products to foreign markets (because they will meet higher SPS standards), increasing private investment in agricultural R&D (because IP is protected), fostering the adoption of more sustainable agricultural practices (because of the introduction and enforcement of environmental standards), and increasing the adoption of GM crops (because biosafety legislation and enforcement are in place).

Accompanying rural investments

Systems-thinking increasingly permeates approaches to economic development, including agricultural development. Criticism of earlier agricultural investments has focused on their tendency to operate as relatively isolated interventions that fail to develop any synergies. The current trend within the World Bank is to formulate bigger and more holistic agricultural development projects with longer time horizons. This module describes examples of the three types of rural investment that have strong synergies with agricultural innovation investments:

- **Rural financing systems.** To eliminate the lack of affordable credit as a constraint on the adoption of agricultural innovations, it is better not to look at credit in isolation, but to take a more holistic approach to financial services, including credit, savings, money transfers, leasing, and insurance. Microfinance initiatives can be an important intermediate step toward a more mature rural financial system. IAP 4 provides an innovative example of microfinance in Andhra Pradesh.
- **Rural infrastructure.** IAP 1 provides an example from Zambia of how investment in improved feeder roads enhances agricultural innovation, value added, and competitiveness.
- **Market institutions and infrastructure.** There is a strong synergy between market development and agricultural innovation, as both tend to take a value chain approach and emphasize the importance of markets and market institutions. Investment opportunities in market development include market infrastructure (such as distribu-

tion and collection points, storage facilities, and market and auction facilities), market institutions (such as supply chain organizations and information systems), and the capacity to explore and develop new markets.

MONITORING AND EVALUATING AN ENABLING ENVIRONMENT FOR AGRICULTURAL INNOVATION

How can a country's progress in creating an enabling environment for agricultural innovation be monitored and evaluated? This section identifies indicators corresponding to the various enabling factors discussed in this module (table 6.1). The indicators can monitor progress through time and, by benchmarking with other countries, give an idea of a country's relative position in establishing an enabling environment. The list of indicators is just an illustration, but a pretty good one to make a start. Other factors and indicators can be added later, and some may not be feasible in all instances because reliable statistical information may be lacking.

Table 6.1	Enabling Environment Factors and Indicators	
Cluster	**Enabling factor**	**Indicator(s)**
Macroeconomic policies	Political and socioeconomic stability	• Political instability index (the *Economist*) or consult www.countryrisk.com for various stability indices
	Favorable macroeconomic policies	• Net taxation of agriculture • Difference between the official and the market exchange rate • Impact of trade agreements on the agricultural sector
	Increased public investment in agriculture	• Agricultural expenditure as a percentage of total government expenditure • Share of public goods in agricultural expenditure
Education	General education	• Literacy rate (urban/rural) • Enrollment in primary education (urban/rural) • Enrollment in secondary education (urban/rural) • Enrollment in higher education (urban/rural) • Programme for International Student Assessment (PISA) scores
	Agricultural education	• Enrollment in agricultural schools at secondary level • Enrollment in on-the-job agricultural training schemes (such as farmer schools, extension courses)
	Agricultural higher education	• Number of agricultural graduates
Innovation policy and governance	A comprehensive national innovation policy in place	• Presence of an innovation policy • Presence and use of innovation policy instruments
	Innovation governance structure in place	• Existence of a governing body at the governmental (highest political) level for STI • Involvement of key stakeholders of the STI system in the governing body (composition of the governing body) • Existence of a national strategy (priorities) for STI • Main activities for the implementation of the national strategy • Intensity of interaction in the STI system vertically and horizontally • Participation and commitment of the private sector in policy preparation and implementation
	General "innovativeness" of a country	• Composite innovation indices such as the World Bank Knowledge Economy Index, the UNCTAD Innovation Capability Index, or the UNDP Technology Achievement Index[a]

(Table continues on the following page)

Table 6.1   Enabling Environment Factors and Indicators (continued)

Cluster	Enabling factor	Indicator(s)
Regulatory reforms	IPR legislation and regulatory regime in place and operating effectively	• Status of IPR legislation (patents, PVRs, trademarks, certification marks, geographic indications) • Capability of the IPR registration system (e.g., average time to complete a registration) • Capability of the legal system to handle IPR disputes • Patent statistics (number of newly registered patents, broken down by local and foreign) • PVR statistics (number of newly registered varieties, broken down by local and foreign) • Use of certification marks and geographic indications
	Biosafety legislation and regulatory regime in place and operating effectively	• Biosafety legislation in place • Biosafety regulatory system in operation • GMO research trials allowed • Introduction of genetically modified crops
	Policy and regulatory frameworks and capacity for managing agricultural health, food safety and associated environmental risks in place and operating effectively	• Legislative and regulatory frameworks upgraded • Institutions operating under clear mandates • Effective mechanisms in place for coordination and collaboration among the entities performing SPS and quality-related functions (including private actors) • Prioritization of investments and short-, medium-, and long-term plans in place to ensure that identified capacity needs (for example, in terms of skills, physical infrastructure, institutional structures, and procedures) are met • Incentives in place to support private sector compliance • Set of sustainable agricultural practices developed and promoted
Accompanying rural investments	Well-functioning rural financial system	• Domestic credit provided by banking sector as percentage of GDP • Agricultural credit as a percentage of total domestic credit
	Good rural infrastructure	• Road density per square kilometer • Percentage of agricultural land under irrigation
	Well-functioning agricultural markets	• Percentage of agricultural production sold in the market • Share of exports in total agricultural production • Presence and strength of supply chain organizations

Source: Author.

(a) UNCTAD = United Nations Conference on Trade and Development; UNDP = United Nations Development Programme.

National Innovation Policy

Johannes Roseboom, Consultant

SYNOPSIS

National innovation policies tend to be overarching, attempting to coordinate a wide range of more specific policies (for science and technology, education, IPRs, the economy and industry, trade, and taxation) and foster optimal conditions for innovation. A national innovation policy (based on NIS ideas and concepts) enables the formulation of an agricultural innovation policy in two ways. First, its economy-wide perspective makes it possible to address issues that should be dealt with at the level of the national economy rather than individual sectors. Second, it promotes innovation system thinking at the sector level. This note describes the steps and tools in this policy-making process.

BACKGROUND AND CONTEXT

The recent rise of innovation policy around the world has been strongly influenced by the NIS school of thinking. NIS thinking has been picked up by national and international policy makers remarkably quickly throughout the world, partly because the OECD was an early promoter of the concept but more importantly because countries are looking for ways to respond to increased global competition. How can a country improve its competitive edge? This question is arguably even more important for the world's poorest countries than for wealthy ones. However, the NIS concept does not provide a simple blueprint for organizing innovation. It is foremost an analytical tool for policy making and planning.

A national innovation policy is not just an extension of the science and technology policy. It is a higher-level policy integrating science and technology, economic, industrial, infrastructure, taxation, trade, labor, and education policies (to name the most relevant). For this reason, it is not limited to one particular ministry or agency. It requires substantial coordination and consensus building among ministries as well as socioeconomic partners such as the business sector, trade unions, and value chain organizations. The mix of policies for innovation depends on the country's political orientation and objectives, and different forms of innovation require different policy instruments and responses (OECD 2010). One developing country with a very articulate innovation policy and strategy is Chile (box 6.3).

ACTIONS NEEDED

The role of an innovation policy is to create the best possible conditions for innovation by developing a range of public goods that are essential for an innovative knowledge economy. The best possible conditions will vary significantly from one country and sector to another, depending on the development phase, economic structure, and national priorities. *Five key functions* have been identified and are recommended as the primary targets of a national innovation policy.[1]

First, organize and implement a national innovation policy. Such a policy should mobilize and engage the various actors and stakeholders within the innovation system, provide guidance regarding their roles and functions within the system, and give a sense of direction, explaining what they want to achieve together. Important steps in this process are described in box 6.4. In most countries, the overall objective of the national innovation policy is to facilitate the transition toward a *knowledge economy*, resulting in increased competitiveness and sustainable economic growth. As noted in the module overview, a national innovation policy should provide an overall framework within which sector-specific and technology-specific innovation policies should fit. This approach may require improved coordination and coherence of policies and different layers of government. The national policy sets out overall priorities across sectors and technologies (and the public resources allocated to them), while the more specific policies set the priorities within a particular sector or technology field.

Box 6.3 Chile's Innovation Strategy

Following a long period of stagnation from the mid-1950s to the mid-1980s, Chile's economy started to take off, and for the past twenty-five years, it has been one of Latin America's better-performing economies. The opposition parties elected to government after 1988 continued the free-market policies introduced by the military junta to a substantial extent but with a greater appreciation of government's role in economic development, including its role in stimulating innovation.

Chile's economy has been booming partly because of high revenues from copper exports. To invest those revenues wisely, the government decided to invest heavily in moving away from a predominantly resource-based economy (agriculture and mining) toward a knowledge-intensive economy. For this purpose, it created a national innovation fund for competitiveness (FIC, Fondo de Innovación para la Competividad), funded by a new tax on mining, in 2005. A newly created national innovation council for competiveness (CNIC, Consejo Nacional de Innovación para la Competividad), in which the various sectors and interest groups are represented, advises FIC on how to allocate its resources, while an interministerial committee on innovation (CMI, Comité de Ministros para la Innovación) is responsible for implementation.

As part of this new initiative, CNIC has formulated a national innovation strategy. After extensive study and consultation, CNIC selected five economic clusters on which to focus science, technology, and innovation (STI) investments: agro-food, aquaculture, mining, tourism, and global services. For each selected cluster, a strategic board with public and private representation has been created to set cluster-specific priorities. The Strategic Board of the Agro-Food Cluster has identified the following subclusters as the most promising for further development and knowledge intensification: fruit, wine, processed food, pigs and poultry, and red meat. These priorities have been passed to the various STI funding agencies, which are organizing calls for proposals for these priorities or giving the selected clusters priority in more generic calls for proposals. Moreover, despite their name, competitive funding schemes are being used to cement stronger links within the innovation system by promoting cross-institutional collaboration between universities and research institutes and by promoting public-private partnerships in the form of "technology consortia." The latter instrument not only cements collaboration between a research agency and the private sector but between companies that share a common technology platform.

Since FIC's creation in 2005, public STI investments in Chile have more than doubled in real terms (reaching US$530 million in 2009). Public STI investments are projected to continue to grow by 10–15 percent per year over the coming ten years.

Parallel to the STI initiative, the Chilean government established a major scholarship scheme (Becas Chile) in 2008, which will allow some 30,000 Chileans to study abroad over the next ten years. The budget for this scheme is some US$6 billion and is also financed out of mining royalties.

Source: Author.

Second, improve the regulatory framework for innovation. Given the many actors within the innovation system and their often conflicting interests, a set of rules and regulations is needed (on dealing with IPRs, fair competition, technical standards, health, and environment, among others) to create a playing field that is transparent and fair.

Third, foster innovation through education. Innovation depends on the level of education in the general population, including the knowledge and skills that people will need in the future and strategies to keep knowledge and skills up to date (in other words, to develop a capacity for lifelong learning). Innovation also depends on the education of science and innovation specialists more specifically, which may involve among other things motivating students to specialize in science.

Fourth, facilitate the creation, exchange, and diffusion of knowledge. This is the core business of an innovation system. Knowledge should not be limited to knowledge generated only by research organizations (and as such codified in scientific publications and patents) but should include the knowledge (a large part of it tacit) accumulated within the economy of a country. It is important to make sure that knowledge (both scientific and industrial) is

Box 6.4   Developing an Innovation Policy

Many of the steps or components of developing an innovation policy do not occur chronologically but simultaneously or in reverse order. They include:

- **Policy analysis.** This step requires a thorough understanding of existing policies and their influence on the innovation system (in other words, their interaction with institutions and actors). A functional analysis may be a useful input in grasping the strengths and weaknesses of the policies in place.

- **Formulating policy advice.** To a large extent, the political context determines whether policy makers are sensitive to evidence and how evidence reaches them. Knowledge of the political context and entry points for evidence and dialogue are essential. Influencing policy through research requires good data as well as credibility of the institution presenting the data. For new ideas to be embodied in policies, it is critical to prove that those ideas provide a solution to a real problem. In presenting such evidence, communications skills are highly important, and using a diversity of communications methods increases the chances of success compared to relying on a single method or pathway. Through links with media,

intermediary organizations, and networks advocating for policy change, policy makers can be pressured from different angles to change policies in a certain direction.

- **Policy making.** Policy makers need to get involved actively in the multistakeholder exchanges and activities that occur to facilitate and realize innovation. When policy makers are immersed in a subject, evidence-based policy making becomes experiential policy making. Policy makers learn, through interaction and engagement with other system actors, how policies influence the system and what changes would be required.

- **Policy implementation.** Often the job ends for policy makers when the policy is written down and made official, yet stating the policy is only the starting point for change, not the end. An inclusive policy-making process makes it more likely that the policy will actually be implemented. When different stakeholders understand the need for policy change, have invested in it, and stand to benefit, there is pressure on the one hand to enforce the policies and on the other hand there is a greater likelihood that stakeholders will abide by the implemented rules and regulations.

Source: Author, drawing on KIT 2011.

---

adequately stored and accessible. An important variable in this context is the quality of a country's ICT infrastructure and the density and quality of its Internet connections. It is also important to stimulate the exchange of knowledge beyond national borders, which may involve measures to improve the language capabilities of knowledge workers, stimulate attendance at international scientific conferences, and create exchange programs and industry-specific study tours.

Fifth, mobilize and allocate resources for innovation activities. Funding of innovation activities can range from fully public to fully private and everything in between, depending on the type of industry and activity. The national innovation policy should: (1) define which innovation activities require public support; (2) define the tax base for public funds (general versus specific taxes; see box 6.5); (3) define

the type of funding instruments to be used (subsidies, tax breaks, patents, procurement, and so on); and (4) prioritize and allocate public resources across the various innovation activities.

These principal functions of an NIS also remain relevant at the AIS level, but many of the policies shaping these functions at the AIS level are formulated as part of the national innovation policy. For example, most investments in education affect all sectors, and the same is true for many regulatory issues, such as IPR legislation and environmental standards. At the same time, the agricultural innovation policy may opt for sector-specific policies if the situation requires. For instance, it may choose to support plant breeders' rights or agricultural advisory services. For this reason, it is important to coordinate the development of an agricultural innovation policy with the development of a

Box 6.5   The Creation of Sectoral Science, Technology, and Innovation Funds in Brazil

While Brazil has invested substantially in science and technology since the 1970s, the economic impact of these investments has been modest and uneven. In the 1990s, the Brazilian science and technology system was criticized for being too science oriented, placing little emphasis on innovation, and lacking stable funding because of economic problems. The Ministry of Science and Technology (MCT, Ministério da Ciência e Tecnologia) introduced major restructuring of the funding of the science and technology system in 1999. It set up sector-specific science, technology, and innovation (STI) funds (in part replacing older funds), which are financed through levies to be negotiated within each sector. In this way a substantial amount of new and more stable "private" resources were mobilized to finance STI of relevance to each specific sector. For sectors in which a levy was not feasible, the government increased the levy on foreign technology transfer payments from 15 percent to 25 percent. The increase finances 4 out of 15 funds.

Of the 15 funds that have been established, 2 are not sector-specific but focus on bottlenecks in the STI system: the university-business fund and the fund for science infrastructure. The infrastructure fund is financed through a 20 percent levy on all of the other sector funds. The sector funds have the status of trust funds and are managed jointly by the academic community, industry, and government. The strong representation of industry in these funds was a particular innovation for the Brazilian STI system.

The sector funds serve four major government objectives: (1) stabilize financial resources for medium- and long-term R&D; (2) improve transparency in funding decisions, merit reviews, and evaluations; (3) reduce regional inequalities; and (4) promote interaction between universities, research institutes, and companies. The selection of strategic sectors, their respective shares of the funds' resources, the blend of basic and applied research, the required overall budget, and sources of support are all jointly decided by the academic community, private sector, and government.

Source: IAC 2003; Roseboom 2004.

national innovation policy and make sure that they are consistent.

POTENTIAL BENEFITS

The development of a national innovation policy usually reflects the priority that a country assigns to knowledge as an important (if not the most important) source of future economic growth, especially as capital accumulation levels off and the transition toward a capital-intensive economy is complete. A national innovation policy helps shift the emphasis in macroeconomic policy from physical capital to human and social capital.

A national innovation policy can also help to create more coherence in a broad range of government policies dealing with issues such as economic development, education, competitiveness, trade, and R&D investment. More coherent policies should help to improve the overall effectiveness of the NIS.

The more prominent role attributed to knowledge in economic development has sparked renewed interest in agricultural innovation and how to improve and modernize it. At the same time, NIS thinking is influencing AIS thinking in important ways:

- Far greater emphasis is placed on private R&D and innovation activities by private firms. Agricultural research and extension were traditionally viewed as government responsibilities. Relatively little attention was given to involving the private sector.
- Education receives greater emphasis as an important enabling factor in agricultural innovation, both on the farm and in research and extension agencies.
- The regulatory framework's importance in shaping innovation is more widely recognized; for instance, biofuel targets play an important role in shaping the biofuel innovation agenda.
- Innovation driven by market demand and market opportunities is given greater emphasis.
- More attention is paid to improving the mobilization, inclusion, and coordination of innovation actors and stakeholders.

POLICY ISSUES

A national innovation policy in a developing country will need to address a number of important policy issues. For example, it will need to define the roles of the public and private sector in innovation; ensure that institutions and incentives are in place to govern and coordinate innovation; determine the priorities for innovation; come to terms with the environmental and social implications of innovation; and acknowledge the informal economy's role in innovation.

Defining public and private roles in innovation

From a market economy perspective, innovation is first and foremost a task of private entrepreneurs. Government policy should focus on creating an enabling environment for private innovation to take place. Some of these policies, such as well-functioning markets, sound corporate governance, and sound financial institutions, may not be specifically aimed at fostering innovation but are nonetheless important. Other policies, such as IPRs, the setting of technological standards, science education, and basic research, may enable private sector innovation more directly. When it comes to innovation, however, a great deal of market or systemic failure requires more direct government intervention or support (Edquist 2001). In primary agriculture, for example, the extreme fragmentation of production into small family farms has traditionally been a legitimate reason for the government to intervene directly.

Governance and coordination

A country's innovation performance depends in part on the strength of the institutional arrangements and incentive structures that govern innovation. Innovative activity is not governed by government alone. Actors from research and the business sector, as well as other stakeholders, play important roles. TN 2 provides a detailed discussion of governance in formulating innovation policy and coordinating innovation.

Making strategic choices

Most countries are too small to excel in all sectors and technologies. One has to be selective and make strategic choices to concentrate innovation investments in specific technology fields and sectors. Spreading resources too thinly will be counterproductive. How to make such strategic choices is a major policy issue and requires sound analysis of the options. Chile offers a good example of an innovation policy that makes clear strategic choices (box 6.3). Many countries avoid

setting these priorities because they are highly sensitive from a political standpoint, but a characteristic shared by successful innovators such as Finland and the Republic of Korea is that they have restructured their economies strategically toward more knowledge-intensive industries (see TN 2).

Environmental sustainability

Many current production and consumption patterns are not sustainable in the long run. They deplete natural resources and are so polluting that they may cause climate change, with far-reaching repercussions for life on earth. Aside from raising overall productivity, innovation must meet increasingly stringent criteria for environmental sustainability and offer green solutions. It is virtually imperative for a national innovation policy to address this overwhelming challenge that faces humankind over the next few generations. Some countries are seeing the environmental crisis as an economic opportunity and positioning themselves as champions of green technologies (a small country like Denmark, for example, is a world leader in windmill technology). In agriculture, reducing greenhouse gas emissions is one of the major environmental challenges, along with developing strategies to cope with the effects of global warming.

Social considerations

The flipside of innovation is creative destruction. Jobs disappear to low-income countries or are replaced by more efficient production methods requiring less labor. Policy measures are needed for people to make the transition to more promising economic activities and new jobs. Education plays a major role in this process, including the concept of lifelong learning. Agriculture is a classic example of how innovation often results in fewer jobs. The exodus of labor from agriculture is characteristic of economic development and coincides with the transition from an agricultural to an industrial economy and from rural to urban life. This transition has never been easy, but it seems to have become even more difficult as innovation in industry has also reduced the demand for labor. Industry's capacity to absorb labor is far lower than it was one hundred or even fifty years ago. China, for example, still has a very large rural labor surplus despite rapid industrialization over the past three decades.

Informal economy

Many developing countries have a very significant informal economy within the overall economy. The informal economy is something of a blind spot for government policies, including innovation policy. Even the most comprehensive concept of NISs has yet to fully address innovation that takes place in the informal sector—yet disregarding the role of the informal sector can produce misleading, asymmetrical, and ineffective innovation strategies (Kraemer-Mbula and Wamae 2010).

LESSONS LEARNED AND RECOMMENDATIONS FOR PRACTITIONERS

A dynamic NIS does not emerge overnight. Such a system is built and continuously improved only through a sustained, long-term commitment by government and the private sector. A national innovation policy should guide this process. In most countries, particularly in the difficult institutional context of developing countries, implementing an innovation policy can be challenging. A long-term strategy to develop a national innovation policy should be inspired by a philosophy of "radical gradualism," which suggests a sequence of finely tuned, small, specific reforms and successful outcomes that paves the way for broader institutional changes (World Bank 2010). One of the first issues to tackle is the fact that in most developing countries the capacity to formulate and implement an innovation policy is usually scarce and must be built over a considerable period. An example of the radical gradualism approach is Argentina's Unleashing Productive Innovation Project, which comprises a wide range of interventions to eliminate critical bottlenecks within Argentina's innovation system (box 6.6).

Box 6.6   Unleashing Argentina's Productive Innovation

The Unleashing Productive Innovation Project is a major World Bank effort (its total budget is US$223 million for five years, of which US$150 million is loan money) to assist Argentina to become more innovative, promote diversification into more knowledge-intensive economic activities, and stimulate economic growth. The project, which strongly reflects a national innovation system approach, consists of the five components:

- *Developing human capital for productive innovation (US$28 million).* This component fills critical human capital gaps in the national innovation system by developing training programs for "technology brokers" and "technology managers" to professionalize and improve innovation processes. It also offers scholarships to pursue studies in information and communication technology (ICT) and reduces Argentina's shortages of qualified personnel.
- *Support for new knowledge-based companies (US$ 54 million).* This component promotes the development of new knowledge-based companies through two complementary activities. The first is a pilot of an early-stage venture capital fund. The second is the creation of a proactive, market-driven incubation cycle—from the initial idea to a commercial project, through early-stage venture capital investment—by establishing "deal flow" promoters that are mainly remunerated on a fee-for-success basis.

- *Fostering sector-specific capacity for productive innovation (US$85 million).* This component will develop critical capacities in three priority areas: biotechnology, nanotechnology, and ICT. Resources will be allocated through competitive funding schemes designed to foster public-private and private-private collaboration. The private business community will play a lead role in these funds, both in specifying the research agenda and funding it.
- *Upgrading research infrastructure (US$36 million).* Funds will be allocated on a competitive basis using predefined selection criteria, such as the extent to which the beneficiary is connected with the productive sector.
- *Strengthening the policy and institutional framework for innovation (US$20 million).* This component will strengthen the policy-making capacity of the Ministry of Science and Technology, strengthen the capacity of the National Agency for the Promotion of Science and Technology in selected areas, and support dissemination of project activities.

Although the project does not target the agricultural sector specifically, the sector can benefit from it in various ways, especially through the biotechnology fund included in component 3. Indirect spillins from the other components are likely as well.

Source: World Bank 2008.
(a) See module 5, TNs 3 and 6, for discussions of how incubators and risk capital are used to support agricultural innovation.

Pay attention to scale and scope

Innovation systems can be considered at different scales, from a national scale to smaller geographic or political units (states, regions, provinces), and with different scopes, from economy-wide to sector- or commodity-specific innovation systems. From the point of view of policies for improving the enabling environment for innovation, the issue of scale and scope is relevant. Policy objectives must be explicit with respect to the scale and scope of the system they intend to influence. At the same time, a balance needs to be struck between centralized and decentralized political decision-making in terms of scale and scope. A national innovation policy should provide the basic architecture for who is responsible for what.

Another challenging issue is where to situate responsibility for innovation policy within the government structure. Many countries have delegated this responsibility to the ministry of science and technology, while others have opted to establish a higher-level entity that brings the various relevant ministries together to coordinate innovation policy (see the Chilean example in box 6.3 and TN 2 on innovation system governance). As noted in the module overview, the latter option seems to be preferred.[2]

A criticism of *national* innovation policies is that they tend to ignore opportunities for *supra-national collaboration* in the innovation sphere. They are often too inward looking and ignore opportunities for regional or international collaboration. Regional economic communities are becoming more active on innovation policy issues, however. They often press hard for product and technology standardization within their communities to create the optimal conditions for a single market. The EU is by far the most advanced regional community in terms of having a regional innovation policy in place (known as the Lisbon Strategy). In addition, various industry-specific or technology-specific innovation platforms in Europe enable European industries to work together on new technologies.

Mobilize a broad spectrum of actors in making innovation policy

An important factor in successfully setting and implementing an innovation policy agenda is the ability to mobilize a broad spectrum of innovation actors. Successful mobilization of these actors depends on factors such as persuasive arguments and incentives, as well as the autonomy of the actors in the institutional landscape, the nature of existing linkages (social capital), and the effectiveness of leadership. The policy agenda for STI is sometimes dominated by narrow scientific elites with considerable influence, particularly in countries with a less mature innovation system (OECD 2010). One way of getting a better overview of the innovation landscape is to conduct a functional analysis of the actors that make up the landscape (box 6.7).

Evaluate and measure innovation performance

Continuous monitoring and evaluation of a country's innovation performance should be an important component of any national innovation policy. Do the various policy instruments and interventions yield the expected results? Benchmarking is a much-used tool at the international level to identify best innovation policy practices, while composite innovation indicators help to monitor innovation performance across countries and through time (box 6.8).

Foster interaction between the national innovation policy and the agricultural innovation policy

The introduction of a national innovation policy often has an important impact on the public agricultural

---

Box 6.7   A Functional Analysis of a National or Sectoral Innovation System

A *functional analysis* is useful to rapidly assess a national or sectoral innovation system.[a] It can help to identify the principal actors within an innovation system and the linkages and interactions between them. Weaknesses identified in such an analysis can form a good starting point for formulating specific innovation

policy interventions. The standard steps in a functional analysis of an innovation system are:

1. *Define the boundaries of the innovation system in focus.* The level of aggregation in a functional appraisal can vary substantially. For example, one

(Box continues on the following page)

Box 6.7 A Functional Analysis of a National or Sectoral Innovation System (continued)

can deal with the business sector as one entity or differentiate it by sector, size, innovation profile, or some other feature. This process will enrich insights into how the innovation system functions and hence help to fine-tune policy interventions.

2. *Identify and engage the principal actors* in each of the stakeholder groups within the innovation system, including (1) the business sector, (2) the government sector (including the principal policy-making, coordinating, financing, and regulatory agencies for science, technology, and innovation), (3) the research sector (research organizations, universities, and others), (4) technology transfer and other intermediary organizations, (5) organized civil society (nongovernmental organizations, consumer groups, trade unions, and the like), and (6) possible foreign innovation partners;

3. *Define the primary functions that the innovation system needs to perform.* The "Actions Needed" section proposes five key functions that an innovation system needs to perform. These functions are not set in stone but can be reformulated to match the specific context to which they apply.

4. *Bring steps two and three together* and map in a matrix format how the different stakeholder groups and their specific actors contribute to the different key functions. This exercise should help identify possible missing actors or weak links between actors or stakeholder groups. Such mapping is best done on the basis of interviews with the various actors involved in the innovation system. An alternative is to make the map based on brainstorming sessions with key experts.

Source: Author.

a. See Paterson, Adam, and Mullen (2003) and Ivanova and Roseboom (2006) for practical examples of applying a functional analysis approach to national innovation systems.

Box 6.8 Benchmarking National Innovation Systems and Policies

Since the mid-1990s, the Organisation for Economic Cooperation and Development (OECD) has conducted an array of studies on national innovation systems and policies, covering its member and nonmember countries (including leading developing countries such as Brazil, Chile, China, and South Africa).[a] The OECD methodology, based on the *Oslo Manual: Guidelines for Collecting and Interpreting Innovation Data* (issued in 1992 and updated in 1997 and 2005), has been copied frequently by other innovation system studies. The latest edition of the *Oslo Manual* includes specific guidelines for the implementation of innovation surveys in developing countries, based largely on experience with the methodology in Latin America.

In 2000, the European Union adopted its "Lisbon Strategy," which aims to "make Europe the most competitive and the most dynamic knowledge-based economy in the world." As part of this strategy, the European Union publishes an annual European Innovation Scoreboard (EIS) to monitor the innovation performance of individual member nations as well as the performance of the European Union in relation to other economies such as the United States, Japan, China, and Brazil. The EIS methodology comprises seven innovation dimensions, grouped into three blocks:

- *Enablers:* Captures the main innovation drivers that are external to the firm and comprises two dimensions: (1) human resources (measures the availability of highly skilled and educated people) and (2) financial resources (measures the availability of finance for innovation projects and the support of governments for innovation activities).
- *Firm activities:* Captures innovation efforts that firms undertake and comprises three dimensions: (1) investment in innovation by firms (multiple

(Box continues on the following page)

variables); (2) linkages and entrepreneurship (captures entrepreneurial efforts and collaboration among innovating firms and also with the public sector); and (3) throughputs (IPR registration and balance of payments regarding technology royalties).

■ *Outputs:* Captures the outputs of firm activities and comprises two dimensions: (1) innovators (measures the number of firms that have introduced innovations onto the market or within their organizations) and (2) economic effects (captures the economic success of innovation in employment, exports, and sales arising from innovation activities).

In addition to EIS, several other international innovation indices are produced, including the World Bank Knowledge Economy Index, the UNCTAD Innovation Capability Index, UNDP Technology Achievement Index, the RAND Science and Technology Capacity Index, the WEF Global Competitiveness Index, the INSEAD Global Innovation Index.[b] The indices use different approaches, but rankings are reasonably stable across indices. Incomplete and poor data cause the rankings of countries at the bottom to be considerably less stable, however. One criticism of current measurement frameworks is that they often fail to measure the social impacts of innovation (on well-being and poverty reduction, for example).

Source: Pro Inno Europe 2010; World Bank 2010; OECD 2010.
(a) The *OECD Reviews of Innovation Policy* comprehensively assesses innovation systems in individual OECD members and nonmembers, focusing on the role of government. The reviews provide recommendations to improve policies affecting innovation performance, including R&D policies. Each review identifies good practices from which other countries can learn (www.oecd.org/sti/innovation/reviews).
(b) UNCTAD = United Nations Conference on Trade and Development; UNDP = United Nations Development Programme; WEF = World Economic Forum; and INSEAD = originally Institut Européen d'Administration des Affaires (European Institute of Business Administration).

research and extension services established decades earlier. These agencies are required to interact more vigorously with STI agencies that have a far wider scope and compete with nonagricultural agencies for resources from STI funding schemes. National innovation policies also tend to introduce new instruments to promote innova-

tion that are not designed specifically for agribusiness but which agribusiness can use. Business incubators, risk capital, technology consortia, technology parks, technology subsidies, and private R&D incentives (subsidies or tax deductions) are examples (many of which are discussed in module 5).

Governance of Innovation Systems

Christopher Palmberg, ADVANSIS Ltd.
Tarmo Lemola, ADVANSIS Ltd.

SYNOPSIS

Investments in developing an NIS should give governance particular attention, especially the systems and practices for setting priorities and agendas, designing and implementing policies, and obtaining knowledge about their impacts. This note provides examples of the roles that innovation policy and its governance have played in the development of innovation systems in Finland, Republic of Korea, and South Africa. Based on these cases, the note identifies governance activities relevant to innovation systems for agriculture in developing countries and discusses the related policy issues, lessons, and recommendations emerging from the case studies. A key issue arising from the cases is that well-functioning innovation systems critically depend on how well governments can bring together and coordinate the activities of the various actors and stakeholders fundamental for advancing science, technology, and innovation in various sectors of the economy.

BACKGROUND AND CONTEXT

Governance concerns the mechanisms by which decisions are made in an organization, whether public, private, or nonprofit. Governance has several dimensions, including power, culture, incentives, leadership, and coordination. In governance of an NIS, the systems and practices for setting priorities and agendas, designing and implementing policies, and obtaining knowledge about their impacts receive special attention (see OECD 2005). A number of factors impinge on the efficiency of the governance of an NIS—in other words, the extent to which policy processes have the greatest effect with a given use of resources (OECD 2010). Evidence indicates that efficient governance depends on certain qualities, including:

- *Legitimacy.* The policy actors and approaches adopted in policy processes have to be widely appropriate and accepted for the tasks at hand.

- *Coherence.* The different strands of innovation policy and associated policy instruments must fit together.
- *Stability.* Innovation requires sufficiently stable framework conditions, institutions, and policy.
- *Ability to adapt.* As the environment for innovation evolves, and innovation evolves along with it, governance actors need to be able to adapt.
- *Ability to steer and give direction.* A related capability is the governance system's ability to provide direction to actors and steer the innovation system as a whole. The ability to provide direction requires commitment and leadership from policy makers at the highest level.

Governance of innovative activity is not provided by government alone. The research and business sectors as well as other stakeholders such as NGOs play important roles in many aspects of the governance of an NIS. For example, a society's accumulated social capital can make an important contribution to innovation by increasing trust among the actors, which makes joint innovation efforts as well as communication and sharing of knowledge between the actors easy and successful.

Innovation system governance at the sectoral level is an important part of overall innovation system governance. In the agricultural sector, the earliest attempts at coordinating AIS were centered on strengthening agricultural research coordination. A number of developing countries have established research governance bodies, but they tend to represent only a narrow range of AIS stakeholders, consisting primarily of ministerial representatives or researchers. They have often lacked a consistent, rigorous process for setting priorities. The current movement to improve the representativeness of these governance bodies and their mode of operation is encouraging, however (for example, seeking to represent a wider range of stakeholders and regions, improving transparency, and using diverse prioritization tools). The overall trend is toward strengthened research governance and multidisciplinary NIS governance (as discussed in this note), wherein agriculture is one sector among many. Some countries have made specific efforts in AIS governance, however. Typically these efforts center on subsectoral governance and coordination—for instance, through commodity boards and subsector networks—rather than on national agriculture/rural innovation governance structures (like Chile's FIA and Australia's Rural Research and Development Council). Module 1 discusses innovation coordination in agriculture in greater detail and provides examples of AIS coordination and governance at the macro, meso, and micro levels.

Although this TN discusses NIS governance, benefits, policy issues, and lessons primarily from developed countries, it can help identify relevant issues and lessons for developing countries and their AISs. Finland, Korea, and South Africa have been chosen as examples because, in different ways, they represent NISs in which government actors and agencies play an important role. They also represent NISs at different phases of development to illustrate governance challenges from different viewpoints. A separate note in this module discusses overall innovation policy issues.

Finland

Finland began to apply the NIS concept before many other countries, and its NIS has a relatively streamlined governance structure, developed in the mid-1980s and early 1990s. The Finnish Funding Agency for Technology and Innovation (Tekes, teknologian ja innovaatioiden kehittämiskeskus), was established in 1983, and R&D programs soon followed. A key characteristic of the Finnish system is that high-level government officials (prime minister, finance minister) as well as representatives from universities, public research organizations, and industry participate in the Research and Innovation Council, which develops national guidelines for innovation. Operational responsibility for policies is delegated to the Ministry of Education and Culture (for basic research), the Ministry of Employment and the Economy (for applied research and the enabling environment for innovation), and other ministries.

A second important characteristic of the Finnish NIS is that the main funding agencies (Academy of Finland for basic research and Tekes for applied research) enjoy considerable autonomy in implementing programs, introducing new policy instruments, and managing these programs and instruments on a day-to-day basis. A third characteristic is the strong tradition of collaboration and coordination throughout the NIS, both across the main ministries and

agencies involved as well as down through the various decision-making levels. There is a strong element of consensus building among the main stakeholders in the design and implementation of policies. Companies and the research community are often involved in policy discussions as experts or through their branch organizations. Innovation policy also explicitly aims to support collaboration and networking between industry, universities, and public research agencies. For example, the R&D programs commissioned by Tekes require collaboration by industry, universities, or public research agencies.

Republic of Korea

Korea's government has taken an active approach to NIS governance, especially since the mid-2000s. As in Finland, in Korea the NIS involves high-level government officials (ministers and other key stakeholders) in designing STI policy through the Presidential Advisory Council for Education, Science, and Technology (with representatives from industry, academia, and research) and the National Science and Technology Council (formed by government ministers). The role of ministries in implementing policy down to the level of individual R&D programs and projects is noteworthy, especially within the Ministry of Education, Science, and Technology (MEST).[1]

Unlike Finland, in Korea the NIS has a complex governance structure. Government science and technology policies have long roots, and the government's overall role has been pronounced. A key challenge for Korea is to govern its rapidly growing portfolio of policy measures (OECD 2009b), and Korea is responding with efforts to improve the coherence of its policies through horizontal coordination (between advisory councils and ministries) and vertical coordination (between ministries and the government research institutes).

A third characteristic of the Korean system is the duality in corporate structures. Large conglomerates or multinationals (*chaebols*, literally "business families") dominate research, development, and industrial transformation, whereas SMEs remain relatively underdeveloped. In this sense, Korea is still a mixture of an advanced and developing country. This duality has crowded out entrepreneurship and may have hampered technology diffusion and knowledge spillovers throughout the system. Especially compared to Finland, collaboration and networking in Korea between companies, universities, and research institutes is less pronounced, though collaboration within chaebols is extensive. A central challenge for the Korean NIS is to encourage more collaboration and networking, both nationally and internationally (OECD and World Bank Institute 2000; OECD 2009b).

South Africa

In the mid-2000s, South Africa became one of the first developing countries to adopt an NIS approach. South Africa is emerging as a global player in STI in certain fields but faces a range of challenges in developing its NIS amid difficult socioeconomic conditions and weak government coordination. Responsibilities for science and technology have been fragmented among numerous ministries, departments, and agencies. Private R&D has been concentrated in a few large, diversified companies with established links to government departments, research organizations, and universities. The innovation system has been virtually disconnected from black communities (Hausman and Klinger 2006; Lingela 2004).

Since 2000, science and technology have been under the purview of the Department of Science and Technology (DST). The Parliamentary Portfolio Committee for Science and Technology oversees DST; the National Advisory Council on Innovation and a large group of stakeholders at the National Science and Technology Forum provides advisory support. Other key STI ministries include the Department of Education, Department of Trade and Industry, and sectoral departments such as minerals and energy, agriculture, water, and forestry. These departments steer their activities through sectoral agencies, foundations, and other funding organizations (OECD 2007b).

South Africa has made remarkable progress in a short period, as evidenced by STI indicators such as a more diversified industrial structure and increasing GDP per capita. Nonetheless, huge social inequalities remain. The limited involvement of the "second economy" of black communities in entrepreneurship and innovation remains a primary characteristic and challenge for the NIS. One source of this problem may be the continued, poor horizontal coordination across the main ministries, agencies, and funders of R&D (OECD 2007b). This lack of overall government coordination is a second characteristic of the South African innovation system.

Limited technology transfer and networking between academia and industry is a third characteristic of the NIS, caused in part by the lack of mental models for how an innovation system functions beyond the public sector. The enabling environment for entrepreneurship is also underdeveloped, as reflected by the limited collaboration between large and small companies, the poor availability of venture capital funding, and an outdated IPR regime.

In 2009, the government established the Technology Innovation Agency (TIA) to improve coordination of innovation funding (Nordling 2009). The new agency is responsible for administering a handful of existing innovation schemes: the Biotechnology Regional Innovation Centers, the Innovation Fund, the National Advanced Manufacturing Technology Strategy, and the Tshumisano Trust.

ACTIVITIES AND CAPABILITIES NEEDED

Good governance is manifested in the degree to which capabilities in the following areas can be developed and supported: perception of and responses to challenges, setting policy priorities and coordinating agendas, implementing and managing policies on a day-to-day basis, and obtaining and processing intelligence.[2] These capabilities are associated with different levels of governance in an innovation system and depend on how interactions and coordination are governed (vertically and horizontally) throughout the system (Nelson 2003; OECD 2007b, 2008, 2009a, 2009b, 2009c). Figure 6.2 depicts typical decision-making levels, key public (or semipublic) organizations, and avenues through which an innovation system can be governed to develop and sustain these capabilities. The figure highlights the key governance capabilities within the institutional and organizational framework of an NIS.

Strengthening policy makers' capacity to perceive and respond to challenges

The ability to perceive and respond to challenges is important for an NIS to be agile and proactive. In other words, these capabilities are vital for developing innovation policy guidelines. These capabilities are *embedded in the NIS as a whole, at all levels of governance,* although councils, advisory committees (consisting of diverse stakeholders), and similar groups subordinate to the government or parliament often play an important role in responding to these challenges by creating a common vision, or consensus, of how to address them.

Finland, Korea, and South Africa illustrate different ways in which the ability to perceive, and respond to challenges plays out in practice. While the Finnish capabilities to perceive challenges are embedded in the NIS in a decentralized way (box 6.9), the Korean innovation system has tended to respond to challenges through a more top-down approach (box 6.10). South Africa's response to the challenge of developing policies to reconfigure the NIS in the years immediately following apartheid can be described as a decentralized as well as top-down NIS (box 6.11). In this case, there was considerable concern about the poor socioeconomic context

Figure 6.2   Typical Governance Structure of a National Innovation System

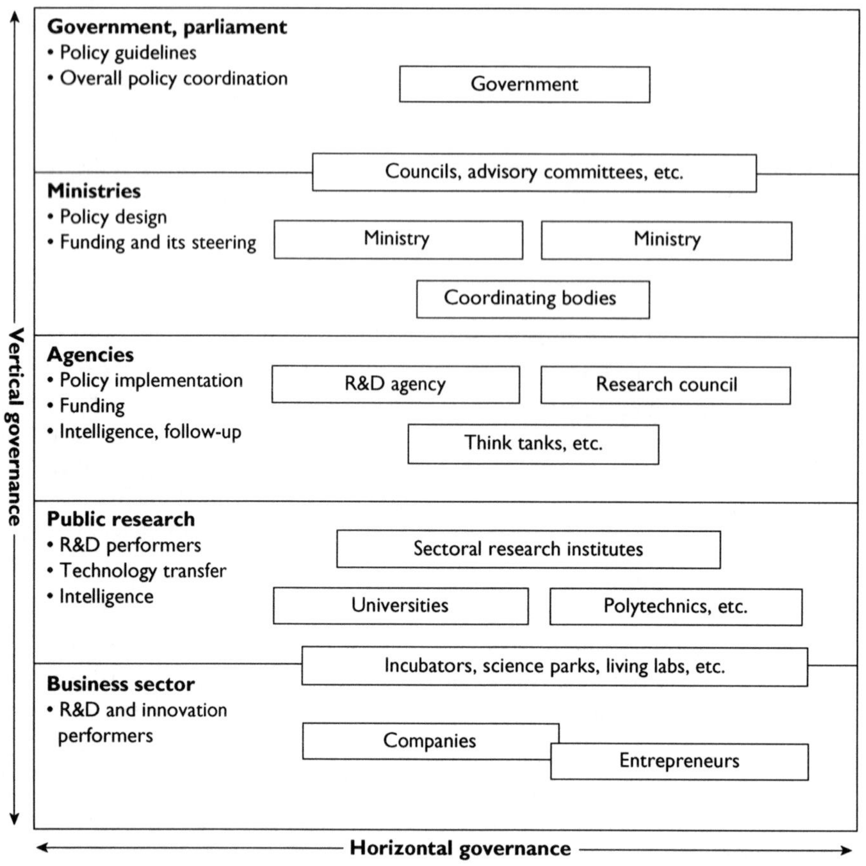

Source: Adapted from OECD 2005.

Box 6.9   Finland Responds to the Challenges of Globalization

The ability of Finland's innovation system to perceive and respond to challenges is best seen in the way that innovation policy reacted to globalization. The impact of globalization was felt most acutely in the business sector, as R&D increasingly moved to foreign locations and price competition became tight, especially in traditional industries. The emergence of countries such as China and India as increasingly competitive locations for manufacturing, research, and development raised concern among labor unions and other national innovation system stakeholders. Public research organizations felt building pressure to compete globally for the best students and become more engaged internationally. In 2004, the government launched a project to assess how globalization would affect various sectors and their employment prospects in Finland and to develop corresponding policies to respond to those challenges by altering the business environment. The final report was based on numerous background studies commissioned from national think tanks and experts, over 20 sectoral dialogues between employers and employee unions, and the work of the high-level steering group appointed by the project. The project was intended to feed into the ongoing, decentralized process to formulate a globalization strategy for Finland which subsequently influenced various areas of policies, such as taxation, R&D programs, and internationalization schemes to support companies.

Source: Prime Minister's Office, http://www.vnk.fi/julkaisukansio/2004/j19-26-osaava-avautuva-uudistuva-suomi/pdf/en.pdf.

Box 6.10   Korea Responds to the Asian Crisis of the Late 1990s

The Korean experience following the Asian financial crisis highlights the importance of capabilities to perceive and act on challenges to innovation at the national level. The crisis caused significant downsizing among large companies, mass layoffs of highly skilled personnel, and large reductions in spending on R&D. Aside from increasing its expenditures on education, the Korean government responded by increasing its R&D budget, to offset the decline in corporate spending. It also used the crisis as an opportunity to develop technology-based small and medium enterprises (SMEs), using the Special Law to Promote Venture Firms enacted in 1998.

Source: OECD 2009b.

A coordinated mix of policy measures was put in place: regulations to improve the environment for venture startups and their growth; government-backed venture funds and tax incentives for investors; and measures to support research. Among other things, these measures fuelled rapid expansion in the number of corporate R&D labs, with SMEs accounting for much of this increase. This success cannot be explained by policy intervention alone, as it was aided by rapid innovations in digital and other technologies, but government action shaped an environment that enabled new businesses to seize emerging opportunities.

Box 6.11   A White Paper and Foresight Exercises Facilitate Changes in South Africa's Innovation Policy

In 1996, a White Paper on Science and Technology laid down the new, post-apartheid government's priorities in science, technology, and innovation. Foresight exercises followed at the end of the 1990s and acknowledged South Africa's many socioeconomic challenges. These combined efforts clarified the challenges to government officials, highlighted weaknesses of the emerging national innovation system, and suggested actions to address these challenges and weaknesses. Human resource issues related to poverty, education,

Source: OECD 2007b.

and absorptive capability were singled out as a key constraint on technological developments and innovation. The preparatory work on the White Paper resulted in a national R&D strategy, endorsed by the government in 2002. It propelled an innovation system approach to the forefront in policy design and highlighted the importance of moving toward an innovation policy with a broad mandate to meet socioeconomic needs through science and technology as well as innovation.

(poverty, segregation, one-sided industrial and company structure) and the narrow science and technology focus of the apartheid regime. These challenges prompted the South African government to adopt a broader and more holistic innovation system approach to policy that could better direct activities toward common socioeconomic goals.

Establishing and/or strengthening capacity in coordination bodies to set policy priorities and coordinate agendas

Capabilities to set policy priorities and coordinate agendas are important to economize on scarce resources (especially

in developing countries) and to align policies with existing structures and framework conditions. These capabilities are usually *embedded in ministries (or department equivalents)*, which typically also design policies and steer funding to sectoral agencies or directly to public research organizations. This level of governance is often vertically linked to the government through various councils and advisory committees.

Ministries also frequently establish *dedicated coordination bodies* to ensure better coordination between ministerial and other agendas, especially in broad technology areas such as nano-, bio-, or environmental technologies. These areas require the involvement of many stakeholders and consultation processes to elicit their views. These coordinating bodies facilitate more horizontal, "'whole-of-government" approaches and policy mixes to respond to an innovation policy agenda that is widening because of globalization, new technologies, and new forms of innovation (open innovation, nontechnical innovation, user-driven innovation, and others) (see discussions in OECD 2005 and EC 2009).

The fact that a high-level policy council plays a central role in research, development, and innovation policy does not mean that the resulting policy favors centralization. For example, the Finnish Research and Innovation Council, chaired by the Prime Minister, does not allocate resources for research, development, and innovation. The Council is very much an advisory body responsible for the strategic development and coordination of Finnish science and technology policy as well as the NIS as a whole. The implemen-

tation of policy (including the allocation of resources) is delegated to various ministries, public funding agencies, and ultimately companies, universities, and public research institutes.

In all three countries, certain organizations play a critical role in addressing challenges proactively by setting priorities and coordinating agendas for action. They are described in box 6.12.

Strengthening the capacity to implement and manage policies on a day-to-day basis

Policy design, prioritization, and agenda setting alone will not respond to socioeconomic needs and deliver innovation and growth; policies must be implemented. Implementa-

---

Box 6.12   Organizations Involved in Prioritizing and Coordinating Policy in Finland, Korea, and South Africa

Research and Innovation Council, Finland. The strategic development and coordination of science, technology, and innovation (STI) policies in Finland are the responsibility of the Science and Technology Policy Council, an advisory body to the government. The composition of this council is distinctive in some respects and underlines its capacity to perceive challenges, draw overall policy guidelines, and facilitate coherence, consensus-building, and coordination throughout the system. It involves a wider range of sectors than similar councils. The chairmanship is held by the Prime Minister, emphasizing its top-level status, and involves key ministers (for employment and the economy, education, and finance, for example). The council also includes representatives from academia, industry, and labor organizations. It dates to 1963, and its mandate for technology was added in 1986.

Ministry of Science and Technology and National Science and Technology Council, Korea. The Ministry of Science and Technology (MoST), which became the Ministry of Education, Science, and Technology (MEST) in 2008, was established in 1967. Its importance grew along with Korea's increasing emphasis on research, development, and innovation in the 1980s and 1990s and the broadening of the innovation policy agenda in the 2000s. It commanded a large budget and had a broad mandate for policy design, coordination, and evaluation of science and technology in Korea, as

well as the formulation of programs and projects. It also promoted public awareness of science and technology.

In the 1980s and 1990s, a range of ministries launched R&D programs, sparking demand for better coordination. The National Science and Technology Council (NSTC), established in 1999 and chaired by the president, has since been Korea's highest decision-making body on STI. As a cross-ministerial body, NSTC has a central role in working across ministries to coordinate the expanding policy priorities and agendas. Its strong links to MEST are underlined by the fact that MEST provides the NSTC with a secretariat. The NSTC's horizontal scope at the sectoral level is strengthened through five subordinate expert committees on key industrial technologies, large-scale technologies, state-led technologies, cutting-edge converging and interdisciplinary technologies, and infrastructure technologies.

Department of Science and Technology, South Africa. The case of South Africa's Department of Science and Technology (DST) is interesting because this department gained responsibility over STI just as South Africa's policy makers endorsed an innovation system approach. The shift toward innovation occurred in response to the enormous socioeconomic challenges of post-apartheid South Africa. Subsequently DST has played an important role in setting priorities and agendas based on white papers and forecast exercises.

Source: Lemola 2002; Dahlman et al. 2006; OECD 2007b, 2009b.

---

tion is an essential element of good NIS governance, but it has often failed owing to competing rationales between ministries, lack of political will and funding, changing external developments (an economic crisis) or other complications (for example, see OECD 2005).

Policy implementation and the management of R&D funding and other schemes are *often delegated to the level of agencies*, for example to R&D agencies (such as Tekes in Finland or TIA in South Africa) and research councils (the Academy of Finland or the Research Council for Fundamental Science and Technology in Korea). Delegation of these responsibilities implies a need either to strengthen the capacities of these agencies or to establish a new agency. It also highlights the need for ministries to strengthen their steering capacity. Delegation of managerial authority is usually accompanied by stronger requirements to report outputs and outcomes and thus increase accountability at lower levels.

The day-to-day management capabilities of NISs are reflected in the routines and procedures that (for example) ministries and agencies use to interact with companies, researchers, and other target groups of R&D programs and policy schemes. These agencies also collect intelligence on technological and market trends to support decision making, as discussed later. Key issues are to avoid unnecessary bureaucracy and red tape, strike a good balance between transparency and secrecy in R&D projects, and ensure policy continuity amid political change and external events.

Aside from vertical coordination of innovation policy, more attention should be paid to horizontal coordination. Horizontal coordination occurs across the boundaries of distinct policy domains and sectors. The development of a horizontal innovation policy involves placing a broader strategic approach above departmental goals by integrating priorities and objectives across various policy sectors. Horizontal governance of innovation policy requires the integration of innovation-oriented thinking into other policy domains and greater attention to interfaces with policy sectors that use and apply science and technology.

The Finnish innovation system offers a good example of the role that agencies such as Tekes can play in implementing policy (box 6.13). In this case, the relatively clear separation between responsibilities for designing innovation policy (occurring at the governmental and ministerial level) and implementing it (occurring at the agency level) has been important for a flexible and proactive innovation policy and for avoiding political deadlocks that block implementation. Overall, this division of labor and the strong vertical and horizontal connections existing throughout the Finnish innovation system have been important preconditions for the relatively short time that elapses between policy design and implementation, which in turn strengthens Finland's capacity to respond quickly to emerging challenges. These preconditions may have been easier to meet in

---

Box 6.13   Tekes as an Implementer of Innovation Policies in Finland

Tekes, founded in 1983, is based in the Ministry of Employment and the Economy. It has relative autonomy to set priorities and agendas in specific technology areas, following guidelines developed at higher levels (the Science and Technology Policy Council and ministries). Tekes' role eventually expanded to include channeling the bulk of public funds for R&D to industry and public research agencies, with the exception of basic research agencies. Its major funding instruments include R&D grants and loans for companies and applied research grants for public agencies. Research grants are typically allocated via technology programs planned and implemented with companies and research institutes. Although the themes of programs are planned with companies, public research organizations, and other agencies, the funding is competitive, and companies must contribute complementary funds (usually around 50 percent). The idea is to stimulate collaboration between program partners and maximize benefits from knowledge spillovers. Each program has a coordinator, a steering group, and a manager from Tekes. Funding for programs ranges from €20–150 million, generally over three to five years. Hundreds of programs have been initiated since 1983; 29 operated in 2009. These programs have played an important role in promoting entrepreneurship, introducing new areas of technology, and renewing industries.

Source: Ylä-Anttila and Palmberg 2007; Tekes (www.tekes.fi).
Note: Tekes = Finnish Funding Agency for Technology and Innovation.

Finland than elsewhere because of its small size and the high level of trust between the main actors in the system.

Establish/strengthen capacities to obtain and analyze intelligence

A well-functioning NIS must have the capabilities and related governance structures to obtain and analyze intelligence on the impacts of innovation policy as well as future technological and market trends. These capabilities relate to technology and innovation studies, development of STI indicators, evaluations of R&D programs, and other types of policy instruments and interventions, as well as technology foresight and assessment. These capabilities are often spread out in the NIS; for example, ministries and agencies typically have their own *research and analysis units* (box 6.14). For the sake of objectivity in impact assessment, however, the most viable arrangement is for independent expert organizations (think tanks, consultancies, public research organizations, universities, and so forth) to gather and analyze intelligence. In the case of public research organizations, the problem may be that many research groups receive R&D funding and

may have vested interests. Transparency and objectivity should be the key criteria in impact assessment.

Capabilities to obtain and analyze intelligence are also often built in collaboration with transnational think tanks such as the World Bank and OECD. Both organizations develop STI indicators and impact assessment methodologies and standards; they also undertake assessments and evaluations of innovation systems.

Evaluations of the inputs, activities, outputs, and impacts of research, development, and innovation are essential to enhance the effectiveness, efficiency, appropriateness, and accountability of policies to foster innovation and improve social welfare (see module 7). For this reason, they are integral to improved innovation intelligence. Aside from improving accountability, the main strength of evaluation may reside in its capacity to provide insight, learning, and understanding.

Evaluation usually includes priority setting, an ex ante impact appraisal, monitoring of progress (interim evaluation), and an ex post evaluation of results and impacts. These cumulative assessments aim to measure performance, support target or performance-based management and

---

Box 6.14   Strategic Intelligence Capabilities and Activities in Finland, Korea, and South Africa

Finland. Tekes monitors results and assesses the impacts of projects it funds. For monitoring, Tekes collects project effectiveness information at the beginning and end of each project and three years after its conclusion. An impact assessment is done to gain feedback on how the project attained its objectives, how effective the project was, and what could be learned from the project to improve Tekes' future operations and strategies. Tekes also follows international comparisons and reports, such as comparisons commissioned by the Organisation for Economic Co-operation and Development, European Union, and others, and conducts peer reviews of innovation activities in various countries.

Korea. The *Korean Institute of Science and Technology Evaluation and Planning (KISTEP)* is the nation's main STI planning agency and supports the Ministry of Education, Science, and Technology's policy planning and coordination. Its specific functions are to formulate, coordinate, and support major science and

technology policies by, for example, forecasting science and technology development trends; analyzing and evaluating science and technology programs by all ministries; conducting research into domestic and overseas research planning, evaluation, and management systems; and disseminating R&D policy information and data.

South Africa. South Africa has also been developing its capacity to undertake policy assessments and analysis. These capabilities have been developed within the main ministries, agencies, and advisory bodies. Of particular importance is the *Centre for Science and Technology and Innovation Indicators (CeSTII)*, which is responsible for national R&D and innovation surveys based on a memorandum of understanding between the Department of Science and Technology and Statistics South Africa in 2004. Several universities also host research groups with a focus on technology and innovation studies.

Source: OECD 2007b, 2009b; Tekes, www.tekes.fi.
Note: Tekes = Finnish Funding Agency for Technology and Innovation.

---

budgeting, enhance accountability and transparency, and improve communication of outcomes to policy and decision makers and sponsors.

In Finland, the evaluation of research, development, and innovation comprises meta-evaluation and system reviews (Ministry of Employment and the Economy 2010), evaluations of scientific and technological fields and programs, and evaluation of universities, research institutes, and other R&D institutions. Using information from evaluations to inform policy has remained a challenge in Finland, however, despite the numerous evaluations undertaken in the past ten to fifteen years.

POTENTIAL BENEFITS

The benefits of an innovation system approach and good governance of an NIS should ultimately be visible at the macroeconomic level through increasing innovation and economic growth. As noted, OECD and others have developed a range of innovation input and output indicators, although the relative role of some factors, such as governance, is virtually impossible to assess through indicators.

The most relevant indicators for measuring the benefits of an innovation system approach capture knowledge flows and collaboration or knowledge distribution throughout the system (under the assumption that they generate innovation and growth). Some of these indicators are available—for example, information on the mobility of researchers and personnel, innovation surveys on R&D collaboration, data on interfirm collaboration, rates of technology diffusion—but it is beyond the scope of this note to apply them to the case study countries.[3] Instead, the experiences of Finland, Korea, and South Africa will be used to highlight some of the more subtle and intangible benefits of an innovation system approach in general.

As emphasized throughout this note, an innovation system approach can *focus the policy debate*—create consensus and a common vision—on issues of key importance for sustaining innovation and growth, especially in response to emerging challenges and in times of crisis. Examples described here include globalization (Finland), economic crises (Finland, Korea), and poverty and segregation (South Africa).

An innovation system approach to policy thinking and analysis can *highlight latent potential for knowledge flows and collaboration* across the various fields of science, technology, and industry and achieve "new combinations" as a source of innovation. To do so, countries will require good capabilities in obtaining and processing intelligence on the structure

and development of different sectors of the economy (as in Finland and Korea).

Similarly, well-governed innovation systems can *bring previously disconnected actors together* and create new nodes and platforms for innovation. Finland and Korea have implemented explicit coordination schemes and policy programs to achieve this goal, such as the Tekes programs. Policy in South Africa has focused on integrating the "second economy" with activities at the traditional core of the innovation system.

The success of an innovation system depends considerably on the extent to which it engages private companies in research, development, and innovation. The *innovation system concept can extend the policy mix from supply-sided schemes* (such as R&D funding) *toward a large array of more demand-oriented schemes* (such as standardization, public procurement, and regulations). (For examples from the three countries discussed here, see Dahlman, Routti, Ylä-Anttila 2006 and OECD 2007b, 2009b). Nonetheless, an important consideration for governance of the innovation system is that a delicate balance must be struck between relying on market forces and more interventionist policies, such as regulations.

Finally, although an NIS generally focuses on developing national innovation capacity, it does not lose sight of the value that the innovation system approach places on knowledge flows and collaboration, including internationally generated knowledge flows and collaboration. *Knowledge flows and collaboration extend beyond national borders, and an innovation system approach can help to identify opportunities and bottlenecks of critical importance* (see Edquist 1997 for a review of innovation system approaches that emphasize the international dimension).

LESSONS LEARNED

The following lessons related to innovation system governance are relevant to developing and sustaining governance in an AIS.

A step-by-step process, building on existing structures and contexts

The development of an innovation system approach to innovation policy may *take significant time* (decades rather than years) *and should be pursued systematically and iteratively* so that emerging challenges and feedback from the research community and private sector can be addressed in a flexible way. Core governance structures for innovation systems often are based on existing policy structures such as ministries, but they also typically involve the establishment of dedicated ministerial departments, councils, agencies, think tanks, and other entities.

Most developing countries have little room to maneuver in research, development, and innovation. Consequently their only strategic choice is to stick to incremental innovations—for example, to improve existing products, services, and processes. All countries will need monitoring and governance arrangements that allow sufficient adaptability to reverse unwise decisions quickly. Countries with relatively small research systems, such as small countries or economies in the initial phases of development or recovery, have a particular need to concentrate their efforts. Many countries have established various prioritization practices in recent years. Korea, for example, uses a mix of instruments for priority setting, including technology foresight and technology road-mapping. These processes are distributed across ministries and agencies and create a diversity of competing priorities and visions (which ideally are reconciled in the national innovation policy). For an example from Thailand of a national innovation council, see the overview in module 1.

An innovation system approach *should acknowledge existing industrial structures* (ICT and Nokia in Finland), company distributions (Korea's chaebols), *and the overall socioeconomic framework* (the lack of involvement of South Africa's "second economy"). Properly applied, the innovation system approach will facilitate collaboration and knowledge flows across actors and stakeholders whose efforts to innovate were previously separate or who were excluded from innovation altogether.

Strong, visible commitment at the highest level

A common feature of countries that have successful research, development, and innovation policies is *strong and visible commitment at the highest political level* to long-term development of financial and human resources for research, development, and innovation. Other key factors are the integration of key ministries (finance, education) in planning and implementation processes, broad-based consensus on the basic elements of research, development, and innovation policy, and wide agreement that investments in research, development, and innovation are needed over the long term. In Finland and Korea, a high-level policy council with representatives from ministries, government, R&D agencies, and the private sector turned out to be an efficient mechanism for overall coordination of research, development, and innovation policies.

Mobilizing actors and resources

For policy to be more relevant and effective, it must embody clear visions, strategies, and priorities. Leadership in the governance of research, development, and innovation are also vital to mobilize actors and resources. Leadership is best undertaken by distinguished individuals (a president, prime minister, minister of finance), ministries, or innovative agencies and enterprises. These leaders have a broader perspective on policy agendas for research, development, and innovation and can help to maintain their coherence.

Coordinating bodies

The *role of coordinating bodies in setting priorities and coordinating agendas is increasingly important* owing to challenges arising from globalization, emerging technologies, new forms of innovation, and a range of global issues such as energy and climate change, poverty, health care, and access to clean water. Coordinating bodies benefit from links to the highest levels of government (vertical coordination) but must also include decision makers and other stakeholders from diverse areas of the economy (horizontal coordination). The councils in Finland and Korea are two examples of such coordinating bodies. Governance of innovative activity is not provided by government or the public sector alone. It is important that *representatives of the private and third sectors* actively participate in formulating and implementing policy through various forms of public-private partnership.

A clear role for high-level councils

High-level councils can and often do play important roles in setting priorities and agendas and as overall policy coordination platforms, but it is evident that simply establishing a council is not enough (OECD 2009a,b,c). Their needs and tasks must be well-defined in the specific context, with attention to the strategic needs of the country's innovation system. The council's composition, too, needs to be considered in view of the strategic tasks. It must be open to newly emerging actors in innovation in the country.

Horizontal coordination

A broader understanding of innovation and innovation policy means that more attention should be paid to *horizontal coordination,* which refers to the crossing of administrative and cultural boundaries between policy domains and sectors.

Autonomy to implement

An innovation system approach can aid both policy design and implementation; policies also need to be implemented to deliver innovation and growth. In particular, *policy implementation may best be facilitated at the level of relatively autonomous agencies rather than ministries and departments where political and other issues may be inhibiting factors* (Tekes is one example; TIA in South Africa may be another).

Transparency

Innovation policies benefit from *transparent schemes and the avoidance of bureaucracy and red tape.* Programs for R&D can be efficient for focusing activities on predefined areas (for example, the specific technology areas represented by Tekes' R&D programs). Care should be taken to include elements of competitive tendering. Policy continuity is also important to stabilize the innovation horizon (of private companies in particular).

Learning and evaluation

The *ability to obtain and analyze intelligence on market and technological developments and trends* is of key importance for a well-functioning, proactive innovation system. Finland, Korea, and (to an increasing extent) South Africa conduct foresight exercises and impact assessments. These capabilities are preferably spread out throughout the innovation system and strengthened through international collaboration and related forums. They should be actively promoted and maintained.

Improved means of evaluating the inputs, activities, outputs, and impacts of research, development, and innovation are needed to manage R&D organizations and instruments and provide important feedback for policy making. The development and implementation of monitoring and evaluation require intervention from the upper levels of innovation policy. Many countries are finding that evaluations of research organizations, research and technology programs, and other policy instruments are an effective and indirect way to control and manage research organizations. Although evaluations are increasingly used to improve the design and implementation of the instruments of research, development, and innovation policy, they are not always readily available or communicated to policy makers at the strategic decision-making level.

RECOMMENDATIONS FOR PRACTITIONERS

Innovation systems emerge gradually and organically if the enabling environment is favorable. Knowledge flows and collaboration cannot be created by policy, but policy can create suitable conditions for them to happen. Good governance is central to the performance of an innovation system, and policies can strongly influence good governance. Practical recommendations for establishing governance structures include the following general and tentative "steps":

1. ***Develop awareness*** of innovation systems concepts and identify good practices in similar sectoral, regional, and national contexts. Engage in international dialogue.
2. ***Communicate the viability and challenges*** of implementing an innovation system approach. Probe the possibilities for seeking, and achieving, consensus and a common vision on key issues.
3. ***Analyze structural and institutional preconditions*** for governance structures related to innovation systems. Involve companies, public research agencies, and other relevant stakeholders (main ministries, regulators, NGOs) in policy design, consultations, and strategizing.
4. ***Consider the suitability of existing institutions*** to handle STI matters. Consider the need for new, STI-dedicated agencies and other institutions.
5. ***Assess the economic, legal, and political viability*** of introducing STI issues and the innovation system concept at various levels of governance. Ensure that mechanisms for priority setting and coordination can be put in place.
6. ***Develop existing institutions*** to support STI or establish new STI institutions if required. Ensure that they have a clear mandate and specific roles to avoid overlap. Be ready to divest obsolete schemes and institutions if necessary to foster the growth of new ones.
7. ***Ensure relative institutional autonomy*** in policy implementation. Ensure that sufficient capabilities and resources are in place for day-to-day management of policy schemes and initiatives, now and in the long run (to ensure policy stability and predictability).
8. ***Implement policy schemes*** (at the agency level or below) and initiatives as considered relevant, based on an assessment of societal needs. Consider which policy mix is most suitable to the context. Ensure that schemes and initiatives are transparent, nonexclusive, and predictable, and support both networking and competition.
9. ***Ensure that institutions and capabilities remain in place*** (compare with the third step) to analyze and assess technological and market trends, as well as to assess the impacts of policy schemes, initiatives, and the innovation system as a whole. Continue to engage in international dialogue.

Managing Intellectual Property to Foster Agricultural Development

Sara Boettiger, University of California, Berkeley
Robert Potter, Consultant
Stanley Kowalski, University of New Hampshire

SYNOPSIS

Over the past decades, consideration of IPRs has become increasingly important in many areas of agricultural development, including foreign direct investment, technology transfer, trade, investment in innovation, access to genetic resources, and the protection of traditional knowledge. The widening role of IPRs in governing the ownership of—and access to—innovation, information, and knowledge makes them particularly critical in ensuring that developing countries benefit from the introduction of new technologies that could radically alter the welfare of the poor. Failing to improve IPR policies and practices to support the needs of developing countries will eliminate significant development opportunities. The discussion in this note moves away from policy prescriptions to focus on investments to improve how IPRs are used in practice in agricultural development. These investments must be seen as complementary to other investments in agricultural development. IPRs are woven into the context of innovation and R&D. They can enable entrepreneurship and allow the leveraging of private resources for resolving the problems of poverty. Conversely, IPRs issues can delay important scientific advancements, deter investment in products for the poor, and impose crippling transaction costs on organizations if the wrong tools are used or tools are badly applied. The central benefit of pursuing the investments outlined in this note is to build into the system a more robust capacity for strategic and flexible use of IPRs tailored to development goals.

BACKGROUND AND CONTEXT FOR ACTION AND INVESTMENT

As public funding for agricultural research has fallen relative to private sector investments, for many countries the era in which agricultural development often depended on public goods and the unchecked sharing of research results has come to an end. Countries have entered a new era in which innovation, R&D, and the sharing of knowledge occur at the same time that global IPRs are on the rise. The current IPR climate demands a nuanced and strategic use of IPRs to enable innovation and support agricultural development, but the shift toward this scenario is proving challenging, and many donors, governments, institutions, and individuals are struggling to respond.

There are good reasons for the hesitant progress in understanding the use of IPRs in agricultural development. First, other priorities upstage IPR issues in environments where limited resources must be allocated across formidable needs. Second, expertise in IPR issues often cuts across many fields, including law, business, science and technology, as well as development policy. Even the range of IPR instruments in agriculture is diverse. In addition to patents, other forms of protection, such as trademarks and geographical indications, can create value in agricultural value chains in developing countries; copyright laws can limit access to agricultural research journals, databases, and software code; PVP certificates, utility models, and sometimes even trade secrets are also relevant to agricultural development (World Bank 2006).[1] Third, the international landscape for IPRs is changing rapidly. Policy makers in developing countries now operate within a system of bilateral, regional, and multilateral treaties that govern a wide range of IPR issues (box 6.15); protection of genetic resources and traditional knowledge (box 6.16); and, in debates about food security and developing countries' capacity to respond to climate change, the increasing importance of the role of IPRs in technology transfer. This note argues, however, that IPR-related investments are

Box 6.15 Beyond TRIPS

The Agreement on Trade-Related Intellectual Property Rights (TRIPS) came into effect in 1995, stipulating that all signatories to the agreement should introduce a minimum amount of legislation to protect IPRs. This international obligation triggered a widespread introduction of IPR legislation in developing countries in recent years, as it became a requirement for entry into the World Trade Organization. More recently, in addition to TRIPS, developing countries operate in a landscape increasingly dominated by preferential trade agreements (PTAs). These agreements often contain obligations relating to domestic intellectual property policies that exceed the minimum standards set forth in TRIPS. Collectively, agreements with intellectual property obligations comprise a landscape referred to as "TRIPS-plus." A recent report estimated that close to 400 PTAs were in force by 2010, governing more than half of global trade. Not only is the number of agreements growing, but IPR provisions are also occurring in increasingly diverse types of agreements, from customs standards to anticounterfeiting agreements.

Source: Heydon and Woolcock 2009; Frankel 2009.

Box 6.16 IPR Issues in Genetic Resources

Ownership of genetic resources and traditional knowledge is an area where IPRs are increasingly considered a serious issue. R&D in crop improvement, for example, depends on the wealth of genetic material held in farmers' fields and national and international gene banks. Both the conservation of genetic resources and access to them are critical for our future capacity to address global food security issues, including drought tolerance, yield improvements, and resistance to diseases and pests.

Maintaining a balance between the preservation of genetic resources and ensuring widespread access depends on finding solutions that can work within a complicated cross-section of national, international, and institutional policies. For example, in 2006 the research centers of the Consultative Group on International Agricultural Research (CGIAR) that maintained ex situ collections of plant genetic resources signed agreements with the Governing Body of the International Treaty on Plant Genetic Resources for Food and Agriculture ("the Treaty"), which placed the collections they hold under the Treaty, and adopted a "Standard Material Transfer Agreement." Exchanges of genetic resources involving the CGIAR centers are now governed by this agreement, which includes IPR obligations.

Further questions over rights to genetic resources are raised by the increased capacity for large-scale DNA sequencing. Currently, many projects to sequence plant genomes promise benefits to agricultural development. The data and associated knowledge hold the potential to assist in breeding for improved yields, disease resistance, and countless other traits. There has been continuing concern, however, in genome sequencing about the optimal use of IPRs that does not impede innovation based on the new data. Some argue that without the ability to patent, the investment in further R&D is not warranted; others argue that allowing proprietary ownership allows for blocking patents that can slow or halt innovation. While it is becoming increasingly difficult to patent DNA sequences in the United States, a large number of patent *applications* still contain claims to sequences in bulk. For example, CAMBIA's analysis indicates that approximately 74 percent of the rice genome is claimed in United States patent applications.

As this brief discussion indicates, future investments in policies and programs involving genetic resources must include considerations of IPRs to support the donors' intended impacts on agricultural development.

Source: CAMBIA ("Mapping of Rice Patents and Patent Applications onto the Rice Genome"); Pollack 2010.
Note: The Standard Material Transfer Agreement can be accessed at this link: http://www.planttreaty.org/smta_en.htm.

critical if developing countries are to benefit sufficiently from advances in STI over the next decades.

MAJOR IPR-RELATED DRIVERS OF DEVELOPMENT

Three main interrelated paths characterize mechanisms through which IPRs impact global poverty: (1) trade and foreign direct investment; (2) national capacity for innovation and development of local commerce; and (3) technology transfer. These paths are discussed in the sections that follow with full recognition of the oversimplification of this framework.[2] Nonetheless, it serves as a useful compass for the analysis that follows.

Trade and foreign direct investment

Box 6.15 showed how trade issues often drive IPR legislation, with varied impacts. Studies of the relationship between IPRs, trade, and foreign direct investment in developing economies have covered a wide range of potential paths of interaction in an attempt to determine whether stronger IPR policies in developing countries are likely to produce benefits for the world's poor. While a significant literature illustrates positive implications of stronger IPR policies on trade and foreign direct investment, there are caveats. Strict enforcement of IPR, for instance, may drive up the costs of imitating or copying inventions, which may reduce growth in very low income countries that rely on these approaches and do not yet have the infrastructure to accept foreign direct investment. Other work has shown that stronger IPR policies can exacerbate income inequalities in developing countries and that the flows from trade and foreign direct investment do not sufficiently impact the very poor (Adams 2008). In short, the empirical work on IPR policies, foreign direct investment, and trade in developing countries leaves unresolved questions about how the poor are affected over time, and debates will continue with further exploration of the issues.[3]

National innovation climate

IPR legislation is one component of the climate for innovation in a country, but legal instruments are not enough on their own to encourage investment in innovation. Without the active involvement of national researchers, there will be little appreciation of the role of IPRs, and thus other investments will do little to encourage innovation. Interventions that focus on protecting inventions in public institutions

can be a way to jump-start a cycle in which increased protection of IP increases awareness of the value of innovation. Box 6.17 shows an approach to increasing locally focused R&D in Botswana, coupled with measures to raise public awareness of new inventions and designs.

Types of IPRs differ in their influence on the innovation climate within low-income countries. Patents can be important IPR instruments for discrete inventions, as seen in box 6.18, which shows how a machine to prevent frost damage in fruit orchards was exported from Chile to important overseas markets. Other forms of IPRs offer greater opportunities to influence the domestic innovation and business development climate, however. Trademarks, geographical indicators, plant breeders' rights, and seed registration laws may garner comparatively less attention than patents in the international press, but as noted, they often have more practical potential to affect agricultural development in low-income countries. Box 6.19 describes how a trademark was initially used to build a brand around Colombian coffee and how geographical indicators have been employed more recently to maintain this brand. The success of this approach led other countries to similarly distinguish their local produce in an international market, such as Pinggu peaches from China. In this case, the agreement between China and the EU on geographical indicators for peaches from this region of China opened an export market for high-quality fruit previously recognized only within China.[4]

Technology transfer

The transfer of technology and knowledge remains perhaps the most influential of the three drivers listed here in terms of IPR investments contributing to poverty reduction. Most well-capitalized engines of innovation are in developed countries, but increasing numbers are found in emerging economies. There is a real need to improve international capacity for agricultural R&D targeted at poverty reduction as well as the flow of knowledge and technologies to benefit developing countries. Whether the "technology" that is transferred refers to a novel plant variety, the tacit knowledge of how to improve a food-processing practice, or an innovative business model for giving smallholder farmers access to microirrigation, making technology and knowledge available to improve the lives of the poor has both direct impacts (for example, by improving health, food security, or access to water and sanitation) and indirect impacts (such as economic development). IPRs are an important factor in public-private partnerships transferring technologies, in the formalization of the knowledge and

Box 6.17 BOTEC Harnesses Innovation in Botswana

Botswana Technology Centre (BOTEC) in Gaborone is a research and technology organization established by the Botswana government in 1979. Operating under the Ministry of Communications, Science, and Technology, BOTEC has pursued the government's policy objective of technology promotion and innovation as a tool for developing the economy and improving the quality of life in Botswana. To strengthen research and development capacity, BOTEC has a longstanding involvement and active participation with a number of local organizations, including the University of Botswana, Botswana Institution of Engineers, Botswana Export Development and Investment Authority, Botswana Innovation Hub, and some nongovernmental organizations.

Botswana's Industrial Property Rights Act (1996) provides a legal framework for the country's innovators to seek intellectual property protection for their intellectual property. BOTEC has worked to improve intellectual property awareness in Botswana to assist inventors to be more creative and benefit from their innovations. BOTEC's intellectual property policy seeks to address a number of issues, including Botswana's increased participation in international treaties related to intellectual property, access to information on inventions related to BOTEC activities, and dealing with new technology that has been transferred to companies. BOTEC initiated the

National Design for Development Awards in 1999 to recognize inventions and innovations that can offer solutions to some of the problems faced by Botswana. As of 2010, three award ceremonies have been hosted. The World Intellectual Property Organization and African Intellectual Property Organization supported the award ceremonies by sponsoring special awards for outstanding innovations in Botswana. BOTEC is contributing to efforts to protect Botswana's traditional knowledge by chairing an Indigenous Knowledge Task Force, which is drafting the indigenous knowledge section for the Industrial Property Rights Act.

BOTEC's solar-powered hearing device was developed through a collaborative scheme with Motse Wa Badiri Camphill, a nongovernmental organization that conducted field tests, raised funds for design improvements, branded the device with the SolarAid name, and took it to market. SolarAid generated considerable interest and was used in many developing countries. BOTEC assisted Motse Wa Badiri Camphill to set up a separate organization, the Godisa Technologies Trust, to develop the promising pilot project into a genuinely successful product. The recharger, now successfully marketed under the SolarAid brand, requires only 6–8 hours of sunlight to maintain a full charge for a week.

Source: Quoted with slight adaptations from WIPO, http://www.wipo.int/ipadvantage/en/details.jsp?id=2623.

---

innovation that lies in the public sector, and in the creation of specific technology transfer offices (TTOs) in such institutions (see module 5, TN 5 for examples of TTOs for individual institutions or a network of institutions). TTOs are one example of policies promoting technology transfer, but they are not the only such policy.

Understanding where to make strategic IPR-related investments, given the diverse pathways of potential impact listed above, requires a closer look at the current context of international IPRs and agricultural development. Investments in this space must take into account (1) the international obligations related to IPR and the rapidly expanding use of IPRs in agriculture and (2) the continuing disparity in capacity between the public sector and the private sector in the strategic use of IPRs. Although biotechnology is playing an increasing role in agricultural development and is one area where the private sector has made large invest-

ments, IPR policies should not be driven by individual technologies. Similarly, the desire to encourage public-private partnerships should not—by itself—drive IPR policies, although clearer understanding of IPRs at both the national and institutional level will help these partnerships flourish.

Disparity in the capacity to manage IP in public and private R&D

Despite increasing opportunities to engage the private sector, the public sector continues to be the primary driver in agricultural R&D for most developing countries. Globally, agricultural investment in the public sector is double that of the private sector, and one-third of the worldwide agricultural R&D budget is spent in developing countries.[5] The lack of capacity for IPR management in public research organizations, and the disparity in IPR management

Box 6.18 Patenting a Chilean Invention to Protect Crops from Frost

In 1991, severe frost decimated Florencio Lazo Barra's fruit orchards. He lost all of his table grape production and 80 percent of his plums. He also incurred high fuel costs for oil burners, which he used in the failed attempt to protect his orchards from frost. After years of experimentation, in 1996 a working prototype of the Lazo Frost Control Machine ("Lazo FCM") was tested successfully. The Lazo FCM is a powerful centrifugal fan with a heater, which is trailed across the field by a tractor. The insertion of a layer of hot air in the cold air mass surrounding the plants protects fruits and vegetables from frost.

Following successful tests with the prototype, Mr. Lazo applied for assistance from Innova Chile, a government agency tasked with promoting innovation. He obtained funding to convert his business idea into reality and produce more machines. Orders from Chilean farmers soon followed, and in 1997, with assistance from an intellectual property expert, he began obtaining a patent abroad. The United States was the first country to grant his invention a patent in 1999. In Chile, patent No. 41776 was granted in 2002 by the

Industrial Property Department. The invention is also protected by patents in Argentina, Australia, China, and the European Union.

In 1998 the Lazo FCM technology was exported to the United States by granting a manufacturing license to Agtec Crop Sprayers (now "Superb Horticulture"), who sold the product under the name "Lazo Frost Dragon." In the first three years, over 500 machines were sold in South America and the United States. In 2000, sales and distribution started in Europe through Agrofrost N.V., a company based in Belgium selling and distributing the machines throughout Europe under the "Lazo Frostbuster" name. More recently, the technology was exported to New Zealand and Australia.

Without the support of Innova Chile, which enabled the inventor to file for patent protection, little of this development would have taken place. Government agencies charged with supporting innovation are often criticized for supporting projects that do not come to fruition, so it is important to recognize cases where they have been successful to balance this impression.

Source: Quoted with slight adaptations from WIPO, http://www.wipo.int/ipadvantage/en/details.jsp?id=2448.

capacity between the public and private sectors, are therefore important considerations for investments in agricultural development.

ACTION AND INVESTMENT NEEDED

This section describes a set of investments related to IPR and agricultural development that can improve activity in this sector. Opportunities exist to improve policies at the institutional level and to develop institutional capacity, as well as to increase knowledge sharing between the public and private sectors and, through these advances, increase capacity for technology transfer.

It is not possible to provide a template for particular laws or IPR regimes that will benefit all countries; *IPR legislation must be tailored to the national context.* This issue is discussed extensively in World Bank (2006), which recommends a dialogue with conscious consideration of needs and priorities prior to enacting IPR legislation for plant breeding. For example, staple crops may be treated differently from crops grown for export. Where a particular

species is considered to be of national importance, exemptions from PVP have been incorporated in some national legislation. Furthermore, as countries develop, needs for protection may change, particularly as they move from industries that exist by copying products produced elsewhere to innovating to develop their own products. IPR regimes continue to evolve even in developed economies and must be flexible enough to cope with changes in national requirements.

Promote the establishment of specific IP policies in public organizations

Establishing institutional policies on the ownership, protection, and dissemination of inventions will have a big impact on enabling technology transfer among public organizations. Institutional IPR polices are critical to the impact of public research, can open an institution to new partners, and create incentives for changing the innovation climate. One of the biggest improvements in technology transfer between public and private organizations in the United

Coffee from Colombia has retained a significant price premium over coffee from Brazil (the world's largest producer of Arabica coffee) for many decades, largely owing to a branding strategy that emphasizes the high quality of Colombia's product. The National Federation of Coffee Growers of Colombia (FNC, *Federación Nacional de Cafeteros de Colombia*) was created in 1927 by Colombian farmers. Today, it represents over half a million coffee growers, the majority of whom are smallholders. The Juan Valdez® brand strategy is an excellent example of continuing creativity in IPR management to promote agricultural development.

Television commercials shown in North America in the 1960s featured Juan Valdez® in the coffee fields with his faithful mule, painstakingly selecting and hand-picking the ripest beans. Consumers began to respond to the message that Colombian beans are grown and harvested with great care, with little help from machines, in ideal climatic conditions with plenty of rain, sun, and fertile volcanic soil. Demand grew. Many coffee roasters began marketing their products as Colombian coffee. A number launched high-end products consisting exclusively of Colombian coffee.

The Juan Valdez® logo was licensed to coffee roasters that used only high-quality Colombian coffee. Not all coffee roasters responded to this initiative, however, and another IPR instrument was included in the strategy: certification. "Colombian" was registered in relation to coffee as a certification mark in North America in the 1980s. The formal standards attached to this certification mark provide a guarantee that the actors in the marketplace are meeting minimum quality standards when selling "Colombian" coffee, thereby protecting its hard-earned reputation. Enforcing and protecting this certification turned out to be expensive but worthwhile, given the premium that Colombian coffee now demands in the market. The continuing expansion of the Juan Valdez® brand also included opening branded coffee shops, which have had varying degrees of success, as well as a partnership with Coca-Cola FoodService to offer a branded liquid coffee system.

"Café de Colombia" was registered as a geographical indicator in Colombia in 2004 and the European Union in 2006. Unlike trademarks and certification marks, geographical indicators are intrinsically linked to attributes and quality standards related to origin. They need to be recognized by governments, so delays can arise in establishing such a system, but the value of these treaties in promoting quality brands is now recognized.

Source: Fridell 2007; March 2007a, 2007b.

States was legislation mandating IPR policies for institutions that receive federal funds. By clarifying ownership of inventions and the responsibility of the institutions to protect them, IPR policies became an integral part of research activities. Without necessarily mandating the use of IPRs through legislative means, in individual organizations the establishment of policies related to ownership and responsibilities for protecting and disseminating inventions will have a big impact on enabling technology transfer.

Create a global corps of trained IPR practitioners

The impact of new IPR legislation in the wake of TRIPS cannot lead to positive cultural shifts in the use of IPRs without sufficient numbers of trained, in-country practitioners. The success of continuing investments in creating patent offices, improving judicial systems, and opening TTOs depends on the quality of the professionals engaged.

Well-trained IP practitioners are critical for a country to represent national interests and negotiate IPR provisions in multiple international forums and for a country to develop national IPR policies that promote development within complex international obligations. Likewise, managers, engineers, and scientists in public and private institutions must be able to understand IP and how to use it if countries are to play an increasingly competitive role in global agricultural development.

Support the creation of TTOs

An effective way to achieve institutional understanding of the value of IPRs may be through the creation of specific TTOs with a mandate to identify and protect innovation use and to use IPRs to promote greater impact of the research and innovations arising within the institution (for example, through licensing technology with other partners). Such offices provide direct opportunities for the professionals targeted in training activities to gain experience and also demonstrate to researchers the value of IPRs in enabling further development of their innovation (for additional information, see module 5, TN 5).

Revisit the importance of nonpatent frameworks and opportunities for agricultural development

Trademarks, PVP, seed laws, and geographical indications are a few of many IPR-related nonpatent instruments that may have an impact on agricultural development.[6] In low-income countries where patenting is limited and mechanisms for enforcing patents are not well developed, these alternative forms of IPRs can be critical instruments in AISs. The PVP system, for instance, granting plant breeders exclusive rights to market new varieties, is implemented in a number of industrialized and developing countries (World Bank 2006). UPOV remains active in promoting harmonization as traditional approaches to PVP are challenged by advances in plant breeding and genetics (Janis and Smith 2007). Trademarks and related brand equity strategies are considered by some scholars to be underutilized as a form of creating value for developing countries' agricultural products. A wide variety of opportunities exist for improving nonpatent, IPR-related instruments at both the policy and practitioner level; for examples, see boxes 6.19 and 6.20.

Encourage donors to require strategic IPR management in development investments

Private foundations, governments, aid agencies, and other donors can influence the outcomes of investments across many fields by instituting internal grant-making requirements that demand a higher level of strategic IPR management. As one example, requiring grantees to provide a plan

---

Box 6.20   Hagar Soya Co., Cambodia: Multiple Benefits from an Innovative Social Business Model

Hagar Soya Co. Limited (HSL) is a small enterprise in Phnom Penh created in 1998 by Hagar, a nongovernmental charity based in Switzerland. In the mid-1990s, Hagar began an income-generation, training, and employment project for abused and abandoned women in postconflict Cambodia. The project led to the incorporation of HSL as one of Hagar's small businesses. HSL's first commercially successful product was a soya milk drink sold under the brand name "So! Soya." The product is nutritious, affordable, and tastes good—all important qualities in helping local children increase their protein intake in a country with very high malnutrition. Following the success of the soya milk drink, the company added more soya-based items to its product line.

Initially, the commercialization of Hagar's soya milk was done in a rather informal way; women from Hagar's programs produced 300 liters of fresh soya milk per day and sold it on the streets of Phnom Penh from push carts. By 2003, HSL was ready for larger-scale production and the "So! Soya" trademark was registered with the intellectual property Department of the Ministry of Commerce of Cambodia. Subsequent HSL products such as "So! Soya kids," "So! Soya Gold," "So! Yo," "So! Yumme," "So! Milk," and "So! Choco" have also been protected by registered trademarks.

The company's intellectual property strategy focuses almost entirely on trademarks and aims at increasing the competitiveness of HSL's products. The company considers trademarks to be effective for preventing unauthorized use of HSL's marks and guarding against counterfeiting. The competitive edge also arises from registering a trademark to protect and increase its value, then publicizing it through a good marketing and business strategy to enlarge the company's market share and stimulate the development of new products. The success of HSL's trademark strategy is reflected in the company's achievements, first, in marketing its brand name through brand development of both the company and its products, and second, in ensuring lasting brand impact through quality products.

The company benefited from Hagar's initial ability to identify the right path to incorporate income-generating activities within a development project, taking into consideration the social needs of Cambodia. HSL is an example of an efficient social enterprise model, which, according to the International Finance Corporation, can be replicated by nongovernmental organizations worldwide.

Source: Quoted with adaptations from WIPO, http://www.wipo.int/ipadvantage/en/details.jsp?id=2563.

---

demonstrating that IPR has been considered in all aspects of the proposed activities will drive demand for building capacity in IPR management, highlight the impact of specific IPR-related policy needs, set important standards, and ultimately benefit donors by reducing the risks of their investments. Furthermore, requiring a consideration of product development beyond the research stage has the advantage of identifying other technologies that may need to be licensed and other partners who will need to be involved to deliver the products to the target population.

IPR-related investments must be made, however, with the recognition that IPR is only one of many factors that can foster or impede technology transfer. While there are key opportunities to address IPR issues in technology transfer, practitioners often find that *risks unrelated to IPR* are more challenging. These risks concern stewardship, products flowing back into commercial markets, and liability issues. It should also be noted that, particularly for technology directed toward rural populations, some form of extension services will be critical for disseminating the technology to the target population (World Bank 2006).

POTENTIAL BENEFITS

Investments in IPRs play a facilitating role, influencing the potential impacts of many other investments in agricultural development. The impacts of improved handling of IPR issues, therefore, are unlikely to be directly measurable. Benefits to R&D activities in the public and private sectors should include improved transfer of technology from public organizations, improved linkages between industry and academic or public research institutions, and improved access to private sector technologies. For example, Unicamp created productive linkages between the university's own R&D and industry once it established a specific TTO (module 5, TN 5).

In individual cases, benefits can often be attributed directly to the particular steps taken to protect the IP within a particular project, such as the patenting of the frost control machine (box 6.18), which allowed the inventor to enter licensing agreements with overseas developers. This connection can also be seen in commercial enterprises, where success is determined by the creation of a particular brand associated with a certain quality of product. Box 6.20 described how a small NGO in Cambodia became a successful enterprise by trademarking its products. Although the success of this enterprise depended on a wide range of factors, trademark protection was an enabling part of the business strategy.

POLICY ISSUES

The World Intellectual Property Organization (WIPO) recognizes a number of policy issues related to IPRs for developing countries and has adopted 45 related recommendations under the WIPO Development Agenda.[7] The recommendations are grouped in the following clusters and cover a number of issues relevant to this discussion, including: Technical Assistance and Capacity Building; Norm-setting, Flexibilities, Public Policy, and Public Domain; Technology Transfer, Information and Communications Technologies (ICTs), and Access to Knowledge; Assessment, Evaluation, and Impact Studies; Institutional Matters, including Mandate and Governance.

Coordination of IPR policies with other innovation policies

Policies seeking to encourage innovation for development are inherently dependent on many other areas of policy and law. Sound policies on education, trade, agricultural input subsidies, farmers' extension services, functioning court systems, and many other elements are integral to the impact of IPR policy on agricultural development (World Bank 2006). Given the interconnectedness of IPR policies with other national policies, and given the wide variety of IPRs affecting agriculture, it is difficult to provide specific policy recommendations. Countries have considerable flexibility (even within TRIPS) to adopt IPR policies that support their own specific needs, and resources exist for them to engage advice on policy changes.

While the appropriate policies will be as diverse as the range of developing countries adopting them, there are common goals for IPR policy supporting agricultural development. These goals support benefits for the poor in access to technology as well as economic development, and they include creating incentives for local innovation, encouraging foreign direct investment, increasing connections between industry and universities or research institutes, facilitating better public-private partnerships, and improving the impact of public agricultural research for the poor.

A functional legal system and extension service

The major precondition for any development of IPRs is a functional legal system under which IPRs and other legally binding agreements, especially contracts, can be enforced. IPRs are a property right, and developing respect for property rights further contributes to social justice and the rule of law in a country (Cavicchi and Kowalski 2007). As noted, extension services are another precondition; introducing technology and innovation into research is pointless if the outcomes are not transferred to farmers and have no impact on poverty (World Bank 2006). Access to improved products is a major equity issue that must be addressed within individual projects, but for products that have a direct commercial value and will be marketed through the private sector, high royalties may have detrimental effects on access. In this case, IPR legislation can include compulsory licensing mechanisms to increase access and/or reduce the market price.

Environmental issues

Given the wide-ranging impact of IPRs on rural development, it is difficult to provide specific policy recommendations with respect to environmental issues. In most cases, access to improved technologies is expected to improve rural productivity. Productivity improvements may have both positive and negative environmental implications, but the major factor in environmental damage is often the lack of better alternatives. If farmers have better alternatives to current practices, they may be able to take better care of their land, use other resources more efficiently, and contribute to greater environmental sustainability.

Roles of public and private sectors

Since the major rationale for a society to develop IPRs is to provide an incentive for individuals and organizations to invest in innovation by increasing the likelihood of a return, IPRs will directly affect the private sector's involvement in agricultural development. IPRs help connect countries to the global innovation marketplace, which includes both private and public actors. At the same time, giving public institutions responsibility to protect their inventions (as well as license them) increases technology transfer. For this reason, there is a major role for the public as well as the private sector in developing an IPR system that is relevant to national needs.

LESSONS LEARNED

As discussed, a wide range of actions and investments can support the management of IPRs to promote agricultural development. The following sections summarize lessons learned over the years as new strategies in IPR management were used to achieve specific socioeconomic goals.

Legislation has to be passed and enforced but is not sufficient on its own

For low-income countries, the impact of IPR policies is often dwarfed by other factors that have a far more direct effect on innovation, including lagging investment in education, institutional and infrastructural challenges, limited access to financing, and the effects of a range of other business development policies. In response to TRIPS, most countries have passed IPR legislation, but few have gone on to create a culture of innovation. To take this next step, the value of protecting and using innovation must be understood within a society. Government support for innovation (box 6.18) that leads to the development of specific products is one way to demonstrate the value of IPRs to a wide cross-section of society.

Investments related to IPRs are focusing on legal systems of developing countries, based on the understanding that the ultimate impact of any IPR depends on how it is enforced. IPR legislation must be supported by well-functioning institutions (courts, patent offices, and the like) if IPRs are to provide any incentive for innovation, but these critical investments in institutional capacity will have more impact if they are designed to support the interactions of institutions and staff with a rapidly changing IPR environment. Examples include investments in improved capacity to negotiate international treaties, increased support for connections to international networks of professionals, and improved access to research and expertise specifically targeted at IPR issues in developing countries.

Managing property rights in public institutions is critical

In agricultural development, public institutions are central to the development and adoption of innovations that will benefit the world's poor. The role of the public sector in agricultural development has shifted considerably over recent decades: grants are for shorter terms and focus more tightly on projects; engagement with the private sector is increasingly a necessity; and organizations operate in a complex web of IP and regulatory law frameworks. The public sector has lagged considerably in understanding how IPR policies and practices affect its role in development goals.

In public organizations, capacity for IPR management is often a low priority due to resource constraints, limitations on available expertise, and a lack of receptivity among some managers to embrace IPRs as an important component of their development work. This lack of capacity can lead to mistakes and missed opportunities in licensing, partnership agreements, and strategic management of innovations. From the donors' perspective, lack of IPR management can lead to delays, potentially reduced impact, or sometimes the halting of a project altogether. Conversely, good capacity for IPR management reduces the risk for donors' R&D investments.

Humanitarian licensing models need more work

Licensing language that supports "humanitarian uses" of proprietary technologies permits them to be used for applications important to international development while preserving the technology owner's commercial market, yet such language is not widely used. Much work remains to be done to move from "model language" to language integrated into working licenses. In addition to IPR provisions, humanitarian use licensing has not yet adequately addressed key issues that repeatedly arise in public-private partnerships, such as liability.

Emerging markets represent a further licensing challenge. So far, many models of humanitarian licenses depend on territorial distinctions. This strategy can leave countries like Brazil, India, and China out of a geographically defined region for humanitarian use, despite the large populations of very poor people in these countries. Legal strategies must be improved to allow the poor in these countries to benefit from technologies that are accessed under humanitarian use licenses.

Flexibility is crucial for future success

One-size-fits-all solutions to IPR management work in limited situations to reduce transaction costs, but over the years it has become clear that most IPR management at the institutional level requires project-specific consideration of the partners, technologies, countries, and many other details. Patent pools, patent commons, clearinghouses,[8] and model licenses do have great value, but the standardized approach must be complemented with (1) the flexibility to modify the IPR strategy and (2) access to resources to support good strategic management. At the policy level, similarly, the complex differences among countries necessitate careful assessment of tailored IPR policy solutions. As noted, the level of development of a particular industry may warrant some kind of special exemption, particularly if the industry is considered of strategic importance. Such exemptions have most often been seen in the pharmaceutical industry, where exemptions from patent protection for certain drugs or even whole classes of drugs have been used to promote a local industry. Often these are temporary measures used to assist in strategic development, because they may be considered to be in breach of trade rules, but where a case can be made for a special need, exemptions are an important option to consider.

RECOMMENDATIONS FOR PRACTITIONERS

The recommendations that follow are intended for policy makers as well as practitioners (researchers, managers, and experts who encounter IP issues at the institutional or project level). The recommendations complement the earlier section on "Actions Needed," which identifies key areas for investment, and should be kept in mind by practitioners as aspects of IPR arise in projects.

Create diverse opportunities for IPR training

Policy makers as well as those at the institutional level can articulate the need to raise awareness of IPR issues across many fields of science and technology. Scientists, engineers, IP managers, government officials, administrators, and many others can benefit from improved understanding of the role of IPRs in agricultural development. The roles that IPRs play, however, and the levels at which they may be encountered are highly diverse, which suggests that a broad range of training options should be considered. For example, box 6.21 details the development of a small enterprise from an NGO-led project to generate income. In this case, IPR awareness training was incorporated into the business planning for the project so that participants would understand the options for protecting any IP. Where a producer organization is involved, such as the Colombia Coffee Federation (box 6.19), the organization's needs may best be served by identifying specific individuals to receive more specialized training in legal aspects of managing IPRs.

In addition, practitioners can work to ensure that training for particular professionals continues—for example, through engagement with an international community. Training within a South–South context can be particularly valuable for professionals to compare the challenges and solutions related to IPRs in developing countries. Where the establishment of technology transfer offices is being considered, exchange programs with existing offices can be highly beneficial and help to forge long-term links between institutions. Box 6.21 includes examples of investments in training IP professionals with funding from national agencies and donors.

Box 6.21   Country and Donor Investments in Intellectual Property Training for Professionals

---

In-depth training. Since its founding in 1998, the World Intellectual Property Organization (WIPO) Academy (http://www.wipo.int/academy/en/) has offered a wide range of courses on intellectual property and intellectual property management. In addition to short workshops, the academy provides key support for degree and certificate programs in many developing countries. For example, in 2010 Zimbabwe's African University (in collaboration with WIPO and the African Intellectual Property Organization) graduated the first class of students with a master's degree in intellectual property.

South–South training. The International Intellectual Property Training Institute (IIPTI, http://iipti.org/), part of the Korean Intellectual Property Office, trains professionals from Malaysia, Vietnam, and other developing countries within the region. India, South Africa, and Brazil play key roles as regional leaders with the capacity to share IPR knowledge specific to the challenges faced by developing countries.

Targeting diverse professionals. Singapore's Mentorship Funding Scheme brings in qualified patent agents from overseas to mentor and train professionals at locally based organizations. The Intellectual Property Office of Singapore (IPOS, www.ipos.gov.sg) funds the costs of the program jointly with local organizations. India's National Institute of Intellectual Property Management (NIIPM, www.ipindia.nic.in), in the Ministry of Commerce and Industry, provides training to R&D managers, scientists, legal professionals, patent agents, researchers, doctors, engineers, and others.

Practical knowledge. As part of its bilateral development assistance, the Swedish International Development Cooperation Agency (SIDA, www.sida.se) offers around 100 Advanced International Training Programmes every year targeted at practitioners in the field. These programs frequently cover various IPR issues, including topics such as genetic resources and IPRs, industrial property in the global economy, and intellectual property for least developed countries. The Public Intellectual Property Resource for Agriculture (PIPRA, www.pipra.org) provides training and educational materials (see the *ipHandbook*, www.iphandbook.org) to developing-country scientists, intellectual property managers, and policy makers, with a focus on providing tools for practicing intellectual property.

Source: Authors.

When developing training programs, it is essential to consider that the trained professionals will need some form of employment in a setting where they can use the skills they have learned. Significant resources have been wasted by training people who have no opportunities to apply their knowledge; these misdirected efforts further erode the impression that IPRs should be taken into account. TTOs provide a focus for training individuals and can also employ them in a role that enables them to maintain their involvement in this field.

Promote collaboration among public and private partners

Practitioners working at the institutional level should seek to ensure that institutional IP policies support partnerships between public and private organizations. Such policies might include, for example, a clear set of principles to govern legal relationships with partners, processes to assess risk in partnerships, transparency mechanisms to enable good governance, clarity around confidentiality issues, and the flexibility for management to implement a strategic IPR management plan. EMBRAPA[9] in Brazil has created such policies, which enabled it to make licensing deals with a number of multinational companies as well as local seed producers and assume a significant role in soybean variety development (Fuck and Banacelli 2009). Policy makers should continue to explore IP policy options through the lens of creating incentives for public and private collaboration in agricultural development.

Balance in-house capacity with prudent use of external services

The IPR management capacity needed to meet the challenges of coming decades does not exist in sufficient depth, even in industrialized countries. In the private sector, due diligence,[10] the negotiation and drafting of agreements, and strategic IPR management are all regular practices (see IAP 2). Universities, nonprofits, governments, international aid agencies, and philanthropic foundations have excellent expertise in IPRs.

Yet the capacity for both strategic management of IPRs and the day-to-day work of IPR practitioners is not as common as it needs to be. In such cases, external sources are commonly employed for a number of specialized tasks, such as legal opinions on freedom to operate and drafting of patent submissions. Practitioners should consider this option when dealing with specific investments that may have detailed requirements for IPRs. The use of external services may be a much more cost-effective option for certain tasks.

One investment discussed earlier was the creation of a TTO in an institution to serve as a focal point for protecting IP as well as for licensing. In module 5, TN 5 addresses the role of TTOs in agricultural development, where the creation of a TTO represents an opportunity to foster an enabling environment for innovation and provide opportunities for training (including raising awareness of IPR among scientists and administrators). For some institutions, the costs of creating and managing a technology transfer office, investing in a portfolio of IPRs, and (importantly) having the resources to enforce those IPRs, may not be feasible, and they will need to explore other options for developing capacity in IPR management and training.

Biosafety Regulatory Systems in the Context of Agricultural Innovation

Morven McLean, ILSI Research Foundation

SYNOPSIS

Biosafety regulation is a multidisciplinary, multifunctional endeavor that should take into account the broader context of agricultural production and innovation. Investments needed to operationalize a biosafety regulatory system should promote interministerial cooperation, sound and pragmatic policy development, scientifically defensible risk assessment and risk management, rational inspection and enforcement activities, and meaningful stakeholder consultation and public participation. Efficiencies can be gained through the cross-utilization of national or regional expertise, regional harmonization, and ensuring that the design of a biosafety regulatory system takes into account programmatic and operational costs, including opportunity costs that may arise from overregulation.

BACKGROUND AND CONTEXT

To date 22 countries have approved genetically engineered (GE) plants for cultivation or consumption (CERA 2010a). In 2010, 148 million hectares (366 million acres) were planted to GE crops, largely soybeans, cotton, maize, and canola (James 2010). Common to all countries where GE crops are cultivated is a system to regulate these products and especially to ensure that they are evaluated with respect to human health and environmental safety (commonly referred to as biosafety) prior to their commercial release.

The regulation of products of agricultural biotechnology, particularly GE crops, has been identified as a constraint to innovation in this sector, largely because of the costs of meeting information and data requirements prescribed by regulatory authorities for assessing the safety of GE plants (Cohen and Paarlberg 2004; Kalaitzandonakes, Alston, and Bradford 2007; Matten, Head, and Quemada

2008) but also because of inadequate capacity to enforce regulatory compliance (Pray et al. 2006).

The converse may be argued as well, however. A well-defined biosafety regulatory system that is consistent in its application (that is, the assessment, decision-making, and enforcement processes are not arbitrary) can be a powerful stimulus for investments in this area. For example, Brazil has seen public and private investments increase in agricultural biotechnology since it passed the Biosafety Law in 2005 (BrBiotec 2010). The new law clarified the regulatory remits of various ministries and clearly defined the roles and responsibilities of the two regulatory authorities, the National Biosafety Council (CNBS, Conselho *Nacional de Biossegurança*) and the National Biosafety Technical Commission (CTNBio; Comissão Técnica Nacional de Biossegurança).

The law ended a five-year moratorium on approvals of GE crops that arose from differences in governmental and judicial interpretation of pre-2005 legislation (Cardoso et al. 2005). The moratorium, in turn, contributed to wide-scale cultivation of illegal (unapproved) GE soybeans. Since 2005, Brazil has approved 20 GE cotton, soybean, and maize lines for commercial cultivation; prior to 2005 it had approved only 1 (CERA 2010a).

A functional biosafety regulatory system is a prerequisite for realizing the benefits that agricultural biotechnology can (and does) provide to poor producers and poor consumers in developing countries (World Bank 2007). Ultimately, environmental and human health protection is the overarching priority of any biosafety regulatory system, and confidence in the decisions that governments make on behalf of the public is a precondition for public acceptance and adoption of agricultural biotechnology products. Strategic investments in programs that foster adaptability, transparency, clarity, and workability in the development and implementation of regulatory systems also foster agricultural innovation.

INVESTMENT NEEDED

Investments in support of biosafety regulation may be needed for any or all stages in the typical progression of events that lead to the development and implementation of a regulatory system. Key issues and policy options for these stages were described in a conceptual framework for biosafety regulation (McLean et al. 2002); World Bank (2003) presented examples for individual countries. In summary, the key stages are:

- *Elaborate a national policy* consistent with other objectives related to economic, social, and rural development, natural resource management, and environmental protection and sustainability.

- *Conduct an assessment and gap analysis* of national development priorities, agricultural policies, existing regulatory regimes, and national and regional scientific and technical means necessary for a biosafety regulatory system to function.

- *Build a strong base of scientific knowledge* in support of the regulatory system and the development of core competencies in biotechnology product evaluation (box 6.22).

- *Develop biosafety regulations to effect specific public policy goals* (as articulated in a national biosafety or even biotechnology strategy).

- *Implement regulations* through the operationalization of the biosafety regulatory system.

- *Address cross-cutting issues* that are common to each stage in the development and implementation of a

---

Box 6.22   Building Human Resource Capacity for Biosafety Risk Assessment

The type of human resource capacity needed to implement a biosafety regulatory system generally, and its risk assessment function specifically, is particular to each country. No standardized lists of human resource requirements specific to individual disciplines exist. It can be instructive, however, to examine how other countries have approached this issue.

In India, the Risk Assessment Unit of the proposed Biotechnology Regulatory Authority of India will be permanently staffed by a multidisciplinary team of scientists responsible for undertaking science-based risk assessments of specific products. The Risk Assessment Unit will comprise thematic cells. The expertise for the two cells pertinent to the regulation of genetically engineered crops is:

- *Core characterization:* Molecular biologist, toxicologist, microbiologist, biochemist, bioinformatics specialist, biostatistician.
- *Plant biotechnology:* Plant physiologist, plant pathologist, entomologist, agronomist, and plant breeder.

In Brazil, the National Biosafety Technical Commission (CTNBio, Comissão Técnica Nacional de Biossegurança) provides technical support and advice to the federal government "in the formulation, updating, and

implementation of the National Biosafety Policy for GMOs and derived products, and for establishing technical safety standards and technical opinions regarding the authorization of activities that involve research and commercial use of GMOs and derived products." CTNBio is comprised of 27 members:

- Twelve specialists (PhDs recommended by scientific organizations).
- Nine government officials appointed by the following agencies: Ministry of Science and Technology; Ministry of Agriculture, Livestock, and Food Supply; Ministry of Health; Ministry of the Environment; Ministry of Development, Industry, and Foreign Trade; Ministry of External Relations; Ministry of Agrarian Development; Ministry of Defense; and Special Office of the President for Aquaculture and Fisheries.
- Six members appointed as follows: one specialist in consumer rights by the Ministry of Justice; one specialist in human health by the Ministry of Health; one specialist in environment by the Ministry of the Environment; one specialist in biotechnology by the Ministry of Agriculture, Livestock, and Food Supply; one specialist in family agriculture by the Ministry of Agrarian Development; one specialist in worker's health by the Ministry of Labor.

Source: DBT 2008; Government of Brazil 2005.

---

national biosafety system, especially the human, financial, and infrastructure resources to: develop and implement a national biosafety system; support the infrastructure required (such as buildings, equipment, and computers); facilitate communication and public participation; train scientific and regulatory personnel; and foster the research required to assure that risk assessments are sound.

POTENTIAL BENEFITS

Investments in support of developing biosafety regulatory capacity have the potential to provide many positive spillovers into related areas. These areas include public agricultural research, extension services, and plant health and quarantine programs.

Private developers of GE crops, particularly multinational companies, are generally disinterested in entering markets, even where there is farmer demand for these crops, unless an operational (and predictable) biosafety regulatory system is in place. More critically, publicly funded and donor-funded initiatives that focus on improving the productivity of staple crops using biotechnology will be unsuccessful unless there is a clear path forward that ensures improved crop varieties will actually move from laboratory to field trials to farmers. (Although when the technology does reach farmers, the impact can be significant; see box 6.23.)

Highly precautionary regulations may be the most significant barrier to innovation in agricultural biotechnology, as they price the technology out of the hands of the public sector and SMEs. These costs include the direct costs of regulatory compliance as well as indirect costs associated with

unanticipated events, such as trade disruptions that can occur as a result of accidental (or sometimes deliberate but illegal) transboundary movement of GE commodities into a jurisdiction where there is no approval for that GE crop or derived food. For example, continued delays in the deployment of pro-vitamin A rice ("Golden Rice") have been attributed exclusively to biosafety regulation by the product developer (Potrykus 2010).

POLICY ISSUES

Key policy considerations include:

- *Coordination of biosafety laws and regulations* with existing legislation related to environmental protection, human health, agricultural production, IP protection, and trade.
- *Interministerial coordination* to ensure that concerns and remits are carefully considered during the establishment of a biosafety regulatory system. Responsibilities and mandates of all involved ministries should be clearly communicated.
- *Multilateral environmental agreements*, particularly the Cartagena Protocol on Biosafety, must be considered during the development or revision of biosafety legislation (box 6.24).
- *Trade:* Biosafety legislation should not promote practices that may be considered or may result in impediments to trade.
- *Resources*—financial, human, and institutional—need to be considered before developing the regulatory system because they can, and should, influence its construction.

Box 6.23   Who Benefits from Agricultural Biotechnology?

It is difficult to quantify the benefits of regulating products of agricultural biotechnology, but the economic impact from commercializing many genetically engineered crops has been studied. Brookes and Barfoot reported that in 2007, the total cost farmers paid for genetically engineered soybean, maize, cotton, and oilseed rape was equal to 24 percent of the technology gains (inclusive of farm income gains plus the cost of the technology payable to the seed supply chain, comprised of sellers of seed to farmers, seed multipliers, plant breeders, distributors, and the providers of

genetically engineered technology). According to this study, farmers in developing countries paid 14 percent of technology gains, whereas farmers in developed countries paid 34 percent of their gains. The higher share of total technology gains accounted for by farm income gains in developing countries relative to the farm income share in developed countries reflected factors such as IPRs in developing countries and the higher average level of farm income gain on a per-hectare basis derived by developing country farmers relative to developed country farmers.

Source: Brookes and Barfoot 2009.

- **Regional coordination and harmonization of elements of the regulatory system** should be considered and/or pursued, as harmonization has the potential to: reduce regulatory disparities between countries; reduce the regulatory burden on national governments and the regulated community; and facilitate trade within region (see IAP 3).

LESSONS LEARNED AND RECOMMENDATIONS FOR PRACTITIONERS

The previous sections have described the often complex interface between agricultural innovation and biosafety regulations. Lessons related to developing and implementing biosafety regulations can be summarized briefly:

- Building capacity to develop and implement biosafety regulatory systems requires a multiyear commitment.
- Interministerial coordination is a prerequisite for successful development of a biosafety regulatory system.
- Investments in biosafety regulatory capacity can be strategically applied to benefit other regulatory programs.
- Biosafety regulatory systems should incorporate provisions for change.
- Investments to develop biosafety regulatory systems should accompany investments in agricultural biotechnology research.

- Biosafety regulation can be rationalized through the promotion and acceptance of international risk assessment standards.

The next sections address each of these points in detail.

Building capacity to develop and implement biosafety regulatory systems requires a multiyear commitment

Workshops, symposia, and conferences can be valuable in raising awareness or catalyzing discussions that may inform the development of strategic programs, but they cannot replace continued and meaningful engagement with those who are tasked with the responsibility of actually developing and implementing the regulatory framework (a task requiring considerable time, coordination, and expertise; see box 6.24 for an example from India). Identifying in-country partners and investing in longer-term capacity building for key individuals, including policy makers and opinion leaders, contributes to systemic versus transient gains. Experience has shown that the willingness of these individuals to understand the impact of, and provide an enabling environment for, (cost)effective biosafety regulation is critical (see box 6.25 for an example from Uganda).

Box 6.24   The Development of Genetically Engineered Food Safety Assessment Guidelines in India

The South Asia Biosafety Program (SABP) has assisted the Governments of Bangladesh and India to further strengthen their institutional governance of biotechnology since 2004. In India, the program started with stakeholder consultations and a gap analysis of the current biosafety regulatory system. The analysis identified the need for comprehensive safety assessment guidelines for foods derived from genetically engineered plants and for technical training in conducting food safety assessments according to international standards.

The Indian Council of Medical Research (ICMR), the technical arm of the Ministry of Health and Family Welfare, in partnership with SABP, undertook a series of activities over the next several years aimed at meeting this need. It began with an international conference on safety assessments for foods derived from genetically

engineered plants. The conference offered an opportunity for stakeholders and technical experts from a number of sectors to exchange experiences and views.

ICMR then hosted a multisectoral stakeholder consultation that achieved consensus on making the safety assessment of genetically engineered foods in India consistent with the internationally accepted Guideline for the Conduct of Food Safety Assessment of Foods Derived from Recombinant-DNA Plants adopted by the Codex Alimentarius in 2003. ICMR formed a drafting committee with representation from several ministries and departments and formulated draft "Guidelines for the Safety Assessment of Foods Derived from Genetically Engineered Plants."

The draft guidelines were circulated to technical experts for input and reviewed by India's Review

(Box continues on the following page)

Committee on Genetic Manipulation and Genetic Engineering Approval Committee before being posted for public comment. Stakeholders' comments were addressed, and the guidelines were reviewed once again by both committees before their final adoption in 2008. The end result is a guidance document that is consistent with internationally accepted practices for assessing the safety of genetically engineered food.

ICMR also collaborated with SABP to conduct technical workshops providing in-depth, hands-on training about key requirements for the safety assessment of foods derived from genetically engineered plants. The training ensured that scientists and regulators, as potential risk assessors and science advisors, understood the concepts and principles of genetically engineered food safety assessment and the methodology outlined in the new guidelines.

From inception to completion, the process of developing new food safety guidelines and ensuring their implementation under existing authority in India took four years. The long-term collaborative relationship between ICMR and SABP contributed to the success of this endeavor. SABP, particularly through strong in-country partnerships, supported ICMR's commitment to developing new guidelines by providing not just technical expertise on food safety assessment, but also institutional support to ICMR and Indian regulatory committees as they took the guidelines through review, adoption, and implementation.

Source: McLean 2010; CERA 2010b.

Uganda has spent almost fifteen years working to develop a functional biosafety regulatory system that will promote an enabling environment for research, development, and deployment of genetically engineered crops. The country was an early recipient of Global Environment Facility support to develop a National Biosafety Framework. The process started in 1998, three years before Uganda ratified the Cartagena Protocol on Biosafety and five years before the Protocol came into force.

Since then, the Ugandan regulatory and science communities have benefitted from significant national and international investments that have supported both human and institutional resource development, such as enhanced technical capacity for product development, management of confined field trials, and premarket risk assessment. The incremental gains achieved through these interventions have been confounded by continued delays in operationalizing the regulatory system, particularly the passage of national biosafety legislation.

Uganda provides an all too common example of a country where innovation in agricultural biotechnology is not necessarily limited by science but by political, social, and market barriers. It is generally accepted that product commercialization will not advance in Uganda until the national Biosafety Bill is promulgated. The process of preparing the Biosafety Bill began in 2003. The Bill was finalized in 2007, approved by the Cabinet in 2008, and currently awaits submission to Parliament. An analysis of the reasons for this protracted process found that a combination of market, policy-political, and sociocultural factors are hindering progress, such as:

- Lack of sustained and coordinated political champions to move the bill forward.
- Lack of clarity among ministries regarding regulatory roles and responsibilities.
- Influence of antibiotechnology organizations.
- Complex and diverse institutional players.
- Poor product development strategies, leading to delays in driving the operationalization of the biosafety regulatory system.

The last bullet may now be a significant catalyst for movement on the Biosafety Bill. Using existing legislation, Uganda has approved confined field trials of genetically engineered cotton, banana, and

(Box continues on the following page)

cassava. These and other pipeline products such as drought-tolerant maize are all considered important for Uganda's agricultural productivity and sustainability. Having farmer-supported, genetically

engineered crops approaching commercialization may be the incentive needed to achieve multistakeholder, and consequently political, support for the Biosafety Bill.

Source: Horna et al. 2012, forthcoming; AATF 2010.

A shortcoming of many capacity-building projects is that they support the drafting of biosafety frameworks, legislation, or related documents but do not provide the follow-on support to finalize, adopt, and then implement the system(s) prescribed in these documents (Chapotin, McLean, and Quemada 2009). For example, 123 countries participated in the Project on Development of National Biosafety Frameworks sponsored by the United Nations Environment Programme and Global Environment Facility (UNEP-GEF). Designed to help countries comply with the Cartagena Protocol, the project was active from 2001 to 2009. Of the 38 African countries that completed their national biosafety frameworks under this project, only three have regulatory systems that can be considered operational: Tanzania and Nigeria have authorized confined field trials (although Tanzania's approvals pre-dated their National Biosafety Frameworks project) and Burkina Faso has assessed and approved a GE plant for commercial release (insect-resistant cotton in 2008). The transition of countries from the framework development projects to the follow-on UNEP-GEF Project on Implementation of National Biosafety Frameworks was limited to 19 countries.

Interventions should be tailored to country needs, but many large capacity-building programs, such as the National Biosafety Frameworks project, implement a common project model. Investments should first support a comprehensive needs assessment and gap analysis to identify and prioritize interventions that will further the operationalization of a functional regulatory system. In addition to evaluating the national situation, it is important also to critically consider capacity building or related initiatives that may be happening regionally or internationally and whether these may assist or constrain follow-on activities. The needs assessment should also take into account the broader context of agricultural production and innovation, because biosafety regulation is but one part of that larger system.

Interministerial coordination is a prerequisite for successful development of a biosafety regulatory system

International support for the establishment of biosafety regulatory systems has favored the creation of new regulatory entities under ministries other than agriculture. Particularly influential in this regard is the Cartagena Protocol. Because of its relationship to the Convention on Biological Diversity, the Protocol has largely been implemented through ministries of environment. Agricultural biotechnology regulation intersects the mandates and interests of multiple ministries, especially agriculture but also ministries of science and technology, environment, health, and trade.

Investments in the development of biosafety regulatory systems should explicitly require meaningful interministerial consultation and a clear delineation of roles and responsibilities between competent authorities. Otherwise, different ministries develop parallel and often redundant or conflicting regulatory requirements that ultimately increase the regulatory burden on product developers. Rational regulation is achievable if the overarching purpose of biosafety regulation (that is, human and environmental safety) drives the development of the regulatory system and is not tied to political or financial gain by specific ministries.

Interministerial coordination, while necessary, is difficult to obtain in practice. As indicated during the 2003 Sub-Regional Workshop for Latin American Countries on the Development of a Regulatory Regime and Administrative Systems, the primary conflict identified for the implementation of national biosafety frameworks was coordination of the administrative tasks and competencies of the institutions involved in them (UNEP 2003a). This issue was also stressed in a similar workshop for Asian countries, where it was noted that "much of the administrative system seemed to be in place in many countries, and that coordination was the major challenge where different agencies were working separately" (UNEP 2003b) (box 6.26 presents an example

Box 6.26 Interministerial Coordination in the Biosafety Regulatory System of Bangladesh

In Bangladesh, the biosafety regulatory system is still in a developmental stage, although institutional procedures cover R&D and the review and approval of foods derived from transgenic plants. The system is based on a National Biosafety Framework document, developed with UNEP-GEF funds in 2004–06, which draws on a set of Biosafety Guidelines initially published by the Ministry of Science and Technology in 1999. With the ratification of the Cartagena Protocol by Bangladesh in 2004, responsibility shifted to the Ministry of Environment and Forests (MoEF), and the Biosafety Guidelines were redrafted to incorporate certain obligations of the Cartagena Protocol. The revised guidelines were published in 2007.

Under the Biosafety Guidelines, the competent authority is the interministerial National Committee on Biosafety (NCB). The subordinate Biosafety Core Committee operates as a scientific review body and so far has been asked by the NCB to provide input into all its decisions. To lend enforcement power to MoEF, a Biosafety Rule has been drafted that incorporates the Biosafety Guidelines and brings them under the formal jurisdiction of the Environment Conservation Act. This Biosafety Rule was prepared by a drafting committee convened by MoEF that sought to proactively include inputs from key ministries. Because of this action, no further government debate is considered necessary for approval.

Guidelines for confined (experimental) field trials of genetically engineered plants have also been prepared through the cooperative efforts of the Department of Environment (DoE in MoEF) and the Bangladesh Agricultural Research Council, Ministry of Agriculture (MoA). The guidelines include procedures for applications, standard operating procedures, and a guide for inspections of confined field trials by officials appointed by MoEF. These guidelines have been approved by the NCB and published as an annex to the Biosafety Guidelines. In 2009 guidelines for genetically engineered food safety assessment were prepared that are consistent with Codex (2003). NCB approved them in 2010, and they will be published as an appendix to the Biosafety Guidelines.

Bangladesh's biosafety regulatory system, while still young, has made significant progress. Confined field trials are now being approved and applications for commercial release are considered imminent. Interministerial cooperation, particularly between DoE of MoEF and the Bangladesh Agricultural Research Council of MoA, has been integral to the success achieved to date.

Source: Author.

---

from Bangladesh). For the majority of countries, both developed and developing, internal coordination between competent authorities remains a significant issue that has yet to be resolved.

Investments in biosafety regulatory capacity can be strategically applied to benefit other regulatory programs

The shared nature of many of the regulatory functions of plant health and quarantine programs and biosafety programs (such as risk assessment, monitoring, and inspection) means that there is an opportunity to apply investments for biosafety regulatory capacity building to strengthen plant health and quarantine systems (and vice versa) so that the objectives of both can be achieved without building redundant administrative and operational services. For example, the Government of Canada recently combined the risk assessment functions for GE plants and plant health into a single Plant and Biotechnology Risk Assessment Unit. This action was taken to align biosafety and phytosanitary risk assessments, leverage complementarities in the scientific expertise required for both, and improve procedural consistencies (P. Macdonald, personal communication).

Investments should strengthen the scientific and knowledge base in ways that will provide benefits that extend beyond biosafety risk assessment and decision making. Many developing countries have only a transient need for biosafety risk assessment per se, because regulatory authorities may receive an application for a field trial or premarket approval only once a year or once every few years. Investments in education and research in the scientific disciplines that support biosafety risk assessment and regulation, especially in the agricultural sciences, will have wide-reaching payoffs, however. Efficiencies can be gained through the cross-utilization of expertise within a country or even through pooling human resources with neighboring countries.

Biosafety regulatory systems should incorporate provisions for change

The regulation of products of modern biotechnology is a relatively new arena for governmental oversight. Advances in biotechnology processes and products, experience gained in regulatory operations over time (both nationally and internationally), the globalization of agricultural trade, and the influence of multilateral agreements and international standard-setting bodies require biosafety regulatory systems to accommodate change (box 6.27). For example, embedding detailed technical provisions about risk assessment into laws versus guidance impedes regulators' ability to accommodate new knowledge or advances in risk assessment approaches, as revising legislation is considerably more burdensome than amending guidance.

Investments to develop biosafety regulatory systems should accompany investments in agricultural biotechnology research

Implementation cannot be meaningfully initiated unless applications related to GE products are ready to "prime the regulatory pump," such as applications for R&D activities in laboratories, field trials of experimental GE products (transgenic plants, insects, or fish, for example) or applications for environmental, food, and/or livestock feed safety assessments prior to marketing a product. Of the 38 African countries mentioned previously, few have substantive public research programs in agricultural biotechnology, and many are not considered priorities for private biotechnology investment.

The lack of substantive private or public R&D, even more than resource constraints, may explain why so few countries have implemented national biosafety frameworks. In effect, there is an absence of demand to drive regulatory development (or reform) forward, and policy makers' attention is redirected to existing priorities (with notable exceptions, as in Burkina Faso; see box 6.28). Another definite requirement is the political will to move the regulatory system forward so that decisions, particularly about product-specific approvals, are actually taken.

Biosafety regulation can be rationalized through the promotion and acceptance of international risk assessment standards

The building of sufficient risk assessment capacity is a particular problem in countries that do not have a base of scientific expertise in biosafety. The development of a regional or subregional approach to risk assessment may be the most practical and cost-effective option in such cases. This approach can be facilitated by the active participation of competent authority representatives in international forums such as the Codex ad hoc Intergovernmental Task Force on Foods Derived from Biotechnology and the OECD Working Group on Harmonization of Regulatory Oversight of Biotechnology, where criteria for risk assessment harmonization are discussed and guidance or standards established. Vietnam developed its own practical approach (in this case to assess risks of GE food), based on a review of risk assessments conducted in other countries (box 6.29).

Rationalization can also be achieved during the design of a biosafety regulatory system. Policy options should be evaluated to take into account not just the government's overarching human health and environmental protection goals but also the costs of sustaining a system that can realistically achieve those goals. These costs include the opportunity costs associated with overregulation. Identifying the funding mechanisms required to sustain a regulatory system can be an effective tool in rationalizing its complexity.

> **Box 6.27   Adaptability in Biosafety Regulation: The Gene Technology Act in Australia**
>
> In 2001, the Gene Technology Act, 2000 introduced a national scheme for the regulation of genetically engineered organisms in Australia. It included a statutory requirement (Section 194) for an independent review of the operation of the act, including the structure of the Office of the Gene Technology Regulator (the OGTR), by the fifth anniversary of the act coming into force. The review was based on issues raised during extensive national public and stakeholder consultations, submissions made in response to the terms of reference for the review, site visits to laboratories and field trials, experience gained by OGTR personnel during the first four years of the act's implementation, international developments in biotechnology, and related reports and literature. The review found that the act's flexibility to deal with changing circumstances and emerging technologies was sufficient but that the act should be reviewed again in five years to ensure that it continues to accommodate emerging trends.
>
> *Source:* GTRS 2006.

Box 6.28   The Approval of Bt Cotton in Burkina Faso

Burkina Faso is sub-Saharan Africa's largest cotton producer. Cotton accounts for 30–50 percent of the country's export earnings and is the main source of foreign exchange. In many rural areas where poverty is high, the sale of cottonseed is the main or only source of cash revenue for Burkinabe farmers. Insect control is a key factor in cotton yield; insect infestations can damage up to 90 percent of the crop. Farmers typically apply 6–8 applications of insecticide per growing season, but yield losses of 30–40 percent persist.

An alternative insect management approach is to plant insect-resistant, transgenic cotton varieties (Bt cotton). Transgenic varieties from the United States were evaluated in confined field trials in Burkina Faso from 2003 to 2005. These Bt varieties had significantly reduced larval populations of cotton bollworm and cotton leafroller, with a commensurate improvement in seed cotton yields and lint quality. After the insect resistance trait was bred into local varieties, further field trials were planted in 2006–07. Precommercial seed production began in 2008, the same year two transgenic

cotton varieties were approved by Burkina Faso's National Biosafety Agency (ANB, Agence Nationale de Biosécurité) for commercial release. Comparisons in 2008 and 2009 showed that Bt cotton yielded 30 percent higher than conventional varieties, and only two insecticide applications were necessary.

Burkina Faso's biosafety regulatory system has developed relatively quickly and smoothly compared to those of other African countries such as Kenya, Uganda, and Nigeria. In 2005 Burkina Faso completed its National Biosafety Framework with resources from the United Nations Environment Programme and Global Environment Facility. In 2006 the ANB was established under Law No. 005-2006 "Pertaining to the security system in regard to biotechnology in Burkina Faso."[a] However, it was the joint commitment of the Ministers of Environment and Agriculture, who publicly championed the economic benefits of Bt cotton to the Bukinabe economy, that effectively catalyzed the rapid operationalization of the ANB, which was achieved in only two years.

Source: Héma et al. 2009; D.J. MacKenzie (personal communication).
(a) Loi N° 005-2006/AN, Portant régime de sécurité en matière de biotechnologie.

Box 6.29   Practical Regulation of Genetically Engineered Foods in Vietnam

In June 2010, the Government of Vietnam issued Decree No. 69/2010/ND-CP on Biosafety for Genetically Modified Organisms, Genetic Specimens, and Products of Genetically Modified Organisms. With respect to the use of genetically engineered organisms as food or animal feed, the Decree permits a written certification of eligibility for use as food if the subject of the application satisfies "either of the following conditions: 1. The dossier of application for a written certification of their eligibility for use as food has been appraised by the Genetically Modified Food Safety Council, which concludes that such genetically

modified organisms have no uncontrollable risks to human health. 2. They have been permitted by at least five (5) developed countries for use as food and no risk has been seen in these countries."

This approach to regulatory approvals is both practical and scientifically defensible. It recognizes that the Vietnamese Ministry of Health considers the biosafety regulatory systems of certain other countries to be consistent with that of Vietnam and that the risk assessment and approvals undertaken by those countries may be considered equivalent to and therefore sufficient to obtain a certificate of eligibility by the Ministry of Health.

Source: Government of Vietnam 2010.

Technical Regulations and Standards

Luz Diaz Rios, Consultant

Johannes Roseboom, Consultant

SYNOPSIS

The current landscape of technical regulations and standards related to agriculture and agrifood is complex and rapidly evolving. Standards represent major challenges for the targeted sectors and industries, yet they also present opportunities to catalyze innovations while achieving public health, trade, environmental, and social objectives. Standards can help to reduce information asymmetries and externalities and promote fair competition. Some agricultural export industries in developing countries have used compliance with standards to gain an important competitive advantage; compliance required not only innovation in production and processing but in collective and organizational behavior. In an evolving landscape of standards, however, individual, one-time innovations offer limited opportunities to leverage long-term benefits. What is required instead is a process of strategic planning, supported by continuous innovation and improvement, to take on new challenges and opportunities as they emerge. Approaches to harmonizing standards across countries or industries can reduce transaction costs by reducing duplicative functions of conformity assessment, including testing and certification. National policy makers need to strike a balance between domestic and international trade interests and, as much as possible, maintain close involvement in regional and international standard-setting efforts.

BACKGROUND AND CONTEXT

A "standard" is a document approved by a recognized body that provides, for common and repeated use, rules, guidelines, or characteristics for products or related processes and production methods. Compliance with standards is not mandatory. "Technical regulations," on the other hand, specify product characteristics or their related processes and production methods, including the applicable administrative provisions, with which compliance is mandatory. Technical regulations include import bans (total or partial), technical specifications (process and product standards), packaging standards, information requirements, and requirements for labeling and claims.

Standards and technical regulations for agriculture and food have become increasingly important in recent decades, but they date to ancient times. Assyrian tablets, for example, describe the method to be used in determining correct weights and measures for food (FAO and WHO 2005). From the late 1800s to early 1900s, countries started to enact national agrifood standards. The ensuing proliferation of requirements complicated the landscape for international trade to such an extent that the first international standards began to be adopted in the early 1900s. Over the course of the century, but especially in the latter half, broader efforts to enact agriculture and food standards at the international level prompted important innovations in the international institutional framework for setting standards (see box 6.30).

The number of agriculture and food-related issues subject to standardization has grown tremendously in the past several decades.[1] This momentum reflects the intensification of regional and global trade and heightened concerns over accompanying threats to food safety and animal and plant health. It also reflects a wider set of innovations in science and technology that permit very sensitive detection and analytical methods, as well as improved knowledge of the quality and associated health hazards of agrifood products. Many standards and regulations relate to naturally occurring hazards, such as foodborne pathogens and toxins, while others have been introduced by innovations in agricultural technologies to increase productivity (such as the use of pesticides, veterinary drugs, and other chemical compounds).

Changes in consumers' concerns and perceptions, as well as pressure from civil society and the enactment of international agreements,[2] have been critical in expanding the range of desirable attributes associated with the quality of

Sanitary and phytosanitary (SPS) measures are taken to protect: (1) human or animal health from risk arising from additives, contaminants, toxins, or disease organisms in food, drink, and feedstuffs; (2) human life from risks associated with diseases carried by plants or animals; (3) animal or plant life from pests, diseases, and disease-causing organisms; and (4) a country from other damage caused by the entry, establishment, or spread of pests.

The need to fight animal diseases (zoonoses) at the global level led to the creation of the **Office International des Epizooties** (OIE) through an international agreement in 1924. An international agreement on plant health was reached in 1952 through the **International Plant Protection Convention** (IPPC). The **Codex Alimentarius Commission** (CAC), focusing on food standards in relation to safety risks, was created in the early 1960s. These international organizations have become even more relevant since the mid-1990s, when they were recognized as the international reference for settling disputes and for international trade under the World Trade Organization (WTO) Agreement for Sanitary and Phytosanitary Measures.[a]

Under the agreement, countries are encouraged to present their concerns to the WTO regarding measures adopted by trade partner countries that do not follow the stated principles. According to WTO, of 312 SPS-related trade concerns raised by countries to the SPS committee over 1995–2010, 28 percent related to food safety, 25 percent to plant health, and 41 percent related to animal health and zoonoses. Animal health concerns mainly included foot-and-mouth disease (24 percent of concerns), transmissible spongiform encephalopathy (35 percent of concerns), and avian influenza.

The Agreement on Technical Barriers to Trade (TBT) deals with product standards. It aims to prevent national or regional technical requirements or standards in general from being used as unjustified barriers to trade. The agreement covers standards relating to all types of products, including industrial and agricultural products. Food standards related to SPS measures are not covered. Codex decisions recognized by the TBT Agreement include those on food labeling, decisions on quality, nutritional requirements, and analytical and sample methods.

The **International Organization of Standardization** (ISO) also enacts international standards; those applicable to agricultural industries and enterprises include standards for quality, safety, and environmental management (series ISO 9000, ISO 22000, and ISO 14000, respectively). The agricultural sector also benefits from standards dealing with conformity assessment that apply across sectors (ISO 17000 series). Other international organizations setting global standards relevant to agriculture include the **International Seed Testing Association** (ISTA) and the **International Federation of Organic Agriculture Movements** (IFOAM).

A plethora of private initiatives also seek to have a global reach. GLOBALG.A.P. enacts standards on good agriculture practice, and the Global Food Safety Initiative (GFSI) focuses on Hazard Analysis and Critical Control Point (HACCP)-based standards with application in agrifood industries. Still other private initiatives apply to particular agricultural subsectors, for example export crops such as coffee, cocoa and tea.

Source: Authors; WTO 2011.

(a) For zoonoses, the International Health Regulations enacted in 2005 are an international legal instrument with the purpose and scope to prevent, protect against, control, and provide a public health response to the international spread of disease in ways that are commensurate with and restricted to public health risks, and which avoid unnecessary interference with international traffic and trade. Another international agreement related to biological risks is the Convention on Biodiversity Cartagena Protocol, discussed in TN 4.

agrifood products. Demands go beyond a product's characteristics (product standards) to include specifications on the conditions under which products are produced and packaged (process standards, which now often include sustainability considerations). Table 6.2 lists examples of the broad range of standards and technical regulations applied to food and agricultural products. The demand for such standards

has induced innovation at many levels in the agricultural sector (box 6.31).

ACTIONS AND INVESTMENTS NEEDED

The capacity of standards and technical regulations to achieve their intended outcomes and also catalyze agricultural

Table 6.2   Examples of Standards and Technical Regulations Applied to Agriculture and Agrifood Products

Food safety	Animal/plant health	Quality or technical attributes	Environment	Social
• Pesticide residue limits • Microbiological standards • Traceability requirements • Hygiene requirements • Vet. Drug residues • Chemical & other contaminants (e.g., mycotoxins) • Code of good agricultural practices	• Plant material quarantine • Pest risk analysis needs • Fumigation requirements & restrictions • Bans/restrictions on antibiotic use in aquaculture • Disease-free areas • Disease surveillance • Restrictions on veterinary drugs • Traceability of animals • Plant material quarantine • Phytosanitary certificates • GMO varietal approval	• Quality grades • General labeling requirements • Packing standards • GMO labeling • Restrictions on animal feed ingredients • Nutritional labeling	• Pesticide use restrictions • Regulations on water/soil contamination codes for organic practices & certification • Protection of specific species • Fish catch restrictions • Regulations on animal waste effluent • Water efficient regulations • Chemical use restrictions • Biosafety regulations (for GMOs) • Codes to limit biodiversity loss	• Monitoring of child labor • Occupational health standards • Animal welfare monitoring • Right to association • Minimum wage

Source: Adapted from Jaffee et al. 2005.

Box 6.31   Standards Induce Innovation throughout the Agriculture Sector

Innovation along agricultural supply chains. The serious effects of mycotoxins on human and animal health following consumption of specific contaminated products (such as groundnuts and maize) have led many countries to enact technical regulations establishing maximum permitted levels of mycotoxins. In sub-Saharan Africa, where the problem is especially serious, numerous collaborative research initiatives have been undertaken to identify cost-effective management options to reduce the threat to trade and human health. Research has emphasized on-farm technologies such as biological control, resistant/tolerant varieties, agronomic practices, cost-effective diagnostic tools, and practices and technologies for drying, storing, and processing food and feed.

Innovation in alternative control methods. Bans on hazardous pesticides and other chemicals for treating pests and diseases are a major incentive for innovations. Methyl bromide, used especially in quarantine operations for controlling pests affecting plants and plant-derived materials, has been recognized as an ozone-depleting substance under the Montreal Protocol. Since 2010 the European Union has banned its use for most purposes, including quarantine and preshipment fumigations, boosting the search for alternative control mechanisms.

Innovation in supply chains. Record-keeping and traceability requirements have been incorporated into public and private standards, leading to innovations in supply chains that include simple tracking methods (pen and paper) as well as more sophisticated systems based on barcodes, radio-frequency identification, wireless sensor networks, and mobile devices and applications.

Innovation in standards themselves. The past two decades have seen the emergence of tremendous innovations in the way standards are developed and implemented. For example, the move toward system approaches to food safety regulation has been influenced by two major developments: (1) the introduction of scientific risk analysis as the basis of establishing food standards and regulatory measures and (2) the

(Box continues on the following page)

adoption of food safety management systems, such as the Hazard Analysis and Critical Control Point (HACCP) system, and the subsequent move from testing end-products to preventive approaches. At the primary production level, HACCP-based approaches are being implemented, complementing a set of preventive measures packaged under good agriculture practices (GAP) programs.

Innovation in certification. The preponderance of system approaches and process standards has fostered the emergence of systems for assessing conformity based on third-party certification. This development opens opportunities for coregulatory approaches by the private and public sector (a combination of legislation and self-regulation by private operators). The movement toward self-regulation in the private sector has been pushed by the incorporation of concepts such as "due diligence" in regulations; due diligence emphasizes the private sector's specific obligations in supplying agrifood products to consumers.

Innovation by private actors. Private "codes of practice" and standards related to sustainability (food safety, environmental and social criteria) are also proliferating, especially in horticultural and export crops (coffee, tea, cocoa, bananas), forestry, aquaculture, and livestock. Tremendous innovations have been put in place by the private sector and NGOs, not only for the development of voluntary standards—with a set of prescribed criteria for ensuring compliance—but also in terms of compliance-related infrastructure (such as the innovative auditing and certification systems described earlier). Innovation has extended to methods for ensuring that certification schemes include farmers of differing capacities. In this regard, the emergence of group certification has been a tremendous innovation, allowing engagement with organized groups of small-scale producers.

Source: Authors.

Note: In the United States, for example, the 2011 Food Safety Modernization Act (FSMA) expands the powers of the Food and Drug Administration; among other provisions, it empowers the FDA to create a system for recognizing bodies that accredit third-party auditors to certify eligible foreign facilities.

---

innovation is influenced by policy and regulatory frameworks and by the mechanisms enabling stakeholders to interact and collaborate to prioritize needs and investments, share costs, and perform specific functions related to SPS and quality. Action and investments are especially important for (1) aligning policy and regulatory frameworks to enable standards to contribute to specific policy goals (such as institutional reform) and (2) enhancing capacities to perform the wide range of roles and functions related to standards.

The alignment of policy and regulatory frameworks

Policy frameworks vary in accordance with specific national or subnational needs and circumstances. To understand how technical regulations and standards can contribute to policy goals, it is essential to clearly define the overarching goals of SPS and quality regulations. The legislative and regulatory process is one of an array of tools that government can use to achieve policy goals, but often it is only in the course of analyzing and discussing concrete legislative actions that outstanding policy questions are identified and resolved. In recent years, government awareness of the importance of SPS and quality issues at the policy level appears to be increasing; for example, many governments have enacted specific food safety or organic production policies.

The alignment and harmonization of policy and legislative frameworks is often the first stage in creating an efficient system for SPS and quality standards. Harmonization addresses the complex, inefficient regulatory frameworks emerging from overlapping institutional roles, identifies outdated regulations and standards, and promotes interagency coordination and communication, among other institutional reforms.

For example, several countries have merged multiple laws related to SPS in new food laws and have updated regulations to reflect new institutional arrangements and competencies. Another trend is to promote integrated policy and regulatory frameworks for managing certain risks together. FAO has developed an integrated "biosecurity approach" for managing biological risks to animal, plant, and human health and life (including associated environmental risk), because they all involve systems and procedures for risk assessment and management, food contamination notification, and exchanging information.[3]

Strengthening institutional arrangements

Once the specific rights and obligations of stakeholders involved in the SPS and quality-related system are defined through the legislative process, the challenge is to create mechanisms enabling the relevant parts of government to collaborate. Major barriers to adopting more effective systems for SPS and quality standards are erected by bureaucratic divisions of responsibility. These barriers can arise from budgetary constraints, unequal institutional capabilities, differing cultures, limited communication of information, the absence of a shared vision, and/or disincentives to working horizontally. Actions and investments to overcome such barriers and achieve greater efficiencies are illustrated in box 6.32.

Enhancing capacities to perform the assigned roles and functions

Along with putting effective policy and regulatory frameworks into place and defining the roles and mechanisms for actors to work together, a third critical area for action and investment is the development and enhancement of the wide range of skills, physical infrastructure, institutional structures, and procedures that ensure that the organizations and individuals can perform SPS and quality-related functions effectively, efficiently, and sustainably. Table 6.3 provides examples of those functions.

Most functions listed in table 6.3 require broader oversight and/or some level of collective action. The foundations of an effective system for SPS and quality standards lie in the broad awareness among stakeholders that standards are

---

Box 6.32 Institutional Arrangements for Improving Systems for SPS and Quality Standards

Develop mechanisms for interagency and stakeholder coordination. Examples include memorandums of understanding among public agencies to clarify roles and responsibilities in specifies areas (such as inspections), the establishment of task forces/working groups to respond to disease outbreaks or emergencies, and identifying liaison staff in each agency to facilitate communication and exchange of information. In many developing countries, task forces have emerged under the leadership of public or private entities, bringing public and private actors together to discuss actions to deal with challenges emerging from SPS and quality-related standards.

Coordinate functions under a lead agency. An example of this approach is ACHIPIA—the Chilean Food

Safety Agency—which defines food safety policy and coordinates the work of institutions with food safety roles.

Merge SPS functions into a single independent agency. An example of this type of arrangement in developing countries is the Belize Agricultural Health Authority (BAHA), established in the early 2000s. BAHA integrates food safety, quarantine, and plant and animal health functions into a single entity.

Consider costs and capacity. Implementation of any of these approaches will involve considerations of cost and capacity. In establishing a new agency, consider the leadership, facilitation, time, and resources required. All options need to be assessed in the context of existing capacities in the public and private sectors, the investments required, and the expected benefits.

Source: Authors.
Note: ACHIPIA = Agencia Chilena para la Calidad e Inocuidad Alimentaria.

Table 6.3 Organizational Functions Related to Sanitary and Phytosanitary (SPS) and Quality Standards

Functions related to SPS and quality standards	
Registering and controlling feed, agrochemicals, veterinary drugs	Developing/maintaining pest- or disease-free areas
Conducting basic research, diagnosis, and analysis	Testing products for residues and contaminants
Accrediting laboratories, veterinarians, and other third-party entities	Establishing/maintaining product traceability
Developing/applying quarantine procedures	Reporting possible hazards to trading partners
Conducting epidemiological surveillance	Providing metrology services
Inspecting/licensing food establishments	Notifying the World Trade Organization and trading partners of new SPS measures
Inspecting and approving consignments for export	Participating in international standard-setting

Source: Adapted from World Bank 2005.

integral to the competitiveness of their country, sector, or firm and that they have particular roles to play in the system (World Bank 2005). For example, educated consumers, entrepreneurs, and others can contribute significantly to setting standards at the national level and push for improvements and efficiencies in the public sector.

It is also through the specific actions of individual producers and processors that compliance with SPS and quality-related standards is achieved. The private sector often invests heavily in compliance with SPS and quality standards (such as the use of HACCP or testing and certification at the farm level). Creating awareness and enabling the private sector to innovate through standards is an important dimension of capacity development. Along with the enactment of standards and regulations, the provision of incentives for private investment can be complementary and serve as a much-needed tool to support innovation. These incentives can take the form of quality promotion policies, national quality awards, national productivity awards, and matching grant programs (to cite some examples).

Given the significant capacities needed to perform SPS and quality-related functions, the investments required to strengthen and develop those capacities can be considerable, particularly in developing countries (box 6.33). The first step in developing this capacity is to identify specific needs. Tools have been developed to support countries in assessing their capacity needs related to standards. For example, FAO has developed guidelines for assessing needs in food safety and biosecurity capacity (FAO 2007a, 2007b). OIE developed the Performance, Vision, and Strategy (PVS) tool as the basis for evaluating performance against international standards published in the Terrestrial Animal Health Code. The World Bank assists countries to perform needs assessments and develop action plans, some of which now include estimates of the costs associated with improving operational capacities (World Bank 2010).

Box 6.33   Actions and Investments for Uganda's Fish Export Industry to Comply with Standards and Technical Regulations

Hazards of a poorly performing regulatory system. Uganda's fish export industry burgeoned in the 1990s, largely because private investments in fish-processing facilities led to strong export performance in European markets. Public investments in food safety policy and regulatory frameworks and enforcement capabilities did not keep pace with private investments in the industry, however. At the end of the 1990s, the weak regulatory system exposed Uganda to three safety-related bans on its fish exports to Europe. Scientific proof that the fish were unsafe never materialized, yet the poor performance of Uganda's public regulatory and monitoring system was used to justify the ban.

Investing and innovating to reposition the industry. Public and private actors made a series of innovations and investments to lift the ban and regain the markets. Innovation and investment were favored by high demand in Europe, technical and financial assistance from development partners, the government's open and decisive leadership; and access to finance for private companies. Specific actions included: (1) streamlining regulations and strengthening the government authority that would implement them; (2) developing a new fishery policy; (3) improving monitoring and inspection systems (drafting inspection manuals and standard operating procedures and training inspectors); (4) initiating regional efforts to harmonize handling procedures in the countries bordering Lake Victoria; (5) upgrading a (small) number of landing sites and plans for upgrading a substantial number of others; (6) upgrading processing plants' procedures and layouts; (7) opening up the U.S. market, which requires HACCP compliance; (8) installing two local laboratories and improving the quality of laboratory services provided to the industry; (9) increasing the number of processing plants and improving export performance; and (10) forming an Association of Quality Assurance Managers to address problems and concerns among industry players.

The fixed investment in upgrading factories, management systems, and other infrastructure between 1997 and 2001 was equivalent to about 6 to percent of the FOB value of exports over that period. The innovations were beyond those required to achieve compliance, such as the adoption of ISO 9000 and even ISO 14000 quality systems. In general, the process enhanced cooperation and relations between the regulatory agency and the industry.

(Box continues on the following page)

Box 6.33   Actions and Investments for Uganda's Fish Export Industry to Comply with Standards and
              Technical Regulations (continued)

Continuously innovating to meet new challenges and opportunities. Despite some lingering food safety issues, the larger challenge for the industry is to deal with the depleted waters and fisheries of Lake Victoria and more general environmental degradation, which have spurred negative campaigns against the industry in Europe. Regulatory controls, complemented by self-regulation and voluntary efforts to gain environmental and sustainable certification, have been adopted to manage market risks. The volume of fish exports to the European Union has not returned to previous levels, partly because of the depletion of fish stocks and competition from other types of white fish from other countries. For the Ugandan fish industry, the capacity to learn from its experience, innovate in response to evolving market demands, and sustain its resource base will be critical to future viability.

Sources: Ponte 2005; Ponte, Kadigi, and Mitullah 2010; Jaffee et al. 2006.

The use of economic analysis to drive policy decisions related to SPS is often emphasized, but the complexity of current methods is driving efforts to find more flexible and practical methodologies. An innovative framework based on multi-criteria decision analysis is being validated by the Standards Trade and Development Facility.[4]

POTENTIAL BENEFITS

In the agriculture and agrifood sectors, standards fulfill a broad range of objectives. A general objective of standardization is to facilitate flows of information between consumers and producers (particularly information on unobservable characteristics, such as the use of GM ingredients) to facilitate trade and spur economic activity. For government, standards allow authorities to achieve several objectives, such as the protection of animal, plant, and human life and health; the protection of the environment; and the incorporation of social and sustainability considerations into agricultural production. Through standards, information imbalances and externalities can be addressed and fair competition promoted.

Compliance with standards is crucial for countries to participate in international trade, because it ensures the compatibility of components and traceability of products and raw materials from different places. Approaches to harmonizing standards between countries and/or industries can reduce transaction costs by reducing duplicative functions of conformity assessment, including testing and certification (Jaffee 2005).

From the perspective of the private sector, standards are a means of transferring technology and diffusing technical information concerning products and processes. They provide incentives to local firms to improve the quality and reliability of their products. They can also be used as a risk management instrument, as a product differentiation tool, or as a cobranding strategy.

Several agricultural export industries in developing countries have used compliance with standards to gain an important competitive advantage. Examples include horticultural industries in Peru (Diaz and O'Brian 2003; Diaz Rios 2007) and Kenya (Jaffee 2003); the groundnut industry in Argentina and Nicaragua (Diaz Rios and Jaffee 2008); and the Brazil nut industry in Bolivia (Coslovsky 2006). In all cases, success required the incorporation of innovations in production and processing but, perhaps most important, in collective and organizational behavior. Examples of collective and organizational innovation include the formation of the Fondation Origine Sénégal—Fruits et Legumes; the collective self-regulation of Bolivia's Brazil nut industry; the collaborative arrangements and interactions between Peru's Commission for Export Promotion (PROMPEX, Comisión Para la Promoción de Exportaciones) and several subsectoral associations.

Clearly the impacts and distributional effects of noncompliance with SPS standards can be devastating for a company or an entire industry.[5] The World Bank (2005) presents several examples of associated distributional effects across agricultural export industries resulting from the imposition of bans or export restrictions following noncompliance with these critical standards. Compliance with standards and the prevention of foodborne illnesses and animal/plant diseases also reinforce a country's reputation as a reliable supplier. Outbreaks can lead to huge costs for governments and the public resulting from diagnosis and treatment of illness, production losses, outbreak investigations, and product tracebacks and recalls.[6]

POLICY ISSUES

Standards and technical regulations are used by governments as a tool to achieve broader policy objectives. The sections that follow discuss the need for a balanced approach to policy and decision making that takes domestic and international trade interests into account. Related issues involve the chaotic proliferation of private standards and their implications for national policy, the strategic uses of standards, and the question of who should provide services related to standards.

Balancing divergent policy goals and dealing with the proliferation of private standards

Policy makers often have to choose between conflicting policy goals with respect to standards and technical regulations. For example, a desire to protect human health may conflict with the desire to facilitate agricultural trade or to develop an industry or sector. The goal of expanding export markets may also conflict with the desire to conserve water or reduce pesticide use (Vapnek and Spreij 2005).

Policy making at the national or local level can be highly influenced by the international environment. Government policies should be consistent with obligations under international agreements as well as with national food security and development goals. It is generally recommended that countries adopt international standards, although their effectiveness depends on their suitability to specific national contexts. The harmonization of regional standards for raw milk in Eastern Africa is one example. Debate revolves around a desire to harmonize with Codex standards, although they do not reflect handling and consumption practices in the region (Jensen, Strychacz, and Keyser 2010).

Trade has become a driving force behind increased public and private investment in SPS and quality systems, but at the same time, many stakeholders are concerned that increasingly stringent trade standards are having adverse effects on the costs to and competitiveness of developing-country suppliers, particularly from LDCs. Consequently, in many countries, compartmentalization of production and adoption of a system of "dual standards"—one focusing on compliance with export market demands and one for local consumption—has been seen as a solution. Another concern is that the heavy emphasis on the trade benefits regarding SPS and quality systems tends to overshadow (at least in the policy discussions) the other benefits of improved standards, such as reduced production losses and improved public health impacts on domestic populations. The challenge for policy makers is to find the right balance between these different interests and options when formulating policies and investment plans.

Another area of concern from a policy perspective is the emergence of private SPS standards. Concerns about their proliferation, prescriptive nature, legitimacy, transparency, potential to undermine public action, as well as their potential economic development impacts, have coalesced around an intense debate within the SPS committee of WTO. Concerns related to the proliferation of private social, environmental and sustainability standards are emerging as well. Discussions in several forums are intensifying over the scope of harmonization and collaboration and the need for a better understanding of intended impacts at the ground level.

Compliance with standards as a strategic issue

Some view the imposition of stricter SPS and quality requirements as a barrier to trade, especially if they entail costly, highly technical requirements or complex administrative procedures. Such requirements erode the competitiveness of industry players and further marginalize small countries, traders, and farmers.

Others view the same standards and requirements playing a catalytic role in innovation and modernization. Demands for compliance with increasingly stringent standards can expose the fragile competitiveness of an industry (or individual players) and the lack of institutional arrangements for collective action and clarify the need for action, as in Uganda's fish industry (box 6.33). This experience illustrates that innovation in response to agricultural standards and regulations is not a one-time event but part of a continual process of anticipating and responding to emerging challenges.

In several cases, industry players and governments have responded effectively to prevailing standards and have consolidated or improved their market position. In some countries, the response has involved a proactive, forward-looking strategy that seeks to reinforce their competitive advantage, as in the groundnut industry in Argentina and the horticultural industries of Peru (Diaz Rios 2007) and Kenya (Jaffee 2003). In other cases, the response has been essentially reactive, seeking to adjust in the face of adverse trade events. (see box 6.34).

The World Bank has advocated for compliance with standards to be viewed as a strategic issue, highlighting the multiple strategic options available to countries (table 6.4).

Box 6.34 Innovating to Quickly Respond to Adverse Trade Events

India's fish and fishery products: An export market lost and regained. In 1997, the European Union banned all fish and fishery products from India due to noncompliance with hygienic standards. The Indian government improved hygiene by requiring measures such as integrating preprocessing operations with processing facilities and imposing strict limits on approved output according to plants' capacities for water, ice making, and effluent treatment. The government implemented programs to support improved hygienic controls in fish processing, including subsidy programs for upgrading processing facilities and training managers and workers throughout the supply chain. Fish exporters acted collectively to establish infrastructure that would link preprocessing units to common water, ice, and effluent facilities. The new facilities include modern laboratories that perform all microbial and chemical tests required by importers. These measures led the European Union to lift the ban on imports.

Peruvian asparagus exports: Success through standards. In 1997, when Spanish health authorities asserted that consumption of canned Peruvian asparagus caused two cases of botulism poisoning, the resulting public scare in European markets created large market losses for Peruvian asparagus exporters. Seeing that even one careless exporter could disrupt the markets, the government and industry decided to take action to bring Peruvian agricultural standards in line with international norms. In 1998, the Peruvian Commission for Export Promotion convinced the asparagus industry to implement the Codex code of practice on food hygiene. Government specialists worked with the companies to ensure proper implementation. In 2001, national fresh asparagus norms were published. They provided a quality and performance baseline for the industry that allowed many firms and farms to generate the necessary skills and experience to gain certification under the stringent international standards.

Source: World Bank 2005.

Table 6.4 Strategic Choices and Responses with Respect to SPS and Quality Standards

Nature of the response	Strategy		
	Exit	**Voice**	**Compliance**
Reactive	Wait for standards and give up	Complain when standards are applied	Wait for standards and then comply
Proactive	Anticipate standards and leave particular markets	Participate in standard creation or negotiate before standards are applied	Participate in standard creation or negotiate before standards are applied
Viability	**Exit**	**Voice**	**Compliance**
Size of firm or industry		++	+
Share of target market	–	++	+
Reputation	–	++	+
Suitability of legal/regulatory framework		++	+
Leadership/coordination within value chain		+	++
Private sector management/technical capacity	+	+	++
Public sector administrative/technical capacities	+	++	++
Clarity of institutional responsibilities		+	+
Geographical/agro-climatic conditions	–/+		–/+
Prevailing challenges	++	–	–/+
Nature of the measure		–/+	–/+

Source: World Bank 2005.

Those options will vary for each country, depending on its economic, political, and social systems and norms, institutional structure, size and location, and so on.

Who provides services related to standards?

Perhaps the most important decision to be made with respect to building capacities related to standards is whether capacity building should be done by the public or private sector. There is certainly an increasing recognition of the critical role that the private sector can play in providing services traditionally viewed as the responsibility of the public sector. In countries where demand for certain standard-related services is high, the private sector may have an opportunity to provide them. Before building, equipping, and maintaining laboratories and other standard compliance-related services, public actors need to consider alternatives. In some instances public authorities have delegated compliance services to private organizations, particularly accreditation, testing, and certification services (for example, public authorities certify compliance on the basis of testing services provided by private laboratories).

LESSONS LEARNED

Standards represent major challenges for developing countries, yet isolated improvements and innovations offer limited opportunities to leverage long-term benefits. A *key lesson* is that countries must be *strategic and proactive.* What is required is a process of strategic planning, supported by continual innovation and improvement, to successfully overcome challenges and take advantage of new opportunities. A proactive stance rests upon public and private awareness of the issues and strong governance.

Quite often, developing countries have a long list of needs for capacity development. Efforts to develop capacity related to standards should aim at maximizing the strategic options available, consider costs/benefits, speed of implementation, sustainability, complementarities between the public and private sectors, and the possibilities for regional collaboration. Certainly one of the "nonregrettable" investments in this domain would be to invest in creating broader public and private awareness of SPS and quality management issues. The sections that follow expand on these points.

Priority setting is essential for effectively managing standard-related challenges and opportunities

Pragmatism is needed when examining the state of a country's SPS and quality-related capacity, and realism is needed to determine the immediate and long-term scope for enhancing that capacity. Prioritization can begin with identifying the most immediate and/or significant risks as well as opportunities for competitive or welfare gains. Policy makers need to weigh the different objectives and their potential distributional impacts, but all too often priorities are driven by the benefits associated with competitive repositioning of industries/sectors or access to remunerative export markets. As challenging as it may be, it is fundamental to consider holistic approaches that merge domestic and trade perspectives, perhaps through strategic prioritization at the national, sectoral, or industry level with stakeholders.

Effective regulatory and voluntary interventions require public and private involvement

The development and enforcement of policies related to standards are enhanced by leveraging support from the private sector and/or creating an enabling environment (incentives) for private investments in capacity related to SPS and quality standards. The conditions for effective coregulatory approaches should be analyzed and explored, as they represent a potential opportunity for public and private collaboration.

Assess the gaps between local and international standards to determine the investments needed to bridge them

From a market perspective, the structure and maturity of an industry should drive the design of public and private interventions related to standards. The first step is to assess the gaps that need to be bridged. The product and the type of market provide a good indicator of the standard-related challenges. Public and private actors will need to make distinct adjustments and investments to meet stricter food safety, quality, and other requirements. Time, significant investments, and incremental upgrades are all needed for an industry to become an effective and competitive supplier in more demanding markets.

Consider the needs of vulnerable groups

New or more stringent standards are likely to pose compliance problems for firms and farms operating under less favorable conditions. An awareness of the distributional effects of standards and their influence on poverty is critical for understanding the strategic choices available to different actors and identifying the most appropriate tools to support them in implementing those choices.

Learn from others, cooperate nationally and regionally, and search for funds to develop capacity

Given the complexity of standards, it is fundamental for policy makers from developing countries to engage in activities where they can influence the setting of public and private standards. Leadership and proactive involvement in initiatives at the regional level are critical. Regional initiatives to harmonize standards addressing common (and cross-border) SPS issues should receive strong consideration from policy makers. Involvement in communities of practice, networks, and forums that promote common learning and information sharing is essential. Examples include the activities undertaken by the Standards Trade and Development Facility, other development partners, and international standard-setting organizations. For voluntary standards in agriculture and agrifood, new spaces for knowledge exchange and learning are emerging, such as the Trade Standards Practitioners Network.

Developing an Enabling Environment to Improve Zambian Smallholders' Agribusiness Skills and Commercial Orientation

Indira Ekanayake, World Bank

SYNOPSIS OF PROJECT DATA

Country: Zambia

Project: Agricultural Development Support Project (ADSP)

Cost: US$37.2 million (total project cost US$39.6 million)

Component cost: Support to Farmers and Agribusiness Enterprises (US$33.2 million); Institutional Development (US$3.9 million); Project Management and Coordination (US$2.6 million)

Dates: FY 2006–14

Contact: Indira Ekanayake, World Bank, Zambia

CONTEXT

Agriculture has become a major driver of growth and a significant source of export earnings and diversification in Zambia as a result of rising mineral prices. Notwithstanding Zambia's abundant and fertile land and water and economic growth (exceeding 5 percent for the past seven to eight years), small-scale farmers have seen little change in their quality of life. Smallholders' productivity is very low compared with that of Zambia's commercial farmers and farmers in other parts of the world. Productivity is partly constrained by the lack of title to land, limited financial resources, and insufficient infrastructure, but much of the problem arises from the unfavorable policy environment for small-scale farmers.

Starting in the 1990s, consecutive investments by the International Development Association (IDA) have sought to raise productivity in Zambian agriculture in line with government strategy to support the commercialization of smallholder agriculture. This strategy aims to reduce poverty by expanding contract farming and outgrower schemes that link smallholders with commercial farmers or agroenterprises.

In 2006, the government initiated the World Bank-funded Agricultural Development Support Project (ADSP). Through support to Zambia's Ministry of Agriculture and Co-operatives (MACO), the ADSP fosters the commercialization of smallholder agriculture by developing a network of competitive value chains in selected high-quality, high-value commodities (such as cotton, horticultural crops, honey, and dairy). Interventions provide better technology (improved seed, microirrigation), strengthen institutions (public-private partnerships, outgrower schemes), and develop well-maintained rural roads in high-potential agricultural areas. The objective is to ensure that the selected value chains operate efficiently to increase value addition, improve smallholders' access to markets, and improve the competitiveness of their agricultural commodities.

PROJECT OBJECTIVES AND DESCRIPTION

As noted, the ADSP was designed to increase the commercialization of smallholder agriculture by improving the productivity, quality, and efficiency of value chains in which smallholders participate. The project funds three investment areas: (1) matching grants to promote innovative agribusiness activities that build synergies to develop value chains, (2) rural feeder roads, and (3) public institutional support for market development.

The Market and Innovation Facility (MIIF) provides matching grants to fund innovative activities in which agribusinesses interact with smallholders or business-oriented farmer groups and cooperatives. The activities match the business development needs of each subsector, emphasize technical assistance, and fall into three categories: (1) technology, training, capacity-building, and agricultural services in production, processing, and marketing in value chain development; (2) information, research, and studies associated with value chain development; and (3) services and capacity-building in business management and development, product promotion, and acquisition of technical and market information.

The Rural Roads Improvement Facility (RRIF) provides resources to rehabilitate and maintain rural and district roads to link selected high-potential agricultural areas to markets as a means of improving incomes and livelihoods. Target roads are in five districts (Choma, Chongwe, Katete, Chipata, and Lundazi) in two provinces (Southern and Eastern). RRIF investment is expected to provide the essential rural road network for improved market access and associated product delivery efficiencies and benefits. The road facility supports the ADSP's general aims, because value chain development is superimposed within the rural road grid. To date, 642 kilometers of critical feeder roads have been rehabilitated (57 percent achievement of the target of 1,129 kilometers).

The Supply Chain Credit Facility (SCCF) was originally designed to provide credit, on a demand-driven basis, for investments to improve the supply chains of existing and emerging outgrower schemes and enable agroenterprises, traders, or nucleus and commercial farmers working with smallholders to finance capital investments, seasonal inputs, and export activities. Following implementation delays, SCCF was modified to improve the productivity of outgrower schemes, scale them up, establish new contract farming enterprises, and upgrade processing and marketing capacity.

Under the project's institutional development component, ADSP builds capacity in selected departments of MACO to provide the core public services for enhancing smallholders' productivity, quality of produce, and access to markets. For example, the project has enabled the Cotton Development Trust (a public-private trust) to provide seed and technical assistance to smallholders and increase its production of foundation seed for cotton through improved irrigation facilities. The project has also helped to build and equip a biotechnology laboratory at the Seed Control and Certification Institute (SCCI) and improve the SPS services of the Zambia Agricultural Research Institute.

The Project has multi-institutional and innovative institutional arrangements for implementation. For example, the National Coordination Office is based in MACO. MIIF is administered by Africare, an international NGO, and coordinated and managed by an independent, outsourced secretariat. Independent technical reviewers assess the technical and financial feasibility of proposed subprojects. A multistakeholder subcommittee of the National Project Steering Committee (with representatives of the Bankers Association of Zambia, the agribusiness sector, MACO, and a member of the secretariat) is responsible for final funding decisions. The project's rural road component is imple-mented by the Road Development Authority (RDA) and the National Road Fund Agency (NRFA). The institutional development component is managed by the respective MACO departments.

INNOVATIVE ELEMENT

The innovative feature of ADSP's design is a demand-driven, value chain approach that facilitates smallholders' participation in key value chains. Innovative features of ADSP's implementation include the demand-led innovation fund, matching investments by agribusiness to finance a sustainable rural road network (crucial for innovation by agribusiness), and the piloting of an improved market information system. Rural road improvements are procured through Output and Performance-based Road Contracting (OPRC). A spatial approach is used to ensure that technological interventions in the selected value chains are compatible with the improved rural road grid.

Under MIIF, matching grants support innovative interventions by agribusiness that add value to agricultural products, improve agricultural productivity, and improve smallholders' links to markets. The MIIF Innovation Categories in agricultural value chains include new products, new technologies or processes, new markets, new strategic partners or organizational arrangements, and new geographical locations.

The innovative element expected of SCCF is that it would enable entrepreneurs to make the capital investments that are vital to stronger and more competitive value chains with or without scaling up while reducing risk absorption.

BENEFITS, IMPACT, AND EXPERIENCE

Value chains strengthened through the project include dairy, cotton, horticultural crops, paprika, honey, biodiesel, and tobacco, among others. Some of the key benefits and outcomes associated with the project are described next.

An innovative matching grant scheme is under way

In its three-plus years of implementation, MIIF has funded 17 subprojects (for which the total budget exceeds US$2.6 million) involving more than 28,800 smallholder beneficiaries. Six additional subprojects are under review, and 20 or more proposals are under development. MIIF subprojects have generated 22 technologies and innovations for a range of value chains, including dairy, groundnuts, honey, biofuels, and fisheries. The grant scheme has leveraged an additional 85.6 percent cofinancing, illustrating the considerable buy-in and commitment by agribusinesses. It is too early to project the outcome of the subproject grants, but initial assessments by beneficiaries have been very positive.

To date, the grant scheme has funded high-quality proposals that are demand led and innovative. One lesson from the experience with MIIF, however, is that it is vital to maintain the number of high-quality subprojects that enter the funding pipeline. Awareness of the facility is spread through continuous publicity; a variety of field days, symposia, and workshops; and word of mouth in the business community. Another lesson is that a favorable external business environment (especially exchange rates for commodity exports and inputs for production and value addition) is essential for strong participation in an innovation grant scheme such as MIIF that attempts to increase competitiveness.

MIIF's implementation has faced several challenges. Initially agribusinesses were reluctant to participate because of their limited awareness of and low interest in the need to innovate for greater competitiveness. The high transaction costs (time, resources) and lack of experience in developing concept notes, proposals, follow-up documentation, and cofinancing commitments also presented a challenge for some participants. The private sector was wary of engaging with what it perceived to be NGO- and government-"driven" activities. Similar issues of limited trust and experience in working with the private sector impeded collaborative arrangements between private and nonprivate actors. Another challenge that must not be underestimated is that the effort involved in working with smallholders in outgrower schemes can limit the private sector's interest in submitting proposals.

Although it is too early to point to specific benefits arising from the project's various kinds of support to specific value chains, the adoption of more productive and favorable technologies has increased. A baseline study in two provinces where rural road work is taking place was completed, and an impact study is being undertaken in the same areas.

Performance-based contracts for rural roads successfully implemented

As noted, the project uses a new method of road contracting called OPRC, in which the contractor rehabilitates the roads under the contract and maintains them for five years. This agreement ensures that project participants in rural areas that are far from markets have consistent access to those markets. Spillover benefits include improved access to health facilities and primary schools. By its third year, the project had rehabilitated 583 rural district and feeder roads in the national road network. It is actively encouraging the use of MIIF grants in contracting for road rehabilitation and maintenance to create synergies between improved crop production and marketing in the value chains. The socioeconomic targets of the OPRCs in selected catchment areas (3,136 households were surveyed as a baseline) are mainly related to process impacts (income-generating opportunities from road rehabilitation), access impacts (associated with providing the road infrastructure), and mobility impacts (on transport services or growth in traffic volumes), but they are still too early to quantify.

An agricultural market information system piloted in an integrated project activity zone

As noted, the project used a spatial approach to target the technology interventions for the selected value chains within the improved rural road grid in Southern Province, where a market information system has also been successfully piloted in three districts. Given the popularity of radio broadcasts of commodity market prices, this program is being scaled up to include all districts in Southern Province and will also be introduced to Eastern Province, where the OPRC rural road work is taking place.

Short-, medium-, and long-term loans to support investment

Loans provided through the SCCF are an important complement to the matching grants provided through MIIF, and access to short-, medium-, and long-term agricultural finance remains critical to the project's success. This aspect of the project has been implemented more slowly than expected, however. Responsibility for implementation has been transferred to the Development Bank of Zambia, where institutional capacity strengthening has been initiated.

Serving the public goods agenda

The outcomes of ADSP far exceed the cost of the public investments. Aside from reinforcing the private sector's capacity to increase the competitiveness of Zambian agriculture and improve smallholders' participation in lucrative value chains, the project strengthens the public goods delivery agenda through targeted institutional development, with long-term benefits for the agricultural sector. Examples of these public goods include wider availability of good quality seed for multiplication by private and public agencies, more skilled human resources in public institutions, and development of the rural road network.

Projects similar to ADSP under way in other parts of the world

Productive partnership projects funded by the World Bank in Colombia and Vietnam also use matching grants to facilitate partnerships and build capacity in value chains. These projects and ADSP are demonstrating the challenges of engaging and retaining the interest of a diverse group of private actors, such as traders, processors, exporters, wholesalers, and retailers. Such projects often require greater attention to entrepreneurial skills than to farming practices.

LESSONS LEARNED AND ISSUES FOR WIDER APPLICATION

To date, the ADSP approach has yielded three *key lessons*. First, the development of market institutions is not in synchrony with development of other parts of the value chains. There is need to consistently identify areas, themes, and issues that can contribute to activities that strengthen value chains. Second, alliances and partnerships for agribusinesses do not "just happen" in projects of this nature. They must be actively facilitated and nurtured and benefit from early technical and financial support. A conducive political economy is essential for success. Third, owing to its demand-driven design, MIIF responded to greater and more varied demand from more diverse businesses than originally anticipated. The focus of the grant scheme became fragmented as a result and increased the administrative burden. Other lessons are discussed in greater detail in the sections that follow.

Engage private sector grantees/actors for dialogue and innovative enterprise development

A project such as ADSP, which seeks to promote innovation in value chains and involve smallholders in commercial agriculture, must engage the private sector when it is first developed and designed. ADSP carefully engaged the private sector as the project was prepared, but a more consistent effort was warranted later, during the project's implementation. Matching grants under ADSP did not automatically strengthen value chains and develop agribusiness. Midway through the project, it was realized that consultative processes (multistakeholder platforms, forums for value chains, sector associations, and field days) were useful instruments to support development of the agricultural sector.

These platforms help to develop a shared understanding of challenges, opportunities, and intervention that may guide the support services and matching grant program. They can also foster collaboration, including partnership between public and private agencies.

Establish a high-caliber secretariat with private sector experience

The secretariat or fund administrator has a key responsibility in implementing a grant scheme. Selection of the grant administrator requires significant effort, and often special capacity building is warranted. Deficiencies in management capacity and leadership could cause delays or even the failure of the scheme. As noted, under ADSP this function was outsourced to an NGO. This option is useful when a project requires autonomy, experience in working with participants at the grassroots level and in decentralized projects, as well as experience with donor requirements (reporting, procurement, and fiduciary issues). NGOs also come with challenges, however, including the potential for greater overhead costs, problems with long-term institutional sustainability, and a greater risk that they will lack business understanding. The essential features for a secretariat to succeed are the available capacity, institutional sustainability, overhead costs, separation of the funding and implementation of the grant fund, potential for political interference, and the interests of the key stakeholders.

Strengthen aspects of the matching grant scheme

The matching grant scheme could be strengthened in a number of ways. The activities and value chains supported by the facility could be adjusted to focus more on high-priority value chains and on moving away from activities involving technology, extension, and studies toward a wider set of business-promoting activities. Stronger, direct communication with actors in the agricultural sector is vital to increase awareness of the facility. The grant application and review process should be streamlined. The MIIF administrator requires greater capacity to interact with private sector stakeholders, train clients, and manage the overall program.

One final lesson from the experience with MIIF is that the grants have been quite useful for building institutional capacity in public organizations at the provincial and district level. In other words, participation in grant schemes that strengthen agribusinesses can benefit not only national goals but provincial and district institutions and economies.

Intellectual Property Management in Livestock Veterinary Medicines for Developing Countries

Josef Geoola, GALVmed

Sara Boettiger, University of California, Berkeley

SYNOPSIS

The Global Alliance for Livestock Veterinary Medicines (GALVmed) is a nonprofit organization that makes livestock vaccines, diagnostics, and medicines accessible and affordable to the very poor. GALVmed coordinates research, development, and deployment (RD&D) among multiple partners, from identifying candidate technologies to manufacturing sustainable supplies of market-ready products. GALVmed uses a wide range of resources to ensure that IP supports innovation for the poor, such as due diligence for accessing upstream technologies, the implementation of IP strategies that work toward development goals, the use of IPRs as incentives to engage partners, and the negotiation of contracts that support the translation of research into products accessible to the poor. GALVmed's IP management system benefits its pro-poor mission by addressing broader issues that prevent innovations from becoming sustainable, market-ready products. Experience with public-private partnerships has taught GALVmed to leverage its interests while providing its partners with the opportunity to achieve their own internal mission.

CONTEXT

The Global Alliance for Livestock Veterinary Medicines (GALVmed, www.galvmed.org) is a nonprofit organization with a mission to make livestock vaccines, diagnostics, and medicines accessible and affordable to the millions for whom livestock is a lifeline. The Bill and Melinda Gates Foundation, the United Kingdom Department for International Development, and the European Commission are major sponsors of GALVmed's work.

The impact of livestock in addressing poverty continues to be underappreciated, particularly livestock's role as living assets for the very poor. Data on the impact of livestock diseases are limited, but four of the many major and unaddressed livestock diseases (East Coast fever, Rift Valley fever, Porcine cysticercosis, and Newcastle disease) cause estimated annual economic losses upwards of US$350 million. Losses on this scale affect the livelihoods of hundreds of millions of poor households in the developing world.

GALVmed currently works on nine disease-control technologies for those four livestock diseases. Many diseases afflicting livestock in developing countries are preventable and well understood from a research perspective. Until recently, however, the developing world has lacked the resources for moving the science out of the lab and into the field to prevent and contain livestock diseases. One reason for this impasse is that disease-preventing and disease-controlling technologies often emerge from R&D in advanced laboratories and are subject to one or more forms of IP protection.

For GALVmed to achieve its mission, the organization must constantly exercise (and review) its IP policies and IP management strategies. Through effective IP policies and management strategies, GALVmed can identify and circumvent IP risks early in the commercialization pathway, therefore avoiding potentially serious and costly downstream impediments to GALVmed projects.

GALVmed is unusual in that it operates across the entire commercialization pathway to make technological solutions accessible to the poor. GALVmed does not have in-house capacity for the research, development, and deployment (RD&D) of products. Instead, its role is to *facilitate* the entire RD&D process, from identifying candidate technologies to manufacturing sustainable supplies of market-ready products. Managing RD&D activities among multiple partners and under pro-poor obligations requires the organization to consider the use of IP strategically to ensure that upstream technologies do ultimately result in downstream products accessible to those who need them most.

By addressing a wide range of IP strategy issues, GALVmed has gained experience that has value for many

The vaccine that GALVmed is currently deploying for East Coast fever has a commercial market, primarily among the Masai in East Africa, and potential for sustainable private sector production and distribution. Protection against East Coast fever adds significant value to Masai calves, and the Masai are willing to pay for the vaccine within a certain price range. With the help of the Public Intellectual Property Resource for Agriculture (PIPRA, www.pipra.org), GALVmed first approached the IP strategy for the vaccine by characterizing the opportunities and risks. The vaccine was nearly ready for the market and would not require substantial further development. PIPRA reviewed the IP in the technology. It determined that the technology and related know-how, although enormous in value, were in the public domain and had no associated IP rights.

The lack of formal IP meant that manufacturers would have less of an incentive to invest in producing the vaccine. GALVmed needed to explore other types of leverage, such as forward market commitments or other assurances of supply channels. Eventually GALVmed learned that deregulation of the vaccine in each country in East Africa was linked to an exclusive marketing authorization that offered some leverage. To create a commercialization strategy for sustainable delivery of the vaccine to East Africa, information on marketing authorizations needed to be integrated with information on the profit incentives of manufacturers and distributors as well as consideration of the transfer of know-how. In summary, even though IP did not play a role in the eventual commercialization strategy, formulation of an IP management strategy was critical

to determine: (1) whether in-licensing was required and which partners might need to be engaged in the process due to IP ownership and (2) what incentives could be derived, either with IP or other levers, to ensure that partners also had incentives to comply with GALVmed's pro-poor obligations.

While commercialization of the East Coast fever vaccine involved private companies as partners in manufacturing and distribution, another vaccine in GALVmed's portfolio, the Porcine cysticercosis vaccine, involves virtually all public partners. In this case, GALVmed recognized that the lack of a private market for the Porcine cysticercosis vaccine (government procurement was anticipated) meant that incentives to engage manufacturers and distributors would need to be different. PIPRA conducted due diligence over relevant technologies and ascertained that, while formal IPRs existed in some countries, it was tangible property rights that would provide GALVmed with both challenges and opportunities in its development of a pro-poor commercialization strategy. GALVmed was then able to employ licensing language to create incentives for partners, whereby a selected partner would gain geographical exclusivity in developing, manufacturing, and distributing the vaccine. As was the case with the East Coast fever vaccine, developing an IP management strategy involved critical due diligence to determine GALVmed's risks and opportunities, and then careful consideration of how to use the available leverage to ensure that partners had incentives that aligned with GALVmed's obligations to deliver products to the very poor.

Source: Authors.

---

organizations that develop technology for the poor. GALVmed has made crucial IP decisions, observed their implications, and employed IP strategies suitable for both public and private partnerships (see box 6.35).

Through broad involvement with the RD&D process, GALVmed addresses IP and contractual challenges, including accessing and transferring proprietarily owned technologies, resolving the distribution of rights, and strategically using IP to promote deployment. The remainder of this profile focuses on the processes and resources

GALVmed has employed to address IP issues, such as due diligence, strategy implementation, and conscious leveraging of IP, as well as some of the challenges involved (for example, negotiating contracts).

GALVMED'S INNOVATIVE APPROACH

As it has grown, GALVmed has developed a systematic approach that anticipates IP hurdles and mitigates IP risks that arise during RD&D (box 6.36). These IP management

As GALVmed has expanded, its needs for managing IP have evolved. During its startup phase, to ensure that IP issues were addressed from the onset of projects with utmost diligence, GALVmed outsourced IP management issues to a group such as PIPRA, with a proven track record and the expertise for managing IP within agriculture. Five years after its founding, GALVmed now manages an ever-growing number of technologies in the RD&D pipeline. The related complex IP challenges demand timely attention and therefore in-house expertise. GALVmed's growing internal capacity for IP management has been achieved through three changes:

■ *Creating a new management role within the organization to deal with IP and agreements.* This role provides for focused, consistent management of the drafting of time-sensitive agreements and delicate negotiations as well as critical accountability for IP management. Moreover, internal expertise allows for IP management strategies that fit the organization's risk tolerance, encompass organizational culture, and can more easily be adapted to changing information of the technical and socioeconomic realities of the RD&D pathway.

■ *Contracting the services of a local attorney from a top-tier law firm to provide weekly and as-needed support in drafting and negotiating complex legal agreements.* A local attorney[a] provides the organization with an external opinion, identifies legal issues

that could be missed internally, and provides insight on regional laws and regulations. The execution of contracts requires expertise in local law, and nonprofits often require legal opinions from local attorneys on risks such as exposure to liability. Most important, a local attorney is essentially local enough to meet individuals in the organization and understand the nuances of issues that would otherwise be missed through a phone call.

■ *Improving utilization of external IP expertise to address the resource gaps that almost always exist internally.* External expertise, in the form of contracted services from organizations or individual consultants, can provide experience-based, impartial advice that would be difficult to gain otherwise. External expertise (in GALVmed's case, from PIPRA) has access to the knowledge and expensive toolsets that small nonprofits may struggle to purchase. These experts have access to a global network of attorneys that can provide regional legal advice that can be valuable, for example, when questions of law arise in countries where GALVmed's partners practice. Lastly, external experts have the latest specialized insight on IP. They are capable of breaking down technologies, conducting highly detailed assessments, acquiring legal insight, and converting a mass of information into one thorough, meaningful report that GALVmed's internal expert can then integrate into a larger commercialization strategy.

Source: Authors.
a. Andy Harris, associate at Maclay Murray & Spens LLP, Edinburgh.

measures are critical to GALVmed's ability to efficiently transform upstream disease-preventing technologies into safe, effective, and accessible downstream products.

The sections that follow provide more detail on GALVmed's four-stage, systematic approach to managing IP. The approach was designed to balance the organization's nonprofit, pro-poor mission with the need to integrate and address a variety of challenges arising throughout the commercialization pathway of the products GALVmed seeks to deliver to the poor.

Stage 1: Technology landscaping

GALVmed's initial step of conducting a technology landscape requires using IP and other sources of information to scout for preexisting and emerging technologies. Technical and scientific value of individual technologies are assessed as well as potential IP risks. In one instance, scientists at GALVmed learned of a number of technically promising, but proprietarily owned, vaccine stabilization technologies. Upon IP review, GALVmed learned of related ongoing patent disputes. The uncertainty and risk associated with these disputes, and the potential impact these risks may have on downstream partners for technology development, were considered in conjunction with technical issues, and the risk was deemed unacceptable. The review of IP issues allowed GALVmed to avoid pursuing a technology that could have potentially led to delays or the expense of late-stage shifts in research strategy. Early identification of technologies that exhibit scientific merit and withstand IP review paves the path to a more resource-efficient commercialization process.

Stage 2: IP due diligence

As candidate technologies are identified from Stage 1, an IP due diligence process is used. This due diligence (or IP auditing) is a resource-intensive process involving in-depth research into the patent landscape surrounding each selected technology (for example, individual investigations of vectors, genes, promoters, markers, and signal sequences of a vaccine). When a patent is particularly important to GALVmed's commercialization strategy or when use of a technology is suspected to infringe existing patents, freedom-to-operate (FTO) assessments may be carried out with the help of attorneys. The information gained from IP due diligence allows GALVmed to identify potential partners, understand in-licensing obligations, and review potential opportunities for the use of IPRs in further development of the technology.

While some large companies employ internal IP legal expertise, it is usually more efficient for small companies and nonprofits to outsource this level of patent landscaping and analysis. For these analyses, GALVmed collaborates with PIPRA. In this stage, GALVmed also incorporates a review of issues of tangible property rights[1] (examining, for instance, material transfer agreements as well as IP licenses); existing claims to both tangible property and IP are mapped to understand the full implications for commercialization. Rights to ownership and the terms of use for technology providers, partners, and GALVmed must be clearly documented for any background (existing) and foreground (future) IP used or generated throughout RD&D. Finally, a review of rights and obligations of relevant existing legal agreements is also conducted at this stage.

The importance of due diligence for RD&D is often underestimated in agricultural development; as a consequence, organizations operate in an environment of uncertainty and risk. Sponsors who invest in organizations like GALVmed are incurring unnecessary risk if they fail to make this type of due diligence a part of their grant-making process. The due diligence task for technology development is undoubtedly complex and requires substantial resources, but there is great value in high-quality IP analysis. Integrating IP analysis with technical information permits decisions to be made based on the evidence and reduces risk.

Stage 3: Technology-specific IP management strategy

The insight gained and information generated through IP due diligence is used for creating a Disease Intellectual Property Plan (DIPP). The DIPP is used to advise GALVmed staff and to address questions from external parties, such as stakeholders, regarding GALVmed's intended IP management strategy for a specific disease-control technology. Aside from presenting the results of the IP due diligence process, DIPPs map the flow of technology from providers to development partners, manufacturers, and so on. This map allows GALVmed to identify the contractual arrangements needed for effectively governing IP transactions between the actors involved in a way that supports pro-poor sustainable delivery of technology. Moreover, by building upon the results of the due diligence process, GALVmed can make informed decisions on critical issues such as ownership and rights allocations as they relate to background and foreground IP.

Issues of ownership and rights allocation are often not straightforward where nonprofit organizations are engaged in technology development. There is, first, the question of whether the coordinating organization should own IP itself. Some would say there is an inherent discord between owning IP and being an "honest broker" that coordinates incentives among partners. However, the ownership of IP allows a facilitator organization to have more leverage in pushing for pro-poor outcomes. GALVmed does not seek to own IP, but it does not rule out the possibility of a future instance in which claiming ownership to IP rights could be critical to achieving the development and deployment of products for the poor.

Stage 4: Contracting

The strategy articulated in a DIPP is ultimately implemented through a set of contracts among partner organizations. Contract drafting and negotiations are among the most challenging and resource-consuming activities that GALVmed undertakes. Some contracts govern straightforward IP transactions. Under other circumstances, contracts need to capture more sophisticated strategies that deal with, for example, issues of pro-poor performance obligations, geographical exclusivities, and activities for which a high degree of uncertainty exists. GALVmed's position as a facilitator in the RD&D process further complicates what might otherwise be a simple contract. As a facilitator, GALVmed engages multiple parties, often playing the role of an intermediary or broker (see module 3, TN 4, for a discussion of innovation brokers). GALVmed must structure contracts to ensure that there is a potential for leveraging to meet pro-poor goals, certain obligations from technology providers are integrated, and an effective recourse process is in place (should obligations be broken) with minimal impact on goals and milestones. In addition, GALVmed must ensure that the expectations of the technology provider and sublicensee are in compliance with one another. It is in GALVmed's interest to release market-ready products as soon as possible. Therefore it becomes GALVmed's responsibility to manage challenging negotiations with all involved parties in a timely and efficient way.

BENEFITS, IMPACT, AND EXPERIENCE

As this profile has illustrated, *IP management processes in the private sector are highly relevant to nonprofits* working to develop technology for the poor. IP management in the private sector minimizes risks and contributes key components to a commercialization strategy that supports the organization's goals. GALVmed, through its systematic approach to IP management, is better able to circumvent and/or minimize IP risks that could adversely affect downstream development and deployment operations (see the sections on IP landscaping and IP due diligence) and can use IP management to support its organizational goals. In the wider scheme of things, GALVmed's IP management system has benefited the organization pro-poor mission by addressing some broader issues that often delayed milestone deliverables, namely, the growth of innovation to sustainable, market-ready products.

One main point highlighted through GALVmed's experience in strategic IP management is that organizational missions and related policies, including IP policies, must be aligned with the ambitions of partners engaged in the RD&D process. Ultimately, the availability of GALVmed's products should not depend on the existence of GALVmed itself. For innovations to become meaningful products with wide adoption, partners, preferably private, must be incentivized to support the existence and availability of a product, throughout and beyond the existence of GALVmed. While GALVmed's facilitation in the development of a vaccine is purely humanitarian, private partners, who are crucial in ensuring that a technology becomes a successful product, are likely to have different ambitions, which must also be considered.

The need to foster stronger public-private partnerships has taught GALVmed to leverage the organization's interests while providing its partners with the opportunity to achieve their own internal mission. This understanding has served GALVmed enormously well while dealing with contentious IP issues and creating conditions for relationships and products conducive to success.

Another benefit GALVmed has enjoyed from its approach to managing IP is the ability to rapidly produce, negotiate, and secure agreements with different partners. The development of core IP principles and more attractive conditions for engaging partners have allowed the organization to significantly increase the rate at which it can negotiate contracts.

LESSONS LEARNED AND ISSUES FOR WIDER APPLICATION

GALVmed's experience in IP management (including its interaction with public and private partners) provides many lessons. A *key lesson* is that superficial surveys of IP are insufficient. All organizations working in the knowledge economy, in the public sector or otherwise, need to proactively address IP matters. Systematic IP management will improve efficient progress, reduce risk, and support the organizational mission, ultimately creating greater impact on livelihoods of the very poor.

The resources needed to implement IP management require organizational decisions to develop certain capacities in-house and determine which elements should be outsourced. Some have suggested that basic understanding of IP and access to patent information (such as information in public patent databases) is sufficient for most public sector operations. As demonstrated here, however, IP issues require significant expertise in analysis and the ability to develop solutions tailored to each project's goals. Public patent data require interpretation, informed analysis, and then translation into a sound IP strategy that serves the organization and its development goals.

GALVmed has found that a hybrid approach to IP capacity building, in which IP expertise is available both in-house and externally, serves the organization best. Internal sources are in closer contact with staff overseeing the RD&D process and can better capture and communicate the organization's needs and wants. External expertise, on the other hand, is impartial, can provide in-depth analysis, and can highlight issues the organization may fail to see internally. External expertise has given GALVmed high-quality analysis and access to top-tier attorneys.

One of the most challenging issues GALVmed has experienced is that of contracting. GALVmed's facilitation role entails the development of multiple contracts and often lengthy negotiations. GALVmed is working toward a new approach that employs significantly simplified contracts fit for multiple purposes. The intention is to decrease the time between drafting and signing contracts, while still effectively integrating the necessary rights and obligations. In addition to benefits for GALVmed, simplified legal contracts benefit developing country partners without good access to legal expertise.

GALVmed has learned that *building in-house capacity to manage IP is only half of the equation.* IP management plays an integral role in achieving a desired result; many related factors, such as business development strategies, go hand-in-hand with IP management practices. Regardless of the diligence GALVmed puts into managing IP, a sustainable endeavor ultimately relies on a partner's ability to interpret GALVmed's knowledge of IP issues and integrate that knowledge into a sound business model for downstream application.

In GALVmed's case, this challenge can prove difficult to meet. The majority of the organization's partners for downstream deployment are from the developing world, and many suffer capacity constraints (either in financial or other resources) or lack experience with IP, complex contracting, the creation of business plans, and other key business tools.

Consequently, GALVmed has recognized that *the second half of the equation for success in commercializing technologies for the poor is to build capacity in its downstream partners.* GALVmed now hires business consultants to work alongside partners to create business plans and strategies that take advantage of the IP knowledge GALVmed holds. In some instances, GALVmed assists its partners by taking the lead in drafting and negotiating complex agreements between partners. This intervention provides the partners with practical experience for dealing with IP issues, while providing GALVmed with the opportunity to impart its knowledge and experience in IP management for pro-poor purposes. *Capacity building on a project-specific basis has made related processes, such as contracting, simpler.* GALVmed can now engage with partners who have a clearer understanding of the needs, steps, risks, costs, and inputs required for a sustainable venture.

In conclusion, GALVmed provides an example of how nonprofits engaged in research, development, and deployment of technologies for the poor can benefit from systematic IP management. IP management plays a key role in reducing risks and improving the organization's capacity to deliver on its mission. Most nonprofits do not have sufficient in-house capacity, and this profile illustrates how the balance of outsourced services and internal capacity can change as an organization grows. Lastly, GALVmed's experience indicates the importance of integrating capacity building in IP management; even where a nonprofit is challenged itself in IP management capacity, there are opportunities to share knowledge and continue to foster improvements in a partner's IP management skills.

Developing a Subregional Approach to Regulating Agricultural Biotechnology in West Africa

Morven McLean, ILSI Research Foundation

SYNOPSIS OF PROJECT DATA

Country: CILSS and ECOWAS member economies in West Africa

Project: Rural Agricultural Income and Sustainable Environment Plus (RAISE Plus) Program: Short-Term Technical Assistance in Biotechnology (STTAB)

Implementing organizations: Michigan State University (MSU), Agriculture and Biotechnology Strategies (AGBIOS), and the Donald Danforth Plant Science Center (DDPSC)

Budget: US$2 million

Date: USAID FY 2006–09

Contact: Dr. Saharah Moon Chapotin, USAID

CONTEXT

In 2004, the Sahel Institute (INSAH, Institut du Sahel) completed a stock-taking exercise in the member countries of the Interstate Committee for Drought Control in the *Sahel* (CILSS, Comité Inter-états de lutte contre la sècheresse au Sahel) plus Ghana to gain a better understanding of the structure of the seed sector in each. During the country consultations, stakeholders provided the following justifications for establishing a subregional regulatory body for conventional and transgenic seed in the Sahel: (1) extending national seed markets that are considered limited; (2) formalizing an ancient transborder seed route; (3) ensuring the quality of the varieties released; and (4) monitoring the release of GE products in particular.

This insight led to the development of the "Framework Convention Introducing a Common Biosafety Regulation for the Prevention of Biotechnological Risks in the CILSS Countries" and the "Framework Convention Instituting Common Regulations for Conventional and Transgenic Seeds in the CILSS Area." The preambles to the conventions recognized both the benefits and potential risks of modern biotechnology. It stated that a subregional approach to biosafety regulation should be undertaken as "each country is neither able to individually take advantage of the known and potential benefits of genetically modified organisms (GMOs), nor cope with their known and potential risks."

In 2005, the Economic Community of West African States (ECOWAS) published an action plan with three operational objectives for the development of biotechnology and biosafety in the subregion, one of which was to develop a subregional approach to biosafety regulation (ECOWAS 2005). The plan was critical of the slow progress in achieving a subregional biosafety framework in West Africa, which it attributed to "an absence of political support in the field of biotechnology and biosafety; lack of communication between stakeholders, even within the same country; lack of coordination between the concerned ministries in the member countries; and poor subregional cooperation on the subject." The subregional approach to biosafety advocated by ECOWAS was to develop and implement a common regulatory framework that would be binding on all ECOWAS member countries.

PROJECT OBJECTIVES AND DESCRIPTION

The primary objective of the Short-Term Technical Assistance in Biotechnology (STTAB) project was to work cooperatively with regulatory officials to develop practical, needs-driven policies, directives, guidance, and review procedures to address the regulation of confined field trials and eventual commercialization of GE crops in West Africa. The project's components are described in the sections that follow.

Technical assistance to INSAH for the review and adoption of technical annexes to the CILSS biosafety convention

The project worked in partnership with INSAH (the technical arm of CILSS) and the West and Central African Council for Agricultural Research and Development (WECARD, referred to more commonly by its French acronym, CORAF)[1] as well as representatives from national environment and agriculture ministries to improve the Framework Convention Instituting Common Regulations for Conventional and Transgenic Seeds in the CILSS Area. During a series of four subregional meetings and with additional bilateral inputs from CILSS country representatives, the CILSS Convention was substantively rewritten in an effort to address the activities of the subregional process consistently and without duplication. The contained, confined, and unconfined uses of GE organisms were clearly differentiated. The regulatory responsibilities for each of these activities were defined. The technical annexes, which describe the technical information required for applications to the regional scientific review panel, were more clearly aligned with the types of applications that will be received in the subregion and with international standards and guidance related to the regulation of GE organisms established by Codex Alimentarius, OECD, and the Cartagena Protocol.

Technical assistance to INSAH to develop and implement an ECOWAS regulation on biosafety

In August 2008, the Experts Group Meeting on ECOWAS Biosafety Regulation, attended by environment and agriculture representatives from 14 ECOWAS countries, concluded with a request to INSAH-CILSS to extend the CILSS Framework Convention to all of the ECOWAS member countries. Building on the STTAB project's support to INSAH for the development of a regional biosafety framework within West Africa, this initiative aimed to extend the CILSS Biosafety Convention under the ECOWAS mandate.

Specifically, the objective was to develop an ECOWAS Regulation governing the importation, development, manufacture, and use of GE organisms and products derived thereof within ECOWAS Member States and to facilitate a consultative process leading to the adoption of the Regulation. The ECOWAS Biosafety Regulation was to be consistent with the spirit of the CILSS Biosafety Convention, incorporating the best elements of that framework, including its technical guidance on risk assessment procedures. The resulting document was "Regulation C/Reg.1/12/08

Establishing a Procedure for the Review and Authorisation of Products of Modern Biotechnology within the ECOWAS."

Technical assistance to enhance the environmental risk assessment capacity of the national biosecurity agency, Burkina Faso

The STTAB project also endeavored to work with national agencies and authorities to build institutional and human resource capacity in risk assessment, risk management, and decision making at the national level. When the project began, Burkina Faso was the only country in West Africa to have approved confined field trials of a GE crop, insect-resistant (Bt) cotton. To approve these trials, Burkina Faso had promulgated biosafety regulations and established ANB, its national biosafety agency reporting to the environment ministry (Ministère de l'Environnement et du Cadre de Vie). While the ANB, which has a legal mandate for the coordination and monitoring of all activities pertaining to the implementation of biosafety in Burkina Faso, was already active in the field, budgetary and technical capacity constraints limited its effectiveness. Preserving and building on the advances in Burkina Faso required building significant and sustainable capacity within the ANB.

INNOVATIVE ELEMENTS

The innovative elements of STTAB were its regional approach to what was initially perceived as a national priority. The approach proved flexible enough to be developed into a novel model for subregional harmonization of biosafety regulations.

Identifying and responding to a national priority with positive regional spillovers

Initially, the STTAB project focused most of its technical capacity-building in Burkina Faso. This strategic decision was based on the fact that: (1) Burkina Faso's government had clearly indicated its support for the commercialization of Bt cotton and, to that end, had made significant steps toward establishing a biosafety regulatory system (see box 6.28 in TN 4 in this module) and (2) farmers expressed significant interest in cultivating Bt cotton, generated by promising results from field trials conducted from 2003 to 2006. Environmental risk assessment training was provided to ANB personnel and other scientists so that a premarket environmental risk assessment of Bt cotton could be undertaken. The assessment was a prerequisite for the decision to approve Bt cotton.

A novel but feasible model for subregional harmonization

Given the ease of transboundary movement of seed between countries in West Africa, the impending commercial authorization of Bt cotton in Burkina Faso was an important catalyst for countries to work toward implementing a subregional approach to biosafety regulation. From prior stock-taking exercises and subregional consultations, it was apparent that the project should direct regional harmonization to the development of a mechanism whereby the science-based risk assessment would be undertaken by a subregional body but all decision-making would remain at the national level. A subregional body responsible for undertaking risk assessments for specific types of applications (such as confined field trials, food safety assessments for GE food, environmental risk assessment of GE plants) and providing scientific opinions to the member countries was considered the most achievable form of harmonization. This model differed from the only other examples of subregional harmonization that have been implemented internationally. In the EU, national decisions about cultivating GE crops are delegated to a subregional body, but this model has been ineffective. In Canada and the United States, harmonization of technical requirements for risk assessment has not resulted in appreciable gains in the efficiency or effectiveness of their representative regulatory systems.

The revised CILSS Convention and follow-on ECOWAS Regulation provide a practical and achievable approach to biosafety regulation in a subregion where national governments have limited scientific resources (human, financial, and institutional) to draw upon. An essential element of this project was to build capacity among the country representatives involved in drafting these documents so that the implications of specific policy choices and regulatory approaches could be considered.

BENEFITS, IMPACT, AND EXPERIENCE

This STTAB project has resulted in both direct and indirect benefits in the subregion. Building the capacity of Burkinabe risk assessors and regulators to undertake the environmental risk assessment of GE cotton was one of the factors contributing to its eventual approval. This effort has strengthened the ANB nationally, promoted its visibility within West Africa as a regional resource for risk assessment

training, and serves as a potential model for other countries in the subregion (or elsewhere in sub-Saharan Africa).

The commercial cultivation of Bt cotton in Burkina Faso contributed to an increase of about 16 percent of overall production in 2009/10. It is anticipated that 95 percent of harvested area (442,900 hectares) in 2010 will be planted to Bt cotton compared to the 2009/10 season (106,000 hectares). This expansion is expected to contribute significantly to national cotton production.

The ECOWAS regulation has not been submitted for approval, so it remains to be seen how implementation will proceed. The West Africa Regional Biosafety Project, launched in June 2009 by the West African Economic and Monetary Union (WAEMU) with funding from UNEP-GEF and the World Bank, has a component to strengthen institutional capacity for preparing regional laws and regulations on biosafety and creating an institutional framework to accompany the dissemination and implementation of the regional biosafety framework in WAEMU countries. A joint CILSS-ECOWAS-WAEMU committee is currently reviewing the ECOWAS Regulation to determine how it may be best incorporated into the WAEMU project. The end result may be that the ECOWAS Regulation will become a joint ECOWAS-WAEMU Regulation.

LESSONS LEARNED AND ISSUES FOR WIDER APPLICATION

The lessons from this experience are summarized in the sections that follow. They focus on the factors that contribute to successful collaboration, including a clear appreciation of the stakeholders involved, the potential incentives for collaboration, and the capacity-building requirements that must be fulfilled if collaboration is to yield useful results.

Understand who the key players are and engage them early in the process

The INSAH-CILSS process that led to the development of the first draft of the Framework Convention was criticized because the Convention was developed by Ministries of Agriculture without representation or input from national biosafety focal points or Ministries of Environment. The process to revise the Convention under the STTAB project deliberately included representation from a broader range of ministries. This more inclusive approach was an important step in correcting the apparent absence of prior interministerial engagement.

Collaboration with like projects should begin early and continue through the life of the project

Deficiencies in cooperation and coordination between the CILSS-ECOWAS initiative to develop a subregional approach to biosafety risk assessment and the West Africa Regional Biosafety Project under WAEMU led to early concerns that two competing approaches to regional biosafety regulation would develop. This concern may have been resolved with the CILSS-ECOWAS-WAEMU committee mentioned previously. Other capacity-building initiatives have also been launched in West Africa since the STTAB project began, notably the African Network of Biosafety Expertise, established by the African Union/New Partnership for Africa's Development (NEPAD) Office of Science and Technology, with a specific mandate to improve technical capacity in biosafety regulation and risk assessment. Collaboration between all of these projects will be essential if subregional harmonization is to be achieved.

Subregional harmonization is unlikely unless there is an imperative for countries to engage meaningfully in the process

In the case of West Africa, the commercial release of Bt cotton in Burkina Faso was a pivotal event. While there had been efforts to promote a subregional approach to biosafety regulation prior to the impending approval of Bt cotton, the expectation that Bt cotton seed would move to other countries within the subregion provided a real-world example of why a subregional approach to risk assessment was desirable and even necessary. Given that most West African countries have very limited capacity in biosafety risk assessment and risk management, a subregional risk assessment of Bt cotton under the process described in the

ECOWAS Regulation would be more efficient and cost-effective than if each country performed its own assessment. It might also help mitigate potential trade disruptions that can occur when trading partners have asynchronous product approvals.

Building national biosafety capacity is necessary for subregional harmonization

It is difficult for policy makers to support efforts to develop subregional approaches to biosafety regulation, let alone determine the appropriate model to advance, unless some national capacity in this area has been achieved. A national government does not need to have established and operationalized a biosafety regulatory system before engaging in such discussions, but it requires at least some expertise in biosafety (or related) regulation and/or risk assessment to ensure that national interests can be met.

Identify how project outcomes can be sustained

Neither the CILSS Convention nor the ECOWAS Regulation identifies provisions for funding the subregional activities described in each (such as convening the subregional scientific panel). Funding for biosafety capacity building in West Africa, including support for the development of national and subregional biosafety regulatory approaches, has come from the EU, United States, and Japanese donor agencies, as well as foundations and international financial institutions such as the Bill and Melinda Gates Foundation, the McKnight Foundation, the Rockefeller Foundation, and the World Bank. Mechanisms for sustainable funding of a subregional biosafety regulatory system by West African governments have not been established.

The Supply Response to New Sources of Demand for Financial and Other Services in Rural Andhra Pradesh

Gunnar Larson, World Bank
Melissa Williams, World Bank

SYNOPSIS

Self-help groups and their federations at the village, subdistrict, and district levels represent a new, self-aware client base for providers of financial and other services. By forming groups that effectively demand services, these clients acquire fundamental financial literacy and other competencies (thrift, savings, inter-lending, bookkeeping, and management skills) that strengthen and sustain their capacity to innovate. Government agencies, NGOs, and private companies have designed products and interventions to answer their demand and fulfill their needs in a number of sectors, including agriculture, finance, nonfarm employment, health, and education. Perhaps the most significant practical lesson from this experience is that stronger institutions for the rural poor enable several positive factors to converge. Public agencies gain a new partner capable of collectively asserting its needs, business gains a promising new market for services, and the wider economy gains a foundation for more pro-poor growth and innovation.

CONTEXT

India is one of the world's fastest-growing economies, yet translating rapid economic growth into reduced poverty remains a persistent challenge, particularly in rural areas. Throughout India, only 23 percent of 200 million rural poor are organized into various forms of groups. Individuals who are not in groups can find it challenging to obtain the credit, other services, and market access that offer the means to increase their incomes. The Government of India estimates that it will need to invest about US$20 billion over the next eight to nine years to tackle poverty but plans to invest just over US$10 billion.[1] Over the same period, the poor are projected to require about US$40 billion–US$50 billion in credit.[2]

Self-help groups (SHGs) are the primary source of credit for the rural poor, including small and marginal farmers. In 2007–08, an estimated US$1.7 billion in credit was disbursed to 1.3 million SHGs, but at this rate of flow, a major credit gap is expected. Per capita credit access is equivalent to US$111, which is less than 40 percent of the average expenditure by small and marginal farmers who cultivate their land (US$286).[3] As a result, countless poor people are left with no recourse other than informal moneylenders, who charge usurious interest rates, sometimes as high as 600 percent annually.

Module 1, IAP 4 described social mobilization among SHGs in Andhra Pradesh to develop a new source of effective demand in that state's rural economy. For rural entrepreneurs and other service providers, the size of this new clientele is sufficient in scale to command substantial attention. As of November 2010, nearly 11 million women had organized themselves into SHGs through Indira Kranthi Patham, creating a new, self-aware client base.[4] The potential returns from serving so vast a population of customers are self-evident.

As noted in the overview of this module, accompanying investments in rural finance show strong synergies with investments in agricultural innovation. SHGs have proven highly effective in bringing rural financial services into areas that are traditionally poorly served. They do so by helping commercial lenders to manage risk through joint liability, which brings tremendous pressure to bear on the respective group members to repay loans on time. The SHG strategy lowers transaction costs and addresses lenders' concerns over the potentially high risks of default in poor, remote rural areas (World Bank 2011). The organization of SHGs into larger aggregates at the village, subdistrict, and district levels was designed intentionally to meet sellers and service providers halfway.

OBJECTIVES AND DESCRIPTION

Aside from eliminating some of the barriers that prevented commercial banks from offering services in rural areas, a major goal of linking organized rural groups to formal credit and other services is to accommodate the constraints typical of SHG members, including time constraints. A premium is placed on convenience and on enabling the individual customer to conduct multiple transactions in a single visit. A closely related goal is to provide them with a relatively complete menu of financial services, including credit, insurance, and instruments for poor households to swap burdensome informal debt obligations for new obligations in the formal sector with more stable and reasonable interest rates (a high priority among the poor in the state). The insurance instruments are designed to protect vulnerable clients from the financial effects of events that often leave people in poverty, including pensions that provide security in old age. Figure 6.3 shows how SHGs and their federations create an enabling environment for innovation by empowering the rural poor to acquire the capacities, services, market access, and social safety nets that pave the way for innovation.

INNOVATIVE ELEMENTS

As the rural poor have organized, saved, accessed credit, and built skills and assets, they have more effectively voiced their demand for goods and services. In response, government agencies, NGOs, and private companies have designed products and interventions to answer their demand and fulfill their needs in a number of sectors, including agriculture, finance, nonfarm employment, health, and education.

In many instances, these service providers use a coproduction model in which the institutions of the poor become agents or franchises of an agency or business to extend its outreach and deliver services more cost-effectively. This practice not only provides services but generates employment within rural areas. In some instances, the Village Organization operates a commodity procurement center where agricultural inputs are sold. The approach builds capacity in the institutions, provides employment, and helps poor clients become more integrated with the value chain (for example, the procurement center will buy their produce and sell them inputs to improve yields in the next cycle).

In other instances, the poor have innovated by developing their own enterprises in response to program-supported activities. Some community members sell biopesticides and biofertilizers to farmers in response to the community-

Figure 6.3   Self-Help Groups Constitute a Rural Institutional Platform That Enables the Rural Poor to Acquire the Capacities, Services, Market Access, and Social Safety Nets That Pave the Way for Innovation

Source: Authors.

managed sustainable agriculture initiative (see module 1, IAP 4). Others provide public services that have not reached their location, such as preschools or nutrition centers for pregnant women and young children. These services are especially important in the tribal areas.

The foundation of this entrepreneurial innovation is access to financial services. These services enable the poor to accumulate assets and create a less risky environment in which they can capitalize on livelihood opportunities.

BENEFITS AND IMPACTS

The benefits and impacts of providing formal financial services to clients previously regarded as too risky to serve have ranged from the tangible benefits that people obtain from the services themselves to less tangible effects such as financial discipline or the sense of security derived from savings and insurance plans. Commercial banks have benefited from innovative business models that make it possible to tap into a vast and underserved rural market. The successes of the program in Andhra Pradesh and other states, and the benefits of the products, services, and new models developed expressly for a large base of very poor clients, inspired the Government of India to establish a National Rural Livelihoods Mission. The Rural Livelihoods Mission will apply the strategies developed through this program at the national level.

Building a bridge to formal credit

The savings, thrift, and inter-lending activities around which SHGs are organized provide members with experience in financial discipline, money management, and in conducting transactions and repaying loans. Over time, these competencies enable people to establish a history of repayment, obtain a credit rating, and then engage with banks or microfinance institutions. As a result, bank lending has increased from Rs 1.97 billion (US$48 million) in 2001–02 to Rs 65 billion (US$1.6 billion) in 2009–10. By early 2010, banks had extended loans of Rs 251 billion (US$6 billion) to SHGs without any collateral.

Total financial inclusion

As banks began to see the rural poor as customers, they altered their business model to accommodate this new source of demand. Rural households generally require working capital to support their current activities, capital to invest in new income-generating activities, and cash to meet basic consumption needs and social obligations, such as health, marriage, and home repair. The banks offered products and services in all these areas, including support for long-term investment in land. Because escaping from debt has been a major priority for many rural households in Andhra Pradesh, banks also arranged debt swaps and provided credit with which to retire costly informal loans. Even better, the new services enabled poor people to avoid the situations that had made them easy prey for informal moneylenders. Lending is based on household investment plans that are vetted by the SHGs and Village Organizations. Community-based recovery mechanisms ensure repayment rates of 95 percent or higher to the banks.

Insurance services to reduce vulnerability

Illness and death can plunge or further entrench a family in poverty. Private companies had often viewed the transaction costs of providing health, disability, and life insurance as prohibitive in rural areas, but community-managed structures dramatically reduce those costs by taking on tasks such as enrolling members and verifying, documenting, and processing claims. In Andhra Pradesh, community resource persons (*bima mithras*) are trained to fulfill these responsibilities on behalf of the Life Insurance Corporation of India (more information on community resource persons appears below). District federations have established call centers and developed a web portal to process transactions. The resource persons and call center make insurance services far more economical to provide and far more accessible to the rural poor, reducing the time to deliver insurance benefits by half.

Throughout Andhra Pradesh, more than 1.5 million SHGs were organized during the first ten years of the Indira Kranthi Patham program. During that period, SHG members accessed more than US$6 billion in credit from commercial banks. More than 11 million members and their families paid for death and disability and health insurance coverage, and over US$100 million worth of claims have been settled. Over 1 million SHG members have a separate health savings account, and as many as 3,000 villages have dedicated health risk funds to mitigate the shocks of health emergencies. More than 3,000 villages have nutrition centers for pregnant and lactating mothers and children under five.

The use of procurement center

Procurement centers operated through Indira Kranthi Patham are an important convening venue for small-scale producers and prospective investors. Small-scale producers, whose sales were previously dispersed widely among informal buyers, command better prices for their produce and buy inputs at lower prices. Procurement centers offer a forum for learning about new crops and growing methods, some of which have been developed locally. Producers are also better positioned to learn about developments in consumer and other markets, because they interact directly with buyers in the value chain. These venues also reduce the costs of outreach to interested investors by assembling a critical mass of potential clients whose purchasing power offsets any discounts they may negotiate. Linking producers to corporate and cooperative partners such as ITC Ltd., APMARKFED, and Olam International became one of the great legacies of Indira Kranthi Patham, and the quest for additional partners continues.

Co-contributory pension scheme

The government of Andhra Pradesh devised a co-contributory pension scheme targeting SHG members over 18, all of them women. The members contribute Re 1 per day, which is matched by the government. The Life Insurance Corporation of India invests the contributions in the market to get higher returns. When the member turns 60 she receives a pension of Rs 500 a month and health insurance coverage. Thus far, about 4.5 million SHG members have individual co-contributory pension accounts, and more than 400,000 are receiving pensions.

Other benefits

In addition to these specific instruments, Village Organizations bundle entitlements from public distribution systems, grain banks, and bulk purchases from the open market in a food security system that benefits as many as three million households. Village Organizations also operate "bridge schools" that offer incentives to ensure high levels of enrollment among girls. Among the more than 600,000 farmers who adopted community-managed sustainable agriculture in its first four years, the use of nonpesticide management caused a dramatic resurgence in local biodiversity in addition to reducing input costs and enabling farmers to escape from debt.

LESSONS LEARNED AND ISSUES FOR WIDER APPLICATION

Some of the most significant practical lessons to emerge from the rural livelihoods program in Andhra Pradesh relate to institutional development. When institutions of the rural poor become strong, they establish the basis for a convergence of factors. Public agencies gain a new partner that articulates the aspirations and concerns of its membership. Private businesses gain clients capable of collectively asserting demand as a market. The wider economy gains a foundation for more inclusive, pro-poor growth and innovation based on improved capacities and access to services. In the case of community-managed sustainable agriculture, stronger institutions for the rural poor create a foundation for dialogue on alternative agriculture in India's semiarid tropics.

The cost advantages of using coproduction models to deliver insurance services and agricultural inputs offset many of the disincentives of investing in rural areas. The organization of a new client base provides businesses with local partners capable of assuming many functions that were once centrally performed. The reduced costs greatly extend the reach of government services and private companies. The transaction costs for poor rural clients also fall dramatically through the use of one-stop shops where they can conveniently conduct multiple transactions. Located close to home, these service points provide complete "end-to-end" financial services including credit, insurance, procurement of inputs, and marketing of produce.

Like most problem-solving efforts, the experience described here has cautionary as well as positive lessons. In seven districts of Andhra Pradesh, the easy availability of credit from commercial banks and microfinance institutions encouraged borrowing well in excess of households' ability to repay, and about 20 percent of participating households began accumulating serious debt. The resulting microfinance crisis points to the ongoing need for building financial literacy among the poor and for discipline in lending. The abusive collection processes employed by some of these institutions emphasizes the need for discretion in selecting which institutions may participate in rural livelihoods programs and to the need for well-defined channels for recourse when borrowers default.

NOTES

Module 6 Overview

1. Of 145 countries for which data were reported for the years 1999/2000 and 2000/01, about one-third reported that more than 95 percent of secondary school students were enrolled in general programs and less than 5 percent in vocational or technical programs. Most European countries reported 20-40 percent enrollment in vocational/technical programs at the secondary level.

2. IPRs: World Trade Organization (WTO) and World Intellectual Property Organization (WIPO); PVRs: International Union for the Protection of New Varieties of Plants (UPOV); Biosafety: Global Environmental Facility (GEF); and SPS standards: Standards and Trade Development Facility (STDF), World Animal Health Organization (OIE), and International Plant Protection Convention (IPPC).

3. Union internationale pour la protection des obtentions végétales, established in 1961.

4. TRIPS is the Agreement on Trade Related Aspects of Intellectual Property Rights, signed in 1994 as part of the Uruguay round of the General Agreement on Tariffs and Trade (GATT). The TRIPS agreement obliges all members of the World Trade Organization (WTO) to have some form of IPR legislation in place. Low-income countries are given additional time to fulfil this obligation but eventually must comply.

5. The Cartagena Protocol, which is part of the Convention on Biological Diversity, was signed in 2000.

Thematic Note 1

1. See, for example, Johnson (2002), Paterson, Adam, and Mullen (2003), Ivanova and Roseboom (2006), Hekkert et al. (2006), and World Bank (2010).

2. Innovation policy calls for a "whole-of-government" approach. It depends on the establishment of efficient government machinery able to ensure the needed coordination. Although its mechanisms must be adapted to existing institutional frameworks and to cultural backgrounds, models that place a powerful coordinating body at the center of government allow innovation policy to have a pervasive influence (World Bank 2010).

Thematic Note 2

1. Formerly MoST, the Ministry of Science and Technology.

2. This taxonomy of key governance capabilities is based on numerous studies undertaken by Advansis. Compare also with findings of the OECD Monitoring and Implementing National Innovation Policies (MONIT) project (OECD 2005).

3. See Benoit (2007) for a good overview of indicators relevant to assessing innovation systems.

Thematic Note 3

1. World Bank (2006) discusses this issue for many highly relevant country-level investments related to plant breeding.

2. For example, it does not adequately emphasize the importance of IPRs in *access and benefit sharing,* in which "access" refers to accessing traditional knowledge and genetic resources, and "sharing" refers to sharing the benefits (commercial and otherwise) arising from the use of traditional knowledge and genetic resource.

3. For a thorough exploration of these issues, see Maskus (2000).

4. See http://www.wipo.int/ipadvantage/en/details.jsp?id =2595.

5. Commission on Intellectual Property Rights (2003).

6. "IPR-related" here refers to a broader definition of IP as creations of the mind or value added by innovative thinking. Even a hybrid plant variety, then, can be considered in some sense a form of IPR in agriculture, because control over the parents prevents others from profiting from the fruits of the breeder's investment.

7. See http://www.wipo.int/export/sites/www/ip-development/en/agenda/recommendations.pdf.

8. Patent pools, patent commons, and clearinghouses are essentially joint marketing systems in which a number of agencies agree to market their IP as a common entity, making it simpler for a licensee to obtain access to a number of different pieces of IP in a single transaction. Patent commons are typically free to access, although this is not always the case.

9. *Empresa* Brasileira de Pesquisa Agropecuária, Brazil's national agricultural research organization.

10. For example, assessing whether the researchers' rights to use technologies, materials, and data are aligned with the project's activities and evaluating potential related risks.

Thematic Note 5

1. The CAC initially formulated international commodity and product standards, but this role has expanded to include commodity-related guidelines and codes of practice; general standards and guidelines on food labeling; general codes and guidelines on food hygiene; guidelines on food safety risk assessment; standards, codes, and guidelines on contaminants in foods; standards, guidelines, and other recommendations on sampling, analysis, inspection, and certification procedures; maximum limits for pesticide

residue; food additives provisions; and maximum limits for veterinary drugs in foods (FAO and WHO 2005).

2. For example, the 1992 Rio Declaration set a foundation not only for government action but for all stakeholders to integrate sustainable development considerations within their consumption and production decisions (Rio Declaration, Principle 8). Other agreements include the Kyoto Protocol (1997) and World Summit on Sustainable Development (2002).

3. For animal diseases, there has been a call for an integrated and global approach under the concept "One World, One Health."

4. STDF is a global program established by FAO, OIE, WHO, WTO, and the World Bank (www.standardsfacility.org), which also offers grants to support specific investment in SPS capacity. The proposed multi-criteria framework aims to enhance the economic efficiency of SPS capacity-building decisions to meet a country's economic development, poverty alleviation, public health, and other objectives (Henson and Masakure 2011).

5. See Calvin (2003) and Calvin et al. (2003) for examples related to raspberries from Guatemala (contaminated with an intestinal parasite) and green onions from Mexico (hepatitis A).

6. In the United States alone, recent data estimate the economic impact of foodborne illness nationwide in US$152 billion annually (Scharff 2010).

Innovative Activity Profile 2

1. Tangible property rights are the set of rights defined by law that relate to a physical object, for example plasmids or vectors.

Innovative Activity Profile 3

1. *Conseil Ouest et Centre Africain* pour la Recherche et le Développement Agricoles.

Innovative Activity Profile 4

1. Including SGSY and NREGS. The Swarnjayanti Gram Swarozgar Yojana (SGSY) is an integrated program for self-employment of the rural poor that seeks to bring poor families above the poverty line by organizing them into SHGs. The National Rural Employment Guarantee Scheme (NREGS) guarantees 100 days of wage employment in every financial year to every household, whose adult members volunteer to do unskilled manual work.

2. Based on SGSY estimates, *swarojgaris* (self-employed persons) were able to leverage US$2.1 from commercial banks for every US$1 of government investment.

3. The US$286 figure is based on NSSO (2003).

4. Andhra Pradesh has 10,978,982 women in 975,362 SHGs, organized into 38,334 village organizations, 1,099 subdistrict organizations, and 22 district organizations, representing all 22 rural districts.

REFERENCES AND FURTHER READING

Module 6 Overview

Daane, J. 2010. "Enhancing Performance of Agricultural Innovation Systems." *Rural Development News* 1/2010: pp. 76-82.

Khandker, S.R., Z. Bakht, and G.B. Koolwal. 2006. "The Poverty Impact of Rural Roads: Evidence from Bangladesh." World Bank Policy Research Working Paper 3875. Washington, DC: World Bank.

OECD (Organisation for Economic Co-operation and Development). 2005. *Governance of Innovation Systems.* Vol. 1: Synthesis Report. Paris.

Walle, D. van de, and D. Cratty. 2002. *Impact Evaluation of a Rural Road Rehabilitation Project.* Washington, DC: World Bank.

Willoughby, C. 2002. Infrastructure and pro-poor growth: Implications of recent research. Unpublished.

World Bank. 2006. *Agricultural Investment Sourcebook.* Washington, DC.

———. 2007a. *World Bank Assistance to Agriculture in Sub-Saharan Africa: An IEG Review.* Washington, DC.

———. 2007b. *World Development Report 2008: Agriculture for Development.* Washington, DC: World Bank.

Thematic Note 1

Edquist, C. 2001. "The Systems of Innovation Approach and Innovation Policy: An Account of the State of the Art." Paper presented at the DRUID Conference, June 12–15, Aalborg, Denmark.

Hall, A., L. Mytelka, and B. Oyeyinka. 2006. "Concepts and Guidelines for Diagnostic Assessments of Agricultural Innovation Capacity." Working Paper No. 17. Maastricht: UNU–MERIT (United Nations University and Maastricht Economic and Social Research Institute on Innovation and Technology).

Hekkert, M.P., R.A.A. Suurs, S.O. Negro, S. Kuhlmann, and R.E.H.M. Smits. 2006. "Functions of Innovation Systems: A New Approach for Analysing Technological Change." *Technological Forecasting and Social Change* 74(4):413–32.

InterAcademy Council (IAC). 2003. *Inventing a Better Future: A Strategy for Building Worldwide Capacities in*

Science and Technology. Amsterdam: InterAcademy Council.

Ivanova, N., and J. Roseboom. February 2006. *A Functional Analysis of the Russian Innovation System: Roles and Responsibilities of Key Stakeholders.* Moscow: Science and Technology Commercialization Project (EuropeAid/115381/C/SV/RU).

Johnson, A. 2002. *Functions in Innovation System Approaches.* Göteborg: Department of Industrial Dynamics, Chalmers University of Technology.

KIT (Royal Tropical Institute). 2011. Dossier: Rural innovation policies, in KIT Information Portal: Rural Innovation Systems. http://portals.kit.nl/Rural_Innovation_Systems and http://www.kit.nl/smartsite.shtml?ch=FAB&id=26844, accessed March 2011.

Kraemer-Mbula, E., and W. Wamae (eds.). 2010. *Innovation and the Development Agenda.* Paris: Organisation for Economic Co-operation and Development (OECD) and International Development Research Centre (IDRC).

OECD (Organisation for Economic Co-operation and Development). 2005. *Oslo Manual: Guidelines for Collecting and Interpreting Innovation Data.* 3rd edition. Paris.

———. 2010. *The OECD Innovation Strategy: Getting a Head Start on Tomorrow.* Paris.

Paterson, A., R. Adam, and J. Mullen. 2003. *The Relevance of the National System of Innovation Approach to Mainstreaming Science and Technology for Development in NEPAD and the AU.* Pretoria: New Partnership for Africa's Development (NEPAD).

Pro Inno Europe. 2010. *European Innovation Scoreboard (EIS) 2009: Comparative Analysis of Innovation Performance.* Brussels: European Union (EU).

Roseboom, J. 2004. *Case Study on the Financing of Agricultural Research in Brazil.* Washington, DC: World Bank.

World Bank. 2008. "Project Appraisal Document on a Proposed Loan in the Amount of US$ 150 million to the Argentine Republic for an Unleashing Productive Innovation Project." Washington, DC.

———. 2010. *Innovation Policy: A Guide for Developing Countries.* Washington, DC.

Thematic Note 2

Benoit, G. 2007. "National Innovation System: The System Approach in Historical Perspective." Working Paper No. 36. Montreal: Project on the History and Sociology of STI Statistics. Canadian Science and Innovation Indicators Consortium, http://www.csiic.ca/PDF/Godin_36.pdf, accessed March 2011.

Dahlman, C.J. Routti, and P. Ylä-Anttila (eds.). 2006. *Finland as a Knowledge Economy: Elements of Success and Lessons Learned.* Washington, DC: World Bank Institute.

EC (European Commission). 2009. *Policy Mixes for R&D in Europe.* Maastricht: UNU-MERIT (United Nations University and Maastricht Economic and Social Research Institute on Innovation and Technology).

Edquist, C. (ed.). 1997. *Systems of Innovation: Technologies, Institutions and Organizations.* Oxon: Francis Pinter Publishers.

Hausmann, R., and B. Klinger. 2006. "South Africa's Export Predicament." CID Working Paper No. 129. Cambridge, MA: Centre for International Development (CID), Harvard University.

Lemola, T. 2002. "Convergence of National Science and Technology Policies: The Case of Finland." *Research Policy* 31(8–9):1481–90.

Lingela, V. 2004. "The Co-evolutionary Framework to Improve Competitiveness in the South African System of Innovation." Paper presented at the Globelics Academy, Lisbon.

Ministry of Employment and the Economy. 2010. *Evaluation of the Finnish National Innovation System.* Helsinki: Taloustieto Oy.

Nelson, R. 2003. *National Innovation Systems: A Comparative Analysis.* Oxford: Oxford University Press.

Nordling, L. 2009. "South African Innovation Agency Takes Shape." SciDevNet, July 21, 2009, http://www.scidev.net/en/news/south-african-innovation-agency-takes-shape.html, accessed March 2011.

OECD (Organisation for Economic Cooperation and Development). 2005. *Governance of Innovation Systems.* Volume 1: Synthesis Report. Paris.

———. 2007a. Chile. *OECD Review of Innovation Policy.* Paris.

———. 2007b. *OECD Reviews of Innovation Policy: South Africa 2007.* Paris: OECD Publishing. http://dx.doi.org/10.1787/9789264038240-en.

———. 2008. China. *OECD Review of Innovation Policy.* Paris.

———. 2009a. "Chile's National Innovation Council for Competitiveness: Interim Assessment and Outlook." Paris.

———. 2009b. *OECD Reviews of Innovation Policy: Korea 2009.* Paris: OECD Publishing. http://dx.doi.org/10.1787/9789264067233-en.

———. 2009c. Mexico. *OECD Review of Innovation Policy.* Paris.

———. 2010. *The OECD Innovation Strategy: Getting a Head Start on Tomorrow*. Paris.

OECD and World Bank Institute. 2000. *Korea and the Knowledge-Based Economy: Making the Transition*. Paris.

Ylä-Anttila, P., and C. Palmberg. 2007. "Economic and Industrial Transformations in Finland." *Journal of Industry, Competition, and Trade* 7(3–4):169–87.

Thematic Note 3

Adams, S. 2008. "Globalization and Income Inequality: Implications for Intellectual Property Rights." *Journal of Policy Modeling* 30(5):725–35.

CAMBIA. n.d. "Mapping of Rice Patents and Patent Applications onto the Rice Genome." Patentlens, http://www .patentlens.net/daisy/RiceGenome/3909.html, accessed September 2010.

Cavicchi, J.R., and S.K. Kowalski. 2007. "Use the Kitchen Door." Op/ed contribution. *National Law Journal* 30(15):23.

Commission on Intellectual Property Rights (UK). 2003. "Agriculture and Genetic Resources." Chapter 3 in *Integrating Intellectual Property Rights and Development Policy*, 3rd ed. London. IPR Commission, http:// www.iprcommission.org/papers/pdfs/final_report/CIPR fullfinal.pdf, accessed March 2011.

Frankel, S. 2009. "Challenging TRIPS-plus Agreements: The Potential Utility of Non-Violation Disputes." *Journal of International Economic Law* 12(4):1023–65.

Fridell, G. 2007. *Fair Trade Coffee: The Prospects and Pitfalls of Market-Driven Social Justice*. Studies in Comparative Political Economy and Public Policy. Toronto: University of Toronto Press.

Fuck, M.P., and M.B.M. Bonacelli. 2009. "Institutions and Technological Learning: Public-Private Linkages in Brazil and Argentina." *Journal of Technology Management and Innovation* 4(2):33–43.

Heydon, K., and W. Woolcock (eds.). 2009. *The Rise of Bilateralism: Comparing American, European and Asian Approaches to Preferential Trade Agreements*. New York: United Nations University Press.

Janis, M.D., and S. Smith. 2007. "Technological Change and the Design of Plant Variety Protection Regimes." *Chicago-Kent Law Review* 82: 1557–615.

Kargbo, A., J. Mao, and C. Wang. 2010. "The Progress and Issues in the Dutch, Chinese, and Kenyan Floriculture Industries. *African Journal of Biotechnology* 9(44):7401–08. Academic Journals, http:// www.academicjournals.org/ AJB/PDF/pdf2010/1Nov/Kargbo%20et%20al.pdf, accessed March 2011.

March, E. 2007a. ". . .And a Tea." *WIPO Magazine* 5/2007.

———. 2007b. "Making the Origin Count: Two Coffees." *WIPO Magazine* 5/2007.

Maskus, K.E. 2000. *Intellectual Property Rights in the Global Economy*. Washington, DC: Institute for International Economics.

Pollack, A. 2010. "Rival Candy Projects Both Parse Cocoa's DNA." *The New York Times*, September 15, 2010.

Sell, S.K. 201. "Cat and Mouse: Forum-Shifting in the Battle over Intellectual Property Enforcement." Unpublished paper. George Washington University, www.gwu .edu/~igis/Sell%20Paper.doc, accessed March 2011.

World Bank. 2006. *Intellectual Property Rights: Designing Regimes to Support Plant Breeding in Developing Countries*. Report No. 33517 GLB. Washington, DC.

Thematic Note 4

AATF (African Agricultural Technology Foundation). 2010. Rationale for a Biosafety Law for Uganda. Policy Brief. http://www.aatf-africa.org/userfiles/WEMA-UG-policy-brief2.pdf, accessed March 2011.

BrBiotec. 2010. "Country Profile: Brazil." Brasil Biotec, http://apps.convention.bio.org/applications/CPA/ ProfileDetailsView.aspx?ccd=bra, accessed March 2011.

Brookes, G., and P. Barfoot. 2009. "Global Impact of Biotech Crops: Income and Production Effects, 1996–2007." *AgBioForum* 12(2):184–208.

Cardoso, T.A.O., M.B.M. Albuquerque Navarro, B.E.C. Soares, F.H. Lima e Silva, S.S. Rocha, and L.M. Oda. 2005. "Memories of Biosafety in Brazil: Lessons to Be Learned." *Applied Biosafety* 10(3):160–68.

CERA (Center for Environmental Risk Assessment). 2010a. GM Crop Database. Center for Environmental Risk Assessment (CERA), ILSI Research Foundation, http://cera-gmc.org/index.php?action=gm_crop_data-base, accessed March 2011.

———. 2010b. The South Asia Biosafety Program. Center for Environmental Risk Assessment (CERA), ILSI Research Foundation, http://cera-gmc.org/index.php? action=s._asia_biosafety_program, accessed March 2011.

Chapotin, S.M., M. McLean, and H. Quemada. 2009. "Biosafety Capacity Building: Lessons Learned from USAID's Global Partnerships." Paper presented at the 13th ICABR Conference on the Emerging Bio-Economy, June 18–20, Ravello. International Consortium on

Agricultural Biotechnology Research, http://www.economia.uniroma2.it/icabr/paper_view.php?id=484&p=5, accessed March 2011.

Cohen, J.I., and R. Paarlberg. 2004. "Unlocking Crop Biotechnology in Developing Countries: A Report from the Field." *World Development* 32 (9):1563–77.

DBT (Department of Biotechnology). 2008. *Establishment Plan for the National Biotechnology Regulatory Authority.* New Delhi: DBT, Ministry of Science and Technology.

EAC (East African Community). 2009. *Biosafety Act, 2009.* Arusha.

Government of Brazil. 2005. *Decreto No. 5.591, de 22 de novembro de 2005.* Brasilia. http://www.planalto.gov.br/ccivil_03/_Ato2004-2006/2005/Decreto/D5591.htm, accessed March 2011.

Government of Vietnam. 2010. Decree No. 69/2010/ND-CP on Biosafety for Genetically Modified Organisms, Genetic Specimens, and Products of Genetically Modified Organisms. http://www.thuvienphapluat.vn/archive/Nghi-dinh/Decree-No-69-2010-ND-CP-on-biosafety-for-genetically-modified-organisms-vb110514t11.aspx, accessed March 2011.

GTRS (Gene Technology Review Secretariat). 2006. "Statutory Review of the Gene Technology Act 2000 and The Gene Technology Agreement." Canberra: GTRS, Department of Health and Ageing.

Héma, O., H.N. Somé, O. Traoré, J. Greenplate, and M. Abdennadher. 2009. "Efficacy of Transgenic Cotton Plant Containing the Cry1Ac and Cry2Ab Genes of *Bacillus thuringiensis* against *Helicoverpa armigera* and *Syllepte derogata* in Cotton Cultivation in Burkina Faso." *Crop Protection* 28(3):205–14.

Horna, D., P. Zambrano, J. Falck-Zepeda, T. Sengooba, G. Gruère, J. Komen, and M. Kyotalimye. 2012 (forthcoming). "Designing an Ex-ante Assessment of GM Technologies to Support Biosafety Regulations and Decision Making: The Case of Cotton in Uganda."

James, C. 2010. "Global Status of Commercialized Biotech/GM Crops: 2010." ISAAA Brief No. 42. Ithaca, NY: International Service for the Acquisition of Agri-biotech Applications (ISAAA).

Kalaitzandonakes, N., J.M. Alston, and K.J. Bradford. 2007. "Compliance Costs for Regulatory Approval of New Biotech Crops." *Nature Biotechnology* 25(5):509–11.

Matten, S.R., G.P. Head, and H.D. Quemada. 2008. "How Government Regulation Can Help or Hinder the Integration of Bt Crops within IPM Programs." *Progress in Biological Control* 5:27–39.

McLean, M.A. 2010. "India's Biosafety System: At Par with the World?" *BiotechNews* 5(2):88–91.

McLean, M.A., R.J. Frederick, P. Traynor, J.I. Cohen, and J. Komen. (2002). "A Conceptual Framework for Implementing Biosafety: Linking Policy, Capacity and Regulation. *ISNAR Briefing Paper* No. 47. IFPRI, Washington, DC.

Potrykus, I. 2010. "Lessons from the 'Humanitarian Golden Rice' Project: Regulation Prevents Development of Public Good Genetically Engineered Crop Products." *New Biotechnology* 27(5):466–72.

Pray, C., B. Ramaswami, J. Huang, R. Hu, P. Bengali, and H. Zhang. 2006. "Costs and Enforcement of Biosafety Regulations in India and China." *International Journal of Technology and Globalisation* 2(1–2):137–57.

UNEP (United Nations Environment Programme). 2003a. *Sub-Regional Workshop for Latin American Countries on: Development of a Regulatory Regime and Administrative Systems.* UNEP, http://www.unep.ch/biosafety/old_site/development/devdocuments/3ChileWebReportEN.pdf, accessed March 2011.

———. 2003b. *Report of the Subregional Workshop for Asian Countries on: the Development of a Regulatory Regime and Administrative Services.* UNEP, http://www.unep.ch/biosafety/old_site/development/devdocuments/shirazreportfinal.pdf, accessed March 2011.

World Bank. 2003. "Biosafety Regulation: A Review of International Approaches." Report No. 26028. Washington, DC.

———. 2007. *World Development Report 2008: Agriculture for Development.* Washington, DC.

Thematic Note 5

Calvin, L. 2003. "Produce, Food Safety, and International Trade: Response to U.S. Foodborne Illness Outbreaks Associated with Imported Produce." In *International Trade and Food Safety: Economic Theory and Case Studies,* edited by J. Buzby. AER-828. Washington, DC: Economic Research Service, United States Department of Agriculture (USDA).

Calvin, L., B. Avendaño, and R. Schwentesius, R. 2003. "The Economics of Food Safety: The Case of Green Onions and Hepatitis A Outbreaks." VGS-305-01. Washington, DC: Economic Research Service, United States Department of Agriculture (USDA).

Coslovsky, S.V. 2006. "How Bolivia's Brazil-Nut Industry Became Competitive in World Markets While Brazil's Fell Behind: Lessons from a Matched Comparison." Department of Urban Studies and Planning Working Paper. Cambridge: Massachusetts Institute of Technology.

Díaz, A., and T. O'Brian. 2004. "Improving Competitiveness and Market Access for Agricultural Exports through the Development and Application of Food Safety and Quality Standards: The Example of Peruvian Asparagus."

Lima: Inter-American Institute for Cooperation on Agriculture (IICA).

Diaz Rios, L. 2007. "Agro-industries Characterization and Appraisal: Asparagus in Peru." Agricultural Management, Marketing, and Finance Working Document No. 23. Rome: Rural Infrastructure and Agro-Industries Division, Food and Agriculture Organization (FAO).

Diaz Rios, L., and S. Jaffee. 2008. "Barrier, Catalyst, or Distraction? Standards, Competitiveness, and Africa's Groundnut Exports to Europe." Agriculture and Rural Development Discussion Paper No. 39. Washington, DC: World Bank.

FAO (Food and Agriculture Organization). 2007a. "FAO Biosecurity Toolkit: Guide to Assess Biosecurity Capacity." Part 2. Rome. ftp://ftp.fao.org/docrep/fao/010/a1140e/a1140e.pdf, accessed August 2011.

———. 2007b. Strengthening National Food Control Systems. A Quick Guide to Assess Capacity Building Needs." Rome. ftp://ftp.fao.org/docrep/fao/010/a1142e/a1142e00.pdf, accessed August 2011.

FAO (Food and Agriculture Organization) and WHO (World Health Organization). 2005. "Understanding the Codex Alimentarius." Revised and updated. Rome.

Henson, S., M. Jensen, S. Jaffee, and L. Diaz Rios. 2010. "Assessing the Demand for Trade-Related Food Safety and Quality Interventions in Agri-Food Chains." Washington, DC: World Bank.

Henson, S., and O. Masakure. 2011. "Establishing Priorities for SPS Capacity-Building: A Guide to Multi-Criteria Decision-Making. Standards and Trade Development Facility." http://www.standardsfacility.org/Files/EconAnalysis/MCDA_FrameworkGuideDraft_20-Jun-11.pdf, accessed August 2011.

Jaffee, S. 2003. "From Challenge to Opportunity: Transforming Kenya's Fresh Vegetable Trade in the Context of Emerging Food Safety and Other Standards in Europe." Washington, DC: World Bank.

Jaffee, S., H. Spencer, M. Sewadeh, P. Pentney, and F. Musonda. 2005. Tanzania's Agro-Food Trade and Emerging Sanitary and Phytosanitary (SPS) Standards: Toward a Strategic Approach and Action Plan. Document prepared as a contribution to the Tanzania Diagnostic Trade Integration Study, March.

Jaffee, S., T. Deeb, T. O'Brien, Y. Strachan, and R. Kiggundu. 2006. Uganda, Standards and Trade: Experience, Capacities, and Priorities. Draft background report for the Uganda Diagnostic Trade Integrated Study, January.

Jaffee, S., S. Henson, and L. Diaz Rios. 2011. "Making the Grade: Smallholder Farmers, Emerging Standards, and Development Assistance Programs in Africa." Report No. 62324-AFR. Washington, DC: World Bank.

Jensen, F.M., N. Strychacz, and J. Keyser. 2010. "Non-Tariff Barriers and Regional Standards in the EAC Dairy Sector." Africa Trade Policy Notes. Washington, DC: World Bank.

Ponte, S. 2005. "Bans, Tests, and Alchemy: Food Safety Standards and the Ugandan Fish Export Industry." Working Paper No. 2005/19. Copenhagen: Danish Institute for International Studies (DIIS).

Ponte, S., R. Kadigi, and M. Mitullah. 2010. When the Market Helps: Standards, Ecolabels and Resource Management Systems in East Africa. Presentation at the SAFE Final Conference, Zanzibar, 31 May-1 June 2010.

Scharff, R.L. 2010. "Health-Related Costs from Foodborne Illness in the United States." Report prepared under the Produce Safety Project at Georgetown University. Produce Safety Project, http://www.producesafetyproject.org/admin/assets/files/Health-Related-Foodborne-Illness-Costs-Report.pdf-1.pdf, accessed September 2011.

Vapnek, J., and M. Spreij. 2005. "Perspectives and Guidelines on Food Legislation, with a New Model Food Law." FAO Legislative Study. Rome: Food and Agriculture Organization (FAO).

World Bank. 2005. "Food Safety and Agricultural Health Standards: Challenges and Opportunities for Developing Country Exports." Washington, DC.

———. 2010. "Operational Costs of Trade-Related Sanitary and Phytosanitary Activities." Washington, DC: World Bank.

WTO (World Trade Organization). 2011. "Specific Trade Concerns." Committee on Sanitary and Phytosanitary Measures. Report G/SPS/GEN/204/Rev.11. Geneva.

Innovative Activity Profile 1

Cristancho, E., I.J. Ekanayake, and W. Janssen. 2007. "Colombia: Lessons from the Competitive Allocation of Research Funding in the Agricultural Sector." *En Breve* 113. [In English and Spanish.] World Bank, http://irispublic.worldbank.org/85257559006C22E9/DOC_VIEWER?ReadForm&I4_KEY=CEE1F9183BCF301C852573B80053A7BB96B7435FBCD5FB548525769A005C5167&I4_DOCID=E70EFADC494EBB71852576A8001A6925&, accessed March 2011.

World Bank. n.d. Implementation Manual: Market Improvement and Innovation Facility, 2009/2010, Zambia Agricultural Development and Support Program. Ministry of Agriculture and Cooperatives, Lusaka.

World Bank. 2010. "Designing and Implementing Agricultural Innovation Funds: Lessons from Competitive Research and Matching Grant Projects." Washington, DC.

Innovative Activity Profile 2

Boettiger, S., and S. Alvarez 2010. "Getting Better Technologies to the Poor: A Landscape of Commercialization Challenges and Opportunities." Berkeley: Public Intellectual Property Resources for Agriculture (PIPRA). http://www.pipra.org/documents/PIPRA_CommercializationReport_photos.pdf, accessed August 2011.

Brown, S. 2009. "Playing Catch-Up with Lifeline Vaccines." Rural 21, http://www.rural21.com/uploads/media/rural_eng_41-42_01.pdf, accessed August 2011.

Innovative Activity Profile 3

CILSS (Comité Inter-états de Lutte contre la Sècheresse au Sahel). 2006. Framework Convention Instituting Common Regulations for Conventional and Transgenic Seeds in the CILSS Area. Bamako.

ECOWAS (Economic Community of West African States). 2005. Action Plan for the Development of Biotechnology and Bio-safety in the ECOWAS Sub-region. Abuja.

———. 2008. Regulation C/Reg.1/12/08 Establishing a Procedure for the Review and Authorisation of Products of Modern Biotechnology within the ECOWAS. Abuja.

Innovative Activity Profile 4

Hayward, N., and A. Brizzi. 2007. "Supporting the 'People Sector': The South Asia Experience in Rural Livelihoods Development—A Summary. *Livelihoods Learning Note* 1 (1). Washington, DC: World Bank. http://www-wds.worldbank.org/external/default/WDSContentServer/WDSP/IB/2007/12/12/000310607_20071212165109/Rendered/PDF/41692optmzd0Le1le1s0sector01PUBLIC1.pdf, accessed March 2011.

Hanumantha Rao, S.V. 2007. "Total Financial Inclusion: A Success Story." *CAB Calling* 31(2):13–15. College of Agricultural Banking, http://www.cab.org.in/CAB%20Calling%20Content/Financial%20Inclusion%20%20A%20Study%20of%20Business%20Correspondents%20in%20Orissa/Total%20Financial%20Inclusion%20-%20A%20Success%20Story.pdf, accessed March 2011.

Mohan, V., M. Takada, V. Kalavakonda, S.S. Banerjee, and P. Shah. 2008. "Community-managed Food Security Enterprises in Andhra Pradesh." *Livelihoods Learning Note* 1(4). Washington, DC: World Bank. http://www-wds.worldbank.org/external/default/WDSContentServer/WDSP/IB/2008/02/27/000310607_20080227114806/Rendered/PDF/427290optmzd0NWP0South0Asia041AP1RCL.pdf, accessed March 2011.

NSSO (National Sample Survey Organization). 2003. "Income, Expenditure, and Productive Assets of Farmer Households." Report No. 497. New Delhi.

Rao, K.P., V. Kalavakonda, S.S. Banerjee, and P. Shah. 2008. "Community-managed Procurement Centers for Small and Marginal Farmers in Andhra Pradesh, India." *Livelihoods Learning Note* 1(2). Washington, DC: World Bank. http://www-wds.worldbank.org/external/default/WDSContentServer/WDSP/IB/2008/02/27/000310607_20080227103401/Rendered/PDF/42721optmzd0WP0South0Asia021AP1MKTG.pdf, accessed March 2011.

Shenoy, M., S. Lakhey, and P. Shah. 2010. "Creating Jobs for Rural Youth in Andhra Pradesh, India: *Livelihoods Learning Note* 2(4). Washington, DC: World Bank. http://siteresources.worldbank.org/INTRURLIV/Resources/LLN-2-4_AP-skills.pdf, accessed March 2011.

Vijay Kumar, T., D.V. Raidu, J. Killi, M. Pillai, P. Shah, V. Kalavadonda, and S. Lakhe. 2009. "Ecologically Sound, Economically Viable: Community Managed Sustainable Agriculture." Washington, DC: World Bank.

World Bank. 2011. "Managing Credit Risks in Financing Agriculture: Lessons from Experiences in Asia and Africa." Washington, DC.

USEFUL WEBLINKS

Thematic Note 4

www.tradestandards.org:	Trade Standards Practitioners Network
www.oie.int:	World Animal Health Organization (OIE)
www.standardsmap.org/en/:	ITC Standard's Mapping Initiative

Thematic Note 5

www.ictsd.org:	International Centre for Trade and Sustainable Development (ICTSD)
www.ifahsec.org:	International Federation of Animal Health (IFAH)
www.ippc.int:	International Plant Protection Convention (IPPC)
www.ipfsaph.org:	International Portal on Food Safety, Animal and Plant Health
www.seedtesting.org:	International Seed Testing Association (ISTA)

www.agrifoodstandards.net: Small Scale Producers and Standards in Agrifood Supply Chains

www.standardsfacility.org: Standards and Trade Development Facility (STDF)

Innovative Activity Profile 5

Society for the Elimination of Rural Poverty (SERP) http://serp.ap.gov.in/SHG/index.jsp.

Assessing, Prioritizing, Monitoring, and Evaluating Agricultural Innovation Systems

OVERVIEW

Helen Hambly Odame, University of Guelph
Andy Hall, LINK Ltd.
Kumuda Dorai, LINK Ltd.

EXECUTIVE SUMMARY

The rationale for investing in assessment, priority setting, monitoring, and evaluation within an innovation system is that practitioners require information for short- and long-term decision making and for managing limited resources effectively within complex, nonlinear processes of technical and institutional change that seek to achieve social, economic, and environmental goals. Change is rarely managed by following a set plan. Instead, information on technical adaptations (*what gets done*) as well as institutional adaptations (policy and organizational changes) involving *how things get done* must be continuously sourced, shared, analyzed and used to inform decision making at multiple levels of the system.

This module examines processes to inform decision making and manage innovation at four generally defined levels of the innovation system for agriculture: policy, investment, organization, and intervention. The module identifies methods relevant at each level for assessing, prioritizing, monitoring, and evaluating innovation processes. For example, at the policy level, international and cross-sectoral comparative analyses may be used (such as benchmarking) and combined with multistakeholder policy dialogues and foresighting studies to inform on future development issues or subsectors. At the organizational level, methods include multi-institutional and organizational performance assessments involving innovation surveys and network mapping. At the level of specific programs or interventions, assessment tools for strengthening the involvement of end-users or farmers in value chains include gender analysis and support for self-organizing networks, among others. Still other methods are relevant at all levels of the innovation system for effective performance management, accompanied by reporting arrangements that ensure accountability.

RATIONALE

In innovation systems, managing depends on information-rich tasks such as assessment, monitoring, and evaluation, which are vital to maintaining learning, performance, and accountability (box 7.1). Decision makers, including investors, public service managers, entrepreneurs, scientists, and primary producers, easily understand the need to identify technical information that can be used to create new products and services. Equally essential—but far more difficult to obtain, analyze, and make accessible—is information about institutional adaptations (organizational and policy changes) that are being made or need to be made within the innovation system before technical innovations can be realized. For this reason, individuals and organizations who are concerned with performance management, learning, and accountability within the AIS will need to learn how to obtain and use information.

Box 7.1   Main Terms Used in This Module

Accountability. A transparent management system that ensures participation and open communication and reporting on results obtained and inputs used to achieve the results.

Assessment. Making decisions about innovation to know *why* investment is necessary to transform modes of production and consumption. Assessment is also done to speed the search function within innovation systems (for example, to identify new, more sustainable, and more equitable routes to value creation in agriculture).

Decision making. Decision making is not centralized but occurs throughout the system. Rather than limiting the focus to any single line of inquiry or information source, interactive learning feeds decision making and requires individuals and groups to be open to different and imaginative ways of thinking as well as to be receptive to new ideas and directions that match the context.

Priorities. The areas to which internal and external resources will be allocated to address problems or take advantage of opportunities.

Performance. Results in the form of productivity (outputs), outcomes, and impact, measured (for example, with indicators using quantitative or qualitative data) in relation to mandate, strategy, objectives, and client needs.

Theory of change or *intervention logic.* The underlying assumptions in an intervention that link intervention inputs with expected outcomes.

Monitoring. Tracking progress in stimulating changes in the policy and institutional environment; internal to an intervention and learning-based.

Evaluation. Performed on behalf of the investor to gauge the effectiveness of the design and execution of the intervention supported by the investor and the observed impacts associated with it. Evaluation is also done to inform the intervention logic of new investments.

Information gathering. The collection, communication, analysis, and reporting of quantitative and qualitative data by various stakeholders, which makes it possible to assess: priorities, capacities, and performance; components within the system, such as institutions, partnerships, and interventions; or the system's resources (human, financial, physical, and other resources).

Source: Authors.

Tasks such as assessing, prioritizing, monitoring, and evaluation will be widely used across the system and by a range of stakeholders. For instance, donors and community leaders may be concerned with tasks such as evaluating social and economic outcomes from past agricultural innovation. Scientists and public planners compile data and analyses of future trends and priorities and monitor indicators of change within the system. Entrepreneurs may track returns to current project investments and farmers may assess value chains of specific interest to them as producers. Efforts are also made to connect all types of decision makers within the system to meet their shared need for information and intelligence about their current context. No single organization or type of organization can provide the knowledge needed to inform policy, set priorities, or propose interventions. Monitoring can facilitate midstream adjustments, while evaluation at the final stages of an intervention enables outcomes and impacts to be thoroughly investigated to inform future investment decisions and negotiation among stakeholders.

Today, the responsibilities of decision making and managing innovations are spread across a range of actors at different levels of the innovation system. It is important to recognize that each of these levels requires different investments, and while coordination may still occur, it is rare for a single, centralized agency to be responsible for assessment, priority setting, monitoring, and evaluation within the AIS. These levels, which are useful for organizing the discussion in the rest of this overview, include the following:

- *Policy level.* At this level, the actors are responsible for creating the enabling environment for innovation (see module 6). At this level, decision making and management emphasize coherence across sectors, scenario development, and benchmarking innovation capacities.

- *Investment level.* At this level, the actors are responsible for designing and prioritizing interventions that support innovation. Decision making and management emphasize evaluating the performance of investments and testing the underlying assumptions that shed light on the "why" question for new investments.

- *Organizational level.* At this level, the actors are responsible for ensuring relevance to the wider innovation system. Decision making and management emphasize assessing organizational performance and understanding the shifting landscape of other organizations and networks in the innovation systems.
- *Intervention level.* At this level, the actors are responsible for managing and implementing innovation processes and resources to achieve desired social and economic goals. Decision making and management emphasize efforts such as monitoring the effectiveness of actions in achieving intervention goals, understanding the dynamic context in which interventions take place and planning around it, and understanding unexpected outcomes.

The "how to" approach for identifying innovation priorities and assessing performance described in this module includes tasks that are familiar to policy makers and managers of science, technology, and innovation, including strategic planning and foresighting, organizational performance assessment, monitoring, evaluation, and impact assessment. These familiar management processes and tools are being adapted to the new demands of a multistakeholder innovation system, however, and being supplemented with interactive learning to gather information on the institutional dimensions of innovation processes and capacities. Examples of these adapted and new tools are discussed throughout this module.

PAST EXPERIENCE

Past agricultural R&D systems traditionally placed great emphasis on gathering information about inputs into or outputs from the agricultural research process, prioritizing where those input investments should be made, listing outputs produced, and gathering information about the outcomes of research that typically combined different outputs and led, for example, to changes in an important agricultural commodity's yield and wider social and economic impact. Investments to gather information on inputs and outputs made it possible to validate technology adoption, investigate the economic surplus generated by research and compute rates of return, and inform economic studies of the contribution of research to impact (Pardey, Alston, and Ruttan 2010).

Historically innovation was oriented toward a linear "technology push" model that focused heavily on R&D, and decision making and management largely were orchestrated by centralized bodies. In agricultural development, national governments predominantly made decisions—constituted as public policies—that directed investment into priority areas (often commodity research programs). In the era that preceded AIS thinking, policy and planning tools commonly resulted in R&D programs and projects that operated with defined resources and specific time scales (table 7.1). Centralized public R&D planning was challenged, however, to respond to constant changes in demand-driven value chains, where decision making and management were decentralized. Entrepreneurs and primary producers became key to processes of assessing existing policy and managing performance within the system. This evolution suggests that innovation systems for agriculture are increasingly more difficult to design, plan, and implement without mechanisms to ensure that rapid adjustments can be made in "game-changing" situations, such as the emergence of new fundamental information, dramatic shifts in resource availability, or demands from new or different stakeholders.

While the basic elements of policy making and planning have not disappeared altogether in an AIS, the need to adapt to uncertainty and appreciate complexity have given rise to greater skepticism about causality and control in policy and planning cycles (Edquist 1997). In an AIS, decision makers are required to be strategic and not just fund and operationalize strategic plans, then monitor and evaluate them. Rather than determining a final set of priorities, planning becomes an iterative process that continuously identifies and (re)prioritizes actions in response to the rapidly changing environment. In an AIS context, the conventional notion of a fixed-purpose or time-bound plan of action must anticipate new information and learning generated through stakeholder interactions, monitoring, and eventually evaluation of results. The move beyond past agricultural policy making and planning methods is most evident when it comes to monitoring and evaluation, which are now understood within the AIS to be distinct tasks with different responsibilities, tools, and time frames (TNs 4 and 5).

KEY POLICY ISSUES

For policy making, priority setting, and performance management in the AIS, key policy issues involve allocating sufficient resources to information gathering and assessments. These activities will help identify opportunities that impinge on the AIS (including those that are nonagricultural) and strategies for ensuring that learning from very localized as well as large-scale interventions can become more useful to individual stakeholders as well as to the AIS as a whole.

Table 7.1 Conventional Agricultural Policy Making and Planning Tools

Tool	Description and time frame	Advantages	Disadvantages and challenges
Strategic planning	Defines direction of change and operational objectives (4–8 years); emphasis on efficiency and relevance.	Creates sense of direction that aligns the organization to its environment.	Operational demands may detract from strategic planning.
Master planning	Define long-term investments and activities (10 years).	Determines major system and organizational activities in relation to financial investment.	Limited with respect to redefining the organization in relation to changes in its environment.
Program planning	Prioritized actions to address constraints (3–5 years).	Systematically analyzes constraints and identifies priorities *before* resource gaps analyzed and projects developed, which is relevant to request funding and negotiate with investors.	Investors may impose new priorities on existing programming, making it difficult to fulfill original program objectives.
Project planning	Define efficient and sufficient set of activities to overcome a "problem" (1–3 years).	Identifies and prepares an integrated plan to resolve a "problem." This integration translates a project idea into a proposal and also ensures that planning integrates monitoring and evaluation operations.	Lack of necessary resources challenges project planning, monitoring, and evaluation. Proposals may have immutable deadlines.
Experiment planning	Develop best option to obtain insight into a scientific question (1 year or less).	Identifies the most efficient and effective option, in the form of research proposals, for achieving research results.	Research ideas often arise and become proposals to respond to investor calls or changing stakeholder demands without sufficient attention to fair and equal treatment within the organization.
Financial planning	Match financial availability to needs (variable).	Develops annual budgets and strategies to identify and develop alternative sources for an organization's core and operational funding.	Needs to ensure a link to project monitoring and evaluation to respond to low-performing activities and rebalance resources with programs.
Human resource and training plans	Efficiently manage and develop human resources (variable).	Analyzes capacity-building needs and prepares to develop knowledge, attitudes, and skills of individuals in an organization.	Needs to ensure a link to project monitoring and evaluation to respond to low-performing individuals, redirect training efforts, and rebalance human resources with programs.
Priority setting	Final stage of planning that defines an investment portfolio that is consistent with national policy and development goals, organizational mission, and program objectives.	Rationalizes investment in relation to limited resources, external demands for transparency, and focus on client needs.	Implementation can deviate from agreed priorities. Priorities set without explicit support of key stakeholders and program staff can lead to conflict in the organization.

Source: Authors; Gijsbers et al. 2000; Alston et al. 1995.

A supportive fiscal environment that values assessing, prioritizing, monitoring, and evaluating AIS and ensures information access

Within public institutions, management tasks associated with information, knowledge, and learning are often squeezed into already tight budgets for financing R&D and innovation (Hall and Learner 2010). Allocating sufficient funding for information management should also ensure access to more and better information. For example, production data in developing countries are notoriously unreliable, and food stock data around the world are highly secretive. Releasing research results or making information such as price and market data widely accessible using a range of communication strategies and media can facilitate access to information that could otherwise be scanty or overwhelming and difficult to sort out. Information accessible to one stakeholder group may not necessarily benefit another, so resources are also needed to assess users' information needs.

It is also appropriate for the public sector to play a key role in mobilizing information for policy, priorities, and performance management in AIS and making this information as widely accessible as possible. Public agencies, such as universities, research institutes, and government departments, can collect, analyze, and communicate information about the AIS, making it available not just to their own networks but to wider professional or producer associations, media agencies, private sector groups, and others. An example of such an initiative is the use of foresighting and scenario planning exercises (TN 3). Furthermore, governments can encourage participation by the private sector, recognizing that proprietary rights may be associated with information sharing among enterprises. Efforts to make as much information publicly available as possible have led to important contributions to technical and institutional innovations (module 5 discusses incentives and resources for innovative partnerships and business development; module 6 covers intellectual property management; and module 1 focuses on coordination and collective action among AIS stakeholders.)

Information on agricultural and nonagricultural opportunities

Although policy making, priority setting, and performance management in an AIS depend on information about interventions with the expected links to agriculture (such as interventions that will raise agricultural productivity), information about the productivity of crucial nonagricultural sectors and linkages to new product markets is also relevant to information-gathering efforts in the AIS. Innovation policy (see TN 1 in module 6) generally needs to include a link between agricultural and nonagricultural economic strategies for resource-poor households—for example, by supporting sustainable agricultural intensification while developing manufacturing and services that will expand nonagricultural employment.

Balancing potentially competing priorities within the wider innovation system requires dedication to information-intensive tasks such as benchmarking. The identification of multipurpose infrastructure or multitasking capacities that are useful to agriculture and transferable to nonagricultural sectors will provide a wider set of options in the innovation system. One example of multipurpose infrastructure is rural infrastructure for information and communication technologies, which can improve flows of agricultural information within and among organizations in addition to performing a multitude of other functions (for an overview, see World Bank 2011).

End-user participation and scaling up information from the local level

Within an AIS, an end-user perspective in processes such as assessment and priority-setting is developed by bringing innovation users' collaboration, behavior, and perceptions of change to bear on the analysis of the system. Baseline data about end users, including farmers, is essential to identify indicators of future performance and impact. Participatory methods of problem analysis such as Most Significant Change (TN 4) or mapping exercises (IAP 2) supplement baseline data and provide end users' critiques of their own situations and past interventions. Such information collection and exchange ensures that all stakeholders are better informed as decision makers by developing intelligence on key information and trends.

Even so, analysis generated about technical and institutional innovations at the end user or local level is often difficult to combine and compare across different temporal, spatial, and need-specific contexts. This requires generic tools for assessing, prioritizing, monitoring, and evaluating innovation processes to be carefully selected and adapted to fit as closely as possible with prevailing social, cultural, political, economic and environmental contexts (see the cases of India in IAP 4 and Chile in IAP 5). For the same reasons, within a country, planning efforts and results from monitoring and evaluating local innovations are not necessarily easily scaled up to constitute national priorities or policy. Policy making and setting priorities will have limited success unless the complex of local circumstances and decision making among stakeholders is taken into account. Rather than trying to gather and compare information about all local interventions, strategies for assessment that compare selected cases of local adaptation and innovation, preferably involving regional or international benchmarking, can be used (see TN 1). The key point is for the end user or local innovation processes not to be overly generalized, and for all stakeholders to be active collaborators within AIS assessment, priority setting, monitoring and evaluation to ensure learning, performance management, and accountability.

NEW DIRECTIONS, PRIORITIES, AND REQUIREMENTS

Innovation systems are developing new management processes which question underlying assumptions and theories of action to reconsider the sustainability of systems that were perhaps no longer effective and relevant in the changed context. Managers within innovation systems now must be equipped to deal with change and not simply administer under changing circumstances. As a result, new emphasis is placed on, for example, defining alternate scenarios and strategies that could influence or create opportunities for technical and institutional innovation. Such methods seek to redefine organizational mandates while ensuring participation and open communication about results obtained from past inputs used to achieve the results. Over the long run, high-performing innovation systems establish internal management processes that can define objectives and indicators of success and can achieve and communicate results while being attentive to resource constraints. In contrast to the past, especially in agricultural R&D, when priority-setting exercises and planning instruments (and their requirements for M&E) were largely mandatory and often static exercises, management tasks within the innovation system are ineffective and irrelevant unless they are established as dynamic processes.

The thematic notes and innovative activity profiles in this module provide more depth on the processes and methods undertaken at these four levels. The overlap among the levels is substantial, but it is useful to highlight the purpose and roles of each level, what is being assessed, and what tools are used for setting innovation priorities and assessing performance (table 7.2).

Examples of key directions for future assessment, priority setting, monitoring, and evaluation at different levels of the AIS include the following:

- **Policy level: Benchmarking innovation capacity.** Managers and investors increasingly need to benchmark the performance of sectors and subsectors in terms of the capacity developed for innovation through innovation system interventions. International investors may also use benchmarking to make international comparisons to track macro-level progress and help target subsectors, sectors, and countries for investment in capacity strengthening or for other investments that are conditional on certain levels of capacity. Methodological challenges need to be addressed, however, in measuring and comparing context-specific and systemic capacities of this sort. (See TNs 1 and 4 in this module and TN 1 in module 6.)

- **Investment program level: Foresighting.** Investment implies committing support to a program of activity to gain a desired return. Projections of future investments needed within an innovation system will benefit from collaborative diagnostic tools such as foresighting. Stakeholder engagement and learning that can lead to technical and institutional changes are facilitated through

Table 7.2 Decision Making and Management Processes and Tools at Different Levels of an Agricultural Innovation System

Level	Stakeholders involved	Key management processes	Tools (related TN/IAP)
Policy	National policy makers, sector committees	– Track progress of the national system and its functions – Coordinate agriculture with other sectors (modules 1 and 6) – Inform global or regional public policy networks – Design an enabling environment (intellectual property, banking, pricing, and tax regimes) (module 6)	– Benchmarking (TN 1) – Innovation surveys (TN 2) – Foresighting and scenario planning (TN 3, IAP 1)
Investment program	Finance ministry, donors, private sector, technical team leaders	– Prioritize and allocate resources – Identify new investment opportunities or bottlenecks – Review effectiveness of past investments – Improve underlying theories of change (intervention logic) of new investments	– Diagnostic studies with a commodity or subsector focus (TN 3, IAP 5) – Benchmarking (TNs 1 and 4) – Evaluation and impact assessment (TN 5)
Organization	Executive officers, board of directors, research organizations, extension organizations	– Assess organizational performance (TN 2) – Set organizational policy and program priorities – Enable organizational and institutional learning and change (module 4, TN 5) – Respond to changing innovation landscape	– Performance indicators (TN 2) – Innovation surveys (TN 2) – Self-organizing networks (IAP 4) – Evaluation and impact assessment (TN 5) – Institutional histories (TN 4) – Network mapping (IAP 2) – Reflexive monitoring in action (TN 4) – Causal process tracing (TN 4)
Intervention	Nongovernmental organizations, private sector, research and extension program leaders, project managers	– Accountability to investors – Managing effectiveness of program/project implementation – Managing innovation processes, including effectiveness of networks, interactions, and ways of working – Testing and reframing theories of change (intervention logic) – Responding to unexpected outcomes – Responding to changing innovation environments	– Participatory impact pathway analysis – Outcome mapping (TN 4) – Gender analysis of value chains (IAP 3) – Rapid appraisal of agricultural knowledge systems (TN 4) – Stakeholder analysis (IAP 2) – Most significant change analysis

Source: Authors.

foresighting processes. Foresighting involves systematically looking at the long-term horizon of science, technology, and institutions within dynamic economic, political, and social contexts to identify strategies that will yield the greatest benefits. (See TN 3 and IAPs 4 and 5.)

- *Organizational level: Institutional assessment.* Investment in an AIS anticipates technical innovation as well as institutional changes involving policy, program, and project implementation (how and when) and resources employed (who, what, and where) to obtain the highest possible potential for impact. Institutional assessments use multiple methods to capture existing and potential changes within and among organizations and their strategic activities. (See TN 2.)

- *Intervention level: Interactive visualization methods for learning, action-oriented planning, monitoring, and evaluation.* Actors and organizations in innovation systems are drawing on information from a wide range of sources, often using multiple methods. Tools that involve interaction, such as outcome mapping or Net-Map, are used to gather and synthesize information and generate new knowledge. These methods generate qualitative data and are used together with more traditional quantitative analysis and diagnostic case studies to enable learning that will improve the prospects for interventions to provide the best possible return on investment. Monitoring allows for mid-stream adjustments that can optimize performance while evaluation leads to better accountability

reporting and negotiation for future decision making. These two crucial innovation management processes can inform one to the other, but they are distinct processes with respect to learning, performance management, and accountability. (See TNs 4 and 5 and IAPs 6 and 7.)

MONITORING AND EVALUATION

The development of an AIS requires continuous decision making and management to obtain critical information to answer critical questions. For example, how have policy and investment in innovation processes achieved or surpassed their objectives? How do stakeholders know that resources for innovation are being allocated and managed effectively? What lessons about institutional adaptations are informing new technical changes? And finally, how can assessments strategically inform future decision making and interventions?

In effect, the act of assessing, prioritizing, monitoring, and evaluating within the AIS is a reflexive practice that catalyzes the experience-based learning that underlies all technical and institutional innovation. Aside from the more detailed discussion of M&E in this module, the notion of revitalizing the way in which monitoring and evaluation are distinctly needed to inform learning, performance management, and accountability is a common thread in all of the themes covered by the modules in this sourcebook.

Assessing Innovation for Prioritizing Investments

Helen Hambly Odame, University of Guelph

SYNOPSIS

From an AIS perspective, approaches to assessment and priority setting must consider both the complexity and unique context of agricultural innovation. The design of agricultural policies and investment programs requires more than analysis of the agricultural sector and its subsectors; it requires information on links with other sectors to which agriculture may contribute or from which agriculture may benefit. Priorities are determined through deeper and wider sets of data, often involving ongoing synthesis through interaction and learning with many stakeholders. Three analytical methods are recommended for assessing innovation system investments and thereby helping to allocate resources for investments: (1) theory of change and flexible planning; (2) measuring the functions of innovation systems; and (3) making comparisons across innovation systems (benchmarking). To develop an effective assessment and priority-setting process for agricultural innovation, public policy makers and donors must insist on the allocation of sufficient resources, perhaps through a unit or platform that would be closely involved with data access, data linkages, and the adoption of interdisciplinary approaches to assessment, monitoring, and evaluation. Stronger individual and organizational capacity is needed to assess change within and across individuals, organizations, and networks or systems. Much work needs to be done to ensure that data collected with stakeholders at the local level feeds into both ex ante and ex post assessment and priority setting. Input from the local level is tremendously important, and participation requires time and commitment to be done well.

BACKGROUND AND CONTEXT

Investing in innovation systems for agriculture has become one of the most important global and national policy efforts of the 21st century,[1] even as most nations face urgent, emergent, and often competing needs for investment in education, health, energy, and telecommunications and transportation infrastructure. Assessment guides decision makers in the allocation of scarce resources (financial, human, and physical) to areas showing the greatest potential for benefit, a process referred to as priority setting. Prior to the application of the innovation systems concept to agriculture, priority setting largely entailed defining a portfolio of public R&D programs and projects that was consistent with national development goals, national agricultural policy, and the objectives of public research institutes and technology transfer services (Contant 2001).

From an AIS perspective, approaches to assessment and priority setting must consider the complexity and unique context of agricultural innovation. The design of agricultural policies and investment programs requires more than the analysis of the agricultural sector and its subsectors; it requires information on links with other sectors to which agriculture may contribute (for example, biofuels for the energy sector) or from which agriculture may benefit (for example, innovations within human health sciences that benefit animal health). Priorities are determined through deeper and wider sets of data, often involving ongoing synthesis through interaction and learning with many stakeholders. As described below, new and modified assessment methods are being used to supplement traditional priority-setting principles and practices in agricultural R&D.[2]

Monitoring and evaluation are crucial for feeding information and intelligence back into assessment and priority setting by, for instance, identifying bottlenecks in the innovation system or recommending new investment opportunities. Instead of relying on linear input-output-impact models, this kind of M&E relies on integrating policy and investment priorities much more closely than before. It uses more flexible forms of planning to feed lessons learned from monitoring—and subsequently from evaluation and impact assessment—back into decision-making processes for investments in innovation.

Regardless of whether the AIS is mature and operating within the context of an established national innovation policy, assessment and priority setting for policy and investment program levels are relevant to any system.[3] Module 6 describes national innovation policy and discusses how the development of innovation priorities across sectors and technologies (and the public resources allocated to them) is coordinated with specific policies and priorities for innovation within a particular sector or technology domain. IAP 4 (India) and IAP 5 (Chile) in this module describe foresighting processes that contribute to overall innovation policy processes, the assessment of policies for innovation, and prioritizing areas for investment. This note focuses on specific methods for exploring the underlying theory of change, measuring the functions of the AIS to identify capacity and resource gaps, and using policy and investment-level benchmarks and indicators to make comparisons across countries and sectors.

ASSESSMENT METHODS

In general, assessment has two major dimensions: ex ante (prior to implementing an investment intervention) and ex post (after implementing an investment intervention) (table 7.3). Each type of assessment seeks to increase the efficiency and effectiveness of past investments while ensuring transparency and accountability for new investments.

Like the other processes discussed in this module—organizational assessment (TN 2), foresighting (TN 3), monitoring (TN 4), and evaluation (TN 5)—in the context of an AIS, the assessment of innovation systems and associated investments recognizes that the dynamics of the innovation system change constantly, that stakeholders throughout the AIS must participate, and that projections of the potential impact of policies and investments are fundamental elements of assessment.

Three analytical methods are recommended for assessing and prioritizing investments in agricultural innovation: (1) understanding the theory of change; (2) measuring the functions of innovation systems; and (3) making comparisons across innovation systems. Each is discussed in the sections that follow.

Theory of change

The process of assessment is one of judging and making decisions. As a lens through which innovation is understood, a theory of change is the set of hypotheses, as developed by stakeholders, of how policy and investment priorities, and the resulting programming and projects, plan to achieve their intended goals and objectives, including social, economic, political, and environmental change objectives. This kind of assessment is sometimes referred to as the *causal model*. The assessment team works with actors and networks having a stake in the innovation process to identify the underlying assumptions, values, and definitions of individuals and organizations. Box 7.2 lists the kinds of questions raised in a theory of change discussion to inform an analytical process that will lead to complementary methods, such as organizational assessment (TN 2), foresighting, building scenarios, and vision statements (TN 3 and IAPs 4 and 5).

Box 7.2	Questions for Assessing the Theory of Change in Innovation Systems

- What is the definition of innovation in the given context?
- How do innovations emerge and develop in the given context?
- Who are the innovation actors (organizations) in the given context?
- What policies and investments exist to support innovation in the given context?
- What policies and investments do not exist to support innovation in the given context?
- How are innovation trends, processes, and products measured in the given context?
- What are the key sources of data on innovation in the given context?

Source: Author.

Table 7.3	Roles of Ex Ante and Ex Post Assessments

Ex ante assessment	Ex post assessment
– Analyzes the likely impact of a proposed intervention. – Identifies the optimal portfolio of investments according to specific programs or projects. – Establishes a baseline or framework to collect and compare information for ex post evaluation.	– Assesses impact. – Supports accountability reporting to investors. – Justifies allocation of resources. – Generates lessons. – Identifies recommendations for future interventions.

Source: Author.

An example of assessing the theory of change is DFID's Research Into Use (RIU) program (see IAP 7), which introduced the method to assess developmental impacts and inform future priorities for investment by DFID and its partners. The RIU assessment explored assumptions of the program and their causal relation to activities, outcomes, outputs, and impact. Information is fed back into policy and investment decision making. Specific attention was given to determining whether the intervention's theory of change was altered during implementation, why this change was important in relation to impact, and why the investors (DFID and its partners) needed to be made aware of the change.

Exploring the theory of change is a relevant activity throughout the management cycle. The theory of change can be developed at the outset of assessing innovation and prioritizing investments; it can later be revisited during monitoring, evaluation, and impact assessment. Analyzing the theory of change contributes information to accountability reporting and is also a useful tool for proposing recommendations for future investment.

Analyzing the theory of change for innovation processes helps to identify expected outputs, outcomes, and impacts for a plan of action. Commissioned surveys or reviews of relevant literature as well as case studies can compile data on conditions of path dependency for technical and institutional innovations. This information may also help to identify stakeholders' participation in identifying complementary and conflicting areas of innovation and document their ideas regarding the theory of change and how it may

differ from policy makers' and investors' conceptions. The process also benefits from participants testing their assumptions against the available data and identifying where data to address those assumptions are missing. The conclusions from these discussions inform decisions on which methods will be used in the next stages of policy and investment planning or evaluation. In Sierra Leone, for example (IAP 1), discussions on the theory of change led to greater clarity in self-organizing networks and their investment processes, including the use of social media for continued discussions and information gathering.

Measuring the functions of an innovation system

Decision-making processes such as ex ante and ex post assessment, the identification of investment priorities, and organizational performance assessment rely on identifying the key functions of innovation systems. Table 7.4 identifies these functions and sources from which data are collected and analyzed.

Case studies

Case studies (including ex ante innovation histories) are especially useful for profiling specific functions of the innovation system or for examining specific sectors or technology domains. Case studies are a powerful assessment tool because innovation systems are highly contextual, and specific technical and institutional innovations are path dependent: Beyond the level of its basic functions, an innovation system

Table 7.4 Ten Functions of Innovation Systems and Related Data Sources

Key functions	Main sources for data or cases
1. Providing R&D and creating new knowledge.	International or government sources, R&D dialogues, end-user surveys.
2. Building competence in the labor force (includes education, training, creation of human capital, production and reproduction of skills, and individual learning).	International or government sources; student performance measures; employer surveys on education, vocational training, and other variables.
3. Forming new product markets.	International, government, and industry sources, including new market surveys.
4. Forming new quality requirements arising from the demand for new products.	International, government, and industry sources, including product surveys or consumer studies.
5. Creating and changing organizations for the development of new fields of innovation.	International, government, and industry sources; policy dialogues; actor-network analysis; innovation surveys.
6. Networking through markets and other mechanisms, including interactive learning among different organizations.	Government and industry sources, policy dialogues, actor-network analysis, innovation surveys.
7. Creating, changing, and abolishing institutions that influence innovating organizations by providing incentives or removing obstacles to innovation.	International, government, and industry sources; policy dialogues; innovation surveys.
8. Incubating activities for new innovating efforts.	Industry, government sources, R&D dialogues, innovation surveys.
9. Financing innovation processes and other activities that can facilitate the commercialization of knowledge and its adoption.	International, government, and industry sources; policy dialogues; investment surveys; and impact assessment.
10. Providing consultancy services relevant to innovation processes.	Government and industry sources, professional association surveys.

Source: Author, based on Edquist 1997.

is difficult to describe in generalities. The socioeconomic, cultural, and geographic contexts tend to define the specific organizations and institutions within an innovation system and the interactions among its actors. Case studies can capture some of this detail, and meta-analysis of case studies can inform comparisons across commodities, institutions or countries (Larsen, Kim, and Theus 2009; Wenninck and Heemskerk 2006; World Bank 2006).

Often case studies begin with a general profile of an innovation system that describes its two basic and related elements: *organizations* and *institutions*. Organizations (TN 2) are the group structures that have staff, facilities, equipment, and funding to conduct activities related to innovation. Examples include public research institutes or advisory services. Institutions are the established formal and informal practices, values, norms, rules, or laws that regulate the relations and interactions between and within organizations. Examples of institutions in an innovation system are tax laws and R&D incentives that encourage innovating organizations. Using tools such as innovation surveys and focus groups, case studies will often elaborate on the range of possible functions performed by a range of possible actors. The identification of key networks or alliances (see IAPs 1 and 4 in module 4) among actors as well as the knowledge base for innovation are also discussed in innovation case studies.

Benchmarking: making comparisons across innovation systems

Benchmarking uses input indicators (investments in the functions of an innovation system) as well as output/outcome indicators (products and processes and trends emerging from an innovation system) to identify disconnects within the system—among institutions and in relation to government policies—that hinder innovation. If indicators are difficult to establish, it may be that the theory of change was not sufficiently developed.

At the policy and investment program level, innovation benchmarks and indicators are used to compare current and historical measures and trends within the same country or to compare the innovative capacity of actors in the system (firms, for example), particularly their relative ability to identify and absorb technologies to enhance innovative performance and overall competitiveness. The Organisation for Economic Co-operation and Development (OECD) has been especially influential in providing guidelines for the systematic collection of innovation data. Table 7.5 summarizes best practices for benchmark data collection (OECD 1997; Bloch 2007; Kraemer-Mbula and Wamae 2010).

Spielman and Birner (2008) identify AIS-oriented benchmarks and indicators. They recognize that agricultural

Table 7.5	Guidelines for Benchmarks and Indicators for Innovation Systems
Key *measurement* considerations	– Interactions among enterprises (e.g., joint research activities; technical collaboration). – Interactions among enterprises, universities, and public research institutes (e.g., joint research, joint patenting, joint publications, network analysis of informal linkages). – Diffusion of information, knowledge, and technology to enterprises (e.g., industry adoption rates for new technologies; consumer data; strengthened competencies; skills training). – Human resource flows (e.g., movement of highly qualified personnel within and between the public and private sectors).
Key *techniques* for assessment	– *Institutional assessments* for analyzing human resource flows; institutional linkages; industrial clusters and innovative firm behavior (TN 2). – *Innovation surveys* question enterprises on their sources of knowledge most relevant to innovation and allow a comparative ranking of sectors/industries and national systems, including regional and global comparisons; surveys used by OECD are typically based on OECD's *Oslo Manual* (first developed in 1990 and currently in its third edition; see OECD 2005). – *Cluster analysis (see also TN 4 in module 5)* focuses on the interactions between particular networks of enterprises across and within specialized sectors and even internationally; tools may include social network analysis and deliberative mapping, among others. – *Evidence-based management* focuses on obtaining the best facts. Even less-than-favorable evidence is accepted during the assessment to "learn by doing." This technique cautions against complacency in achieving benchmarks set by comparing one country to another, perhaps within the same region, and by avoiding dialogue about mistakes, risks, and uncertainties.
Key *investment* and *financing* considerations	– Upgrading competencies in individual organizations; incentives and recognition for innovative work in individual enterprises and clusters. – Investing in higher education; R&D in public institutions; access to public research. – Private-public research initiatives; improved intellectual property and knowledge management; investing in information and communication technology. – Competitive or matching grants or low-interest loans; science/trade fairs; new business incubators. – Facilitate the national system to engage in international knowledge flows, including knowledge acquired abroad as capital or intermediate goods; foreign direct investment; purchases of foreign patents/licenses; establish learning alliances; trade in services such as technical consultancies; aid-for-trade; internationally coauthored publications and R&D.

Source: Author; OECD 1997, 2005.

benchmarks and indicators have to be attentive not only to measures of sectoral performance but to processes that underlie sectoral performance and the contribution of innovation to performance. Table 7.6 presents examples of the wider scope of benchmarks and indicators for assessing an AIS at the level of policy and program investments.

POTENTIAL BENEFITS

Assessment and associated priority setting direct scarce resources towards the best possible return on innovation activities. Assessment compiles a vast range of data on key functions in the system. These data include both quantitative and qualitative information supporting the analysis of policy effectiveness, enabling adjustments in the resource allocation as needed, and ultimately enabling the nonlinear feedback loops expected of planning, monitoring, and evaluation in complex adaptive systems (Pant 2010).

Opening up discussions with a wide range of stakeholders on the underlying theory of change catalyzes a process of planning strategically and transparently, versus mechanistically setting out a strategic plan (TN 3). It improves the chances that stakeholders will reach consensus on the action plan. With this method, questions about the objectives of the innovation system, the resources available, and alternatives for allocating those resources are addressed. Specific strategies to bridge the gaps within and among the functions of innovation systems can be developed, such as strategies for empowering people to innovate, unleashing innovation in firms, creating and applying knowledge, applying innovation to address key development challenges, or improving policies influencing the governance of innovation systems. Benchmarks and indicators provide a means for comparative assessments of

countries, sectors, institutions, and commodities. Together, these methods support the ongoing design, monitoring, and evaluation that enable technical and institutional innovations.

POLICY ISSUES

Assessing and setting priorities for innovation involves information and learning-intensive processes. These processes constitute an organized way of thinking about how and why innovation takes place and how the key functions of an innovation system can be identified, understood, and assessed. Approaches to assessing innovation and prioritizing investments are still evolving, given the relatively recent application of the innovation systems concept to agricultural development. It is clear, however, that two important policy issues for assessing innovation and prioritizing innovation investments include appropriate financing and organization of priority setting within the public sector and the need for inclusiveness in assessment and priority-setting processes.

Supporting public sector innovation assessment and priority setting

To develop an effective assessment and priority-setting process for agricultural innovation, public policy makers and donors must insist on the allocation of sufficient time, human resources, and funding. Such a process would gain greater visibility if a unit or platform responsible for this analysis were established within the national system (see TNs 1 and 2 in module 6). This platform or unit, and assessment teams linked to it, would be closely involved with sourcing data, ensuring access to the data, and adopting

Table 7.6 Examples of Classic Indicators for Policy and Program Investments Compared with Indicators for AIS Policy and Program Investments

Classical indicators	AIS indicators
Agricultural GDP and GDP growth rate	Share of farmers who have tried/adopted some new agricultural production practice (e.g., new crop variety or livestock breed)[a]
Total agricultural factor productivity	Share of farmers who have tried/adopted some new agricultural marketing practice (e.g., pre-production contracts, collective marketing)[a]
Yields per hectare of major food staple and high-value crops/livestock	Share of farmers who have tried/adopted some new natural resource management technique (e.g., conservation tillage, soil erosion controls, water harvesting)[a]
Share of cultivable land under modern varieties	Share of agricultural firms that have tried/introduced some new product or process innovation

Source: Spielman and Birner 2008.
a. Includes use of sex-disaggregated data and disaggregated by income group (poverty).

interdisciplinary approaches to assessment, monitoring, and evaluation. International bodies such as OECD and the World Bank can enable national systems to share methodologies and knowledge. Regional forums are providing a similar opportunity (CPR, CRISP, and LINK 2011; SWAC 2005).

Policies to support inclusive priority setting

To ensure that priority setting for innovation actually happens and that it yields the widest possible range of benefits, policy makers must promote mechanisms to help farmers and other actors articulate demands, build public/private sector dialogue, and increase accountability to the local level (Sulaiman 2009; Spielman and Lynam 2010). For instance, AIS benchmarks and indicators can integrate the analysis of poverty and gender issues (Spielman and Birner 2008). Decentralized methods for assessing innovation and investment priority setting are being explored by OECD, among others (box 7.3) (see also module 1).

Box 7.3 The Local Economic and Employment Development Project

National systems of innovation are sometimes challenged by consultation at the local level when assessing institutional capacities, negotiating targets, and establishing multistakeholder processes such as boards and panels, which allow a wider group of actors to have input into national programs. Such efforts are vital, however, for ensuring incremental responsibility and decentralized decision making that strengthen local economies.

The Local Economic and Employment Development (LEED) Project (supported by the Organisation for Economic Co-operation and Development) has tracked experiences in national systems of innovation with multistakeholder processes at the local level. LEED has found that flexible funding schemes and special funds for local innovation help to build stakeholder involvement and that multistakeholder action learning enables the innovation process to continue after external funding ends.

Source: Potter 2008.

The approaches described here continue to develop. The following lessons and recommendations have emerged so far:

- ***Recognize supportive norms for assessment and priority setting.*** Assessment that involves interactions among multiple stakeholders (such as learning alliances or communities of interest or practice) requires spending time and material resources on creating environments in which values and attitudes that complement the innovation process can thrive. Incentives and recognition activities as well as communication strategies are often useful to convey innovation values and attitudes. Creating an environment for presenting, discussing, and reviewing priorities is integral to the process of innovation itself.

- ***When selecting tools for assessing innovation and prioritizing investment in AISs, determine which tools are already being used and why.*** Introduce new methods such as theory of change, analysis of innovation functions, and benchmarking on an experimental basis. Gaps in data will become apparent; they should be addressed through an action plan to collect the missing data or by focusing on the data that are available (a larger assessment suffering from a lack of data would be ineffective). When selecting tools, use experience developed by stakeholders within the system and seek recommendations from communities of practice.

- ***Develop competencies for assessment and priority setting.*** Capacity for assessing innovation and prioritizing investments is considered essential for the future of global agriculture (G20 2011b). Priority assessment of commodity innovations has received methodological attention, but technical skills for econometric analysis and participatory research can be lacking (Raitzer and Norton 2009). Human capacities are best built through effective curriculums and learning in action. The teaching of assessment and priority-setting methods cannot follow a blueprint, because valuable learning comes from working within different contexts to identify relevant concepts, definitions, and methods. Collaborative projects for sharing tools and using assessment and priority-setting methods in AISs are encouraged. In this respect, competencies such as willingness to communicate, learn, and interact with multiple stakeholders will be essential. In the end, such efforts are more likely to create engaged individuals, organizations, and networks.

- *Continue to strengthen the integration of the end-user/ farmer perspective in assessment and priority-setting processes.* The Agricultural Science and Technology Innovation systems work by CTA, KIT, CABI, and the Free University of Amsterdam is a case in point. It integrates different perspectives into a more comprehensive framework that encompasses end-user/farmer-led innovation processes, the private sector's role in value chain development, and the facilitating role of public research and extension organizations. In low-income countries, end-users may not be literate or may be marginalized in other ways. For contextual reasons, they may be difficult to engage in participatory processes. Much work needs to be done to ensure that data collected with stakeholders at the local level feeds into both ex ante and ex post assessment and priority-setting activities.

Methods for Organizational Assessments in Agricultural Innovation Systems

Regina Birner, University of Hohenheim

SYNOPSIS

This note presents methods that can be used to assess individual organizations within the innovation system, such as agricultural research and extension organizations. These assessments are not only useful diagnostic tools for planning AIS interventions but important components of monitoring, evaluating, and assessing the impact of AIS interventions. If time and resources are limited, assessments can be based on secondary data and expert interviews. More detailed assessments may involve surveys among staff of organizations in the AIS, farm household surveys, and participatory methods. From an implementation and policy perspective, it is important to create demand for assessment data to achieve sustainability.

BACKGROUND AND CONTEXT FOR INVESTMENT

Investments to improve the overall performance of an AIS often include components to improve the functioning of individual organizations, often in conjunction with components to improve the coordination between organizations and to create an enabling environment for them to innovate. For planning, managing, and evaluating investment projects to support agricultural innovation, it is essential to have diagnostic tools for assessing the organizations within the system. This note describes such tools and methods (methods for assessing an AIS in its entirety are discussed in TN 1).

Assessments can support AIS investment projects in the following ways:

- *Diagnostic assessments (ex ante).* In the planning phase of development interventions, assessments can be used to identify the strengths and weaknesses of the organizations involved in an AIS, to identify entry points for reforms, and to assess the feasibility and expected costs and benefits of planned investments.

- *Monitoring.* During the execution of development interventions, assessments are needed to monitor changes over time. In particular, assessments can help to reveal whether organizational reforms are proceeding as intended.

- *Evaluation and impact assessment.* After completing an intervention, such as a reform of a country's research organization, assessments are required to evaluate the effectiveness of the investment, to quantify its impact, and to derive lessons for future interventions.

Since the AIS approach is comparatively new, project managers face a lack of clarity about the methods that can be used for these purposes. Existing methods for organizational assessments can be adjusted, however, to take the role of organizations within the innovation system into account. This note describes a set of assessment methods and gives specific advice on how they can be used in assessing investment projects that reflect an AIS approach.

THE ASSESSMENT FRAMEWORK, OBJECTIVES, AND REFERENCE SITUATION

Since an AIS is a theoretical construct, it is important to use an *assessment framework* that defines the elements and relations of the AIS in which a given organization is to be assessed. Several assessment frameworks are available (see TN 1). This note uses the framework developed for the World Bank by Spielman and Birner (2008) for illustration, because it identifies the types of organizations involved in an AIS and their relations (figure 7.1). Although this note focuses on one framework, the methods discussed can be applied to other assessment frameworks.

In the context of investment projects, assessments may be carried out (1) at the level of the AIS as a whole, (2) at the level of the innovation system for specific commodities or value chains, or (3) at the level of different organizations

Figure 7.1 Conceptual Diagram of a National Agricultural Innovation System

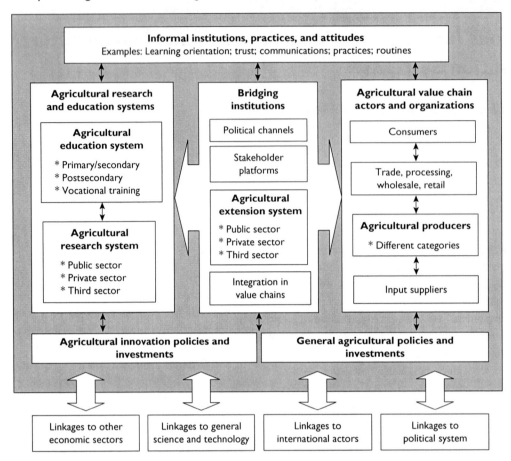

Source: Spielman and Birner 2008.

within the innovation system (organizational assessments). In conducting an assessment at the organizational level, a primary task is to *clarify the objectives of the assessment.* Box 7.4 presents a number of questions related to the performance of an organization within an AIS. The questions illustrate the range of objectives that an organizational assessment may need to consider to gain a comprehensive understanding of that performance.

In addition to clarifying the objectives of an assessment, it is also important to identify who will use the results. Assessments may be carried out by organizations that fund, or intend to fund, organizations within the innovation system. However, assessment results need to be fed back to the management of the organizations involved to stimulate institutional learning and change. Moreover, as discussed below, assessments benefit from the involvement of staff members as well as users of innovation-related services.

Another primary task is to *select the standard or reference situation* against which an organization's performance

within the innovation system can be assessed. The following options may be considered:

- **Benchmarking.** One approach is to compare the organizations within the system in a particular country with those of other countries, typically countries in the same region or countries that are otherwise comparable. If quantitative indicators are used, this approach is known as "benchmarking."
- **Changes over time.** Another approach (which can be combined with benchmarking) is to compare the performance of the organization at different points in time and determine whether it improved or deteriorated.
- **Policy goals, organizational objectives, or standards.** A third approach is to compare organizations against goals that have been set by policy makers, by managers of the respective organization, or by another entity, such as a donor organization or certification or accreditation agency.

Box 7.4   Questions That Illustrate the
           Range of Objectives in an
           Organizational Assessment

A number of questions illustrate the varied objectives of an assessment to understand how well a certain organization—for example, the public research institute responsible for a certain commodity—performs within an AIS:

- What internal management factors influence the organization's performance?
- Which factors in the organization's environment influence its performance?
- What are the most promising types of interventions to improve the organization's performance and its contribution to the innovation system?
- Is the evolving nature of the agricultural sector placing new demands on the organization?
- What is the role of the organization within the innovation system? Is this role still relevant? How may it need to change?
- Are the organization's patterns of partnership and collaboration sufficient to maintain its relevance in the evolving agricultural sector? What new linkages might be required?
- What learning-based mechanisms for enhancing organizational performance are in place?
- How can the organization's contribution to the overall innovation system be improved?

Source: Author.

SELECTING ASSESSMENT METHODS AND OBTAINING DATA

Planners, analysts, researchers, and evaluators can use a variety of methods to assess organizations within an innovation system. Assessments may be based on quantitative or qualitative methods and may be derived from different disciplinary backgrounds, such as economics and its branches (public economics and New Institutional Economics, for example), organizational sociology, public administration, and political science. The assessment can call upon existing data and statistics to develop a picture of the national context and the main contours of the innovation system in which the organization is situated. Data may also be available on the organization's performance in relation to specific indicators, such as the number of staff publications,

quantities of inputs distributed, numbers of trainees or students receiving instruction, and so on. The assessment will need to generate much of its own information, however, and much of it will be qualitative. Examples include information on the organization's patterns and strength of collaboration with other organizations in the AIS, the relevance of the organization's roles, and the existence of learning-based performance management arrangements.

Criteria for selecting assessment methods

Before describing assessment methods in detail, it is useful to consider the criteria that influence the choice of an appropriate method:

- *The scope of the assessment.* Even though the assessment is looking at the performance of an individual organization, from an innovation systems perspective it is important to take into account specific linkages and coordination mechanisms.
- *The existing data and knowledge.* The data and knowledge about an organization that are available (and accessible) have a large influence on the choice of the assessment method and the amount of primary data that must be collected. Organizations in the innovation system for agriculture, such as extension organizations, typically have their own reporting systems. It is helpful to examine whether the data generated from such reporting systems will be useful for the assessment.
- *The time and resources available.* The choice of an assessment method is also determined by the time and the resources available for the assessment. Ideally, there should be a match between the purpose of the assessment—for example, to determine the level of the planned investment—and the time and resources available for ex ante, ongoing, and ex post assessments.

The role of preliminary AIS assessments

Before embarking on an organizational assessment, it is also useful to develop an understanding of the AIS in which the organization operates. It is important to keep these preliminary assessments of the AIS (mapping the AIS and reviewing expert opinion about the AIS) in proportion to the main task that lies ahead, which is the organizational assessment. These system assessments provide context, "locate" an organization within the wider innovation system, and highlight its relationships throughout the system. A further function of these system assessments is that they engage system stakeholders in a dialogue about the role and performance of a specific organization with the innovation system.

MAPPING THE AIS. Even if an assessment ultimately focuses on only one organization within an AIS, an organizational mapping of the AIS as a whole is useful to gain a clear understanding of the environment in which the focus organization operates (for example, it can help to answer some of the questions raised in box 7.4). The outcome of organizational mapping is a diagram that displays the essential organizations in the innovation system and their relations to each other. The mapping process may take different forms, but it needs to be based on an assessment framework. If time and resources are limited, the major organization within the AIS usually can be identified by compiling information from the documents and literature available and interviewing experts based on a semistructured questionnaire.

For example, if the goal is to map the national innovation system for dairy production and the assessment framework displayed in figure 7.1 is used, the analyst will have to compile information on the following questions:

- Which research organizations deal with dairy production?
- Which education organizations provide training for dairy production at different levels (diploma, graduate, postgraduate)?
- Which extension organizations provide advice on dairy production?
- Who are the major players in the value chain (dairy processing companies, for example)?
- Which organizations of dairy farmers, such as dairy cooperatives, exist?

The analyst can construct a diagram based on the information collected and use it to collect further information on innovation system actors and organizations, especially their roles and interactions. Some mapping techniques make it possible to visualize the innovation system during the interview process; for an example, see the description of Net-Map in IAP 2.

CONDUCTING EXPERT SURVEYS. The collection of information from experts is useful for gathering valuable information about an innovation system in a comparatively short period. It allows the analyst to draw on the comprehensive knowledge gained by professionals who have long experience in the AIS. The list of experts to interview can be derived during a mapping exercise (discussed

previously) and by using the "snowball system"—that is, by asking respondents to identify other persons who should be interviewed.

Information can be collected from experts in different ways. One alternative is to conduct semistructured interviews, using an interview guideline. It is also possible to use a questionnaire for an expert survey and ask respondents to score the various actors in the innovation system on a scale (of one to four, for example) regarding their effectiveness, responsiveness, accountability, organizational performance, and other criteria. Even though the scoring results will reflect a subjective assessment of the actors involved, this approach yields useful information. Similarly, governance indicators are often based on expert assessments, and if a standardized approach is applied, such data can be used as indicators to monitor changes over time or make comparisons across countries. For an example of how an expert survey was used in Ethiopia, see Spielman and Kelemework (2009).

METHODS FOR ASSESSING ORGANIZATIONS WITHIN AN INNOVATION SYSTEM

This section introduces methods that can be used to assess a specific organization within an AIS, such as an agricultural research institute, agricultural training center, or agricultural extension organization. The methods include staff and farm household surveys as well as methods derived from business administration to assess organizational performance.

Organizational performance assessment

The business administration literature describes a wide range of methods that organizations can use to assess and manage their performance. One approach that is particularly relevant for innovation systems is the Organizational Performance Assessment (OPAS), developed for agricultural research institutes by the former International Service for National Agricultural Research (ISNAR) (Peterson, Gijsbers, and Wilks 2003).

OPAS was first tested in 1996–97 at the research institutes of the Council for Scientific and Industrial Research (CSIR) in Ghana and later adapted and used by national research organizations in Benin (Institut National des Recherches Agricoles du Bénin) and Uganda (National Agricultural Research Organisation) (Peterson, Gijsbers, and Wilks 2003, 8).

In OPAS, organizational performance is defined as "the ability of an organization to use its resources efficiently and to produce outputs that are consistent with its objectives and relevant for its users" (Peterson, Gijsbers, and Wilks 2003, 1). Box 7.5 displays the elements of OPAS and explains the relations between them.

OPAS has been designed for assessments conducted by managers and staff of the respective research organization with the assistance of one or two external facilitators. The assessment has two major elements: an output assessment and a management assessment. Each component follows a number of clearly defined steps. Scores (which may be weighted according to organizational priorities) are applied, making it possible to monitor progress over time. An assessment of the output trends of research organizations shows that the organization considerably increased its public services such as dissemination and training events but that its technology output fell (Peterson, Gijsbers, and Wilks 2003, 18).

Box 7.5   Elements of the Organizational Performance Assessment

Agricultural research organizations use **resources** and inputs (funds, personnel, equipment, and facilities) to undertake their **research operations** in order to produce **outputs** (agricultural technologies and services) for the benefit of farmers, agro-industries, and other users. The **outcomes** (or consequences) of adopting or applying these outputs are measured by their effects, positive or negative, on such factors as production costs, yields, and use of natural resources. In this sequence of events, which is illustrated in the upper part of the diagram, **performance assessment and** **feedback mechanisms** are required at different levels to ensure that research organizations plan their resources efficiently and produce relevant and useful outputs.... An underlying assumption in organizational performance is driven by a number of **critical management factors**, as indicated in the lower part of the diagram. Through a **periodic assessment** of these factors, managers can determine if appropriate mechanisms and procedures are in place and functioning, and can take steps to correct management deficiencies that contribute to poor (or lower) organizational performance.

Figure B7.5   Diagram of Organizational Performance Assessment

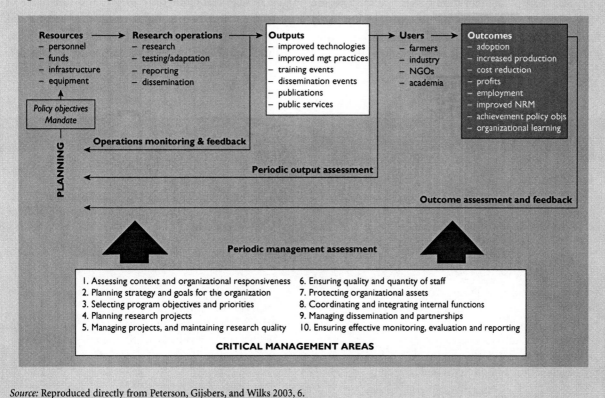

Source: Reproduced directly from Peterson, Gijsbers, and Wilks 2003, 6.

Box 7.6 displays sample questions for human resource management (management area 6 in figure B7.5 in box 7.5). As in the output assessment, scores can be applied to each question, which makes it possible to monitor changes over time.

From an innovation systems perspective, it is useful to expand the OPAS approach and include indicators that capture the relation of the organization to be evaluated with other organizations in the AIS, both at the level of the output assessment and the level of the management assessment. For example, at the output level, research organizations may include scores for collaboration with agricultural extension and education organizations, and vice versa. At the management level, critical management area 9 already refers to "managing dissemination and partnerships" (critical management areas are listed in figure B7.5, box 7.5). Indicators in this area may be expanded to include all types of partners in the innovation system, as indicated in figure 7.1. Such data could then also be used as measurable indicators of the performance of different AIS members. These indicators could also be considered in external evaluations, which may increase the incentives for organizations not only to improve their individual performance but to improve their performance as a member of a wider innovation system.

To reflect more of an innovation systems perspective, the OPAS can also be modified in the area of learning-based performance management. Successful organizations continuously update and reframe their relationships with the rest of the system and the competencies, roles, and ways of working that the wider system demands. Organizations use a suite of methods, referred to as "institutional and organizational learning" (see also module 1 and TN 4 in module 4) to enable this continuous adaptation and updating to take place (box 7.7).

Surveys among the staff of organizations

Another organizational assessment method, which can be combined with OPAS, is a survey of an organization's staff members. These surveys are particularly useful in organizations that have large numbers of field staff, such as public agricultural extension systems. They can provide in-depth information about the organization's capacity and staff incentives, but they need to be carefully planned, as they require genuine support from management. As in other surveys, the anonymity of the respondents has to be ensured, and interviewers have to be careful to create an atmosphere in which respondents are willing to talk freely about their assessments, especially when sensitive issues are raised, such as issues of political interference. Moreover, it is important to pre-test the survey instrument with a group of respondents who reflect the diversity of the AIS. This step is important not only to test the suitability of the instrument but also to build confidence among staff. In designing the questionnaire, the trade-off between simplicity and capturing all relevant details must be considered.

Box 7.6 Sample Question Set for Assessing Human Resource Management Performance

- To what extent does the organization maintain and update staff information (e.g., biodata, publications, projects)?
- To what extent does the organization plan and update its staffing, recruitment, and training requirements?
- How effectively are staffing, recruitment, and training plans linked to program and project needs?
- How effective are selection procedures (for management, scientific, and support posts) in terms of objectivity and transparency?
- To what extent is training based on merit and on organization and program objectives?

- How effective are mechanisms to promote a good working environment and high staff morale?
- How effective is the performance-evaluation process for research staff?
- How effective is the performance-evaluation process for nonresearch (management, administrative, and support) staff?
- How effective are reward and sanction processes, in terms of motivating staff?
- How effectively does the organization compete with the private sector in providing salaries and benefits that attract and retain quality staff?

Source: Reproduced directly from Peterson, Gijsbers, and Wilks 2003, 22.

Box 7.7 Organizational Learning and Institutional Change

Institutional or organizational learning is the deliberate and ongoing process in which information from research and evaluation activities and outcomes feeds into a reflective analysis of what has worked and not worked in an institution. In turn, the lessons from such reflective analysis inform decisions about future directions for the organization. Leadership, incentives, resources, and flexibility within the organization's routines are required for this process to work.

The Institutional Learning and Change initiative defines a **learning organization** as an organization with a culture that supports this kind of analysis and change. The term "institution" is used instead of "organization" when referring to the learning process that takes place across organizations and among a diverse set of people involved in research and evaluation activities. Shambu Prasad, Laxmi, and Wani discuss an "unusual coalition" between an international research center (the International Crops Research Institute for the Semi-Arid Tropics) and a private Indian donor (Tata), in which the research center established new institutional mechanisms, both internal and external. The internal mechanisms involved new ways of organizing work among site coordinators and activity coordinators who have to seek input from each other. External mechanisms included a new steering committee as well as state and district committees with multiple stakeholders who were engaged in the use of new tools such as actor-linkage mapping. Module 3 examines additional cases of organizational learning and institutional change.

Source: Author; Watts et al. 2003; Shambu Prasad, Laxmi, and Wani 2006.

In designing staff surveys, it is also useful to take gender into account. For example, a survey may include specific questions on career opportunities and constraints for female staff. Moreover, the data for male and female respondents can be analyzed separately. Box 7.8 describes the constraints to agricultural innovation that were identified in a survey of agricultural extension agents in six districts of Ghana. (See IAP 3 for an example from Peru.)

Farm household surveys

For organizations that deal directly with farmers, such as agricultural extension organizations, surveys among farmers—the clients of the organization—are essential for an ultimate assessment of organizational performance. At the same time, farm household surveys can provide important information about the performance of the AIS as a whole, which means that they can also form an important component of a system-level assessment. Farm household surveys are the most expensive and time-consuming approach to collecting data about agricultural innovation, but they provide particularly relevant information, especially if secondary data on farm households that capture aspects of agricultural innovation are not available. Box 7.8 describes how this assessment method was used in Ghana.

If a survey that includes agricultural households is planned for another purpose, it may be possible to include questions on the performance of the organization to be evaluated and on other aspects of agricultural innovation. If a survey is planned specifically to collect information on the assessment of an organization in the AIS, it will be useful to include questions on outcome indicators, such as the adoption of innovations, as well as questions on household access to the services provided by the organization (such as extension services) as well as household satisfaction with those services. It may also be useful to include information on other aspects of the innovation system, such as access to agricultural inputs and complementary services as well as marketing opportunities. It will often be useful to collect such data separately from male-headed households, female-headed households, and female spouses in male-headed households.

When farm household surveys are not possible and secondary data are limited, Participatory or Rapid Rural Appraisal methods will be useful, since assessments of organizations in the AIS, or of the system as a whole, should take the farmers' perspective into account.

POTENTIAL BENEFITS

The key challenge facing agricultural research institutes, development organizations, and enterprises is to maintain

Box 7.8 Using Staff Surveys in Assessing Agricultural Extension Services in Six Districts of Ghana

To assess the performance of the organizations providing agricultural extension services in Ghana, a team from the University of Ghana–Legon and the International Food Policy Research Institute carried out an assessment that involved surveys among agricultural extension providers (70 interviewees) as well as agricultural household heads (1,168) and their spouses (613).

The assessment showed that the public extension service remained the main provider of extension information (an important finding, given the crucial role of advisory services in agricultural innovation). Only one respondent had received a visit from a nongovernment organization providing extension services, and nongovernmental organizations organized only 4 percent of group meetings on extension.

The assessment also showed that female household heads as well as female spouses in male-headed households had very low access to agricultural extension services even though women play an important role in Ghana's smallholder-based agriculture, and even though Ghana has a special program for Women in Agricultural Development.

The survey also showed that less than 12 percent of household heads and less than 6 percent of spouses had adopted a new technology in the previous two years. The implication is that considerable constraints prevent male and female farmers from innovating.

The survey among agricultural extension agents revealed that female extension agents were more effective in reaching female farmers than male extension agents, but only 14 percent of extension agents were women. Extension agents as a group identified the lack of transport and access to credit as major constraints on farmers (see the figure). An inadequate number of extension staff was seen as the least important constraint.

The assessment revealed management problems as well. Extension agents had limited incentives to perform, priorities and targets were not set, and training opportunities were limited. The assessment identified entry points for interventions to improve the performance of this key organization in Ghana's agricultural innovation system.

Source: World Bank and IFPRI 2010.

their relevance and performance in the rapidly evolving agricultural sector. Organizational assessments provide a way of investigating the capacities and outcomes of an organization and identifying ways of strengthening different aspects of capacity to improve relevance and performance. These actions improve the effectiveness of both public and private investments in the development of organizations within an AIS. If similar assessment tools are used in different countries, additional benefits can be realized from cross-country comparisons and benchmarking.

POLICY ISSUES

Policy issues related to organizational assessments range from ensuring that assessments are properly resourced, that local capacity to conduct assessments is developed and sustained, that assessments reflect environmental and social considerations, and that wide support develops for using the results to improve performance. Policy responses to these concerns include the following:

- *Commit resources to organizational assessment.* The benefits of organizational assessments are realized only if they are conducted on a regular basis to maintain an organization's relevance in the AIS and to monitor progress over time. Conducting assessments regularly can be a considerable challenge, especially if organizations rely on external donors to fund this activity.
- *Build local ownership for assessment.* The sustainability of an assessment regime can be improved by generating buy-in from local organizations, such as the ministries in charge of agriculture, science, and technology.
- *Build local capacity for assessment.* The sustainability of an assessment regime can also be improved by building and institutionalizing local capacity for conducting organizational assessments. It may also be useful to involve regional networks of agricultural research organizations, such as the Forum for Agricultural Research in Africa (FARA), or regional economic communities, such as the Economic Community of West African States (ECOWAS), in the use of assessment

tools, especially if benchmarking approaches are used (see the final point).

- **_Emphasize social and environmental considerations in designing assessment methods._** For example, assessment methods can be designed to capture the extent to which individual organizations in the system are biased towards large-scale farmers. As shown in box 7.8, assessment methods are also useful to assess the gender dimensions of the innovation system.
- **_Create regional guidelines._** International and regional organizations may take on the role of developing guidelines for harmonizing assessments and publishing results, especially if benchmarking approaches are used.

IMPLEMENTATION ISSUES

To some extent, implementation issues for organizational assessments reflect the policy issues just mentioned:

- **_Timeliness._** It is important to conduct organizational assessments regularly and in a timely fashion, as they are a key mechanism for organizations to maintain their relevance within an innovation system. It is preferable to use frequent organizational assessments to encourage continuous incremental change rather than to rely on infrequent major organizational reforms (see, for example, the CGIAR as an example of major reform; www.cgiar.org).

- **_Resource implications._** As with all assessments, in organizational assessments carefully consider the resources and capacity required to undertake the assessment. Assessments involving household surveys are particularly resource-intensive. Depending on the country and region, the costs may range from US$25 to US$120 per household. Yet certain types of data about an organization's performance in the innovation system can be collected only through surveys, such as data on male and female farmers' access to extension services.
- **_Inclusiveness._** To create "buy-in," it is important to generate "demand" for assessment data at the organizational, national, and regional level. It may be useful, for example, to work with parliamentary committees in charge of agriculture or science and technology. In democratic systems, such committees may have considerable latitude to use assessment tools to hold the government accountable for the performance of an innovation system and its organizations. Likewise, it may be useful to involve farmer organizations, which can also play an important role in creating accountability.
- **_Choosing local partners for assessments._** As indicated, it is essential to build local capacity for assessments. Potential partners may include analytical units within the ministries in charge of agriculture or science and technology, university departments that work in this field, as well as think tanks and local consulting companies.

Foresighting Investments in Agricultural Innovation

Anthony Clayton, University of the West Indies

SYNOPSIS

Individuals, firms, and governments have a common interest in gaining a more accurate sense of the future to identify potential achievements, successes, and failures; discern new opportunities; or reduce risk. New, future-oriented evaluation methods are needed to complement current methods, which are largely ex post—in other words, they account for past outputs, outcomes, and impact. Given the impossibility of seeing into the future, the only solution is to gather and analyze information to think about and prepare for the future. Tools such as foresighting can make an important contribution to this process by clarifying a country's position with regard to strengths, weaknesses, threats, challenges, and opportunities, focusing attention on longer-term issues (including difficult institutional and political issues), and securing a sufficient level of commitment from stakeholders to enable the necessary processes of reform, restructuring, transformation, and change.

BACKGROUND AND CONTEXT

Most decisions are based on implicit assumptions about the future. People depositing funds with a bank assume that they will be able to withdraw their capital when it is required. Investors fund a new business venture because they anticipate profits. When farmers expect good demand for their crop, they may plant an extra field. Assumptions such as these are both rational and functional; few decisions can be made without assuming that a significant degree of stability and continuity will prevail. Because it is also true that these assumptions sometimes prove incorrect, it is important to assess the balance between risk and reward involved in each decision and to gauge the reliability of the individuals and the strength and trustworthiness of the institutions on which the plan depends.

Individuals, firms, and governments therefore have a common interest in gaining a more accurate sense of the future, either to improve their chances of making a successful investment or to reduce the risk of loss. The problem is that the future for any given sector, nation, or individual will be determined by a complex combination of interacting variables that cannot be anticipated with precision. The response is to identify better ways to think about and prepare for the future (Postrel 1998). Foresighting is a means of gathering and using information to think strategically about the future, including the future of agriculture (de Lattre-Gasquet 2006). Foresighting assumes that the future is not predetermined or even predictable but that it will be influenced by choices made today.

FORESIGHTING TOOLS AND APPLICATIONS

Foresighting is a strategy for change, rather than a strategic plan, although it often feeds into a strategic planning process. Foresighting is both a process and a set of tools for managing and communicating knowledge, setting priorities, coordinating goals, and encouraging innovation in science and technology. It involves an iterative and interactive process of systematically exploring possible future economic and social dynamics, including factors such as science, technology, institutions, environment, and development options. The aim is to identify strategic areas of investment that will yield the greatest economic and/or social benefits (Rutten 2001). From an innovation systems perspective, organizations pursue foresighting exercises to (Popper et al. 2007): encourage strategic and future-oriented thinking; support innovation strategies and priority setting; identify research/investment opportunities; generate visions and images of the future; cope with "grand challenges" facing the economy, society, and environment; and promote public debate and trigger necessary actions.

Box 7.9 Foresighting to Transform Ireland's Agrifood Sector (Teagasc 2030)

Teagasc, established in 1988, is the national body providing integrated research, advisory, and training services to Ireland's agriculture, food industry, and rural communities. The Teagasc 2030 foresight exercise (launched in 2006, ending in 2008 with an international foresight conference) sought to establish a broadly shared vision for the Irish agrifood industry and rural economy in 2030. That vision would enable Teagasc to meet science and technology needs in the short, medium, and long term.

The foresighting process. Teagasc 2030 was overseen by a steering committee of national and international representatives from government, industry, and universities. The committee was assisted by a foresight panel of experts from Teagasc, other government departments, state bodies, universities, farming and rural organizations, and food industries. A foresight working group from Teagasc's research, advisory, and training directorates completed background papers on drivers of change and possible future scenarios and handled day-to-day running of the project. The knowledge base was developed in workshops, other events, and consultations with stakeholders and international experts.

The scenarios. Involving elements of imagination but drawing on likely developments in the agrifood sector, rural economy, and world over the next 20 years, Teagasc 2030 developed five scenarios and described their effects on the organization:

1. *Ireland—The Food Island.* The value-added food sector in 2030 has many new elements such as convergence of the food and pharmaceutical industries.
2. *Globally Competitive Farming.* In 2030, a diversified agrifood sector competitively produces milk, beef/sheep, and tillage crops.

3. *Energy Squeeze Fuels Agriculture.* With oil production declining steadily, agriculture is vital to global food and energy security.
4. *A European Agriculture.* This scenario addresses the socioeconomic aspects of the rural economy and the consequences of a European economy partly isolated from the rest of the world by tariffs and restricted trade.
5. *Sustainable and Rural.* Sustainability, climate change, and environmental security have precedence: The 2030 bioeconomy delivers competitive agri-environmental products and services.

Lessons and response. Based on the scenarios, strategic and operational responses emerged:

- *For the sector:* Promote knowledge generation and dissemination, learning, and problem solving. Policy drivers at the European Union and national level create new markets and opportunities.
- *For farms, firms, and policy makers:* Adapt quickly to changing circumstances, generate added value, and support innovation.
- *For Teagasc:* Address the new challenges and needs facing the Irish agrifood knowledge system. Excellence in supporting science-based innovation will depend on building organizational capabilities in leadership, partnership, and accountability. Teagasc will establish a new technology transfer service for food companies, strengthen investment in biosciences, enhance the depth of its scientific effort, and continue upgrading its educational programs to the highest international standards.

Source: Teagasc, www.teagasc.ie.

The next sections of this note describe elements of the foresighting process, how to select an appropriate foresighting tool, the use of complementary tools, and the development of alternative scenarios. They conclude with examples of how these processes unfolded in agricultural foresighting in Ireland (box 7.9) and Jamaica (box 7.10).

Foresighting process

Foresighting is often spearheaded by individual organizations or collaborating groups who have a stake within the innovation system (for example, by strategic partnerships or innovation councils, among others). The organization

Box 7.10 Foresighting for Jamaica's Sugar Industry

Objective and process. Jamaica's sugar industry was the largest employer of agricultural labor and used the most (and much of the best) arable land, but it depended on a European Union trade regime that was being phased out. Jamaican sugar was not competitive and would have to undergo profound restructuring to survive. A research program at the University of the West Indies attempted to address these issues through linked foresight exercises involving government, academics, and industry from 2003 to 2010. Following a 2003 Delphi study that identified drivers of change and a 2004 foresight workshop to identify land-use scenarios, by 2009 public and private partners had developed an integrated assessment that generated more accurate predictions than official forecasts. An integrated policy development project in 2010–11 identified options for improved policies, using geographic information system mapping as the basis for a national spatial plan. The cost of the seven years of foresighting was an estimated US$150,000.

The scenarios. In early 2008, a diversification plan involving privatization and the manufacture of ethanol presented three scenarios:

1. **Diversification would succeed.** The area used to produce cane for ethanol would increase significantly. The contribution to mitigating climate change would offset probable negative consequences for river and coastal water quality.
2. **Diversification would fail.** Much land would become available for other purposes (housing, tourism, forestry) or revert to scrub. Some options would be environmentally positive, but much

would depend on how the change process was managed.
3. **Multi-objective optimization** demonstrated a possible solution that could achieve a range of developmental goals. Extensive, low-value agriculture would shift to intensive, high-value agriculture, increasing revenue, profits, and skill transfer while reducing environmental impact.

Outcomes and lessons. Following the withdrawal of Brazil's plan to invest in Jamaica's ethanol industry and an even more acute budgetary crisis in 2010–11, the Government of Jamaica chose to sell the sugar industry to a foreign company. This outcome illustrates how pressing short-term needs, in conjunction with international uncertainties and domestic politics, often entail the loss of strategic direction and potential future prosperity.

The foresighting exercises resulted in three significant advances. First, the identification of future scenarios helped to establish that better future outcomes were possible. Second, the process helped to identify the weaknesses and vulnerabilities in current policies. Third, the dialogue between officials, academics, and industry representatives was genuinely useful. Yet the exercises could not resolve deep core weaknesses in governance. Most decisions are still taken on a sectoral basis without considering cross-sectoral effects. Ministry officials participated in the debates, but not ministers. No sufficiently powerful group exists to address the large disparities in wealth, influence, and power in Jamaica and arrive at an optimal outcome. Vested interests continue to dominate the informal networks where key decisions are made.

Source: Author.

initiating the foresighting exercise does not necessarily facilitate the process. A common procedure is to establish a steering committee for the exercise and a foresight working group to manage implementation. This approach was used for a foresighting exercise for agriculture in Ireland (box 7.9). For the Jamaican sugar industry, a team of foresight experts was contracted to help users define the steps and tools in a foresighting exercise to meet the needs of different users (in general, users might include firms, industry associations, multinationals, or governments).

Rutten (2000) explains that there are no blueprints for organizing a foresighting exercise. The process itself has four key stages, however:

1. Bring together a wide range of information resources and key stakeholders to discuss and define the core objective(s) of the foresighting exercise.
2. Identify and engage a wider group of stakeholders (typically involving both public and private sector organizations) in the process.

3. Establish a foresighting task force (or a steering committee with an implementing working group) with key stakeholders represented, ideally with the support of a foresight expert.
4. Design and develop the foresighting exercise. Select the tools for the strategic analyses, and set the timelines for completion.

This process is not a linear series of steps. At each of these four stages, new information may be injected, additional stakeholders may become involved (they can sometimes include criminal or vested interests), and organizational changes may affect the level of human and financial resources allocated to the exercise. The exercise will need to remain sufficiently flexible to respond to these challenges and changes while remaining productive. The outputs of the process should include the following:

- **Determining** the economic, social, environmental, and institutional factors that might encourage or inhibit innovation (either in a particular country, or among firms, or in government).
- **Clarifying** the position of a country (or firms or government) with regard to its strengths, weaknesses, threats, challenges, and opportunities, usually by focusing attention on the longer-term issues.
- **Securing** a sufficient level of commitment from stakeholders to enable the necessary processes of strategic reform, restructuring, transformation, and change.

Selecting the appropriate foresighting tool

As summarized in table 7.7, three key strategic planning tools are used to develop future-oriented analyses in the foresighting process:

- **Technology roadmapping** identifies key trends in the market and clarifies those trends and their relation to organizational goals; then technological and managerial decision-making occur to achieve the preferred future.
- **Delphi studies** review significant trends in relevant areas (such as emerging scientific and technological opportunities, needs for education and training, and so forth) and identify the most plausible outcomes, plus any associated threats and opportunities.
- **Foresighting and backcasting exercises** identify organizational and institutional drivers of change plus their

interactions, clarify "known unknowns," assess vulnerability to events, identify possible outcomes, backcast to the present day, and build a strategy for managing change.

Examples of online toolkits for foresighting are listed in the "resources" section of this note, following the references.

Complementary tools in the foresighting process

Aside from the three major foresighting tools, complementary methods of information gathering can be used in the process, depending on timelines and tasks assigned within the task force or to the foresight expert:

- **A horizon scan** examines the external environment for potential threats and opportunities or early signs of disruptive technological change.
- **A decision tree** is developed by constructing a logical sequence of pertinent questions, such as "If this plan fails, what are our other options?"
- **User requirements capture.** When a particular group (such as consultants or a local elite) has all of the relevant information, they may be able to control the agenda and determine the answers. A user requirements capture process helps to forestall elite capture and information asymmetries. The information base for all users is developed, displayed, discussed, and modified in an iterative process, often using graphic-rich software that allows data to be overlaid in layers (a GIS is one example).

Foresighting and defining alternative scenarios

Scenarios (different possible visions of the future) can be used to formulate long-term policy, institutional strategy, and research programs (Johnson and Paez 2000). In a foresight exercise, a process of defining and describing scenarios is used to explore the way that choices made today will lead to alternative futures. Scenarios are usually encapsulated in brief, illustrative descriptions of possible future states of a system. A common technique is to develop a small set of alternative scenarios, which is helpful for imagining, structuring, and analyzing different possible futures. Scenarios can range from the probable (most likely) future to possible best-case and worst-case futures. The storylines are discussed widely and critiqued.[1] Ideally, like the overall foresighting

Table 7.7 Foresight Tools

Characteristic	Foresight tool		
	Roadmapping	Delphi	Foresight/backcasting
Purpose	Identify key trends in market, clarify organizational goals, make technological and managerial choices to achieve preferred future	Review significant trends in relevant areas and identify most likely outcome, plus any associated threats and opportunities; can be more accurate than individual consultation and more reliable than statistical groups made up of noninteracting individuals whose judgments are aggregated	Identify drivers of change plus interactions, identify "known unknowns," assess vulnerability to events, identify possible outcomes, backcast to present day, build strategy for managing change
Typical user	Individual firms	Industry associations	Large multinational corporations, governments
Typical time horizon examined (years)	1–10	5–20	10–50+
People involved and numbers	Experts and decision makers; 10–20	Experts (different disciplines); 30–100+	Experts and stakeholders; 20–50
Type of process	1 day introduction, 1 day run-time, monitoring and implementation	Coordinated, iterative discussion of expert judgments on defined issues; responses are kept anonymous but fed back to group, clarified, ranked and reevaluated in a minimum of 3 rounds, 3 months' preparation, 3 months' run-time	Facilitated discussion, present "what-if" challenges and counterfactuals, 3 months' preparation, 2 days' run-time; a large national exercise will require 1 year's preparation, 2 years' run-time
Role of foresight expert	Introduce concept	Coordinate, derive scenarios	Facilitate, challenge, manage process
Key challenges	Getting experts and decision makers to engage, overcoming organizational inertia	Identifying and recruiting the panels, managing the process	Identifying and recruiting key stakeholders, facilitating the process, maintaining momentum
Key questions to guide information gathering/analysis	– What are our core technologies? – Could we use them more effectively? – Could an innovation make our technology obsolete? How quickly could we adapt? – What are the trends in our sector? Could a new competitor or a new market emerge? – What is our competitive advantage? – What are the priorities for maintaining, upgrading, or replacing our core technologies? – What resources will be required to update our technologies and upgrade our skills?	– What is the issue to be addressed? – What is important about the issue? – What are possible future scenarios for this issue? – What views create tensions on the issue?	– What are the key assumptions underlying this plan? – What is the "worst case" situation? – What if this plan doesn't work? – What is the contingency plan?
Examples	Institute of Grocery Distribution (UK) used roadmapping to identify food production issues that could affect the food chain and to facilitate a more considered introduction of new technologies in the future on behalf of its members, who were major retailers and food and agricultural research institutes (IGD 2003)	Rikkonen, Kaivo-oja, and Aakkula (2006) described the use of Delphi expert panels in the scenario-based strategic planning of agriculture in Finland; Stewman and Lincoln (1981) conducted a Delphi study on the likely timeframe for expected breakthroughs in the biological sciences from recombinant DNA (RDNA) research, with emphasis on basic knowledge and three applied areas: agriculture, industry, and medicine	De Lattre-Gasquet (2006) examined three case studies on the use of foresighting in agricultural research and development priorities: one at the commodity level (cocoa), one at the level of a national system (Dutch agriculture), and one at the level of an organization (the International Food Policy Research Institute's 2020 Vision)

Source: Author.

exercise, scenarios are updated with further assessments of the environment, drivers of change, and likely interactions between system variables in the progression from current conditions to a future state. Boxes 7.9 and 7.10 provide examples of scenarios developed within wider foresighting in Ireland and Jamaica; IAPs 4 and 5 provide examples from India and Chile.

POTENTIAL BENEFITS

Attempts to prepare for the future can generate a range of benefits. The general advantages of a strategic planning process are that it can help any organization assess its threats and opportunities, clarify the issues, determine its priorities, and integrate all of the important variables into a single coherent plan. This is true irrespective of the quality of the plan that results. A strategic planning process also helps to identify any weaknesses (such as insufficient capital, inadequate technical capacity, inappropriate management, or gaps in the supply chain) that will have to be addressed and thereby clarifies the priorities for borrowing and investment, hiring and firing, and R&D.

The advantage of future-oriented strategic planning is that it facilitates more objective consideration of today's real problems. The foresight exercises may depoliticize and decontextualize the problems, which then allow the problems to be addressed more realistically. Thinking about forces that will shape the future, such as demographic trends, can indicate the future pattern of market demand (for example, for food, energy, water, and housing) and the need for investments in new technologies to meet that demand (see, for example, Clayton and Staple-Ebanks 2002). A foresight approach, using techniques such as scenario planning and integrated assessment to incorporate social, economic, and environmental factors, can ultimately identify and present the best available, optimal, or even win-win outcomes, which different organizations and interest groups may find acceptable. Finally, many planning exercises are flawed because of information asymmetries. A possible solution to this problem is a user requirements capture process, which can be built into a strategic planning exercise. This process involves two or three iterations in which users are shown possible outcomes, give their feedback, and refine the model.

POLICY ISSUES

The chief policy issues related to foresighting are related to reducing the risk of failed policies; addressing cultural,

institutional, and political barriers; providing sufficient resources to implement the results of foresighting exercises; and ensuring that policy change can be sustained.

Reduce the risk of costly policy failures

For poor and developing countries, the primary policy issue in foresighting and strategic planning is that these exercises can reduce the risk of policy failures with high human and developmental costs. One of the most important differences between rich and poor countries is that the cost of a policy failure in a rich country is usually in terms of its opportunity cost—the loss of the wealth that might have otherwise been generated. Although poor and developing countries have a greater need to prepare for the future, most have just a small fraction of the capacity (the skills, knowledge base, access to information networks, and so on) that rich countries deploy on such exercises. For this reason, it is exceptionally important for poor countries to use policy tools that are effective, robust, and cost-effective.

Address barriers to progress and provide sufficient resources for foresighting

Strategies for change, including the use of foresighting, usually have to address issues such as institutional culture and local politics. As the example of the Jamaican sugar industry suggests, it is often important to address the political, cultural, or economic factors and institutional policies that seriously impede progress, because development trajectories can become locked in by the real or perceived cost of developing alternatives, the reluctance to write off sunk expenditures, or a lack of relevant skills.

Managing in the present while preparing for the future often involves gathering new sources of information, establishing where and when change is needed, building a consensus, and mobilizing people and institutions around the new strategy. Commitment, time, and financial resources are essential to support foresighting, especially the implementation phase.

Link foresighting with institutional change to ensure sustainability

A policy issue especially relevant to innovation systems is that innovation cannot usually be imposed. A solution that is not widely understood, or which cannot be easily assimilated or at least accommodated by local cultural and political systems, generally will not gain wide support and will fail once external funding and/or pressure are withdrawn. Foresighting exercises can inform innovation at the policy and investment levels, but to ensure sustainability, they must be linked to assessment and change processes within the organizations involved in the innovation system (see TN 2). Ultimately, the dynamics of innovation and change cannot be understood without a broad concept of knowledge, including competencies and capabilities, practices and routines, and meanings, beliefs, and perceptions (Williams and Markusson 2002). A foresight exercise takes all of these factors into account to map out a development plan which may be visionary but is also practical and realistic, so that it ensures immediate and long-term benefits for the economy and society.

LESSONS LEARNED AND RECOMMENDATIONS

Foresighting cannot, by itself, solve all problems. Other factors must be addressed, as seen in the example from Jamaica and discussed throughout this sourcebook. These factors include the role of multistakeholder collaboration for good governance (modules 1 and 6); sensible macroeconomic and regulatory policies (module 6); investments in education, advisory services, and research capacity (modules 2, 3, and 4); and an innovative, entrepreneurial private sector (module 5). All have an important role to play in encouraging technological dynamism, investment, and associated processes of economic diversification and growth, as well as a move into higher-value products and services that meet market demands.

Experience with foresighting exercises suggests that foresighting is a best practice for formulating a long-term strategic plan that enables reform and encourages innovation. The foresighting process and associated tools provide a structure for assessing factors that are internal (internal strengths and weaknesses of the science, technology, production, and institutional base) and external (changing global markets and other critical factors in the external environment). A foresighting exercise also lays the groundwork for a review process to anticipate key market opportunities by updating findings with information on new and emerging technologies in conjunction with an analysis of the current restructuring of key sectors of the global economy in relation to science, technology, and production. The identification of such market opportunities could, with good management, create demand for a process of institutional

and wider economic restructuring, provide the economic impetus to support a widening skill base, attract and retain human and financial capital, and make a decisive move along the value chain, thereby escaping from low-growth, low-margin markets.

Key recommendations for practitioners include the following:

- *Many policy advisors, planners, and donor agencies find it difficult to talk openly about political problems* that can undermine innovation processes, such as vested interests and corruption. They often prefer to talk about technical issues as if they could be addressed in isolation. Transparency is sacrificed and the exercise loses credibility with stakeholders. In fact, the real impediments to progress are often the political, cultural, or economic factors that determine which solutions are adopted and supported and how and when this occurs. It is useless to present an analysis of technological choices as if the final decision is value-free. The only way to map out a potentially viable solution is to engage with and understand the political and cultural issues.

- *For these reasons, it is extremely important that participants in foresighting exercises understand that a systematic examination of the possibility of failure and an honest discussion of the pattern of previous failures are the only ways to improve the chances of success.* Well-organized feedback to the client and participants enables the process to tolerate the dissent, complexity, and uncertainty that is typical of strategic analyses.

- *Foresighting is a process, not a one-time activity.* The time required may have to be extended in light of new information or significant changes in the external environment (such as a recession). A national process can take several years and cycle over a longer period. Regular interaction with participants at all stages of the process is needed to ensure that the exercise does not suffer from stagnation or "groupthink" as well as to ensure that new stakeholders are identified, recruited, and fully involved.

- *Ideally a foresighting exercise should be commissioned formally* so that it is clearly accountable to client demand. It is recommended that a foresighting expert be contracted as a facilitator, but the expert does not replace the need for a steering committee and working group to implement the exercise.

Monitoring Agricultural Innovation System Interventions

Andy Hall, LINK Ltd.

Kumuda Dorai, LINK Ltd.

Trish Kammili, National Institute for Agricultural Research, France (INRA)

SYNOPSIS

This note deals with strategies for monitoring AIS interventions. Innovation system interventions make explicit assumptions about the nonlinearity of change and innovation in their design, and in doing so, they place specific demands on monitoring arrangements. These assumptions hinge on the recognition that innovation usually involves simultaneous technical adaptation and changes in the way things are done—in other words, institutional (and policy) adaptation—and that the final impacts will occur only when institutional adaptation has been achieved. Process-oriented monitoring methods that can cope with learning-based interventions have been around for some time; innovation system interventions should rediscover and adapt these methods rather than reinvent them. These types of monitoring methods include Outcome Mapping, Rapid Appraisal of Agricultural Knowledge Systems (RAAKS), the Most Significant Change (MSC) approach, and Participatory Impact Pathway Analysis.

BACKGROUND AND CONTEXT

In common with all interventions, good practice in agricultural innovation interventions requires effective performance management accompanied by reporting arrangements that ensure accountability. This thematic note deals with monitoring—the effective management of performance by implementers so that they can achieve desired outcomes and report their progress to investors. A separate thematic note (TN 5) deals with evaluation—the assessment of impacts and the generation of lessons for future interventions by investors and planners. Traditionally, monitoring systems in interventions have focused on tracking performance against a set of milestones agreed with the investor at the intervention's inception. In reality, agricultural development interventions—particularly those related to innovation

systems—are rarely this simple, and the underlying assumptions often prove to be unrealistic. Monitoring practice has for some time recognized this nonlinearity of agricultural and other development interventions, accepting that outcomes and impacts are achieved following constant iterations of an approach based on experience emerging from the intervention itself. This awareness is evident in the range of learning-based interventions that have emerged in recent years, such as adaptive collaborative management techniques in natural resource management (Colfer 2005; Guijt 2007) or techniques such as project Outcome Mapping (described in box 7.12 later in this note).

Innovation system interventions make explicit assumptions about the nonlinearity of change and innovation in their design and, in doing so, place specific demands on monitoring arrangements. These assumptions hinge on the recognition that innovation usually involves simultaneous technical adaptation and changes in the way things are done—institutional (and policy) adaptation. A related assumption is that final impacts such as changes in yield, incomes, food availability, or environmental sustainability will occur only when institutional adaptation has been achieved. The intervention logic of innovation system interventions is that while technological adaptations have the potential for immediate impacts, institutional and policy adaptations strengthen capacities for innovation that remain and continue to develop beyond the life of an intervention. These capacities lay the foundation for future technical adaptations and lead to social and economic impacts.

This focus on institutional adaptation highlights the need for those implementing an intervention to monitor how effectively their actions stimulate new ways of doing things. Often this kind of change involves stimulating the adaptation of informal institutions—for example, by developing links between research, enterprise, development, and (sometimes) policy players and then finding ways in which these networks can work effectively. Sometimes it involves stimulating changes in formal institutions, such as land tenure arrangements, regulatory regimes, pricing policies, or the roles of certain organizations. This process, in turn, requires interventions to facilitate negotiations about change between different stakeholders. Both types of institutional change need to be tracked.

Because the environments in which institutional changes must take place have highly specific features, the process of stimulating change can rarely follow a set plan. Instead an experimental approach is needed. This means that interventions must be learning based and reflexive, which is why monitoring is so critical: It assumes the role of helping to determine whether the intervention's chosen approach is resulting in desirable outcomes and whether the intervention needs to respond to other (often unexpected) changes in its environment.

Since the nature of institutional bottlenecks may become apparent only during the course of an intervention, indicators of performance will need to be developed on a case-by-case basis and constantly revised. Recording unexpected institutional changes is an additional way of alerting an intervention's investor that progress is being made.

Interventions also will need to monitor progress toward final social, economic, and environmental impacts. Such monitoring implies a set of assumptions about the relationship between these institutional outcomes and final impacts. Interventions need to test these assumptions by monitoring outcomes farther down the results chain toward final impact to verify whether the institutional changes that have been enacted are likely to lay the foundation for those final impacts. This monitoring provides information that can feed into adjustments in the intervention; it also acts as a means of reporting on progress and remaining accountable to investors.

The monitoring of innovation system interventions also emphasizes the need to make information accessible to all stakeholders involved. Information collection approaches will need to be inclusive and transparent, and information management systems must provide open access to all, not just those involved in designing monitoring arrangements.

INVESTMENT NEEDED

Practitioners have become better equipped to manage the performance of learning-based, institutional adaptation processes and the complex impact chains that are likely to be encountered in innovation system interventions. The challenge of monitoring innovation system interventions is therefore to learn how current good practice and principles

are best deployed to meet the demands of this type of intervention. Process-oriented monitoring methods have been around for some time, and innovation system interventions might need to rediscover and adapt those methods rather than reinvent them. Boxes 7.11 and 7.12 illustrate two such methods: Rapid Appraisal of Agricultural Knowledge Systems (RAAKS) and Outcome Mapping.

Principles for selecting monitoring methods

Several principles from good practice stand out and make a useful guide for selecting monitoring methods from among the many that are being adapted to the learning orientation of innovation system interventions. These principles are particularly useful at a time when experience of monitoring innovation system interventions remains limited. They include the following:

- *Balance accountability and learning.* Aside from collecting the data required for reporting to the investor, implementers will need to collect monitoring data that will help them manage the performance of their intervention. All of the methods summarized in table 7.8 support learning-based monitoring for interventions.
- *Make assumptions explicit, and revisit theories of change.* The greater learning orientation of monitoring implies a need to make assumptions about change processes explicit in planning interventions and a need to revisit and test those assumptions. Previously practitioners tended to regard such assumptions as a given. By exploring and responding to the validity of the assumptions on which an intervention is based, however, implementers can improve the intervention's effectiveness. All of the learning-based approaches in table 7.8 involve testing underlying assumptions. If an intervention's theory of change must be altered, it is important that the investor be made aware of the change in accountability reporting.
- *Incorporate different stakeholders' perspectives.* The shift to address the question of how things happened and to track unexpected outcomes requires a much stronger emphasis on widening the scope of participation in monitoring. Wider participation helps to capture the perspectives of the poor (and other stakeholders) on the actual effects of an intervention (social, economic, and institutional). It also makes it possible to develop a fuller understanding of the process through which those effects came about. Different stakeholders may have different interpretations of cause and effect. It is now good

Box 7.11   Rapid Appraisal of Agricultural Knowledge Systems

A Rapid Appraisal of Agricultural Knowledge Systems (RAAKS) focuses primarily on knowledge and information systems. The appraisal is a structured inquiry into the social organization of innovation, based on the inputs of those involved: the way the actors behave, how they interact and form networks, how they go about cooperating and communicating, what stimulates them to learn, and what blocks them from learning. Actors gain a shared understanding of their performance as innovators—their perceptions, judgments, understanding, and capacity to take decisions and act—and learn to contribute more effectively to innovation.

The main elements of RAAKS are as follows:

- **Strategic diagnosis.** Joint definition of useful strategies through an appraisal of opportunities and obstacles.
- **Creative tension.** Compare and contrast the multiple findings that represent the various analytical perspectives.
- **Design of solutions.** Participants are encouraged to analyze, interpret, and, based on these steps, design potentially useful solutions.

Source: Kammili 2011; Salomon and Engel 1997.

Strengths

- Provides insights into the social organization of innovation and people's values, motivations, and reactions.
- Improves the generation, exchange, and use of knowledge and information for innovation.
- Builds capacity among the actors involved by making them conscious of their performance as innovators.

Weaknesses

- A complex methodology with a series of steps, exercises, and tools to be implemented.
- A strong focus on rural activities; does not consider the wider setting of the innovation system (actors other than those involved in activities in the rural domain).

Best use or application

- Together, actors develop a common understanding of their performance as innovators.

---

practice in monitoring to recognize these divergent perceptions. This principle extends to the need to draw on perspectives from a wider range of stakeholders in developing the theory of change that will guide an intervention's implementation and learning. Box 7.13 illustrates one approach for widening participation in monitoring; others are included in table 7.8.

- **Mixed methods.** Expanding views of monitoring require quantitative methods (to measure outcomes) to be combined with qualitative methods (to understand and learn from institutional and process changes). To understand institutional and process changes, and to establish their causal links to outcomes and impact, monitoring will need to place much greater emphasis on qualitative methods such as Innovation and Institutional Histories (box 7.14) or Causal Process Tracing (table 7.8). Rather than measuring levels of income and social variables, it may be more appropriate to use proxy indicators of changes that will lead to these impacts in the future (for example,

changes in yields or quantities of fertilizer sold), although these indicators contain their own assumptions about the causal chain to impact. Data collection techniques for tracking outcomes include small sample surveys, participatory appraisal techniques, and longitudinal household case studies. It is important to keep these activities in proportion to the task of managing the intervention, however. Costly and time-consuming baseline surveys do not lend themselves to learning-based interventions for the simple reason that they do not generate information quickly enough to inform how an intervention is managed (see the examples from IAPs 6 and 7).

Available monitoring methods

Table 7.8 presents the strengths and weaknesses of a range of monitoring methods that have relevance to innovation system interventions. These methods have a number of

Box 7.12   Outcome Mapping

Outcome Mapping reflects the idea that development is done by and for people. The central concept of Outcome Mapping is that development is achieved through changes in the behavior, actions, relationships, and activities of people, groups, and organizations with which an intervention works directly (the "boundary partners"). The originality of this approach stems from the fact that there is a clear shift from measuring the outputs of an intervention (poverty alleviation, reduced conflict, and so forth) toward trying to assess changes in behaviors, relationships, and actions of the people and the organizations directly involved. By emphasizing behavioral change, Outcome Mapping aims to assess "contributions" to impacts rather than claim "attribution" for impacts.

The boundary partners are identified, as are strategies for equipping them with tools and resources so that they can contribute to the development process. An intervention thus facilitates changes but does not cause or control them directly. Outcome Mapping maps how an intervention influences the roles partners play in development through a set of graduated indicators of changed behavior. It monitors and evaluates three elements of the intervention: behavioral changes, the strategies used by the intervention to stimulate change among the partners, and how the intervention functions as an organizational unit. Through these three elements, Outcome Mapping unites process and outcome monitoring and evaluation.

Source: Kammili 2011; Earl, Carden, and Smutylo 2001.

Strengths

- A robust methodology that can be adapted to a wide range of contexts. Outcome Mapping's very flexible approach allows it to be used as a planning or replanning tool and at the beginning or midway through an intervention. Outcome Mapping can also be used as a monitoring approach throughout an intervention or as a framework for evaluation.
- It complements standard approaches and thus can be used in combination with other methodologies.
- It unites process and outcome evaluation.
- Monitoring provides the space for critical self-reflection and learning.

Weaknesses

- Most assessment data are generated by the intervention, raising the question of whether the data are objective (for example, failures may be whitewashed).
- Uncertainties about combining Outcome Mapping data with more quantitative data.
- Lack of clarity about how to deal with and integrate new boundary partners.

Best use or application

- Satisfies the need for accountability as well as learning about the change process.

---

characteristics, discussed below, that make them different from conventional milestone-based monitoring techniques. The discussion provides a flavor of the new approaches to monitoring that will be seen in the coming years in innovation system interventions.

The value added by the new monitoring techniques reviewed in table 7.8, compared to conventional milestone-based techniques, may be summarized as follows:

1. *Explanatory.* A focus on reconstructing events in an attempt to understand why a particular course of action led to the outcomes observed or failed to achieve expected outcomes. This understanding is important for innovation system interventions, in which experimentation and action learning are the main routes to success.
2. *Inquisitive.* Recognition of the importance of unexpected outcomes and the need to record and learn from them. This perspective is particularly useful in innovation system interventions, because their process-driven nature can lead to unanticipated outcomes that have significance.
3. *Communicative and accessible.* Monitoring approaches as ways of sharing results and lessons and building a joint understanding of events that have taken place. Making information accessible to all stakeholders is important in innovation systems, because it is a way that organizations learn and improve their performance.

Table 7.8 Overview of Methods for Monitoring AIS Interventions

Method	Purpose/use	Strengths	Weaknesses	Major references
Innovation and Institutional Histories	– Understand past innovation processes and identify institutional factors that foster or hinder innovation. – Forge a shared vision of the future among stakeholders.	– Fosters discussion among stakeholders and leads to reflection and learning. – Helps build a shared vision of the future.	– Written accounts are based on recollections and could be biased. – Need skilled facilitators to help collect and analyze the information. – Usually takes a very long time for significant lessons to emerge.	Douthwaite and Ashby (2005); Shambu Prasad, Hall, and Thummuru (2006)
Participatory Impact Pathway Analysis	– Guide project management, especially in complex situations where innovation is seen as emerging from a network.	– Involves staff and key stakeholders of an intervention in constructing impact pathways. – Includes both (1) causal chain of activities, outputs, and outcomes that shed light on how an intervention achieves its goals and (2) network maps that show evolving relationships between participants. – Underlines the fact that innovations emerge from a network and not a linear "pipeline." – Promotes reflection, self-evaluation, and learning. – Provides a framework for carrying out action-research.	– A relatively new approach that needs to be further implemented to gauge its weaknesses.	Douthwaite et al. (2003)
Causal Process Tracing	– Well suited for complex and long-term interventions with systems learning goals.	– Places data and theory in close proximity. One quickly sees what works and what does not in an intervention's lifetime.	– Regarded as not very strong for wider generalization but more suited to narrow specification of the reach of causal propositions. – Takes a great deal of time. – Not conducive to parsimonious theory and leads to partial, middle-range theory. It is easy to miss causal complexity. – Easy to lose sight of the broader context.	George and Bennett (2005); CoS-SIS (2009); Walters and Vayda (2009)
Reflexive Monitoring in Action	– Best suited for long-term interventions with systems learning built into their mandates.	– Mechanisms built into the intervention permit all participants to contribute to learning by reflecting on the relationships between key aspects and ambitions of the intervention as well as the practices and institutions in which they are embedded. – Monitoring is integral to the intervention, so insights gained are built into and experimented with in new activities. – Encourages investigators to look for creative solutions. – Reflexive monitoring in an intervention ensures that those involved develop new ways of working to keep up with changes in the intervention's institutional context.	– Works in theory for long-term interventions oriented to systems learning, but most development interventions do not have the luxury of long-term learning as the sole goal and need to demonstrate (developmental) impacts throughout the life of the intervention. – A coherent set of tools and principles, but in essence the approach is still being developed and not in widespread use, so experiences from the field are few.	van Mierlo et al. (2010)

(Table continues on the following page)

Table 7.8 Overview of Methods for Monitoring AIS Interventions (continued)

Method	Purpose/use	Strengths	Weaknesses	Major references
Appreciative Inquiry	– Identify positive changes and look for unexpected outcomes.	– Fosters learning from past and contemporary situations. – Opens up the possibility of looking at different things in new places/ways. – Interviews allow for deep connections, unexpected learning, and a sense of empowerment.	– Like Innovation and Institutional Histories, this method can suffer from bias or incorrect reporting.	Biggs (2006); Acosta and Douthwaite (2005); Hall, Sulaiman, Bezkorowajnyj (2007)
Outcome Mapping	– Satisfy the need for accountability as well as learning about the process of change. – Especially useful for assessing what causes change in behavior, relationships, activities, or actions of the people, groups, and organizations with whom an intervention works directly. – Well suited for complex and long-term aspects of interventions with outcomes that are intertwined and difficult to segregate.	– A robust methodology that can be adapted to a wide range of contexts. Its flexible approach allows it to be used as a planning or replanning tool, at the beginning or midway through an intervention. It can also be used as a monitoring approach throughout an intervention or as a framework for evaluation. – It can complement standard approaches and can be used in combination with other methodologies. – Unites process and outcome evaluation. – Monitoring provides the required space for critical self-reflection and learning.	– Does not replace but complements *logical framework analysis.* – Most data is self-assessment data generated by the intervention, which raises the question of objectivity. – Unclear how to combine resulting information with more quantitative data and how to deal with and integrate new boundary partners.	Smutylo (2005); Earl, Carden, and Smutylo (2001); IDRC n.d.
Most Significant Change	– Make sense of an intervention's impact and foster learning. – With the help of all primary stakeholders, identify the most significant changes that have occurred as a result of an intervention.	– Enhances capacities of stakeholders (at the organizational and individual levels) and fosters learning. – Helps identify unexpected changes or outcomes. – Large amounts of information are processed, from which negative and positive changes are deduced. – More accessible than traditional M&E techniques; no specific skills required to participate.	– A subjective expression of the values and concerns of the stakeholders designated to select the stories.	ECDPM (2006); IFAD (2002); Davies (1996); Davies and Dart (2005)

			Source
Rapid Appraisal of Agricultural Knowledge Systems	– Help actors as a group to understand their performance as innovators.	– Provides insights into the social organization of innovation and people's values, motivations, and reactions. – Improves the generation, exchange, and utilization of knowledge and information for innovation. – Enhances capacity building of the actors involved by making them conscious of their performance as innovators.	– Complex methodology with a series of steps, exercises, and tools to implement. – Strong focus on rural activities does not consider the wider setting of the innovation system (actors other than those involved in activities in the rural domain).	Salomon and Engel (1997); ECDPM (2006)
Stories and Narratives	– Shed light on the changes that have occurred at the individual, organizational, or institutional level. – Uncover intangible factors (qualities, values, culture, and so forth) that determine the organization's character.	– Exchanging stories builds trust between participants; in some cases may lead to an environment conducive to learning and stimulate change. – An effective way to deal with passions and emotions of individuals involved. – Has the potential to stimulate change if told correctly. – Individual focus can translate into organizational development. – Not hierarchical.	– Biased; depends on the perspective of the person telling the story.	Asif (2005)
Performance Indicators	– Used to assess innovations for which cause and effect are known and can be linked through predetermined performance indicators.	– Effective means to measure progress towards fixed objectives. – Facilitates benchmarking comparisons over time.	– Definition of indicators guarantees success. When indicators are defined poorly, they are not good measures of effectiveness. – Predetermined indicators do not allow for measuring unexpected changes. – A risk that the intervention will need too many indicators; data for some indicators may be inaccessible or costly and impractical to measure.	World Bank (2004)

Source: Authors.

Most Significant Change (MSC) is a form of participatory monitoring and evaluation that involves many participants in an intervention to decide what kinds of changes need to be recorded and to analyze the information collected, which can be used to assess the intervention's performance and impact. The approach focuses on collecting significant change stories that emerge from the field. Significant changes can include changes in people's lives and participation levels as well as changes in the sustainability of people's institutions and their activities.

Together, stakeholders decide what is going to be monitored. MSC process managers identify broad domains of change that they assume to be important and that should be evaluated. These domains of change are deliberately wide and inclusive. Stakeholders identify significant changes in a particular domain of change and justify why they think these changes are the most significant. The stories are analyzed by stakeholders at every level (field, organization, investor, and so on). This approach is a fairly simple way to make sense of a large amount of information. The central aspect of the technique is not the stories themselves but the deliberations and dialogues surrounding their selection. If implemented successfully, MSC causes whole teams of people to focus their attention on the intervention's impact.

Source: Kammili 2011; Davies 1996; Davies and Dart 2005.

Strengths

- Enhances capacities of the stakeholders (at organizational and individual levels) and fosters learning.
- Helps identify unexpected changes or outcomes.
- Processes large amounts of information, from which negative and positive changes are deduced.
- More accessible than traditional techniques for monitoring and evaluation, and requires no specific skills to participate (everyone can tell a story).

Weaknesses

- The approach is a subjective expression of the values and concerns of the stakeholders designated to select the stories.

Best use or application

- To make sense of an intervention's impact and foster learning.
- To identify the most significant changes that have occurred as a result of an intervention, with the help of all primary stakeholders.

4. **Inclusive.** A focus on the inclusiveness of the monitoring process. Inclusiveness can help diffuse tensions around the change process associated with innovation system interventions. It is also a way of helping to build the linkages that these types of interventions need.

5. **Rapid.** These methods generate information quickly. This consideration is important for innovation system interventions, which must be nimble in responding to unfolding events.

6. **Nonexpert/open access.** Methods are designed to be used by all those involved in interventions—that is, for self-assessment—rather than by monitoring experts. These methods are also designed to promote access to the information generated. Open access is important in innovation system interventions, because managing performance and achieving results is the responsibility of all those involved.

7. **Tailor-made.** A number of methods involve the use of institutional change indicators. The emphasis is on developing these indicators based on the nature of the intervention being monitored. Approaches accommodate the fact that indicators of performance are a moving target and need to be revised constantly. Tailoring indicators in this way is particularly suitable to the learning-based characteristic of innovation system interventions, in which the specific nature of the institutional change being sought rarely can be predicted, aside from generic terms such as the degree of participation, the strengthening of links between stakeholders, or the inclusiveness of decision-making processes.

Box 7.14   Innovation and Institutional Histories

The Institutional History method helps people involved in the innovation process construct a shared understanding of how innovation has occurred. Institutional histories are narratives written jointly by people who have been involved in an innovation. The history records changes in institutional arrangements (new ways of working) that evolve over time and facilitate the achievement of goals.

This approach highlights the importance of institutional innovations. The main idea behind these histories is to introduce institutional factors into the legitimate narrative of success and failure in research organizations. Histories can be written by using interviews to construct a timeline, gain a clear understanding of roles and relationships, inquire into what triggers or hinders successful innovations, and reflect on failures. Lessons drawn from the analysis can be used to improve performance.

The dialogue that is promoted between the actors during the preparation of institutional histories can promote learning and capacity building. The conclusions drawn can be used in subsequent planning and help to formulate a shared vision that can catalyze

Source: Kammili 2011.

change. These experiences can then be scaled out by disseminating the findings of the innovation process.

Strengths

- Fosters discussion among stakeholders and leads to reflection and learning.
- Helps build a shared vision of the future.

Weaknesses

- Written accounts are based on recollections and could be biased.
- Skilled facilitators are needed to help assemble and analyze the information.
- It usually takes a long time for significant lessons to emerge.

Best use or application

- To understand past innovation processes and identify institutional factors that foster or hinder innovation.
- To forge a shared vision of the future among stakeholders.

POTENTIAL BENEFITS OF MONITORING APPROACHES

In common with all interventions, well-executed monitoring arrangements are central to the performance of innovation system interventions. Investments in developing such arrangements will *strengthen the effectiveness of investments in achieving developmental goals* and, in doing so, *improve value for money.*

POLICY ISSUES RELATED TO MONITORING

A number of policy issues are important for ensuring that monitoring delivers some of the benefits that have just been described.

A primary concern is to *improve the demarcation of responsibilities for monitoring and evaluation.* A lack of clarity on the part of investors as well as intervention implementers about the different purposes of monitoring and

evaluation and about who is responsible for each task causes monitoring systems to underperform and results in badly executed evaluations (see IAPs 6 and 7). These wasted resources lead to ineffective interventions and prevent investors from learning vital lessons for designing future interventions. This issue is addressed easily by simply clarifying roles, responsibilities, and time frames for the separate tasks of monitoring and evaluation.

Strengthen the capacity to implement learning-based performance management approaches. Within the agricultural research and innovation profession such expertise is limited. Although these skills are better developed in the general development and rural development communities, particularly in the nongovernmental sector, training in these methods is needed urgently, because monitoring is the lynchpin of effective innovation system interventions.

As a central performance management tool, *monitoring must be integrated and mainstreamed in innovation system interventions* rather than exist outside of them. Isolating monitoring as a specialist domain simply to satisfy the investor, without all the intervention's personnel taking performance management seriously, defeats the objective of an innovation system intervention.

Investors will need to modify their expectation of accountability reporting, accepting the process nature of interventions and becoming more comfortable with institutional change outcomes. This changed perspective places additional responsibility on the investor to ensure that impact evaluation is undertaken in a timely fashion and in a way that recognizes the nature of these interventions.

LESSONS AND RECOMMENDATIONS FOR PRACTITIONERS

Innovation system interventions demand an expanded suite of monitoring arrangements that respond to the learning-based nature of these interventions and their primary focus on using institutional adaptation as the foundation for future impacts. A wide range of tools and approaches is available; the challenge is to know which to select and how

to adapt them to best match the performance management demands of innovation system interventions. Practical lessons for doing so include:

- **Negotiate accountability reporting with the investor.** Learning-based monitoring generates mainly qualitative information on processes and institutional arrangements, including unexpected outcomes. Investors may be unfamiliar with this kind of information in accountability reporting, so it is important to discuss reporting expectations beforehand.
- **Ensure that the indicators of institutional change are specific to the particular intervention and revised as needed.** Generic institutional change indicators, such as the degree of participation or the strengthening of links between stakeholders, can act as guidelines for categories of institutional change that are likely to occur. In managing the performance of an individual intervention, however, more case-specific indicators are needed. Since the nature of institutional bottlenecks may become apparent only as the intervention unfolds, performance

---

Box 7.15 Developing Institutional Change Indicators

Institutional change encompasses a very wide range of changes, from new ways of doing things to formal policy changes. Developing indicators can be difficult, because decisions need to be made about what types of institutional change are important to help understand the progress of a particular innovation system intervention. At the same time, it is important to capture the range of changes that an intervention is helping to stimulate, some of which may not be expected. This last objective is important when reporting outcomes to investors. The DFID-funded Research Into Use program (see www.researchintouse.com) faced this challenge. The program recognized that institutional change would be the main route through which it would achieve long-term impacts, but it had no systematic mechanism for capturing evidence of those changes, nor did it have an accessible way of reporting institutional changes to its investor other than through lengthy case studies. For this reason, the program's Central Research Team developed an inventory of all institutional changes observed (IAP 7) and categorized them as follows:

- Changes in the poverty relevance of actions and interventions.
- Market-related institutional change.
- New types of organizations playing new types of roles.
- Old types of organizations playing new roles.
- New forms of rural credit.
- Changes in agricultural research practice.
- Changes in the policy-making process.
- New network configurations.
- Formal policy changes.
- Changes in donor practice.

This practice helped identify categories of institutional change where limited progress was being made—changes in donor practice and changes in the policy-making process. It also helped to improve communication of the program's progress in stimulating institutional change, which had previously been difficult for an external audience to see.

Source: Authors and Adwera et al., forthcoming 2012.

---

indicators will need to be revised constantly (box 7.15 illustrates institutional change indicators developed in an innovation system intervention).

■ *Collect data on outcomes in a timely way to contribute to the intervention's learning cycles, with due attention to the approach and scale of data collection.* Quantitative methods are likely to play a role in collecting outcome or proxy outcome information to understand the effectiveness of process changes brought about by an intervention. The approach and scale of this data collection need to be in proportion to implementers' need to manage the performance of the intervention and ensure that it is on track to achieve its agreed outcomes. Implementers should guard against collecting impact evaluation data. This time-consuming activity will not necessarily provide data at the appropriate time to manage an intervention successfully.

■ *Revisit milestones and expected outcomes.* Learning-based interventions may evolve. Their evolution will lead to unexpected outcomes and modified theories of change. It is important to make the investor aware of these changes and negotiate how new milestones and outcomes will be reported.

Evaluating Agricultural Innovation System Interventions

Andy Hall, LINK Ltd.

Kumuda Dorai, LINK Ltd.

Trish Kammili, National Institute for Agricultural Research, France (INRA)

SYNOPSIS

The evaluation of innovation system interventions is linked to the monitoring of interventions but is a separate function. Evaluation is usually performed on behalf of the investor by independent evaluators and not by those responsible for implementing the intervention. A well-designed evaluation is particularly important for innovation system interventions, as their process-driven nature means that the result chains are complex, dynamic, and not amenable to simple linear and anecdotal evaluation. Great care is needed in making judgments about the relationship between the effectiveness of the design and execution of the intervention and the observed impacts associated with it. Evaluation good practice is better equipped than ever to deal with this complexity. This note outlines key good practice principles relevant to evaluating innovation system interventions: a stronger learning orientation in evaluation, the use of counterfactuals, the use of mixed (quantitative and qualitative) methods, and incorporating the perspectives of different stakeholders. The example of Theory-Based Impact Evaluation (TBIE) illustrates how these principles work together. Experience with applying evaluation approaches and principles to agricultural innovation system interventions is currently in its infancy. A policy priority is to develop the capacity for this type of evaluation, which will help to improve the long-term effectiveness of innovation system interventions.

BACKGROUND AND CONTEXT

The effectiveness of innovation system interventions lies in their outcomes and impacts being investigated thoroughly and the resulting lessons applied to future investments. This process contributes to accountability as well as future investment performance. This note focuses on the evaluation of innovation system interventions.

Monitoring—the effective management of performance by implementers to achieve desired outcomes and report progress—is covered in TN 4.

Three critical features of AIS interventions influence how they are evaluated. First, they focus on strengthening capacity; second, they use a learning-based process; and third, they require a distinction to be made between impact, shared impacts, unexpected impacts, and unrelated impacts. Each of these issues is discussed next.

A focus on strengthening capacity

An innovation system intervention is an investment to improve how change and innovation take place. Although such an investment is made ultimately to achieve certain social, economic, or environmental impacts, the immediate outcome is improved ways of doing things—usually referred to as institutional change. For example, the National Agricultural Innovation Project in India established research, development, and private sector consortia around selected themes to introduce a new way of working that would achieve wide-scale impact in the future (see module 4, IAP 2). An innovation systems research project addressing fodder scarcity in West Africa and India (the Fodder Innovation Project, described in IAP 6) experimented with ways to induce institutional change that would enable innovation leading to social and economic impacts.

A learning-based intervention process

Improvements in how change and innovation take place will depend on the specific circumstances of each intervention. For this reason, ways of achieving those improvements are hard to design in advance. Innovation system interventions address this issue by adopting a learning-based approach, in which the intervention continuously tests the assumptions on which it is based (its intervention logic). Over the course of the intervention, investors' expectations of what the intervention is going to achieve, and how it will do so, often need to evolve significantly. For example, an intervention in Sierra Leone (IAP 1) designed to help put agricultural research results into use began by trying to establish innovation platforms around commodity chains. It soon found that the main bottlenecks to research use and innovation were policy issues in the enabling environment for innovation.

Distinguishing impact, shared impacts, unexpected impacts, and unrelated impacts

By definition, innovation system interventions operate within an often complex web of activity. Much of this wider set of events is beyond the control and influence of the intervention, but the performance of the intervention itself is often greatly influenced by this context. While the immediate outcomes of an intervention—for example, the establishment of innovation platforms (see module 4, particularly TN 1)—can be evaluated easily, it is more difficult to attribute impacts to those platforms. The causal chain from intervention to impact is often complex, may take time to mature, and might be geographically removed from the point of intervention. As a result, tracing and attributing impact are difficult.

In India, for example, the Fodder Innovation Project (see IAP 6) introduced an institutional change, in which an NGO helped the National Department of Animal Husbandry use its resources to hold livestock health camps in villages. Can the resulting impact be attributed solely to the intervention that supported the NGO to explore institutional change? In Nigeria, the Fodder Innovation Project led to an unexpected outcome: An NGO and the local livestock research institute collaborated on animal disease surveillance, an activity that was outside the scope of the project's mandate to focus on fodder but was likely to lead to important future impacts. Finally, since complex impact chains can often mask underlying causal processes, there is the danger that impacts may be attributed to an intervention when in fact they arose from unrelated events.

INVESTMENT NEEDED

Viewed from the perspective of traditional approaches to assessing the impacts of agricultural research investments (see box 7.16), designing approaches for evaluating innovation system interventions seems daunting. Traditional approaches relied on (1) technology adoption studies, (2) investigation of economic surplus generated by research and computation of rates of return, and (3) economic studies of the contribution of research to impact. While these approaches are powerful (see Evenson, Waggoner, and Ruttan 1979; Pardey and Beintema 2001; Alston et al. 1995), a persistent critique is that they have weak diagnostic value. Their inability to elucidate underlying causal processes and account for institutional change are particular weaknesses (Hall et al. 2003; Horton and Mackay 2003; Watts et al. 2003).

In contrast, the wider development evaluation tradition is well equipped to deal with the investigation of causal links between dynamic theories of change and impacts of the sort likely to be encountered in innovation system interventions (see, for example, the guidance on impact evaluations in NONIE, Leeuw, and Vaessen 2009). The challenge is mainly to understand how current good practice in the wider development evaluation community can be more widely deployed in the evaluation of innovation system interventions. Of particular relevance is the recent focus on evaluations that link the assessment of outcomes and impacts (what were the end results) with learning (what processes and practices brought about those results) (Savedoff, Levine, and Birdsall 2006; White 2009a).

Principles for evaluation

Emerging from this trend is a set of principles that practitioners can draw upon in evaluating innovation system interventions. These are now general principles for all types of evaluations.

- *Situational responsiveness influences the design of the evaluation.* Referred to as "situational responsiveness," the key principle involves matching the design to the needs, constraints, and opportunities of the particular situation rather than one particular method. The application of this principle rests on understanding the characteristics of an intervention or parts of it and determining where the underlying change processes are simple, complicated, or complex.
- *Seek a stronger learning orientation in impact evaluation.* A number of new initiatives—notably the International Initiative for Impact Evaluation (3IE) but also others—have stressed the need to combine accountability and learning objectives. Learning is important for identifying what worked or did not work and why, and this information is valuable for designing future investments. While this perspective emphasizes the need for

Ex post impact assessment

- Although it addresses the accountability imperative for funders of interventions, it does not help in drawing the institutional lessons required to understand the innovation process. Does not cover noneconomic dimensions.
- Difficult to develop credible counterfactuals and establish definitive causalities.
- Focuses on intended positive results and frequently ignores unexpected and negative results.
- Not appropriate in more complex areas such as natural resource management, policy and biodiversity research, and training and capacity building.

Randomized control trials

- Weak in external validity (or generalizability) and in identifying the mechanisms responsible for differences observed in the experimental and control situations.
- Rarely appropriate in complex situations where outcomes arise as result of interactions of multiple factors that cannot be "controlled."
- Limited in their ability to deal with emerging and unanticipated outcomes.

Source: Kammili 2011.

Cost-benefit analysis

- Calls upon significant financial and human resources.
- Benefits are estimated in advance based on assumptions that may not always be correct.
- Not all costs and benefits can be quantified (social and environmental costs/benefits, for example); results obtained do not reflect all benefits.
- Items included in the analysis reflect the bias of whoever performs the analysis; coverage and quality thus vary greatly.
- Given its complexity, involves only economists and project designers and does not engage other primary stakeholders.

Economic surplus approach and rate of return studies

- Requires substantial resources for collecting, processing, and interpreting technical and economic data.
- Simplistic assumptions about lags, costs, and supply shifts have biased rates of return (usually upwards).
- Not suitable for ranking noncommodity research such as socioeconomic and interdisciplinary research.

---

rigor in measuring impact, it also explicitly acknowledges the need to test interventions' assumptions and theories of change (White 2009a). For details, see the example of Theory-Based Impact Evaluation in box 7.19.

- *Make assumptions explicit and revisit theories of change.* Adopting a greater learning orientation in evaluation means making assumptions about change processes explicit when planning interventions and revisiting and testing those assumptions at the time of evaluation. By exploring the validity of the assumptions, evaluators can learn critical lessons for designing future interventions (box 7.19).
- *Use counterfactuals in impact evaluation.* A stronger learning orientation has also been accompanied by demands for greater rigor in the quantitative methods used to measure impacts, as well as for methods that better establish what would have happened without a par-

ticular intervention—the counterfactual. The best way to achieve this goal is still a point of debate (see box 7.17 for details on the challenges involved). Box 7.18 provides an example of Propensity Score Matching, one approach to developing a counterfactual case.

- *Mixed methods.* Quantitative methods (to measure impacts) clearly are central to investigating impacts, but they must be complemented by a range of qualitative methods (to understand and learn from institutional and process changes) that can help to understand the context in which the intervention took place and the process to which the intervention contributed that brought about those impacts. Practitioners can draw upon a very wide array of qualitative methods, and the combination of methods selected depends considerably on the nature of the intervention and the precise demands of evaluation functions. (For a discussion of

Box 7.17 Challenges of Developing Counterfactuals

The origins of counterfactuals can be traced to scientific experiments conducted with a control, the classic example being the randomized control trials used in drug testing. This approach poses ethical and operational problems that make it impractical for development interventions.

Another approach, "before" and "after" comparisons, struggles to account for changes that would have happened despite the intervention. Variants of this approach have been developed, however, which compare the effects of an intervention that starts in different places at different times. "With" and "with-out" comparisons appear more promising for capturing the counterfactual case, particularly when used in conjunction with statistical techniques that account for differences between the starting conditions in comparator sites.

The establishment of counterfactuals remains an evolving science for the evaluation community. Despite the challenges, mechanisms for establishing what would have happened in the absence of an intervention should be built into the overall design of interventions and the monitoring and evaluation arrangements that are put into place for them.

Source: Authors.

Box 7.18 Propensity Score Matching

Propensity Score Matching is a tool for identifying a suitable group with which the recipients of an intervention (the treatment group) can be compared. Evaluators find a comparison group comprising individuals who did not, in fact, receive the intervention but who, given their observable characteristics, had the same probability of receiving it as individuals in the treatment group. The intervention's impact is the difference in outcomes between the treatment and comparison group.

Source: World Bank 2004.

mixed methods in Theory-Based Impact Evaluation, see box 7.19.)

■ *Incorporate different stakeholders' perspectives.* Greater participation is needed to capture the perspectives of the poor (and other stakeholders) on what were the actual results of the intervention (social, economic, and institutional). Similarly, wider participation is needed to more fully understand the process through which these results came about. Different stakeholders may have different interpretations of cause and effect. It is now good practice in evaluation to recognize these divergent perceptions. This principle extends to the need to draw on perspectives from a wider range of stakeholders in investigating the intervention's theory of change.

Evaluation approaches

Table 7.9 presents a comparative overview of evaluation approaches appropriate to the demands of innovation system interventions. In different ways and with different emphases, these methods use the principles discussed in this note. They should be viewed as a menu of approaches which practitioners can draw upon to ensure that evaluations achieve the correct balance between learning and accountability. There are a number of points that need to be highlighted about these methods. Only the first method mentioned, Theory-Based Impact Evaluation, explicitly makes provisions for measuring outcomes and investigating underlying process. In reality, Theory-Based Impact Evaluation is an evaluation framework rather than a specific method, and it relies on a suite of qualitative tools.

The other methods outlined in table 7.9 are specific tools for qualitative investigation of what happened and what processes lead to the outcomes observed. While these methods for learning lessons are powerful, practitioners must recognize that there is an appetite among investors—public and private—for the quantification of outcomes and impacts. The nascent approaches to evaluating innovation system interventions will need to satisfy this demand.

The least developed of the methods presented in table 7.9 is benchmarking of innovation capacity. An important area of methodological development is to find measures of

Box 7.19   Theory-Based Impact Evaluation

Theory-based Impact Evaluation (TBIE) is advocated for understanding why an intervention has or has not had an impact. The approach calls for examining and mapping the causal chain of an intervention—from inputs to outcomes and impact—to test the underlying assumptions and shed light on the "why" question. TBIE involves six steps:

1. *Map the causal chain.* The causal chain links inputs to outcomes and impacts. It is the intervention's theory of change, which explains how the intervention is expected to have its intended impact. Testing assumptions is central to a theory-based approach. One criticism of the causal chain approach is that it is static and does not account for an intervention's ability to adapt and evolve—an important consideration for innovation system interventions. This criticism is addressed in TBIE by suggesting that the intervention theory should reflect the new design, and the evaluation should document the learning process that resulted in the new design. In this way, the intervention theory remains dynamic. Another criticism is that it misses unintended consequences, which is addressed by identifying those consequences through a careful application of intervention theory, fed by thorough preliminary fieldwork, and using the resulting information to develop new theories of change.

2. *Understand the context.* Clearly, understanding the context is crucial to understanding its impact and in designing the evaluation, as context influences how the causal chain plays out.

3. *Anticipate heterogeneity.* Understanding the context also makes it possible to design the evaluation to anticipate possible variation in impacts (arising from how the intervention is designed, the characteristics of the beneficiaries, varied socioeconomic settings, and so on).

4. *Conduct a rigorous evaluation of impact using a credible counterfactual.* The appropriate counterfactual is most usually defined with reference to a control group, which has to be identified in a way that avoids selection bias, meaning the use of either experimental or quasi-experimental approaches. Panel data help to strengthen the design, so baselines are encouraged. Where they are not available, they might be recreated using existing data sets or recall. In addition to selection bias, important issues to

consider in the design are the possibility of spillover effects (the control is affected by the intervention) and contagion or contamination (the control is affected by other interventions).

5. *Conduct a rigorous factual analysis.* The counterfactual analysis of impact needs to be supplemented by rigorous factual analysis of various kinds, given that many links in the causal chain are based on factual analysis. Targeting analysis is the most common form of factual analysis: Who benefits from the intervention? To the extent that there is a defined target group, then what is the extent of the targeting errors? Such errors can be quantified and their source identified. Factual analysis often highlights a crucial break in the causal chain and explains low impact.

6. *Use mixed methods.* A major step toward mixed methods is to increase the use of rigorous quantitative methods in qualitative studies (quantitative analysis informed by qualitative insight) or the use of qualitative data in quantitative studies. Without qualitative methods, the danger is that researchers will conduct impact studies with no exposure at all to the intervention.

Strengths

- The combination of counterfactual impact estimates with detailed exploration of causal links and theories of change has the potential to provide compelling lessons and evidence.
- Lessons gleaned from TBIE are valuable in trying to understand what works in development.
- Far more valuable lessons for policy are obtained through the insights on what doesn't work in interventions.

Weaknesses

- Not yet in widespread use, so experiences are limited.
- Construction of a robust counterfactual can be challenging.
- Data-intensive.
- Expensive, although good value for money if well executed.

Best use or application

- An evaluation tool best suited to investigating impacts and undertaking policy learning in large-scale interventions.

Source: White 2009b and authors.

innovation system performance at the macro level that rely on indicators of system behavior and functions rather than on input indicators such as research spending.

POTENTIAL BENEFITS

Effective evaluation is central to improving the performance of investments in current and future innovation system interventions. Well-designed evaluations of innovation system interventions are particularly important because their process-driven nature means that the result chains are complex, dynamic, and not amenable to simple, linear, anecdotal evaluation. Great care is needed in making judgments about the relationship between the effectiveness of an intervention's design and execution and the observed impacts associated with the intervention. Consequently the evaluation of innovation system interventions is likely to be expensive but will be an important investment in strengthening the long-term performance of these interventions.

POLICY ISSUES

Policy issues related to the evaluation of innovation system interventions are partly but not entirely similar to those for monitoring (see TN 4).

One similarity is the need to *build capacity in evaluating innovation system interventions*. Experience with these sorts of evaluations remains limited, because innovation system interventions are relatively recent. Nor are many investors familiar with the evaluation principles and approaches applicable to those interventions. Professional evaluators from the wider development community do have experience in applying these principles, but there is a need to develop a new cadre of evaluators with experience in applying them to AIS interventions.

A second policy priority is to *clearly demarcate responsibilities for monitoring and evaluation*. Investors need to make a very clear distinction between evaluation and monitoring and not be tempted to include the collection of impact evaluation data within the intervention. This point is particularly important for innovation system interventions, because the ultimate impacts may not emerge until some time after the intervention ends.

Third, *sufficient resources must be available to use a combination of evaluation methods*. The evaluation of innovation system interventions requires greater rigor in measuring impacts, investigating result chains, and testing the validity of theories of change. Inevitably, evaluations will take longer and cost more. Investors need to plan ahead and put sufficient money aside for evaluation when planning new innovation system interventions.

Finally, over and above the evaluation of individual interventions, investors will increasingly need to *benchmark innovation capacity developed in sectors and subsectors through innovation system interventions*. International investors may also wish to use benchmarking to make international comparisons. These comparisons are needed to track macro-level progress and to help target subsectors, sectors, and countries for investments to strengthen innovation capacity or for other investments that require certain levels of innovation capacity as a precondition. A number of methodological challenges in measuring and comparing context-specific and systemic capacities of this sort remain to be addressed, however.

LESSONS AND RECOMMENDATIONS FOR PRACTITIONERS

The main lessons and recommendations for practitioners mirror the principles of good practice outlined earlier. A description of Theory-Based Impact Evaluation (box 7.19) illustrates how these good practice principles and the methods associated with them work together to ensure that the evaluation of innovation system interventions contains accountability and learning dimensions.

The evaluation of innovation system interventions should give particular attention to the following:

- *Timing of evaluations.* While many innovation system interventions will have quick wins, most impacts will emerge only much later as institutional change kicks in, creating new capacities for innovation. While process and institutional change evaluation can take place shortly after the completion of an intervention, impact evaluation can take place only after sufficient time has passed, often 3–5 years later.
- *Looking widely for impact.* The evolving nature of interventions means that outcomes and impacts can be unpredictable, both in the types of impact observed and their geographical and social location. Evaluation needs to be sensitive to this unpredictability.
- *Investigating unexpected outcomes.* Since innovation system interventions operate in dynamic environments, changes and unexpected outcomes can occur that have significance for impacts or can provide opportunities for new investments. These unexpected outcomes need to be recorded and investigated for any lessons they provide.

Table 7.9 Evaluation Approaches Relevant to Innovation System Interventions

Method	Purpose/use	Strengths	Weaknesses	Major references
Theory-based Impact Evaluation	– Helps in understanding why a program has or has not had impact. – Best suited to investigating impacts and undertaking policy learning in large-scale investments.	– A combination of counterfactual-based impact estimates with a detailed exploration of causal links and theories of change has the potential to provide compelling lessons and evidence. – Lessons gleaned from such an exercise are valuable in trying to understand what works in development. – Offers far more valuable lessons for policy through its insights on what doesn't work in development.	– Not yet in widespread use, so experience limited. – Construction of a robust counterfactual can be challenging. – Data-intensive. – Expensive, though good value for money if well executed.	White (2009b)
Innovation and Institutional Histories	– Understand past innovation processes and identify institutional factors that foster or hinder innovation. – Forge a shared vision of the future among stakeholders.	– Fosters discussion among stakeholders and leads to reflection and learning. – Helps build a shared vision of the future.	– Written accounts are based on recollections and could be biased. – Need skilled facilitators to help collect and analyze the information. – Usually takes a very long time for significant lessons to emerge.	Douthwaite and Ashby (2005); Shambu Prasad, Hall, and Thummuru (2006)
Participatory Impact Pathway Analysis	– Guide project management, especially in complex situations where innovation is seen as emerging from a network.	– Involves intervention staff and key stakeholders in constructing impact pathways. – Includes both (1) a causal chain of activities, outputs, and outcomes that sheds light on how an intervention achieves its goals and (2) network maps that show evolving relationships between participants. – Underlines the fact that innovations emerge from a network and not a linear "pipeline." – Promotes reflection, self-evaluation, and learning. – Provides a framework for carrying out action-research.	– A relatively new approach that needs to be further implemented to gauge its weaknesses.	Douthwaite et al. (2003)
Causal Process Tracing	– Well suited for complex, long-term interventions with systems learning goals.	– Places data and theory in close proximity. One quickly sees what works and what does not in an intervention's lifetime.	– Regarded as not very strong for wider generalization but more suited to narrow specification of the reach of causal propositions. – Takes a great deal of time. – Not conducive to parsimonious theory and leads to partial, middle-range theory. It is easy to miss causal complexity. – Easy to lose sight of the broader context.	George and Bennett (2005); CoS-SIS (2009); Walters and Vayda (2009)

Most Significant Change	– Make sense of an intervention's impact and foster learning. – With the help of all primary stakeholders, identify the most significant changes that have occurred as a result of an intervention.	– Enhances capacities of stakeholders (at the organizational and individual levels) and fosters learning. – Helps identify unexpected changes or outcomes. – Large amounts of information are processed, from which negative and positive changes are deduced. – More accessible than traditional M&E techniques; no specific skills required to participate.	– A subjective expression of the values and concerns of the stakeholders designated to select the stories.	ECDPM (2006); IFAD (2002); Davies and Dart (2005)
Stories and Narratives	– Shed light on the changes that have occurred at the individual, organizational, or institutional level. – Uncover intangible factors (qualities, values, culture, and so forth) that determine the organization's character.	– Exchanging stories builds trust between participants; in some cases may lead to an environment conducive to learning and stimulate change. – An effective way to deal with passions and emotions of individuals involved. – Has the potential to stimulate change if told correctly. – Individual focus can translate into organizational development. – Not hierarchical.	– Biased; depends on the perspective of the person telling the story.	Asif (2005)
Benchmarking Innovation Capacity	– More than just evaluating individual interventions. Can help investors benchmark the performance of sectors and subsectors in terms of capacity built for innovation.	– Can be useful in tracking macro-level progress and to help target subsectors, sectors, and countries for investment in capacity strengthening or for investments that require certain levels of capacity as a precondition.	– Methodologies still being developed.	CPR, CRISP, and LINK (2008); Kraemer-Mbula (2012 forthcoming); Spielman and Birner (2008)

Source: Authors.

■ *Acting on evaluation lessons.* Investors must be willing to act upon the lessons that emerge from evaluations of their innovation system interventions. This can be easier said than done when lessons point investors away from traditional approaches and the interests of stakeholders associated with them. One approach that investors increasingly use to bolster their confidence in emerging lessons is to use systematic reviews of similar projects to test key assumptions about suggested ways forward.

The principles outlined in this module and the evaluation approaches summarized in table 7.9 are well known to professional evaluators in the wider development evaluation community. What is important is that investors commission evaluations that embody these principles and select evaluators with experience in applying them. In the short term, they probably must look beyond the pool of evaluators who have undertaken traditional assessments of the impact of agricultural research investments.

Self-Organizing Networks in Policy and Planning: Experience from Sierra Leone's Partnership for Agricultural Innovation and Development

Steen Joffe, Innodev Ltd.
David Suale, Research Into Use (RIU)
Adolphus Johnson, Research Into Use (RIU)

SYNOPSIS OF PROJECT DATA

Project name: Sierra Leone Partnership for Agricultural Innovation and Development (SL-PAID)

Established: July 2008 by Memorandum of Association, in a process facilitated by the DFID-financed Research Into Use Programme

Incorporated: August 2008 in Sierra Leone as Company Limited by Guarantee

Launched: January 2009

BACKGROUND AND CONTEXT

For Sierra Leone, the period of rapid change following the disastrous 1991–2002 war has provided opportunities to rebuild and reinvent institutional frameworks supporting agriculture. The country's new vision for agriculture requires the sector to shift toward a pluralistic and competitive rural service economy, with agribusiness acting as an engine of socioeconomic growth and development. For that vision to become a reality, a stream of new policy, process, and technological innovations must be created. Although inside or outside of Sierra Leone there is no shortage of knowledge relevant to Sierra Leone's current needs, this knowledge is not yet used on any scale for innovation, because the conditions to use it are lacking.

In a well-functioning AIS, knowledge flows between all the principle domains, underpinning myriad investment decisions and behavior changes that collectively drive productivity and growth. In Sierra Leone, as in many low-income development contexts, these flows and interactions around problems and opportunities are weak and have historically taken place within a limited institutional context.

Many research agencies in sub-Saharan Africa have evolved little over recent decades, are institutionally hidebound, and have little accountability to other elements of the innovation system. Priority-setting mechanisms that shape investments in the AIS still tend to be formal, generally top-down exercises. They lack the sort of broad-based, "many-to-many" problem-solving exchanges that are essential drivers of innovation.

PROJECT OBJECTIVES AND DESCRIPTION

Within the Research Into Use Programme (IAP 7), a country strategy team worked with local stakeholders in Sierra Leone to develop an innovative strategy for an autonomous, self-organizing group to develop as a platform for improving policy and practice related to agricultural innovation (Joffe et al. 2008). This open, inclusive network would be a medium for creating new productive alliances and creating value.

As a first step, the team worked through a local secretariat to bring key actors together in a series of workshops, ensuring participation across the innovation system. Participants included farmers, farmer-based organizations, and representatives of rural communities; agribusiness and market actors (processors, wholesalers, retailers, input companies, equipment suppliers, and financial service providers); knowledge intermediaries, including technical advisory and business development services; communications services and the media; knowledge generators (research, education, and other widely used sources); and policy makers, decision makers, and regulators with influence over "framework conditions."

The participants used innovation system mapping techniques to visualize the main elements and actors in the AIS and the strength of the linkages and knowledge flows between them. More specifically, they assessed and mapped knowledge flows between elements of the system and the factors influencing these flows in relation to rural and non-rural livelihoods and routes out of poverty. They also identified key drivers of change in the context for agricultural policies and programs and determined where investments and other economic activities were likely to create new challenges and demands for knowledge. This process required participants to develop an initial map of the key institutional elements and organizations relating to the AIS, which was transposed into a more formalized innovation system domain structure. Next, the participants worked in groups to map linkages between those domains. The subsequent analysis and discussion highlighted a number of findings:

- **Weak or ineffective knowledge flows from markets** into the small-farm sector resulted in asymmetries that led to exploitive behavior.
- **Very unidirectional and supply-driven knowledge flows from intermediaries and from policy processes** into the small-farm sector led to coordination problems and high transaction costs (see the next point).
- **Formal policy processes were poorly linked with other key innovation system elements.** Weaknesses at the center and coordination problems limited the public sector's effectiveness.
- **Weak links between financial institutions and other elements of the AIS** meant that formal lenders lacked information to support credit flows to rural entrepreneurs.
- **The research system had poor links with all other actors in the AIS.** It did not deliver knowledge oriented towards value addition in the market chain, and a lack of trust persisted between researchers and actors in the production and agribusiness "side" of the sector.

As a result of the workshops and related meetings, a core group of actors decided to organize and form a partnership. Through flexible alliances, members would engage in innovative activities and build a better business environment to foster and scale out those activities. The Partnership for Agricultural Innovation and Development (PAID), established in Freetown in 2008, now operates autonomously as a "partnership of service-providers," open to all who subscribe to its vision and mission.

Under a Memorandum of Association, PAID has a general assembly of all members, a board (elected by the members and mandated to manage the affairs of the partnership), and a secretariat (responsible for day-to-day operations, coordination, and administration). In summary, the association's defining features are that it is self-governing, membership-based and has a broad, representative membership; it is financed through subscriptions; it receives broad support from the Government of Sierra Leone and principal agencies; and it is embedded in key policy forums.

BENEFITS, IMPACT, AND EXPERIENCE TO DATE

PAID is a young organization but already operates successfully in a number of key respects. PAID members have formulated and voted on appropriate governance structures, rules, norms, and the interaction mechanisms required for PAID to operate effectively as a vehicle for delivering innovations in the agricultural sector. Under rules established by the membership, "formal" interaction in PAID is collective and consensual. Strategic decisions are made or endorsed by the general membership at the annual general meeting. Operational decisions are taken at the district level and at events held at the platform level.

The organization has also launched two fast-track, commodity-based innovation platforms. These initial platforms were selected based on information from field assessments by or with the participation of PAID member organizations:

- **Solar drying in fruit and other horticultural value chains.** Twenty solar drying units operate in communities in four districts in the Northern Province; early reports indicate that they are used successfully.
- **Poultry feed production and marketing.** Maize is being grown on 110 acres (46 hectares) across locations in Bo, Kenema, and Kailahun Districts for a seed multiplication drive and eventual use for feed in intensive poultry production systems.

The platforms have spun off active partnerships with other actors who want to adopt platform technology, including CARE International, MADAM-Sierra Leone, the Sierra Leone Centre for Agribusiness Development, and the Nehemiah Project. To guide future priorities for developing platforms, PAID has documented a more structured approach consisting of open calls for proposals, concept screening, investment events, and small grants for opportunity development.

Membership in PAID makes it easier for people and organizations to obtain information to guide choices, gain support for their decisions, and build alliances. Typically, information flows informally among the wide range of stakeholders present at key events and meetings, including stakeholders from the Ministry of Agriculture, Fisheries, and Forestry, other central and district government agencies, FAO, DFID, the Sierra Leone Agricultural Research Institute, banks, farmer organizations, civil society and nongovernmental organizations, producers, processors, traders, and transporters.

PAID also has a place on the Agriculture Advisory Group and Technical Committee, a formal, policy-making standing committee that was involved in developing Sierra Leone's Comprehensive Africa Agriculture Development Programme compact and National Sustainable Agricultural Development Plan, linked to the national poverty reduction strategy. PAID has also brought together actors in extension to form the Sierra Leone Forum for Agricultural Advisory Services. Stakeholders in this process identified gaps in extension services; as a result, a policy on extension in Sierra Leone is being developed. Finally, PAID members in Bo District raised concern over the activity of a fraudulent actor in their region. This concern was channeled to the secretariat and brought to the notice of ministry officials. The situation was corrected.

INNOVATIVE ELEMENT: SOCIAL BUSINESS NETWORKS FOR INNOVATION SYSTEMS

From a public policy perspective, the means to foster innovation often focus on improving linkages and flows of information between actors and interests. For this reason, governments offer incentives for innovative businesses to locate in clusters (see TN 4 in module 5), where their close association can generate a variety of spillover benefits and externalities. Governments also subsidize interactions by organizing events and meetings around key policy areas in which the state wishes to drive innovation that the market alone may not deliver.

These strategies are valid but require complementary approaches to enable the "conversation" that will support policy and practice. The new frontier for agencies wishing to support the AIS—rather than supporting priority-setting processes alone—is to foster self-organizing social business networks (see TN 2 in module 1). Such networks respond to the interests of a diverse group of stakeholders and can operate flexibly and dynamically in light of demands and opportunities identified by their membership.

This general model is well established as a way to organize innovation in industry, where social business design approaches, increasingly underpinned by new social media tools and engagement strategies, are used to harness

collective intelligence around investment decisions. Within the social business paradigm, boundaries between public, private, and third sector roles begin to break down for practical purposes. Similarly, boundaries between policy and practice and between planning and delivery are no longer necessarily institutionally or organizationally distinct; these activities are undertaken instead by individuals and groups applying creative and flexible solutions to common problems across organizational boundaries, representing particular interests and competencies in pursuit of value-creating opportunities. These approaches are already driving transformational change in more advanced economies and may be equally transformational in developing country contexts.

LESSONS AND ISSUES FOR WIDER APPLICATION

Within the PAID network, a number of early decisions were highly influential in *building social capital and reducing transaction costs*.

Network membership and the role of facilitation

Members decided that the partnership would be autonomous and establish its own secretariat, membership would be open to those who subscribed to the partnership's vision, and organizations would be represented by senior representatives with decision-making power. Membership also would be open to service providers of various kinds, in agriculture and agribusiness. An elected executive would represent the partnership between general meetings, a general code of conduct would be drafted and agreed, and the network would be financially independent and sustainable. In addition to charging registration and annual subscription fees, the network would raise financing independently and develop revenues from innovative activities. Revenue would not only sustain the partnership but serve as an incentive for membership.

The network was able to come to these decisions because of the *initial investment by key actors in facilitating a core understanding and common vision of the network's role*. Members' early involvement in decisions on structure and governance increased confidence and promoted engagement. Higher levels of trust and reduced transaction costs have enabled PAID to rely more on signals through the internal "knowledge market" and less on formal priority-setting mechanisms.

Existing sector frameworks are still needed

Networks for innovation are not a replacement for or an alternative to existing sector-wide and/or vertical subsector frameworks. Such frameworks are still needed to provide the formal process and democratic accountability around policy and practice. Networks like PAID bridge such frameworks, providing an opportunity for more flexible and less formal interaction within a different institutional paradigm.

For networks like PAID, *engagement with government is important but should be balanced to avoid prejudicing the network's independence and openness in decision making.* Attaining such a balance can be tricky. For example, Rwanda's National Innovation Coalition is restricted to a few public sector bodies. Plans to establish another separate and more open agricultural innovation network attracted no high-level support and were never implemented. In Bolivia, the Natural Resources Information and Knowledge Network (SICTAF, Sistema de Información y Conocimiento Tecnológico Agropecuario y Forestal) explored establishing itself as a self-governing, socially inclusive knowledge network. The responsible ministry was unable to engage productively with such a network other than as its apex institution, however; it never accepted that the network could operate within a self-defined governance framework. This lack of formal support was a key reason why the initiative foundered.

One lesson from this diverse experience is that *the open, "flat," nonhierarchical structure that favors innovation is not always consistent with established ways of doing government business.* Indeed, where coalitions are facilitated to promote AIS they can end up reinforcing existing networks rather than bringing in new voices and influences on policy.

Financial sustainability

As always, financial sustainability remains a major consideration. The facilitation, coordination, and "market research" underpinning the creation of PAID in Sierra Leone were financed by DFID, along with the core administration and personnel costs. *This initial subsidy is very likely to be a generic requirement for such networks, but it should be provided only in the context of a strategy and path towards financial independence.* As noted, PAID's member organizations pay a registration fee and annual subscription (approximately US$30 and US$150, respectively) to support core network costs.

Improving linkages and working relationships between research and the business sector

Many networks dissolve because members perceive little real value in participating. A key lesson from Sierra Leone and elsewhere is to *provide pathways to see initiatives through and gain the rewards of participation, both professional and financial.* In practical terms, this means that the network should offer a framework for adaptive research, enterprise development, and scaling out innovations. "Pull" mechanisms such as innovation-financing events and competitions and related processes will help to seed ideas and new enterprises out of the network. For one of its members, the Sierra Leone Agricultural Research Institute, PAID acts as a new framework for *improving linkages and working relationships between research and the business sector.* The network also serves as a channel to bring technologies already developed by the institute into use and for the institute to "plug in" to the new priorities emerging from innovation platforms.

PAID's continuing heavy reliance on transmitting information directly through meetings and field operations raises costs and is one area where improvement is needed. *The value of face-to-face interaction will never be replaced, but social networks for AIS can and should develop strategies to apply mobile and web-based social media.* These technologies permit information and knowledge to be exchanged at a low cost, both "internally" and with other stakeholders and influencers locally and globally. PAID is exploring these avenues through the African Forum for Agricultural Advisory Services, which is piloting the Innodev platform (www.innodev.org) in Sierra Leone and Uganda to support problem solving through networks, enable groups to form around opportunities, and attract support for entrepreneurs through mentors and investment partners.

Using Net-Map to Assess and Improve Agricultural Innovation Systems

Eva Schiffer, Consultant

SYNOPSIS

Practitioners require methods that capture the complexity of an AIS and structure it in a way that allows actors to use the detailed information rapidly. Net-Map is a participatory influence network mapping method based on social network analysis and power mapping. This pen-and-paper method helps those involved in or observing agricultural innovation to determine and discuss who the actors are, how they are linked, how influential they are, what their goals are, and what the crucial bottlenecks and opportunities are. Net-Map is useful for understanding complex, dynamic situations in which multiple actors influence each other and the outcome. It can be used for an initial assessment of an innovation system and can also help to monitor the innovation system's development over time.

BACKGROUND AND CONTEXT: REQUIREMENTS FOR AN AIS ASSESSMENT TOOL

The AIS approach, instead of focusing on specific actors, appreciates that an innovation involves multiple partners who have formal and informal ties, have different goals, and use their influence in various ways to further or block the innovation. Practitioners increasingly require methods that capture this complexity and structure it in a way that allows actors to use the detailed information rapidly.

AIS assessment tools ideally need to capture a specific range of complex data and to do so under particular logistic conditions. The following data are needed to understand an AIS:

- Who are all the actors involved (impacting on and being impacted by the innovation, formally and informally involved, supportive and unsupportive of the innovation)?

- How do these actors interact (including formal and informal links, material flows such as funding or seed, nonmaterial flows such as ideas, policy pressure)?

- What are their goals with regard to a specific innovation or the general innovativeness of the system (are they supportive, unsupportive, or neutral)?

- How strongly do they influence the innovation system's ability to innovate?

- What are the crucial strength and weaknesses of the innovation network? Where are bottlenecks and coalitions? What links are missing? What strategies are successful?

- How does the innovation system change over time?

With regard to the logistics of the method needed, it is crucial for it to be straightforward and easy for people to apply in the field; provide results quickly; allow for exploring and understanding systems with many unknowns; structure the complexity but leave room for in-depth explanations; support users in developing strategies for improving the AIS; and collect data that are comparable between sectors, countries, and over time.

INNOVATIVE ELEMENT

Net-Map, a participatory social network mapping approach (Schiffer and Hauck 2010), is based on social network analysis (Hanneman and Riddle 2005), power mapping (Schiffer 2007), stakeholder analysis (Grimble and Wellard 1997), and participatory action-research (Kindon, Pain, and Kesby 2007). This method helps those involved in or observing agricultural innovation to determine and discuss who the actors are, how they are linked, how influential they are, what their goals are, and what the crucial bottlenecks and opportunities are (Schiffer and Hauck 2010; http://netmap.wordpress.com).

HOW NET-MAP WORKS

Net-Map can be used in planning, implementing, monitoring, and evaluating interventions. The following step-by-step description briefly explains how it works.

1: Getting started

Net-Map is a pen-and-paper method that involves drawing networks together with participants (individuals or groups) to capture their complex knowledge of a system and make implicit or tacit knowledge explicit. Net-Map can be used as a tool by external actors, such as donors or researchers, to acquire a better understanding of the situation and monitor its development. It may also be used internally (for example, by an implementing NGO or ministry) to help decision makers and implementers improve stakeholders' involvement, strategic planning, monitoring, and evaluation. Net-Map sessions are typically facilitated by a trained Net-Map practitioner.

Before starting the activity, it is important to determine its framework and goal: Net-Map can be used as a one-off activity for planning or as a startup tool to get an activity on track. For monitoring and evaluation, a baseline Net-Map at the beginning of the intervention and one or more follow-up Net-Maps are recommended (for example, after one, three, or five years). It is possible to do a string of individual interviews or one (or more) group meetings.

The first step is to develop the overall question. It normally has the format: "Who influences XY?" XY can be specific ("Who influences farmers' adoption of this new rice variety in this area within the next five years?") or more general ("Who influences the innovativeness of the agricultural sector in this country?").

Often the more specific questions provide more specific and therefore useful answers. For example, one might learn that network structures that encourage the adoption of a new rice variety might be similar to those for other crops.

2: Who is involved?

The people attending the session normally consist of the host (the person/organization who is looking for answers), the facilitator (expert in the Net-Map method, neutral in the content question), participants from different areas of the innovation system, and a note-taker. Choosing the right participants is crucial, because the knowledge of the people interviewed is the core source of information. For example,

a broad set of perspectives could be gained by including people from: the private sector, NGOs, donor agencies, government, and civil society (farmers, consumers); the national, regional, district, and local level; different ethnic groups, nationals and foreigners, different ages and genders; and agriculture, trade, finance, and industry. The ideal group is between 6 and 12 people. Larger groups should be split into (equally diverse) subgroups.

The participants are asked to name all actors (individuals, groups, organizations) involved. Actors include not only those who are involved in formal decision making but everyone who can influence or is influenced by the issue.

3: How are they linked?

A link is something that flows from one actor to another (like money) or connects two actors (like friendship). Typical links in an innovation system are flows of money, ideas, innovative products, political pressure, and formal lines of command.

4: How strong is their influence?

This question focuses on how strongly the different actors can influence the specific issue at hand (not in the country at large)—for example, "How strongly can this actor influence whether farmers use this new rice variety?" Actors' influence is defined as their ability to achieve their goals in a social setting, despite resistance (Weber 1922).

The level of influence is represented by an "influence tower" (using some small, stackable objects). The greater the influence, the higher the tower. The influence tower is used to assess an actor's actual influence on a given issue; the actor's influence can be based on a number of attributes, such as money, formal position, persuasiveness, informal ties, and so on. The influence tower *does not* measure the actor's formal position or how influential the actor should be.

5: What are their goals?

The next step focuses on understanding the actors' goals. In some cases it makes sense to ask who actively supports the innovation or innovativeness, who is passive, and who actively hinders it. In other cases, actors might follow two competing philosophies or goals.

6: Discussion (what does this mean)?

In this step, the map is drawn, and any issues that came up with the mapping are discussed. The discussion can include looking at bottlenecks, conflicts, coalitions, future strategies, missing actors, or links that should be developed in the future. It is not important to reach agreement on every point but to explore and understand different points of view, why people hold them, and how these different views can affect the innovation system. For example, if a representative from the agriculture ministry and one from the environmental protection agency disagree on a crucial issue, that information, in and of itself, can be important for shaping a future strategy of engagement with these two agencies.

7: Results

The Net-Map session yields the map and discussion notes. The network map can be entered into social network analysis software (such as Visualyzer™ or UCINET™), which produces a computerized network picture and also allows for some quantitative analysis, such as identifying bottlenecks or boundary spanners. While the network structure provides the bones, the discussion adds the meat, giving concrete information about how and why the network performs or fails.

Less tangible (but sometimes even more relevant) outcomes of a Net-Map session are the learning and energy shared by the people attending. Participants regularly report that they have gained enthusiasm for a common cause, have resolved misunderstandings, and have a clearer vision and shared strategy after attending Net-Map sessions. The intangible effects are especially powerful if participants discover blind spots together (see box 7.21 in the next section) or if a diverse group develops a common understanding. To make the most of these process results, it is crucial that the host is seriously committed to using them and continuing to collaborate with the participants.

BENEFITS, IMPACT, AND TWO CASE STUDIES

Since its development in 2007, Net-Map has been used in a variety of ways, within and beyond agriculture, in Africa, Asia, Europe, and the United States (for detailed case studies and methodological development, see http://netmap.wordpress.com). The uses have been as diverse as the following:

- Developing benchmarks and indicators for chicken and maize innovation systems in Ethiopia (box 7.20); see also Spielman and Birner (2008); Spielman and Kelemework (2009).
- Assessing communication channels concerning avian influenza in Ghana (box 7.21), Ethiopia, and Nigeria; see also Schiffer, Narrod, and von Grebmer (2008).

- Engaging stakeholders in Nigeria, Ethiopia, and Uganda in developing bisosafety legislation under the International Food Policy Research Institute's (IFPRI's) Program for Biosafety Systems; see http://programs.ifpri.org/pbs/).
- Understanding and improving regional water governance in northern Ghana; see Schiffer and Hauck (2010).
- Understanding fisheries management in small reservoirs in northern Ghana; see Hauck and Youkhana (2008). The Net-Map exercise revealed that overlapping governance systems (traditional and modern, top-down and bottom-up) were one reason for unsustainable management practices and poor enforcement of rules.
- Increasing the impact of agricultural research on policy making in Malawi and Nigeria; see Aberman et al. (2010). By looking at concrete case studies (such as studies of fertilizer policy), this project aims to understand when and how research can enter policy-making processes. Follow-up Net-Map sessions in Malawi will track changes over time.

Typically, the goals of a Net-Map intervention are twofold—to understand and to improve a situation. Two case studies provide more detail on how Net-Map was used in analyzing an innovation system in Ethiopia (box 7.20) and developing strategies to prevent the spread of avian influenza in Ghana (box 7.21). General lessons from the use of Net-Map are provided in the concluding section.

LESSONS LEARNED AND ISSUES FOR WIDER APPLICATION

Net-Map is a useful tool for understanding complex, dynamic situations in which multiple actors influence each other and the outcome. It can be used for an initial assessment of an innovation system in a country or sector and can also help to monitor the innovation system's development over time.

A Net-Map facilitator needs to be good at working with groups and individuals, giving them room to express themselves but also guiding them when the discussion goes off on a tangent. It helps if the facilitator is able to think in structures and discover patterns in complex maps. Prior knowledge of social network analysis is a plus but not necessary.

The Net-Map steps are normally taught in a learning-by-doing approach. After a brief (one-hour) session, new Net-Map facilitators are able to draw their first Net-Map on an issue of their choice. In five to eight days, with the help of an experienced Net-Map practitioner, a new Net-Map

Box 7.20 Net-Mapping a Poultry Innovation System in Ethiopia

The Debre Zeit-Mojo corridor in the Addis Ababa market shed is an exceptionally innovative area for poultry, where new breeds and methods are embraced much more rapidly and widely than in other areas of the country. Could Net-Map discover the network conditions that encouraged agricultural innovation? What lessons could be learned for other geographical areas and other agricultural products?

Net-Maps were drawn with researchers, extension agents, and poultry farmers. The links that were mapped included: production inputs and equipment (embodied knowledge); knowledge and information (disembodied knowledge); credit and financial services; regulatory oversight; and coordination and cooperation.

Through this process, participants discovered an innovation cluster of strongly interlinked private and public sector actors who had facilitated the development of a small commercial poultry farm sector. The analysis showed that this innovation cluster was inextricably bound to the specific location (the Addis market shed, with a market for white-fleshed chicken) and the collaboration between large-scale poultry producers and an agricultural research center. This context-specificity meant that the Debre Zeit-Mojo innovation cluster did not deliver a blueprint for poultry innovation systems in more remote areas of the country. Further research, for example through Net-Maps of poultry systems in less-privileged areas, would be needed to understand how innovation could be fostered under different conditions.

This experience shows how Net-Map can tease out which innovation conditions are specific to a given context and which can be transferred as general lessons for other areas. The method also helped colleagues who had worked in the country for a long time to see aspects of the innovation system of which they were unaware.

Source: Author; Spielman and Kelemework 2009.

Box 7.21 Net-Mapping to Reduce the Risk of Avian Influenza in Ghana

Net-Map was used in kick-off workshops for a project on pro-poor strategies to reduce the risk of avian influenza (http://www.hpai-research.net). Stakeholders from different areas of poultry production, marketing, and government oversight mapped all of the actors involved, focusing on two links: (1) flows of information about suspicious bird deaths and (2) flows of intervention if avian influenza was confirmed.

Group mapping allowed participants to exchange knowledge about this network and highlight specific bottlenecks. In Ghana, mapping revealed critical issues that had not been clear to the participants or researchers beforehand. The Net-Mapping session indicated that if there was an outbreak on a small farm, considerable information would be exchanged at the village level (including among teachers, opinion leaders, and other actors unrelated to the poultry subsector). Only one actor, however, bridged the gap between the community actors and district administrators: the animal health technician. The relatively low number of animal health technicians in the system increased the risk that reports of suspicious bird deaths would be delayed.

An even more crucial insight was related to the neglect of market actors in avian flu compensation schemes. Farmers were compensated for every bird culled by the government in an outbreak, but no such compensation was available to live bird traders. Participants diagnosed a potential corruption hot spot at the national border: Suspicious bird deaths in a trader's flock would give the trader strong incentives to bribe border veterinarians, cross to the neighboring country, sell the birds, and leave as soon as possible. This kind of activity sets the scene for a regional pandemic. The network figure shown here represents flows of information about suspicious bird deaths in Ghana, indicating the corruption hotspot at the border.

Source: Author.

Note: More information including illustrative Net-Map examples can be found at http://netmap.wordpress.com.

intervention can be developed with a team of facilitators with no prior training in the method. The basic process is taught; the proposed question and links are pretested with a number of interview partners; the questions are adjusted; and the new facilitators learn how to enter the data. A particular challenge in every Net-Map intervention is to ask the right general question.

Following this preparation, the group of facilitators will either invite participants to a group mapping session or conduct a series of individual interviews. Group mapping sessions are especially powerful for getting consensus and buy-in, developing strategic plans, and getting answers rapidly, without much additional analysis. If possible, plan one full day for a group session to allow for discussion and avoid rushing participants. It is possible to do a group Net-Map in half a day, however, and make it part of a bigger

event, such as an inception workshop or annual planning meeting.

In some cases, however, individual interviews are more convenient. Actors may be geographically spread out or otherwise difficult to reach; interview partners may speak more truthfully about sensitive issues, especially if there is a great power difference between stakeholders or a history of conflict.

As noted, it is beneficial to have a time series; for example, Net-Mapping could be done at the beginning of a project, halfway through, and at the end. During each session, discuss what is useful, identify any underutilized opportunities, and identify bottlenecks. Develop strategies accordingly and use the next mapping session to see how the network changed, which strategies were successful, and what still needs to happen.

Gender Analysis for the Assessment of Innovation Processes: The Case of Papa Andina in Peru

Silvia Sarapura, University of Guelph

SYNOPSIS OF PROJECT DATA

Project name: Papa Andina

Country/region: Papa Andina works through a range of strategic local partners in each country: the PROINPA Foundation (Bolivia); the National Potato Program, INIAP (Ecuador); and the INCOPA Project (Peru)[1]

Starting date: Papa Andina (1998); Peru PMCA (2001)

Closing date: Ongoing

Project financing: Initially Swiss Agency for Development and Cooperation; also New Zealand Aid Programme, McKnight Foundation

Implementing agency: Partnership Program hosted by the International Potato Center (CIP)

Website: http://www.papandina.org/

CONTEXT

Across the Andean region, small-scale farmers face the challenge of gaining access to dynamic new markets for high-value produce while remaining resilient amid the forces of climate change and globalization. The Papa Andina regional initiative, anchored in the International Potato Center (CIP), promotes innovation that leads to the development of market niches and value addition, particularly for the native potatoes grown by poor smallholders in Bolivia, Ecuador, and Peru (box 7.22). The assessment of gender issues plays a critical role in Papa Andina's two principal approaches to engage market chain actors: the Participatory Market Chain Approach (PMCA) and stakeholder platforms (see also TN 1 and IAP 1 in module 4):

Box 7.22   New Market Niches and Value Addition for Small-Scale Growers of Native Potatoes in the Andes

Papa Andina led to the creation of T'ikapapa, the first commercial brand that supports the sale of native potatoes under strict quality standards. T'ikapapa connects small-scale potato farmers in the Andes with high-value niche markets in urban centers, exports its products to other countries within the region such as Venezuela, explores potential European markets for high-end potato products, and partners with an increasing numbers of nongovernmental organizations and private operations to further promote native crops.

Source: Author.

- **The PMCA** is based on the participatory approach to stakeholder collaboration in agricultural R&D known as Rapid Appraisal of Agricultural Knowledge Systems (RAAKS; see also box 7.11) (Engel and Salomon 2003). The PMCA fosters commercial, technological, and institutional innovation through a three-step process that builds interest, trust, and collaboration among participants, improves farmers' links to markets, and stimulates pro-poor innovation.

- **Stakeholder platforms** (see also TN 2 in module 1) are spaces and events where public and private stakeholders interact, share reciprocal interests, build trust, and join in common initiatives. Often such platforms are developed as a result of PMCA and continue after the approach has been implemented; in other cases, the PMCA works through platforms that already exist.

Both the PMCA and stakeholder platforms facilitate the articulation of demand and supply for innovation-linked services and reduce transaction costs in marketing the produce of many small farmers (Bernet et al. 2008). In the Andes, PMCA has been validated in two complete cycles, both in Peru and Bolivia (2003–04). The method has been shared with other organizations in these countries, which has led to further testing. In Peru, the Intermediate Technology Development Group, an international NGO, subsequently used the method in the cheese, coffee, and cacao subsectors. Starting in 2005, PMCA was introduced and tested in potato, sweet potato, and vegetable commodity chains in Uganda.

OBJECTIVES AND DESCRIPTION

A key feature of Papa Andina is that it brings together many participants in the AIS, including smallholders, market agents, and agricultural service providers, many of whom did not know one another or who actively distrusted one another, and helps to identify new opportunities for all of these stakeholders to collaborate and innovate. Papa Andina recognizes that gender analysis and female farmers' active involvement in assessing innovation processes and systems are central to developing sustainable, profitable agricultural market chains that are well integrated into the wider innovation system. In turn, this system-level integration is important for gender equality and the empowerment of resource-poor women and their families.

Each phase of the PMCA incorporates specific gender-related assessments and activities (table 7.10). Flexibility in the duration of each phase and in the use of specific tools (quantitative surveys, focus groups, and so forth) is necessary (Bernet et al. 2008).

INNOVATIVE ELEMENT

From a gender perspective, Papa Andina has three innovative elements. *The first innovative element* is that the PMCA and stakeholder platforms enable women to share their findings and customs with other members of the AIS through events and activities that highlight women's knowledge of genetic diversity. When women participate in events such as family competitions, their roles in the farming household, the wider community, the market chain, and the AIS are recognized and reinforced (box 7.23).

The *second innovative element* is that the empowerment of women farmers has resulted in systemic changes.

Through the PMCA, women's involvement and the involvement of different groups of women are systematized in the following ways:

- ***Representation.*** Smallholders, female and male, representing their communities at events return to their communities and share their findings and innovative ideas.
- ***Replication.*** Initial farmers, now acting as representative farmers, work with R&D partners to replicate knowledge-sharing events and activities with more farmers in their area who grow native potatoes. For example, a woman farmer in Puno shared information with representatives of 12 communities in the Lake Titicaca basin. In this way, innovative ideas for making coffee from dried potato and adding value to freeze-dried potato products spread to at least 10,000 farmers in those areas.
- ***Communication and recognition.*** Native potato product ideas and technologies were also shared between women farmers in Peru and women's groups and R&D institutions in Uganda, Bolivia, and Ecuador (Horton 2008; Kaganzi et al. 2009).

The *third innovative element* is that Papa Andina purposefully demonstrated the value of women's involvement in the AIS. The initiative showed that it is possible to involve resource-poor women farmers as key stakeholders in the potato value chain; the participating R&D institutions demonstrated the value added by gender analysis and investing in women's innovation; and the donor agencies played an important role in establishing the need for gender assessment and the integrated involvement of women farmers in R&D as key stakeholders.

BENEFITS, IMPACT, AND EXPERIENCE

A number of gender-related benefits, impacts, and experiences are linked to each of the three phases of the PCMA and to the stakeholder platforms. In *phases 1 and 2*, experiences with gender assessment and gender-related activities in organizing the PMCA and stakeholder platforms have shown how to foster the organization of female and male farmer groups based on common interests and resources. Organizing enables farmer groups to consider the economic feasibility of production and marketing issues beyond the household level. The groups can build their human and social capital to access platforms where support is available from R&D and government institutions as well

Table 7.10 Phases of the Participatory Market Chain Approach and Gender Assessment and Related Activities in Each Phase

Phase	Overall activity[a]	Gender-related assessment and activity
Stakeholders identified Phase 1: 2–4 months	– Get to know the market chain actors and other stakeholders—their activities, interests, ideas, problems, and so forth: Step 1: Conduct a 3- to 6-week rapid assessment of the market chain and identify key stakeholders. Step 2: Hold a one-day workshop to define impact groups using the impact filter. Step 3: Hold a final event for phase 1 to share information and secure stakeholders' continued involvement.	– Integrate gender sensitivity training into R&D organizations as they begin their stakeholder identification activities. – Include women farmers as a stakeholder group in the rapid assessment of the market chain. R&D partners, including investors, reinforce the need to address women's specific needs in PMCA. – Initiate family and community competitions for innovation to recognize women's contribution to the value chain (box 7.23). – Among the stakeholders, identify women's groups and male and female leaders who support gender equity and empowerment; encourage them to highlight or discuss issues and benefits for women farmers in workshop events.
Stakeholder platforms engaged Phase 2: 3–5 months	– In a participatory manner, analyze potential business opportunities; work in thematic groups of 10–20 persons; establish new stakeholder platforms or strengthen existing stakeholder platforms; R&D organization involved provides facilitator to assist groups. – Using the following tools, each thematic group analyzes potential business opportunities: rapid market appraisal; quantitative market survey; focus groups.	– Platforms bring together female and male small-scale farmers from different communities in the region, market agents, and agricultural service providers to share findings and customs, with support from R&D institutions. Many of these stakeholders will be unfamiliar with each other. – Identify and involve NGOs engaged in related gender analysis and women's empowerment programs. Their involvement may be the key to the success of these platforms. – Gender equity (participation of women representing different ages, classes, and ethnic groups) is included in the platforms and in the selection criteria for the thematic groups.
Implementation of joint market innovations Phase 3: 4–6 months	– Implement joint market innovations: work in thematic groups of 10–20 persons; R&D organization involved provides facilitator to assist groups. – Each thematic group uses marketing concept development and business plan to test or implement, monitor, and evaluate their innovations. If necessary, phase 2 activities can be revised (for instance, by adding focus groups to clarify consumer preferences).	– Analyze continued knowledge sharing by women farmers and gender roles and relations within the stakeholder platforms for further technical and institutional innovation at the national and international levels. – Women and men continue to participate in fairs and events outside their communities (regional and national) to demonstrate their knowledge and stimulate participation in stakeholder platforms. – R&D partners monitor/evaluate how individual women farmers have gained confidence to join new and extended networks and to exchange varieties cultivated in other areas of the Andes. – R&D partners monitor/evaluate how male and female farmers have gained individual and collective capacities and skills for communication, negotiation, facilitation, and teamwork. – Encourage ongoing discussion within the project of how market chains empower disadvantaged farmers who otherwise have little opportunity to participate and make decisions. – Encourage ongoing discussion within the project of how women have the chance to interact with other market chain actors and professionals from R&D organizations, thereby increasing their access to knowledge, innovation, contacts, and self-development.

Source: Author.

a. The activities in each phase of the PMCA (described in detail in the "User Guide"; see Bernet, Thiele, and Zschocke 2006) occur consecutively over 9–15 months.

Box 7.23   Innovation Fairs to Assess and Recognize Women's Contributions to Market Chains and the Agricultural Innovation System

Every family and community in the high Andes has developed its own varieties of native potato. Seed of native potato varieties is usually obtained by inheritance, barter, or as a gift. The PMCA partners support local, provincial, regional, and even national fairs—public events where farmers (men and women) have a chance to demonstrate the varieties they prefer to select, store, cultivate, harvest, process, consume, and market. These fairs are opportunities for communities and farmers to highlight the enormous diversity of potatoes they use and explain how they have managed this native potato biodiversity over time. The participants may exchange seed or buy tubers from one another at these fairs.

Fairs represent an excellent opportunity for farmers to obtain information from one another as well as from R&D partners. In most cases, wives accompany their husbands to the fairs, because women are the farm household members with the best knowledge of the morphological and qualitative characteristics of each potato variety. Family collections can be extensive: A small-scale farming family at one fair presented more than 600 varieties. Women farmers report that the fairs enable them to feel rewarded and recognized for their efforts in preserving and maintaining the extraordinary biodiversity of native potatoes.

Source: Author.

as NGOs. This support can also entail technology transfer to farmers and opportunities to fine-tune technologies to specific conditions.

In *phases 2 and 3*, thematic groups use communication and collaboration to address and break down traditional gender roles, divisions of labor, and power relations. Recognizing women's role in the selective breeding of native potato varieties in different ecosystems and their detailed knowledge of different potato phenotypes helps to counteract gender bias. Communication activities, including the innovation fairs, focus on how Andean women have cultivated native potatoes. These activities enable women to

bring their large store of knowledge to bear on the innovation process for native potato.

In recent years, women farmers in some regions of Peru have established profitable businesses supplying native potatoes to national and/or international markets. Messages about women's advancement in marketing chains and innovations have been highlighted in public-private R&D partnerships and corporate social responsibility commitments involving such companies as Pepsi-Co and its subsidiary, Frito Lay.[2] New products marketed by some companies have used the image of an award-winning female farmer. These examples have been reported to the author as motivating female producers to participate in the native potato market chain.

LESSONS AND ISSUES FOR WIDER APPLICATION

Several gender-related lessons have emerged from Papa Andina. Donor priorities were an important contextual consideration for incorporating gender assessment in the native potato innovation system. Donor agencies' initial proposal development and planning criteria for gender, empowerment, and working with NGOs stimulated the requirements for gender assessment and the integrated involvement of women farmers in R&D as key stakeholders. As a result, "researchers and NGOs that have worked with Papa Andina are more aware of gender issues and the need to achieve impact at farmer level" (Devaux et al. 2010).

In some cases, the benefits of traditional and newly developed innovations generated by the stakeholder platforms remain highly localized. For example, with support from USAID, one farming community sold a local variety of potato known as "Capiro" to Frito Lay to produce potato chips for the domestic market (the company had previously imported potatoes from Colombia). Farmers earned more than US$1.6 million in sales, but this success cannot be replicated easily because the domestic market for snack foods is limited. Farmers are also cautioned not to regard this success story as an inducement to grow just one variety of potato. The maintenance of potato diversity remains central to the innovation system and its stakeholder platforms. Although female farmers, especially indigenous women farmers, have brought a wealth of experience to market chains and agricultural innovation, women farmers often struggle to ensure that their knowledge benefits themselves, their families, and their communities. Investment strategies that establish networks of information and knowledge sharing can increase the impact of locally developed and innovative practices and strengthen the abilities of women and their communities to meet their agricultural and economic needs in a culturally appropriate and environmentally sensitive manner.

Despite women's critical role in the potato market chain, subsistence production, in which women are usually involved, receives less institutional support than cash crop production. The number of female extension officers in public extension systems is very limited (although the only NGO working in the high Andes, Fovida, provides a few female agents). As a result, resource-poor women farmers are less likely than their male counterparts to receive agricultural extension services. Forming links to NGOs within phases 2 and 3 of the PMCA is important to strengthening the innovation system in this regard.

Aside from these relatively specific lessons, Papa Andina offers a number of more general considerations about the successful integration of women into any AIS:

- *Assess the entire system and individual agrifood value chains using a mainstreaming approach that includes the use of gender analysis to recognize women's role and gender relations in production and decision making.* Indispensable tools for gender analysis in innovation assessment are gender-disaggregated data; analysis of women's and men's access to resources such as labor, land, capital, and knowledge; and the engagement of women in capacity-building activities.
- *Ensure that the full range of women's and men's activities, resources, and benefits* is reflected in the assessment

of the innovation system and the continuing activities of the stakeholder platforms.

- *Through networking provided by the stakeholder platforms, identify suitable technological and institutional innovations.* In particular, review the suitability of technologies or institutional arrangements available in other market chains that have become successful and sustainable for women farmers.
- *Identify and respond to socioeconomic factors that may affect the adoption of proposed technological or institutional innovations* (for example, security of resources; tenurial arrangements for land or water; access to inputs such as credit, seed, and fertilizer; and membership in producer groups). Identify activities that are particularly time- and/or energy-consuming for women and address them with targeted investments and supporting interventions.
- *Increase and sustain the supply of information, technologies, and facilities that women may fail to access because of social exclusion* (examples include market information, transport, appropriate tools and equipment, and so forth).

Papa Andina illustrates the centrality of gender issues in sustainable and inclusive agricultural development and the effectiveness of the AIS as a whole. Gender assessment and strategies to ensure the participation of women in value chains are important tools to identify the strengths and diversity of actors in innovation systems. R&D institutions play an especially important role in ensuring that innovation benefits small-scale male and female farmers.

Scenario Planning to Guide Long-Term Investments in Agricultural Science and Technology in India

Riikka Rajalahti, World Bank

SYNOPSIS

An Indian Council for Agricultural Research (ICAR) and World Bank team engaged in scenario planning from September 2004 to June 2006 to assess critical policy and institutional challenges for agriculture and corresponding reforms that would enable the research system to meet them. ICAR management evaluated the likely benefits and impact of alternative reform scenarios and determined which specific reforms to support through the National Agricultural Innovation Project (NAIP). This process increased the government's ownership of the reforms and its commitment to implement them. This profile summarizes key elements of the process and its findings. The cost of the scenario planning in India involved several components: external facilitators (US$100,000); preparatory studies (US$30,000); workshops (US$60,000); peer reviewing (US$5,000); and dissemination (US$20,000). Funding came from the budget for preparing NAIP (US$155,000) and a US$60,000 grant from the World Bank's Agricultural and Rural Development Department (ARD) to support knowledge generation. The investment of staff time was also substantial for ICAR and the World Bank, on the order of 30 weeks for each institution.

BACKGROUND AND CONTEXT

To address the challenges facing agriculture in India and consolidate the gains under the completed National Agricultural Technology Project, the Government of India and the World Bank agreed to undertake a new National Agricultural Innovation Project (NAIP) (see IAP 2 in module 4). In preparing the project, it became clear that many uncertainties faced agriculture and agricultural science and technology in India. For example, how would global warming affect the production characteristics of Indian agriculture?

How would the technology system embrace the growing importance of the private sector? Would Indian agriculture remain competitive in the global marketplace? What would be the fate of the small-scale farmer?

Scenario planning may help address such questions

Scenario planning is a structured process of thinking about and anticipating the future that helps to break the mindset that the future will be a continuation of the past (van der Heijden 1996). It entails the development and collective analysis of a set of scenarios, which are narratives of alternative environments that show how different interpretations of driving forces can lead to different plausible futures (Ogilvy and Schwartz 1998; van der Heijden 1996).

PROJECT OBJECTIVES AND DESCRIPTION

Scenario planning was used to explore the uncertainties surrounding Indian agriculture and identify the key decisions that would need to be taken to ensure that India's agricultural technology system was prepared for the future. The assessment included a wide range of stakeholders and enabled participants to develop a shared perspective on a future that was not necessarily a continuation of the past. The scenario development and analysis were conducted in parallel with the design of NAIP.

For these parallel efforts to succeed, they required the participation of high-level officials, farm leaders, senior leaders from the public and the private sectors, NGO leaders, donor representatives, experts on agricultural development, and some "remarkable people" (a term used in the scenario planning literature to describe lateral thinkers). The process was managed jointly by the regionally and centrally based staff of the World Bank's ARD. An Indian co-leader was invited, and experienced scenario planning experts facilitated the process, which was organized around seven information-gathering and knowledge-sharing steps:

1. ***Identify driving forces for future change,*** taking into consideration political conditions, economic developments, social developments, environmental trends, and technological changes.
2. ***Identify predetermined factors.*** Which future developments will take place in any scenario?
3. ***Identify critical uncertainties***—in other words, critical areas in which the future is uncertain.
4. ***Develop scenario plots.*** A scenario is defined by a combination of two critical uncertainties, drawn out and shown as axes on which the scenarios are plotted. Then a comprehensive description of how the future will look under this scenario is developed. These futures must be plausible.
5. ***Consult with those having relevant expertise.*** The scenarios are presented to a large number of people who have relevant expertise; their comments are collected and incorporated in the scenarios. Consultation helps to identify knowledge gaps and guides decisions on whether and what additional knowledge must be gathered.
6. ***Assess the implications of different scenarios.*** The best possible responses of the client organizations to each of the plausible future scenarios are assessed.
7. ***Compare possible responses to the scenarios.*** Two elements in the comparison require special attention. First, there are those actions that can be found in all responses and tend to be low risk. Second, there are the responses that differ strongly among scenarios. Responses in these fields may require further assessment to understand how the impact of change on these variables can be managed.

As hoped, the scenario project co-evolved with the NAIP project (table 7.11), enabling NAIP to benefit from the understanding emerging through the scenario work. In this way, the NAIP model was tested in various "environmental" conditions specified by the scenarios—a process sometimes referred to as "wind tunneling."

The main steps included the following:

1. ***An initial workshop*** at World Bank headquarters in Washington, DC, to introduce the scenario planning concept and process and to receive wider buy-in among staff for the process that would unfold (figure 7.3).
2. ***Interviews with "remarkable people"*** to explore the issues and concerns for future agricultural development in India (van der Heijden 1996; box 7.24).
3. ***A workshop*** in India to launch the process and obtain input from participants.
4. ***A scenario analysis and design workshop*** to identify the key scenarios that would be developed (following the steps described earlier to identify the critical elements of each scenario: driving forces, predetermined factors, and main uncertainties). The scenario plots (figure 7.3) had two main dimensions. The first was *economic management*, which could be strongly market based and liberalized but also more government controlled and centrally led. The second was the *social fabric of the countryside and the country in general,* which

Table 7.11   Timing for Preparing the National Agricultural Innovation Project (NAIP) in Relation to Scenario Development

Timing	NAIP	Scenario project
September 2004		Internal planning workshop in the World Bank
April 2005	Start of project preparation	Scenario agenda workshop at Indian Council for Agricultural Research
July 2005	Draft Project Appraisal Document (PAD)	Scenario building workshop; development of first-generation scenarios
August 2005	Polished PAD	Research on the validity of first-generation scenarios
October 2005	Quality enhancement review	
December 2005	Project appraisal	
February 2005	Negotiations	Second-generation scenarios concluded and circulated for comments
April and June 2006	Board approval	Workshops in India and at the World Bank on the implications of the scenarios
July 2006	Project becomes effective and implementation begins	

Source: Rajalahti et al. 2006.

Figure 7.3   Final Scenario Plots Developed during the Scenario Design and Consultation Processes

The Four Scenarios

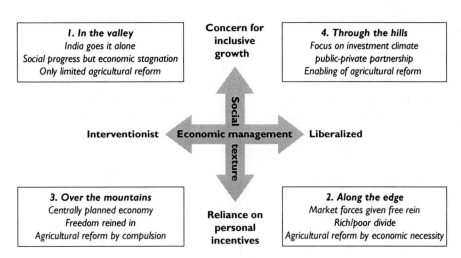

Source: Rajalahti et al. 2006.

could be strong, with rural people well organized in villages that are able to take care of their problems, or weak, in which case the poor would be more marginalized. Using these two dimensions as the axes of a 2 x 2 matrix, 4 combinations emerged that can serve as perspectives on the future of Indian development. One of the combinations introduced a third dimension of rapid global warming.

5. *Finalization and presentation of the scenario storylines*, including the development of full scenario stories (see Rajalahti et al. 2006) and their validation.

6. *Scenario analysis results workshops* were organized to define the way forward in relation to NAIP and ICAR. The specific goal was to identify how the scenarios could help to identify which critical decisions needed to be made to maximize the future role and impact of India's technology system.

INNOVATIVE ELEMENT

Scenario planning was done to build consensus and elicit outside-the-box thinking among diverse stakeholders that traditionally did not engage with each other in science and policy discussions. The results were used for designing a long-term investment project in science and technology.

BENEFITS TO NAIP AND IMPLICATIONS FOR SCIENCE, TECHNOLOGY, AND INNOVATION

Scenario analysis contributed to the design of NAIP in many ways. The process truly engaged people in thinking outside their everyday domains and resulted in four very different but plausible scenarios. The process led ICAR to think about the issues beyond its own technical competence and to strengthen the realization that the world it serves requires new approaches, including social organization and institutional innovation.

The scenarios were considered very useful for envisioning long-term science and technology needs. Scenario development clearly revealed two major needs: to work on institutional arrangements for R&D (farmer organizations, sector boards, cooperatives) and fully explore the potential of nonfarm rural employment.

The scenario process also helped the project design team to define the scope of NAIP's components, particularly the institutional development needed for the AIS to evolve. It highlighted the importance of enhancing the capacity for dialogue and interacting with other stakeholders in the innovation system. Flexibility, rather than the pursuit of one reform strategy, was considered a key trait for a successful organization in a rapidly changing world.

The client organizations used the national scenarios to strengthen their visioning capacity and strategy development

> **Box 7.24** Key Issues Raised by "Remarkable People"—Including Opinion Leaders and Policy Makers—in the Scenario Development Process, India
>
> The key question posed to the interviewees was, "When thinking of the future of Indian agriculture, what keeps you awake at night?" Four main themes emerged from these interviews:
>
> - Will there be enough water for future generations? How can water be managed sustainably?
> - What will drive Indian agriculture in the future: government or the market? What is the right balance?
> - How will rural communities change? How fast will rural–urban migration proceed, and what is the future of small-scale farming?
> - How can rural stakeholders voice their views—women, farmers, the private sector?
>
> *Source:* Rajalahti et al. 2006.

at the level of specific regions and products, such as rice, dairy products, and medicinal plants. For this purpose, groups of stakeholders were asked to develop the national scenarios for the product or region of their interest.

The national competitive fund for research consortiums, managed by ICAR, subsequently was aligned with the issues identified by the scenarios. These consortiums have been the main means of reforming India's agricultural research system and enabling it to move toward a more demand-driven, multistakeholder approach in addressing innovation needs. See module 4, IAP 2.

LESSONS LEARNED AND ISSUES FOR WIDER APPLICATION

Scenarios provided a neutral space for building consensus about critical decisions surrounding the future role and impact of India's technology system. The following recommendations, suggested adjustments, and limitations should be kept in mind by those engaged in a similar exercise:

- *Adapt scenario analysis to specific planning tasks*, such as the development of a country assistance strategy, sector strategy, project, or regional plans. Applying the analysis to larger, global issues is far more challenging.
- *Implement the scenario process ahead of project preparation* because scenario analysis requires a significant time commitment, particularly for consultation and validation.
- *Allocate sufficient time and resources* for clients to understand and come to own the process.
- *Form a multidisciplinary scenario team, led by an experienced scenario leader(s).*
- *Draw participants from many disciplines and representing a range of views* (India's scenario-building process, for example, included people from outside the agricultural sector). It is also essential to include participants representing the groups that the process aims to influence.
- *Pay close attention to the following:* the need for a full-time manager to oversee the process; the availability of research capacity with adequate resources; the need to manage and guide research performed by third-party institutes; the coordination required to operate a virtual team over long distances; and managing political sensitivities.

A Vision for Agriculture in Chile in 2030 and the Implications for Its Innovation System

Willem Janssen, World Bank
Ariani Wartenberg, World Bank

SYNOPSIS

In 2009, Chile's Ministry of Agriculture commissioned the World Bank to identify the long-term changes required for its AIS to be more effective. The World Bank collaborated with Chile's Foundation for Agricultural Innovation (FIA) to design a participatory process combining an analysis of major trends with input from opinion leaders, sector representatives, and others. The two major outcomes were (1) a vision statement that expresses Chile's agricultural potential and aspirations and (2) an action plan outlining changes for the innovation system to achieve that vision. The Ministry of Agriculture financed the study through a fee-based service agreement with the World Bank. The service cost US$250,000 (US$150,000 to develop the vision and US$100,000 to prepare the implications for the innovation system). FIA invested US$175,000 in the study and made a staff member available full time for the two years that the study required.

CONTEXT

Chile is a leading player in regional and global agricultural markets and regards itself as a food and forest powerhouse. Despite substantial development in the agricultural sector, agricultural growth has leveled off in Chile over the past decade, signaling that Chile's innovation system was more effective in the past and that changes are needed to forestall future negative growth.

Changes in the innovation system should anticipate the challenges of the future in addition to those that are apparent at present, because innovation is usually a slow process. Many years are likely to intervene between an initial idea or finding and its widespread application. Potential challenges can be related to any number of variables: domestic and international market factors, climatic and production conditions, competition with other sectors, social unrest, access to information and technology, logistics, and trade agreements, to mention just a few. It may be preferable for a country to overestimate such threats and be overly prepared than to underestimate them and be marginalized. Investments in innovation that respond to those challenges are a key ingredient in ensuring the future global competitiveness of Chilean agriculture and in meeting the increasingly sophisticated domestic demand for agricultural products.

This profile describes collaboration between Chile's Fundación para la Innovación Agraria (FIA, Foundation for Agricultural Innovation) and the World Bank to develop a vision for Chilean agriculture in 2030 and a corresponding action plan for the innovation system to realize that vision. For additional information on Chile's agricultural technology consortiums, see module 4, IAP 3; for a discussion of FIA and its activities, see module 1, IAP 3.

PROJECT OBJECTIVES AND DESCRIPTION

The project had two interlinked objectives. The first objective was to identify the main opportunities and challenges that Chile needs to address if it wishes to reinvigorate agricultural growth and propose a vision for Chilean agriculture toward 2030. The second was to identify the adjustments required for AIS to contribute effectively to realizing this vision. These objectives would be achieved through a process combining analysis, multidisciplinary consultation at different phases of the analysis, and synthesis of the results. The process is detailed in the sections that follow.

Developing scenarios and building the vision

Four *driver studies* were commissioned to identify key trends and driving forces of change that could be to be used for defining and building the scenarios. The studies focused on markets and trade, rural policies and rural development, natural resource management and climate change, and science and technology. The team produced short summaries of the studies and shared them at the first workshop. Information on *seven subsectors (clusters)* (fresh fruit, processed food, wine, native forestry, dairy, red meats, and cereals) contributed to building scenarios and discussing their implications. Further input came from *interviews* with 11 opinion leaders and "remarkable people" (lateral thinkers who could bring alternative perspectives to the dialogue) representing a wide range of views from the public and private sector, as well as academia and civil society. These views helped in drawing implications and identifying important elements of the vision for 2030 (box 7.25).

A subsequent *scenario building workshop,* facilitated by FIA and World Bank staff and external consultants in December 2010, convened 24 experts from the public and private sectors, academia, and civil society. Four scenarios were outlined in a series of plenary sessions and smaller working groups. The core team then *consolidated the scenarios,* developing comprehensive descriptions that were submitted for *validation* to the original workshop participants, the scenario team, other experts in Chile (more than 70 people), and five external peer reviewers. The team incorporated the feedback, emphasizing scenario 2 ("Terra Calida") and consolidating scenarios 1 and 3 in a "Business as Usual" scenario (figure 7.4).

Action planning and dissemination of results

The core team combined the analysis and proposals from the position papers into an *action plan.* Box 7.26 summarizes the main thematic recommendations. In the final step, *consultation and dissemination,* the action plan was widely shared in the country through presentations, press releases, and web publishing. Feedback from these events concluded the process and initiated the *preparation of budget proposals.*

INNOVATIVE ELEMENT

The Chile exercise recognized the uncertainties of the future, used this insight to understand the basic long-term competitive advantages of its agricultural sector, and drew the implications for change. Rather than diagnosing the present situation and proposing remedies, the study formulated future ambitions and specified the action required to get there. The proposed changes to the innovation system were (rightly) perceived as the next step forward and not as a recognition of past failure.

BENEFITS TO CHILE'S AIS

The benefits of the approach described here derive from its constructive, creative, and collaborative features and from the comparative strengths of FIA (local and national expertise) and the World Bank (international experience). The involvement of former presidents, cabinet ministers, journalists, scientists, businesspeople, individual farmers, farmer organizations, and many other individuals was highly productive. The discussion was conducted in simple, frank language. Participants shared and developed considerable knowledge and linked it to a specific plan for action. The plan recognized that Chile has sophisticated ambitions in high-value markets and is not satisfied with imitating wealthier countries.

Somewhat by chance, the approach proved politically robust: The work was started by a left-leaning government and finished by right-leaning government. Nor did the second biggest earthquake in the history of the world derail the study.

The forward-looking nature of the study motivated strong participation and interest in its results. By looking far into the future, participants ensure that several issues that had been forgotten or considered out of bounds (technology transfer, the role of the ministry, the importance of qualified human resources) regained relevance and received attention.

LESSONS LEARNED AND ISSUES
FOR WIDER APPLICATION

Briefly, lessons from this process include the following:

■ *The local partnership was key to success.* Vision building is a very labor- and communication-intensive process.

> ### Box 7.25   The Vision for Chilean Agriculture in 2030
>
> In 2030 Chile is a quality producer of a range of food and fiber products. Its international image is marked by the diversity that its geography allows it to produce. The sector has an emphasis on environmental sustainability and wholesomeness, valued by both domestic and international consumers. Through the application of ICT, investments in agricultural technology and the training of its labor force, Chile has been able to develop profitable value chains, well integrated from production to final markets, and able to remunerate its participants at comparable levels to the rest of the Chilean economy.
>
> *Source:* World Bank 2011b.

Figure 7.4 Scenarios for Chile's Agricultural Innovation System

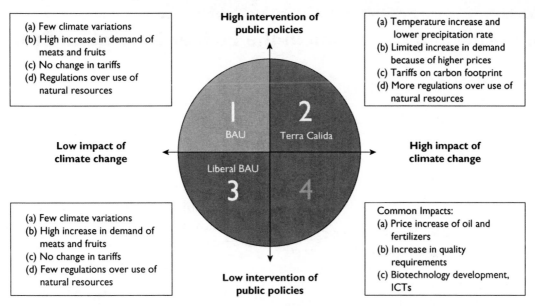

(a) Few climate variations
(b) High increase in demand of meats and fruits
(c) No change in tariffs
(d) Regulations over use of natural resources

High intervention of public policies

(a) Temperature increase and lower precipitation rate
(b) Limited increase in demand because of higher prices
(c) Tariffs on carbon footprint
(d) More regulations over use of natural resources

Low impact of climate change

1 BAU

2 Terra Calida

Liberal BAU

3

4

High impact of climate change

(a) Few climate variations
(b) High increase in demand of meats and fruits
(c) No change in tariffs
(d) Few regulations over use of natural resources

Low intervention of public policies

Common Impacts:
(a) Price increase of oil and fertilizers
(b) Increase in quality requirements
(c) Biotechnology development, ICTs

Source: Authors.
Note: BAU = Business as Usual; ICTs = information and communication technologies; "Terra Calida" is a reference to the effects of global warming.

Box 7.26 Summary of Action Plan Recommendations in Six Thematic Areas

- *Genetic improvement.* Multidisciplinary teams; biotechnology tools; intellectual property and patents.
- *Farm management.* Management of natural resources and water; information and communication technologies; ecological inputs.
- *Harvest and postharvest.* Proposals from competitive funds; shared funding between government and private sector.
- *Standards and quality.* Private sector-led expansion of Chile GAP (good agricultural practice standards); benchmarking of standards with importers.
- *Qualified human resources.* Collaboration with Becas Chile (a national scholarship program); international exchange networks.
- *Labor resources.* Basic and vocational education in rural areas; monitoring system.

Source: World Bank 2011a.

The many consultation and dissemination sessions could have been organized only by the national partner. Identifying resource people (for background studies and work days) requires in-depth understanding of the national setting. The phone must be answered if someone (a journalist, entrepreneur, student) has questions.

- *The development of a vision that expresses an ambition made the study interesting to the political players in the sector.* Politicians cannot easily sell the need for more institutional integration or long-term research, but they can piggyback those measures on the ambition for higher farm incomes or less pollution.

- *The step from analysis to vision was smaller than expected.* The 10 workdays brought out similar ideas across subsectors and remarkable agreement on how to pursue them. The main difference is that vision cannot be based on evidence alone; it requires some "structured dreaming."

- *The use of simple language allowed everybody to contribute and to understand the goals that were being pursued.* This frank approach creates much wider acceptance and better feedback.

- *Development of the vision created the room for change needed to implement the action plan.* If this willingness to consider change is sustained during the implementation of the action plan, it will constitute a further achievement of the process described here.

Monitoring and Evaluation in the Fodder Innovation Project

Andy Hall, LINK Ltd.[1]

SYNOPSIS

The rationale of the Fodder Innovation Project (FIP), implemented in India and Nigeria in 2007–10, was that persistent fodder shortages arose from a scarcity of fodder innovation capacity, not of technology. A diagnosis of the limitations of networks and institutional arrangements associated with specific fodder-related themes was done to design activities to address those limitations. The project's monitoring system was designed to include a baseline survey of households, a map of current innovation capacity, and an actor linkage matrix and scoreboard to track institutional change. The difficulties encountered with all of these methods provide some cautionary lessons. First, an essential foundation for M&E within any innovation system project is to determine whether the expected outcomes are developmental or institutional, because this orientation directly influences the kind of M&E approach required. Second, data requirements and tools for monitoring must be adapted to the realities of short projects. Third, collaborative development of tools for monitoring institutional change helps to ensure that they are appropriate and owned by the partners who need to use them. Finally, the evaluation of innovation system projects such as FIP, in which socioeconomic impacts may become apparent some time after the project ends, should include and adequately fund strategies for learning how the project led to impacts.

CONTEXT

The innovation system perspective in the Fodder Innovation Project (FIP)[2] built on lessons from an earlier project (2004–07) that indicated the limitations of a technology-transfer approach in addressing fodder scarcity. The project was implemented in India and Nigeria in collaboration with five key partner organizations (KPOs), which formed the nucleus for stimulating change in their local innovation environments. The KPOs were all NGOs, with one exception: the animal husbandry department of an agricultural university (see module 4, IAP 1).

PROJECT OBJECTIVES AND DESCRIPTION

The project's rationale was that the persistent fodder shortage suffered by many poor livestock keepers did not arise from a scarcity of technology itself but from a scarcity of fodder innovation capacity (Hall, Sulaiman, and Bezkorowajnyj 2008). Fodder innovation capacity was defined as the networks of organizations that mobilized ideas and resources and the institutional settings that framed the relationships involved in innovation. The project used an approach inspired by action-research to explore how these networks and institutional settings could be strengthened, with the expectation that they would drive fodder innovation.

The project's implementation strategy was for the project management team to help the KPOs select fodder-related themes (challenges but also opportunities, such as new markets for milk), diagnose the limitations of current networks and institutional arrangements associated with these themes, and design activities to address those limitations. The project management team then helped the KPOs monitor the effectiveness of their efforts in stimulating institutional changes and, where needed, helped them alter their plans in light of information revealed by monitoring. The project hired two research fellows to conduct diagnostic studies, investigate which activities and processes were enabling the development of fodder innovation capacity, and identify any resulting developmental outcomes.

After three years (of which the first year was spent identifying suitable KPOs and introducing them to the project's rationale), some immediate outcomes became evident in the project sites. They included more efficient veterinary and input service delivery systems; changing collaborative practices of actors; changing institutional arrangements to make additional fodder produced available to women, the landless, and poor livestock-keeping households; evidence of demand being generated for fodder varieties and other livestock-related knowledge and technologies; and the KPOs institutionalizing and mainstreaming their approach in their other activities or across different organizations.

INNOVATIVE ELEMENTS OF MONITORING AND EVALUATING THE FODDER INNOVATION PROJECT

Because FIP was inspired by an action-research approach, the iteration and recasting of activities based on their relative effectiveness in strengthening networks and institutional arrangements were regarded as central to exploring how to develop innovation capacity. A monitoring system was a critical element of this iterative process, but the issue of how to design a monitoring system for FIP was contested within the project's management team.

Some felt that the project was a development project and that monitoring should therefore focus on what they viewed as the project's major deliverable: household-level outcomes. This view was reinforced by the donor's desire to see tangible results in areas planted to new fodder species and related live-stock feeding practices adopted by poor households. Others regarded the project as a research project. Their view was that any developmental outcome would be on a relatively small scale and at best a byproduct of an experiment attempting to understand how to facilitate institutional change associated with the development of innovation capacity.

Given the strongly held and often opposing views about the types of outcomes against which the project should monitor its performance, the team decided to proceed on two tracks. The first was an impact assessment exercise; the second was to monitor changes in the institutional environment.

Impact assessment

The impact assessment involved designing and conducting a household survey to collect baseline information about animal feeding practices, cropping patterns, and household incomes. Statistical expertise was brought into the project to help design the survey instrument and develop an appropriate sampling approach. The design anticipated repeating the survey at the end of the project to judge impacts.

The designers of the impact assessment recognized the need for a counterfactual, although the means of developing one was debated considerably. Finding comparator "without" sites was problematic, so the designers adopted a "before-and-after" approach. Seventeen locations were selected from the five project sites in India and Nigeria and 2,047 households interviewed. Data collection began six months into the project, took over a year to complete, and the initial analysis of the baseline data was incomplete at the end of the project. The survey was not repeated at the end of the project to develop the "after" scenario, as time and resources were not available.

Monitoring institutional change

To monitor changes in the institutional environment, the project team planned to develop an institutional baseline. The baseline would have two functions. A diagnostic function—identifying institutional issues that needed attention—would form the basis for the initial set of project interventions. A monitoring function would track progress in facilitating institutional change.

To collect this baseline information, FIP investigated fodder innovation capacity using a methodology developed by the World Bank (Hall, Mytelka, and Oyelaran-Oyeyinka 2006; World Bank 2006). The methodology involved exploring four main elements of innovation capacity: (1) actors and their roles; (2) patterns of interaction among the actors; (3) the institutions (rules that govern interactions); and (4) the enabling policy environment. Using this framework, project staff struggled to develop sufficiently detailed accounts of the institutional dimension of fodder innovation capacity to inform implementation design. They tended to develop more macro-level or generic accounts of the weaknesses in innovation capacity—the missing links between research and development actors—with a view to publishing them as academic papers. It should be stressed that the weakness was not in the method but in the guidance given to those who were expected to use it.

Acknowledging that the diagnostic and institutional baselines were not helping to develop action plans, project staff held workshops with the implementing partners to diagnose critical issues and develop plans to proceed. These workshops relied on implementers' knowledge of their own operating environments; the role of the project was to facilitate them to identify key bottlenecks that needed to be addressed in this environment. The question of how to monitor institutional change remained open. The project recognized that monitoring was the key to iterative learning and thus a critical part of the experiment to investigate how to facilitate fodder innovation capacity.

The project also recognized that the primary institutional dimension of capacity that needed to be addressed across all project initiatives was the pattern and quality of linkages between players associated with fodder innovation. The project selected two tools to discern these patterns. The first was an actor linkage matrix (Biggs and Matseart 1999). This tool draws up a list of organizations on the axes of a matrix. In a workshop setting, the matrix can be used to map patterns, linkages, and collaboration and identify which missing links and relationships could be formed for the change process to work more effectively. The second tool was a scoreboard developed by the project to help partners assess qualitative changes in the nature of relationships. For each relationship identified in the actor linkage matrix, the KPOs used the scoreboard parameters to specify the quality of these relationships. The project team, partners, and other stakeholders would conduct this scoring exercise periodically. The parameters were (1) embracing the project approach; (2) openness and flexibility; (3) level of joint actions; (4) the perceived value of interaction; (5) punctuality and commitment; and (6) use of own resources for project activities. Table 7.12 depicts a partial scoreboard.

EXPERIENCE TO DATE WITH MONITORING INSTITUTIONAL CHANGE

The project management team felt that these tools were well conceived, yet the KPOs found them cumbersome, particularly the scoreboard. KPOs felt that the scoreboard could not be used in a participatory way with their partners and stakeholders, because it raised sensitive issues that could undermine the relationship-building process (table 7.12 compares externally and internally generated scores). A more worrying trend quickly became apparent: The KPOs were using the actor linkage matrix and scoreboard mainly to report back to project management rather than for their own learning.

In fact, the KPOs were all well embedded in the prevailing institutional context. They had their own informal ways of analyzing that context and identifying which partners and stakeholders they needed to draw in to support their intervention. They were well aware of the sorts of incentives that different partners and stakeholders would require to participate in particular project activities. When they were given project resources to focus on this networking, they were able to use their pragmatic knowledge of the context and make good progress.

The single remaining monitoring issue was that the project still had to collect evidence that the interventions it put into place led to institutional changes, which led in turn to changes in livestock feeding practices. The project team searched for examples of institutional change and developed accounts of how they took place and their outcomes. They learned that many of the institutional changes taking place related only indirectly to fodder use. Instead, they concerned innovations in output markets for milk, disease surveillance, veterinary services, and conflicts over access to land and grazing areas. Project partners had felt that these issues needed to be addressed as a precondition for fodder innovation.

Evaluation in FIP was external, independent, and commissioned by the donor. The approach was an output-to-purpose review based on FIP's logical framework. While this exercise was adequate for accountability, it was not clear that it generated project-level lessons that could be used in future interventions.

LESSONS LEARNED AND ISSUES FOR WIDER APPLICATION

The experience with M&E in FIP yielded cautionary lessons about mixed messages on expected outcomes and the risks of using data-intensive monitoring methods in short projects. Greater collaboration in designing monitoring tools and a greater emphasis on using evaluations as learning opportunities would also have been valuable.

- *Clarify the nature of the interventions and expected outcomes.* An action-research project of this type—and most innovation system interventions are going have an action-learning orientation—runs the risk of sending mixed messages to the donor and project partners about the nature of the outcomes it is expected to deliver. These mixed messages directly affect the monitoring strategies adopted and set up internal conflicts about whether developmental impacts or institutional changes are to be monitored. Getting agreement on these issues right from the start is an essential foundation for the M&E system within any innovation system project.
- *Tailor data requirements and tools for monitoring impact to the realities of short projects.* Resources spent on an elaborate baseline for a short project of this type were probably misspent. The project did have a responsibility to track its contribution to developmental outcomes, but the team should have selected an approach that was more appropriate for generating data that the implementers could use in designing, redesigning, and managing the intervention. Such an approach would

| Table 7.12 | Partial Results of an Externally and Internally Generated Scoreboard for Monitoring and Learning in the Fodder Innovation Project |

	1a		1b		2a		2b		3a		3b	
	JDPC/ILRI		JDPC/ILRI		Farmer Group		Farmer Group		OSADEP		OSADEP	
	EGSB	IGSB	EGSB	IGSB	EGSB	IGSB	EGSB	IGSB	EGSB	IGSB	EGSB	IGSB
Parameter	AS	AS	AN	AN	AS	AS	AN	AN	AS	AS	AN	AN
Embracing the project approach, openness, and flexibility (institutional)	3	3	6	6	0	1	5	4	0	0	5	3
Level of joint actions and sharing resources Interaction (5 marks)	4	6	8	8	2	4	5	6	2	4	6	6
Value of interaction (5 marks)	4	6	8	8	2	4	4	8	2	2	6	4
Sense of belonging as an actor (individual)	2	3	7	7	2	2	6	6	2	1	4	4
Forum as platform for mutual benefits	2	4	7	6	2	3	4	6	2	1	5	4
Punctuality and commitment of actors	3	5	7	7	2	2	6	7	2	1	6	4
Use of own resources (institutional)	3	4	8	6	5	2	8	7	0	1	2	3
Misunderstanding among actors (drawn internally for internal use)	*	0	*	0	*	0	*	1	*	0	*	0

Source: Author.

Note: Scores are 0 = lowest mark, 10 = highest mark; * = not determined; – = does not yet exist; AS = at start (January 2008); AN = at now (July 2009); EGSB= externally generated scoreboard; IGSB = internally generated scoreboard; JDPC = Justice, Development, and Peace Commission; ILRI = International Livestock Research Institute; OSADEP = Osun State Agricultural Development Programme.

almost certainly have involved rapid, qualitative appraisal methods. Another valuable approach would have been to use Causal Process Tracing to help unravel the underlying institutional (and other) causes of outcomes observed. This approach would have been particularly important in FIP, as many of the outcomes recorded were either unexpected or tangential to the original ambition to reduce fodder scarcity.

■ *Collaboratively develop tools for monitoring institutional change.* The institutional monitoring tools for FIP were well conceived but failed largely because they were "expert"-driven and not appropriate to or owned by the partners who needed to use them. A more useful approach would have been to develop monitoring tools collaboratively with each partner, adapting existing institutional-learning tools and principles to each partner's specific management needs. In hindsight, the scoreboard parameters appear ambiguous and poorly framed.

■ *Link impact measurement to learning.* The evaluation of FIP could have paid much more attention to learning how a project like this could lead to impacts. The baseline survey conducted for FIP could be valuable for this purpose: Repeating the survey some years after the project's end would provide valuable lessons about impact as well as underlying processes of change. Donors could consider providing projects with resources to commission baseline surveys, and donors could commission impact assessment and evaluations to be done after the project's end. This approach would be particularly useful for innovation system projects in which large-scale socioeconomic impacts are evident only some time after the project is implemented. Caution would be necessary, however, to (1) ensure that an adequate counterfactual analysis could be done; (2) capture unintended outcomes that may take place away from the original project site; and (3) fully interrogate theories of change and investigate causal links to impacts observed.

Monitoring and Evaluation in the Research Into Use Program

Andy Hall, LINK Ltd.
Kumuda Dorai, LINK Ltd.

SYNOPSIS

The Research Into Use (RIU) program, designed with an innovation systems perspective, emphasized strengthening networks and partnerships and also gave prominence to private sector and enterprise perspectives to drive research into use. The program pursued explicit learning objectives, including drawing key policy lessons about better strategies for putting agricultural research into use. In RIU, the challenge for M&E lay in the need to track developmental as well as institutional outcomes and the limited experience, confidence, and consensus in the use of methods that could address those two outcomes together. A key lesson from RIU's experience with M&E is that it is critical to maintain the distinction between monitoring and evaluation and to separate the timing and responsibilities for these two functions. Monitoring progress toward institutional and developmental targets is challenging; programs should have specific M&E expertise to help design integrated monitoring strategies for each of their interventions (not to collect data for M&E experts to analyze). Donors should be realistic about the type and scale of outcomes likely to be evident in the action-to-impact results chain during the life of an intervention. The final impact needs to be explored after the program has come to an end, especially for a program such as RIU, with its emphasis on achieving impact by stimulating institutional and policy change. Sensitive management of the interaction between evaluators and programs is needed to deal with the tensions between accountability and learning. This point is particularly important for innovation system interventions, because theories of change are multidimensional, evolve, and are often difficult to articulate.

CONTEXT

A series of reviews funded by DFID indicated that investments in agricultural research often delivered excellent research findings but that the findings produced more limited social and economic impacts than expected. In July 2006, DFID established a five-year flagship program, Research Into Use (RIU). As its name implies, the program's fundamental purpose was to make better use of agricultural research.

OBJECTIVES AND DESCRIPTION

Implemented in South Asia and Africa with a budget of US$50 million, RIU drew inspiration from the innovation systems perspective. It emphasized driving research into use by strengthening networks and partnerships and giving private sector and enterprise perspectives more prominence. It also pursued explicit learning objectives for internal purposes and external policy audiences (for details, see www.research-intouse.org and Hall, Dijkman, and Sulaiman 2010).

RIU had three main elements:

- *The Asia Challenge Fund (ACF)* supported 15 consortiums of research and development partners to scale out previously developed technologies. Projects were located in India, Bangladesh, and Nepal. The logic was that the main task in putting research into use was the promotion of technologies and other research products. The technologies dealt with in this way included new crop varieties developed through client-oriented breeding, fish fingerling production techniques, crab and seaweed production, and new management and analytical techniques, including participatory floodplain management and participatory market chain analysis. Over time, many of the consortiums recognized that their main task was not to promote technology per se but to marshal the different players around existing value chains or to develop new ones. Often this work involved bringing additional partners with entrepreneurial expertise into the consortiums. While the Asia Challenge Fund projects certainly yielded direct developmental outcomes, they were most noteworthy for the extensive institutional changes they stimulated.

- *The Africa Country Programmes (ACPs)* were established in Malawi, Nigeria, Rwanda, Sierra Leone, Tanzania, and Zambia with the explicit agenda of brokering new clusters of organizations around selected R&D themes. This agenda was based on the recognition that developmental and market-based opportunities often arise and can use research expertise and findings, but institutional inertia often prevents an appropriate mix of organizations, knowledge, and resources from assembling to innovate in response to such opportunities. The ACPs used innovation platforms and other dialogue mechanisms as starting points to identify opportunities and help organizations link with each other. Some of the initiatives brokered in this way included a smallholder indigenous poultry value chain in Tanzania, an improved fish fingerling supply chain in Malawi, and an integrated livestock fodder and vet service arrangement in Nigeria. With the ACPs' broad, opportunity-driven agenda, many unexpected adaptations took place as organizations in the consortiums found new ways of working with each other, such as new financing mechanisms, new roles for research partners, and new ways of influencing policy. Existing research products (and research expertise) were put into use in these initiatives, and their developmental outcomes were recorded, but the main outcomes from the ACPs were institutional.

- *The Best Bets*. RIU envisaged that it would identify Best Bet technologies for scaling up, but quickly it shifted to identifying best bet business models and unique consortiums that successfully combined enterprise principles (specifically, a focus on the poor as a market for products and services) with science-Based innovation. The approach proved useful for tapping the ability of entrepreneurs with social credentials to marshal research and other knowledge, resources, and partners to create business innovations that addressed issues as diverse as sleeping sickness control, farm input supplies, and biological control of an aggressive parasitic weed (Striga). Support for these businesses created capacity for continuous innovation around the themes covered. Having focused on supporting existing enterprise-like organizations and consortiums, the Best Bets were better placed than other RIU interventions to achieve direct developmental outcomes. Institutional change was also anticipated, however, as many of the organizations involved were encouraged to assume new roles in the innovation process.

RIU incorporated two further elements: a communications and a research function. The research team, distributed across Asia and Africa, was mandated to draw key policy lessons to inform national and particularly international development investors about better strategies for putting agricultural research into use.

INNOVATIVE ELEMENT: MONITORING AND EVALUATING DEVELOPMENTAL AND INSTITUTIONAL CHANGE

RIU set targets for development (outcomes measured in terms of benefits to poor people) and institutional and policy change (outcomes measured in terms of changes in key stakeholders' behavior in the innovation process and changes in policies that shape the national and international innovation environment). The emphasis given to these outcomes shifted in the latter part of the project, when a 70 percent impact weighting was assigned to institutional and policy outcomes.

The challenge for RIU was to monitor institutional and policy changes and direct developmental outcomes, even though it was recognized that developmental outcomes would not arise on a significant scale until after the project ended. The task was even more challenging because the interventions evolved a great deal to reflect the emerging understanding of how to put research into use and to take advantage of emerging opportunities to do so.

Part of the challenge lay in the tension over whether RIU was a developmental or institutional change initiative. Management and staff changes brought differing views on this point, which had implications for how M&E was addressed. A consensus eventually emerged that RIU needed to track developmental as well as institutional outcomes. Because RIU was implemented before there was much experience, confidence, or consensus in the use of methods that could address those two outcomes together, the M&E task was largely exploring new ground. The next sections describe the resulting experience and learning.

EXPERIENCE

Following advice from the donor, RIU initially set aside one-third of its budget for the combined task of monitoring impact and learning (MIL). A specialist group was brought in to design and implement an M&E plan, which had two notable elements. The first was a baseline survey for the ACF projects and ACPs to conduct (the Best Bet projects had not yet started). The second element was that projects were asked to record the events unfolding around their interventions. The data were sent to the UK-based MIL group for analysis.

A new management team and M&E strategy

A midterm and subsequent technical review of RIU were critical of the MIL approach, noting that the baseline exercise was particularly burdensome and provided no clear feedback to the ACF or ACPs. Following these reviews, a new management team was brought into RIU at the beginning of year four of RIU's five-year program. The MIL group disbanded, a new M&E strategy was put into place, and ACF projects and ACPs became responsible for monitoring their progress as they saw fit.

Some projects continued baseline and follow-up studies and tracking. For example, a project promoting varieties developed through client-oriented breeding found those studies useful for targeting, because they identified agroclimatic zones where adoption patterns indicated that the new varieties performed well and were acceptable to farmers. Others found the approach less useful or did not see it as a priority and stopped.

Impact assessment

The new M&E strategy specified that RIU would appoint an independent team to assess impact. Initially RIU struggled to find a suitable evaluator. An evaluator was finally identified and appointed early in 2010, when RIU had about 15 months left to run.

The impact assessment team (as RIU described it) would assemble evidence about outcomes to substantiate lessons that the research team was developing as well as to report to the donor for accountability. Responding to the donor's demands, however, the impact team assumed a more broadly conceived evaluative role, exploring the effectiveness of project cycle management and reporting its findings to the donor.

The impact team also developed a learning approach, drawing on the Theory-Based Impact Evaluation methods developed by Howard White and the 3IE group to assess developmental impacts, explore RIU's theory of change, and revisit its assumptions (for a brief description of the methods, see box 7.19 in TN 5; see also White 2009a, 2009b). Information was collected through household surveys in selected countries and extensive interviews were conducted with RIU staff and stakeholders in RIU focus countries. Box 7.27 summarizes key elements of the evaluation framework.

Problems encountered

The impact team's dual responsibilities for accountability (judging the effectiveness of RIU's implementation) and learning (helping RIU to understand its impact over time)

Box 7.27   Key Elements of the Framework Used to Evaluate Research Into Use

The impact assessment team developed a series of questions to examine the theory of change embedded in the interventions of Research Into Use (RIU). The following are the main categories of questions posed:

- *Overarching question.* Has the underlying theory of change—that "new forms of partnership will lead to innovation (which in turn will contribute to poverty reduction and economic growth)"—been shown to be appropriate?
- *Relevance.* Given its theory of change, was RIU's design appropriate to explore how to put research into use? Was the program's design appropriate to its ambition to impact on poor people?

- *Efficiency.* To what extent was the RIU's information management system (including the M&E system) fit-for-purpose? How did RIU assess the progress of innovations and their contribution (both positive and negative) to building knowledge and addressing market failures?
- *Effectiveness.* What partnership arrangements were most effective in understanding and addressing the barriers to innovation, both nationally and locally, and why? What partnership arrangements are effectively ensuring that the innovation process focuses on the issues of gender and social exclusion? Was the research monitoring system effective?
- *Impact.* To what extent has RIU impacted poor people?

Source: Adapted from RIU project document.

Box 7.28   Framework for Tracking Institutional Change

Research Into Use (RIU) recognized that it needed to track institutional change, but what sort of institutional change should be tracked? Given that innovation is embedded in a very wide range of relationships in economic systems, the range of institutions that are important in the innovation process is likely to be equally large and varied. Some areas of expected institutional change are very obvious—for example, changes in research practice or changes in patterns of partnership—but because RIU operated in complex development arenas, some institutional changes would be difficult to predict from theory alone.

The broad categories of change listed below were identified through a rapid inventory of institutional changes observed in association with RIU's activities. Institutional changes were defined as things that were being done differently as well as changes in formal policies and rules. The institutional changes in the inventory were sorted into groups to arrive at broad categories of institutional change. Illustrative indicators of each type of change were developed, and this framework was used for deeper investigation and documentation of institutional changes through case studies and writeshops.

- New ways of financing rural innovation.
- New, poverty-relevant ways of working or organizing things.
- Market-related institutional changes.
- Existing types of organizations playing new roles.
- New types of organizations playing new roles.
- Changes in research practice.
- Changes in the policy formulation space/process.
- Effects on donor/government investment behavior.
- New network configurations.
- Formal policy changes.

Source: Adwera et al. 2011.

were managed insensitively. The accountability function was perceived as a policing exercise and tended to impede the learning function, preventing the sharing of information and perspectives. Frequent changes in the impact team, including its leader, exacerbated this problem.

The evaluators and RIU disagreed about the RIU's theory of change. Evaluators articulated it as "partnerships lead to innovation," whereas RIU articulated it as "institutional and policy change will enable innovation." The evaluators found that it was too early to collect the impact data needed to satisfy the Theory-Based Impact Evaluation approach that inspired the design of the evaluation. In other words, an impact evaluation was premature.

A mechanism for systematically capturing change

A more positive result of this experience was that the impact team identified evidence that institutional change was occurring as a result of RIU's efforts. The team also called attention to the fact that RIU lacked a mechanism for systematically capturing this information and using it in dialogue with policy makers and others to leverage wider policy and institutional change.

RIU responded to these findings in a number of ways. It changed its quarterly reporting formats to include institutional change issues. It developed a framework to categorize and track an expanding range of different types of institutional change (box 7.28; Adwera et al. 2011). Institutional histories of the ACPs were commissioned to develop a deeper understanding of how they promoted innovation (box 7.29). Finally, writeshops helped staff implementing interventions to record institutional changes and unexpected outcomes and use the writeups to engage other stakeholders.

LESSONS LEARNED: WHAT COULD HAVE BEEN DONE DIFFERENTLY?

By the end of RIU's initial five-year lifecycle in June 2011, the impact team had not yet reported its findings (as of this writing, RIU has been extended to June 2012, partly to complete the impact evaluations). Even so, from the work completed so far, a number of lessons related to M&E stand out:

- *Separate responsibilities for M&E.* It is critical to maintain the distinction between monitoring and evaluation and to separate the timing and responsibilities for these two functions. RIU started off collecting its own impact

Box 7.29   Rationale and Approach for Innovation Studies Based on Institutional Histories
of Africa Country Programmes

The Africa Country Programmes (ACPs) of Research Into Use (RIU) used innovation platforms as one means of enabling innovation. RIU commissioned institutional histories of the ACPs to understand the specific details of how the various innovation platforms were designed and functioned in each setting and to learn how each ACP functioned as a broking or intermediary organization within the wider innovation and development landscape. The decision to use institutional histories reflected the fact that the arrangements and approaches used in each program evolved significantly. All programs took advantage of a range of opportunities; some approaches were less effective than others, but all were instructive. The resulting institutional histories contributed to innovation studies with the following elements:

■ A short institutional history of the evolution of RIU, with a strong focus on understanding the changing

prominence of core concepts and the way this evolution played out in RIU's strategy, with particular emphasis on the ACPs.

■ A detailed institutional history of the ACPs, emphasizing how they organized their work, learned along the way, and evolved in response to the evolution of RIU and the local development, political, and institutional environment.

■ A detailed account (in accessible language) of the nature, role, and function of the intermediary/ brokering task, including the innovation platforms, to explain what brokering involves.

■ Based on those accounts, develop guiding principles for designing a program enabling intermediary agencies/brokers to catalyze innovation and put research into use.

Source: RIU project document.

data (an evaluation function) and, in the process, impeded progress on the action part of its interventions. The expectations of the donor were also important: Donors should be realistic about the type and scale of outcomes likely to be evident in the action-to-impact results chain during the life of an intervention. This statement does not mean that interventions should not track their performance; it means that tracking and managing performance will require data different from the data needed for assessing an intervention's final impact. A suite of well-executed qualitative methods and rapid quantitative surveys would have been much more useful to the ACF projects and ACPs as a way of monitoring and generating feedback on the effectiveness of their actions.

■ *Time the evaluation carefully.* The impact team initiated its activities prematurely. It could be argued that the effectiveness of program cycle management could be investigated only when the program was still on the ground, but the final impact needs to be explored after the program has come to an end. The nature of RIU, with its emphasis on achieving impact by stimulating institutional and policy change, suggests that this time lag is particularly important. Since this route to impact is complex, chains of causation will need to be explored care-

fully. In this sense, the baseline studies may yet prove valuable for post-program evaluation, although a more appropriate approach would have been for RIU to commission the design and execution of the baselines independently of the interventions. Then the evaluators could have repeated the surveys after the program ended.

■ *Develop a systematic monitoring plan.* Monitoring progress toward institutional and developmental targets is challenging and requires technical backstopping so that projects can perform this function as an integral part of their management. Those who need to act on the information generated should have ownership of the monitoring role: Responsibility should lie with the individual projects rather than the central program. Programs should have specific M&E expertise to help design integrated monitoring strategies for others to use rather than to collect data for the M&E expert to analyze. A very large suite of techniques is available for exploring institutional change and understanding relationships between process and outcome (TN 4). The key is to have expertise that is sufficiently familiar with this suite of methods to adapt principles and tools to the specific monitoring needs of each project in the larger program.

- *Manage tensions between accountability and learning.* Sensitive management of the interaction between evaluators and programs is needed to deal with the tensions between accountability and learning. This point is particularly important for innovation system interventions, because theories of change are multidimensional and evolving and appear difficult for programs such as RIU to articulate. One approach—viewed as a good practice in the evaluation community—is to employ conversational rather than interrogative information collection techniques with program staff and stakeholders.

NOTES

Thematic Note 1

1. See, for example, the G20 press release on its September 2011 Ministerial Meeting on Development (G20 2011a) and progress on the Global Conferences on Agricultural Research for Development process (http://gcardblog.word-press.com/; FAO 2011).

2. For further information on conventional methods of priority setting and investment in agricultural R&D see Tabor, Janssen, and Bruneau (1998), Contant (2001), and Alston, Norton, and Pardey (1995).

3. World Bank (2007) identifies three main contexts for innovation: (1) agriculture-based countries (mainly in sub-Saharan Africa) where farmers lack access to well-functioning agricultural markets; (2) transforming countries (mainly in South Asia, East Asia and the Pacific, the Middle East, and North Africa) where agricultural markets are developing and some farmers gain from good connections to markets; and (3) mature innovation countries (most countries in Latin America and the Caribbean and many in Europe and Central Asia) where agricultural markets function relatively efficiently and farmers are effective market players.

Thematic Note 3

1. For example, see the discussion of scenarios developed under India's National Agricultural Innovation Project on the future of agriculture in India (http://www.naip.icar.org .in/workshops2.htm).

Innovative Activity Profile 3

1. Fundación PROINPA (Promoción e Investigación de Productos Andinos) (www.proinpa.org/); Programa Nacional de Raíces y Tubérculos rubro Papa (PNRT-Papa), Instituto Nacional Autónomo de Investigaciones Agropecuarias (INIAP) (www.iniap-ecuador.gov.ec/); and Innovación tecnológica y competitividad de la papa en Per (INCOPA) (www.cipotato.org/papandina/incopa/incopa.htm).

2. See "A Quest for the Perfect Potato" (*Newsweek* 2008), Mapstone (2010), and the August 2010 speech by Pepsi-Co Chairperson and CEO Ms. Indra Nooys (http://www.pepsico .com/assets/speeches/IndraNooyiPeruReception-2010.pdf).

Innovative Activity Profile 6

1. The author led the research in the Fodder Innovation Project.

2. Funded by DFID between 2007 and 2010, RIU was led by the International Livestock Research Institute (ILRI) in collaboration with United Nations University-Maastricht Economics and Social Research Institute on Innovation and Technology (UNU-MERIT), the International Center for Crop Research in the Semi-Arid Tropics (ICRISAT), and the International Institute of Tropical Agriculture (IITA).

Innovative Activity Profile 7

1. The author was head of RIU's Central Research Team.

REFERENCES AND FURTHER READING

Module 7 Overview

Alston, J. M., G. W. Norton, and P. G. Pardey. 1995. *Science Under Scarcity: Principles and Practice for Agricultural Research Evaluation and Priority Setting.* Ithaca: Cornell.

Edquist, C. 1997. *Systems of Innovation, Institutions, and Organizations.* London: Pinter Publishers.

Gijsbers, G., W. Janssen, H. Hambly Odame, and G. Meijerink. 2000. *Planning Agricultural Research: A Sourcebook.* The Hague: International Service for National Agricultural Research (ISNAR).

Hall, B. H., and J. Learner. 2010. "The Financing of R&D and Innovation." In *Handbook of the Economics of Innovation,* edited by B. H. Hall and N. Rosenberg. Amsterdam: Elsevier. Pp. 610–38.

Hall, B. H., and N. Rosenberg. 2010. "Introduction to the Handbook." In *Handbook of the Economics of Innovation,* edited by B. H. Hall and N. Rosenberg. Amsterdam: Elsevier. Pp. 3–9.

OECD (Organisation for Economic Co-operation and Development). 1986. "Glossary of Terms Used in Evaluation." In *Methods and Procedures in Aid Evaluation.* Paris.

———. 1991. *Principles for Evaluation of Development Assistance.* Paris.

———. 2000. Glossary of Evaluation and Results-Based Management (RBM) Terms. Paris.

Pardey, P. G., J. M. Alston, and V. Ruttan. 2010. "The Economics of Innovation and Technical Change in Agriculture." In *Handbook of the Economics of Innovation,* Vol. 2, edited by B. H. Hall and N. Rosenberg. Amsterdam: Elsevier. Pp. 939–84.

World Bank. 2011. *Information and Communication Technologies for Agriculture e-Sourcebook.* Washington, DC.

Thematic Note 1

Alston, J. M., G. W. Norton, and P. G. Pardey. 1995. *Science under Scarcity: Principles and Practice for Agricultural*

Research Evaluation and Priority Setting. Ithaca: Cornell.

Bloch, C. 2007. "Assessing Recent Developments in Innovation Measurement": The Third Edition of the *Oslo Manual Science and Public Policy* 34 (1): 23–34.

Contant, R. 2001. "Priority Setting." In *Planning Agricultural Research: A Sourcebook*, edited by G. Gijsbers, W. Janssen, H. Hambly Odame, and G. Meijerink. The Hague: International Service for National Agricultural Research (ISNAR).

CPR, CRISP, and LINK (Centre for Policy Research, Centre for Research on Innovation and Science Policy, and the Learning, INnovation, Knowledge initiative). 2011. *South Asia Rural Innovation Capacity Benchmarking Workshop, August 19–20, 2011*. New Delhi: CPR.

Edquist, C. 1997. *Systems of Innovation, Institutions, and Organizations*. London: Pinter Publishers.

FAO (Food and Agriculture Organization). 2011. "The GCARD Road Map: Transforming Agricultural Research for Development Systems for Global Impact." Rome. http://www.fao.org/docs/eims/upload//294891/GCARD%20Road%20Map.pdf, accessed September 2011.

G20. 2011a. Ministerial Meeting on Development: Communiqué. 23 September 2011, Washington, DC. http://www.g20.org/Documents2011/09/Ministerial%20Declaration-final.pdf, accessed September 2011.

———. 2011b. "Concept Paper Prepared by Brazil, Canada, France, Japan, CGIAR, FAO, GFAR and the WB." Conference on Agricultural Research for Development: Promoting Scientific Partnerships for Food Security, Montpellier, September 12–13. http://www.egfar.org/egfar/digitalAssets/4797_Concept_Paper_G20_conference_on_ARD.pdf, accessed October 2011.

Gijsbers, G., W. Janssen, H. Hambly Odame, and G. Meijerink. 2001. *Planning Agricultural Research: A Sourcebook*. The Hague: International Service for National Agricultural Research (ISNAR).

Hall, A., L. K. Mytelka, and B. Oyelaran-Oyeyinka. 2006. "Concepts and Guidelines for Diagnostic Assessments of Agricultural Innovation Capacity." UNU-MERIT Working Paper No. 2006-017. Maastricht: United Nations University–Maastricht Economics and Social Research Institute on Innovation and Technology (UNU-MERIT).

Kraemer-Mbula, E., and W. Wamae, (eds.) 2010. *Innovation and the Development Agenda*. Paris: Organisation for Economic Co-Operation and Development (OECD) and International Development Research Centre (IDRC).

Larsen, K., R. Kim, and F. Theus. 2009. "Agribusiness and Innovation Systems in Africa." Washington, DC: World Bank.

Pant, L. P. 2010. "Assessing Innovations in International Research and Development Practice." UNU-MERIT Working Paper No. 2010-043. Maastricht: United Nations University–Maastricht Economics and Social Research Institute on Innovation and Technology (UNU-MERIT).

OECD (Organisation for Economic Co-operation and Development). 1997. *National Innovation Systems*. Paris.

———. 2005. *Oslo Manual: Guidelines for Collecting and Interpreting Innovation Data*. 3rd edition. Paris.

———. 2010. *Measuring Innovation: A New Perspective*. Paris.

Potter, J. (ed.) 2008. *Making Local Strategies Work: Building the Evidence Base*. Paris: Organisation for Economic Co-operation and Development (OECD).

Raitzer, D. A., and G. W. Norton. 2009. *Prioritizing Research for Development: Experiences and Lessons*. Wallingford, UK: CAB International.

Spielman, D. J., and R. Birner. 2008. "How Innovative Is Your Agriculture? Using Innovation Indicators and Benchmarks to Strengthen National Agricultural Innovation Systems." ARD Discussion Paper No. 41. Washington, DC: World Bank.

Spielman, D. J., and J. Lynam. 2010. "Design Elements for Agricultural Research within an Agricultural Innovation Systems Framework: Evidence from and for Africa." In *Proceedings of the Second Science with Africa Conference 2010*. United Nations Economic Commission for Africa, http://www.uneca.org/sciencewithafrica/Resources.html, accessed October 2011.

Sulaiman, R. 2009. "Farmer First or Still Last? Uneven Institutional Development in the Indian Agricultural Innovation System." Paper presented to Farmer First Revisited: Farmer Participatory Research and Development Twenty Years On, Institute of Development Studies, University of Sussex, December 12–14, 2007.

SWAC (Sahel and West Africa Club). 2005. "The Family Economy and Agricultural Innovation in West Africa: Towards New Partnerships." Report No. SAH/D(2005)550. Paris: Organisation for Economic Co-operation and Development (OECD).

Tabor, S., W. Janssen, and H. Bruneau. 1998. *Financing Agricultural Research: A Sourcebook*. The Hague: International Service for National Agricultural Research (ISNAR).

Wennink, B., and W. Heemskerk (eds.) 2006. "Farmers Organizations and Agricultural Innovation: Case Studies from Benin, Rwanda, and Tanzania." Bulletin No. 374. Amsterdam: Royal Tropical Institute (KIT).

World Bank. 2006. "Enhancing Agricultural Innovation: How to Go Beyond the Strengthening of Research Systems." Washington, DC.

———. 2007. *World Development Report 2008: Agriculture for Development.* Washington, DC.

Thematic Note 2

Akramov, K. T. 2009. "Decentralization, Agricultural Services, and Determinants of Input Use in Nigeria." IFPRI Discussion Paper No. 941. Washington, DC: International Food Policy Research Institute (IFPRI).

Peterson, W., G. Gijsbers, and M. Wilks. 2003. "An Organizational Performance Assessment System for Agricultural Research Organizations: Concepts, Methods, and Procedures." ISNAR Research Management Guidelines No. 7. The Hague: International Service for National Agricultural Research (ISNAR).

Shambu Prasad, C., T. Laxmi, and S. P. Wani. 2006. "Institutional Learning and Change (ILAC) at ICRISAT: A Case Study of the Tata-ICRISAT Project." Global Theme on Agroecosystems Report No. 19. Patancheru: International Crops Research Institute for the Semi-Arid Tropics (ICRISAT).

Spielman, D., and R. Birner. 2008. "How Innovative Is Your Agriculture? Using Innovation Indicators and Benchmarks to Strengthen National Agricultural Innovation Systems." Agriculture and Rural Development Discussion Paper No. 41. Washington, DC: World Bank.

Spielman, D., and D. Kelemework. 2009. "Measuring and Benchmarking Agricultural Innovation System Properties and Performance: Illustrations from Ethiopia and Vietnam." IFPRI Discussion Paper No. 851. Washington, DC: International Food Policy Research Institute (IFPRI).

Watts, J., R. Mackay, D. Horton, A. Hall, B. Douthwaite, R. Chambers, and A. Acosta. 2003. "Institutional Learning and Change: An Introduction." ISNAR Discussion Paper No. 03–10. The Hague: International Service for National Agricultural Research (ISNAR).

World Bank and IFPRI (International Food Policy Research Institute). 2010. "Gender and Governance in Rural Services: Insights from India, Ghana, and Ethiopia." Washington, DC.

Thematic Note 3

Clayton, A., K. K'nIfe, and A. Spencer. 2009. "Integrated Assessment of Trade-Related Policies on Biological Diversity in the Agricultural Sector in Jamaica: Transition Strategies for the Sugar Industry in Jamaica." Nairobi: United Nations Environment Programme (UNEP) and National Environmental and Planning Agency of Jamaica. http://www.unep.ch/etb/initiatives/pdf/Final%20Study%20Jamaica%2012%202009.pdf, accessed September 2011.

Clayton, A., and C. Staple-Ebanks. 2002. "Nutraceuticals and Functional Foods: A New Development Opportunity for Jamaica: Market-Scoping Study." Technical report for the National Commission on Science and Technology. Kingston: Environmental Foundation of Jamaica (EFJ).

de Lattre-Gasquet, M. 2006. "The Use of Foresight in Agricultural Research." In *Science and Technology Policy for Development: Dialogues at the Interface*, edited by L. Box and R. Engelhard. London: Anthem.

Georghiou, L. 1996. "The UK Technology Foresight Programme." *Futures* 28 (4): 359–77.

IGD (Institute of Grocery Distribution). 2003. "Future Foods for Well-Being: An Expert Panel's View of the Next 25 Years." Watford, UK: IGD.

Johnson, B., and M. L. d'Apice Paez. 2000. "Alternative Scenarios for Agricultural Research." In *Planning Agricultural Research: A Sourcebook*, edited by G. Gijsbers, W. Janssen, H. Hambly Odame, and G. Meijerink. The Hague: International Service for National Agricultural Research (ISNAR).

Miles, I. 1997. "Technology Foresight: Implications for Social Science." Centre for Research on Innovation and Competition Working Paper No. 3. Manchester: University of Manchester.

Popper, A., M. Keenan, I. Miles, M. Butter, and G. Sainz de la Fuenta. 2007. *Global Foresight Outlook 2007.* European Foresight Monitoring Network/European Foresight Network, http://www.foresight-network.eu/files/reports/efmn_mapping_2007.pdf, accessed October 2011.

Postrel, V. 1998. *The Future and Its Enemies.* New York: Simon and Schuster.

Rajalahti, R., K. van der Heijden, W. Janssen, and E. Pehu. 2006. "Scenario Planning to Guide Long-Term Investments in Agricultural Science and Technology: Theory and Practice from a Case Study on India." Agriculture and Rural Development Discussion Paper No. 29. Washington, DC: World Bank.

Rikkonen, P., J. Kaivo-oja, and J. Aakkula. 2006. "Delphi Expert Panels in the Scenario-based Strategic Planning of Agriculture." *Foresight* 8 (1): 66–81.

Rutten, H. W. 2001. "Science and Technology Foresight." In *Planning Agricultural Research: A Sourcebook,* edited by G. Gijsbers, W. Janssen, H. Hambly Odame, and G. Meijerink. The Hague: International Service for National Agricultural Research (ISNAR).

Shepherd R. 2010. "Societal Attitudes to Different Food Production Models: Biotechnology, GM, Organic, and Extensification." Science Review No. 12, Foresight Project on Global Food and Farming Futures. London: Government Office for Science.

Stewman, S., and D. Lincoln. 1981. "Recombinant DNA Breakthroughs in Agriculture, Industry, and Medicine: A Delphi Study." *Futures* 13 (2): 128–40.

Wehrmeyer, W., A. Clayton, and K. Lum (eds.). 2002. "Foresighting for Development." *Greener Management International* 37.

Williams, R., and N. Markusson. 2002. "Knowledge and Environmental Innovations." Paper presented at the 1st BLUEPRINT workshop, January 23–24.

Thematic Note 4

Acosta, A., and B. Douthwaite. 2005. "Appreciative Inquiry: An Approach for Learning and Change Based on Our Own Best Practices." ILAC Brief No. 6. Rome: Institutional Learning and Change (ILAC) Initiative.

Adwera, A., J. Dijkman, K. Dorai, A. Hall, C. Kilelu, A. Kingiri, E. Madzudzo, H. Ojha, T. S. V. Reddy, R. Sulaiman V, and U. Ugbe. 2011 (forthcoming). "Institutional Change and Innovation: A Framework and Preliminary Analysis of RIU." RIU Discussion Paper. UK: Research Into Use (RIU).

Asif, M. 2005. "Listening to the People in Poverty Project: A Manual for Life History Collection." Unpublished.

Biggs, S. D. 2006. "Learning from the Positive to Reduce Rural Poverty: Institutional Innovations in Agricultural and Natural Resources Research and Development." Paper prepared for the Impact Assessment Workshop organized by the CGIAR System-Wide Program on Participatory Research and Gender Analysis for Technology Development and Institutional Innovation and the International Maize and Wheat Improvement Center (CIMMYT), October 19–21, 2005, Texcoco, Mexico.

Colfer, C. J. P. (ed.) 2005. *The Complex Forest: Communities, Uncertainty, and Adaptive Collaborative Management.* Washington, DC: Resources for the Future.

CoS-SIS (Convergence of Science–Strengthening Agricultural Innovation Systems). CoS-SIS. 2009. "Revised Proposal for CoS-SIS Research Design, March 2, 2009." Draft Working Paper No. 3. CoS-SIS, http://cos-sis.org/pdf/CoS-SIS%20VERSION%20Res%20Design%20(March%202009).pdf, accessed September 2011.

Davies, R. J. 1996. "An Evolutionary Approach to Facilitating Organizational Learning: An Experiment by the Christian Commission for Development in Bangladesh." Monitoring and Development News, http://www.mande.co.uk/docs/ccdb.htm, accessed September 2011.

Davies, R., and J. Dart. 2005. "The 'Most Significant Change' (MSC) Technique: A Guide to Its Use." *Monitoring and Development News,* www.mande.co.uk/docs/MSCGuide.pdf, accessed September 2011.

Douthwaite, B., and J. Ashby. 2005. "Innovation Histories: A Method from Learning from Experience." ILAC Brief No. 5. Rome: Institutional Learning and Change (ILAC) Initiative.

Douthwaite, B., T. Kuby, E. van de Fliert, and S. Schulz. 2003. "Impact Pathway Evaluation: An Approach for Achieving and Attributing Impact in Complex Systems." *Agricultural Systems* 78: 243–65.

Earl, S., F. Carden, and T. Smutylo. 2001. "Outcome Mapping: Building Learning and Reflection into Development Programs," Ottawa: International Development Research Centre (IDRC). www.idrc.ca/en/ev-9330-201-1-DO_TOPIC.html, accessed September 2011.

ECDPM (European Center for Development Policy Management). 2006. "Draft Study on Capacity Change and Performance: Mapping of Approaches towards M&E of Capacity and Capacity Development." Brussels.

George, A., and A. Bennett. 2005. *Case Studies and Theory Development in the Social Sciences.* Oxford: Oxford University Press.

Guijt, I. (ed.) 2007. "Negotiated Learning: Collaborative Monitoring for Forest Resource Management." Washington, DC: Resources for the Future.

Hall, A., R. Sulaiman V, and P. Bezkorowajnyj. 2007. "Reframing Technical Change: Livestock Fodder Scarcity Revisited as Innovation Capacity Scarcity." Nairobi: International Livestock Research Institute (ILRI), International Crops Research Institute for the Semi-Arid Tropics (ICISAT), Consultative Group on International Agricultural Research (CGIAR) Systemwide Livestock Programme, International Institute of Tropical Agriculture (IITA), and United Nations University–Maastricht Economics and Social Research Institute on Innovation and Technology (UNU-MERIT).

IDRC (International Development Research Centre). "Evaluation for Development." http://www.idrc.ca/en/ev-26586-201-1-DO_TOPIC.html, accessed March 2009.

IFAD (International Fund for Agricultural Development). 2002. "Managing for Impact in Rural Development: A Guide for Project M&E." Rome.

Kammili, T. 2011. "A Briefing Paper on Monitoring and Evaluation Practice for Rural/Agricultural Innovation: How Do You Measure the Impact of Innovation Initiatives?" LINK Policy Resources on Rural Innovation. Hyderabad: Learning, Innovation, and Knowledge (LINK).

Patton, M. Q. 2008. "State of the Art in Measuring Development Assistance." Presentation at the World Bank Independent Evaluation Group Conference, Measuring Development Effectiveness: Progress and Constraints, April 10, Washington, DC.

Salomon, M., and P. Engel. 1997. "Facilitating Innovation for Development: A RAAKS Resource Box." Amsterdam: Royal Tropical Institute (KIT).

Shambu Prasad, C., A. Hall, and L. Thummuru. 2006. "Engaging Scientists through Institutional Histories." ILAC Brief No. 14. Rome: Institutional Learning and Change (ILAC) Initiative.

Smutylo, T. 2005. "Outcome Mapping: A Method for Tracking Behavioural Changes in Development Programs." ILAC Brief No. 7. Rome: Institutional Learning and Change (ILAC) Initiative.

van Mierlo, B., B. Regeer, M. van Amstel, M. Arkesteijn, V. Beekman, J. Bunders, T. de Cock Buning, B. Elzen, A. Hoes, and C. Leeuwis. 2010. "Reflexive Monitoring in Action: A Guide for Monitoring Systems Innovation Projects. Wageningen: Communication and Innovation Studies, Wageningen University and Research Centre (WUR), Athena Institute, and VU University Amsterdam.

Walters, B. B., and A. P. Vayda. 2009. "Event Ecology, Causal Historical Analysis, and Human-Environment Research." *Annals of the Association of American Geographers* 99 (3): 534–53.

Winter, R. 1989. *Learning from Experience: Principles and Practice in Action Research*. Philadelphia: Falmer Press.

World Bank. 2004. *Monitoring and Evaluation: Some Tools, Methods and Approaches*. Washington, DC.

Thematic Note 5

Alston, J. M., G. W. Norton. and P. G. Pardey. 1995. *Science under Scarcity: Principles and Practice for Agricultural Research Evaluation and Priority Setting*. Ithaca: Cornell University Press.

Asif, M. 2005. "Listening to the People in Poverty: A Manual for Life History Collection." Unpublished, http://www.dgroups.org/groups/pelican/docs/LPP_2005.doc?ois=no.

CoS-SIS (Convergence of Science–Strengthening Agricultural Innovation Systems). 2009. "Revised Proposal for Cos-SIS Research Design, March 2, 2009." Draft Working Paper No. 3. CoS-SIS, http://cos-sis.org/pdf/CoS-SIS%20VERSION%20Res%20Design%20(March%202009).pdf, accessed September 2011.

CPR, CRISP, and LINK (Centre for Policy Research, Centre for Research on Innovation and Science Policy, and Learning, Innovation, and Knowledge). 2008. "Report on CPR-CRISP-LINK Workshop on Benchmarking Rural Innovation Capacity, August 2008." Ottawa: International Development Research Centre (IDRC).

Davies, R., and J. Dart. 2005. "The 'Most Significant Change' (MSC) Technique: A Guide to Its Use." Monitoring and Development News, www.mande.co.uk/docs/MSCGuide.pdf, accessed September 2011.

Douthwaite, B., and J. Ashby. 2005. "Innovation Histories: A Method from Learning from Experience." ILAC Brief No. 5. Rome: Institutional Learning and Change (ILAC) Initiative.

Douthwaite, B., T. Kuby, E. van de Fliert, and S. Schulz. 2003. "Impact Pathway Evaluation: An Approach for Achieving and Attributing Impact in Complex Systems." *Agricultural Systems* 78: 243–65.

ECDPM (European Center for Development Policy Management). 2006. "Draft Study on Capacity Change and Performance: Mapping of Approaches towards M&E of Capacity and Capacity Development." Brussels.

Evenson, R. E., P. Waggoner, and V. W. Ruttan. 1979. *Economic Benefits from Research: An Example from Agriculture*. New Haven: Economic Growth Center, Yale University.

George, A., and A. Bennett. 2005. *Case Studies and Theory Development in the Social Sciences*. Oxford: Oxford University Press.

Hall, A., R. Sulaiman, N. Clark, and B. Yoganand. 2003. "From Measuring Impact to Learning Institutional Lessons: An Innovation Systems Perspective on Improving the Management of International Agricultural Research." *Agricultural Systems* 78 (2): 213–41.

Horton, D., and R. Mackay. 2003. "Using Evaluation to Enhance Institutional Learning and Change: Recent Experiences with Agricultural Research and Development." *Agricultural Systems* 78 (2): 127–42.

Horton, D., P. Ballantyne, W. Peterson, B. Uribe, D. Gapasin, and K. Sheridan (eds.) 1993. *Monitoring and Evaluating Agricultural Research, A Sourcebook*. Wallingford: CAB International and International Service for National Agricultural Research (ISNAR).

IFAD (International Fund for Agricultural Development). 2002. "Managing for Impact in Rural Development: A Guide for Project M&E." Rome.

ISNAR (International Service for National Agricultural Research). 2003. "Monitoring, Evaluation, and Impact Assessment of R&D Investments in Agriculture." The Hague.

Kammili, T. 2011. "A Briefing Paper on Monitoring and Evaluation Practice for Rural/Agricultural Innovation: How Do You Measure the Impact of Innovation Initiatives?" LINK Policy Resources on Rural Innovation. Hyderabad: Learning, Innovation, and Knowledge (LINK).

Kelly, T., J. Ryan, and H. Gregersen. 2008. "Enhancing Ex Post Impact Assessment of Agricultural Research: The CGIAR Experience." *Research Evaluation* 17 (3): 201–12.

Kraemer-Mbula, E. 2011 (forthcoming). "Rethinking the Benchmarking of Agricultural and Rural Innovation." LINK Policy Resources on Rural Innovation. Hyderabad: Learning, Innovation, and Knowledge (LINK).

Maredia, M., D. Byerlee, and J. Anderson. 2000. "Ex Post Evaluations of Economic Impacts of Agricultural Research Programs: A Tour of Good Practice." Paper presented at the workshop on The Future of Impact Assessment in the CGIAR: Needs, Constraints, and Options, May 3–5, Rome.

NONIE (Network of Networks on Impact Evaluation), F. Leeuw, and J. Vaessen. 2009. "Impact Evaluations and Development: NONIE Guidance on Impact Evaluation." Washington, DC: NONIE. http://siteresources.worldbank.org/EXTOED/Resources/nonie_guidance.pdf, accessed September 2011.

Pardey, P. G., and N. M. Beintema. 2001. "Slow Magic: Agricultural R&D a Century after Mendel." IFPRI Food Policy Report. Washington, DC: International Food Policy Research Institute (IFPRI).

Patton, M. Q. 2008. "State of the Art in Measuring Development Assistance." Presentation at the World Bank Independent Evaluation Group Conference, Measuring Development Effectiveness: Progress and Constraints, April 10, Washington, DC.

Savedoff, W., R. Levine, and N. Birdsall. 2006. "When Will We Ever Learn? Improving Lives through Impact Evaluation." Report of the Evaluation Gap Working Group. Washington, DC: Center for Global Development (CGD).

Shambu Prasad, C., A. Hall, and L. Thummuru. 2006. "Engaging Scientists through Institutional Histories." ILAC Brief No. 14. Rome: Institutional Learning and Change (ILAC) Initiative.

Spielman, D., and R. Birner. 2008. "How Innovative Is Your Agriculture? Using Innovation Indicators and Benchmarks to Strengthen National Agricultural Innovation Systems." Agriculture and Rural Development Discussion Paper No. 41. Washington, DC: World Bank.

Walters, B. B., and A. P. Vayda. 2009. "Event Ecology, Causal Historical Analysis, and Human-Environment Research." *Annals of the Association of American Geographers* 99 (3): 534–53.

Watson, J. D. (ed.) 2003. "International Conference on Impacts of Agricultural Research and Development: Why Has Impact Assessment Research Not Made More of a Difference?" Mexico DF: International Maize and Wheat Improvement Center (CIMMYT).

Watts, J., D. Horton, B. Douthwaite, R. La Rovere. G. Thiele, S. Prasad, and C. Staver. 2007. "Transforming Impact Assessment: Beginning the Quiet Revolution of Institutional Learning and Change." *Experimental Agriculture* 44: 21–35.

Watts J., R. Mackay, D. Horton A. Hall, B. Douthwaite, R. Chambers, and A. Acosta. 2003. "Institutional Learning and Change: An Introduction." ISNAR Discussion Paper No. 03–10. The Hague: International Service for National Agricultural Research (ISNAR).

White, H. 2009a. "Some Reflections on Current Debates in Impact Evaluation." 3ie Working Paper No. 1. New Delhi: International Initiative for Impact Evaluation.

White, H. 2009b. "Theory-based Impact Evaluation: Principles and Practice." 3ie Working Paper No. 3. New Delhi: International Initiative for Impact Evaluation.

World Bank. 2004. *Monitoring and Evaluation: Some Tools, Methods, and Approaches.* Washington, DC.

Innovative Activity Profile 1

Arnold, E., and M. Bell. 2001. "Some New Ideas about Research for Development." In *Danida, Partnerships at the Leading Edge: A Danish View for Knowledge, Research, and Development.* Copenhagen: The Commission on Development-related Research Funded by Danida, Ministry of Foreign Affairs.

Hinchcliffe, D. 2010. "The Social Enterprise: A Case for Disruptive Transformation." Dachis Group, http://www.dachisgroup.com/2010/04/a-case-for-disruptive-transformation/, accessed September 2011.

Joffe, S., D. Suale, H. Jalloh, F. M'Cormack, and F. Johnston. 2008. "RIU Strategy for Sierra Leone." Edinburgh: Research Into Use (RIU). http://www.dfid.gov.uk/r4d/PDF/Outputs/ResearchIntoUse/riu08sl-countrystrategy.pdf, accessed September 2011.

Kanyarukiga, S., A. Barnett, V. Rusharza, C. Zaninka, D. Cownie, S. Joffe, V. Mwesigwa, J. Espasa, and M-A. Kamikazi. 2007. "RIU Country Strategy in Rwanda." Edinburgh: Research Into Use (RIU). http://www.researchintouse.com/resources/riu07rw-countrystrategy.pdf, accessed September 2011.

Maconachie, R. 2008. "New Agricultural Frontiers in Post-conflict Sierra Leone? Exploring Institutional Challenges for Wetland Management in the Eastern Province." Working paper. Manchester: University of Manchester.

"National Innovation Coalition," Research Into Use, http://www.researchintouse.com/programmes/riu-sierra-leone/riu-sl31natcoalition.html, accessed September 2011. Information on PAID in Sierra Leone.

"RIU Sierra Leone." Research Into Use, http://www.researchintouse.com/programmes/riu-sierra-leone/index.html, accessed September 2011. Background on RIU program in Sierra Leone.

"The Birth of the John Lewis State: Public Sector Workers Urged to Set Up Co-operatives," *The Daily Mail*, November 18, 2010. http://www.dailymail.co.uk/news/article-1330742/John-Lewis-state-Public-sector-workers-urged-set-operatives.html?ito=feeds-newsxml, accessed September 2011.

Innovative Activity Profile 2

Aberman, N., E. Schiffer, M. Johnson, and V. Oboh. 2010. "Mapping the Policy Process in Nigeria: Examining Linkages between Research and Policy." IFPRI Discussion Paper No. 1000. Washington, DC: International Food Policy Research Institute (IFPRI).

Grimble, R., and K. Wellard. 1997. "Stakeholder Methodologies in Natural Resource Management: A Review of Principles, Contexts, Experiences, and Opportunities." *Agricultural Systems* 55 (2): 173–93.

Hanneman, R. A., and M. Riddle. 2005. *Introduction to Social Network Methods.* Riverside: University of California, Riverside. http://faculty.ucr.edu/~hanneman/, accessed September 2011.

Hauck, J., and E. Youkhana. 2008. "Claims and Realities of Community-based Water Resource Management: A Case Study of Rural Fisheries in Ghana." In *Natural Resources in Ghana,* edited by D. M. Nanang and T. K. Nunifu. Hauppauge, NY: Nova Science.

Kindon, S., R. Pain, and M. Kesby (eds.). 2007. *Participatory and Action Research Approaches and Methods: Connecting People, Participation, and Place.* New York: Routledge.

Schiffer, E. 2007. "The Power Mapping Tool: A Method for the Empirical Research of Power Relations." IFPRI Discussion Paper No. 703. Washington, DC: International Food Policy Research Institute (IFPRI).

Schiffer, E., and J. Hauck. 2010. "Net-Map: Collecting Social Network Data and Facilitating Network Learning through Participatory Influence Network Mapping." *Field Methods* 22 (3): 231–49.

Schiffer, E., C. Narrod, and K. von Grebmer. 2008. "The Role of Information Networks in Communicating and Responding to HPAI Outbreaks." HPAI Research Brief No. 5. Pro-Poor HPAI Risk Reduction, http://www.hpairesearch.net/docs/Research_briefs/IFPRI-ILRI/IFPRI_ILRI_rbr05.pdf, accessed September 2011.

Spielman, D., and R. Birner. 2008. "How Innovative Is Your Agriculture? Using Innovation Indicators and Benchmarks to Strengthen National Agricultural Innovation Systems." Agriculture and Rural Development Discussion Paper No. 41. Washington, DC: World Bank.

Spielman, D., and D. Kelemework. 2009. "Measuring and Benchmarking Agricultural Innovation System Properties and Performance: Illustrations from Ethiopia and Vietnam." IFPRI Discussion Paper No. 851. Washington, DC: International Food Policy Research Institute (IFPRI).

Weber, M. 1922. *Economy and Society: An Outline of Interpretive Sociology.* 1978 ed. Berkley: University of California Press.

Innovative Activity Profile 3

"A Quest for the Perfect Potato." 2008. *Newsweek,* July 25. Daily Beast, http://www.thedailybeast.com/newsweek/2008/07/25/a-quest-for-the-perfect-potato.html, accessed September 2011.

Alderman, H., J. Hoddinott, L. Haddad, and C. Udry. 2003. "Gender Differentials in Farm Productivity." FCND Discussion Paper No. 6. Washington, DC: International Food Policy Research Institute (IFPRI).

Bebbington, A., and E. Rotondo. 2010. "Informe de la evaluación externa de la fase 3 de Papa Andina." Lima: International Potato Center (CIP) and Papa Andina.

Bernet, T., A. Devaux, G. Thiele, G. López, C. Velasco, K. Manrique, and M. Ordinola. 2008. "The Participatory Market Chain Approach: Stimulating Pro-poor Market-Chain Innovation." Rome: Institutional Learning and Change (ILAC).

Bernet, T., G. Thiele, and T. Zschocke. 2006. "Participatory Market Chain Approach: User Guide." Lima: International Potato Center (CIP) and Papa Andina. http://www.cipotato.org/publications/pdf/003296.pdf, accessed September 2011.

Devaux, A., J. Andrade-Piedra, D. Horton, M. Ordinola, G. Thiele, A. Thomann, and C. Velasco. 2010. "Brokering Innovation for Sustainable Development: The Papa Andina Case." ILAC Working Paper No. 12. Rome: Institutional Learning and Change (ILAC).

Devaux, A., D. Horton, C. Velasco, G. Thiele, G. López, T. Bernet, I. Reinoso, and M. Ordinola. 2009. "Collective Action for Market Chain Innovation in the Andes." *Food Policy* 34: 31–38.

Devaux, A., C. Velasco, G. López,, T. Bernet, M. Ordinola, H. Pico, G. Thiele, and D. Horton. 2007. "Collective Action for Innovation and Small Farmer Market Access: The Papa Andina Experience." CAPRi Working Paper No. 68. Washington, DC: International Food Policy Research Institute (IFPRI).

Engel, P., and M. Salomon. 2003. "Facilitating Innovation for Development: A RAAKS Resource Box." Amsterdam: Royal Topical Institute (KIT).

Horton, D. 2008. "Facilitating Pro-poor Market Chain Innovation: An Assessment of the Participatory Market Chain Approach in Uganda." Social Sciences Working Paper No.2008-1. Lima: International Potato Center (CIP).

Horton, D., B. Akello, L. Aliguma, T. Bernet, T. Devaux, B. Lemaga, D. Magala, S. Mayanja, I. Sekitto, G. Thiele, and C. Velasco. 2010. "Developing Capacity for Agricultural Market Chain Innovation: Experience with the 'PMCA' in Uganda." *Journal of International Development* 22 (3): 367–89.

Kaganzi, B., S. Ferris, J. Barham, A. Abenayko, P. Sanginga, and J. Njuki. 2009. "Sustaining Linkages to High-Value Markets through Collective Action in Uganda." *Food Policy* 34: 23–30.

Mapstone, N. 2010. "Old Roots Tap New Customers." *Financial Times*, September 27. http://www.ft.com/intl/cms/s/0/fd19a3fa-ca7b-11df-a860-00144feab49a.html#axzz1YoAqM4QP, accessed September 2011.

Practical Action. 2010. Lessons and insights in Participatory Market Chain Analysis (PMCA): An action-research of PMCA applications in Bangladesh, Sri Lanka, Sudan, Peru, and Zimbabwe. Unpublished. http://practicalaction.org/docs/ia2/DFID-PCMA_report_rev6.pdf, accessed September 2011.

Research Into Use. 2010. "New Market Chain Approach Yields Fast Results: Participatory Market Chain Approach." RIU, http://www.researchintouse.com/nrk/RIUinfo/PF/CPH01.htm#L5, accessed March 2011.

World Bank. 2008. *Gender in Agriculture Sourcebook*. Washington, DC: Food and Agriculture Organization (FAO), World Bank, and International Fund for Agricultural Development (IFAD).

Innovative Activity Profile 4

Ogilvy, J., and P. Schwartz. 1998. "Plotting Your Scenarios: Global Business Network." In *Learning from the Future*, edited by L. Fahey and R. Randall. Chichester, UK: John Wiley.

Rajalahti, R., W. Janssen, and E. Pehu. 2007. "Scenario Planning to Guide Long-term Investments in Agricultural Science and Technology: Theory and Practice from a Case Study on India." Agriculture and Rural Development Note No. 28. Washington, DC: World Bank.

Rajalahti, R., K. van der Heijden, W. Janssen, and E. Pehu. 2006. "Scenario Planning to Guide Long-term Investments in Agricultural Science and Technology: Theory and Practice from a Case Study on India." Agriculture and Rural Development Discussion Paper No. 29. Washington, DC: World Bank.

Rajalahti, R., J. Woelcke, and E. Pehu. 2005. "Development of Research Systems to Support the Changing Agricultural Sector." Proceedings. Agriculture and Rural Development Discussion Paper No. 14. Washington, DC: World Bank.

van der Heijden, K. 1996. *Scenarios: The Art of Strategic Conversation.* Chichester, UK: John Wiley.

———. 2005. *Scenarios: The Art of Strategic Conversation.* 2nd edition. Chichester, UK: John Wiley.

World Bank. 2006. "Enhancing Agricultural Innovation: How to Go Beyond Strengthening Research Systems." Washington, DC.

Innovative Activity Profile 5

World Bank. 2011a. "Chile's Agricultural Innovation System: An Action Plan towards 2030." Unpublished internal report, Washington, DC.

———. 2011b. "Towards a Vision for Agricultural Innovation in Chile in 2030." Unpublished internal report, Washington, DC.

Innovative Activity Profile 6

Biggs, S. D., and H. Matseart. 1999. "An Actor-Oriented Approach for Strengthening Research and Development Capabilities in Natural Resource Systems." *Public Administration and Development* 19: 231–62.

Hall, A., L. K. Mytelka, and B. Oyelaran-Oyeyinka. 2006. "Concepts and Guidelines for Diagnostic Assessments of Agricultural Innovation Capacity." UNU-MERIT Working Paper No. 2006-017. Maastricht: United Nations University–Maastricht Economics and Social Research Institute on Innovation and Technology (UNU-MERIT).

Hall, A., R. Sulaiman V., and P. Bezkorowajnyj. 2008. "Reframing Technical Change: Livestock Fodder Scarcity Revisited as Innovation Capacity Scarcity. Part 2: A Framework for Analysis." UNU-MERIT Working Paper No. 2008-003. Maastricht: United Nations University–Maastricht Economics and Social Research Institute on Innovation and Technology (UNU-MERIT).

World Bank. 2006. "Enhancing Agricultural Innovation: How to Go Beyond the Strengthening of Research Systems." Washington, DC.

Innovative Activity Profile 7

Adwera, A., J. Dijkman, K. Dorai, A. Hall, C. Kilelu, A. Kingiri, E. Madzudzo, H. Ojha, T. S. V. Reddy, R. Sulaiman V, and U. Ugbe. 2012 (forthcoming). "Institutional Change and Innovation: A Framework and Preliminary Analysis of RIU." RIU Discussion Paper. UK: Research Into Use (RIU).

Hall, A., J. Dijkman, and R. Sulaiman V. 2010. "Research Into Use: Investigating the Relationship between Agricultural Research and Innovation." RIU Discussion Paper No. 2010-01. Edinburgh: Research Into Use (RIU). http://www.researchintouse.com/learning/learning40discussionpaper01.html, accessed September 2011.

White, H. 2009a. "Some Reflections on Current Debates in Impact Evaluation." 3ie Working Paper No. 1. New Delhi: International Initiative for Impact Evaluation.

White, H. 2009b. "Theory-based Impact Evaluation: Principles and Practice." 3ie Working Paper No. 3. New Delhi: International Initiative for Impact Evaluation.

RESOURCES

Module 7 Overview

The CGIAR's **Institutional Learning and Change Program** (http://www.cgiar-ilac.org) brings together a group of national and international partners who are committed to strengthening the contributions of collaborative applied R&D programs to pro-poor agricultural innovation.

Knowledge Management for Development (http://www.km4dev.org/) is an internet portal and discussion group on knowledge sharing and knowledge issues, including reflective management, action-learning, and network development.

At the United Nations University–Maastricht Economic and Social Research and Training Centre, **LINK (Learning, INnovation, Knowledge)** (http://www.innovationstudies.org/) provides policy-relevant resources on innovation for a new rural economy through concepts, lessons, and guidelines.

The **Pelican Initiative** (http://dgroups.org/Community .aspx?c=3c4b8b5b-d151-4c38-9e7b-7a8a1a456f20) is an online community of practice for monitoring and evaluation (M&E) and participatory inquiry.

ODI's **Research and Policy in Development (RAPID)** program (http://www.odi.org.uk/Rapid/Index.html) aims to improve the uses of research and evidence in development policy and practice through research, advice, and debate.

Thematic Note 3

The **European Foresight Platform** (http://www.foresight-platform.eu/), financed by the European Commission DG Research, is part of a series of initiatives to provide a knowledge sharing platform for policy makers in the EU.

The **Foresight for Development** initiative (http://www .foresightfordevelopment.org/) is piloted in Africa by the South Africa Node of the Millennium Project, with support from the Rockefeller Foundation.

The **Global Food and Farming Futures 2011 Foresight Project** from the UK Departments for Environment, Food and Rural Affairs (Defra) and International Development (DFID) involved around 400 leading experts and stakeholders from about 35 countries across the world. More than 100 peer-reviewed evidence papers commissioned by the project are available at http://www .bis.gov.uk/foresight/our-work/projects/published-projects/global-food-and-farming-futures.

iKnow (http://wiwe.iknowfutures.eu/) is one of six Blue Sky foresight research projects funded by the European Commission's Seventh Framework Programme for Research and Technology Development (FP7) under the Socio-economic Sciences and Humanities (SSH) theme. The project aims to interconnect knowledge on issues and developments potentially shaking or shaping the future of STI in Europe and the world.

Millennia2015 Foresighting Women's Life in the Knowledge Society (http://www.millennia2015.org/) examines issues that will strongly influence women's life in the knowledge society in every country in the future. It also examines responsibilities with regard to how these issues evolve at a global level. The foresighting exercise has developed a methodology and is analyzing 37 variables to build an action plan to empower women.

Science and Technology Foresighting (http://www.tech-foresight.ca/tools.html) is an online community of practice offering tools, templates, and links to various foresighting and scenario planning initiatives around the world.

Glossary

Action-research. An iterative process of diagnosis, planning, action, evaluation, and reflection.

Adaptive management. "A structured, iterative process of optimal decision making in the face of uncertainty, with an aim to reducing uncertainty over time via system monitoring" (http://en.wikipedia.org/wiki/Adaptive_management). Adaptive management functions as a tool for learning as well as change within a system: As new information emerges, the system uses it (actively or passively) to change its operations and improve outcomes over the long term. The management process includes present and future stakeholders, bases iterative decision making on the results of monitoring (learning), and regards uncertainty as a means of improving understanding.

Advisory services. Agricultural advisory services can be defined as the entire set of organizations that support and facilitate people engaged in agricultural production to solve problems and to obtain information, skills, and technologies to improve their livelihoods and well-being. See also *agricultural extension.*

Agricultural education and learning (AEL). A variation on *agricultural education and training* that reflects a more student-centered approach to formal programs.

Agricultural education and training (AET). Organized programs and activities that serve the need for information, knowledge, and skills among those who work in various parts of the agriculture sector and the broader rural space. An AET system typically consists of tertiary educational institutions (agricultural universities or faculties and col-leges of agriculture within comprehensive universities) in addition to the polytechnics, institutes, or colleges that prepare technicians at the diploma level (postsecondary, sub-degree level). This second category of education, often termed *agricultural technical–vocational education and training (ATVET)*, *technical–vocational education and training (TVET)*, or *vocational education and training (VET)*, prepares technicians in a variety of specializations in agriculture subsectors. Some secondary schools offer agriculture as an elective. Agricultural training, frequently in training centers or training institutes, is offered to public employees as in-service training and/or to farmers as farmer training. Although the various elements in the AET delivery chain are often referred to collectively as a "system," they do not necessarily form a robust system in which communication and feedback flow between institutions and allow for continuous improvements. Many countries divide responsibility for AET between the ministries of agriculture and education.

Agricultural extension. The entire set of organizations that support and facilitate people engaged in agricultural production to solve problems and to obtain information, skills, and technologies to improve their livelihoods and well-being.

Agricultural knowledge and information system (AKIS). A system that links people and institutions to promote mutual learning and generate, share, and utilize agriculture-related technology, knowledge, and information. The system integrates farmers, agricultural educators, researchers, and extensionists to harness knowledge and information from various sources for improved livelihoods. Farmers are at the heart of this knowledge triangle.

Agricultural innovation system (AIS). A network of organizations, enterprises, and individuals focused on bringing new products, new processes, and new forms of organization into economic use, together with the institutions and policies that affect their behavior and performance.

Agricultural technical–vocational education and training (ATVET). See *agricultural education and training.*

Business development services (BDS). A range of nonfinancial services provided by public and private suppliers to entrepreneurs, who use them to operate more efficiently and expand their businesses. BDSs can include training, consultancy, and advisory services, marketing assistance, information, technology development and transfer, and business linkage promotion.

Business incubators. Programs designed to accelerate the successful development of entrepreneurial activities through an array of business support resources and services, developed and orchestrated by incubator management and offered both in the incubator and through its network of contacts. Incubators vary in the way they deliver their services, in their organizational structure, and in the types of clients they serve.

Central nodes. In the terminology of social network analysis, well-connected partners who pull promising new entrants into networks and collaborate with a wide assortment of partners, exposing them to more experiences, different competencies, and added opportunities. By linking clusters of network actors, the central nodes facilitate flows of information and resources. *Innovation brokers* are particularly prepared to become central nodes.

Change agent. See *innovation broker.*

Commercialization or *commercial services.* A focus in the agricultural research and extension system on commercial crops or the provision of services for specific cash crops such as tobacco or cotton.

Commodity extension. An extension system focused on one cash crop, for which advice and inputs are provided by one institution.

Competitive research grants (CRGs). Grants that fund research based on national competition and scientific peer review. Transparent procedures are used to select the proposals that will receive funding, based on rigorous criteria. Well-designed grants can bring greater contestability to the innovation process; the funding may not necessarily flow to the traditional recipients. Grants can promote research partnerships, leverage research resources, and help to develop a more efficient, demand-driven, and pluralistic research system by involving clients in setting priorities and financing, executing, and evaluating research. See *matching grants.*

Contracting. When one organization contracts with another to perform a function or service. *Contracting in* occurs when research, extension, and other government agricultural services provide services for an external group, such as a nongovernmental organization or a private organization. *Contracting out* is when government hires an external organization to perform services.

Cooperative extension. An extension model developed in the U.S. university system, in which universities, linked to the extension system, produce research results that are disseminated to farmers.

Cost sharing. When users of services pay at least a partial amount of the cost.

Decentralization. Moving responsibility and/or funding of public governmental services to local administrative levels such as districts. Decentralization can take many forms, such as *deconcentration* (accountability remains within the Department of Agriculture), *devolution* to local governments (accountability is transferred from central to locally elected governments), or *delegation* to semiautonomous services.

Demand articulation. Identifying the needs of different user groups for the knowledge and information produced by agricultural research organizations or the services (credit, advice) provided by other types of organizations.

Downstream research. Usually refers to research that adapts a technology (agronomic practice, new variety) to a particular country's or locality's needs. Sometimes called *applied* or *adaptive research.*

Economies of scale. In agricultural research, economies of scale occur when the cost per unit of research output falls with the number of units of output produced, usually through better use of major fixed investment or specialized skills.

Economies of scope. In agricultural research, economies of scope occur when the cost of a desired output falls with an increase in the number of different research outputs being produced. These gains occur when there is cross-commodity or cross-disciplinary learning as an external economy.

Embedded services. Companies provide information with the inputs they sell or other products they market.

Enabling environment. The environment (political, regulatory, institutional, economic, and social) that supports, promotes, and sustains a given outcome. For agricultural innovation, an enabling environment comprises those factors that influence agricultural innovation positively but are controlled by policy domains other than agricultural innovation policy itself.

Farm business schools. Schools that facilitate learning on production, management, business finance, and marketing skills. Farm business schools and cooperatives have an important learning role in promoting entrepreneurship among farmers, but initially they require external facilitation.

Farmer organization or *producer organization.* An organization constituted by farmers who seek solutions to production or commercial problems. Some agricultural services focus on providing extension, conducting research, or offering other services through these organizations.

Farmer field school. A participatory method of learning, technology development, and dissemination based on adult-learning principles such as experiential learning. Typically groups of 20–25 farmers meet weekly in an informal setting on their farms with a facilitator. The defining characteristics of farmer field schools include discovery learning, farmer experimentation, and group action. This interactive, practical training method empowers farmers to be their own technical experts on major aspects of their farming systems. Farmers are facilitated to conduct their own research, diagnose and test problems, devise solutions, and disseminate their learning to others.

Farmer field school networks. Networks of informal or formal groupings with a common interest that draw their membership from all the farmer field schools within a given geographic or administrative boundary.

Farming systems research and extension (FSRE). System of research and extension that is focused on understanding the farming systems of small-scale farmers through applied, multidisciplinary, on-farm, farmer-centered research.

Fee for service. The provision of services for a cost by government, nongovernmental, or private organizations.

Genetically engineered. A genetically engineered or *modified* organism in which the genetic material has been transformed using the techniques of genetic engineering. These techniques combine DNA molecules from different sources into one molecule to create a new set of genes. This recombined DNA is then transferred into an organism, giving it modified or novel genes. Transgenic organisms, a subset of genetically modified organisms, carry DNA that originated in a different species. Examples include cotton that has been genetically transformed to resist a particular herbicide. Many countries strictly control the production, use, export, and import of genetically modified plants and animals. (Based on http://en.wiki pedia.org/wiki/Genetically_modified_organism.)

Going concern. An enterprise that is expected to generate sufficient revenues and manage its resources in a manner that will allow it to remain in business for an indefinite period.

Governance. The systems and practices that governments use to set priorities and agendas, design and implement policies, and obtain knowledge about their impacts. Also refers to any systems and practices performing the same function at the subnational level (provincial research institutes, input subsidy programs) and in smaller institutions (farmer organizations, irrigation schemes).

Green revolution. The enormous increases in cereal production resulting from the adoption of high-yielding wheat, maize, and rice varieties, grown under irrigation with fertilizer and other inputs, across large areas of Asia and Latin America from the 1960s to 1980s.

Group-based approach. Extension programs that work with farmer groups or other common interest groups.

Industry clusters. Agglomerations of strongly interdependent firms (including specialized suppliers) linked to each other in a value-adding production chain, service providers, and associated institutions in a particular field. Some clusters encompass strategic alliances with universities, research institutes, knowledge-intensive business services, bridging institutions (brokers, consultants), and customers. Cluster-based approaches for business development and innovation have increased agricultural productivity, innovation, and business formation.

Information and communications technology (ICT). The wide and growing array of modern communications technology such as the Internet, e-mail, electronic databases, mobile phones and telephones, computers, personal digital devices, radio-frequency infrared devices, and the related infrastructure to support it (wireless networks, fiber-optic cable, and so on).

Innovation. An *invention* that is used for the first time in a product that reaches the market or produces a change in a social process. An innovation that is well known elsewhere may still be regarded as an innovation if it is new locally.

Innovation brokers. Teams of specialists that combine a strong background in science with knowledge of business and commercialization and/or the creation of innovation networks. Innovation brokers are also known as *change agents* or *technology brokers.*

Innovation capabilities. The skills to build and integrate internal and external resources to address problems or take advantage of opportunities. Innovation capabilities depend not only on innovative individuals but also on internal features of an organization, especially incentives, cultures, organizational spaces for experimentation, coordinating structures, and collective action.

Innovation-led growth. Growth based on innovative technologies, processes, products, markets, or organizational arrangements rather than on large additional uses of natural resources.

Innovation network. A diverse group of actors that voluntarily contribute knowledge and other resources (such as money, equipment, and land) to develop jointly or improve a social or economic process or product. Innovation networks are a special form of organization with a nonhierarchical structure, a collaboration-based culture, consensus-based coordination (because members are free to leave the network at any time), usually no legal personality (especially in their early stages), and often relatively fuzzy objectives (such as improving the management of natural resources). They evolve with market opportunities and the technologies they develop. Innovation networks differ from *farmer organizations* in that farmer organizations have a homogeneous membership and more formal, stable relations. Innovation networks differ from *value chains* in that the latter are more stable, are focused on delivering a product or service, and are coordinated by a central actor. Innovation networks are also known as *innovation platforms.*

Innovation platform. See *innovation network.*

Intellectual property rights (IPRs). Intellectual property law grants owners of intellectual property (creations of the mind) certain exclusive rights to a variety of intangible assets, such as musical, literary, and artistic works; discoveries and inventions; and words, phrases, symbols, and designs. Common types of intellectual property include copyrights, trademarks, patents, industrial design rights, and trade secrets. See *tangible property rights.* (Based on http://en.wikipedia.org/wiki/Intellectual_property.)

Intervention logic. The underlying assumptions in an intervention that link intervention inputs with expected outcomes.

Invention. The creation of knowledge. An invention becomes an *innovation* only when it is first used in a product that reaches the market or produces a change in a social process.

Learning alliance. A process-driven approach that facilitates the development of shared knowledge among different actors. Learning alliances contribute to improved development outcomes because lessons are more quickly identified and learned and because stronger links among research organizations and other actors in the AIS improve the focus on research and development practices.

Local agribusiness development services. Services that improve the performance of a small-scale enterprise oriented to agricultural production, be it individual or cooperative, in accessing markets, financial services, and enhanced agribusiness environments. Examples of these services include training and advisory services, market information services, technology, and business linkage information. Also referred to as *value-chain oriented services, market-oriented agricultural advisory services,* and *marketing extension.*

Marketing chain. Modern marketing chains for agricultural crops have many features of buyer-driven *value chains*: an actor close to consumers (usually a supermarket or broker) dominates, organizing many producers and intermediaries, deciding who participates in the chain, overseeing all the links from the farm to the shelf, defining the nature of the interactions and commercial conditions, and setting quality and safety standards. Other important features of modern marketing chains are that they focus on marketing specific products (such as vegetables, fruits, meat), access to the chain is highly restricted, verbal contracts based on trust are common but informal transactions rare, and technologies are generated mostly in developed countries and imposed by the leading agent. Only farmers with strong capabilities for innovation (especially entrepreneurship, physical and financial resources, and social capital) can survive in the highly competitive environment of modern marketing chains.

Matching grants. The matching of funds from the granting organization (usually a public agency) with funds from the beneficiary. Matching grants increasingly promote near-market technology generation, technology transfer and adoption, private economic activity, and overall innovation, often by including multiple stakeholders. By focusing greater attention on demand and use from the very beginning, basically by attracting users of technologies and knowledge in partnerships (and requiring a matching commitment), matching grants may be more effective than *competitive research grants* at enhancing the use of technology and knowledge by farmers and other entrepreneurs.

National agricultural research system (NARS). The entities responsible within a given country for organizing, coordinating, or executing research that contributes explicitly to the development of the country's agriculture and maintenance of its natural resource base.

National innovation policy. Overarching policies coordinating a wide spectrum of policy domains—science and technology policy, education policy, economic policy, industrial policy, infrastructure policy, taxation policy, and justice policy, among others—in such a way that together they create an environment that enables and stimulates innovation. Sector-specific innovation policies (such as a policy for agricultural innovation) replicate the national innovation policy's overarching and coordinating nature but have considerably less political clout to influence policies outside their domains.

Niche markets. A small, specific, and well-defined subset of the market on which a particular product focuses. Examples include markets for fair trade, organic, or other products certified to have particular qualities; small-scale growers of particular kinds of produce purchased directly by consumers in a nearby city; a new market for a traditional product (quinoa, amaranth, and acai are three of many kinds), or a new product derived from a traditional crop. Some niches have more demanding quality and commercial requirements than others, and farmers' participation in these markets depends on whether they can meet those requirements. Although niche markets have had important impacts on local communities, they cannot expand beyond a certain size without becoming commoditized. For this reason, they can make only a limited contribution to alleviating poverty. Niche markets can be considered a form of innovation network.

Nongovernmental organization (NGO) extension. Extension systems run by NGOs, often in a project mode and focused on participatory methods.

No-till agriculture. An agronomic practice in which crops are planted in previously unprepared soil by opening a narrow slot or trench of the smallest width and depth needed to obtain proper coverage of the seed. Conventional tillage practices involve multiple tractor passes to accomplish plowing, harrowing, planking, and seeding operations; no-till requires only one or two passes for spraying herbicide and seeding. In addition to reducing the number of operations, no-till requires less-powerful tractors and reduces equipment depreciation. While no-till principles are the same everywhere—entailing minimal soil disturbance, keeping soil covered, and using crop rotations—the actual packages differ greatly by location.

Organization. A group of actors that collaborate over a sustained period. An organization can be either formal or informal. Collaboration may take different forms, including frequent exchanges of information, joint priority setting for policies and programs, and joint implementation of innovation projects.

Organizational capabilities for innovation. The *abilities of the organization's members* and the *organization's key characteristics.* Organizational abilities for innovation include maintaining specialized knowledge, creativity, and commitment to the organization; developing a long-term vision for the organization; absorbing information generated by other agents (also called the absorptive capacity); creating new knowledge; and using this knowledge to develop innovations that address commercial, social, organizational, or technological needs or opportunities. An organization's key characteristics include its culture, governance, and communications routines (whether they are hierarchical or allow individual exploration of opportunities and horizontal communication); learning routines (the heuristics and methods used for collectively accepting new ideas and procedures); the propensity to interact and cooperate with other actors in the AIS; and the availability of resources for the development of innovations (capital and specialized assets).

Organizational interface. Modalities that help to transform knowledge and information produced by research organizations into socially and economically relevant goods and services. Examples include *innovation platforms, value-chain approaches,* and *public-private partnerships.*

Outgrower. A farmer operating under a formal or informal agreement (often a contract) to grow produce for a commercial agricultural enterprise (for example, a sugarcane processor) or a large-scale farmer. Outgrowers may receive credit and advice from the processor.

Participatory or demand-driven approaches. Method of research and/or extension focused on bottom-up approaches and empowerment of clientele. These approaches include methods such as farmer field schools and farmer research groups.

Pluralistic extension. Extension system based on multiple service providers, including public, private, and civil society organizations, in which the focus is often on demand-driven, participatory approaches.

Privatization. Full transfer of ownership (usually by sale) from government to a private entity.

Privatized research or extension services. Services run for profit, not necessarily for cash crops only.

Producer organizations. See *farmer organization.*

Public-private partnerships. At least one public and one private organization share resources, knowledge, and risks to achieve a match of interests and jointly deliver products and services. In agricultural research, PPPs bring together partners with different skills and knowledge to contribute jointly to the generation, adaptation, and/or diffusion of an innovation. Usually the partnership agreement is in the form of a contract that establishes each partner's commitments and the distribution of benefits. PPPs in agricultural research can be set up not only to generate knowledge via research but also to foster the diffusion and application of knowledge among private actors (agribusiness, farmers) and public actors (universities, research institutes, and extension agencies).

Qualitative growth. Growth associated with a range of additional public goods that especially reduce extreme poverty, provide food security, narrow structural inequalities, protect the environment, or sustain the growth process itself.

Risk capital. Money explicitly available for investment into a high-risk business or a security of some type—typically those that are not publicly traded on any national stock exchange. In this sourcebook, risk capital refers to investment in a company or project at an early or high-risk stage. Private investors are the major sources of risk capital; public investment cannot meet the needs, although it can stimulate and leverage private investment in the sector. See *venture capital.*

Rural productive alliance. An economic agreement between formally organized producers and at least one buyer, which specifies product characteristics (such as size and varieties to be produced); quantity to be produced or bought; production modalities (such as how a product will be delivered, by whom, and when, as well as grading and packing requirements); payment modalities and price determination criteria; and the buyer's contribution (such as technical assistance, specific inputs, and arrangements for input reimbursement—for example, at the time of sale).

Sanitary and phytosanitary (SPS) protection. Measures, including regulations and agreements, to protect: (1) human or animal health from risk arising from additives, contaminants, toxins, or disease organisms in food, drink, and feedstuffs; (2) human life from risks associated with diseases carried by plants or animals; (3) animal or plant life from pests, diseases, and disease-causing organisms; and (4) a country from other damage caused by the entry, establishment, or spread of pests. Such measures include national control of contaminants, pests, and diseases (vaccination programs, limits on pesticide residues in food) as well as international controls to prevent their inadvertent spread (for example, the rejection of insect-infested food shipments that pose a risk to domestic food production). See *standards* and *technical regulations.*

Social capital. The institutions, relationships, and norms that shape the quality and quantity of a society's social interactions. Increasing evidence shows that social cohesion is critical for societies to prosper economically and for development to be sustainable. A narrow view of social capital regards it as a set of horizontal associations between people, consisting of social networks and associated norms that have an effect on community productivity and well-being. Social networks can increase productivity by reducing the costs of doing business. Social capital facilitates coordination and cooperation. This quality is strong within mature groups with strong internal institutions, intragroup trust, altruistic behavior, membership in other groups, and ties to external service providers.

Standard. A document approved by a recognized body that provides, for common and repeated use, rules, guidelines, or characteristics for products or related processes and production methods, with which compliance is not mandatory. See *technical regulations.*

Strategic alliances. Partnerships that usually involve long-term cooperation (10 or more years), multinational companies, or groups of companies. Examples include the development and introduction of minimum social and environmental standards for agricultural or forestry products, fair trade arrangements, and similar ambitious programs.

Tangible property rights. The set of rights defined by law that relate to a physical object, for example plasmids or vectors. See *intellectual property rights.*

Technical regulations. Regulations that specify product characteristics or their related processes and production methods, including the applicable administrative provisions, with which compliance is mandatory. Technical regulations include import bans (total or partial), technical specifications (process and product standards), packaging standards, information requirements, and requirements for labeling and claims. See *standards.*

Technology broker. See *innovation broker.*

Theory of change. The underlying assumptions in an intervention that link intervention inputs with expected outcomes.

Traditional extension system (general extension). Extension focused on increasing agricultural productivity, run by central government, using a top-down approach and often emphasizing the transfer of technology.

Training and visit (T&V). A system of extension management with a focus on improving technical knowledge of extension agents and regular visits to farms.

Transfer of technology (TOT). Programs focused on disseminating information and new technologies. Such programs often include an integrated approach in which technology is pushed as a package deal with the requisite institutional support, such as credit and fertilizer facilities.

Value chain. The set of linked activities pursued by the different actors that a firm organizes to produce and market a product. See also *marketing chain.*

Value-chain approach. Attention to improving efficiency along the value chain for a particular agricultural commodity, often through applied agricultural research integrated with institutional innovations in farmer organization and marketing.

Venture capital. Venture capital is a form of private equity provided for early-stage and more mature companies with substantial market potential. Returns on venture capital investment are from a trade sale (sale to, or merger with, another company) or an initial public offering in which the company becomes authorized to sell its stock to the general public on a stock exchange. Venture capital funds will not only provide money but will mentor their investee firms. See *risk capital.*

Vocational education and training (VET). See *agricultural education and training.*

AUTHORS AND THEIR AFFILIATIONS

Nuria Ackermann, United Nations Industrial Development Organization (UNIDO)

John Allgood, International Fertilizer Development Center (IFDC)

S. Aravazhi, International Crops Research Institute for the Semi-Arid Tropics (ICRISAT)

Seth Ayers, World Bank Institute, and formerly with infoDev

Peter Ballantyne, International Livestock Research Institute (ILRI)

Alan B. Bennett, University of California, Davis

Rupert Best, Catholic Relief Services (CRS)

Regina Birner, University of Hohenheim

Sara Boettiger, Public Intellectual Property Resource for Agriculture (PIPRA), University of California, Berkeley

Arnoud R. Braun, Farmer Field School Foundation, Wageningen University

Alistair Brett, Consultant

Judy Chambers, Program for Biosafety Systems, International Food Policy Research Institute (IFPRI)

Anthony Clayton, University of the West Indies, Jamaica

Marie-Hélène Collion, Latin America and Caribbean Region, Agriculture and Rural Development, World Bank

Kristin Davis, International Food Policy Research Institute (IFPRI) and Global Forum for Rural Advisory Services (GFRAS)

Luz Diaz Rios, Consultant

Kumuda Dorai, Programme Officer, LINK Ltd. (Link Ltd. started as a UNU-MERIT/FAO initiative in 2005)

Deborah Duveskog, Consultant

Javier Ekboir, Institutional Learning and Change Initiative (ILAC, www.cgiar-ilac.org)

Howard Elliott, Consultant

Wyn Ellis, Consultant

Josef Ernstberger, Consultant

Michelle Friedman, World Bank

Josef Geoola, GALVmed

Peter Gildemacher, Royal Tropical Institute (KIT)

María Verónica Gottret, Tropical Agriculture Research and Education Center (CATIE)

Andy Hall, LINK Ltd. (Link Ltd. started as a UNU-MERIT/FAO initiative in 2005)

Helen Hambly Odame, Assistant Professor, University of Guelph

Frank Hartwich, United Nations Industrial Development Organization (UNIDO)

Willem Heemskerk, Royal Tropical Institute (KIT)

Douglas Horton, Consultant

Indira Ekanayake, Africa Agriculture and Development, World Bank

Steen Joffee, Director, Innodev U.K.

Adolphus J. Johnson, Ministry of Agriculture, Forestry and Food Security, Sierra Leone

Trish Kammili, French National Institute for Agricultural Research (INRA)

S.M. Karuppanchetty, International Crops Research Institute for the Semi-Arid Tropics (ICRISAT)

Godrick Khisa, Farmer Field School Foundation Promotion Services, Wageningen University

Laurens Klerkx, Communications and Innovation Studies, Wageningen University

Stanley Kowalski, University of New Hampshire School of Law, Concord

Patti Kristjanson, World Agroforestry Centre (CIFOR)

Anton Krone, SaveAct, Prolinnova South Africa

Gunnar Larson, Agriculture and Rural Development Department, World Bank

Ninatubu Lema, NARS Tanzania

Tarmo Lemola, ADVANSIS Ltd.

David Lugg, Food and Agriculture Organization of the United Nations (FAO)

Mark Lundy, Decision and Policy Analysis Program, CIAT

John Lynam, Consultant

Charles J. Maguire, Consultant

Vijay Mahajan, BASIX Group, India

Morven McLean, ILSI Research Foundation

Mohinder S. Mudahar, Consultant

Bernardo Ospina Patiño, Latin American and Caribbean Consortium to Support Cassava Research and Development (CLAYUCA)

Christopher Palmberg, ADVANSIS Ltd.

Andrea Pape-Christiansen, Consultant

Robert Potter, Robert Potter Consulting

John Preissing, Food and Agriculture Organization of the United Nations (FAO)

Ranjitha Puskur, International Livestock Research institute (ILRI)

Catherine Ragasa, International Food Policy Research Institute (IFPRI)

Riikka Rajalahti, Agriculture and Rural Development Department, World Bank

Johannes Roseboom, Innovation Policy Consultancy

Silvia Sarapura, University of Guelph

Eva Schiffer, Consultant

Kiran K. Sharma, International Crops Research Institute for the Semi-Arid Tropics (ICRISAT)

David J. Spielman, International Food Policy Research Institute (IFPRI)

David S. Suale, Ministry of Agriculture, Forestry and Food Security, Sierra Leone

Rasheed Sulaiman V, Centre for Research on Innovation and Science Policy (CRISP), India

Florian Theus, World Bank Institute

Bernard Triomphe, Agricultural Research for Development, France (CIRAD)

Klaus Urban, Food and Agriculture Organization of the United Nations (FAO)

K. Vasumathi, BASIX Group, India

Rodrigo Vega Alarcón, Consultant, formerly with FIA

Laurens van Veldhuizen, Prolinnova International Secretariat, ETC Foundation

Ann Waters-Bayer, Prolinnova International Secretariat, ETC Foundation

Melissa Williams, South Asia Agriculture and Rural Development, World Bank

Mariana Wongtschowski, Royal Tropical Institute (KIT)

Douglas Zhihua Zeng, Africa Finance and Private Sector Development, World Bank

INDEX

Boxes, figures, and tables are indicated by b, f, and t following the page number.

capacity building (*continued*)
  extension and advisory services, 10
  farmer organizations and, 61, 64
  for governance, 473–76
  higher education students and, 133*b*
  innovation brokers and, 226
  innovation systems and, 9–10
  for intellectual property (IP) management, 518*b*
  local business development services and, 211–12
  for national coordination, 35, 36*b*
  organizational change and, 316–17, 318*b*
  for pluralistic extension systems, 198–200
  for public-private partnerships, 379
Casas Agrárias (Mozambique), 208*b*
Cassava research networks, 90–94, 92*b*. See also CLAYUCA
catalytic agents, 46, 50. *See also* business incubators; innovation
  brokers
Catholic Relief Services (CRS), 347, 348*b*
CATIE, 344–49. *See also* Learning Alliances
CBR (community-based research), 157–59, 159*b*
CCARDESA (Center for Coordination of Agricultural Research
  and Development in Southern Africa), 307*b*
CEDAC (Cambodian Center for Study and Development in
  Agriculture), 440*b*
CEGE (Center for Entrepreneurial Management)
  (Nicaragua), 432*b*
CENICAÑA (Colombian Sugarcane Research Center), 66
Center for Community-Based Research (Thailand), 158
Center for Coordination of Agricultural Research and
  Development in Southern Africa (CCARDESA), 307*b*
Center for Entrepreneurial Management (CEGE)
  (Nicaragua), 432*b*
Central Africa, regional research coordination in, 299*b*.
  See also specific countries
Central American Learning Alliance, 283, 344, 347
CFC (Common Fund for Commodities, Kenya), 222*b*
CGIAR. *See* Consultative Group on International Agricultural
  Research
Cheetah Network (Mali), 209*b*
Chiang Mai University (Thailand), 126, 133, 157–59, 159*b*
Chile
  agricultural innovation council in, 23
  Agricultural Technology Consortium model in, 338–43, 339*b*,
    342–43*b*
  assessment in, 607–9, 608–9*b*, 609*f*
  coordination organizations in, 36, 80–83
  intellectual property management in, 484*b*
  national innovation policy in, 461*b*
  no-till agriculture in, 72
  olive oil industry in, 82*b*
  research and development tax incentives in, 365*b*
China
  business incubators in, 392*b*
  education and training in, 129, 138
  higher education reform in, 157–59, 159*b*
  intellectual property management in, 409*b*, 426–29
  no-till agriculture in, 70, 72, 73*t*
  public-private partnerships in, 362, 375*b*, 377*b*, 380*b*

public-private research partnerships in, 90
tertiary education reform in, 124–25*b*
value chains in, 54*b*
Chinese University of Technology, 409*b*
CIAT (International Center for Tropical Agriculture), 17*b*, 91–93,
  347, 348*b*
  CLAYUCA, 90–94
  Learning Alliances, 344–49
CIMMYT (International Maize and Wheat Improvement
  Center), 47*b*, 50
CIP (International Potato Center), 45*b*, 222*b*
civil society. *See also* nongovernmental organizations (NGOs);
  specific organizations
  cluster-based business development and, 431*b*
  coordination and collective action role, 28, 31
  education and, 138
  innovation networks and, 48, 51
  marketing chains, 57
  research and, 273, 371*b*
Clayton, Anthony, 562
CLAYUCA (Latin American and Caribbean Consortium to
  Support Cassava Research and Development), 90–94, 92*b*
Client-Oriented Research and Development Management
  Approach (CORDEMA), 318*b*, 322
climate change, 12*b*, 188. *See also* environmental issues
cluster-based business development, 368, 396–405
  capacity building and, 431, 433–34
  infrastructure and, 397
  innovative activity profile, 430–34, 431–33*b*
  institutional framework and, 398–99, 398*f*
  investment needs, 397–99
  investment rationale, 396–97
  lessons learned, 402–5, 434
  policy issues, 400
  potential benefits, 399–400, 431–34
  public-private partnerships and, 404*b*
  recommendations, 402–5
  regulatory environment and, 397–98
  value chains vs., 397
CMSA (community-managed sustainable agriculture), 85, 86–89,
  87–88*b*
codesigned innovations, 308–15
  investment context, 308–9, 309*t*
  investment needs, 309–13, 310*b*, 310*t*, 312*b*
  lessons learned, 314–15
  partnerships for, 282–83, 295
  policy issues, 314
  potential benefits, 314
COFUPRO (Coordinadora Nacional de las Fundaciones Produce,
  Mexico), 75–77, 78*b*
collective action. *See* coordination and collective action
colleges and universities. *See* tertiary education and training
Collion, Marie-Hélène, 95
Colombia
  agricultural innovation systems in, 3*b*
  business incubators in, 395*b*
  farmer organizations in, 67
  intellectual property rights in, 485*b*

International Crops Research Institute for Semi-Arid Tropics (ICRISAT), 285*b*, 393*b*, 421–25, 422*f*, 423–24*b*, 425*t*
International Development Agency (Canada), 299*b*
International Development Enterprises (IDE), 224*b*
International Development Research Centre (IDRC), 17*b*, 344
International Fertilizer Development Center (IFDC), 232–33, 233*b*
International Food Policy Research Institute, 181*b*
International Institute of Rural Reconstruction (IIRR), 207
International Institute of Tropical Agriculture–FOODNET, 348*b*
International Labour Organization (ILO), 111, 375*b*
International Livestock Research Institute (ILRI), 44, 223*b*, 299*b*, 326–30, 327*b*
International Maize and Wheat Improvement Center (CIMMYT), 47*b*, 50
International Potato Center (CIP), 45*b*, 222*b*
International Service for National Agricultural Research (ISNAR), 78*b*
internship programs, 153, 161
intervention logic, defined, 540*b*
IP. *See* intellectual property management
iPark (business incubator), 392
IPB (Institut Pertanian Bogor, Indonesia), 127*b*, 133
Ireland
    education and training in, 126
    foresighting investments in, 563*b*
    scenarios, 563*b*
    Teagasc, 536*b*
IRR (internal rate of return), 418
ISNAR (International Service for National Agricultural Research), 78*b*

J

Jamaica
    foresighting investments in, 564*b*
    public-private research partnerships in, 385*b*
Janssen, Willem, 607
Jimma University College of Agriculture and Veterinary Medicine (Ethiopia), 163
Joffe, Steen, 589
Johnson, Adolphus, 589
joint experimentation, 311
joint venture, 79*b*, 81*b*, 141–42, 195, 286, 341

K

Kammili, Trish, 569, 580
KAPP (Kenya Agricultural Productivity Project), 238*b*, 269
KARI (Kenya Agricultural Research Institute), 222*b*, 269
Karuppanchetty, S. M., 421
Kazakhstan, competitive grant funding in, 436
KDGCBP (Kenya Dairy Goat and Capacity Building Project), 185*b*
Kelemework, D., 275
Kellogg Commission on the Future of State and Land-Grant Universities, 127–28
Kenya
    agrodealer development in, 234*b*
    business development services in, 201, 207

cluster-based business development in, 401*b*
coordination organizations in, 24*b*, 64*b*
extension and advisory services in, 222*b*, 238*b*
Farmer Field Schools in, 236, 238, 238*b*
farmer organizations in, 62*b*
gender-inclusive education in, 109*b*
innovation brokers in, 222, 222*b*, 225
research and development in, 271*b*, 327*b*
value chains in, 54*b*
Kenya Agricultural Productivity Project (KAPP), 238*b*, 269
Kenya Agricultural Research Institute (KARI), 222*b*, 269
Kenya Dairy Goat and Capacity Building Project (KDGCBP), 185*b*
Kenya Flower Council, 24*b*
Kenya Good Agricultural Practices (Kenya-GAP), 24*b*
Kenya Horticulture Council, 24*b*
Kenya Tea Board, 24, 64*b*
Kenya Tea Development Agency (KTDA), 53, 60, 64*b*, 66–67
Kerala Horticultural Development Programme (KHDP, India), 215*b*, 217*b*, 219*b*
Khisa, Godrick, 236
Klerkx, Laurens, 221
Kline, S. J., 268
knowledge transfer, 232, 233*b*, 235*b*. *See also* education and training; technology transfer
Korea, Republic of
    governance of innovation systems in, 470–71, 473*b*
    policy coordination organizations in, 474*b*
    strategic intelligence capabilities in, 476*b*
Kowalski, Stanley, 480
Kristjanson, Patti, 326
Krone, Anton, 435
KTDA. *See* Kenya Tea Development Agency
Kwadaso Agricultural College (Ghana), 156

L

Labor standards, 375*b*
Lake Victoria Environmental Management Program, 404*b*
Lambayeque Institute for Agricultural Development (Peru), 242
Land Care movement (Philippines), 270
Land O'Lakes, 165, 166, 167
Larson, Gunnar, 84, 526
Latin America and the Caribbean. *See also specific countries*
    competitive research grants in, 386*b*
    farmer organizations in, 65–66
    public-private research partnerships in, 90–94, 294
    regional research coordination in, 297–98, 298*b*
    research and development in, 263
    science and technology funds in, 40
Latin American and Caribbean Consortium to Support Cassava Research and Development (CLAYUCA), 90–94, 92*b*
Latin American Fund for Irrigated Rice (FLAR), 298*b*
LBDSs. *See* local business development services
learning alliances, 17*b*, 273, 344–49, 345*b*, 346*t*, 348*b*
learning organizations, 559*b*
LEED (Local Economic and Employment Development) Project, 551*b*

Peking University (China), 409*b*

PepsiCo, 249*b*

performance, defined, 540*b*. *See also* monitoring and evaluation (M&E)

performance-based contracts, 514

Peru
  agricultural innovation systems in, 3*b*
  extension and advisory services in, 240–45
  gender and program assessment in, 598–602, 600*t*
  innovation networks in, 45*b*
  public-private research partnerships in, 90, 385*b*
  research and development in, 283
  research funding in, 320–21*b*

Philippines
  public-private research partnerships in, 274*b*
  research and development in, 270

Pingali, P., 300

plant variety rights (PVRs), 453*b*

pluralistic extension systems, 191, 194–203, 196*t*
  capacity building for, 198–200
  environmental issues, 201
  equity issues, 201
  gender issues, 201
  institutional framework and, 201
  investment context, 195–97
  investment needs, 197–200, 199*t*
  lessons learned, 202
  policy issues, 200–201
  potential benefits, 200
  public and private sector roles, 201
  recommendations, 203
  sustainability, 200

PMCA (participatory market chain approach), 313*b*

policy issues
  agricultural research linkages, 286–87
  assessment, 541–43, 542*t*, 550–51
  biosafety regulatory systems, 494–95
  business incubators, 391
  cluster-based business development, 400, 431*b*
  codesigned innovations, 314
  coordination and collective action, 28–29, 37
  curriculum reform, 134
  education and training, 112–13
  enabling environment for innovation, 450–52, 450*f*, 455–56, 457
  evaluation, 585
  extension and advisory services, 187–91
  "extension-plus" approach, 216–18
  foresighting investments, 567–68
  innovation brokers, 226
  innovation funds, 384–85
  innovation networks, 47–49
  intellectual property (IP) management, 410, 411–12, 487–88
  local business development services (LBDSs), 208–10
  marketing chains, 56–57
  monitoring and evaluation (M&E), 541–43, 542*t*, 577–78, 585
  national innovation policy, 464–65
  organizational assessment, 560–61
  organizational change, 323–24
  partnerships, 369–71
  pluralistic extension systems, 200–201
  public-private partnerships, 293–94, 376
  regional research, 300–302
  research, 269–70
  risk capital investments, 418–19
  technician development, 138–39
  Technology Transfer Offices, 410, 411–12
  tertiary education and training, 128
  value chains, 56–57

Polytechnic University of Bobo-Dioulasso (Burkina Faso), 156

Potter, Robert, 480

poverty reduction
  coordination and collective action role in, 17–18
  enabling environment for innovation and, 456

Preissing, John, 240

priorities, defined, 540*b*. *See also* agenda setting

private sector. *See also* public-private partnerships
  coordination and collective action, 28
  extension and advisory services, 181*b*, 187, 189, 190*t*, 195, 196*t*, 202
  "extension-plus" approach, 218
  innovation networks and, 48–49, 49*b*, 51
  in-service training capacity, 142
  intellectual property (IP) management, 483–84, 488, 517*b*
  local business development services (LBDSs), 205–7, 205*f*, 209–10
  marketing chains, 52, 57
  pluralistic extension systems, 201
  research financing, 5*b*, 267, 271–72
  role of, 12*b*
  value chains, 57, 93

PROCISOR, 298*b*

Produce Foundations (Mexico), 23, 31, 60, 75–79

Producer organization, 5*b*, 6*b*, 24, 31, 62*b*, 95–99, 220*t*, 222*b*, 241, 244, 281*b*, 342*b*, 385*b*, 431*b*, 433*b*, 489. *See also* farmer organization
  defined, 631

Productive Partnerships Project (Colombia), 385*b*

product marketing companies, 23

PROfarm courses, 137*b*

Programa Nacional de Extensão Agrária (PRONEA, Mozambique), 184, 200

PROINPA Foundation (Bolivia), 45*b*

project-based funding, 40, 41*b*, 42*t*

project management, 142–43

Prolinnova network, 60

Propensity Score Matching, 583*b*

public-private partnerships, 290*b*, 290*t*, 374–80
  business development services, 208, 212
  capacity building for, 27*b*, 379
  cluster-based business development and, 404*b*
  contracts for, 378–79
  for coordination and collective action, 90–94, 92*b*
  education and training, 143
  enabling environment for innovation and, 93, 456
  exit strategies, 369*b*, 379

ECO-AUDIT
Environmental Benefits Statement

The World Bank is committed to preserving endangered forests and natural resources. The Office of the Publisher has chosen to print *Agricultural Innovation Systems: An Investment Sourcebook* on recycled paper with 50 percent post-consumer waste, in accordance with the recommended standards for paper usage set by the Green Press Initiative, a nonprofit program supporting publishers in using fiber that is not sourced from endangered forests. For more information, visit www.greenpressinitiative.org.

Saved:
- 53 trees
- 22 million BTU of total energy
- 5,369 pounds of net greenhouse gases (CO_2 equivalent)
- 24,213 gallons of waste water
- 1,534 pounds of solid waste